Atomic Masses of the Elements
Based on the 2005 IUPAC Table of Atomic Masses

Name	Symbol	Atomic Number	Atomic Mass	Name	Symbol	Atomic Number	Atomic Mass
Actinium*	Ac	89	227	Meitnerium*	Mt	109	268
Aluminum	Al	13	26.981538	Mendelevium*	Md	101	258
Americium*	Am	95	243	Mercury	Hg	80	200.59
Antimony	Sb	51	121.760	Molybdenum	Mo	42	95.94
Argon	Ar	18	39.948	Neodymium	Nd	60	144.24
Arsenic	As	33	74.92160	Neon	Ne	10	20.1797
Astatine*	At	85	210	Neptunium*	Np	93	237
Barium	Ba	56	137.327	Nickel	Ni	28	58.6934
Berkelium*	Bk	97	247	Niobium	Nb	41	92.90638
Beryllium	Be	4	9.012182	Nitrogen	N	7	14.00674
Bismuth	Bi	83	208.98038	Nobelium*	No	102	259
Bohrium*	Bh	107	264	Osmium	Os	76	190.23
Boron	B	5	10.811	Oxygen	O	8	15.9994
Bromine	Br	35	79.904	Palladium	Pd	46	106.42
Cadmium	Cd	48	112.411	Phosphorus	P	15	30.973762
Calcium	Ca	20	40.078	Platinum	Pt	78	195.078
Californium*	Cf	98	251	Plutonium*	Pu	94	244
Carbon	C	6	12.0107	Polonium*	Po	84	209
Cerium	Ce	58	140.116	Potassium	K	19	39.0983
Cesium	Cs	55	132.90545	Praseodymium	Pr	59	140.90765
Chlorine	Cl	17	35.4527	Promethium*	Pm	61	145
Chromium	Cr	24	51.9961	Protactinium	Pa	91	231.03588
Cobalt	Co	27	58.933200	Radium*	Ra	88	226
Copper	Cu	29	63.546	Radon*	Rn	86	222
Curium*	Cm	96	247	Rhenium	Re	75	186.207
Darmstadtium*	Ds	110	271	Rhodium	Rh	45	102.90550
Dubnium*	Db	105	262	Roentgenium*	Rg	111	272
Dysprosium	Dy	66	162.500	Rubidium	Rb	37	85.4678
Einsteinium*	Es	99	252	Ruthenium	Ru	44	101.07
Erbium	Er	68	167.26	Rutherfordium*	Rf	104	261
Europium	Eu	63	151.964	Samarium	Sm	62	150.36
Fermium*	Fm	100	257	Scandium	Sc	21	44.955910
Fluorine	F	9	18.9984032	Seaborgium*	Sg	106	266
Francium*	Fr	87	233	Selenium	Se	34	78.96
Gadolinium	Gd	64	157.25	Silicon	Si	14	28.0855
Gallium	Ga	31	69.723	Silver	Ag	47	107.8682
Germanium	Ge	32	72.61	Sodium	Na	11	22.989770
Gold	Au	79	196.96655	Strontium	Sr	38	87.62
Hafnium	Hf	72	178.49	Sulfur	S	16	32.066
Hassium*	Hs	108	277	Tantalum	Ta	73	180.9479
Helium	He	2	4.002602	Technetium*	Tc	43	98
Holmium	Ho	67	164.93032	Tellurium	Te	52	127.60
Hydrogen	H	1	1.00794	Terbium	Tb	65	158.92534
Indium	In	49	114.818	Thallium	Tl	81	204.3833
Iodine	I	53	126.90447	Thorium	Th	90	232.0381
Iridium	Ir	77	192.217	Thulium	Tm	69	168.93421
Iron	Fe	26	55.845	Tin	Sn	50	118.710
Krypton	Kr	36	83.80	Titanium	Ti	22	47.867
Lanthanum	La	57	138.9055	Tungsten	W	74	183.94
Lawrencium*	Lr	103	262	Uranium	U	92	238.0289
Lead	Pb	82	207.2	Vanadium	V	23	50.9415
Lithium	Li	3	6.941	Xenon	Xe	54	131.29
Lutetium	Lu	71	174.967	Ytterbium	Yb	70	173.04
Magnesium	Mg	12	24.3050	Yttrium	Y	39	88.90585
Manganese	Mn	25	54.938049	Zinc	Zn	30	65.39
				Zirconium	Zr	40	91.224

* This element has no stable isotopes. The atomic mass given is that of the isotope with the longest known half-life.

Foundations of College Chemistry

TWELFTH EDITION

Morris Hein
Mount San Antonio College

Susan Arena
University of Illinois, Urbana-Champaign

John Wiley & Sons, Inc

Executive Editor: *Deborah Edson*
Project Editor: *Jennifer Yee*
Marketing Manager: *Amanda Wygal*
Senior Production Editor: *Sandra Dumas*
Designer: *Hope Miller*
Senior Media Editor: *Tom Kulesa*
Photo Editor: *Elle Wagner*
Production Management: *Ingrao Associates*
Cover Photo: *Digital Vision/Getty Images*

This book was set in 10/12 Minister Light by Preparé Inc. and printed and bound by Quebecor/Versailles. The cover was printed by Phoenix Color Corporation.

This book is printed on acid free paper. ∞

To order books or for customer service please call 1-800-CALL WILEY (225-5945).

Hein, Morris
Foundations of college chemistry, 12th ed.

ISBN-13 978-0-471-74153-4
ISBN-10 0-471-74153-1

Printed in the United States of America
10 9 8 7 6 5 4 3 2

An Education isn't how much you have committed to memory, or even how much you know. It's being able to differentiate between what you do know and what you don't. It's knowing where to go to find out what you need to know; and it's knowing how to use the information once you get it.

William Feather

For Our Future Generations

Dillon Brayden Prum
Great-grandson of Morris Hein

Jessica, Joshua, Moriah and David Hendrie
Lucia Cavaluzzi

Alexander Wettergren
Grandchildren of Susan Arena

ABOUT THE AUTHORS

Morris Hein is professor emeritus of chemistry at Mt. San Antonio College, where he regularly taught the preparatory chemistry course and organic chemistry. He is the original author of *Foundations of College Chemistry* and his name has become synonymous with clarity, meticulous accuracy, and a step-by-step approach that students can follow. Over the years, more than three million students have learned chemistry using a text by Morris Hein. In addition to *Foundations of College Chemistry*, Twelfth Edition, he is co-author of *Introduction to General, Organic, and Biochemistry*, Eighth Edition, and *Introduction to Organic and Biochemistry*. He is also co-author of *Foundations of Chemistry in the Laboratory*, Twelfth Edition, and *Introduction to General, Organic and Biochemistry in the Laboratory*, Eighth Edition.

Susan Arena earned a BS and MA in Chemistry at California State University-Fullerton. She has taught science and mathematics at all levels, including middle school, high school, community college, and university. At the University of Illinois she developed a program for increasing the retention of minorities and women in science and engineering. This program focused on using active learning and peer teaching to encourage students to excel in the sciences. She has coordinated and led workshops and progrms for science teachers from elementary through college levels that encourage and support active learning and creative science teaching techniques. For several years she was director of an Institute for Chemical Education (ICE) field center in Southern California. Susan enjoys reading, knitting, traveling, classic cars, and gardening in her spare time when she is not playing with her grandchildren.

This new Twelfth Edition of *Foundations of College Chemistry* presents chemistry as a modern, vital subject and is designed to make introductory chemistry accessible to all beginning students. The central focus is the same as it has been from the first edition: to make chemistry interesting and understandable to students and teach them the problem-solving skills they will need. In preparing this new edition we considered the comments and suggestions of students and instructors to design a revision that builds on the strengths of previous editions. We have especially tried to relate chemistry to the real lives of our students as we develop the principles that form the foundation for the further study of chemistry, and to provide them with problem-solving skills and practice needed in their future studies.

Course Options *Foundations of College Chemistry*, 12th Edition, is intended for students who have never taken a chemistry course or those who have had a significant interruption in their studies but plan to continue with the general chemistry sequence. Since its inception this book has helped define the preparatory chemistry course and has since developed a much wider audience. In addition to preparatory chemistry, our text is used extensively in one-semester general purpose courses (such as those for applied health fields) and in courses for nonscience majors.

Alternate Editions For the convenience of instructors and to accommodate the various lengths of academic terms, two versions of this book are available. *Foundations of College Chemistry*, 12th Edition, includes 20 chapters and is our main text. *Foundations of College Chemistry*, Alternate 12th Edition, provides a shorter, 17 chapter text in paperback with the same material, but without the nuclear, organic, and biochemistry chapters.

Development of Problem–Solving Skills

We all want our students to develop real skills in solving problems. We believe that a key to the success of this text is the fact that our problem-solving approach works for students. It is a step-by-step process that teaches the use of units and shows the change from one unit to the next. Students learn concepts most easily in a step-by-step process. In this edition we continue to use examples to incorporate fundamental mathematical skills, scientific notation, and significant figures. Painstaking care has been taken to show each step in the problem-solving process and to give *alternative methods for solutions* where appropriate. These alternative methods give students flexibility in choosing the one that works best for them. We continue to use four significant figures for

Problem-Solving Steps

Practice Problems

atomic and molar masses for consistency and for rounding off answers appropriately. We have been meticulous in providing answers, correctly rounded, for students who have difficulty with mathematics.

Fostering Student Skills *Attitude* plays a critical role in problem solving. We encourage students to learn that a systematic approach to solving problems is better than simple memorization. Throughout the book we encourage students to begin by writing down the facts or data given in a problem and to think their way through the problem to an answer, which is then checked to see if it makes sense. Once we have laid the foundations of concepts, we highlight the steps so students can locate them easily. Important rules and equations are highlighted for emphasis and ready reference.

Student Practice Practice problems follow the examples in the text, with answers provided at the end of the chapter. The end of each chapter begins with a new *Chapter Review* and *Review Questions* section, which help students review key terms and concepts, as well as material presented in tables and figures. This is followed by *Paired Exercises*, covering concepts and numerical exercises, where two similar exercises are presented side by side. The final section, *Additional Exercises*, includes further practice problems presented in a more random order. In our new edition we have changed a significant number of exercises per chapter, and added new "Challenge Exercises." Throughout, challenging questions and exercises are denoted with an asterisk. Answers for selected questions and exercises appear in Appendix VI, and answers for *Putting It Together* review exercises appear in Appendix VII.

Emphasis on Real-World Aspects

We continue to emphasize the less theoretical aspects of chemistry early in the book, leaving the more abstract theory for later. This sequence seems especially appropriate in a course where students are encountering chemistry for the very first time. Atoms, molecules, and reactions are all an integral part of the chemical nature of matter. A sound understanding of these topics allows the student to develop a basic understanding of chemical properties and vocabulary.

Chapters 1 through 3 present the basic mathematics and the language of chemistry, including an explanation of the metric system and significant figures. In Chapter 4 we present chemical properties—the ability of a substance to form new substances. Then, in Chapter 5, students encounter the history and language of basic atomic theory.

We continue to present new material at a level appropriate for the beginning student by emphasizing nomenclature, composition of compounds, and reactions in Chapters 6 through 9 before moving into the details of modern atomic theory. Some applications of the Periodic Table are shown in early chapters and discussed in detail in Chapters 10 and 11. Students gain confidence in their own ability to identify and work with chemicals in the laboratory before tackling the molecular models of matter. As practicing chemists we have little difficulty connecting molecular models and chemical properties. Students, especially those with no prior chemistry background, may not share this ability to connect the molecular models and the macroscopic properties of matter. Those instructors who feel it is essential to teach atomic theory and bonding early in the course can cover Chapters 10 and 11 immediately following Chapter 5.

New to This Edition

In this Twelfth Edition we have tried to build on the strengths of the previous editions. We have worked to update the language and to provide students with clear explanations for concepts and useful ways to solve chemistry problems. We continually strive to keep the material at the same level so students can easily read and use the text to learn chemistry. Some specific changes in the text are highlighted below:

- **Chapter 1** has been rewritten to include a discussion of matter and physical and chemical changes. This new material engages the students immediately into the vocabulary of chemistry and the particulate nature of matter. We have added new end-of-chapter review material and exercises so that students immediately begin to review and practice what they are learning.
- **Chapter 3** has been revised for a greater emphasis on classification of matter and to include an introduction to the periodic table. This introductory material supports laboratory work and presents an overview of how the periodic table provides the big picture for how the elements are related.
- **Chapter 13** has been reorganized to move the discussion of hydrogen bonding and hydrates forward to complete the discussion of liquids. The information on water has been collected in a new section called Water, a Unique Liquid and serves to discuss water and its properties as an example of a unique liquid.
- **Molecular art**: Learning chemistry requires the ability to connect the macroscopic world of everyday life to the microscopic world of atoms and molecules. In this edition we have added molecular art to macroscopic pictures to emphasize this connection.
- **Chapter Reviews** have been reformatted to include key terms and to review each section in a bulleted format similar to what a student might use to review the material. We have included summary art from the chapter in this section to aid visual learners. *Review Questions* are included in the *Chapter Reviews* to encourage students to restate concepts from the chapter and to apply *Tables* and *Figures* to new situations.

- **Design and Illustration Program:** This edition has a fresh new design, enhanced by an updated art and photo program. New photos appear throughout.

- **Problem Solving:** Because the development of problem-solving skills is essential to learning chemistry, in this new design the sections on problem solving are highlighted throughout for easy student reference.

- **New Exercises:** 8 to 12 *new* end of chapter exercises have been added in most chapters. *Challenge Exercises* have also been added to most chapters.

- **Math Skills Learning Aids: A Review of Mathematics** is provided as Appendix I for students who need assistance.

- **New Book Companion Web Site:** This site supplies a number of helpful resources for students and instructors, including **Interactive Learningware problems**, **Online Quizzing**, and **Power Point Lecture Slides.**

- **Each chapter opens** with a *color photograph* relating the chapter to our daily lives. A *chapter outline* assists students in viewing the topics covered in the chapter, and the *introductory paragraph* further connects the chapter topic to real life.

- **Chemistry in Action** sections, a number new to this edition, show the impact of chemistry in a variety of practical applications. These essays cover a range of relevant topics and introduce experimental information on new chemical discoveries and applications.

- **Important statements**, equations, and laws are highlighted for emphasis. A **Glossary** is provided to help students review key terms, with section numbers given for each term to guide the student to the contextual definition. The margins of the glossary pages are color tinted to provide ready access. Glossary terms are also printed in bold type in the text.

- **End of chapter exercises**, including 8 to 12 new exercises in most chapters, provide practice and review of the chapter material. *Paired exercises* present two parallel exercises, side by side, so the student can solve one problem, check the answer in Appendix VI, and use the same problem-solving skills with the second exercise. **Additional exercises** are provided at the end of most chapters, arranged in random order, to encourage students to review the chapter material. **Answers to selected questions and exercises** are given in Appendix VI.

- **Putting It Together** review sections appear after every 2 to 4 chapters and include additional conceptually oriented exercises for effective self-review. These review sections are printed on colored pages for easy reference and provide students with a helpful summary of the preceding material. **Answers to Putting It Together** review sections are given in Appendix VII.

- Directions on using a calculator to solve problems are given in Appendix II, **Using a Scientific Calculator**.

- **Units of measurement** are shown in table format in Appendix IV and in the endpapers.

Learning Aids

To help the beginning student gain the confidence necessary to master technical material we have refined and enhanced a series of learning aids:

- Important **terms** are set off in bold type where they are defined, and are printed in blue in the margin. Most glossary terms are also defined in the glossary.

- Worked **examples** show students the how of problem solving before they are asked to tackle problems on their own.
- **Practice problems** permit immediate reinforcement of a skill shown in the example problems. Answers are provided at the end of the chapter to encourage students to check their problem solving immediately.
- **Marginal notations** help students understand basic concepts and problem-solving techniques. These are printed in green ink to clearly distinguish them from text and vocabulary terms.

Learning Aids: Math Skills For students who may need help with the mathematical aspects of chemistry, the following learning aids are available:

- A **Review of Mathematics**, covering the basic functions, is provided in Appendix I.
- **Math Survival Guide: Tips and Tricks for Science Students**, 2nd Edition, by Jeffrey R. Appling and Jean C. Richardson, a brief paperback summary of basic skills that can be packaged with the text, provides an excellent resource for students who need help with the mathematical aspects of chemistry.

Supplements Package and Companion Web Site

For the Student Study Guide by Rachael Henriques Porter is a self-study guide for students. For each chapter, the **Study Guide** includes a self-evaluation section with student exercises, a summary of chapter concepts, one or more "challenge problems," and answers and solutions to all **Study Guide** exercises.

Solutions Manual by Morris Hein, Susan Arena, and Kathy Mitchell includes answers and solutions to all end-of-chapter questions and exercises.

Math Survival Guide: Tips and Tricks for Science Students, 2nd Edition, a paperback summary of basic skills with practice exercises in every chapter.

Foundations of Chemistry in the Laboratory, 12th Edition, by Morris Hein, Judith N. Peisen, Leo R. Best, and Robert L. Miner includes 28 experiments for a laboratory program that may accompany the lecture course. Featuring updated information on waste disposal and emphasizing safe laboratory procedures, the lab manual also includes study aids and exercises.

Student Companion Web Site The Student Companion Web Site includes a number of helpful tools and materials to enhance student learning:

- **Interactive Learningware**: a step-by-step problem-solving tutorial program that guides students through over 40 interactive problems.
- The website also features **Online Quizzing** for additional practice.

For the Instructor Test Bank, by Anthony Stellato. Includes chapter tests with additional test questions, and answers to all test questions.

Computerized Test Bank. The test bank contains true-false, multiple-choice, and open-ended questions, and is available in two formats.

WileyPLUS is a powerful online tool that provides instructors with an integrated suite of resources, including an online version of the text, in one easy-to-use website. Organized around the essential activities you perform in class, *WileyPLUS* allows you to create class presentations, assign homework and quizzes for automatic grading, and track student progress.

For the Laboratory **Foundations of Chemistry in the Laboratory, 12th Edition, by Morris Hein, Judith Peisen, Leo Best, and Robert Miner**, has been completely updated and revised to reflect the most current terminology and environmental standards. Instructors can customize their own lab manual to meet the distinct needs of their laboratory by selecting from any of the 28 experiments, adding their own experiments or exercises.

Instructor's Manual for Foundations of Chemistry in the Laboratory, 12th Edition, includes valuable and important information on the management of the lab, evaluation of experiments, notes for individual experiments, a list of reagents needed, and answer keys to each experiment's report form and to all exercises.

Instructor's Companion Web Site Instructors have access to all student Web Site features and the following:

- **Digital Image Library**: images from the text are available online in JPEG format. Instructors may use these to customize their presentations and to provide additional visual support for quizzes and exams.
- **Test Bank**: contains true-false, multiple-choice, and free-response questions.
- **Power Point Lecture Slides**: created by Eugene Passer, these slides contain lecture outlines and key topics from each chapter of the text, along with supporting artwork and figures from the text.

Acknowledgments

Books are the result of a collaborative effort of many talented and dedicated people. We are grateful for the many helpful comments from colleagues and students who, over the years, have made this book possible. We hope they will continue to share their ideas for change with us, either directly or through our publisher.

We are especially thankful for the support, friendship, and constant encouragement of our spouses, Edna and Steve, who have endured many lost weekends and been patient and understanding through the long hours of this process. Their optimism and good humor have given us a sense of balance and emotional stability.

Our sincere appreciation goes to the following reviewers who were kind enough to read and give their professional comments.

For the 12th Edition:

Eric Antonelli
Ocean County College

Yelda Hangun Balkir
California University of Pennsylvania

Godfrey Barlatt
Atlantic Cape Community College

Frank Cange
Southwestern Illinois College

Jing-Yi Chin
Suffolk County Community College

Douglas Cody
Nassau Community College

Brahmadeo Dewprashad
Borough of Manhattan Community College

Saul Grunfeld
Oakland Community College

Kathleen Harter
Community College of Philadelphia

Tom Higgins
Harold Washington College

Peter Krieger
Palm Beach Community College

Juan Lozano
University of Wisconsin–Rock County

Kathy Mitchell
St. Petersburg College

Kathy Nebril
Wilbur Wright College

James Pazun
Pfeiffer University

Anthony Stellato
Suffolk County Community College

John Stenger-Smith
Cerro Coso Community College

Kaveh Zarrabi
Community College of Southern Nevada

Morris Hein and Susan Arena

BRIEF CONTENTS

1 An Introduction to Chemistry 1

2 Standards for Measurement 14

3 Elements and Compounds 47

4 Properties of Matter 65

5 Early Atomic Theory and Structure 84

6 Nomenclature of Inorganic Compounds 103

7 Quantitative Composition of Compounds 127

8 Chemical Equations 150

9 Calculations from Chemical Equations 176

10 Modern Atomic Theory and the Periodic Table 203

11 Chemical Bonds: The Formation of Compounds from Atoms 224

12 The Gaseous State of Matter 265

13 Properties of Liquids 307

14 Solutions 336

15 Acids, Bases, and Salts 373

16 Chemical Equilibrium 405

17 Oxidation–Reduction 437

18 Nuclear Chemistry 468

19 Introduction to Organic Chemistry 495

20 Introduction to Biochemistry 541

Appendixes A–1

Glossary G–1

Photo Credits PC–1

Index I–1

CONTENTS

1 An Introduction to Chemistry   **1**

1.1 Why Study Chemistry?   2
1.2 The Nature of Chemistry   2
1.3 Thinking Like a Chemist   3

CHEMISTRY IN ACTION
Clearing the Fog   4

1.4 A Scientific Approach
to Problem Solving   4
1.5 The Scientific Method   5
1.6 The Particulate Nature of Matter   6
1.7 Physical States of Matter   6
1.8 Classifying Matter   8
Review   11
Review Questions   12
Paired Exercises, Additional Exercises   12

2 Standards for Measurement   **14**

2.1 Scientific Notation   15
2.2 Measurement and Uncertainty   16
2.3 Significant Figures   18
2.4 Significant Figures in Calculations   20
2.5 The Metric System   23
2.6 Problem Solving   24

CHEMISTRY IN ACTION
Keeping Track of Units   29

2.7 Measuring Mass and Volume   29

CHEMISTRY IN ACTION
Setting Standards   33

2.8 Measurement of Temperature   33

CHEMISTRY IN ACTION
**Taking the Temperature
of Old Faithful**   36

2.9 Density   37
Review   41
Review Questions   42

Paired Exercises, Additional Exercises
Challenge Exercises   42
Answers to Practice Exercises   46

3 Elements and Compounds   **47**

3.1 Elements   48
3.2 Distribution of Elements   49
3.3 Names of the Elements   49
3.4 Symbols of the Elements   51
3.5 Introduction to the Periodic Table   52
3.6 Elements in Their Natural States   54

CHEMISTRY IN ACTION
**Using Oxygen to Revive
Damaged Art**   55

3.7 Elements That Exist as
Diatomic Molecules   55
3.8 Compounds   56

CHEMISTRY IN ACTION
Cars: Is Hydrogen the Answer?   58

3.9 Chemical Formulas   59
Review   61
Review Questions   61
Paired Exercises, Additional Exercises
Challenge Exercises   62
Answers to Practice Exercises   64

4 Properties of Matter   **65**

4.1 Properties of Substances   66

CHEMISTRY IN ACTION
Making Money   67

4.2 Physical Changes   67
4.3 Chemical Changes   68
4.4 Conservation of Mass   70
4.5 Energy   71
4.6 Heat: Quantitative Measurement   72

4.7 Energy in Chemical Changes 74

CHEMISTRY IN ACTION
Popping Popcorn 75

4.8 Conservation of Energy 75
Review 76
Review Questions 77
Paired Exercises, Additional Exercises
Challenge Exercises 77
Answers to Practice Exercises 80

PUTTING IT TOGETHER
Review for Chapters 1–4 81

5 Early Atomic Theory and Structure 84

5.1 Early Thoughts 85
5.2 Dalton's Model of the Atom 85
5.3 Composition of Compounds 86
5.4 The Nature of Electric Charge 87
5.5 Discovery of Ions 88
5.6 Subatomic Parts of the Atom 88
5.7 The Nuclear Atom 91
5.8 Isotopes of the Elements 94

CHEMISTRY IN ACTION
Isotope Detectives 95

5.9 Atomic Mass 95
Review 98
Review Questions 99
Paired Exercises, Additional Exercises
Challenge Exercises 100
Answers to Practice Exercises 102

6 Nomenclature of Inorganic Compounds 103

6.1 Common and Systematic Names 104
6.2 Elements and Ions 105

CHEMISTRY IN ACTION
What's in a Name? 106

6.3 Writing Formulas from Names
of Ionic Compounds 109
6.4 Naming Binary Compounds 111

6.5 Naming Compounds Containing
Polyatomic Ions 116
6.6 Acids 118
Review 120
Review Questions 121
Paired Exercises, Additional Exercises
Challenge Exercises 122
Answers to Practice Exercises 124

PUTTING IT TOGETHER
Review for Chapters 5–6 125

7 Quantitative Composition of Compounds 127

7.1 The Mole 128
7.2 Molar Mass of Compounds 132
7.3 Percent Composition of Compounds 136

CHEMISTRY IN ACTION
Vanishing Coins? 139

7.4 Empirical Formula versus
Molecular Formula 139
7.5 Calculating Empirical Formulas 140
7.6 Calculating the Molecular Formula
from the Empirical Formula 143
Review 145
Review Questions 145
Paired Exercises, Additional Exercises
Challenge Exercises 146
Answers to Practice Exercises 149

8 Chemical Equations 150

8.1 The Chemical Equation 151
8.2 Writing and Balancing
Chemical Equations 152
8.3 Information in a
Chemical Equation 156

CHEMISTRY IN ACTION
CO Poisoning—A Silent Killer 157

8.4 Types of Chemical Equations 159
8.5 Heat in Chemical Reactions 165
8.6 Global Warming:
The GreenhouseEffect 168

Review   170
Review Questions   171
Paired Exercises, Additional Exercises
Challenge Exercises   172
Answers to Practice Exercises   175

9 Calculations from Chemical Equations   176

9.1   A Short Review   177
9.2   Introduction to Stoichiometry   178
9.3   Mole-Mole Calculations   180
9.4   Mole-Mass Calculations   182
9.5   Mass-Mass Calculations   184

CHEMISTRY IN ACTION
A Shrinking Technology   186

9.6   Limiting-Reactant and
      Yield Calculations   186
      Review   192
      Review Questions   193
      Paired Exercises, Additional Exercises
      Challenge Exercises   193
      Answers to Practice Exercises   198

PUTTING IT TOGETHER
Review for Chapters 7–9   199

10 Modern Atomic Theory and the Periodic Table   203

10.1   A Brief History   204
10.2   Electromagnetic Radiation   204

CHEMISTRY IN ACTION
You Light Up My Life   205

10.3   The Bohr Atom   206
10.4   Energy Levels of Electrons   207

CHEMISTRY IN ACTION
Atomic Clocks   210

10.5   Atomic Structures of
       the First 18 Elements   210
10.6   Electron Structures and
       the Periodic Table   214

Review   219
Review Questions   220
Paired Exercises, Additional Exercises
Challenge Exercises   220
Answers to Practice Exercises   223

11 Chemical Bonds: The Formation of Compounds from Atoms   224

11.1   Periodic Trends in Atomic Properties   225
11.2   Lewis Structures of Atoms   228
11.3   The Ionic Bond: Transfer of Electrons
        from One Atom to Another   230
11.4   Predicting Formulas
        of Ionic Compounds   236
11.5   The Covalent Bond:
        Sharing Electrons   238
11.6   Electronegativity   240

CHEMISTRY IN ACTION
Trans-forming Fats   242

CHEMISTRY IN ACTION
Goal! A Spherical Molecule   243

11.7   Lewis Structures of Compounds   244
11.8   Complex Lewis Structures   247

CHEMISTRY IN ACTION
Cleaner Showers
through Chemistry   248

11.9   Compounds Containing
        Polyatomic Ions   249
11.10   Molecular Shape   250

CHEMISTRY IN ACTION
Strong Enough to Stop
a Bullet?   251

11.11   The Valence Shell Electron
        Pair Repulsion (VSEPR) Model   251
        Review   255
        Review Questions   257
        Paired Exercises, Additional Exercises
        Challenge Exercises   257
        Answers to Practice Exercises   261

PUTTING IT TOGETHER
Review for Chapters 10–11   262

12 The Gaseous State of Matter 265
12.1 General Properties 266
12.2 The Kinetic-Molecular Theory 266
12.3 Measurement of Pressure of Gases 268
12.4 Dependence of Pressure on Number of Molecules and Temperature 270
12.5 Boyle's Law 271
12.6 Charles' Law 275
12.7 Gay-Lussac's Law 279
12.8 Combined Gas Laws 280
12.9 Dalton's Law of Partial Pressures 283

CHEMISTRY IN ACTION
Messenger Molecules 284

12.10 Avogadro's Law 285
12.11 Mole–Mass–Volume Relationships of Gases 287
12.12 Density of Gases 288
12.13 Ideal Gas Law 290
12.14 Gas Stoichiometry 292
12.15 Real Gases 295

CHEMISTRY IN ACTION
Wetland Blanket for Global Warming 297

Review 298
Review Questions 300
Paired Exercises, Additional Exercises Challenge Exercises 301
Answers to Practice Exercises 306

13 Properties of Liquids 307
13.1 What Is a Liquid? 308
13.2 Evaporation 309
13.3 Vapor Pressure 309
13.4 Surface Tension 311
13.5 Boiling Point 312
13.6 Freezing Point or Melting Point 314
13.7 Changes of State 314
13.8 The Hydrogen Bond 316

CHEMISTRY IN ACTION
How Sweet It Is! 317

13.9 Hydrates 319

CHEMISTRY IN ACTION
Hot Ice—It's a Gas! 320

13.10 Water, a Unique Liquid 322
13.11 Water Purification 326
Review 330
Review Questions 331
Paired Exercises, Additional Exercises Challenge Exercises 332
Answers to Practice Exercises 335

14 Solutions 336
14.1 General Properties of Solutions 337

CHEMISTRY IN ACTION
Exploding Silicon 338

14.2 Solubility 338
14.3 Factors Related to Solubility 340
14.4 Rate of Dissolving Solids 344
14.5 Solutions: A Reaction Medium 345
14.6 Concentration of Solutions 346
14.7 Colligative Properties of Solutions 355

CHEMISTRY IN ACTION
The Scoop on Ice Cream 359

14.8 Osmosis and Osmotic Pressure 360
Review 362
Review Questions 363
Paired Exercises, Additional Exercises Challenge Exercises 365
Answers to Practice Exercises 368

PUTTING IT TOGETHER
Review for Chapters 12–14 369

15 Acids, Bases, and Salts 373
15.1 Acids and Bases 374
15.2 Reactions of Acids 377

15.3   Reactions of Bases   378

CHEMISTRY IN ACTION
A Cool Fizz   379

15.4   Salts   379
15.5   Electrolytes and Nonelectrolytes   380
15.6   Dissociation and Ionization of
        Electrolytes   381
15.7   Strong and Weak Electrolytes   383
15.8   Ionization of Water   386
15.9   Introduction to pH   386

CHEMISTRY IN ACTION
In the Pink—
A Sign of Corrosion   389

15.10   Neutralization   389
15.11   Writing Net Ionic Equations   392
15.12   Acid Rain   395
15.13   Colloids   396

CHEMISTRY IN ACTION
Foam Cars—
Wave of the Future?   397

15.14   Properties of Colloids   397
15.15   Applications of Colloidal Properties   398
        Review   399
        Review Questions   401
        Paired Exercises, Additional Exercises
        Challenge Exercises   401
        Answers to Practice Exercises   404

16 Chemical Equilibrium   405

16.1   Reversible Reactions   406
16.2   Rates of Reaction   407
16.3   Chemical Equilibrium   408

CHEMISTRY IN ACTION
New Ways in Fighting Cavities
and Avoiding the Drill   409

16.4   Le Châtelier's Principle   410
16.5   Effect of Concentration
        on Equilibrium   410
16.6   Effect of Volume on Equilibrium   413
16.7   Effect of Temperature
        on Equilibrium   414

16.8   Effect of Catalysts on
        Equilibrium   415
16.9   Equilibrium Constants   416
16.10   Ion Product Constant for Water   418
16.11   Ionization Constants   420
16.12   Solubility Product Constant   423
16.13   Acid–Base Properties of Salts   426
16.14   Buffer Solutions:
        The Control of pH   426

CHEMISTRY IN ACTION
Exchange of Oxygen
and Carbon Dioxide
in the Blood   427

        Review   429
        Review Questions   430
        Paired Exercises, Additional Exercises
        Challenge Exercises   431
        Answers to Practice Exercises   436

17 Oxidation-Reduction   437

17.1   Oxidation Number   438
17.2   Oxidation–Reduction   441
17.3   Balancing Oxidation–Reduction
        Equations   442
17.4   Balancing Ionic Redox Equations   446

CHEMISTRY IN ACTION
Sensitive Sunglasses   447

17.5   Activity Series of Metals   450
17.6   Electrolytic and Voltaic Cells   452

CHEMISTRY IN ACTION
Superbattery Uses
Hungry Iron Ions   457

        Review   458
        Review Questions   459
        Paired Exercises, Additional Exercises
        Challenge Exercises   460
        Answers to Practice Exercises   463

PUTTING IT TOGETHER
Review for Chapters 15–17   464

18 **Nuclear Chemistry**   **468**

18.1   Discovery of Radioactivity   469
18.2   Natural Radioactivity   470
18.3   Alpha Particles, Beta Particles,
       and Gamma Rays   473
18.4   Radioactive Disintegration Series   476
18.5   Transmutation of Elements   477
18.6   Artificial Radioactivity   478
18.7   Measurement of Radioactivity   478
18.8   Nuclear Fission   480

CHEMISTRY IN ACTION
**Does Your Food Glow
in the Dark?**   482

18.9   Nuclear Power   482
18.10   The Atomic Bomb   484
18.11   Nuclear Fusion   485
18.12   Mass-Energy Relationship
        in Nuclear Reactions   486
18.13   Transuranium Elements   487
18.14   Biological Effects of Radiation   488

CHEMISTRY IN ACTION
**A Window into Living
Organisms**   489

Review   490
Review Questions   491
Paired Exercises, Additional Exercises
Challenge Exercises   492
Answers to Practice Exercises   494

19 **Introduction to
Organic Chemistry**   **495**

19.1   The Beginnings of Organic Chemistry   496
19.2   The Carbon Atom   497
19.3   Hydrocarbons   498

CHEMISTRY IN ACTION
**Biodiesel: Today's
Alternative Fuel**   499

19.4   Alkanes   500
19.5   Structural Formulas and Isomerism   501
19.6   Naming Alkanes   504

19.7   Alkenes and Alkynes   508
19.8   Naming Alkenes and Alkynes   510
19.9   Reactions of Alkenes   511
19.10   Aromatic Hydrocarbons   512
19.11   Naming Aromatic Compounds   513
19.12   Hydrocarbon Derivatives   516
19.13   Alkyl Halides   517
19.14   Alcohols   518
19.15   Naming Alcohols   521
19.16   Ethers   522
19.17   Aldehydes and Ketones   524
19.18   Naming Aldehydes and Ketones   525
19.19   Carboxylic Acids   527
19.20   Esters   528
19.21   Polymers—Macromolecules   530

CHEMISTRY IN ACTION
Getting Clothes CO_2 Clean!   532

Review   532
Review Questions   535
Paired Exercises, Additional Exercises
Challenge Exercises   535
Answers to Practice Exercises   540

20 **Introduction
to Biochemistry**   **541**

20.1   Chemistry in Living Organisms   542
20.2   Carbohydrates   542
20.3   Lipids   547
20.4   Amino Acids and Proteins   550

CHEMISTRY IN ACTION
The Taste of Umami   557

20.5   Enzymes   557
20.6   Nucleic Acids   559
20.7   DNA and Genetics   563
       Review   564
       Review Questions   566
       Answers to Practice Exercises   568

PUTTING IT TOGETHER
Review for Chapters 18–20   569

Appendixes

I. Mathematical Review   A-1
II. Using a Scientific Calculator   A–11
III. Units of Measurement   A–16
IV. Vapor Pressure of Water
at Various Temperatures   A–17
V. Solubility Table   A–18
VI. Answers to Selected Questions
and Exercises   A–19
VII. Answers to Putting It Together
Review Exercises   A–40

Glossary   **G–1**

Photo Credits   **PC–1**

Index   **I–1**

THE WILEY BICENTENNIAL—KNOWLEDGE FOR GENERATIONS

Each generation has its unique needs and aspirations. When Charles Wiley first opened his small printing shop in lower Manhattan in 1807, it was a generation of boundless potential searching for an identity. And we were there, helping to define a new American literary tradition. Over half a century later, in the midst of the Second Industrial Revolution, it was a generation focused on building the future. Once again, we were there, supplying the critical scientific, technical, and engineering knowledge that helped frame the world. Throughout the 20th Century, and into the new millennium, nations began to reach out beyond their own borders and a new international community was born. Wiley was there, expanding its operations around the world to enable a global exchange of ideas, opinions, and know-how.

For 200 years, Wiley has been an integral part of each generation's journey, enabling the flow of information and understanding necessary to meet their needs and fulfill their aspirations. Today, bold new technologies are changing the way we live and learn. Wiley will be there, providing you the must-have knowledge you need to imagine new worlds, new possibilities, and new opportunities.

Generations come and go, but you can always count on Wiley to provide you the knowledge you need, when and where you need it!

PRESIDENT AND CHIEF EXECUTIVE OFFICER   **CHAIRMAN OF THE BOARD**

An Introduction to Chemistry

The colors, fragrances, and textures of a tulip garden are all the results of chemistry.

Chapter Outline

1.1 Why Study Chemistry?

1.2 The Nature of Chemistry

1.3 Thinking Like a Chemist

1.4 A Scientific Approach to Problem Solving

1.5 The Scientific Method

1.6 The Particulate Nature of Matter

1.7 Physical States of Matter

1.8 Classifying Matter

Do you know how the battery in your car works to start your engine? Have you ever wondered how a tiny seedling can grow into a corn stalk taller than you in just one season? Perhaps you have been mesmerized by the flames in your fireplace on a romantic evening as they change color and form. And think of your relief when you dropped a container and found that it was plastic, not glass. These phenomena are the result of chemistry that occurs all around us, all the time. Chemical changes bring us beautiful colors, warmth, light, and products to make our lives function more smoothly. Understanding, explaining, and using the diversity of materials we find around us is what chemistry is all about.

1.1 Why Study Chemistry?

A knowledge of chemistry is useful to virtually everyone—we see chemistry occurring around us every day. An understanding of chemistry is useful to doctors, health care professionals, attorneys, homemakers, businesspeople, firefighters, and environmentalists just to name a few. Even if you're not planning to work in any of these fields, chemistry is important and is used by us every day. Learning about the benefits and risks associated with chemicals will help you to be an informed citizen, able to make intelligent choices concerning the world around you. Studying chemistry teaches you to solve problems and communicate with others in an organized and logical manner. These skills will be helpful in college and throughout your career.

1.2 The Nature of Chemistry

chemistry

Key words are highlighted in bold and color in the margin to alert you to new terms defined in the text.

What is chemistry? One dictionary gives this definition: "**Chemistry** is the science of the composition, structure, properties, and reactions of matter, especially of atomic and molecular systems." Another, somewhat simpler definition is "Chemistry is the science dealing with the composition of *matter* and the changes in composition that matter undergoes." Neither of these definitions is entirely adequate. Chemistry and physics form a fundamental branch of knowledge. Chemistry is also closely related to biology, not only because living organisms are made of material substances but also because life itself is essentially a complicated system of interrelated chemical processes.

The scope of chemistry is extremely broad. It includes the whole universe and everything, animate and inanimate, in it. Chemistry is concerned with the composition and changes in the composition of matter and also with the energy and energy changes associated with matter. Through chemistry we seek to learn and to understand the general principles that govern the behavior of all matter.

The chemist, like other scientists, observes nature and attempts to understand its secrets: What makes a tulip red? Why is sugar sweet? What is occurring when iron rusts? Why is carbon monoxide poisonous? Problems such as these—some of which have been solved, some of which are still to be solved—are all part of what we call chemistry.

A chemist may interpret natural phenomena, devise experiments that reveal the composition and structure of complex substances, study methods for improving natural processes, or synthesize substances. Ultimately, the efforts of successful chemists advance the frontiers of knowledge and at the same time contribute to the well-being of humanity.

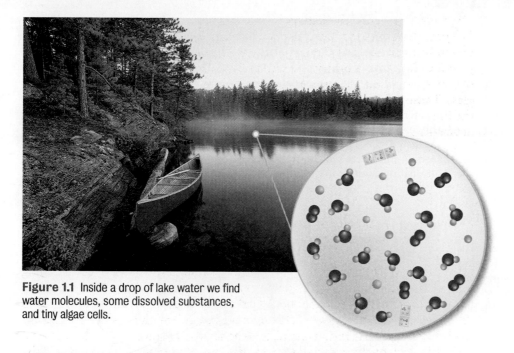

Figure 1.1 Inside a drop of lake water we find water molecules, some dissolved substances, and tiny algae cells.

1.3   Thinking Like a Chemist

Chemists take a special view of things in order to understand the nature of the chemical changes taking place. Chemists "look inside" everyday objects to see how the basic components are behaving. To understand this approach, let's consider a lake. When we view the lake from a distance, we get an overall picture of the water and shoreline. This overall view is called the *macroscopic* picture.

As we approach the lake we begin to see more details—rocks, sandy beach, plants submerged in the water, and aquatic life. We get more and more curious. What makes the rocks and sand? What kind of organisms live in the water? How do plants survive underwater? What lies hidden in the water? We can use a microscope to learn the answers to some of these questions. Within the water and the plants, we can see single cells and inside them organelles working to keep the organisms alive. For answers to other questions, we need to go even further inside the lake. A drop of lake water can itself become a mysterious and fascinating *microscopic* picture full of molecules and motion. (See Figure 1.1). A chemist looks into the world of atoms and molecules and their motions. Chemistry makes the connection between the *microscopic* world of molecules and the *macroscopic* world of everyday objects.

Think about the water in the lake. On the surface it has beauty and colors, and it gently laps the shore of the lake. What is the microscopic nature of water? It is composed of tiny molecules represented as

In this case H represents a hydrogen atom and O an oxygen atom. The water molecule is represented by H_2O since it is made up of two hydrogen atoms and one oxygen atom.

If only someone could find a way to keep that annoying fog off mirrors in the bathroom on a more permanent basis. Michael Rabner, a materials scientist at MIT, just might have figured out how to clear the fog from glass surfaces. He found his inspiration in the beautiful lotus plant. It turns out that lotus leaves repel water so well that when raindrops hit the leaf, the drops remain spherical. He began exploring the possibility of trying to make a coating that did the same thing. Fogging occurs when water drops from the air condense on a cool surface. The drops scatter light, creating the fogging. The majority of antifogging sprays are polymers that flatten the drops so they do not scatter light. Unfortunately, they wear off the surface quickly.

Rubner and his colleagues found a nanocoating (a very thin coating) that had the opposite effect of the lotus leaf. Their coating is extremely water loving, or superhydrophilic. It is composed of tiny, hydrophilic glass particles that are packed together very irregularly. This leaves tiny pockets between the particles that can fill with water, spreading the droplets into a film that doesn't scatter light.

An additional bonus of the glass and air coating is that it prevents glare. Nearly 100% of light travels through this nanocoat, compared to 92% on untreated glass. The coating is made from 7-nanometer layers of polymer alternating with layers of glass particles. The layers are constructed so that the glass particles don't pack together well, producing a Swiss cheese structure. To increase the durability of the coating, it is heated to 500°C to fuse the glass particles and disintegrate the polymer. This makes the coating scratch-resistant and also makes the antifog property last for more than a year.

A lotus flower seen through nanocoated glass.

1.4 A Scientific Approach to Problem Solving

One of the most common and important things we do every day is to solve problems. For example,

- You have two exams and a laboratory report due on Monday. How should you divide your time?
- You leave for school and learn from the radio that there is a big accident on the freeway. What is your fastest alternate route to avoid the traffic problem?
- You need to buy groceries, mail some packages, attend your child's soccer game, and pick up the dry cleaning. What is the most efficient sequence of events?

We all face these kind of problems and decisions. A logical approach can be useful for solving daily problems:

1. Define the problem. We first need to recognize we have a problem and state it clearly, including all the known information. When we do this in science, we call it *making an observation*.
2. Propose possible solutions to the problem. In science this is called *making a hypothesis*.

3. Decide which is the best way to proceed or solve the problem. In daily life we use our memory of past experiences to help us. In the world of science we *perform an experiment*.

Using a scientific approach to problem solving is worthwhile. It helps in all parts of your life whether you plan to be a scientist, doctor, businessperson, or writer.

1.5   The Scientific Method

Chemists work together and also with other scientists to solve problems. As scientists conduct studies they ask many questions, and their questions often lead in directions that are not part of the original problem. The amazing developments from chemistry and technology usually involve what we call the **scientific method**, which can generally be described as follows:

1. **Collect the facts or data** that are relevant to the problem or question at hand. This is usually done by planned experimentation. The data are then analyzed to find trends or regularities that are pertinent to the problem.
2. **Formulate a hypothesis** that will account for the data and that can be tested by further experimentation.
3. **Plan and do additional experiments to test the hypothesis**.
4. **Modify the hypothesis** as necessary so that it is compatible with all the pertinent data.

Confusion sometimes arises regarding the exact meanings of the words *hypothesis*, *theory*, and *law*. A **hypothesis** is a tentative explanation of certain facts that provides a basis for further experimentation. A well-established hypothesis is often called a **theory** or model. Thus a theory is an explanation of the general principles of certain phenomena with considerable evidence or facts to support it. Hypotheses and theories explain natural phenomena, whereas **scientific laws** are simple statements of natural phenomena to which no exceptions are known under the given conditions.

These four steps are a broad outline of the general procedure that is followed in most scientific work, but they are not a "recipe" for doing chemistry or any other science (Figure 1.2). Chemistry is an experimental science, however, and much of its progress has been due to application of the scientific method through systematic research.

Scientists employ the scientific method every day in their laboratory work.

scientific method

hypothesis

theory

scientific laws

Chemists continue to explore the effects of nicotine on the smoker in order to help those who want to quit smoking.

Figure 1.2
The scientific method.

We study many theories and laws in chemistry; this makes our task as students easier because theories and laws summarize important aspects of the sciences. Certain theories and models advanced by great scientists in the past have since been substantially altered and modified. Such changes do not mean that the discoveries of the past are any less significant. Modification of existing theories and models in the light of new experimental evidence is essential to the growth and evolution of scientific knowledge. Science is dynamic.

1.6   The Particulate Nature of Matter

matter

The entire universe consists of matter and energy. Every day we come into contact with countless kinds of matter. Air, food, water, rocks, soil, glass, and this book are all different types of matter. Broadly defined, **matter** is *anything* that has mass and occupies space.

Matter may be quite invisible. For example, if an apparently empty test tube is submerged mouth downward in a beaker of water, the water rises only slightly into the tube. The water cannot rise further because the tube is filled with invisible matter: air (see Figure 1.3).

To the macroscopic eye, matter appears to be continuous and unbroken. We are impressed by the great diversity of matter. Given its many forms, it is difficult to believe that on a microscopic level all of matter is composed of discrete, tiny, fundamental particles called *atoms*. It is truly amazing to understand that the fundamental particles in ice cream are very similar to the particles in air that we breathe. Matter is actually discontinuous and is composed of discrete, tiny particles called *atoms*.

Figure 1.3 An apparently empty test tube is submerged, mouth downward, in water. Only a small volume of water rises into the tube, which is actually filled with invisible matter—air.

1.7   Physical States of Matter

solid

Matter exists in three physical states: solid, liquid, and gas (see Figure 1.4). A **solid** has a definite shape and volume, with particles that cling rigidly to one another. The shape of a solid can be independent of its container. In Figure 1.4a we see water in its solid form. Another example, a crystal of sulfur, has the same shape and volume whether it is placed in a beaker or simply laid on a glass plate.

Most commonly occurring solids, such as salt, sugar, quartz, and metals, are *crystalline*. The particles that form crystalline materials exist in regular, repeating, three-dimensional, geometric patterns (see Figure 1.5). Some solids such as plastics, glass, and gels do not have any regular, internal geometric pattern. Such

amorphous
liquid

solids are called **amorphous** solids. (*Amorphous* means "without shape or form").

A **liquid** has a definite volume but not a definite shape, with particles that stick firmly but not rigidly. Although the particles are held together by strong attractive forces and are in close contact with one another, they are able to move freely. Particle mobility gives a liquid fluidity and causes it to take the shape of the container in which it is stored. Note how water looks as a liquid in Figure 1.4b.

gas

A **gas** has indefinite volume and no fixed shape, with particles that move independently of one another. Particles in the gaseous state have gained enough energy to overcome the attractive forces that held them together as liquids or solids. A gas presses continuously in all directions on the walls of any container.

Solid (Ice)   Liquid (Water)   Gas (Steam)
(a)   (b)   (c)

Figure 1.4 The three states of matter. (a) Solid—water molecules are held together rigidly and are very close to each other. (b) Liquid—water molecules are close together but are free to move around and slide over each other. (c) Gas—water molecules are far apart and move freely and randomly.

Because of this quality, a gas completely fills a container. The particles of a gas are relatively far apart compared with those of solids and liquids. The actual volume of the gas particles is very small compared with the volume of the space occupied by the gas. Observe the large space between the water molecules in Figure 1.4c compared to ice and liquid water. A gas therefore may be compressed into a very small volume or expanded almost indefinitely. Liquids cannot be compressed to any great extent, and solids are even less compressible than liquids.

Figure 1.5 A large crystal of table salt. A salt crystal is composed of a three-dimensional array of particles.

If a bottle of ammonia solution is opened in one corner of the laboratory, we can soon smell its familiar odor in all parts of the room. The ammonia gas escaping from the solution demonstrates that gaseous particles move freely and rapidly and tend to permeate the entire area into which they are released.

Although matter is discontinuous, attractive forces exist that hold the particles together and give matter its appearance of continuity. These attractive forces are strongest in solids, giving them rigidity; they are weaker in liquids, but still strong enough to hold liquids to definite volumes. In gases, the attractive forces are so weak that the particles of a gas are practically independent of one another. Table 1.1 lists common materials that exist as solids, liquids, and gases. Table 1.2 compares the properties of solids, liquids, and gases.

Table 1.1 Common Materials in the Solid, Liquid, and Gaseous States of Matter

Solids	Liquids	Gases
Aluminum	Alcohol	Acetylene
Copper	Blood	Air
Gold	Gasoline	Butane
Polyethylene	Honey	Carbon dioxide
Salt	Mercury	Chlorine
Sand	Oil	Helium
Steel	Vinegar	Methane
Sulfur	Water	Oxygen

Table 1.2 Physical Properties of Solids, Liquids, and Gases

State	Shape	Volume	Particles	Compressibility
Solid	Definite	Definite	Rigidly clinging; tightly packed	Very slight
Liquid	Indefinite	Definite	Mobile; adhering	Slight
Gas	Indefinite	Indefinite	Independent of each other and relatively far apart	High

1.8 Classifying Matter

substance

The term *matter* refers to all materials that make up the universe. Many thousands of distinct kinds of matter exist. A **substance** is a particular kind of matter with a definite, fixed composition. Sometimes known as *pure substances*, substances are either elements or compounds. Familiar examples of elements are copper, gold, and oxygen. Familiar compounds are salt, sugar, and water. We'll discuss elements and compounds in more detail in Chapter 3.

(a)   (b)

(a) Water is the liquid in the beaker, and the white solid in the spoon is sugar.
(b) Sugar can be dissolved in the water to produce a solution.

We classify a sample of matter as either *homogeneous* or *heterogeneous* by examining it. **Homogeneous** matter is uniform in appearance and has the same properties throughout. Matter consisting of two or more physically distinct phases is **heterogeneous**. A **phase** is a homogeneous part of a system separated from other parts by physical boundaries. A **system** is simply the body of matter under consideration. Whenever we have a system in which visible boundaries exist between the parts or components, that system has more than one phase and is heterogeneous. It does not matter whether these components are in the solid, liquid, or gaseous states.

homogeneous

heterogeneous
phase
system

A pure substance may exist as different phases in a heterogeneous system. Ice floating in water, for example, is a two-phase system made up of solid water and liquid water. The water in each phase is homogeneous in composition, but because two phases are present, the system is heterogeneous.

A **mixture** is a material containing two or more substances and can be either heterogeneous or homogeneous. Mixtures are variable in composition. If we add a spoonful of sugar to a glass of water, a heterogeneous mixture is formed immediately. The two phases are a solid (sugar) and a liquid (water). But upon stirring, the sugar dissolves to form a homogeneous mixture or solution. Both substances are still present: All parts of the solution are sweet and wet. The proportions of sugar and water can be varied simply by adding more sugar and stirring to dissolve. Solutions do not have to be liquid. For example, air is a homogeneous mixture of gases. Solid solutions also exist. Brass is a homogeneous solution of copper and zinc.

mixture

Many substances do not form homogeneous mixtures. If we mix sugar and fine white sand, a heterogeneous mixture is formed. Careful examination may be needed to decide that the mixture is heterogeneous because the two phases (sugar and sand) are both white solids. Ordinary matter exists mostly as mixtures.

Flowcharts can help you to visualize the connections between concepts.

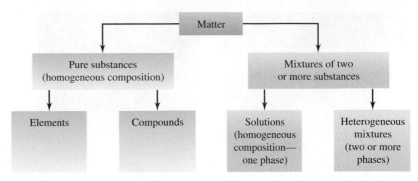

Figure 1.6 Classification of matter. A pure substance is always homogeneous in composition, whereas a mixture always contains two or more substances and may be either homogeneous or heterogeneous.

If we examine soil, granite, iron ore, or other naturally occurring mineral deposits, we find them to be heterogeneous mixtures. Figure 1.6 illustrates the relationships of substances and mixtures.

Distinguishing Mixtures from Pure Substances

Single substances—elements or compounds—seldom occur naturally in a pure state. Air is a mixture of gases; seawater is a mixture of a variety of dissolved minerals; ordinary soil is a complex mixture of minerals and various organic materials.

How is a mixture distinguished from a pure substance? A mixture always contains two or more substances that can be present in varying concentrations. Let's consider two examples.

Homogeneous Mixture Homogeneous mixtures (solutions) containing either 5% or 10% salt in water can be prepared simply by mixing the correct amounts of salt and water. These mixtures can be separated by boiling away the water, leaving the salt as a residue.

Heterogeneous Mixture The composition of a heterogeneous mixture of sulfur crystals and iron filings can be varied by merely blending in either more sulfur or more iron filings. This mixture can be separated physically by using a magnet to attract the iron.

Granite is a heterogeneous mixture.

(a)   (b)

(a) When iron and sulfur exist as pure substances, only the iron is attracted to a magnet.
(b) A mixture of iron and sulfur can be separated by using the difference in magnetic attraction.

Chapter 1 Review

1.1 Why Study Chemistry?

- Chemistry is important to everyone because chemistry occurs all around us in our daily lives.

1.2 The Nature of Chemistry

KEY TERM

Chemistry

- Chemistry is the science dealing with matter and the changes in composition that matter undergoes.
- Chemists seek to understand the general principles governing the behavior of all matter.

1.3 Thinking Like a Chemist

- Chemistry "looks inside" ordinary objects to study how their components behave.
- Chemistry connects the macroscopic and microscopic worlds.

1.4 A Scientific Approach to Problem Solving

- Scientific thinking helps us solve problems in our daily lives.
- General steps for solving problems include:
  - Defining the problem
  - Proposing possible solutions
  - Solving the problem

1.5 The Scientific Method

KEY TERMS

Scientific method
Hypothesis
Theory
Scientific laws

- The scientific method is a procedure for processing information in which we:
  - Collect the facts
  - Formulate a hypothesis
  - Plan and do experiments
  - Modify the hypothesis if necessary

1.6 The Particulate Nature of Matter

KEY TERM

Matter

- Matter is anything with the following two characteristics:
  - Has mass
  - Occupies space
- On the macroscopic level matter is continuous.
- On the microscopic level matter is discontinuous and composed of atoms.

1.7 Physical States of Matter

KEY TERMS

Solid
Amorphous
Liquid
Gas

- Solid—rigid substance with a definite shape
- Liquid—fluid substance with a definite volume that takes the shape of its container
- Gas—takes the shape and volume of its container

1.8 Classifying Matter

KEY TERMS

Substance
Homogeneous
Heterogeneous
Phase
System
Mixture

- Matter can be classified as a pure substance or a mixture.
- A mixture has variable composition:
  - Homogeneous mixtures have the same properties throughout.
  - Heterogeneous mixtures have different properties in different parts of the system.
- A pure substance always has the same composition.
  - There are two types of pure substances
    - Elements
    - Compounds

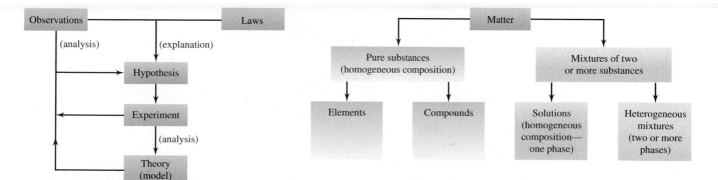

Review Questions

All questions with blue numbers have answers in the appendix of the text.

1. Explain the difference between
   (a) a hypothesis and a theory
   (b) a theory and a scientific law
2. Define a phase.
3. How many phases are present in the graduated cylinder?

4. Consider each of the following statements and determine whether it represents an observation, a hypothesis, a theory, or a scientific law:
   (a) The battery in my watch must be dead since it is no longer keeping time.
   (b) My computer must have a virus since it is not working properly.
   (c) The air feels cool.
   (d) The candle burns more brightly in pure oxygen than in air because oxygen supports combustion.
   (e) My sister wears red quite often.
   (f) A pure substance has a definite, fixed composition.

5. List four different substances in each of the three states of matter.
6. In terms of the properties of the ultimate particles of a substance, explain
   (a) why a solid has a definite shape but a liquid does not
   (b) why a liquid has a definite volume but a gas does not
   (c) why a gas can be compressed rather easily but a solid cannot be compressed appreciably
7. What evidence can you find in Figure 1.3 that gases occupy space?
8. Which liquids listed in Table 1.1 are not mixtures?
9. Which of the gases listed in Table 1.1 are not pure substances?
10. When the stopper is removed from a partly filled bottle containing solid and liquid acetic acid at 16.7°C, a strong vinegarlike odor is noticeable immediately. How many acetic acid phases must be present in the bottle? Explain.
11. Is the system enclosed in the bottle in Question 10 homogeneous or heterogeneous? Explain.
12. Is a system that contains only one substance necessarily homogeneous? Explain.
13. Is a system that contains two or more substances necessarily heterogeneous? Explain.
14. Distinguish between homogeneous and heterogeneous mixtures.
15. Which of the following are pure substances?
    (a) sugar
    (b) sand
    (c) gold
    (d) maple syrup
    (e) eggs

Paired Exercises

All exercises with blue numbers have answers in the appendix of the text.

1. Refer to the illustration and determine which state(s) of matter are present.

2. Refer to the illustration and determine which states(s) of matter are present.

3. Look at the photo and determine whether it represents a homogeneous or heterogeneous mixture.

4. Look at the maple leaf below and determine whether it represents a homogeneous or heterogeneous mixture.

5. For each of the following mixtures, state whether it is homogeneous or heterogeneous:
(a) tap water
(b) carbonated beverage
(c) oil and vinegar salad dressing
(d) people in a football stadium

6. For each of the following mixtures, state whether it is homogeneous or heterogeneous:
(a) stainless steel
(b) motor oil
(c) soil
(d) a tree

Additional Exercises

All exercises with blue numbers have answers in the appendix of the text.

7. At home, check your kitchen and bathroom cabinets for five different substances, then read the labels and list the first ingredient of each.

8. During the first week of a new semester, consider that you have enrolled in five different classes, each of which meets for 3 hours per week. For every 1 hour that is spent in class, a minimum of 1 hour is required outside of class to complete assignments and study for exams. You also work 20 hours per week, and it takes you 1 hour to drive to the job site and back home. On Friday nights, you socialize with your friends. You are fairly certain that you will be able to successfully complete the semester with good grades. Show how the steps in the scientific method can help you predict the outcome of the semester.

CHAPTER 2

Standards for Measurement

Different-sized and -calibrated containers are needed to measure quantities of liquids accurately.

Chapter Outline

2.1 Scientific Notation

2.2 Measurement and Uncertainty

2.3 Significant Figures

2.4 Significant Figures in Calculations

2.5 The Metric System

2.6 Problem Solving

2.7 Measuring Mass and Volume

2.8 Measurement of Temperature

2.9 Density

Doing an experiment in chemistry is very much like cooking a meal in the kitchen. It's important to know the ingredients and the amounts of each in order to have a tasty product. Working on your car requires specific tools, in exact sizes. Buying new carpeting or draperies is an exercise in precise and accurate measurement for a good fit. A small difference in the concentration or amount of medication a pharmacist gives you may have significant effects on your well-being. As we saw in Chapter 1, observation is an important part of the scientific process. Observations can be *qualitative* (the substance is a blue solid) or *quantitative* (the mass of the substance is 4.7 grams). A quantitative observation is called a **measurement**. Both a number and a unit are required for a measurement. For example, at home in your kitchen measuring 3 flour is not possible. We need to know the unit on the 3. It is cups, grams, or tablespoons? A measurement of 3 cups tells us both the amount and the size of the measurement. In this chapter we will discuss measurements and the rules for calculations using these measurements.

measurement

2.1 Scientific Notation

Scientists often use numbers that are very large or very small in measurements. For example, the Earth's age is estimated to be about 4,500,000 (4.5 billion) years. Numbers like these are bulky to write, so to make them more compact scientists use powers of 10. Writing a number as the product of a number between 1 and 10 multiplied by 10 raised to some power is called **scientific notation.**

scientific notation

    To learn how to write a number in scientific notation, let's consider the number 2468. To write this number in scientific notation:

1. Move the decimal point in the original number so that it is located after the first nonzero digit.

   $2468 \rightarrow 2.468$ (decimal moves three places to the left)

2. Multiply this new number by 10 raised to the proper exponent (power). The proper exponent is equal to the number of places that the decimal point was moved.

   2.468×10^3

3. The sign on the exponent indicates the direction the decimal was moved.

   moved right $\rightarrow$ negative exponent

   moved left $\rightarrow$ positive exponent

Examples show you problem-solving techniques in a step-by-step form. Study each one and then try the Practice Exercises.

Write 5283 in scientific notation.

5283.   Place the decimal between the 5 and the 2. Since the decimal
3   was moved three places to the left, the power of 10 will be 3 and the number 5.283 is multiplied by 10^3.

5.283×10^3

Example 2.1

SOLUTION

Scientific notation is a useful way to write very large numbers such as the distance between Earth and the moon, or very small numbers such as the length of these *E. coli* bacteria (shown here as a colored scanning electron micrograph ×30,000).

Example 2.2   Write 4,500,000,000 in scientific notation (two significant figures).

SOLUTION   4 500 000 000.   Place the decimal between the 4 and the 5. Since the decimal
                       9   was moved nine places to the left, the power of 10 will be 9
                                      and the number 4.5 is multiplied by 10^9.

4.5×10^9

Example 2.3   Write 0.000123 in scientific notation.

SOLUTION   0.000123   Place the decimal between the 1 and the 2. Since the decimal was
                    4   moved four places to the right, the power of 10 will be -4 and the
                             number 1.23 is multiplied by 10^{-4}.

1.23×10^{-4}

Practice 2.1

Write the following numbers in scientific notation:

(a) 1200 (four digits)   (c) 0.0468
(b) 6,600,000 (two digits)   (d) 0.00003

2.2   Measurement and Uncertainty

To understand chemistry, it is necessary to set up and solve problems. Problem solving requires an understanding of the mathematical operations used to manipulate numbers. Measurements are made in an experiment, and chemists use these data to calculate the extent of the physical and chemical changes occurring in the substances that are being studied. By appropriate calculations an experiment's results can be compared with those of other experiments and summarized in ways that are meaningful.

A measurement is expressed by a numerical value together with a unit of that measurement. For example,

$$\overset{\displaystyle\frown \text{numerical value} \frown}{70.0 \text{ kilograms} = 1\underset{\underset{\displaystyle\smile\text{— unit —}\smile}{}}{54} \text{ pounds}}$$

Whenever a measurement is made with an instrument such as a thermometer or ruler, an estimate is required. We can illustrate this by measuring temperature. Suppose we measure temperature on a thermometer calibrated in degrees and observe that the mercury stops between 21 and 22 (see Figure 2.1a). We then know that the temperature is at least 21 degrees and is less than 22 degrees. To express the temperature with greater precision, we estimate that the mercury is about two-tenths the distance between 21 and 22. The temperature is therefore 21.2 degrees. The last digit (2) has some uncertainty because it is an estimated value. Because the last digit has some uncertainty (we made a visual estimate), it may be different when another person makes the same measurement. If three more people make this same reading, the results might be

Person	Measurement
1	21.2
2	21.3
3	21.1
4	21.2

Notice that the first two digits of the measurements did not change (they are *certain*). The last digit in these measurements is *uncertain*. The custom in science is to record all of the certain digits and the first uncertain number.

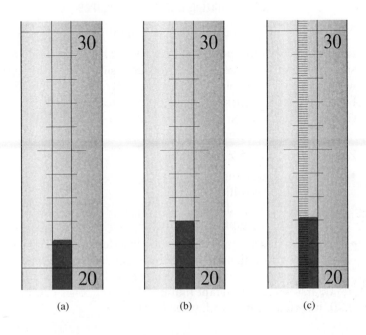

(a)   (b)   (c)

Figure 2.1
Measuring temperature (°C) with various degrees of precision.

Numbers obtained from a measurement are never exact values. They always have some degree of uncertainty due to the limitations of the measuring instrument and the skill of the individual making the measurement. The numerical value recorded for a measurement should give some indication of its reliability (precision). To express maximum precision, this number should contain all the digits that are known plus one digit that is estimated. This last estimated digit introduces some uncertainty. Because of this uncertainty, every number that expresses a measurement can have only a limited number of digits. These

significant figures   digits, used to express a measured quantity, are known as **significant figures**.

Now let's return to our temperature measurements. In Figure 2.1a the temperature is recorded as 21.2 degrees and is said to have three significant figures. If the mercury stopped exactly on the 22 (Figure 2.1b), the temperature would be recorded as 22.0 degrees. The zero is used to indicate that the temperature was estimated to a precision of one-tenth degree. Finally, look at Figure 2.1c. On this thermometer, the temperature is recorded as 22.11°C (four significant figures). Since the thermometer is calibrated to tenths of a degree, the first estimated digit is the hundredths.

2.3   Significant Figures

Since all measurements involve uncertainty, we must use the proper number of significant figures in each measurement. In chemistry we frequently do calculations involving measurements, so we must understand what happens when we do arithmetic on numbers containing uncertainties. We'll learn several rules for doing these calculations and figuring out how many significant figures to have in the result. You will need to follow these rules throughout the calculations in this text.

The first thing we need to learn is how to determine how many significant figures are in a number.

Rules for Counting Significant Figures

Rule 1   *Nonzero digits.* All nonzero digits are significant.

Rule 2   *Exact numbers.* Some numbers are exact and have an infinite number of significant figures. Exact numbers occur in simple counting operations; when you count 25 dollars, you have exactly 25 dollars. Defined numbers, such as 12 inches in 1 foot, 60 minutes in 1 hour, and 100 centimeters in 1 meter, are also considered to be exact numbers. Exact numbers have no uncertainty.

Rules for significant figures should be memorized for use throughout the text.

Rule 3   *Zeros.* A zero is *significant* when it is

- between nonzero digits:
  205 has three significant figures
  2.05 has three significant figures
  61.09 has four significant figures

- at the end of a number that includes a decimal point:
  0.500 has three significant figures (5, 0, 0)
  25.160 has five significant figures (2, 5, 1, 6, 0)
  3.00 has three significant figures (3, 0, 0)
  20. has two significant figures (2, 0)

A zero is *not significant* when it is

- before the first nonzero digit. These zeros are used to locate a decimal point:
  0.0025 has two significant figures (2, 5)
  0.0108 has three significant figures (1, 0, 8)
- at the end of a number without a decimal point:
  1000 has one significant figure (1)
  590 has two significant figures (5, 9)

One way of indicating that these zeros are significant is to write the number using scientific notation. Thus if the value 1000 has been determined to four significant figures, it is written as 1.000×10^3. If 590 has only two significant figures, it is written as 5.9×10^2.

Practice 2.2 _____

How many significant figures are in each of these measurements?

(a) 4.5 inches
(b) 3.025 feet
(c) 125.0 meters
(d) 0.001 mile
(e) 25.0 grams
(f) 12.20 liters
(g) 100,000 people
(h) 205 birds

Answers to Practice Exercises are found at the end of each chapter.

Rounding Off Numbers

When we do calculations on a calculator, we often obtain answers that have more digits than are justified. It is therefore necessary to drop the excess digits in order to express the answer with the proper number of significant figures. When digits are dropped from a number, the value of the last digit retained is determined by a process known as **rounding off numbers**. Two rules will be used in this book for rounding off numbers:

rounding off numbers

Rules for Rounding Off

Rule 1 When the first digit after those you want to retain is 4 or less, that digit and all others to its right are dropped. The last digit retained is not changed. The following examples are rounded off to four digits:

Not all schools use the same rules for rounding. Check with your instructor for variations in these rules.

$$74.693 = 74.69 \qquad 1.00629 = 1.006$$
This digit is dropped.   These two digits are dropped.

Rule 2 When the first digit after those you want to retain is 5 or greater, that digit and all others to the right are dropped and the last digit retained is increased by one. These examples are rounded off to four digits:

$$1.026868 = 1.027 \qquad 18.02500 = 18.03$$
These three digits are dropped.   These three digits are dropped.
This digit is changed to 7.   This digit is changed to 3.

$$12.899 = 12.90$$
This digit is dropped.
These two digits are changed to 90.

Practice 2.3

Round off these numbers to the number of significant figures indicated:

(a) 42.246 (four)   (d) 0.08965 (two)
(b) 88.015 (three)   (e) 225.3 (three)
(c) 0.08965 (three) (f) 14.150 (three)

2.4 Significant Figures in Calculations

The results of a calculation based on measurements cannot be more precise than the least precise measurement.

Use your calculator to check your work in the examples. Compare your results to be sure you understand the mathematics.

Multiplication or Division

In calculations involving multiplication or division, the answer must contain the same number of significant figures as in the measurement that has the least number of significant figures. Consider the following examples:

Example 2.4 $(190.6)(2.3) = 438.38$

SOLUTION   The value 438.38 was obtained with a calculator. The answer should have two significant figures because 2.3, the number with the fewest significant figures, has only two significant figures.

```
              ┌──── Round off this digit to 4.
      │   ┌──── Drop these three digits.
      ↓  ┌─┐
   438.38
   ↑
   └──────── Move the decimal 2 places to the left to express in scientific notation.
```

The correct answer is 440 or 4.4×10^2.

Example 2.5 $\dfrac{(13.59)(6.3)}{12} = 7.13475$

SOLUTION   The value 7.13475 was obtained with a calculator. The answer should contain two significant figures because 6.3 and 12 each have only two significant figures.

```
           ┌──── Drop these four digits.
       ↓  ┌───┐
   7.13475
      └──────── This digit remains the same.
```

The correct answer is 7.1.

Practice 2.4

(a) (134 in.)(25 in.) = ?

(b) $\dfrac{213 \text{ miles}}{4.20 \text{ hours}}$ = ?

(c) $\dfrac{(2.2)(273)}{760}$ = ?

(d) 0.0321×42 = ?

(e) $\dfrac{0.0450}{0.00220}$ = ?

(f) $\dfrac{1.280}{0.345}$ = ?

Addition or Subtraction

The results of an addition or a subtraction must be expressed to the same precision as the least precise measurement. This means the result must be rounded to the same number of decimal places as the value with the fewest decimal places (blue line in examples).

Add 125.17, 129, and 52.2.

```
125.|17
129.|
 52.|2
306.|37
```

The number with the least precision is 129. Therefore the answer is rounded off to the nearest unit: 306.

Example 2.6

SOLUTION

Subtract 14.1 from 132.56.

```
  132.5|6
−  14.1|
  118.4|6
```

14.1 is the number with the least precision. Therefore the answer is rounded off to the nearest tenth: 118.5.

Example 2.7

SOLUTION

Subtract 120 from 1587.

```
  158|7
−  12|0
  146|7
```

120 is the number with least precision. The zero is not considered significant; therefore the answer must be rounded to the nearest ten: 1470 or 1.47×10^3.

Example 2.8

SOLUTION

Example 2.9   Add 5672 and 0.00063.

SOLUTION

$$
\begin{array}{r}
5672 \\
+\quad 0.\,|\,00063 \\
\hline
5672.\,|\,00063
\end{array}
$$

Note: When a very small number is added to a large number, the result is simply the original number.

The number with least precision is 5672. So the answer is rounded off to the nearest unit: 5672.

Example 2.10 $\dfrac{1.039 - 1.020}{1.039} = ?$

SOLUTION $\dfrac{1.039 - 1.020}{1.039} = 0.018286814$

The value 0.018286814 was obtained with a calculator. When the subtraction in the numerator is done,

$$1.039 - 1.020 = 0.019$$

the number of significant figures changes from four to two. Therefore the answer should contain two significant figures after the division is carried out:

Drop these six digits.

0.018286814

This digit remains the same.

The correct answer is 0.018, or 1.8×10^{-2}.

Practice 2.5

How many significant figures should the answer in each of these calculations contain?

(a) (14.0)(5.2)

(b) (0.1682)(8.2)

(c) $\dfrac{(160)(33)}{4}$

(d) 8.2 + 0.125

(e) 119.1 − 3.44

(f) $\dfrac{94.5}{1.2}$

(g) 1200 + 6.34

(h) 1.6 + 23 − 0.005

If you need to brush up on your math skills, refer to the "Mathematical Review" in Appendix I.

Additional material on mathematical operations is given in Appendix I, "Mathematical Review." Study any portions that are not familiar to you. You may need to do this at various times during the course when additional knowledge of mathematical operations is required.

2.5   The Metric System

The **metric system**, or **International System** (**SI**, from *Système International*), is a decimal system of units for measurements of mass, length, time, and other physical quantities. Built around a set of standard units, the metric system uses factors of 10 to express larger or smaller numbers of these units. To express quantities that are larger or smaller than the standard units, prefixes are added to the names of the units. These prefixes represent multiples of 10, making the metric system a decimal system of measurements. Table 2.1 shows the names, symbols, and numerical values of the common prefixes. Some examples of the more commonly used prefixes are

metric system or SI

$$1 \, kilo\text{meter} = 1000 \text{ meters}$$
$$1 \, kilo\text{gram} = 1000 \text{ grams}$$
$$1 \, milli\text{meter} = 0.001 \text{ meter}$$
$$1 \, micro\text{second} = 0.000001 \text{ second}$$

The prefixes most commonly used in chemistry are shown in bold.

Table 2.1 Common Prefixes and Numerical Values for SI Units

Prefix	Symbol	Numerical value	Power of 10 equivalent
giga	G	1,000,000,000	10^9
mega	M	1,000,000	10^6
kilo	k	1,000	10^3
hecto	h	100	10^2
deka	da	10	10^1
—	—	1	10^0
deci	d	0.1	10^{-1}
centi	c	0.01	10^{-2}
milli	m	0.001	10^{-3}
micro	μ	0.000001	10^{-6}
nano	n	0.000000001	10^{-9}
pico	p	0.000000000001	10^{-12}
femto	f	0.000000000000001	10^{-15}

The common standard units in the International System, their abbreviations, and the quantities they measure are given in Table 2.2. Other units are derived from these units. The metric system, or International System, is currently used by most of the countries in the world, not only in scientific and technical work, but also in commerce and industry.

Most products today list both systems of measurement on their labels.

Table 2.2 International System's Standard Units of Measurement

Quantity	Name of unit	Abbreviation
Length	meter	m
Mass	kilogram	kg
Temperature	kelvin	K
Time	second	s
Amount of substance	mole	mol
Electric current	ampere	A
Luminous intensity	candela	cd

Measurement of Length

meter (m)

The standard unit of length in the metric system is the **meter (m)**. When the metric system was first introduced in the 1790s, the meter was defined as one ten-millionth of the distance from the equator to the North Pole, measured along the meridian passing through Dunkirk, France. The latest definition describes a meter as the distance that light travels in a vacuum during 1/299,792,458 of a second.

A meter is 39.37 inches, a little longer than 1 yard. One meter equals 10 decimeters, 100 centimeters, or 1000 millimeters (see Figure 2.2). A kilometer contains 1000 meters. Table 2.3 shows the relationships of these units.

The nanometer (10^{-9} m) is used extensively in expressing the wavelength of light, as well as in atomic dimensions. See the inside back cover for a complete table of common conversions.

Common length relationships:

$1\,m = 10^6\,\mu m = 10^{10}\,\text{Å}$
$\qquad = 100\,cm = 1000\,mm$
$1\,cm = 10\,mm = 0.01\,m$
$1\,in. = 2.54\,cm$
$1\,mile = 1.609\,km$

See inside back cover for a table of conversions.

Table 2.3 Metric Units of Length

Unit	Abbreviation	Meter equivalent	Exponential equivalent
kilometer	km	1000 m	10^3 m
meter	m	1 m	10^0 m
decimeter	dm	0.1 m	10^{-1} m
centimeter	cm	0.01 m	10^{-2} m
millimeter	mm	0.001 m	10^{-3} m
micrometer	μm	0.000001 m	10^{-6} m
nanometer	nm	0.000000001 m	10^{-9} m
angstrom	Å	0.0000000001 m	10^{-10} m

Figure 2.2
Comparison of the metric and American systems of length measurement: 2.54 cm = 1 in.

2.6 Problem Solving

Many chemical principles can be illustrated mathematically. Learning how to set up and solve numerical problems in a systematic fashion is *essential* in the study of chemistry.

See Appendix II for help in using a scientific calculator.

Usually a problem can be solved by several methods. But in all methods it is best to use a systematic, orderly approach. The *dimensional analysis method* is emphasized in this book because it

- provides a systematic, straightforward way to set up problems.
- gives a clear understanding of the principles involved.
- trains you to organize and evaluate data.
- helps to identify errors, since unwanted units are not eliminated if the setup of the problem is incorrect.

The basic steps for solving problems are

Steps for solving problems are highlighted in color for easy reference.

Step 1	**Read.** Read the problem carefully. Determine what is known and what is to be solved for, and write them down. It is important to label *all* factors and measurements with the proper units.
Step 2	**Plan.** Determine which principles are involved and which unit relationships are needed to solve the problem. You may need to refer to tables for needed data.
Step 3	**Set up.** Set up the problem in a neat, organized, and logical fashion, making sure that unwanted units cancel. Use sample problems in the text as guides for setting up the problem.
Step 4	**Calculate.** Proceed with the necessary mathematical operations. Make certain that your answer contains the proper number of significant figures.
Step 5	**Check.** Check the answer to see if it is reasonable.

A few more words about problem solving: Don't allow any formal method of problem solving to limit your use of common sense and intuition. If a problem is clear to you and its solution seems simpler by another method, by all means use it. But in the long run you should be able to solve many otherwise difficult problems by using dimensional analysis.

Dimensional analysis converts one unit to another unit by using conversion factors.

$$\text{unit}_1 \times \text{conversion factor} = \text{unit}_2$$

Important equations and statements are boxed or highlighted in color.

If you want to know how many millimeters are in 2.5 meters, you need to convert meters (m) to millimeters (mm). Start by writing

$$\text{m} \times \text{conversion factor} = \text{mm}$$

This conversion factor must accomplish two things: It must cancel (or eliminate) meters *and* it must introduce millimeters—the unit wanted in the answer. Such a conversion factor will be in fractional form and have meters in the denominator and millimeters in the numerator:

$$\cancel{\text{m}} \times \frac{\text{mm}}{\cancel{\text{m}}} = \text{mm}$$

We know that multiplying a measurement by 1 does not change its value. Since our conversion factors both equal 1, we can multiply the measurement by the appropriate one to convert units.

We know that $1\,\text{m} = 1000\,\text{mm}$. From this relationship we can write two conversion factors—by dividing both sides of the equation by the same quantity:

$$\frac{1\,\text{m}}{1000\,\text{mm}} = \frac{1000\,\text{mm}}{1000\,\text{mm}} \longrightarrow \frac{1\,\text{m}}{1000\,\text{mm}} = 1$$

or

$$\frac{1\,\text{m}}{1\,\text{m}} = \frac{1000\,\text{mm}}{1\,\text{m}} \longrightarrow 1 = \frac{1000\,\text{mm}}{1\,\text{m}}$$

Therefore the two conversion factors are

$$\frac{1\,\text{m}}{1000\,\text{mm}} \quad \text{and} \quad \frac{1000\,\text{mm}}{1\,\text{m}}$$

Choosing the conversion factor $\dfrac{1000\,\text{mm}}{1\,\text{m}}$, we can set up the calculation for the conversion of 2.5 m to millimeters:

$$(2.5\,\cancel{\text{m}})\left(\frac{1000\,\text{mm}}{1\,\cancel{\text{m}}}\right) = 2500\,\text{mm} \quad \text{or} \quad 2.5 \times 10^{3}\,\text{mm}$$

(two significant figures)

Note that, in making this calculation, units are treated as numbers; meters in the numerator are canceled by meters in the denominator.

Now suppose you need to change 215 centimeters to meters. We start with

$$\text{cm} \times \text{conversion factor} = \text{m}$$

The conversion factor must have centimeters in the denominator and meters in the numerator:

$$\cancel{\text{cm}} \times \frac{\text{m}}{\cancel{\text{cm}}} = \text{m}$$

From the relationship $100\,\text{cm} = 1\,\text{m}$, we can write a factor that will accomplish this conversion:

$$\frac{1\,\text{m}}{100\,\text{cm}}$$

Now set up the calculation using all the data given:

$$(215\,\cancel{\text{cm}})\left(\frac{1\,\text{m}}{100\,\cancel{\text{cm}}}\right) = 2.15\,\text{m}$$

Units are emphasized in problems by using color and flow diagrams to help visualize the steps in the process.

Some problems require a series of conversions to reach the correct units in the answer. For example, suppose we want to know the number of seconds in 1 day. We need to convert from the unit of days to seconds in this manner:

$$\text{day} \rightarrow \text{hours} \rightarrow \text{minutes} \rightarrow \text{seconds}$$

This sequence requires three conversion factors, one for each step. We convert days to hours (hr), hours to minutes (min), and minutes to seconds (s). The conversions can be done individually or in a continuous sequence:

$$\cancel{\text{day}} \times \frac{\text{hr}}{\cancel{\text{day}}} \longrightarrow \cancel{\text{hr}} \times \frac{\text{min}}{\cancel{\text{hr}}} \longrightarrow \cancel{\text{min}} \times \frac{\text{s}}{\cancel{\text{min}}} = \text{s}$$

$$\cancel{\text{day}} \times \frac{\cancel{\text{hr}}}{\cancel{\text{day}}} \times \frac{\cancel{\text{min}}}{\cancel{\text{hr}}} \times \frac{\text{s}}{\cancel{\text{min}}} = \text{s}$$

Inserting the proper factors, we calculate the number of seconds in 1 day to be

$$(1 \, day)\left(\frac{24 \, hr}{1 \, day}\right)\left(\frac{60 \, min}{1 \, hr}\right)\left(\frac{60 \, s}{1 \, min}\right) = 86{,}400. \, s$$

All five digits in 86,400 are significant, since all the factors in the calculation are exact numbers.

Label all factors with the proper units.

The dimensional analysis used in the preceding examples shows how unit conversion factors are derived and used in calculations. As you become more proficient, you can save steps by writing the factors directly in the calculation. The following examples show the conversion from American units to metric units.

How many centimeters are in 2.00 ft?

Example 2.11

SOLUTION

The stepwise conversion of units from feet to centimeters may be done in this sequence. Convert feet to inches; then convert inches to centimeters:

$$ft \rightarrow in. \rightarrow cm$$

The conversion factors needed are

$$\frac{12 \, in.}{1 \, ft} \quad \text{and} \quad \frac{2.54 \, cm}{1 \, in.}$$

$$(2.00 \, ft)\left(\frac{12 \, in.}{1 \, ft}\right) = 24.0 \, in.$$

$$(24.0 \, in.)\left(\frac{2.54 \, cm}{1 \, in.}\right) = 61.0 \, cm$$

Since 1 ft and 12 in. are exact numbers, the number of significant figures allowed in the answer is three, based on the number 2.00.

How many meters are in a 100.-yd football field?

Example 2.12

SOLUTION

The stepwise conversion of units from yards to meters may be done by this sequence, using the proper conversion factors:

$$yd \rightarrow ft \rightarrow in. \rightarrow cm \rightarrow m$$

$$(100. \, yd)\left(\frac{3 \, ft}{1 \, yd}\right) = 300. \, ft$$

$$(300. \, ft)\left(\frac{12 \, in.}{1 \, ft}\right) = 3600 \, in.$$

$$(3600 \, in.)\left(\frac{2.54 \, cm}{1 \, in.}\right) = 9144 \, cm$$

$$(9144 \, cm)\left(\frac{1 \, m}{100 \, cm}\right) = 91.4 \, m \quad \text{(three significant figures)}$$

If you need help in doing the calculation all at once on your calculator, see Appendix II.

Examples 2.11 and 2.12 may be solved using a linear expression and writing down conversion factors in succession. This method often saves one or two calculation steps and allows numerical values to be reduced to simpler terms, leading to simpler calculations. The single linear expressions for Examples 2.11 and 2.12 are

$$(2.00 \text{ ft})\left(\frac{12 \text{ in.}}{1 \text{ ft}}\right)\left(\frac{2.54 \text{ cm}}{1 \text{ in.}}\right) = 61.0 \text{ cm}$$

$$(100. \text{ yd})\left(\frac{3 \text{ ft}}{1 \text{ yd}}\right)\left(\frac{12 \text{ in.}}{1 \text{ ft}}\right)\left(\frac{2.54 \text{ cm}}{1 \text{ in.}}\right)\left(\frac{1 \text{ m}}{100 \text{ cm}}\right) = 91.4 \text{ m}$$

Using the units alone (Example 2.12), we see that the stepwise cancellation proceeds in succession until the desired unit is reached:

$$\text{yd} \times \frac{\text{ft}}{\text{yd}} \times \frac{\text{in.}}{\text{ft}} \times \frac{\text{cm}}{\text{in.}} \times \frac{\text{m}}{\text{cm}} = \text{m}$$

Practice 2.6

(a) How many meters are in 10.5 miles?
(b) What is the area of a 6.0-in. × 9.0-in. rectangle in square meters?

Example 2.13   How many cubic centimeters (cm^3) are in a box that measures 2.20 in. by 4.00 in. by 6.00 in.?

SOLUTION   First we need to determine the volume of the box in cubic inches (in.^3) by multiplying the length × width × height:

$$(2.20 \text{ in.})(4.00 \text{ in.})(6.00 \text{ in.}) = 52.8 \text{ in.}^3$$

Now we convert in.^3 to cm^3 by using the inches and centimeters relationship (1 in. = 2.54 cm) three times:

$$\text{in.}^3 \times \frac{\text{cm}}{\text{in.}} \times \frac{\text{cm}}{\text{in.}} \times \frac{\text{cm}}{\text{in.}} = \text{cm}^3$$

$$(52.8 \text{ in.}^3)\left(\frac{2.54 \text{ cm}}{1 \text{ in.}}\right)\left(\frac{2.54 \text{ cm}}{1 \text{ in.}}\right)\left(\frac{2.54 \text{ cm}}{1 \text{ in.}}\right) = 865 \text{ cm}^3$$

Example 2.14   A driver of a car is obeying the speed limit of 55 miles per hour. How fast is the car traveling in kilometers per second?

SOLUTION   Three conversions are needed to solve this problem:

$$\text{mi} \rightarrow \text{km}$$
$$\text{hr} \rightarrow \text{min} \rightarrow \text{s}$$

To convert $\text{mi} \rightarrow \text{km}$,

$$\left(\frac{55 \text{ mi}}{\text{hr}}\right)\left(\frac{1.609 \text{ km}}{1 \text{ mi}}\right) = 88 \frac{\text{km}}{\text{hr}}$$

Next we must convert $\text{hr} \rightarrow \text{min} \rightarrow \text{s}$. Notice that hours is in the denominator of our quantity, so the conversion factor must have hours in the numerator:

$$\left(\frac{88 \text{ km}}{\text{hr}}\right)\left(\frac{1 \text{ hr}}{60 \text{ min}}\right)\left(\frac{1 \text{ min}}{60 \text{ s}}\right) = 0.024 \frac{\text{km}}{\text{s}}$$

Why do scientists worry about units? The National Aeronautics and Space Administration (NASA) recently was reminded of just why keeping track of units is so important. In 1999 a $125 million satellite was lost in the atmosphere of Mars because scientists made some improper assumptions about units. NASA's scientists at the Jet Propulsion Lab (JPL) in Pasadena, California, received thrust data from the satellite's manufacturer, Lockheed Martin Aeronautics in Denver, Colorado. Unfortunately, the Denver scientists used American units in their measurements and the JPL scientists assumed the units were metric. This mistake caused the satellite to fall 100 km lower into the Mars atmosphere than planned. The spacecraft burned up from the friction with the atmosphere.

Measuring and using units correctly is very important. In fact, it can be critical, as we have just seen. For example, a Canadian jet almost crashed when the tanks were filled with 22,300 pounds (instead of kilograms) of fuel. Calculations for distance were based on

kg, and the jet almost ran out of fuel before landing at the destination. Correct units are also important to ensure perfect fits for household purchases such as drapes, carpet, or appliances. Be sure to pay attention to units in both your chemistry problems and in everyday life!

Mars climate orbiter.

Practice 2.7 _____

How many cubic meters of air are in a room measuring 8 ft × 10 ft × 12 ft?

2.7 Measuring Mass and Volume

Although we often use mass and weight interchangeably in our everyday lives, they have quite different meanings in chemistry. In science we define the **mass** of an object as the amount of matter in the object. Mass is measured on an instrument called a balance. The **weight** of an object is a measure of the effect of gravity on the object. Weight is determined by using an instrument called a scale, which measures force against a spring. This means mass is independent of the location of an object, but weight is not. In this text we will use the term mass for all of our metric mass measurements. The gram is a unit of mass measurement, but it is a tiny amount of mass; for instance, a U.S. nickel has a mass of about 5 grams.

mass

weight

kilogram Therefore the *standard unit* of mass in the SI system is the **kilogram** (equal to 1000 g). The amount of mass in a kilogram is defined by international agreement as exactly equal to the mass of a platinum-iridium cylinder (international prototype kilogram) kept in a vault at Sèvres, France. Comparing this unit of mass to 1 lb (16 oz), we find that 1 kg is equal to 2.205 lb. A pound is equal to 453.6 g (0.4536 kg). The same prefixes used in length measurement are used to indicate larger and smaller gram units (see Table 2.4). A balance is used to measure mass. Two examples of balances are shown in Figure 2.3.

Table 2.4 Metric Units of Mass

Unit	Abbreviation	Gram equivalent	Exponential equivalent
kilogram	kg	1000 g	10^3 g
gram	g	1 g	10^0 g
decigram	dg	0.1 g	10^{-1} g
centigram	cg	0.01 g	10^{-2} g
milligram	mg	0.001 g	10^{-3} g
microgram	μg	0.000001 g	10^{-6} g

Common mass relationships:

1 g = 1000 mg
1 kg = 1000 g
1 kg = 2.205 lb
1 lb = 453.6 g

To change grams to milligrams, we use the conversion factor 1000 mg/g. The setup for converting 25 g to milligrams is

$$(25 \cancel{g})\left(\frac{1000 \text{ mg}}{1 \cancel{g}}\right) = 25{,}000 \text{ mg} \qquad (2.5 \times 10^4 \text{ mg})$$

(a)

(b)

Figure 2.3
(a) Digital electronic, top-loading balance with a precision of 0.001 g (1 mg); and (b) a digital electronic analytical balance with a precision of 0.0001 g (0.1 mg).

Note that multiplying a number by 1000 is the same as multiplying the number by 10^3 and can be done simply by moving the decimal point three places to the right:

$$(6.428)(1000) = 6428 \qquad (6.\underset{3}{\underbrace{428}})$$

To change milligrams to grams, we use the conversion factor 1 g/1000 mg. For example, we convert 155 mg to grams as follows:

$$(155 \ \cancel{mg})\left(\frac{1 \ g}{1000 \ \cancel{mg}}\right) = 0.155 \ g$$

Mass conversions from American to metric units are shown in Examples 2.15 and 2.16.

Example 2.15

A 1.50-lb package of baking soda contains how many grams?

SOLUTION

We are solving for the number of grams equivalent to 1.50 lb. Since 1 lb = 453.6 g, the conversion factor is 453.6 g/lb:

$$(1.50 \ \cancel{lb})\left(\frac{453.6 \ g}{1 \ \cancel{lb}}\right) = 680. \ g$$

Example 2.16

Suppose four ostrich feathers weigh 1.00 lb. Assuming that each feather is equal in mass, how many milligrams does a single feather weigh?

SOLUTION

The unit conversion in this problem is from 1.00 lb/4 feathers to milligrams per feather. Since the unit *feathers* occurs in the denominator of both the starting unit and the desired unit, the unit conversions are

$$lb \rightarrow g \rightarrow mg$$

$$\left(\frac{1.00 \ \cancel{lb}}{4 \ feathers}\right)\left(\frac{453.6 \ \cancel{g}}{1 \ \cancel{lb}}\right)\left(\frac{1000 \ mg}{1 \ \cancel{g}}\right) = \frac{113,400 \ mg}{feather}(1.13 \times 10^5 \ mg/feather)$$

Practice 2.8

You are traveling in Europe and wake up one morning to find your mass is 75.0 kg. Determine the American equivalent (in pounds) to see whether you need to go on a diet before you return home.

Practice 2.9

A tennis ball has a mass of 65 g. Determine the American equivalent in pounds.

Volume, as used here, is the amount of space occupied by matter. The SI unit of volume is the *cubic meter* (m^3). However, the liter (pronounced *lee-ter* and abbreviated L) and the milliliter (abbreviated mL) are the standard units of volume used in most chemical laboratories. A **liter** is usually defined as 1 cubic decimeter (1 kg) of water at 4°C.

volume

liter

Common volume relationships:
1 L = 1000 mL = 1000 cm³
1 mL = 1 cm³
1 L = 1.057 qt
946.1 mL = 1 qt

The most common instruments or equipment for measuring liquids are the graduated cylinder, volumetric flask, buret, pipet, and syringe, which are illustrated in Figure 2.4. These pieces are usually made of glass and are available in various sizes.

The volume of a cubic or rectangular container can be determined by multiplying its length × width × height. Thus a box 10 cm on each side has a volume of (10 cm)(10 cm)(10 cm) = 1000 cm³. Let's try some examples.

Graduated   Volumetric   Buret   Pipet   Syringe
cylinder   flask

Figure 2.4
Calibrated glassware for measuring the volume of liquids.

Example 2.17   How many milliliters are contained in 3.5 liters?

SOLUTION   The conversion factor to change liters to milliliters is 1000 mL/L:

$$(3.5 \text{ L})\left(\frac{1000 \text{ mL}}{\text{L}}\right) = 3500 \text{ mL} \qquad (3.5 \times 10^3 \text{ mL})$$

Liters may be changed to milliliters by moving the decimal point three places to the right and changing the units to milliliters:

1.500 L = 1500. mL

Example 2.18   How many cubic centimeters are in a cube that is 11.1 inches on a side?

SOLUTION   First we change inches to centimeters; our conversion factor is 2.54 cm/in.:

$$(11.1 \text{ in.})\left(\frac{2.54 \text{ cm}}{1 \text{ in.}}\right) = 28.2 \text{ cm on a side}$$

Then determine volume (length × width × height):

$$(28.2 \text{ cm})(28.2 \text{ cm})(28.2 \text{ cm}) = 22{,}426 \text{ cm}^3 \qquad (2.24 \times 10^4 \text{ cm}^3)$$

The kilogram is the standard base unit for mass in the SI system. But who decides just what a kilogram is? Before 1880, a kilogram was defined as the mass of a cubic decimeter of water. But this is difficult to reproduce very accurately since impurities and air are dissolved in water. The density of water also changes with temperature, leading to inaccuracy. So in 1885 a Pt-Ir cylinder was made with a mass of exactly 1 kilogram. This cylinder is kept in a vault at Sèvres, France, outside Paris, along with six copies. In 1889 the kilogram was determined to be the base unit of mass for the metric system and has been weighed against its copies three times in the past 100 years (1890, 1948, 1992) to make sure that it is accurate. The trouble with all this is that the cylinder itself can vary. Experts in the science of measurement want to change the definition of the kilogram to link it to a property of matter that does not change.

The kilogram is the only one of seven SI base units defined in terms of a physical object instead of a property of matter. The meter (redefined in 1983) is defined in terms of the speed of light (1 m = distance light travels in $\frac{1}{299,792,458}$ s). The second is defined in terms of the natural vibration of the cesium atom. Unfortunately, three of the SI base units (mole, ampere, and candela) depend on the definition of the kilogram. If the mass of the kilogram is uncertain, then all of these units are uncertain.

How else could we define a kilogram? Scientists have several ideas. They propose to fix the kilogram to a set value of a physical constant (which does not change). These constants connect experiment to theory. Right now they are continuing to struggle to find a new way to determine the mass of the kilogram. By 2007 they hope to have a new definition of the kilogram based on constants and accurately reproducible anywhere in the world.

Kilogram artifact in Sevres, France.

Practice 2.10_____
A bottle of excellent chianti holds 750. mL. What is its volume in quarts?

Practice 2.11_____
Milk is often purchased by the half gallon. Determine the number of liters equal to this amount.

When doing problems with multiple steps, you should round only at the end of the problem. We are rounding at the end of each step in example problems to illustrate the proper significant figures.

2.8 Measurement of Temperature

Heat is a form of energy associated with the motion of small particles of matter. The term *heat* refers to the quantity of energy within a system or to a quantity of energy added to or taken away from a system. *System*, as used here, simply refers to the entity that is being heated or cooled. Depending on the amount of heat energy present, a given system is said to be hot or cold. **Temperature** is a measure of the intensity of heat, or how hot a system is, regardless of its size. Heat always flows from a region of higher temperature to one of lower temperature. The SI unit of temperature is the kelvin. The common laboratory instrument for measuring temperature is a thermometer (see Figure 2.5).

heat

temperature

Figure 2.5
Comparison of Celsius, Kelvin, and
Fahrenheit temperature scales.

The temperature of a system can be expressed by several different scales. Three commonly used temperature scales are Celsius (pronounced *sell-see-us*), Kelvin (also called absolute), and Fahrenheit. The unit of temperature on the Celsius and Fahrenheit scales is called a *degree*, but the size of the Celsius and the Fahrenheit degree is not the same. The symbol for the Celsius and Fahrenheit degrees is °, and it is placed as a superscript after the number and before the symbol for the scale. Thus, 10.0°C means 10.0 *degrees Celsius*. The degree sign is not used with Kelvin temperatures:

degrees Celsius = °C

Kelvin (absolute) = K

degrees Fahrenheit = °F

On the Celsius scale the interval between the freezing and boiling temperatures of water is divided into 100 equal parts, or degrees. The freezing point of water is assigned a temperature of 0°C and the boiling point of water a temperature of 100°C. The Kelvin temperature scale is known as the absolute temperature scale, because 0 K is the lowest temperature theoretically attainable. The Kelvin zero is 273.15 kelvins below the Celsius zero. A kelvin is equal in size to a Celsius degree. The freezing point of water on the Kelvin scale is 273.15 K. The Fahrenheit scale has 180 degrees between the freezing and boiling temperatures of water. On this scale the freezing point of water is 32°F and the boiling point is 212°F:

$$0°C \cong 273\,K \cong 32°F \qquad 100°C \cong 373\,K \cong 212°F$$

The three scales are compared in Figure 2.5. Although absolute zero (0 K) is the lower limit of temperature on these scales, temperature has no upper limit. Temperatures of several million degrees are known to exist in the sun and in other stars.

By examining Figure 2.5, we can see that there are 100 Celsius degrees and 100 kelvins between the freezing and boiling points of water, but there are 180 Fahrenheit degrees between these two temperatures. Hence, the size of the Celsius degree and the kelvin are the same, but 1 Celsius degree is equal to 1.8 Fahrenheit degrees.

$$\frac{180}{100} = 1.8$$

From these data, mathematical formulas have been derived to convert a temperature on one scale to the corresponding temperature on another scale:

$$K = °C + 273.15$$
$$°F = (1.8 \times °C) + 32$$
$$°C = \frac{°F - 32}{1.8}$$

The temperature at which table salt (sodium chloride) melts is 800.°C. What is this temperature on the Kelvin and Fahrenheit scales?

Example 2.19

SOLUTION

To calculate K from °C, we use the formula

$$K = °C + 273.15$$
$$K = 800.°C + 273.15 = 1073 \, K$$

To calculate °F from °C, we use the formula

$$°F = (1.8 \times °C) + 32$$
$$°F = (1.8)(800.°C) + 32$$
$$°F = 1440 + 32 = 1472°F$$

Summarizing our calculations, we see that

$$800.°C = 1073 \, K = 1472°F$$

Remember that the original measurement of 800.°C was to the units place, so the converted temperature is also to the units place.

The temperature on December 1 in Honolulu, Hawaii, was 110.°F, a new record. Convert this temperature to °C.

Example 2.20

SOLUTION

We use the formula

$$°C = \frac{°F - 32}{1.8}$$
$$°C = \frac{110. - 32}{1.8} = \frac{78}{1.8} = 43°C$$

What temperature on the Fahrenheit scale corresponds to −8.0°C? (Notice the negative sign in this problem.)

Example 2.21

SOLUTION

$$°F = (1.8 \times °C) + 32$$
$$°F = (1.8)(-8.0) + 32 = -14 + 32$$
$$°F = 18°F$$

If you have ever struggled to take the temperature of a sick child, imagine the difficulty in taking the temperature of a geyser. Such are the tasks scientists set for themselves! In 1984, James A. Westphal and Susan W. Keiffer, geologists from the California Institute of Technology, measured the temperature and pressure inside Old Faithful during several eruptions in order to learn more about how a geyser functions. The measurements, taken at eight depths along the upper part of the geyser, were so varied and complicated that the researchers returned to Yellowstone in 1992 to further investigate Old Faithful's structure and functioning. To see what happened between eruptions, Westphal and Keiffer lowered an insulated 2-inch video camera into the geyser. Keiffer had assumed that the vent (opening in the ground) was a uniform vertical tube, but this is not the case. Instead, the geyser appears to be an east–west crack in the Earth that extends downward at least 14 m. In some places it is over 1.8 m wide, and in other places it narrows to less than 15 cm. The walls of the vent contain many cracks, allowing water to enter at several depths. The complicated nature of the temperature data is explained by these cracks. Cool water enters the vent at depths of 5.5 m and 7.5 m. Superheated water and steam blast into the vent 14 m underground. According to Westphal, temperature increases of up to 130°C at the beginning of an eruption suggest that water and steam also surge into the vent from deeper geothermal sources.

During the first 20–30 seconds of an eruption, steam and boiling water shoot through the narrowest part of the vent at near the speed of sound. The narrow tube limits the rate at which the water can shoot from the geyser. When the pressure falls below a critical value, the process slows and Old Faithful begins to quiet down again.

The frequency of Old Faithful's eruptions are not on a precise schedule, but vary from 45 to 105 minutes— the average is about 79 minutes. The variations in time between eruptions depend on the amount of boiling water left in the fissure. Westphal says, "There's no real pattern except that a short eruption is always followed by a long one." Measurements of temperatures inside Old Faithful have given scientists a better understanding of what causes a geyser to erupt.

Old Faithful eruptions shoot into the air an average of 130 feet.

Temperatures used throughout this book are in degrees Celsius (°C) unless specified otherwise. The temperature after conversion should be expressed to the same precision as the original measurement.

Practice 2.12

Helium boils at 4 K. Convert this temperature to °C and then to °F.

Practice 2.13

"Normal" human body temperature is 98.6°F. Convert this to °C and K.

2.9 Density

Density (d) is the ratio of the mass of a substance to the volume occupied by that mass; it is the mass per unit of volume and is given by the equation

density

$$d = \frac{\text{mass}}{\text{volume}}$$

Density is a physical characteristic of a substance and may be used as an aid to its identification. When the density of a solid or a liquid is given, the mass is usually expressed in grams and the volume in milliliters or cubic centimeters.

$$d = \frac{\text{mass}}{\text{volume}} = \frac{g}{mL} \quad \text{or} \quad d = \frac{g}{cm^3}$$

Since the volume of a substance (especially liquids and gases) varies with temperature, it is important to state the temperature along with the density. For example, the volume of 1.0000 g of water at 4°C is 1.0000 mL; at 20°C, it is 1.0018 mL; and at 80°C, it is 1.0290 mL. Density therefore also varies with temperature.

The density of water at 4°C is 1.0000 g/mL, but at 80°C the density of water is 0.9718 g/mL:

$$d^{4°C} = \frac{1.0000 \text{ g}}{1.0000 \text{ mL}} = 1.0000 \text{ g/mL}$$

$$d^{80°C} = \frac{1.0000 \text{ g}}{1.0290 \text{ mL}} = 0.97182 \text{ g/mL}$$

The density of iron at 20°C is 7.86 g/mL.

$$d^{20°C} = \frac{7.86 \text{ g}}{1.00 \text{ mL}} = 7.86 \text{ g/mL}$$

The densities of a variety of materials are compared in Figure 2.6.

Densities for liquids and solids are usually represented in terms of grams per milliliter (g/mL) or grams per cubic centimeter (g/cm^3). The density of gases, however, is expressed in terms of grams per liter (g/L). Unless otherwise stated, gas densities are given for 0°C and 1 atmosphere pressure (discussed further in Chapter 13). Table 2.5 lists the densities of some common materials.

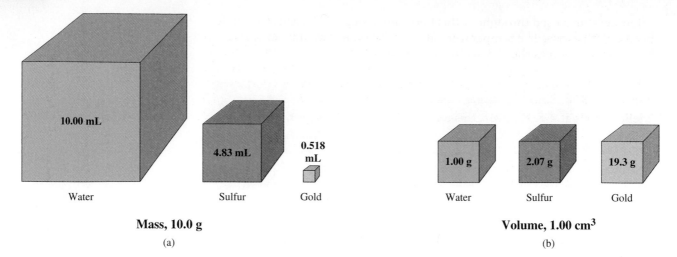

Figure 2.6
(a) Comparison of the volumes of equal masses (10.0 g) of water, sulfur, and gold.
(b) Comparison of the masses of equal volumes (1.00 cm³) of water, sulfur, and gold.
(Water is at 4°C; the two solids, at 20°C.)

Table 2.5 Densities of Some Selected Materials

Liquids and solids		Gases	
Substance	Density (g/mL at 20°C)	Substance	Density (g/L at 0°C)
Wood (Douglas fir)	0.512	Hydrogen	0.090
Ethyl alcohol	0.789	Helium	0.178
Vegetable oil	0.91	Methane	0.714
Water (4°C)	**1.000***	Ammonia	0.771
Sugar	1.59	Neon	0.90
Glycerin	1.26	Carbon monoxide	1.25
Karo syrup	1.37	Nitrogen	1.251
Magnesium	1.74	**Air**	**1.293***
Sulfuric acid	1.84	Oxygen	1.429
Sulfur	2.07	Hydrogen chloride	1.63
Salt	2.16	Argon	1.78
Aluminum	2.70	Carbon dioxide	1.963
Silver	10.5	Chlorine	3.17
Lead	11.34		
Mercury	13.55		
Gold	19.3		

*For comparing densities, the density of water is the reference for solids and liquids; air is the reference for gases.

When an insoluble solid object is dropped into water, it will sink or float, depending on its density. If the object is less dense than water, it will float, displacing a *mass* of water equal to the mass of the object. If the object is more dense than water, it will sink, displacing a *volume* of water equal to the volume of the object. This information can be used to determine the volume (and density) of irregularly shaped objects.

The **specific gravity** (sp gr) of a substance is the ratio of the density of that substance to the density of another substance, usually water at 4°C. Specific gravity has no units because the density units cancel. The specific gravity tells us how many times as heavy a liquid, a solid, or a gas is as compared to the reference material. Since the density of water at 4°C is 1.00 g/mL, the specific gravity of a solid or liquid is the same as its density in g/mL without the units.

specific gravity

$$sp\ gr = \frac{\text{density of a liquid or solid}}{\text{density of water}}$$

Sample calculations of density problems follow.

What is the density of a mineral if 427 g of the mineral occupy a volume of 35.0 mL?

Example 2.22

We need to solve for density, so we start by writing the formula for calculating density:

SOLUTION

$$d = \frac{\text{mass}}{\text{volume}}$$

Then we substitute the data given in the problem into the equation and solve:

mass = 427 g   volume = 35.0 mL

$$d = \frac{\text{mass}}{\text{volume}} = \frac{427\ g}{35.0\ mL} = 12.2\ g/mL$$

The density of gold is 19.3 g/mL. What is the mass of 25.0 mL of gold?

Example 2.23

There are two ways to solve this problem: (1) Solve the density equation for mass, and then substitute the density and volume data into the new equation and calculate. (2) Solve by dimensional analysis.

SOLUTION

Method 1 (a) Solve the density equation for mass:

$$d = \frac{\text{mass}}{\text{volume}} \qquad d \times \text{volume} = \text{mass}$$

(b) Substitute the data and calculate.

$$\text{mass} = \left(\frac{19.3\ g}{mL}\right)(25.0\ mL) = 483\ g$$

When alternative methods of solution are available, more than one is shown in the example. Choose the method you are most comfortable with.

Method 2 Dimensional analysis: Use density as a conversion factor, converting

$$mL \rightarrow g$$

The conversion of units is

$$mL \times \frac{g}{mL} = g$$

$$(25.0\ mL)\left(\frac{19.3\ g}{mL}\right) = 483\ g$$

Example 2.24   Calculate the volume (in mL) of 100. g of ethyl alcohol.

SOLUTION   From Table 2.5 we see that the density of ethyl alcohol is 0.789 g/mL. This density also means that 1 mL of the alcohol has a mass of 0.789 g (1 mL/0.789 g).

Method 1   Solve the density equation for volume and then substitute the data in the new equation:

$$d = \frac{mass}{volume}$$

$$volume = \frac{mass}{d}$$

$$volume = \frac{100. \text{ g}}{0.789 \text{ g/mL}} = 127 \text{ mL}$$

Method 2   Dimensional analysis. For a conversion factor, we can use either

$$\frac{\text{g}}{\text{mL}} \quad \text{or} \quad \frac{\text{mL}}{\text{g}}$$

In this case the conversion is from g → mL, so we use mL/g. Substituting the data,

$$100. \text{ g} \times \frac{1 \text{ mL}}{0.789 \text{ g}} = 127 \text{ mL of ethyl alcohol}$$

Example 2.25   The water level in a graduated cylinder stands at 20.0 mL before and at 26.2 mL after a 16.74-g metal bolt is submerged in the water. (a) What is the volume of the bolt? (b) What is the density of the bolt?

SOLUTION   (a) The bolt will displace a volume of water equal to the volume of the bolt. Thus, the increase in volume is the volume of the bolt:

$$\begin{aligned}
26.2 \text{ mL} &= \text{volume of water plus bolt} \\
-20.0 \text{ mL} &= \text{volume of water} \\
\hline
6.2 \text{ mL} &= \text{volume of bolt}
\end{aligned}$$

(b) $d = \dfrac{\text{mass of bolt}}{\text{volume of bolt}} = \dfrac{16.74 \text{ g}}{6.2 \text{ mL}} = 2.7 \text{ g/mL}$

Practice 2.14

Pure silver has a density of 10.5 g/mL. A ring sold as pure silver has a mass of 18.7 g. When it is placed in a graduated cylinder, the water level rises 2.0 mL. Determine whether the ring is actually pure silver or whether the customer should contact the Better Business Bureau.

Practice 2.15

The water level in a metric measuring cup is 0.75 L before the addition of 150. g of shortening. The water level after submerging the shortening is 0.92 L. Determine the density of the shortening.

Chapter 2 Review

2.1 Scientific Notation
KEY TERMS
Measurement
Scientific notation

- Quantitative observations consist of a number and a unit and are called measurements.
- Very large and very small numbers can be represented compactly by using scientific notation:
  - The number is represented as a decimal between 1 and 10 and is multiplied by 10 raised to the appropriate exponent.
  - The sign on the exponent is determined by the direction the decimal point is moved in the original number.

2.2 Measurement and Uncertainty
KEY TERM
Significant figures

- All measurements reflect some amount of uncertainty, which is indicated by the number of significant figures in the measurement.
- The significant figures include all those known with certainty plus one estimated digit.

2.3 Significant Figures
KEY TERM
Rounding off numbers

- Rules exist for counting significant figures in a measurement:
  - Nonzero numbers are always significant.
  - Exact numbers have an infinite number of significant figures.
  - The significance of a zero in a measurement is determined by its position within the number.
- Rules exist for rounding off the result of a calculation to the correct number of significant figures.
  - If the first number after the one you want to retain is 4 or less, that digit and all those after it are dropped.
  - If the first number after the one you want to retain is 5 or greater, that digit and all those after it are dropped and the last digit retained is increased by one.

2.4 Significant Figures in Calculations
- The results of a calculation cannot be more precise than the least precise measurement.
- Rules exist for determining the correct number of significant figures in the result of a calculation.

2.5 The Metric System
KEY TERMS
Metric system (SI)
Meter (m)

- The metric system uses factors of 10 and a set of standard units for measurements.

- Length in the metric system is measured by the standard unit of the meter.

2.6 Problem Solving
- Dimensional analysis is a common method used to solve problems:
  - Conversion factors are used to convert one unit into another:

  $$\text{unit}_1 \times \text{conversion factor} = \text{unit}_2$$

- The basic steps for solving problems are:
  - Read the problem and write down the known information.
  - Plan how you will solve the problem by deciding what principles or equations are needed.
  - Set up the calculation using dimensional analysis, making sure all the units cancel except the unit required in the answer.
  - Calculate the result and round to the correct number of significant figures.
  - Check the answer to determine whether it is reasonable.

2.7 Measuring Mass and Volume
KEY TERMS
Mass
Weight
Kilogram (kg)
Volume
Liter (L)

- The standard unit for mass in the metric system is the kilogram. In chemistry we often use the gram instead, as we tend to work in smaller quantities.
- Volume is the amount of space occupied by matter.
- The standard unit for volume is the cubic meter. In chemistry we usually use the volume unit of the liter or the milliliter.

2.8 Measurement of Temperature
KEY TERMS
Heat
Temperature

- There are three commonly used temperature scales: Fahrenheit, Celsius, and Kelvin.
- We can convert among the temperature scales by using mathematical formulas:
  - $K = {}^\circ C + 273.15$
  - ${}^\circ F = (1.8 \times {}^\circ C) + 32$
  - ${}^\circ C = \dfrac{{}^\circ F - 32}{1.8}$

2.9 Density
KEY TERMS
Density
Specific gravity

- The density of a substance is the amount of matter (mass) in a given volume of the substance:

$$d = \frac{mass}{volume}$$

- Specific gravity is the ratio of the density of a substance to the density of another reference substance (usually water).

Review Questions

All questions with blue numbers have answers in the appendix of the text.

1. How many centimeters make up 1 km? (Table 2.3)
2. What is the metric equivalent of 3 in.? (Figure 2.3)
3. Why is the neck of a 100-mL volumetric flask narrower than the top of a 100-mL graduated cylinder? (Figure 2.4)
4. Describe the order of the following substances (top to bottom) if these three substances were placed in a 100-mL graduated cylinder: 25 mL glycerin, 25 mL mercury, and a cube of magnesium 2.0 cm on an edge. (Table 2.5)
5. Arrange these materials in order of increasing density: salt, vegetable oil, lead, and ethyl alcohol. (Table 2.5)
6. Ice floats in vegetable oil and sinks in ethyl alcohol. The density of ice must lie between what numerical values? (Table 2.5)

7. Distinguish between heat and temperature.
8. Distinguish between density and specific gravity.
9. State the rules used in this text for rounding off numbers.
10. Compare the number of degrees between the freezing point of water and its boiling point on the Fahrenheit, Kelvin, and Celsius temperature scales. (Figure 2.5)
11. Why does an astronaut weigh more on Earth than in space when his or her mass remains the same in both places?
12. Ice floats in water, yet ice is simply frozen water. If the density of water is 1.0 g/mL, how is this possible?

Paired Exercises

All exercises with blue numbers have answers in the appendix of the text.

1. What is the numerical meaning of each of the following?
   (a) kilogram
   (b) centimeter
   (c) microliter
   (d) millimeter
   (e) deciliter

2. What is the correct unit of measurement for each of the following?
   (a) 1000 meters
   (b) 0.1 gram
   (c) 0.000001 liter
   (d) 0.01 meter
   (e) 0.001 liter

3. State the abbreviation for each of the following units:
   (a) gram
   (b) microgram
   (c) centimeter
   (d) micrometer
   (e) milliliter
   (f) deciliter

4. State the abbreviation for each of the following units:
   (a) milligram
   (b) kilogram
   (c) meter
   (d) nanometer
   (e) angstrom
   (f) microliter

Significant Figures, Rounding, Exponential Notation

5. Are the zeros in these numbers significant?
   (a) 503
   (b) 0.007
   (c) 4200
   (d) 3.0030
   (e) 100.00
   (f) 8.00×10^2

6. Are the zeros significant in these numbers?
   (a) 63,000
   (b) 6.004
   (c) 0.00543
   (d) 8.3090
   (e) 60.
   (f) 5.0×10^{-4}

7. How many significant figures are in each of the following numbers?
   (a) 0.025
   (b) 22.4
   (c) 0.0404
   (d) 5.50×10^3

8. State the number of significant figures in each of the following numbers:
   (a) 40.0
   (b) 0.081
   (c) 129,042
   (d) 4.090×10^{-3}

9. Round each of the following numbers to three significant figures:
   (a) 93.246
   (b) 0.02857
   (c) 4.644
   (d) 34.250

10. Round each of the following numbers to three significant figures:
    (a) 8.8726
    (b) 21.25
    (c) 129.509
    (d) 1.995×10^6

11. Express each of the following numbers in exponential notation:
    (a) 2,900,000   (c) 0.00840
    (b) 0.587   (d) 0.0000055

12. Write each of the following numbers in exponential notation:
    (a) 0.0456   (c) 40.30
    (b) 4082.2   (d) 12,000,000

13. Solve the following problems, stating answers to the proper number of significant figures:
    (a) $12.62 + 1.5 + 0.25 = ?$
    (b) $(2.25 \times 10^3)(4.80 \times 10^4) = ?$
    (c) $\dfrac{(452)(6.2)}{14.3} = ?$
    (d) $(0.0394)(12.8) = ?$
    (e) $\dfrac{0.4278}{59.6} = ?$
    (f) $10.4 + 3.75(1.5 \times 10^4) = ?$

14. Evaluate each of the following expressions. State the answer to the proper number of significant figures:
    (a) $15.2 - 2.75 + 15.67$
    (b) $(4.68)(12.5)$
    (c) $\dfrac{182.6}{4.6}$
    (d) $1986 + 23.84 + 0.012$
    (e) $\dfrac{29.3}{(284)(415)}$
    (f) $(2.92 \times 10^{-3})(6.14 \times 10^5)$

15. Change these fractions into decimals. Express each answer to three significant figures:
    (a) $\dfrac{5}{6}$ (c) $\dfrac{12}{16}$
    (b) $\dfrac{3}{7}$ (d) $\dfrac{9}{18}$

16. Change each of the following decimals to fractions in lowest terms:
    (a) 0.25   (c) 1.67
    (b) 0.625   (d) 0.8888

17. Solve each of these equations for x:
    (a) $3.42x = 6.5$
    (b) $\dfrac{x}{12.3} = 7.05$
    (c) $\dfrac{0.525}{x} = 0.25$

18. Solve each equation for the variable:
    (a) $x = \dfrac{212 - 32}{1.8}$
    (b) $8.9\dfrac{g}{mL} = \dfrac{40.90\ g}{x}$
    (c) $72°F = 1.8x + 32$

Unit Conversions

19. Complete the following metric conversions using the correct number of significant figures:
    (a) 28.0 cm to m   (e) 6.8×10^4 mg to kg
    (b) 1000. m to km   (f) 8.54 g to kg
    (c) 9.28 cm to mm   (g) 25.0 mL to L
    (d) 10.68 g to mg   (h) 22.4 L to μL

20. Complete the following metric conversions using the correct number of significant figures:
    (a) 4.5 cm to Å   (e) 0.65 kg to mg
    (b) 12 nm to cm   (f) 5.5 kg to g
    (c) 8.0 km to mm   (g) 0.468 L to mL
    (d) 164 mg to g   (h) 9.0 μL to mL

21. Complete the following American/metric conversions using the correct number of significant figures:
    (a) 42.2 in. to cm   (d) 42.8 kg to lb
    (b) 0.64 m to in.   (e) 3.5 qt to mL
    (c) 2.00 in.² to cm²   (f) 20.0 L to gal

22. Make the following conversions using the correct number of significant figures:
    (a) 35.6 m to ft   (d) 95 lb to g
    (b) 16.5 km to mi   (e) 20.0 gal to L
    (c) 4.5 in.³ to mm³   (f) 4.5×10^4 ft³ to m³

23. After you have worked out at the gym on a stationary bike for 45 minutes, the distance gauge indicates that you have traveled 15.2 miles. What was your rate in km/hr?

24. A competitive college runner ran a 5-K (5.0-km) race in 15 minutes, 23 seconds. What was her pace in miles per hour?

25. A pharmacy technician is asked to prepare an antibiotic IV that will contain 500. mg of cephalosporin for every 100 mL of normal saline solution. The total volume of saline solution will be 1 L. How many grams of the cephalosporin will be needed for this IV?

26. An extra-strength aspirin tablet contains 0.500 gram of the active ingredient, acetylsalicylic acid. Aspirin strength used to be measured in grains. If 1 grain = 60 mg, how many grains of the active ingredient are in 1 tablet? (Report your answer to 3 significant figures.)

27. The velocity of light is 1.86×10^8 mi/hr. The distance of Mercury from the sun is approximately 5.78×10^7 km. How many minutes will it take for light from the sun to travel to Mercury?

28. How many days would it take to lose 25 lb if the average weight loss is 1.0 kg per week?

29. A personal trainer uses calipers on a client to determine his percent body fat. After taking the necessary measurements, the personal trainer determines that the client's body contains 11.2% fat by mass (11.2 lb of fat per 100 lb of body mass). If the client weighs 225 lb, how many kg of fat does he have?

30. The weight of a diamond is measured in carats. How many pounds does a 5.75-carat diamond weigh? 1 carat = 200. mg

31. A competitive high school swimmer takes 52 seconds to swim 100. yards. What is his rate in m/min?

32. In 2005, Jarno Trulli was the Pole Winner of the US Grand Prix Race with a speed of 133 miles per hour. What was his speed in cm/s?

33. If gasoline is $2.50 per gallon for regular grade, how much will it cost a commuter to drive to and from work each day if the round-trip distance is 45.5 miles and his car averages 11 km/L?

34. The price of gold varies greatly and has been as high as $875 per ounce. What is the value of 250 g of gold at $559 per ounce? Gold is priced by troy ounces (14.58 troy ounces = 1 lb).

35. An adult ruby-throated hummingbird has an average mass of 3.2 g, while an adult California condor may attain a mass of 21 lb. How many hummingbirds would it take to equal the mass of one condor?

36. A very strong camel can carry 990 lb. If one straw weighs 1.5 grams, how many straws can the camel carry without breaking his back?

37. Assuming that there are 20. drops in 1.0 mL, how many drops are in 1.0 gallon?

38. How many liters of oil are in a 42-gal barrel of oil?

39. Calculate the number of milliliters of water in a cubic foot of water.

40. Oil spreads in a thin layer on water called an "oil slick." How much area in m^2 will 200 cm^3 of oil cover if it forms a layer 0.5 nm thick?

41. A textbook is 27 cm long, 21 cm wide, and 4.4 cm thick. What is the volume in:
(a) cubic centimeters?
(b) liters?
(c) cubic inches?

42. An aquarium measures 16 in. $\times$ 8 in. $\times$ 10 in. How many liters of water does it hold? How many gallons?

43. A toddler in Italy visits the family doctor. The nurse takes the child's temperature, which reads 38.8°C. (a) Convert this temperature to °F. (b) If 98.6°F is considered normal, does the child have a fever?

44. Driving to the grocery store, you notice the temperature is 45°C. Determine what this temperature is on the Fahrenheit scale and what season of the year it might be.

45. Make the following conversions and include an equation for each one:
(a) 162°F to °C   (c) −18°C to °F
(b) 0.0°F to K   (d) 212 K to °C

46. Make the following conversions and include an equation for each one:
(a) 32°C to °F   (c) 273°C to K
(b) −8.6°F to °C   (d) 100 K to °F

*47. At what temperature are the Fahrenheit and Celsius temperatures exactly equal?

*48. At what temperature are Fahrenheit and Celsius temperatures the same in value but opposite in sign?

49. The normal boiling point of ethyl alcohol is 173.3°F. What is this temperature in °C?

50. The normal boiling point of oxygen, O_2, is −183°C. What is this temperature in °F?

51. Calculate the density of a liquid if 50.00 mL of the liquid have a mass of 78.26 g.

52. A 12.8-mL sample of bromine has a mass of 39.9 g. What is the density of bromine?

53. When a 32.7-g piece of chromium metal was placed into a graduated cylinder containing 25.0 mL of water, the water level rose to 29.6 mL. Calculate the density of the chromium.

54. An empty graduated cylinder has a mass of 42.817 g. When filled with 50.0 mL of an unknown liquid, it has a mass of 106.773 g. What is the density of the liquid?

55. Concentrated hydrochloric acid has a density of 1.19 g/mL. Calculate the mass of 250.0 mL of this acid.

56. What mass of mercury (density 13.6 g/mL) will occupy a volume of 25.0 mL?

Additional Exercises

All exercises with blue *numbers have answers in the appendix of the text.*

57. You have measured out 10.0123576 grams of NaCl. What amount should you report if the precision of the balance you used was
   (a) + or − 0.01 g?
   (b) + or − 0.001 g?
   (c) + or − 0.0001 g?

58. Suppose you want to add 100 mL of solvent to a reaction flask. Which piece of glassware shown in Figure 2.4 would be the **best** choice for accomplishing this task and why?

59. A reaction requires 21.5 g of $CHCl_3$. No balance is available, so it will have to be measured by volume. How many mL of $CHCl_3$ need to be taken? (Density of $CHCl_3$ is 1.484 g/mL.)

60. A 25.27-g sample of pure sodium was prepared for an experiment. How many mL of sodium is this? (Density of sodium is 0.97 g/mL.)

61. One liter of homogenized whole milk has a mass of 1032 g. What is the density of the milk in grams per milliliter? in kilograms per liter?

62. The volume of blood plasma in adults is 3.1 L. Its density is 1.03 g/cm^3. Approximately how many pounds of blood plasma are there in your body?

63. The dashed lane markers on an interstate highway are 2.5 ft long and 4.0 in. wide. One (1.0) qt of paint covers 43 ft^2. How many dashed lane markers can be painted with 15 gal of paint?

64. Will a hollow cube with sides of length 0.50 m hold 8.5 L of solution? Depending on your answer, how much additional solution would be required to fill the container or how many times would the container need to be filled to measure the 8.5 L?

65. The accepted toxic dose of mercury is 300 μg/day. Dental offices sometimes contain as much as 180 μg of mercury per cubic meter of air. If a nurse working in the office ingests 2×10^4 L of air per day, is he or she at risk for mercury poisoning?

66. Which is the higher temperature, 4.5°F or −15°C? Show your calculation.

67. According to the National Heart, Lung, and Blood Institute, LDL-cholesterol levels of less than 130 mg of LDL-cholesterol per deciliter of blood are desirable for heart health in humans. On the average, a human has 4.7 L of whole blood. What is the maximum number of grams of LDL-cholesterol that a human should have?

68. Pure water at 4.0°C has a density of 1.00 g/mL. How many pounds does 2.50 gallons of water weigh?

69. The height of a horse is measured in hands (1 hand = exactly 4 inches). How many meters is a horse that measures 14.2 hands?

70. Camels have been reported to drink as much as 22.5 gallons of water in 12 hours. How many liters can they drink in 30. days?

71. You are given three cubes, A, B, and C; one is magnesium, one is aluminum, and the third is silver. All three cubes have the same mass, but cube A has a volume of 25.9 mL, cube B has a volume of 16.7 mL, and cube C has a volume of 4.29 mL. Identify cubes A, B, and C.

72. How many cubic centimeters are in 1.00 cubic inch?

* 73. A cube of aluminum has a mass of 500. g. What will be the mass of a cube of gold of the same dimensions?

74. A 25.0-mL sample of water at 90°C has a mass of 24.12 g. Calculate the density of water at this temperature.

75. The mass of an empty container is 88.25 g. The mass of the container when filled with a liquid (d = 1.25 g/mL) is 150.50 g. What is the volume of the container?

76. Which liquid will occupy the greater volume, 50 g of water or 50 g of ethyl alcohol? Explain.

77. The Sacagawea gold-colored dollar coin has a mass of 8.1 g and contains 3.5% manganese. What is its mass in ounces (1 lb = 16 oz) and how many ounces of Mn are in this coin?

78. The density of sulfuric acid is 1.84 g/mL. What volume of this acid will weigh 100. g?

79. The density of palladium at 20°C is 12.0 g/mL and at 1550°C the density is 11.0 g/mL. What is the change in volume (in mL) of 1.00 kg Pd in going from 20°C to 1550°C?

* 80. As a solid substance is heated, its volume increases but its mass remains the same. Sketch a graph of density versus temperature showing the trend you expect. Briefly explain.

81. A gold bullion dealer advertised a bar of pure gold for sale. The gold bar had a mass of 3300 g and measured 2.00 cm × 15.0 cm × 6.00 cm. Was the bar pure gold? Show evidence for your answer.

82. A 35.0-mL sample of ethyl alcohol (density 0.789 g/mL) is added to a graduated cylinder that has a mass of 49.28 g. What will be the mass of the cylinder plus the alcohol?

83. The largest nugget of gold on record was found in 1872 in New South Wales, Australia, and had a mass of 93.3 kg. Assuming the nugget is pure gold, what is its volume in cubic centimeters? What is it worth by today's standards if gold is $559/oz? (14.58 troy oz = 1 lb)

Challenge Exercises

All exercises with blue numbers have answers in the appendix of the text.

*84. Your boss found a piece of metal in the lab and wants you to determine what the metal is. She's pretty sure that the metal is either of lead, aluminum, or silver. The lab bench has a balance and a 100-mL graduated cylinder with 50 mL of water in it. You decide to weigh the metal and find that it has a mass of 20.25 g. After dropping the piece of metal into the graduated cylinder containing water, you notice that the volume increased to 57.5 mL. Identify the metal.

*85. Forgetful Freddie placed 25.0 mL of a liquid in a graduated cylinder with a mass of 89.450 g when empty. When Freddie placed a metal slug with a mass of 15.454 g into the cylinder, the volume rose to 30.7 mL. Freddie was asked to calculate the density of the liquid and of the metal slug from his data, but he forgot to obtain the mass of the liquid. He was told that if he found the mass of the cylinder containing the liquid and the slug, he would have enough data for the calculations. He did so and found its mass to be 125.934 g. Calculate the density of the liquid and of the metal slug.

Answers to Practice Exercises

2.1 (a) $1200 = 1.200 \times 10^3$
(left means positive exponent)
(b) $6,600,000 = 6.6 \times 10^6$
(left means positive exponent)
(c) $0.0468 = 4.68 \times 10^{-2}$
(right means negative exponent)
(d) $0.00003 = 3 \times 10^{-5}$
(right means negative exponent)

2.2 (a) 2; (b) 4; (c) 4; (d) 1; (e) 3; (f) 4; (g) 1; (h) 3

2.3 (a) 42.25 (Rule 2); (b) 88.0 (Rule 1); (c) 0.0897 (Rule 2);
(d) 0.090 (Rule 2); (e) 225 (Rule 1); (f) 14.2 (Rule 2)

2.4 (a) $3350 \text{ in.}^2 = 3.4 \times 10^3 \text{ in.}^2$; (b) 50.7 mi/hr; (c) 0.79;
(d) 1.3; (e) 20.5; (f) 3.71

2.5 (a) 2; (b) 2; (c) 1; (d) 2; (e) 4; (f) 2; (g) 2; (h) 2

2.6 (a) 1.69×10^4 m; (b) 3.5×10^{-2} m^2

2.7 30 m^3 or 3×10^1 m^3

2.8 165 lb

2.9 0.14 lb

2.10 0.793 qt

2.11 1.89 L (the number of significant figures is arbitrary)

2.12 $-269°C$, $-452°F$

2.13 $37.0°C$, 310.2 K

2.14 The density is 9.35 g/mL; therefore the ring is not pure silver. The density of silver is 10.5 g/mL.

2.15 0.88 g/mL

Elements and Compounds

Flower varieties are often classified by color in the nursery.

Chapter Outline

3.1 Elements

3.2 Distribution of Elements

3.3 Names of the Elements

3.4 Symbols of the Elements

3.5 Introduction to the Periodic Table

3.6 Elements in Their Natural States

3.7 Elements That Exist as Diatomic Molecules

3.8 Compounds

3.9 Chemical Formulas

In Chapter 1 we learned that matter can be divided into the broad categories of pure substances and mixtures. We further learned that pure substances are either elements or compounds. The chemical elements are very important to us in our daily lives. In tiny amounts they play a large role in our health and metabolism. Metallic elements are used for the skin of airplanes, buildings, and sculpture. In this chapter we explore the nature of the chemical elements and begin to learn how chemists classify them.

3.1   Elements

element

See the periodic table on the inside front cover.

All words in English are formed from an alphabet consisting of only 26 letters. All known substances on Earth—and most probably in the universe, too—are formed from a sort of "chemical alphabet" consisting of over 100 known elements. An **element** is a fundamental or elementary substance that cannot be broken down by chemical means to simpler substances. Elements are the building blocks of all substances. The elements are numbered in order of increasing complexity beginning with hydrogen, number 1. Of the first 92 elements, 88 are known to occur in nature. The other four—technetium (43), promethium (61), astatine (85), and francium (87)—either do not occur in nature or have only transitory existences during radioactive decay. With the exception of number 94, plutonium, elements above number 92 are not known to occur naturally but have been synthesized, usually in very small quantities, in laboratories. The discovery of trace amounts of element 94 (plutonium) in nature has been reported. No elements other than those on Earth have been detected on other bodies in the universe.

Most substances can be decomposed into two or more simpler substances. Water can be decomposed into hydrogen and oxygen. Sugar can be decomposed into carbon, hydrogen, and oxygen. Table salt is easily decomposed into sodium and chlorine. An element, however, cannot be decomposed into simpler substances by ordinary chemical changes.

If we could take a small piece of an element, say copper, and divide it and subdivide it into smaller and smaller particles, we would finally come to a single unit of copper that we could no longer divide and still have copper (see Figure 3.1).

Figure 3.1
The surface of a penny is made up of tiny identical copper atoms packed tightly together.

This smallest particle of an element that can exist is called an **atom**, which is also the smallest unit of an element that can enter into a chemical reaction. Atoms are made up of still smaller subatomic particles. However, these subatomic particles (described in Chapter 5) do not have the properties of elements.

atom

3.2   Distribution of Elements

Elements are distributed unequally in nature, as shown in Figure 3.2. At normal room temperature two of the elements, bromine and mercury, are liquids. Eleven elements—hydrogen, nitrogen, oxygen, fluorine, chlorine, helium, neon, argon, krypton, xenon, and radon—are gases. All the other elements are solids.

Ten elements make up about 99% of the mass of the Earth's crust, seawater, and atmosphere. Oxygen, the most abundant of these, constitutes about 50% of this mass. The distribution of the elements shown in Figure 3.2 includes the Earth's crust to a depth of about 10 miles, the oceans, fresh water, and the atmosphere but does not include the mantle and core of the Earth, which are believed to consist of metallic iron and nickel. Because the atmosphere contains relatively little matter, its inclusion has almost no effect on the distribution. But the inclusion of fresh and salt water does have an appreciable effect since water contains about 11.2% hydrogen. Nearly all of the 0.9% hydrogen shown in the second graph in Figure 3.2 is from water.

The average distribution of the elements in the human body is shown in Figure 3.2c. Note again the high percentage of oxygen.

3.3   Names of the Elements

The names of the elements come to us from various sources. Many are derived from early Greek, Latin, or German words that describe some property of the element. For example, iodine is taken from the Greek word *iodes,* meaning

(a)   (b)

(a) Bromine (Br_2) and (b) mercury (Hg) are elements that are liquid at room temperature.

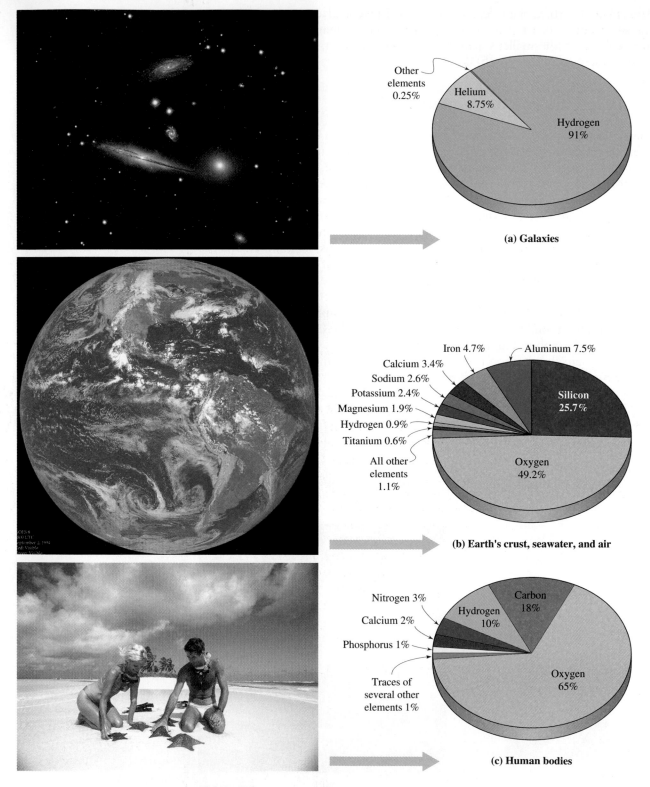

Figure 3.2
Distribution of the common elements in nature.

"violetlike," and iodine is certainly violet in the vapor state. The name of the metal bismuth originates from the German words *weisse masse,* which means "white mass." Miners called it *wismat*; it was later changed to *bismat,* and finally to bismuth. Some elements are named for the location of their discovery—for example, germanium, discovered in 1886 by a German chemist. Others are named in commemoration of famous scientists, such as einsteinium and curium, named for Albert Einstein and Marie Curie, respectively.

3.4 Symbols of the Elements

We all recognize Mr., N.Y., and Ave. as abbreviations for mister, New York, and avenue. In a like manner, each element also has an abbreviation; these are called **symbols** of the elements. Fourteen elements have a single letter as their symbol, and the rest have two letters. A symbol stands for the element itself, for one atom of the element, and (as we shall see later) for a particular quantity of the element.

symbol

Rules governing symbols of elements are as follows:

1. Symbols have either one or two letters.

2. If one letter is used, it is capitalized.

3. If two letters are used, only the first is capitalized.

Examples: Iodine I   Barium Ba

The symbols and names of all the elements are given in the table on the inside front cover of this book. Table 3.1 lists the more commonly used elements and their symbols. Examine Table 3.1 carefully and you will note that most of

Table 3.1 Symbols of the Most Common Elements

Element	Symbol	Element	Symbol	Element	Symbol
Aluminum	Al	Gold	Au	Platinum	Pt
Antimony	Sb	Helium	He	Plutonium	Pu
Argon	Ar	Hydrogen	H	Potassium	K
Arsenic	As	Iodine	I	Radium	Ra
Barium	Ba	Iron	Fe	Silicon	Si
Bismuth	Bi	Lead	Pb	Silver	Ag
Boron	B	Lithium	Li	Sodium	Na
Bromine	Br	Magnesium	Mg	Strontium	Sr
Cadmium	Cd	Manganese	Mn	Sulfur	S
Calcium	Ca	Mercury	Hg	Tin	Sn
Carbon	C	Neon	Ne	Titanium	Ti
Chlorine	Cl	Nickel	Ni	Tungsten	W
Chromium	Cr	Nitrogen	N	Uranium	U
Cobalt	Co	Oxygen	O	Xenon	Xe
Copper	Cu	Palladium	Pd	Zinc	Zn
Fluorine	F	Phosphorus	P		

Iodine in its elemental form is a bluish-black crystal. The symbol for iodine is I.

The metal sodium is soft enough to cut with a knife. The symbol for sodium is Na.

The colors of these varieties of quartz are the result of the presence of different metallic elements in the samples.

the symbols start with the same letter as the name of the element that is represented. A number of symbols, however, appear to have no connection with the names of the elements they represent (see Table 3.2). These symbols have been carried over from earlier names (usually in Latin) of the elements and are so firmly implanted in the literature that their use is continued today.

Special care must be taken in writing symbols. Capitalize only the first letter, and use a lowercase second letter if needed. This is important. For example, consider Co, the symbol for the element cobalt. If you write CO (capital C and capital O), you will have written the two elements carbon and oxygen (the *formula* for carbon monoxide), *not* the single element cobalt. Also, make sure that you write the letters distinctly; otherwise, Co (for cobalt) may be misread as Ca (for calcium).

Table 3.2 Symbols of the Elements Derived from Early Names*

Present name	Symbol	Former name
Antimony	Sb	Stibium
Copper	Cu	Cuprum
Gold	Au	Aurum
Iron	Fe	Ferrum
Lead	Pb	Plumbum
Mercury	Hg	Hydrargyrum
Potassium	K	Kalium
Silver	Ag	Argentum
Sodium	Na	Natrium
Tin	Sn	Stannum
Tungsten	W	Wolfram

*These symbols are in use today even though they do not correspond to the current name of the element.

Knowledge of symbols is essential for writing chemical formulas and equations, and will be needed in the remainder of this book and in any future chemistry courses you may take. One way to learn the symbols is to practice a few minutes a day by making flash cards of names and symbols and then practicing daily. Initially it is a good plan to learn the symbols of the most common elements shown in Table 3.1.

3.5 Introduction to the Periodic Table

Almost all chemistry classrooms have a chart called the *periodic table* hanging on the wall. It shows all the chemical elements and contains a great deal of useful information about them. As we continue our study of chemistry, we will learn much more about the periodic table. For now let's begin with the basics.

A simple version of the periodic table is shown in Table 3.3. Notice that in each box there is the symbol for the element and, above it, a number called the *atomic number*. For example nitrogen is $\begin{smallmatrix}7\\N\end{smallmatrix}$ and gold is $\begin{smallmatrix}79\\Au\end{smallmatrix}$.

The elements are placed in the table in order of increasing atomic number in a particular arrangement designed by Dimitri Mendeleev in 1869. His arrangement organizes the elements with similar chemical properties in columns called families or **groups**. An example of this is the column

group

| 2
He |
| 10
Ne |
| 18
Ar |
| 36
Kr |
| 54
Xe |
| 86
Rn |

These elements are all gases and nonreactive. The group is called the **noble gases**. Other groups with special names are the **alkali metals** (under the 1A on the table), **alkaline earth metals** (Group 2A), and **halogens** (Group 7A).

noble gases
alkali metals
alkaline earth metals
halogens

The tall columns of the periodic table (1A–7A and the noble gases) are known as the **representative elements**. Those elements in the center section of the periodic table and called **transition elements**.

representative elements
transition elements

Metals, Nonmetals, and Metalloids

The elements can be classified as metals, nonmetals, and metalloids. Most of the elements are metals. We are familiar with them because of their widespread use in tools, construction materials, automobiles, and so on. But nonmetals are equally useful in our everyday life as major components of clothing, food, fuel, glass, plastics, and wood. Metalloids are often used in the electronics industry.

metal

The **metals** are solids at room temperature (mercury is an exception). They have high luster, are good conductors of heat and electricity, are *malleable* (can be rolled or hammered into sheets), and are *ductile* (can be drawn into wires). Most metals have a high melting point and high density. Familiar metals are aluminum, chromium, copper, gold, iron, lead, magnesium, mercury, nickel, platinum, silver, tin, and zinc. Less familiar but still important metals are calcium, cobalt, potassium, sodium, uranium, and titanium.

Metals have little tendency to combine with each other to form compounds. But many metals readily combine with nonmetals such as chlorine, oxygen, and sulfur to form compounds such as metallic chlorides, oxides, and sulfides. In nature, minerals are composed of the more reactive metals combined with other elements. A few of the less reactive metals such as copper, gold, and silver are sometimes found in a native, or free, state. Metals are often mixed with one another to form homogeneous mixtures of solids called alloys. Some examples are brass, bronze, steel, and coinage metals.

Tiny computer chips contain silicon, a metalloid.

Nonmetals, unlike metals, are not lustrous, have relatively low melting points and densities, and are generally poor conductors of heat and electricity.

nonmetal

Carbon, phosphorus, sulfur, selenium, and iodine are solids; bromine is a liquid; the rest of the nonmetals are gases. Common nonmetals found uncombined in nature are carbon (graphite and diamond), nitrogen, oxygen, sulfur, and the noble gases (helium, neon, argon, krypton, xenon, and radon).

Nonmetals combine with one another to form molecular compounds such as carbon dioxide (CO_2), methane (CH_4), butane (C_4H_{10}), and sulfur dioxide (SO_2). Fluorine, the most reactive nonmetal, combines readily with almost all other elements.

metalloid

Several elements (boron, silicon, germanium, arsenic, antimony, tellurium, and polonium) are classified as **metalloids** and have properties that are intermediate between those of metals and those of nonmetals. The intermediate position of these elements is shown in Table 3.3. Certain metalloids, such as boron, silicon, and germanium, are the raw materials for the semiconductor devices that make the electronics industry possible.

Table 3.3 The Periodic Table

1A												3A	4A	5A	6A	7A	Noble gases
1 H	2A																2 He
3 Li	4 Be											5 B	6 C	7 N	8 O	9 F	10 Ne
11 Na	12 Mg											13 Al	14 Si	15 P	16 S	17 Cl	18 Ar
19 K	20 Ca	21 Sc	22 Ti	23 V	24 Cr	25 Mn	26 Fe	27 Co	28 Ni	29 Cu	30 Zn	31 Ga	32 Ge	33 As	34 Se	35 Br	36 Kr
37 Rb	38 Sr	39 Y	40 Zr	41 Nb	42 Mo	43 Tc	44 Ru	45 Rh	46 Pd	47 Ag	48 Cd	49 In	50 Sn	51 Sb	52 Te	53 I	54 Xe
55 Cs	56 Ba	57 La*	72 Hf	73 Ta	74 W	75 Re	76 Os	77 Ir	78 Pt	79 Au	80 Hg	81 Tl	82 Pb	83 Bi	84 Po	85 At	86 Rn
87 Fr	88 Ra	89 Ac†	104 Rf	105 Db	106 Sg	107 Bh	108 Hs	109 Mt	110 Ds	111 Rg							

Metals
Metalloids
Nonmetals

	58 Ce	59 Pr	60 Nd	61 Pm	62 Sm	63 Eu	64 Gd	65 Tb	66 Dy	67 Ho	68 Er	69 Tm	70 Yb	71 Lu
*	90 Th	91 Pa	92 U	93 Np	94 Pu	95 Am	96 Cm	97 Bk	98 Cf	99 Es	100 Fm	101 Md	102 No	103 Lr

3.6   Elements in Their Natural States

Most substances around us are mixtures or compounds. Elements tend to be reactive, and they combine with other elements to form compounds. It is rare to find elements in nature in pure form. There are some exceptions, however. Gold, for example, can be found as nuggets. Silver and platinum can also be found in nature in pure form. In fact, these metals are sometimes called the

Treating damaged works of art is a tricky and time-consuming process that often produces mixed results. Now NASA has found a way to help bring damaged paintings back to life. NASA was breaking down oxygen molecules into oxygen atoms (a process that happens in nature in the upper atmosphere). The researchers were using the atomic oxygen to test the durability of satellite materials when they discovered that atomic oxygen could remove organic materials from the surface of objects without damaging the object. Bad news for satellite parts (which needed their coatings) but great news for the art world.

The chief conservator at the Cleveland Museum of Art tried the atomic oxygen treatment on two paintings damaged in a church fire in Cleveland. Although the paintings were not extremely valuable (and so were good subjects for an experiment), all other attempts to restore them had failed. Atomic oxygen proved to work wonders, and the soot and char came off to reveal the image below it. Since the treatment is a gas, the underlying layers were not harmed. The treatment doesn't work on everything and won't replace other techniques altogether, but the conservator was impressed enough to continue to work with NASA on the process.

A small, portable atomic oxygen unit has also been built to treat artworks that have been damaged in a small area (by grafitti, etc.). A beam of atomic oxygen is applied directly to the damaged area to "spot clean" a piece of art. This technique was used to clean a lipstick mark from Andy Warhol's painting *Bathtub* (1961). A cosmetic party had been held at the Andy Warhol Museum in 1997. Free lipstick samples had been distributed at the party, and one reveler decided to kiss the painting! The painting was not varnished, so the lipstick stuck fast to the paint. After a day of treatment with the atomic oxygen beam, the lipstick mark was gone. Space technology can now be used to revive and restore art!

This famous Andy Warhol painting was spot cleaned by using a beam of atomic oxygen.

noble metals since they have a low reactivity. The noble gases are also not reactive and can be found in nature in uncombined form. Helium gas, for example, consists of tiny helium atoms moving independently.

Air can also be divided into its component gases. It is mainly composed of nitrogen and oxygen gases. But when we "look inside" these gases, we find tiny molecules (N_2 and O_2) instead of independent atoms like we see in the noble gases.

Nitrogen and oxygen gases are composed of molecules (N_2, ●●) and (O_2, ●●).

3.7 Elements That Exist as Diatomic Molecules

Diatomic molecules each contain exactly two atoms (alike or different). Seven elements in their uncombined state are **diatomic molecules**. Their symbols, formulas, and brief descriptions are listed in Table 3.4. Whether found free in nature or prepared in the laboratory, the molecules of these elements always contain two atoms. The formulas of the free elements are therefore always written to show this molecular composition: H_2, N_2, O_2, F_2, Cl_2, Br_2, and I_2.

diatomic molecules

Table 3.4 Elements That Exist as Diatomic Molecules

Element	Symbol	Molecular formula	Normal state
Hydrogen	H	H_2	Colorless gas
Nitrogen	N	N_2	Colorless gas
Oxygen	O	O_2	Colorless gas
Fluorine	F	F_2	Pale yellow gas
Chlorine	Cl	Cl_2	Yellow-green gas
Bromine	Br	Br_2	Reddish-brown liquid
Iodine	I	I_2	Bluish-black solid

Container of hydrogen molecules.

H_2 (blue) and O_2 (red) molecule.

It is important to see that symbols can designate either an atom or a molecule of an element. Consider hydrogen and oxygen. Hydrogen gas is present in volcanic gases and can be prepared by many chemical reactions. Regardless of their source, all samples of free hydrogen gas consist of diatomic molecules.

Free hydrogen is designated by the formula H_2, which also expresses its composition. Oxygen makes up about 21% by volume of the air that we breathe. This free oxygen is constantly being replenished by photosynthesis; it can also be prepared in the laboratory by several reactions. The majority of free oxygen is diatomic and is designated by the formula O_2. Now consider water, a compound designated by the formula H_2O (sometimes HOH). Water contains neither free hydrogen (H_2) nor free oxygen (O_2). The H_2 part of the formula H_2O simply indicates that two atoms of hydrogen are combined with one atom of oxygen to form water.

Symbols are used to designate elements, show the composition of molecules of elements, and give the elemental composition of compounds.

Practice 3.1

Identify the physical state of each of the following elements at room temperature (20°C):

H, Na, Ca, N, S, Fe, Cl, Br, Ne, Hg

Hint: You may need to use a resource (such as the Internet or a chemical handbook) to assist you.

Practice 3.2

Identify each of the following elements as a nonmetal, metal, or metalloid:

Na, F, Cr, Mo, Kr, Si, Cu, Sb, I, S

3.8   Compounds

compound   A **compound** is a distinct substance that contains two or more elements chemically combined in a definite proportion by mass. Compounds, unlike elements, can be decomposed chemically into simpler substances—that is, into simpler compounds and/or elements. Atoms of the elements in a compound are combined in whole-number ratios, never as fractional parts. Compounds fall into two general types, *molecular* and *ionic*. Figure 3.3 illustrates the classification of compounds.

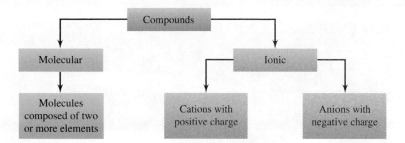

A **molecule** is the smallest uncharged individual unit of a compound formed by the union of two or more atoms. Water is a typical molecular compound. If we divide a drop of water into smaller and smaller particles, we finally obtain a single molecule of water consisting of two hydrogen atoms bonded to one oxygen atom, as shown in Figure 3.4a. This molecule is the ultimate particle of water; it cannot be further subdivided without destroying the water molecule and forming hydrogen and oxygen.

An **ion** is a positively or negatively charged atom or group of atoms. An ionic compound is held together by attractive forces that exist between positively and negatively charged ions. A positively charged ion is called a **cation** (pronounced *cat-eye-on*); a negatively charged ion is called an **anion** (pronounced *an-eye-on*).

Sodium chloride is a typical ionic compound. The ultimate particles of sodium chloride are positively charged sodium ions and negatively charged chloride ions, shown in Figure 3.4b. Sodium chloride is held together in a crystalline structure by the attractive forces existing between these oppositely charged ions. Although ionic compounds consist of large aggregates of cations and anions, their formulas are normally represented by the simplest possible ratio of the atoms in the compound. For example, in sodium chloride the ratio is one sodium ion to one chlorine ion, so the formula is NaCl.

There are more than 11 million known registered compounds, with no end in sight as to the number that will be prepared in the future. Each compound is unique and has characteristic properties. Let's consider two compounds, water and sodium chloride, in some detail. Water is a colorless, odorless, tasteless liquid that can be changed to a solid (ice) at 0°C and to a gas (steam) at 100°C. Composed of two atoms of hydrogen and one atom of oxygen per molecule, water is 11.2% hydrogen and 88.8% oxygen by mass. Water reacts chemically with sodium to produce hydrogen gas and sodium hydroxide, with lime to produce calcium hydroxide, and with sulfur trioxide to produce sulfuric acid. When water is decomposed, it forms hydrogen and oxygen molecules (see Figure 3.5). No other compound has all these exact physical and chemical properties; they are characteristic of water alone.

molecule

(a) H_2O

(b) NaCl

Figure 3.4
Representation of molecular and ionic (nonmolecular) compounds. (a) Two hydrogen atoms combined with an oxygen atom to form a molecule of water. (b) A positively charged sodium ion and a negatively charged chloride ion form the compound sodium chloride.

ion
cation
anion

Water molecules $\longrightarrow$ Oxygen molecules + Hydrogen molecules

Figure 3.5
A representation of the decomposition of water into oxygen and hydrogen molecules.

The increasing cost of petroleum and the increased pollution in our cities have sparked a race to find alternative fuels for our cars. Manufacturers agree that the best choice for automobiles of the future is hydrogen fuel cells. The greatest controversy now is how to obtain and deliver the hydrogen required for the fuel cells.

General Motors scientists are working to get hydrogen by "cracking," or re-forming, gasoline. Basically, larger hydrocarbon molecules are cracked, or broken down, to produce hydrogen and smaller hydrocarbons. This could be accomplished right onboard the vehicle. In fact, a Chevrolet S-10 truck has been designed as a fuel cell/battery hybrid. The fuel cell gets hydrogen from an onboard re-former, but the re-former and fuel cell equipment take up half the truck's bed. After adding batteries (for acceleration), the total weight of the truck is about 6300 pounds! The good news is the fuel economy is 40 miles per gallon with a range of 525 miles. This combined with the possibility of re-forming the gasoline or natural gas into hydrogen at gas stations (once enough hybrid vehicles are

Chrysler Town and Country Natrium fuel cell vehicle.

on the road) give General Motors optimism to continue development.

DaimlerChrysler is also working on fuel cells using another process. The company's engineers plan to bypass fossil fuels altogether. In March 2002, the firm unveiled the Natrium fuel cell Town and Country minivan prototype. This electric van is fueled by hydrogen obtained from a noncombustible aqueous solution of sodium borohydride. The liquid is carried in a large fuel tank and

is re-formed onboard to release hydrogen to the fuel cells. Since Natrium uses a nonfossil fuel and has a range of 300 miles without taking up passenger and cargo space, DaimlerChrysler plans to continue development in this direction. The biggest hurdle to overcome is the lack of an economic means to get the fuel from a borax mine to a filling station. Hopefully, one or both of these methods will produce economical alternatives to our present-day automobiles.

Sodium chloride is a colorless crystalline substance with a ratio of one atom of sodium to one atom of chlorine. Its composition by mass is 39.3% sodium and 60.7% chlorine. It does not conduct electricity in its solid state; it dissolves in water to produce a solution that conducts electricity. When a current is passed through molten sodium chloride, solid sodium and gaseous chlorine are produced (see Figure 3.6). These specific properties belong to sodium

Sodium chloride ⟶ Sodium metal + Chlorine gas

(a)   (b)   (c)

Figure 3.6
When sodium chloride (a) is decomposed, it forms sodium metal (b) and chlorine gas (c).

chloride and to no other substance. Thus, a compound may be identified and distinguished from all other compounds by its characteristic properties. We consider these chemical properties further in Chapter 4.

3.9 Chemical Formulas

Chemical formulas are used as abbreviations for compounds. A **chemical formula** shows the symbols and the ratio of the atoms of the elements in a compound. Sodium chloride contains one atom of sodium per atom of chlorine; its formula is NaCl. The formula for water is H_2O; it shows that a molecule of water contains two atoms of hydrogen and one atom of oxygen.

chemical formula

The formula of a compound tells us which elements it is composed of and how many atoms of each element are present in a formula unit. For example, a unit of sulfuric acid is composed of two atoms of hydrogen, one atom of sulfur, and four atoms of oxygen. We could express this compound as HHSOOOO, but this is cumbersome, so we write H_2SO_4 instead. The formula may be expressed verbally as "H-two-S-O-four." Numbers that appear partially below the line and to the right of a symbol of an element are called **subscripts**. Thus, the 2 and the 4 in H_2SO_4 are subscripts (See Figure 3.7). Characteristics of chemical formulas are as follows:

subscript

1. The formula of a compound contains the symbols of all the elements in the compound.
2. When the formula contains one atom of an element, the symbol of that element represents that one atom. The number 1 is not used as a subscript to indicate one atom of an element.

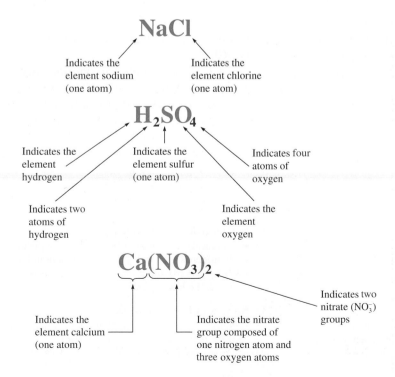

Figure 3.7
Explanation of the formulas NaCl, H_2SO_4, and $Ca(NO_3)_2$.

3. When the formula contains more than one atom of an element, the number of atoms is indicated by a subscript written to the right of the symbol of that atom. For example, the 2 in H_2O indicates two atoms of H in the formula.

4. When the formula contains more than one of a group of atoms that occurs as a unit, parentheses are placed around the group, and the number of units of the group is indicated by a subscript placed to the right of the parentheses. Consider the nitrate group, NO_3^-. The formula for sodium nitrate, $NaNO_3$, has only one nitrate group, so no parentheses are needed. Calcium nitrate, $Ca(NO_3)_2$, has two nitrate groups, as indicated by the use of parentheses and the subscript 2. $Ca(NO_3)_2$ has a total of nine atoms: one Ca, two N, and six O atoms. The formula $Ca(NO_3)_2$ is read as "C-A [pause] N-O-three taken twice."

5. Formulas written as H_2O, H_2SO_4, $Ca(NO_3)_2$, and $C_{12}H_{22}O_{11}$ show only the number and kind of each atom contained in the compound; they do not show the arrangement of the atoms in the compound or how they are chemically bonded to one another.

Practice 3.3

How would you read these formulas aloud?
(a) KBr   (b) $PbCl_2$ (c) $CaCO_3$ (d) $Mg(OH)_2$

Example 3.1   Write formulas for the following compounds; the atomic composition is given. (a) hydrogen chloride: 1 atom hydrogen + 1 atom chlorine; (b) methane: 1 atom carbon + 4 atoms hydrogen; (c) glucose: 6 atoms carbon + 12 atoms hydrogen + 6 atoms oxygen.

SOLUTION   (a) First write the symbols of the atoms in the formula: H Cl. Since the ratio of atoms is one to one, we bring the symbols together to give the formula for hydrogen chloride as HCl.

(b) Write the symbols of the atoms: C H. Now bring the symbols together and place a subscript 4 after the hydrogen atom. The formula is CH_4.

(c) Write the symbols of the atoms: C H O. Now write the formula, bringing together the symbols followed by the correct subscripts according to the data given (six C, twelve H, six O). The formula is $C_6H_{12}O_6$.

Chapter 3 Review

3.1   Elements

KEY TERMS

Element
Atom

- All matter consists of about 100 elements.
- An element is a fundamental chemical substance.
- The smallest particle of an element is an atom.
- Elements cannot be broken down by chemical means to a simpler substance.

3.2   Distribution of Elements

- Chemical elements are not distributed equally in nature.
- Hydrogen is the most abundant element in the universe.
- Oxygen is the most abundant element on the Earth and in the human body.

3.3   Names of the Elements

- Names for the chemical elements come from a variety of sources, including Latin, location of discovery, and famous scientists.

3.4 Symbols of the Elements

KEY TERM

Symbols

- Rules for writing symbols for the elements are:
  - One or two letters
  - If one letter, use a capital
  - If two letters, only the first is a capital

3.5 Introduction to the Periodic Table

KEY TERMS

Group	Representative elements
Noble gases	Transition elements
Alkali metals	Metals
Alkaline earth metal	Nonmetals
Halogens	Metalloids

- The periodic table was designed by Dimitri Mendeleev and arranges the elements according to their atomic numbers and in groups by their chemical properties.
- Elements can be classified as representative or as transition elements.
- Elements can also be classified by special groups with similar chemical properties. Such groups include the noble gases, alkali metals, alkaline earth metals, and halogens.
- Elements can be classified as metals, nonmetals, or metalloids:
  - Most elements are metals.
  - Metals have the following properties:
    - High luster
    - Good conductor of heat and electricity
    - Malleable
  - Nonmetals have the following properties:
    - Not lustrous
    - Poor conductors of heat and electricity

3.6 Elements in Their Natural States

- Most elements are found in nature combined with other elements.
- Elements that are found in uncombined form in nature include gold, silver, copper, and platinum as well as the noble gases (He, Ne, Ar, Kr, Xe, Rn).

3.7 Elements That Exist as Diatomic Molecules

KEY TERM

Diatomic molecules

- Diatomic molecules contain exactly two atoms (alike or different).
- Seven elements exist as diatomic molecules—H_2, N_2, O_2, F_2, Cl_2, Br_2, and I_2.

3.8 Compounds

KEY TERMS

Compound
Molecule
Ion
Cation
Anion

- A compound is a substance that contains two or more elements chemically combined in a definite proportion by mass.
- There are two general types of compounds:
  - Molecular—formed of individual molecules composed of atoms
  - Ionic—formed from ions that are either positive or negative
    - Cation—positively charged ion
    - Anion—negatively charged ion

3.9 Chemical Formulas

KEY TERMS

Chemical formula
Subscript

- A chemical formula shows the symbols and the ratios of atoms for the elements in a chemical compound.
- Characteristics of chemical formulas include:
  - It contains symbols of all elements in the compound.
  - The symbol represents one atom of the element.
  - If more than one atom of an element is present, the number of atoms is indicated by a subscript.
  - Parentheses are used to show multiple groups of atoms occurring as a unit in the compound.
  - A formula does not show the arrangement of the atoms in the compound.

Review Questions

All questions with blue numbers have answers in the appendix of the text.

1. Are there more atoms of silicon or hydrogen in the Earth's crust, seawater, and atmosphere? Use Figure 3.2 and the fact that the mass of a silicon atom is about 28 times that of a hydrogen atom.

2. Give the symbols for each of the following elements:
   - (a) silver
   - (b) oxygen
   - (c) hydrogen
   - (d) carbon
   - (e) iron
   - (f) nitrogen
   - (g) magnesium
   - (h) potassium

3. Give the names for each of these elements:
   - (a) Na
   - (b) F
   - (c) Ni
   - (d) Zn
   - (e) Ne
   - (f) He
   - (g) Ca
   - (h) Cl

4. What does the symbol of an element stand for?

5. Write down what you believe to be the symbols for the elements phosphorus, aluminum, hydrogen, potassium, magnesium, sodium, nitrogen, nickel, and silver. Check yourself by looking up the correct symbols in Table 3.1.

6. Interpret the difference in meanings for each of these pairs: (a) Si and SI (b) Pb and PB (c) 4 P and P_4

7. List six elements and their symbols in which the first letter of the symbol is different from that of the name. (Table 3.2)

8. Write the names and symbols for the 14 elements that have only one letter as their symbol. (See periodic table on inside front cover.)

9. Distinguish between an element and a compound.

10. How many metals are there? nonmetals? metalloids? (Table 3.3)

11. Of the ten most abundant elements in the Earth's crust, seawater, and atmosphere, how many are metals? nonmetals? metalloids? (Figure 3.2)

12. Of the six most abundant elements in the human body, how many are metals? nonmetals? metalloids? (Figure 3.2)

13. Why is the symbol for gold Au rather than G or Go?

14. Give the names of (a) the solid diatomic nonmetal and (b) the liquid diatomic nonmetal. (Table 3.4)

15. Distinguish between a compound and a mixture.

16. What are the two general types of compounds? How do they differ from each other?

17. What is the basis for distinguishing one compound from another?

18. What is the major difference between a cation and an anion?

19. Write the names and formulas of the elements that exist as diatomic molecules. (Table 3.4)

Paired Exercises

All exercises with blue numbers have answers in the appendix of the text.

1. Which of the following are diatomic molecules?
   (a) Cl_2 (e) N_2
   (b) CO   (f) HBr
   (c) CCl_4 (g) ClO_2
   (d) NO_2 (h) H_2O

2. Which of the following are diatomic molecules?
   (a) CO_2 (e) CS_2
   (b) H_2S (f) NO
   (c) H_2 (g) ClF
   (d) P_4

3. What elements are present in each compound?
   (a) potassium iodide KI
   (b) sodium carbonate Na_2CO_3
   (c) aluminum oxide Al_2O_3
   (d) calcium bromide $CaBr_2$
   (e) acetic acid $HC_2H_3O_2$

4. What elements are present in each compound?
   (a) magnesium bromide $MgBr_2$
   (b) carbon tetrachloride CCl_4
   (c) nitric acid HNO_3
   (d) barium sulfate $BaSO_4$
   (e) aluminum phosphate $AlPO_4$

5. Write the formula for each compound (the composition is given after each name):
   (a) zinc oxide   1 atom Zn, 1 atom O
   (b) potassium chlorate   1 atom K, 1 atom Cl, 3 atoms O
   (c) sodium hydroxide   1 atom Na, 1 atom O, 1 atom H
   (d) ethyl alcohol   2 atoms C, 6 atoms H, 1 atom O

6. Write the formula for each compound (the composition is given after each name):
   (a) aluminum bromide   1 atom Al, 3 atoms Br
   (b) calcium fluoride   1 atom Ca, 2 atoms F
   (c) lead(II) chromate 1 atom Pb, 1 atom Cr, 4 atoms O
   (d) benzene   6 atoms C, 6 atoms H

7. Explain the meaning of each symbol and number in these formulas:
   (a) H_2O
   (b) Na_2SO_4
   (c) $HC_2H_3O_2$

8. Explain the meaning of each symbol and number in these formulas:
   (a) $AlBr_3$
   (b) $Ni(NO_3)_2$
   (c) $C_{12}H_{22}O_{11}$ (sucrose)

9. How many atoms are represented in each formula?
   (a) KF   (d) $NaC_2H_3O_2$
   (b) $CaCO_3$ (e) $(NH_4)_2C_2O_4$
   (c) $K_2Cr_2O_7$

10. How many atoms are represented in each formula?
    (a) NaCl   (d) CCl_2F_2 (Freon)
    (b) N_2 (e) $Al_2(SO_4)_3$
    (c) $Ba(ClO_3)_2$

11. How many atoms of oxygen are represented in each formula?
    (a) H_2O (d) $Fe(OH)_3$
    (b) $CuSO_4$ (e) $Al(ClO_3)_3$
    (c) H_2O_2

12. How many atoms of hydrogen are represented in each formula?
    (a) H_2 (d) $HC_2H_3O_2$
    (b) $Ba(C_2H_3O_2)_2$ (e) $(NH_4)_2Cr_2O_7$
    (c) $C_6H_{12}O_6$

13. Identify each of the following as a pure substance or a mixture:
    (a) a bottle of Gatorade (d) coffee
    (b) aluminum   (e) fluorine gas
    (c) a piece of bread   (f) potassium carbonate

14. Identify each of the following as a pure substance or a mixture:
    (a) the human body   (d) copper
    (b) bottled spring water (e) Mountain Dew
    (c) aluminum oxide   (f) oak wood

15. For Question 13, state whether each pure substance is an element or a compound.

16. For Question 14, state whether each pure substance is an element or a compound.

17. Classify each of the following as an element, compound, or mixture:

18. Classify each of the following as an element, compound, or mixture:

(a)   (c)

(b)

(a)

(b)   (c)

19. Classify each material as an element, compound, or mixture:
    (a) xenon   (c) crude oil
    (b) sugar   (d) nitric acid

20. Classify each material as an element, compound, or mixture:
    (a) carbon monoxide (c) mouthwash
    (b) iced tea   (d) nickel

21. Is there a pattern to the location of the gaseous elements on the periodic table? If so, describe it.

22. Is there a pattern to the location of the liquid elements on the periodic table? If so, describe it.

23. What percent of the first 36 elements on the periodic table are metals?

24. What percent of the first 36 elements on the periodic table are solids at room temperature?

Additional Exercises

All exercises with blue numbers have answers in the appendix of the text.

25. Consider a homogeneous mixture of salt dissolved in water. What method could you use to separate the two components of the mixture?

26. Consider a heterogeneous mixture of golf balls, tennis balls, and footballs. What method could you use to separate the three components of the mixture?

27. You accidentally poured salt into your large grind pepper shaker. This is the only pepper that you have left and you need it to cook a meal, but you don't want the salt to be mixed in. How can you successfully separate these two components?

28. On the periodic table at the front of this book, do you notice anything about the atoms that make up the following ionic compounds: NaCl, KI, and $MgBr_2$? (*Hint:* Look at the position of the atoms in the given compounds on the periodic table.)

29. How many total atoms are present in each of the following compounds?
  (a) CO
  (b) BF_3
  (c) HNO_3
  (d) $KMnO_4$
  (e) $Ca(NO_3)_2$
  (f) $Fe_3(PO_4)_2$

30. The formula for vitamin B_{12} is $C_{63}H_{88}CoN_{14}O_{14}P$.
  (a) How many atoms make up one molecule of vitamin B_{12}?
  (b) What percentage of the total atoms are carbon?
  (c) What fraction of the total atoms are metallic?

*31. It has been estimated that there is 4×10^{-4} mg of gold per liter of seawater. At a price of \$19.40/g, what would be the value of the gold in $1 \text{ km}^3 (1 \times 10^{15} \text{ cm}^3)$ of the ocean?

32. Calcium dihydrogen phosphate is an important fertilizer. How many atoms of hydrogen are there in ten formula units of $Ca(H_2PO_4)_2$?

33. How many total atoms are there in one molecule of $C_{145}H_{293}O_{168}$?

34. Name the following:
  (a) three elements, all metals, beginning with the letter M
  (b) four elements, all solid nonmetals
  (c) five elements, all solids in the first five rows of the periodic table, whose symbols start with letters different from the element name.

35. How would you separate a mixture of sugar and sand and isolate each in its pure form?

36. How many total atoms are there in seven dozen molecules of nitric acid, HNO_3?

37. Make a graph using the data below. Plot the density of air in grams per liter along the *x*-axis and temperature along the *y*-axis.

Temperature (°C)	Density (g/L)
0	1.29
10	1.25
20	1.20
40	1.14
80	1.07

  (a) What is the relationship between density and temperature according to your graph?
  (b) From your plot, find the density of air at these temperatures:
    5°C   25°C   70°C

38. These formulas look similar but represent different things.
$$8 \text{ S} \qquad S_8$$
Compare and contrast them. How are they alike? How are they different?

39. Write formulas for the following compounds that a colleague read to you:
  (a) NA-CL
  (b) H2-S-O4
  (c) K2-O
  (d) Fe2-S3
  (e) K3-P-O4
  (f) CA (pause) CN taken twice
  (g) C6-H12-O6
  (h) C2-H5 (pause) OH
  (i) CR (pause) NO3 taken three times

40. The abundance of iodine in seawater is 5.0×10^{-8} % by mass. How many kilograms of seawater must be treated to obtain 1.0 g iodine?

Challenge Exercises

All exercises with blue numbers have answers in the appendix of the text.

*41. Write the chemical formulas of the neutral compounds that would result if the following ions were combined. Use the Charges of Common Ions table on the inside back cover of the book to help you.
  (a) ammonium ion and chloride ion
  (b) hydrogen ion and hydrogen sulfate ion
  (c) magnesium ion and iodide ion
  (d) iron(II) ion and fluoride ion
  (e) lead(II) ion and phosphate ion
  (f) aluminum ion and oxide ion

*42. Write formulas of all the compounds that will form between the first five of the Group 1A and 2A metals and the oxide ion.

Answers to Practice Exercises

3.1
gases	H, N, Cl, Ne
liquids	Br, Hg
solids	Na, Cu, S, Fe

3.2
nonmetal	F, Kr, I, S
metal	Na, Cr, Mo, Cu
metalloid	Si, Sb

3.3 (a) K–BR
  (b) P–B–CL–2
  (c) CA (pause) CO–3
  (d) MG (pause) OH taken twice

Properties of Matter

Iron can be melted at very high temperatures and then cast into a variety of shapes.

Chapter Outline

4.1 Properties of Substances

4.2 Physical Changes

4.3 Chemical Changes

4.4 Conservation of Mass

4.5 Energy

4.6 Heat: Quantitative Measurement

4.7 Energy in Chemical Changes

4.8 Conservation of Energy

The world we live in is a kaleidoscope of sights, sounds, smells, and tastes. Our senses help us to describe these objects in our lives. For example, the smell of freshly baked cinnamon rolls creates a mouthwatering desire to gobble down a sample. Just as sights, sounds, smells, and tastes form the properties of the objects around us, each substance in chemistry has its own unique properties that allow us to identify it and predict its interactions.

These interactions produce both physical and chemical changes. When you eat an apple, the ultimate metabolic result is carbon dioxide and water. These same products are achieved by burning logs. Not only does a chemical change occur in these cases, but an energy change occurs as well. Some reactions release energy (as does the apple or the log) whereas others require energy, such as the production of steel or the melting of ice. Over 90% of our current energy comes from chemical reactions.

4.1   Properties of Substances

properties

How do we recognize substances? Each substance has a set of **properties** that are characteristic of that substance and give it a unique identity. Properties— the "personality traits" of substances—are classified as either physical or chemical. **Physical properties** are the inherent characteristics of a substance that can be determined without altering its composition; they are associated with its physical existence. Common physical properties include color, taste, odor, state of matter (solid, liquid, or gas), density, melting point, and boiling point. **Chemical properties** describe the ability of a substance to form new substances, either by reaction with other substances or by decomposition.

physical properties

chemical properties

Let's consider a few of the physical and chemical properties of chlorine. Physically, chlorine is a gas at room temperature about 2.4 times heavier than air. It is yellowish-green in color and has a disagreeable odor. Chemically, chlorine will not burn but will support the combustion of certain other substances. It can be used as a bleaching agent, as a disinfectant for water, and in many chlorinated substances such as refrigerants and insecticides. When chlorine combines with the metal sodium, it forms a salt called sodium chloride. These properties, among many others, help us characterize and identify chlorine.

Substances, then, are recognized and differentiated by their properties. Table 4.1 lists four substances and several of their common physical properties. Information about physical properties, such as that given in Table 4.1,

Table 4.1 Physical Properties of Chlorine, Water, Sugar, and Acetic Acid

Substance	Color	Odor	Taste	Physical state	Melting point (°C)	Boiling point (°C)
Chlorine	Yellowish-green	Sharp, suffocating	Sharp, sour	Gas (20°C)	−101.6	−34.6
Water	Colorless	Odorless	Tasteless	Liquid	0.0	100.0
Sugar	White	Odorless	Sweet	Solid	—	Decomposes 170–186
Acetic acid	Colorless	Like vinegar	Sour	Liquid	16.7	118.0

Chemists are heavily involved in the manufacture of our currency. In fact, in a very real way, the money industry depends on chemistry and finding substances with the correct properties. The most common paper currency in the United States is the dollar bill. Chemistry is used to form the ink and paper and in processes used to defeat counterfeiters. The ink used on currency has to do a variety of things. It must be just the right consistency to fill the fine lines of the printing plate and release onto the paper without smearing. The ink must dry almost immediately, since the sheets of currency fly out of the press and into stacks 10,000 sheets tall. The pressure at the bottom of the stack is large, and the ink must not stick to the back of the sheet above it.

The security of our currency also depends on substances in the ink, which is optically variable. This color-changing ink shifts from green to black depending on how the bill is tilted. The ink used for the numbers in the lower right corner on the front of $10, $20, $50, and $100 bills shows this color change.

Once the currency is printed, it has to undergo durability tests. The bills are soaked in different solvents for

Dollar bill.

24 hours to make sure that they can stand dry cleaning and chemical exposure to household items such as gasoline. Then the bills must pass a washing-machine test to be sure that, for example, your $20 bill is still intact in your pocket after the washer is through with it. Paper currency is a blend of 75% cotton and 25% linen. Last, the bills must pass the crumple test, in which they are rolled up, put in a metal tube, crushed by a plunger, removed, and flattened. This process is repeated up to 50 times.

Once a bill is in circulation, Federal Reserve banks screen it using light to measure wear. If the bill gets too dirty, it is shredded and sent to a landfill. The average $1 bill stays in circulation for only 18 months. No wonder Congress decided in 1997 to revive the dollar coin by creating a new dollar coin to succeed the Susan B. Anthony dollar.

Currency in stacks.

is available in handbooks of chemistry and physics. Scientists don't pretend to know all the answers or to remember voluminous amounts of data, but it is important for them to know where to look for data in the literature and on the Internet.

Many chemists have reference books such as *Handbook of Chemistry and Physics* **to use as a resource.**

No two substances have identical physical and chemical properties.

4.2 Physical Changes

Matter can undergo two types of changes, physical and chemical. **Physical changes** are changes in physical properties (such as size, shape, and density) or changes in the state of matter without an accompanying change in composition. The changing of ice into water and water into steam are physical changes from one state of matter into another. No new substances are formed in these physical changes.

physical change

When a clean platinum wire is heated in a burner flame, the appearance of the platinum changes from silvery metallic to glowing red. This change is physical because the platinum can be restored to its original metallic appearance by cooling and, more importantly, because the composition of the platinum is not changed by heating and cooling.

4.3   Chemical Changes

chemical change

In a **chemical change**, new substances are formed that have different properties and composition from the original material. The new substances need not resemble the original material in any way.

When a clean copper wire is heated in a burner flame, the appearance of the copper changes from coppery metallic to glowing red. Unlike the platinum wire, the copper wire is not restored to its original appearance by cooling but instead becomes a black material. This black material is a new substance called copper(II) oxide. It was formed by chemical change when copper combined with oxygen in the air during the heating process. The unheated wire was essentially 100% copper, but the copper(II) oxide is 79.9% copper and 20.1% oxygen. One gram of copper will yield 1.251 g of copper(II) oxide (see Figure 4.1). The platinum is changed only physically when heated, but the copper is changed both physically and chemically when heated.

When 1.000 g of copper reacts with oxygen to yield 1.251 g of copper(II) oxide, the copper must have combined with 0.251 g of oxygen. The percentage of copper and oxygen can be calculated from these data [the copper and oxygen each being a percent of the total mass of copper(II) oxide]:

$$1.000 \text{ g copper} + 0.251 \text{ g oxygen} \rightarrow 1.251 \text{ g copper(II) oxide}$$

$$\frac{1.000 \text{ g copper}}{1.251 \text{ g copper(II) oxide}} \times 100 = 79.94\% \text{ copper}$$

$$\frac{0.251 \text{ g oxygen}}{1.251 \text{ g copper(II) oxide}} \times 100 = 20.1\% \text{ oxygen}$$

Figure 4.1
Chemical change: Forming of copper(II) oxide from copper and oxygen:

(a) Before heating, the wire is 100% copper (1.000 g)
(b) Copper and oxygen from the air combine chemically when the wire is heated.
(c) After heating, the wire is black copper(II) oxide (79.9% copper, 20.1% oxygen) (1.251 g).

(a)   (b)   (c)

Figure 4.2
Electrolysis of water produces
hydrogen gas (on the right) and
oxygen gas (on the left). Note
the ratio of the gases is 2:1.

Water can be decomposed chemically into hydrogen and oxygen. This is usually accomplished by passing electricity through the water in a process called
electrolysis. Hydrogen collects at one electrode while oxygen collects at the other
(see Figure 4.2). The composition and the physical appearance of the hydrogen
and the oxygen are quite different from that of water. They are both colorless
gases, but each behaves differently when a burning splint is placed into the sample: The hydrogen explodes with a pop while the flame brightens considerably
in the oxygen (oxygen supports and intensifies the combustion of the wood).
From these observations, we conclude that a chemical change has taken place.

Chemists have devised a shorthand method for expressing chemical changes
in the form of **chemical equations**. The two previous examples of chemical
changes can be represented by the following molecular representations, word
and symbol equations.

chemical equations

Reaction 1		
Type of equation	**Reactants**	**Products**
Word	water $\xrightarrow{\text{electrical energy}}$	hydrogen + oxygen
Molecular	$\xrightarrow{\text{electrical energy}}$	+
Symbol (formula)	$2\,H_2O \xrightarrow{\text{electrical energy}}$	$2\,H_2$ + O_2

Reaction 2				
Type of equation	Reactants			Products
Word	copper + oxygen		$\xrightarrow{\Delta}$	copper(II) oxide
Molecular	+		$\xrightarrow{\Delta}$	
Symbol (formula)	2 Cu + O_2		$\xrightarrow{\Delta}$	2 CuO

In reaction 1 water decomposes into hydrogen and oxygen when electrolyzed. In reaction 2 copper plus oxygen when heated produce copper(II) oxide. The arrow means "produces," and it points to the products. The Greek letter delta (Δ) represents heat. The starting substances (water, copper, and oxygen) are called the **reactants** and the substances produced [hydrogen, oxygen, and copper(II) oxide] are called the **products**. We will learn more about writing chemical equations in later chapters.

reactants
products

Physical change usually accompanies a chemical change. Table 4.2 lists some common physical and chemical changes; note that wherever a chemical change occurs, a physical change also occurs. However, wherever a physical change is listed, only a physical change occurs.

4.4   Conservation of Mass

law of conservation of mass

The **law of conservation of mass** states that no change is observed in the total mass of the substances involved in a chemical change. This law, tested by extensive laboratory experimentation, is the basis for the quantitative mass relationships among reactants and products.

The decomposition of water into hydrogen and oxygen illustrates this law. Thus, 100.0 g of water decompose into 11.2 g of hydrogen and 88.8 g of oxygen:

water $\longrightarrow$ hydrogen + oxygen
100.0 g   11.2 g   88.8 g

100.0 g $\longrightarrow$ 100.0 g
reactant   products

In a chemical reaction

mass of reactants = mass of products

Table 4.2 Physical or Chemical Changes of Some Common Processes

Process taking place	Type of change	Accompanying observations
Rusting of iron	Chemical	Shiny, bright metal changes to reddish-brown rust.
Boiling of water	Physical	Liquid changes to vapor.
Burning of sulfur in air	Chemical	Yellow, solid sulfur changes to gaseous, choking sulfur dioxide.
Boiling an egg	Chemical	Liquid white and yolk change to solids.
Combustion of gasoline	Chemical	Liquid gasoline burns to gaseous carbon monoxide, carbon dioxide, and water.
Digestion of food	Chemical	Food changes to liquid nutrients and partially solid wastes.
Sawing of wood	Physical	Smaller pieces of wood and sawdust are made from a larger piece of wood.
Burning of wood	Chemical	Wood burns to ashes, gaseous carbon dioxide, and water.
Heating of glass	Physical	Solid becomes pliable during heating, and the glass may change its shape.

Sawing wood is a physical charge.

Combustion of gasoline is a chemical charge.

4.5 Energy

From the early discovery that fire can warm us and cook our food to our discovery that nuclear reactors can be used to produce vast amounts of controlled energy, our technical progress has been directed by our ability to produce, harness, and utilize energy. **Energy** is the capacity of matter to do work. Energy exists in many forms; some of the more familiar forms are mechanical, chemical, electrical, heat, nuclear, and radiant or light energy. Matter can have both potential and kinetic energy.

Potential energy (PE) is stored energy, or energy that an object possesses due to its relative position. For example, a ball located 20 ft above the ground has more potential energy than when located 10 ft above the ground and will bounce higher when allowed to fall. Water backed up behind a dam represents potential energy that can be converted into useful work in the form of electrical or mechanical energy. Gasoline is a source of chemical potential energy. When gasoline burns (combines with oxygen), the heat released is associated with a decrease in potential energy. The new substances formed by burning have less chemical potential energy than the gasoline and oxygen.

Kinetic energy (KE) is energy that matter possesses due to its motion. When the water behind the dam is released and allowed to flow, its potential energy is changed into kinetic energy, which can be used to drive generators and produce electricity. Moving bodies possess kinetic energy. We all know the results when two moving vehicles collide: Their kinetic energy is expended in the crash that occurs. The pressure exerted by a confined gas is due to the kinetic energy of rapidly moving gas particles.

energy

potential energy

kinetic energy

The mechanical energy of falling water is converted to electrical energy at the hydroelectric plant at Niagara Falls.

joule
calorie

1 kJ = 1000 J
1 kcal = 1000 cal = 1 Cal

specific heat

Energy can be converted from one form to another form. Some kinds of energy can be converted to other forms easily and efficiently. For example, mechanical energy can be converted to electrical energy with an electric generator at better than 90% efficiency. On the other hand, solar energy has thus far been directly converted to electrical energy at an efficiency of only about 15%. In chemistry, energy is most frequently expressed as heat.

4.6   Heat: Quantitative Measurement

The SI-derived unit for energy is the joule (pronounced *jool*, rhyming with *tool*, and abbreviated J). Another unit for heat energy, which has been used for many years, is the calorie (abbreviated cal). The relationship between joules and calories is

$$4.184\,J = 1\,cal \quad (exactly)$$

To give you some idea of the magnitude of these heat units, 4.184 **joules**, or 1 **calorie**, is the quantity of heat energy required to change the temperature of 1 gram of water by 1°C, usually measured from 14.5°C to 15.5°C.

Since joules and calories are rather small units, kilojoules (kJ) and kilocalories (kcal) are used to express heat energy in many chemical processes. The kilocalorie is also known as the nutritional, or large, Calorie (spelled with a capital C and abbreviated Cal). In this book, heat energy will be expressed in joules.

The difference in the meanings of the terms *heat* and *temperature* can be seen by this example: Visualize two beakers, A and B. Beaker A contains 100 g of water at 20°C, and beaker B contains 200 g of water also at 20°C. The beakers are heated until the temperature of the water in each reaches 30°C. The temperature of the water in the beakers was raised by exactly the same amount, 10°C. But twice as much heat (8368 J) was required to raise the temperature of the water in beaker B as was required in beaker A (4184 J).

In the middle of the 18th century, Joseph Black, a Scottish chemist, was experimenting with the heating of elements. He heated and cooled equal masses of iron and lead through the same temperature range. Black noted that much more heat was needed for the iron than for the lead. He had discovered a fundamental property of matter—namely, that every substance has a characteristic heat capacity. Heat capacities may be compared in terms of specific heats. The **specific heat** of a substance is the quantity of heat (lost or gained) required to change the temperature of 1 g of that substance by 1°C. It follows then that the specific heat of liquid water is 4.184 J/g°C (1.000 cal/g°C). The specific heat of water is high compared with that of most substances. Aluminum and copper, for example, have specific heats of 0.900 J/g°C and 0.385 J/g°C, respectively (see Table 4.3). The relation of mass, specific heat, temperature change (Δt), and quantity of heat lost or gained by a system is expressed by this general equation:

$$\left(\begin{array}{c}\textbf{mass of}\\\textbf{substance}\end{array}\right)\left(\begin{array}{c}\textbf{specific heat}\\\textbf{of substance}\end{array}\right)(\Delta t) = \textbf{heat}$$

Thus the amount of heat needed to raise the temperature of 200. g of water by 10.0°C can be calculated as follows:

$$(200.\,g)\left(\frac{4.184\,J}{g\,°C}\right)(10.0\,°C) = 8.37 \times 10^3\,J$$

Table 4.3 Specific Heat of Selected Substances

Substance	Specific heat (J/g°C)	Specific heat (cal/g°C)
Water	4.184	1.000
Ethyl alcohol	2.138	0.511
Ice	2.059	0.492
Aluminum	0.900	0.215
Iron	0.473	0.113
Copper	0.385	0.0921
Gold	0.131	0.0312
Lead	0.128	0.0305

Calculate the specific heat of a solid in J/g°C and cal/g°C if 1638 J raise the temperature of 125 g of the solid from 25.0°C to 52.6°C.

Example 4.1

Substitute the data into the equation (mass)(specific heat)(Δt) = heat and solve for specific heat:

SOLUTION

$$\text{specific heat} = \frac{\text{heat}}{g \times \Delta t}$$

$$\text{heat} = 1638\,J \qquad \text{mass} = 125\,g \qquad \Delta t = 52.6°C - 25.0°C = 27.6°C$$

$$\text{specific heat} = \frac{1638\,J}{(125\,g)(27.6°C)} = 0.475\,J/g°C$$

Now convert joules to calories using 1.000 cal/4.184 J:

$$\text{specific heat} = \left(\frac{0.475\,\cancel{J}}{g°C}\right)\left(\frac{1.000\,cal}{4.184\,\cancel{J}}\right) = 0.114\,cal/g°C$$

A sample of a metal with a mass of 212 g is heated to 125.0°C and then dropped into 375 g water at 24.0°C. If the final temperature of the water is 34.2°C, what is the specific heat of the metal? (Assume no heat losses to the surroundings.)

Example 4.2

When the metal enters the water, it begins to cool, losing heat to the water. At the same time, the temperature of the water rises. This process continues until the temperature of the metal and the temperature of the water are equal, at which point (34.2°C) no net flow of heat occurs.

SOLUTION

The heat lost or gained by a system is given by (mass)(specific heat)(Δt) = energy change. First calculate the heat gained by the water, and then calculate the specific heat of the metal:

$$\text{temperature rise of the water} \qquad (\Delta t) = 34.2°C - 24.0°C = 10.2°C$$

$$\text{heat gained by the water} = (375\,\cancel{g})\left(\frac{4.184\,J}{\cancel{g}\cancel{C}}\right)(10.2\,\cancel{°C}) = 1.60 \times 10^4\,J$$

The metal dropped into the water must have a final temperature the same as the water (34.2°C):

$$\text{temperature drop by the metal} \qquad (\Delta t) = 125.0°C - 34.2°C = 90.8°C$$

$$\text{heat lost by the metal} = \text{heat gained by the water} = 1.60 \times 10^4\,J$$

To determine the specific heat of the metal, we rearrange the equation (mass)(specific heat)(Δt) = heat to obtain

$$\text{specific heat} = \frac{\text{heat}}{(\text{mass})(\Delta t)}$$

$$\text{specific heat of the metal} = \frac{1.60 \times 10^4 \, \text{J}}{(212 \, \text{g})(90.8°\text{C})} = 0.831 \, \text{J/g°C}$$

Practice 4.1 _____

Calculate the quantity of energy needed to heat 8.0 g of water from 42.0°C to 45.0°C.

Practice 4.2 _____

A 110.0-g sample of metal at 55.5°C raises the temperature of 150.0 g of water from 23.0°C to 25.5°C. Determine the specific heat of the metal in J/g°C.

4.7 Energy in Chemical Changes

In all chemical changes, matter either absorbs or releases energy. Chemical changes can produce different forms of energy. For example, electrical energy to start automobiles is produced by chemical changes in the lead storage battery. Light energy is produced by the chemical change that occurs in a light stick. Heat and light are released from the combustion of fuels. All the energy needed for our life processes—breathing, muscle contraction, blood circulation, and so on—is produced by chemical changes occurring within the cells of our bodies.

A rain forest in Marajo Island, Amazon estuary, Brazil.

Rapid combustion provides the energy for lifting the space shuttle.

W hat's the secret to getting all those kernels in a bag of microwave popcorn to pop? Food scientists have looked at all sorts of things to improve the poppability of popcorn. In every kernel of popcorn there is starch and a little moisture. As the kernel heats up, the water vaporizes and the pressure builds up in the kernel. Finally, the kernel bursts and the starch expands into the white foam of popped corn.

Researchers in Indiana tested 14 types of popcorn to determine the most poppable. They found a range of 4–47% of unpopped kernels and determined that the kernels that held mois-ture the best produced the fewest un-popped kernels. Further analysis led them to the conclusion that the deter-mining factor was the pericarp.

The pericarp is the hard casing on the popcorn kernel. It is made of long polymers of cellulose. When the kernel is heated up, the long chains of cellu-lose line up and make a strong crys-talline structure. This strong structure keeps the water vapor inside the ker-nel, allowing the pressure to rise high enough to pop the kernel.

Getting the most pops comes from the molecular changes resulting from heating the kernel. Energy and chem-istry combine to produce a tasty snack!

Popcorn.

Conversely, energy is used to cause chemical changes. For example, a chem-ical change occurs in the electroplating of metals when electrical energy is passed through a salt solution in which the metal is submerged. A chemical change occurs when radiant energy from the sun is used by green plants in the process of photosynthesis. And as we saw, a chemical change occurs when elec-tricity is used to decompose water into hydrogen and oxygen. Chemical changes are often used primarily to produce energy rather than to produce new substances. The heat or thrust generated by the combustion of fuels is more important than the new substances formed.

4.8 Conservation of Energy

An energy transformation occurs whenever a chemical change occurs (see Figure 4.3). If energy is absorbed during the change, the products will have more chemical potential energy than the reactants. Conversely, if energy is given off in

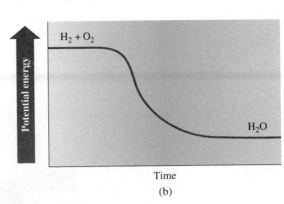

Figure 4.3
(a) In electrolysis of water, energy is absorbed by the system, so the products H_2 and O_2 have a higher potential energy.
(b) When hydrogen is burned (in O_2), energy is released and the product (H_2O) has lower potential energy.

a chemical change, the products will have less chemical potential energy than the reactants. Water can be decomposed in an electrolytic cell by absorbing electrical energy. The products, hydrogen and oxygen, have a greater chemical potential energy level than that of water (see Figure 4.3a). This potential energy is released in the form of heat and light when the hydrogen and oxygen are burned to form water again (see Figure 4.3b). Thus, energy can be changed from one form to another or from one substance to another, and therefore is not lost.

The energy changes occurring in many systems have been thoroughly studied. No system has ever been found to acquire energy except at the expense of energy possessed by another system. This is the **law of conservation of energy**: Energy can be neither created nor destroyed, though it can be transformed from one form to another.

law of conservation of energy

Chapter 4 Review

4.1   Properties of Substances

KEY TERMS

Properties
Physical properties
Chemical properties

- Each substance has a set of properties that are characteristic of the substance and give it a unique identity.
- Physical properties are inherent in the substance and can be determined without altering the composition of the substance:
  - Color
  - Taste
  - Odor
  - State of matter
  - Density
  - Melting point
  - Boiling point
- Chemical properties describe the ability of the substance to interact with other substances to form different substances.

4.2   Physical Changes

KEY TERM

Physical changes

- A physical change is a change in the physical properties or a change in the state of matter for a substance without altering the composition of the substance.

4.3   Chemical Changes

KEY TERMS

Chemical change
Chemical equations
Reactants
Products

- In a chemical change different substances are formed that have different properties and composition from the original material.
- Chemical changes can be represented by chemical equations:

- Word equation
- Molecular equation
- Symbol (formula) equation

4.4   Conservation of Mass

KEY TERM

Law of conservation of mass

- The law of conservation of mass states that matter is neither gained nor lost during a chemical reaction.

4.5   Energy

KEY TERMS

Energy
Potential energy
Kinetic energy

- Energy is the capacity to do work.
- Potential energy is energy that results from position or is stored within a substance.
- Kinetic energy is the energy matter possesses as a result of its motion.
- Energy can be converted from one form to another:
  - Common forms of energy:
    - Mechanical
    - Chemical
    - Electrical
    - Heat
    - Nuclear
    - Light
- In chemistry, energy is most frequently expressed as heat.

4.6   Heat: Quantitative Measurement

KEY TERMS

Joule
Calorie
Specific heat

- The SI unit for heat is the joule
  $4.184 \, J = 1 \, cal$

- The calorie is defined as the amount of heat required to change the temperature of 1 gram of water 1°C.
- Every substance has a characteristic heat capacity:
  - The specific heat of a substance is a measure of its heat capacity.
  - The specific heat is the quantity of heat required to change the temperature of 1 gram of the substance by 1°C.
- Heat lost or gained by a system can be calculated by

heat = (mass of substance) (specific heat of substance) (Δ*t*).

4.7 Energy in Chemical Changes
- In all chemical changes matter either absorbs or releases energy.

- Energy can be used to cause chemical change.
- Chemical changes are often used to produce energy rather than to produce new substances.

4.8 Conservation of Energy
KEY TERM
Law of conservation of energy

- Energy can neither be created or destroyed. It can be transformed from one form to another.
- The energy released or absorbed in a chemical reaction can be summarized in graphical form.

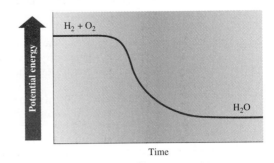

Review Questions
All questions with blue numbers have answers in the appendix of the text.

1. In what physical state does acetic acid exist at 393 K? (Table 4.1)

2. In what physical state does chlorine exist at −65°C? (Table 4.1)

3. What evidence of chemical change is visible when electricity is run through water? (Figure 4.2)

4. What is the fundamental difference between a chemical change and a physical change?

5. Distinguish between potential and kinetic energy.

6. Calculate the boiling point of acetic acid in
   (a) kelvins
   (b) degrees Fahrenheit (Table 4.1)

7. Calculate the melting point of acetic acid in (Table 4.1)
   (a) degrees Fahrenheit
   (b) kelvins

8. When water boils, bubbles form. What is inside those bubbles?

9. Is dissolving salt in water a physical or chemical change and why? How could you verify your answer?

Paired Exercises
All exercises with blue numbers have answers in the appendix of the text.

1. Determine whether each of the following represents a physical property or a chemical property:
   (a) Vinegar has a pungent odor.
   (b) Carbon cannot be decomposed.
   (c) Sulfur is a bright yellow.
   (d) Sodium chloride is a crystalline solid.
   (e) Water does not burn.
   (f) Mercury is a liquid at 25°C.
   (g) Oxygen is not combustible.
   (h) Aluminum combines with oxygen to form a protective oxide coating.

2. Determine whether each of the following represents a physical property or a chemical property:
   (a) Fluorine gas has a greenish-yellow tint.
   (b) The density of water at 4°C is 1.000 g/mL.
   (c) Hydrogen gas is very flammable.
   (d) Aluminum is a solid at 25°C.
   (e) Water is colorless and odorless.
   (f) Lemon juice tastes sour.
   (g) Gold does not tarnish.
   (h) Copper cannot be decomposed.

3. Cite the evidence that indicates that only physical changes occur when a platinum wire is heated in a Bunsen burner flame.

4. Cite the evidence that indicates that both physical and chemical changes occur when a copper wire is heated in a Bunsen burner flame.

5. Identify the reactants and products when a copper wire is heated in air in a Bunsen burner flame.

6. Identify the reactants and products for the electrolysis of water.

7. State whether each of the following represents a chemical change or merely a physical change:
   (a) A steak is cooked on a grill until well done.
   (b) In the lab, students firepolish the end of a glass rod. The jagged edge of the glass has become smooth.
   (c) Chlorine bleach is used to remove a coffee stain on a white lab coat.
   (d) When two clear and colorless aqueous salt solutions are mixed together, the solution turns cloudy and yellow.
   (e) One gram of an orange crystalline solid is heated in a test tube, producing a green powdery solid whose volume is 10 times the volume of the original substance.
   (f) In the lab, a student cuts a 20-cm strip of magnesium metal into 1-cm pieces.

8. State whether each of the following represents a chemical change or merely a physical change:
   (a) A few grams of sucrose (table sugar) are placed in a small beaker of deionized water; the sugar crystals "disappear," and the liquid in the beaker remains clear and colorless.
   (b) A copper statue, over time, turns green.
   (c) When a teaspoon of baking soda (sodium bicarbonate) is placed into a few ounces of vinegar (acetic acid), volumes of bubbles (effervescence) are produced.
   (d) When a few grams of a blue crystalline solid are placed into a beaker of deionized water, the crystals "disappear" and the liquid becomes clear and blue in color.
   (e) In the lab, a student mixes 2 mL of sodium hydroxide with 2 mL of hydrochloric acid in a test tube. He notices that the test tube has become very warm to the touch.
   (f) A woman visits a hairdresser and has her hair colored a darker shade of brown. After several weeks the hair, even though washed several times, has not changed back to the original color.

9. Are the following examples of potential energy or kinetic energy?
   (a) a ball sitting on the top of a hill
   (b) a stretched rubber band
   (c) a lit match
   (d) diving into a pool
   (e) a person rolling down a hill

10. Are the following examples of potential energy or kinetic energy?
    (a) runners poised at a starting line
    (b) a stretched arrow bow
    (c) running water
    (d) rubbing your hands together
    (e) a swing when it is poised at its highest position

11. What happens to the kinetic energy of a speeding car when the car is braked to a stop?

12. What energy transformation is responsible for the fiery reentry of the space shuttle into Earth's atmosphere?

13. Indicate with a plus sign (+) any of these processes that require energy, and a negative sign (−) any that release energy:
    (a) melting ice
    (b) relaxing a taut rubber band
    (c) a rocket launching
    (d) striking a match
    (e) a Slinky toy (spring) "walking" down stairs

14. Indicate with a plus sign (+) any of these processes that require energy, and a negative sign (−) any that release energy:
    (a) boiling water
    (b) releasing a balloon full of air with the neck open
    (c) a race car crashing into a wall
    (d) cooking a potato in a microwave oven
    (e) ice cream freezing in an ice cream maker

15. How many joules of energy are required to raise the temperature of 75 g of water from 20.0°C to 70.0°C?

16. How many joules of energy are required to raise the temperature of 65 g of iron from 25°C to 95°C?

17. How many joules of heat are required to heat 25.0 g of ethyl alcohol from the prevailing room temperature, 22.5°C, to its boiling point, 78.5°C?

18. How many joules of heat are required to heat 35.0 g of isopropyl alcohol from the prevailing room temperature, 21.2°C, to its boiling point, 82.4°C? (The specific heat of isopropyl alcohol is 2.604 J/g°C.)

19. A 250.0-g metal bar requires 5.866 kJ to change its temperature from 22°C to 100.0°C. What is the specific heat of the metal?

20. A 1.00-kg sample of antimony absorbed 30.7 kJ, thus raising the temperature of the antimony from 20.0°C to its melting point (630.0°C). Calculate the specific heat of antimony.

21. A 325.0-g piece of gold at 427°C is dropped into 200.0 g of water at 22.0°C. (The specific heat of gold is 0.131 J/g°C.) Calculate the final temperature of the mixture. (Assume no heat loss to the surroundings.)

22. A 500.0-g iron bar at 212°C is placed in 2.0 L of water at 24.0°C. What will be the change in temperature of the water? (Assume no heat is lost to the surroundings.)

Additional Exercises

All exercises with blue numbers have answers in the appendix of the text.

23. A 110.0-g sample of a gray-colored, unknown, pure metal was heated to 92.0°C and put into a coffee-cup calorimeter containing 75.0 g of water at 21.0°C. When the heated metal was put into the water, the temperature of the water rose to a final temperature of 24.2°C. The specific heat of water is 4.184 J. (a) What is the specific heat of the metal. (b) Is it possible that the metal is either iron or lead? Explain.

24. A 40.0-g sample of an unknown, yellowish-brown, pure metal was heated to 62.0°C and put into a coffee-cup calorimeter containing 85.0 g of water at 19.2°C. The water was heated by the hot metal to a temperature of 21.0°C. (a) What is the specific heat of the metal? (b) Is it possible that the metal is gold? Explain.

*25. You learned about the law of conservation of mass in this chapter. Suppose you burn a piece of wood that has a mass of 20 kg. You then determine the mass of the ashes remaining after the wood has been burned. Its mass is less than 20 kg. According to the law of conservation of mass, the mass before reaction should equal the mass after reaction (in this case burning) is over. What happened?

26. A sample of metal with a mass 250. g is heated to 130.°C and dropped into 425 g of water at 26.0°C. The final temperature of the water is 38.4°C. What is the specific heat capacity of the metal if 1.2×10^3 J are lost to the environment?

27. 18 g of oxygen are reacted with sufficient nitrogen to make 56 g of nitrogen dioxide according the following reaction:

$$2\,O_2(g) + N_2(g) \longrightarrow 2\,NO_2(g)$$

If the reaction goes to completion, how much nitrogen reacted?

28. The specific heat of zinc is 0.096 cal/g°C. Determine the energy required to raise the temperature of 250.0 g of zinc from room temperature (24°C) to 150.0°C.

29. If 40.0 kJ of energy are absorbed by 500.0 g of water at 10.0°C, what is the final temperature of the water?

*30. The heat of combustion of a sample of coal is 5500 cal/g. What quantity of this coal must be burned to heat 500.0 g of water from 20.0°C to 90.0°C?

31. One gram of anthracite coal gives off 7000. cal when burned. How many joules is this? How many grams of anthracite are required to raise the temperature of 4.0 L of water from 20.0°C to 100.0°C?

(For Questions 32 and 33, you may need to refer to Table 4.3)

32. A 100.0-g sample of copper is heated from 10.0°C to 100.0°C.
    (a) Determine the number of calories needed. (The specific heat of copper is 0.0921 cal/g°C.)
    (b) The same amount of heat is added to 100.0 g of Al at 10.0°C. (The specific heat of Al is 0.215 cal/g°C.) Which metal gets hotter, the copper or the aluminum?

33. A 500.0-g piece of iron is heated in a flame and dropped into 400.0 g of water at 10.0°C. The temperature of the water rises to 90.0°C. How hot was the iron when it was first removed from the flame? (The specific heat of iron is 0.473 J/g°C.)

*34. A 20.0-g piece of metal at 203°C is dropped into 100.0 g of water at 25.0°C. The water temperature rises to 29.0°C. Calculate the specific heat of the metal (J/g°C). Assume that all of the heat lost by the metal is transferred to the water and no heat is lost to the surroundings.

*35. Assuming no heat loss by the system, what will be the final temperature when 50.0 g of water at 10.0°C are mixed with 10.0 g of water at 50.0°C?

36. Three 500.0-g pans of iron, aluminum, and copper are each used to fry an egg. Which pan fries the egg (105°C) the quickest? Explain.

*37. At 6:00 P.M., you put a 300.0-g copper pan containing 800.0 g of water (all at room temperature, which is 25°C) on the stove. The stove supplies 628 J/s. When will the water reach the boiling point? (Assume no heat loss.)

38. Why does blowing gently across the surface of a cup of hot coffee help to cool it? Why does inserting a spoon into the coffee do the same thing?

39. If you are boiling some potatoes in a pot of water, will they cook faster if the water is boiling vigorously than if the water is only gently boiling? Explain your reasoning.

40. Homogenized whole milk contains 4% butterfat by volume. How many milliliters of fat are there in a glass (250 mL) of milk? How many grams of butterfat ($d = 0.8$ g/mL) are in this glass of milk?

41. A 100.0-mL volume of mercury (density = 13.6 g/mL) is put into a container along with 100.0 g of sulfur. The two substances react when heated and result in 1460 g of a dark, solid matter. Is that material an element or a compound? Explain. How many grams of mercury were in the container? How does this support the law of conservation of matter?

Challenge Exercises

All exercises with blue numbers have answers in the appendix of the text.

*42. Suppose a ball is sitting at the top of a hill. At this point, the ball has potential energy. The ball rolls down the hill, so the potential energy is converted into kinetic energy. When the ball reaches the bottom of the hill, it goes halfway up the hill on the other side and stops. If energy is supposed to be conserved, then why doesn't the ball go up the other hill to the same level as it started from?

43. From the following molecular picture determine whether a physical or a chemical change has occurred. Justify your answer.

44. From the following illustration (a) describe the change that has occurred and (b) determine whether the change was a physical change or a chemical change. Justify your answer.

45. (a) Describe the change(s) that you see in the following illustration. (b) Was this a physical or a chemical change?

Answers to Practice Exercises

4.1 1.0×10^2 J = 24 cal

4.2   0.477 J/g°C

PUTTING IT TOGETHER   Review for Chapters 1–4

Answers for Putting It Together Reviews are found in Appendix VII.

Multiple Choice:

Choose the correct answer to each of the following.

1. 1.00 cm is equal to how many meters?
   (a) 2.54   (b) 100.   (c) 10.0   (d) 0.0100

2. 1.00 cm is equal to how many inches?
   (a) 0.394   (b) 0.10   (c) 12   (d) 2.54

3. 4.50 ft is how many centimeters?
   (a) 11.4   (b) 21.3   (c) 454   (d) 137

4. The number 0.0048 contains how many significant figures?
   (a) 1   (b) 2   (c) 3   (d) 4

5. Express 0.00382 in scientific notation.
   (a) 3.82×10^3 (c) 3.82×10^{-2}
   (b) 3.8×10^{-3} (d) 3.82×10^{-3}

6. 42.0°C is equivalent to
   (a) 273 K   (c) 108°F
   (b) 5.55°F   (d) 53.3°F

7. 267°F is equivalent to
   (a) 404 K   (c) 540 K
   (b) 116°C   (d) 389 K

8. An object has a mass of 62 g and a volume of 4.6 mL. Its density is
   (a) 0.074 mL/g   (c) 7.4 g/mL
   (b) 285 g/mL   (d) 13 g/mL

9. The mass of a block is 9.43 g and its density is 2.35 g/mL. The volume of the block is
   (a) 4.01 mL   (c) 22.2 mL
   (b) 0.249 mL   (d) 2.49 mL

10. The density of copper is 8.92 g/mL. The mass of a piece of copper that has a volume of 9.5 mL is
    (a) 2.6 g   (c) 0.94 g
    (b) 85 g   (d) 1.1 g

11. An empty graduated cylinder has a mass of 54.772 g. When filled with 50.0 mL of an unknown liquid, it has a mass of 101.074 g. The density of the liquid is
    (a) 0.926 g/mL   (c) 2.02 g/mL
    (b) 1.08 g/mL   (d) 1.85 g/mL

12. The conversion factor to change grams to milligrams is
    (a) $\dfrac{100 \text{ mg}}{1 \text{ g}}$ (c) $\dfrac{1 \text{ g}}{1000 \text{ mg}}$
    (b) $\dfrac{1 \text{ g}}{100 \text{ mg}}$ (d) $\dfrac{1000 \text{ mg}}{1 \text{ g}}$

13. What Fahrenheit temperature is twice the Celsius temperature?
    (a) 64°F   (c) 200°F
    (b) 320°F   (d) 546°F

14. A gold alloy has a density of 12.41 g/mL and contains 75.0% gold by mass. The volume of this alloy that can be made from 255 g of pure gold is
    (a) 4.22×10^3 mL   (c) 27.4 mL
    (b) 2.37×10^3 mL   (d) 20.5 mL

15. A lead cylinder ($V = \pi r^2 h$) has radius 12.0 cm and length 44.0 cm and a density of 11.4 g/mL. The mass of the cylinder is
    (a) 2.27×10^5 g   (c) 1.78×10^3 g
    (b) 1.89×10^5 g   (d) 3.50×10^5 g

16. The following units can all be used for density *except*
    (a) g/cm^3 (b) kg/m^3 (c) g/L   (d) kg/m^2

17. 37.4 cm $\times$ 2.2 cm equals
    (a) 82.28 cm^2 (c) 82 cm^2
    (b) 82.3 cm^2 (d) 82.2 cm^2

18. The following elements are among the five most abundant by mass in the Earth's crust, seawater, and atmosphere *except*
    (a) oxygen   (c) silicon
    (b) hydrogen   (d) aluminum

19. Which of the following is a compound?
    (a) lead   (c) potassium
    (b) wood   (d) water

20. Which of the following is a mixture?
    (a) water   (c) wood
    (b) chromium   (d) sulfur

21. How many atoms are represented in the formula Na_2CrO_4?
    (a) 3   (b) (5)   (c) (7)   (d) 8

22. Which of the following is a characteristic of metals?
    (a) ductile   (c) extremely strong
    (b) easily shattered   (d) dull

23. Which of the following is a characteristic of nonmetals?
    (a) always a gas
    (b) poor conductor of electricity
    (c) shiny
    (d) combines only with metals

24. When a pure substance was analyzed, it was found to contain carbon and chlorine. This substance must be classified as
    (a) an element
    (b) a mixture
    (c) a compound
    (d) both a mixture and a compound

25. Chromium, fluorine, and magnesium have the symbols
    (a) Ch, F, Ma   (c) Cr, F, Mg
    (b) Cr, Fl, Mg   (d) Cr, F, Ma

26. Sodium, carbon, and sulfur have the symbols
    (a) Na, C, S   (c) Na, Ca, Su
    (b) So, C, Su   (d) So, Ca, Su

27. Coffee is an example of
    (a) an element   (c) a homogeneous mixture
    (b) a compound   (d) a heterogeneous mixture

28. The number of oxygen atoms in $Al(C_2H_3O_2)_3$ is
    (a) 2   (b) 3   (c) 5   (d) 6

29. Which of the following is a mixture?
    (a) water   (c) sugar solution
    (b) iron(II) oxide   (d) iodine

30. Which is the most compact state of matter?
    (a) solid   (c) gas
    (b) liquid   (d) amorphous

31. Which is not characteristic of a solution?
    (a) a homogeneous mixture
    (b) a heterogeneous mixture
    (c) one that has two or more substances
    (d) one that has a variable composition

32. A chemical formula is a combination of
    (a) symbols   (c) elements
    (b) atoms   (d) compounds

33. The number of nonmetal atoms in $Al_2(SO_3)_3$ is
    (a) 5   (b) 7   (c) 12   (d) 14

34. Which of the following is not a physical property?
    (a) boiling point   (c) bleaching action
    (b) physical state   (d) color

35. Which of the following is a physical change?
    (a) A piece of sulfur is burned.
    (b) A firecracker explodes.
    (c) A rubber band is stretched.
    (d) A nail rusts.

36. Which of the following is a chemical change?
    (a) Water evaporates.
    (b) Ice melts.
    (c) Rocks are ground to sand.
    (d) A penny tarnishes.

37. When 9.44 g of calcium are heated in air, 13.22 g of calcium oxide are formed. The percent by mass of oxygen in the compound is
    (a) 28.6%   (b) 40.0%   (c) 71.4%   (d) 13.2%

38. Barium iodide, BaI_2, contains 35.1% barium by mass. An 8.50-g sample of barium iodide contains what mass of iodine?
    (a) 5.52 g   (b) 2.98 g   (c) 3.51 g   (d) 6.49 g

39. Mercury(II) sulfide, HgS, contains 86.2% mercury by mass. The mass of HgS that can be made from 30.0 g of mercury is
    (a) 2586 g   (b) 2.87 g   (c) 25.9 g   (d) 34.8 g

40. The changing of liquid water to ice is known as a
    (a) chemical change
    (b) heterogeneous change
    (c) homogeneous change
    (d) physical change

41. Which of the following does not represent a chemical change?
    (a) heating of copper in air
    (b) combustion of gasoline
    (c) cooling of red-hot iron
    (d) digestion of food

42. Heating 30. g of water from 20. °C to 50. °C requires
    (a) 30. cal   (c) 3.8×10^3 J
    (b) 50. cal   (d) 6.3×10^3 J

43. The specific heat of aluminum is 0.900 J/g°C. How many joules of energy are required to raise the temperature of 20.0 g of Al from 10.0°C to 15.0°C?
    (a) 79 J   (b) 90. J   (c) 100. J   (d) 112 J

44. A 100.-g iron ball (specific heat = 0.473 J/g°C) is heated to 125°C and is placed in a calorimeter holding 200. g of water at 25.0°C. What will be the highest temperature reached by the water?
    (a) 43.7°C   (c) 65.3°C
    (b) 30.4°C   (d) 35.4°C

45. Which has the highest specific heat?
    (a) ice   (b) lead   (c) water   (d) aluminum

46. When 20.0 g of mercury are heated from 10.0°C to 20.0°C, 27.6 J of energy are absorbed. What is the specific heat of mercury?
    (a) 0.726 J/g°C   (c) 2.76 J/g°C
    (b) 0.138 J/g°C   (d) no correct answer given

47. Changing hydrogen and oxygen into water is a
    (a) physical change
    (b) chemical change
    (c) conservation reaction
    (d) no correct answer given

Free Response Questions:

Answer each of the following. Be sure to include your work and explanations in a clear, logical form.

1. You decide to go sailing in the tropics with some friends. Once there, you listen to the marine forecast, which predicts in-shore wave heights of 1.5 meters, offshore wave heights of 4 meters, and temperature of 27°C. Your friend the captain is unfamiliar with the metric system and he needs to know whether it is safe for your small boat and if it will be warm. He asks you to convert the measurements to feet and Fahrenheit, respectively.

2. Jane is melting butter in a copper pot on the stove. If she knows how much heat her stove releases per minute, what other measurements does she need to determine how much heat the butter absorbed? She assumes that the stove does not lose any heat to the surroundings.

3. Julius decided to heat 75 g $CaCO_3$ to determine how much carbon dioxide is produced. (*Note:* When $CaCO_3$ is heated, it produces CaO and carbon dioxide.) He collected the carbon dioxide in a balloon. Julius found the mass of the CaO remaining was 42 g. If 44 g of carbon dioxide take up 24 dm^3 of space, how many *liters* of gas were trapped in the balloon?

Use these pictures to answer Question 4.

(1)   (2)   (3)   (4)

4. (a) Which picture best describes a homogeneous mixture?
   (b) How would you classify the contents of the other containers?
   (c) Which picture contains a compound? Explain how you made your choice.

Use these pictures to answer Question 5.

(1)   (2)   (3)   (4)

5. (a) Which picture best represents fluorine gas? Why?
   (b) Which other elements could that picture also represent?
   (c) Which of the pictures could represent SO_3 gas?

6. Sue and Tim each left a one-quart bowl outside one night. Sue's bowl was full of water and covered while Tim's was empty and open. The next day there was a huge snowstorm that filled Tim's bowl with snow. The temperature that night went down to 12°F.
   (a) Which bowl would require less energy to bring its contents to room temperature (25°C)? Why?
   (b) What temperature change (°C) is required to warm the bowls to 25°C?

(c) How much heat (in kJ) is required to raise the temperature of the contents of Sue's bowl to 0°C (without converting the ice to water)?
(d) Did the water in Sue's bowl undergo chemical or physical changes or both?

7. One cup of Raisin Bran provides 60.% of the U.S. recommended daily allowance (RDA) of iron.
   (a) If the cereal provides 11 mg of iron, what is the U.S. RDA for Fe?
   (b) When the iron in the cereal is extracted, it is found to be the pure element. What is the volume of iron in a cup of the cereal?

8. 16% of the U.S. RDA for Ca is 162 mg.
   (a) What mass of calcium phosphate, $Ca_3(PO_4)_2$, provides 162 mg of calcium?
   (b) Is $Ca_3(PO_4)_2$ an element, mixture, or compound?
   (c) Milk is a good source of calcium in the diet. If 120 mL of skim milk provide 13% of the U.S. RDA for Ca, how many cups of milk should you drink per day if that is your only source of calcium?

9. Absent-minded Alfred put down a bottle containing silver on the table. When he went to retrieve it, he realized he had forgotten to label the bottle. Unfortunately there were two full bottles of the same size side-by-side. Alfred realized he had placed a bottle of mercury on the same table last week. State two ways Alfred can determine which bottle contains silver without opening the bottles.

10. Suppose 25 g of solid sulfur and 35 g of oxygen gas are placed in a sealed container.
    (a) Does the container hold a mixture or a compound?
    (b) After heating, the container was weighed. From a comparison of the total mass before heating to the total mass after heating, can you tell whether a reaction took place? Explain.
    (c) After the container is heated, all the contents are gaseous. Has the density of the container including its contents changed? Explain.

CHAPTER 5

Early Atomic Theory and Structure

Multiple lightning strikes over the desert in Arizona. Lightning occurs when electrons move to neutralize a charge difference between the clouds and the Earth.

Chapter Outline

5.1 Early Thoughts

5.2 Dalton's Model of the Atom

5.3 Composition of Compounds

5.4 The Nature of Electric Charge

5.5 Discovery of Ions

5.6 Subatomic Parts of the Atom

5.7 The Nuclear Atom

5.8 Isotopes of the Elements

5.9 Atomic Mass

Pure substances are classified as elements or compounds, but just what makes a substance possess its unique properties? How small a piece of salt will still taste salty? Carbon dioxide puts out fires, is used by plants to produce oxygen, and forms dry ice when solidified. But how small a mass of this material still behaves like carbon dioxide? Substances are in their simplest identifiable form at the atomic, ionic, or molecular level. Further division produces a loss of characteristic properties.

What particles lie within an atom or ion? How are these tiny particles alike? How do they differ? How far can we continue to divide them? Alchemists began the quest, early chemists laid the foundation, and modern chemists continue to build and expand on models of the atom.

5.1   Early Thoughts

The structure of matter has long intrigued and engaged us. The earliest models of the atom were developed by the ancient Greek philosophers. About 440 B.C. Empedocles stated that all matter was composed of four "elements"—earth, air, water, and fire. Democritus (about 470–370 B.C.) thought that all forms of matter were composed of tiny indivisible particles, which he called atoms, derived from the Greek word *atomos*, meaning "indivisible." He held that atoms were in constant motion and that they combined with one another in various ways. This hypothesis was not based on scientific observations. Shortly thereafter, Aristotle (384–322 B.C.) opposed the theory of Democritus and instead endorsed and advanced the Empedoclean theory. So strong was the influence of Aristotle that his theory dominated the thinking of scientists and philosophers until the beginning of the 17th century.

5.2   Dalton's Model of the Atom

More than 2000 years after Democritus, the English schoolmaster John Dalton (1766–1844) revived the concept of atoms and proposed an atomic model based on facts and experimental evidence (Figure 5.1). His theory, described in a series of papers published from 1803 to 1810, rested on the idea of a different kind of atom for each element. The essence of **Dalton's atomic model** may be summed up as follows:

Dalton's atomic model

1. Elements are composed of minute, indivisible particles called atoms.
2. Atoms of the same element are alike in mass and size.
3. Atoms of different elements have different masses and sizes.
4. Chemical compounds are formed by the union of two or more atoms of different elements.
5. Atoms combine to form compounds in simple numerical ratios, such as one to one, one to two, two to three, and so on.
6. Atoms of two elements may combine in different ratios to form more than one compound.

Dalton's atomic model stands as a landmark in the development of chemistry. The major premises of his model are still valid, but some of his statements must be modified or qualified because later investigations have shown that (1) atoms are composed of subatomic particles; (2) not all the atoms of a specific element have the same mass; and (3) atoms, under special circumstances, can be decomposed.

(a)

(b)   (c)

Figure 5.1
(a) Dalton's atoms were individual particles, the atoms of each element being alike in mass and size but different in mass and size from other elements. (b) and (c) Dalton's atoms combine in specific ratios to form compounds.

In chemistry we use models (theories) such as Dalton's atomic model to explain the behavior of atoms, molecules, and compounds. Models are modified to explain new information. We frequently learn the most about a system when our models (theories) fail. That is the time when we must rethink our explanation and determine whether we need to modify our model or propose a new or different model to explain the behavior.

5.3 Composition of Compounds

A large number of experiments extending over a long period have established the fact that a particular compound always contains the same elements in the same proportions by mass. For example, water always contains 11.2% hydrogen and 88.8% oxygen by mass (see Figure 5.1b). The fact that water contains hydrogen and oxygen in this particular ratio does not mean that hydrogen and oxygen cannot combine in some other ratio but rather that a compound with a different ratio would not be water. In fact, hydrogen peroxide is made up of two atoms of hydrogen and two atoms of oxygen per molecule and contains 5.9% hydrogen and 94.1% oxygen by mass; its properties are markedly different from those of water (see Figure 5.1c).

	Water	**Hydrogen peroxide**
	11.2% H	5.9% H
	88.8% O	94.1% O
Atomic composition	2 H + 1 O	2 H + 2 O

natural law
law of definite composition
We often summarize our general observations regarding nature into a statement called a **natural law**. In the case of the composition of a compound, we use the **law of definite composition**, which states that a compound always contains two or more elements chemically combined in a definite proportion by mass.

Let's consider two elements, oxygen and hydrogen, that form more than one compound. In water, 8.0 g of oxygen are present for each gram of hydrogen. In hydrogen peroxide, 16.0 g of oxygen are present for each gram of hydrogen. The masses of oxygen are in the ratio of small whole numbers, $16:8$ or $2:1$. Hydrogen peroxide has twice as much oxygen (by mass) as does water. Using Dalton's atomic model, we deduce that hydrogen peroxide has twice as many oxygen atoms per hydrogen atom as water. In fact, we now write the formulas for water as H_2O and for hydrogen peroxide as H_2O_2. See Figure 5.1b and c.

The **law of multiple proportions** states atoms of two or more elements may combine in different ratios to produce more than one compound.

law of multiple proportions

Some examples of the law of multiple proportions are given in Table 5.1. The reliability of this law and the law of definite composition is the cornerstone of the science of chemistry. In essence, these laws state that (1) the composition of a particular substance will always be the same no matter what its origin or how it is formed, and (2) the composition of different compounds formed from the same elements will always be unique.

Table 5.1 Selected Compounds Showing Elements That Combine to Give More Than One Compound

Compound	Formula	Percent composition
Copper(I) chloride	CuCl	64.2% Cu, 35.8% Cl
Copper(II) chloride	$CuCl_2$	47.3% Cu, 52.7% Cl
Methane	CH_4	74.9% C, 25.1% H
Octane	C_8H_{18}	85.6% C, 14.4% H
Methyl alcohol	CH_4O	37.5% C, 12.6% H, 49.9% O
Ethyl alcohol	C_2H_6O	52.1% C, 13.1% H, 34.7% O
Glucose	$C_6H_{12}O_6$	40.0% C, 6.7% H, 53.3% O

You need to recognize the difference between a *law* and a *model (theory)*. A law is a summary of observed behavior. A model (theory) is an attempt to explain the observed behavior. This means that laws remain constant—that is, they do not undergo modification—while theories (models) sometimes fail and are modified or discarded over time.

5.4 The Nature of Electric Charge

You've probably received a shock after walking across a carpeted area on a dry day. You may have also experienced the static associated with combing your hair and have had your clothing cling to you. These phenomena result from an accumulation of *electric charge*. This charge may be transferred from one object to another. The properties of electric charge are as follows:

1. Charge may be of two types, positive and negative.
2. Unlike charges attract (positive attracts negative), and like charges repel (negative repels negative and positive repels positive).

3. Charge may be transferred from one object to another, by contact or induction.
4. The less the distance between two charges, the greater the force of attraction between unlike charges (or repulsion between identical charges). The force of attraction (F) can be expressed using the following equation:

$$F = \frac{kq_1q_2}{r^2}$$

where q_1 and q_2 are the charges, r is the distance between the charges, and k is a constant.

5.5   Discovery of Ions

When ions are present in a solution of salt water and an electric current is passed through the solution, the light bulb glows.

English scientist Michael Faraday (1791–1867) made the discovery that certain substances when dissolved in water conduct an electric current. He also noticed that certain compounds decompose into their elements when an electric current is passed through the compound. Atoms of some elements are attracted to the positive electrode, while atoms of other elements are attracted to the negative electrode. Faraday concluded that these atoms are electrically charged. He called them *ions* after the Greek word meaning "wanderer."

Any moving charge is an electric current. The electrical charge must travel through a substance known as a conducting medium. The most familiar conducting media are metals formed into wires.

The Swedish scientist Svante Arrhenius (1859–1927) extended Faraday's work. Arrhenius reasoned that an ion is an atom (or a group of atoms) carrying a positive or negative charge. When a compound such as sodium chloride (NaCl) is melted, it conducts electricity. Water is unnecessary. Arrhenius's explanation of this conductivity was that upon melting, the sodium chloride dissociates, or breaks up, into charged ions, Na^+ and Cl^-. The Na^+ ions move toward the negative electrode (cathode), whereas the Cl^- ions migrate toward the positive electrode (anode). Thus positive ions are called *cations*, and negative ions are called *anions*.

From Faraday's and Arrhenius's work with ions, Irish physicist G. J. Stoney (1826–1911) realized there must be some fundamental unit of electricity associated with atoms. He named this unit the electron in 1891. Unfortunately, he had no means of supporting his idea with experimental proof. Evidence remained elusive until 1897, when English physicist J. J. Thomson (1856–1940) was able to show experimentally the existence of the electron.

5.6   Subatomic Parts of the Atom

The concept of the atom—a particle so small that until recently it could not be seen even with the most powerful microscope—and the subsequent determination of its structure stand among the greatest creative intellectual human achievements.

Any visible quantity of an element contains a vast number of identical atoms. But when we refer to an atom of an element, we isolate a single atom from the multitude in order to present the element in its simplest form. Figure 5.2 shows individual atoms as we can see them today.

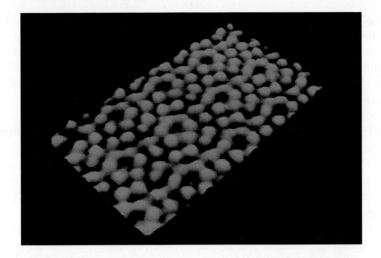

Figure 5.2
A scanning tunneling microscope shows an array of copper atoms.

What is this tiny particle we call the atom? The diameter of a single atom ranges from 0.1 to 0.5 nanometer (1 nm = 1×10^{-9} m). Hydrogen, the smallest atom, has a diameter of about 0.1 nm. To arrive at some idea of how small an atom is, consider this dot (•), which has a diameter of about 1 mm, or 1×10^{6} nm. It would take 10 million hydrogen atoms to form a line of atoms across this dot. As inconceivably small as atoms are, they contain even smaller particles, the **subatomic particles**, including electrons, protons, and neutrons.

Subatomic particles

The development of atomic theory was helped in large part by the invention of new instruments. For example, the Crookes tube, developed by Sir William Crookes (1832–1919) in 1875, opened the door to the subatomic structure of the atom (Figure 5.3). The emissions generated in a Crookes tube are called *cathode rays*. J. J. Thomson demonstrated in 1897 that cathode rays (1) travel in straight lines, (2) are negative in charge, (3) are deflected by electric and magnetic fields, (4) produce sharp shadows, and (5) are capable of moving a small paddle wheel. This was the experimental discovery of the fundamental unit of charge—the electron.

(a)

(b)

(c)

Figure 5.3
Crookes tube. Emissions generated in Crookes tube (a) travel in straight lines and are negative in charge, (b) are deflected by a magnetic field, and (c) provide a sharp shadow of the cross in the center of the tube.

electron   The **electron** (e⁻) is a particle with a negative electrical charge and a mass of 9.110×10^{-28} g. This mass is 1/1837 the mass of a hydrogen atom. Although the actual charge of an electron is known, its value is too cumbersome for practical use and has therefore been assigned a relative electrical charge of -1. The size of an electron has not been determined exactly, but its diameter is believed to be less than 10^{-12} cm.

Protons were first observed by German physicist Eugen Goldstein (1850–1930) in 1886. However, it was Thomson who discovered the nature of the proton. He showed that the proton is a particle, and he calculated its mass

proton   to be about 1837 times that of an electron. The **proton** (p) is a particle with actual mass of 1.673×10^{-24} g. Its relative charge ($+1$) is equal in magnitude, but opposite in sign, to the charge on the electron. The mass of a proton is only very slightly less than that of a hydrogen atom.

Thomson had shown that atoms contain both negatively and positively charged particles. Clearly, the Dalton model of the atom was no longer acceptable. Atoms are not indivisible but are instead composed of smaller parts. Thomson proposed a new model of the atom.

Thomson model of   In the **Thomson model of the atom**, the electrons are negatively charged
the atom   particles embedded in the atomic sphere. Since atoms are electrically neutral, the sphere also contains an equal number of protons, or positive charges. A neutral atom could become an ion by gaining or losing electrons.

Positive ions were explained by assuming that the neutral atom loses electrons. An atom with a net charge of $+1$ (for example, Na⁺ or Li⁺) has lost one electron. An atom with a net charge of $+3$ (for example, Al³⁺) has lost three electrons (Figure 5.4a).

Negative ions were explained by assuming that additional electrons can be added to atoms. A net charge of -1 (for example, Cl⁻ or F⁻) is produced by the addition of one electron. A net charge of -2 (for example, O²⁻ or S²⁻) requires the addition of two electrons (Figure 5.4b).

(a)   (b)

Figure 5.4
(a) When one or more electrons are lost from an atom, a cation is formed. (b) When one or more electrons are added to a neutral atom, an anion is formed.

The third major subatomic particle was discovered in 1932 by James Chad- wick (1891–1974). This particle, the **neutron** (n), has neither a positive nor a negative charge and has an actual mass (1.675×10^{-24} g) which is only very slightly greater than that of a proton. The properties of these three subatomic particles are summarized in Table 5.2.

neutron

Nearly all the ordinary chemical properties of matter can be explained in terms of atoms consisting of electrons, protons, and neutrons. The discussion of atomic structure that follows is based on the assumption that atoms contain only these principal subatomic particles. Many other subatomic particles, such as mesons, positrons, neutrinos, and antiprotons, have been discovered, but it is not yet clear whether all these particles are actually present in the atom or whether they are produced by reactions occurring within the nucleus. The fields of atomic and high-energy physics have produced a long list of subatomic par- ticles. Descriptions of the properties of many of these particles are to be found in physics textbooks.

Table 5.2 Electrical Charge and Relative Mass of Electrons, Protons, and Neutrons

Particle	Symbol	Relative electrical charge	Actual mass (g)
Electron	e^-	−1	9.110×10^{-28}
Proton	p	+1	1.673×10^{-24}
Neutron	n	0	1.675×10^{-24}

The mass of a helium atom is 6.65×10^{-24} g. How many atoms are in a 4.0-g sample of helium?

Example 5.1

$$(4.0 \text{ g})\left(\frac{1 \text{ atom He}}{6.65 \times 10^{-24} \text{ g}}\right) = 6.0 \times 10^{23} \text{ atoms He}$$

SOLUTION

Practice 5.1

The mass of an atom of hydrogen is 1.673×10^{-24} g. How many atoms are in a 10.0-g sample of hydrogen?

5.7 The Nuclear Atom

The discovery that positively charged particles are present in atoms came soon after the discovery of radioactivity by Henri Becquerel (1852–1908) in 1896. Radioactive elements spontaneously emit alpha particles, beta particles, and gamma rays from their nuclei (see Chapter 18).

By 1907 Ernest Rutherford (1871–1937) had established that the positive- ly charged alpha particles emitted by certain radioactive elements are ions of the element helium. Rutherford used these alpha particles to establish the

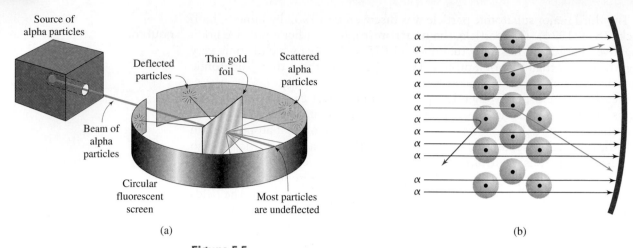

Figure 5.5
(a) Rutherford's experiment on alpha-particle scattering, where positive alpha particles (α), emanating from a radioactive source, were directed at a thin gold foil. (b) Deflection (red) and scattering (blue) of the positive alpha particles by the positive nuclei of the gold atoms.

nuclear nature of atoms. In experiments performed in 1911, he directed a stream of positively charged helium ions (alpha particles) at a very thin sheet of gold foil (about 1000 atoms thick). See Figure 5.5a. He observed that most of the alpha particles passed through the foil with little or no deflection; but a few of the particles were deflected at large angles, and occasionally one even bounced back from the foil (Figure 5.5b). It was known that like charges repel each other and that an electron with a mass of 1/1837 amu could not possibly have an appreciable effect on the path of a 4-amu alpha particle, which is about 7350 times more massive than an electron. Rutherford therefore reasoned that each gold atom must contain a positively charged mass occupying a relatively tiny volume and that, when an alpha particle approaches close enough to this positive mass, it is deflected. Rutherford spoke of this positively charged mass as the *nucleus* of the atom. Because alpha particles have relatively high masses, the extent of the deflections (some actually bounced back) indicated to Rutherford that the nucleus is very heavy and dense. (The density of the nucleus of a hydrogen atom is about 10^{12} g/cm^3—about 1 trillion times the density of water.) Because most of the alpha particles passed through the thousand or so gold atoms without any apparent deflection, he further concluded that most of an atom consists of empty space.

When we speak of the mass of an atom, we are referring primarily to the mass of the nucleus. The nucleus contains all the protons and neutrons, which represent more than 99.9% of the total mass of any atom (see Table 5.1). By way of illustration, the largest number of electrons known to exist in an atom is 111. The mass of even 111 electrons is only about 1/17 of the mass of a single proton or neutron. The mass of an atom therefore is primarily determined by the combined masses of its protons and neutrons.

General Arrangement of Subatomic Particles

The alpha-particle scattering experiments of Rutherford established that the atom contains a dense, positively charged nucleus. The later work of Chadwick demonstrated that the atom contains neutrons, which are particles with mass,

Figure 5.6
In the nuclear model of the atom, protons (p) and neutrons (n) are
located in the nucleus. The electrons are found in the remainder of the
atom (which is mostly empty space because electrons are very tiny).

but no charge. Rutherford also noted that light, negatively charged electrons
are present and offset the positive charges in the nucleus. Based on this
experimental evidence, a model of the atom and the location of its subatomic
particles was devised in which each atom consists of a **nucleus** surrounded by   **nucleus**
electrons (see Figure 5.6). The nucleus contains protons and neutrons but does
not contain electrons. In a neutral atom the positive charge of the nucleus
(due to protons) is exactly offset by the negative electrons. Because the charge
of an electron is equal to, but of opposite sign than, the charge of a proton, a
neutral atom must contain exactly the same number of electrons as protons.
However, this model of atomic structure provides no information on the
arrangement of electrons within the atom.

A neutral atom contains the same number of protons and electrons.

Atomic Numbers of the Elements

The **atomic number** of an element is the number of protons in the nucleus of   **atomic number**
an atom of that element. The atomic number determines the identity of an
atom. For example, every atom with an atomic number of 1 is a hydrogen atom;
it contains one proton in its nucleus. Every atom with an atomic number of 6
is a carbon atom; it contains 6 protons in its nucleus. Every atom with an atomic number of 92 is a uranium atom; it contains 92 protons in its nucleus. The
atomic number tells us not only the number of positive charges in the nucleus
but also the number of electrons in the neutral atom, since a neutral atom contains the same number of electrons and protons.

atomic number = number of protons in the nucleus

You don't need to memorize the atomic numbers of the elements because
a periodic table is usually provided in texts, in laboratories, and on examina-
tions. The atomic numbers of all elements are shown in the periodic table on
the inside front cover of this book and are also listed in the table of atomic
masses on the inside front endpapers.

Carbon, shown here as a beautiful diamond, has six protons and six electrons in each atom.

5.8   Isotopes of the Elements

Shortly after Rutherford's conception of the nuclear atom, experiments were performed to determine the masses of individual atoms. These experiments showed that the masses of nearly all atoms were greater than could be accounted for by simply adding up the masses of all the protons and electrons that were known to be present in an atom. This fact led to the concept of the neutron, a particle with no charge but with a mass about the same as that of a proton. Because this particle has no charge, it was very difficult to detect, and the existence of the neutron was not proven experimentally until 1932. All atomic nuclei except that of the simplest hydrogen atom contain neutrons.

All atoms of a given element have the same number of protons. Experimental evidence has shown that, in most cases, all atoms of a given element do not have identical masses. This is because atoms of the same element may have different numbers of neutrons in their nuclei.

isotopes   Atoms of an element having the same atomic number but different atomic masses are called **isotopes** of that element. Atoms of the various isotopes of an element therefore have the same number of protons and electrons but different numbers of neutrons.

Three isotopes of hydrogen (atomic number 1) are known. Each has one proton in the nucleus and one electron. The first isotope (protium), without a neutron, has a mass number of 1; the second isotope (deuterium), with one neutron in the nucleus, has a mass number of 2; the third isotope (tritium), with two neutrons, has a mass number of 3 (see Figure 5.7).

The three isotopes of hydrogen may be represented by the symbols $_1^1H$, $_1^2H$, $_1^3H$, indicating an atomic number of 1 and mass numbers of 1, 2, and 3, respectively. This method of representing atoms is called *isotopic notation*. The subscript (Z) is **mass number**   the atomic number; the superscript (A) is the **mass number**, which is the sum of the number of protons and the number of neutrons in the nucleus. The hydrogen isotopes may also be referred to as hydrogen-1, hydrogen-2, and hydrogen-3.

The mass number of an element is the sum of the protons and neutrons in the nucleus.

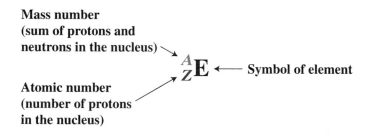

Mass number
(sum of protons and
neutrons in the nucleus)

$_Z^A E$ ←—— Symbol of element

Atomic number
(number of protons
in the nucleus)

Figure 5.7
The isotopes of hydrogen. The number of protons (purple) and neutrons (blue) are shown within the nucleus. The electron (e^-) exists outside the nucleus.

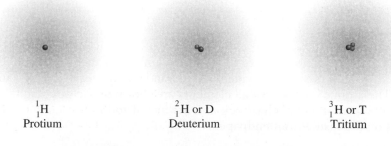

$_1^1H$
Protium

$_1^2H$ or D
Deuterium

$_1^3H$ or T
Tritium

Scientists are learning to use isotopes to determine the origin of drugs and gems. It turns out that isotope ratios similar to those used in carbon dating can also identify the source of cocaine or the birthplace of emeralds.

Researchers with the Drug Enforcement Agency (DEA) have created a database of the origin of coca leaves that pinpoints the origin of the leaves with a 90% accuracy. Cocaine keeps a chemical signature of the environment where it grew. Isotopes of carbon and nitrogen are found in a particular ratio based on climatic conditions in the growing region. These ratios correctly identified the source of 90% of the samples tested, according to James Ehleringer of the University of Utah, Salt Lake City. This new method can trace drugs a step further back than current techniques, which mainly look at chemicals introduced by processing practices in different locations. This could aid in tracking the original exporters and stopping production at the source.

It turns out that a similar isotopic analysis of oxygen has led researchers in France to be able to track the birthplace of emeralds. Very high quality emeralds have few inclusions (microscopic

An uncut (left) and cut (right) emerald from Brazil.

cavities). Gemologists use these inclusions and the material trapped in them to identify the source of the gem. High-quality gems can now also be identified by using an oxygen isotope ratio. These tests use an ion microscope that blasts a few atoms from the gems' surface (with virtually undetectable damage). The tiny sample is analyzed for its oxygen isotope ratio and then compared to a database

from emerald mines around the world. Using the information, gemologists can determine the mine from which the emerald was born. Since emeralds from Colombian mines are valued much more highly than those from other countries, this technique can be used to help collectors know just what they are paying for, as well as to identify the history of treasured emeralds.

Most of the elements occur in nature as mixtures of isotopes. However, not all isotopes are stable; some are radioactive and are continuously decomposing to form other elements. For example, of the seven known isotopes of carbon, only two, carbon-12 and carbon-13, are stable. Of the seven known isotopes of oxygen, only three—$^{16}_{8}O$, $^{17}_{8}O$, and $^{18}_{8}O$—are stable. Of the fifteen known isotopes of arsenic, $^{75}_{33}As$ is the only one that is stable.

Practice 5.2 _____

How many protons, neutrons, and electrons are in each of the three stable isotopes of oxygen?

5.9 Atomic Mass

The mass of a single atom is far too small to measure on a balance, but fairly precise determinations of the masses of individual atoms can be made with an instrument called a *mass spectrometer* (see Figure 5.8). The mass of a single

Figure 5.8
A modern mass spectrometer. A beam of positive ions is produced from the sample as it enters the chamber. The positive ions are then accelerated as they pass through slits in an electric field. When the ions enter the magnetic field, they are deflected differently, depending on mass and charge. The ions are then detected at the end of the tube. From intensity and position of the lines on the mass spectrogram, the different isotopes of the elements and their relative amounts can be determined.

atomic mass unit

$1\,amu = 1.6606 \times 10^{-24}\,g$

The copper used in casting the Liberty bell contains a mixture of the isotopes of copper.

hydrogen atom is 1.673×10^{-24} g. However, it is neither convenient nor practical to compare the actual masses of atoms expressed in grams; therefore, a table of relative atomic masses using *atomic mass units* was devised. (The term *atomic weight* is sometimes used instead of *atomic mass*.) The carbon isotope having six protons and six neutrons and designated carbon-12, or $^{12}_{6}C$, was chosen as the standard for atomic masses. This reference isotope was assigned a value of exactly 12 atomic mass units (amu). Thus, 1 **atomic mass unit** is defined as equal to exactly 1/12 of the mass of a carbon-12 atom. The actual mass of a carbon-12 atom is 1.9927×10^{-23} g, and that of one atomic mass unit is 1.6606×10^{-24} g. In the table of atomic masses, all elements then have values that are relative to the mass assigned to the reference isotope, carbon-12.

A table of atomic masses is given on the inside front cover of this book. Hydrogen atoms, with a mass of about 1/12 that of a carbon atom, have an average atomic mass of 1.00797 amu on this relative scale. Magnesium atoms, which are about twice as heavy as carbon, have an average mass of 24.305 amu. The average atomic mass of oxygen is 15.9994 amu.

Since most elements occur as mixtures of isotopes with different masses, the atomic mass determined for an element represents the average relative mass of all the naturally occurring isotopes of that element. The atomic masses of the individual isotopes are approximately whole numbers, because the relative masses of the protons and neutrons are approximately 1.0 amu each. Yet we find that the atomic masses given for many of the elements deviate considerably from whole numbers.

For example, the atomic mass of rubidium is 85.4678 amu, that of copper is 63.546 amu, and that of magnesium is 24.305 amu. The deviation of an atomic mass from a whole number is due mainly to the unequal occurrence of the various isotopes of an element. The two principal isotopes of copper are $^{63}_{29}Cu$ and $^{65}_{29}Cu$. It is apparent that copper-63 atoms are the more abundant isotope, since the atomic mass of copper, 63.546 amu, is closer to 63 than to 65 amu (see Figure 5.9). The actual values of the copper isotopes observed by mass spectra determination are shown in the following table:

Isotope	Isotopic mass (amu)	Abundance (%)	Average atomic mass (amu)
$^{63}_{29}Cu$	62.9298	69.09	
			63.55
$^{65}_{29}Cu$	64.9278	30.91	

Figure 5.9
A typical reading from a mass spectrometer. The two principal isotopes of copper are shown with the abundance (%) given.

The average atomic mass can be calculated by multiplying the atomic mass of each isotope by the fraction of each isotope present and adding the results. The calculation for copper is

$$(62.9298 \text{ amu})(0.6909) = 43.48 \text{ amu}$$
$$(64.9278 \text{ amu})(0.3091) = \underline{20.07 \text{ amu}}$$
$$63.55 \text{ amu}$$

> The average atomic mass is a weighted average of the masses of all the isotopes present in the sample.

The **atomic mass** of an element is the average relative mass of the isotopes of that element compared to the atomic mass of carbon-12 (exactly 12.0000 . . . amu).

> atomic mass

The relationship between mass number and atomic number is such that if we subtract the atomic number from the mass number of a given isotope, we obtain the number of neutrons in the nucleus of an atom of that isotope. Table 5.3 shows this method of determining the number of neutrons. For example, the fluorine atom ($^{19}_{9}F$), atomic number 9, having a mass of 19 amu, contains 10 neutrons:

mass number	−	atomic number	=	number of neutrons
19	−	9	=	10

The atomic masses given in the table on the front endpapers of this book are values accepted by international agreement. You need not memorize atomic masses. In the calculations in this book, the use of atomic masses to four significant figures will give results of sufficient accuracy. (See periodic table.)

> Use four significant figures for atomic masses in this text.

Table 5.3 Determination of the Number of Neutrons in an Atom by Subtracting Atomic Number from Mass Number

	Hydrogen ($^{1}_{1}H$)	Oxygen ($^{16}_{8}O$)	Sulfur ($^{32}_{16}S$)	Fluorine ($^{19}_{9}F$)	Iron ($^{56}_{26}Fe$)
Mass number	1	16	32	19	56
Atomic number	(−)1	(−)8	(−)16	(−)9	(−)26
Number of neutrons	0	8	16	10	30

Example 5.2   How many protons, neutrons, and electrons are found in an atom of $^{14}_{6}C$?

SOLUTION   The element is carbon, atomic number 6. The number of protons or electrons equals the atomic number and is 6. The number of neutrons is determined by subtracting the atomic number from the mass number: $14 - 6 = 8$.

Practice 5.3 _____

How many protons, neutrons, and electrons are in each of these isotopes?
(a) $^{16}_{8}O$,   (b) $^{80}_{35}Br$,   (c) $^{235}_{92}U$,   (d) $^{64}_{29}Cu$

Practice 5.4 _____

What is the atomic number and the mass number of the elements that contain
(a) 9 electrons   (b) 24 protons and 28 neutrons   (c) $^{197}_{79}X$

What are the names of these elements?

Example 5.3   Chlorine is found in nature as two isotopes, $^{37}_{17}Cl$ (24.47%) and $^{35}_{17}Cl$ (75.53%). The atomic masses are 36.96590 and 34.96885 amu, respectively. Determine the average atomic mass of chlorine.

SOLUTION   Multiply each mass by its percentage and add the results to find the average:

$$(0.2447)(36.96590 \text{ amu}) + (0.7553)(34.96885 \text{ amu})$$
$$= 35.4575 \text{ amu}$$
$$= 35.46 \text{ amu (4 significant figures)}$$

Practice 5.5 _____

Silver occurs as two isotopes with atomic masses 106.9041 and 108.9047 amu, respectively. The first isotope represents 51.82% and the second 48.18%. Determine the average atomic mass of silver.

Chapter 5 Review

5.1 Early Thoughts
- Greek model of matter:
  - Four elements—earth, air, water, fire
  - Democritus—atoms (indivisible particles) make up matter
  - Aristotle—opposed atomic ideas

5.2 Dalton's Model of the Atom

KEY TERM

Dalton's atomic model

- Summary of Dalton's model of the atom:
  - Elements are composed of atoms.
  - Atoms of the same element are alike (mass and size).
  - Atoms of different elements are different in mass and size.
  - Compounds form by the union of two or more atoms of different elements.
  - Atoms form compounds in simple numerical ratios.
  - Atoms of 2 elements may combine in different ratios to form different compounds.

5.3 Composition of Compounds

KEY TERMS

Natural law
Law of definite composition
Law of multiple proportions

- The law of definite composition states that a compound always contains two or more elements combined in a definite proportion by mass.
- The law of multiple proportions states that atoms of two or more elements may combine in different ratios to form more than one compound.

5.4 The Nature of Electric Charge
- The properties of electric charge:
  - Charges are one of two types—positive or negative.
  - Unlike charges attract and like charges repel.
  - Charge is transferred from one object to another by contact or by induction.
  - The force of attraction between charges is expressed by

$$F = \frac{kq_1q_2}{r^2}.$$

5.5 Discovery of Ions
- Michael Faraday discovered electrically charged ions.
- Svante Arrhenius explained that conductivity results from the dissociation of compounds into ions:
  - Cation—positive charge—attracted to negative electrode (cathode)
  - Anion—negative charge—attracted to positive electrode (anode)

5.6 Subatomic Parts of the Atom
KEY TERMS
Subatomic particles
Electron
Proton
Thomson model of the atom
Neutron

- Atoms contain smaller subatomic particles:
  - Electron—negative charge, 1/1837 mass of proton
  - Proton—positive charge, 1.673×10^{-24} g
  - Neutron—no charge, 1.675×10^{-24} g
- Thomson model of the atom:
  - Negative electrons are embedded in a positive atomic sphere.
  - Number of protons = number of electrons in a neutral atom.
  - Ions are formed by losing or gaining electrons.

5.7 The Nuclear Atom
KEY TERMS
Nucleus
Atomic number

- Rutherford gold foil experiment modified the Thomson model to a nuclear model of the atom:
  - Atoms are composed of a nucleus containing protons and/or neutrons surrounded by electrons, which occupy mostly empty space.
  - Neutral atoms contain equal numbers of protons and electrons.

5.8 Isotopes of the Elements
KEY TERMS
Isotopes
Mass number

- The mass number of an element is the sum of the protons and neutrons in the nucleus.

Mass number (sum of protons and neutrons in the nucleus) $\searrow$

$^A_Z E$ $\longleftarrow$ **Symbol of element**

Atomic number (number of protons in the nucleus) $\nearrow$

5.9 Atomic Mass
KEY TERMS
Atomic mass unit
Atomic mass

- The average atomic mass is a weighted average of the masses of all the isotopes present in the sample.
- The number of neutrons in an atom is determined by

mass number − atomic number = number of neutrons
A − Z = neutrons

Review Questions
All questions with blue numbers have answers in the appendix of the text.

1. What are the atomic numbers of (a) copper, (b) nitrogen, (c) phosphorus, (d) radium, and (e) zinc?
2. A neutron is approximately how many times heavier than an electron?
3. From the chemist's point of view, what are the essential differences among a proton, a neutron, and an electron?
4. Distinguish between an atom and an ion.
5. What letters are used to designate atomic number and mass number in isotopic notation of atoms?
6. In what ways are isotopes alike? In what ways are they different?

Paired Exercises

All exercises with blue numbers have answers in the appendix of the text.

1. In Section 5.3, there is a statement about the composition of water. It says that water (H_2O) contains 8 grams of oxygen for every 1 gram of hydrogen. Show why this statement is true.

2. In Section 5.3, there is a statement about the composition of hydrogen peroxide. It says that hydrogen peroxide (H_2O_2) contains 16 grams of oxygen for every 1 gram of hydrogen. Show why this statement is true.

3. Explain why, in Rutherford's experiments, some alpha particles were scattered at large angles by the gold foil or even bounced back.

4. What experimental evidence led Rutherford to conclude the following?
   (a) The nucleus of the atom contains most of the atomic mass.
   (b) The nucleus of the atom is positively charged.
   (c) The atom consists of mostly empty space.

5. Describe the general arrangement of subatomic particles in the atom.

6. What part of the atom contains practically all its mass?

7. What contribution did these scientists make to atomic models of the atom?
   (a) Dalton
   (b) Thomson
   (c) Rutherford

8. Consider the following models of the atom: (a) Dalton, (b) Thomson, (c) Rutherford. How does the location of the electrons in an atom vary? How does the location of the atom's positive matter compare?

9. Explain why the atomic masses of elements are not whole numbers.

10. Is the isotopic mass of a given isotope ever an exact whole number? Is it always? (Consider the masses of $^{12}_{6}C$ and $^{63}_{29}Cu$.)

11. What special names are given to the isotopes of hydrogen?

12. List the similarities and differences in the three isotopes of hydrogen.

13. What is the nuclear composition of the five naturally occurring isotopes of germanium having mass numbers 70, 72, 73, 74, and 76?

14. What is the nuclear composition of the five naturally occurring isotopes of zinc having mass numbers 64, 66, 67, 68, and 70?

15. Write isotopic notation symbols for each of the following:
    (a) $Z = 29$, $A = 65$
    (b) $Z = 20$, $A = 45$
    (c) $Z = 36$, $A = 84$

16. Write isotopic notation symbols for each of the following:
    (a) $Z = 47$, $A = 109$
    (b) $Z = 8$, $A = 18$
    (c) $Z = 26$, $A = 57$

17. Give the nuclear composition and isotopic notation for
    (a) an atom containing 27 protons, 32 neutrons, and 27 electrons
    (b) an atom containing 15 protons, 16 neutrons, and 15 electrons
    (c) an atom containing 110 neutrons, 74 electrons, and 74 protons
    (d) an atom containing 92 electrons, 143 neutrons, and 92 protons

18. Give the nuclear composition and isotopic notation for
    (a) an atom containing 12 protons, 13 neutrons, and 12 electrons
    (b) an atom containing 51 neutrons, 40 electrons, and 40 protons
    (c) an atom containing 50 protons, 50 electrons, and 72 neutrons
    (d) an atom containing 122 neutrons, 80 protons, and 80 electrons

19. An unknown element contains 24 protons, 21 electrons, and has mass number 54. Answer the following questions:
    (a) What is the atomic number of this element?
    (b) What is the symbol of the element?
    (c) How many neutrons does it contain?

20. An unknown element contains 35 protons, 36 electrons, and has mass number 80. Answer the following questions:
    (a) What is the atomic number of this element?
    (b) What is the symbol of the element?
    (c) How many neutrons does it contain?

21. Naturally occurring lead exists as four stable isotopes: ^{204}Pb with a mass of 203.973 amu (1.480%); ^{206}Pb, 205.974 amu (23.60%); ^{207}Pb, 206.9759 amu (22.60%); and ^{208}Pb, 207.9766 amu (52.30%). Calculate the average atomic mass of lead.

22. Naturally occurring magnesium consists of three stable isotopes: ^{24}Mg, 23.985 amu (78.99%); ^{25}Mg, 24.986 amu (10.00%); and ^{26}Mg, 25.983 amu (11.01%). Calculate the average atomic mass of magnesium.

23. 68.9257 amu is the mass of 60.4% of the atoms of an element with only two naturally occurring isotopes. The atomic mass of the other isotope is 70.9249 amu. Determine the average atomic mass of the element. Identify the element.

24. A sample of enriched lithium contains 30.00% ^{6}Li (6.015 amu) and 70.00% ^{7}Li (7.016 amu). What is the average atomic mass of the sample?

25. An average dimension for the radius of an atom is 1.0×10^{-8} cm, and the average radius of the nucleus is 1.0×10^{-13} cm. Determine the ratio of atomic volume to nuclear volume. Assume that the atom is spherical [$V = (4/3)\pi r^3$ for a sphere].

*26. An aluminum atom has an average diameter of about 3.0×10^{-8} cm. The nucleus has a diameter of about 2.0×10^{-13} cm. Calculate the ratio of the atom's diameter to its nucleus.

Additional Exercises

All exercises with blue numbers have answers in the appendix of the text.

27. What experimental evidence supports these statements?
(a) The nucleus of an atom is small.
(b) The atom consists of both positive and negative charges.
(c) The nucleus of the atom is positive.

28. What is the relationship between the following two atoms:
(a) one atom with 10 protons, 10 neutrons, and 10 electrons; and another atom with 10 protons, 11 neutrons, and 10 electrons
(b) one atom with 10 protons, 11 neutrons, and 10 electrons; and another atom with 11 protons, 10 neutrons, and 11 electrons

29. The radius of a carbon atom in many compounds is 0.77×10^{-8} cm. If the radius of a Styrofoam ball used to represent the carbon atom in a molecular model is 1.5 cm, how much of an enlargement is this?

30. How is it possible for there to be more than one kind of atom of the same element?

31. Which element contains the largest number of neutrons per atom: ^{210}Bi, ^{210}Po, ^{210}At, or ^{211}At?

*32. An unknown element Q has two known isotopes: ^{60}Q and ^{63}Q. If the average atomic mass is 61.5 amu, what are the relative percentages of the isotopes?

*33. The actual mass of one atom of an unknown isotope is 2.18×10^{-22} g. Calculate the atomic mass of this isotope.

34. The mass of an atom of argon is 6.63×10^{-24} g. How many atoms are in a 40.0-g sample of argon?

35. Using the periodic table inside the front cover of the book, determine which of the first 20 elements have isotopes that you would expect to have the same number of protons, neutrons, and electrons.

36. Complete the following table with the appropriate data for each isotope given (all are neutral atoms):

Element	Symbol	Atomic number	Mass number	Number of protons	Number of neutrons	Number of electrons
	^{36}Cl					
GOLD			197			
		56			79	
					20	18
			58	28		

37. Complete the following table with the appropriate data for each isotope given (all are neutral atoms):

Element	Symbol	Atomic number	Mass number	Number of protons	Number of neutrons	Number of electrons
	^{134}Xe					
SILVER			107			
				9		
		92			143	92
			41	19		

38. Draw diagrams similar to those shown in Figure 5.4 for the following ions:
(a) F^- (b) Pb^{2+} (c) S^{2-} (d) Al^{3+}

39. Draw pictures similar to those shown in Figure 5.7 for the following isotopes of oxygen:
(a) $^{16}_8O$ (b) $^{17}_8O$ (c) $^{18}_8O$

40. What percent of the total mass of one atom of each of the following elements comes from electrons?
(a) iron (mass of one atom = 9.274×10^{-23} g)
(b) nitrogen (mass of one atom = 2.326×10^{-23} g)
(c) carbon (mass of one atom = 1.994×10^{-23} g)
(d) potassium (mass of one atom = 6.493×10^{-23} g)

41. What percent of the total mass of one atom of each of the following elements comes from protons?
(a) sodium (mass of one atom = 3.818×10^{-23} g)
(b) oxygen (mass of one atom = 2.657×10^{-23} g)
(c) mercury (mass of one atom = 3.331×10^{-22} g)
(d) fluorine (mass of one atom = 3.155×10^{-23} g)

42. Figure 5.6 is a representation of the nuclear model of the atom. The area surrounding the nucleus is labeled as the electron region. What is the electron region?

Challenge Exercise

All exercises with blue numbers have answers in the appendix of the text.

***43.** You have discovered a new element and are trying to determine where on the periodic table it would fit. You decide to do a mass spectrometer analysis of the sample and discover that it contains three isotopes with masses of 270.51 amu, 271.23 amu, and 269.14 amu and relative abundances of 34.07%, 55.12%, and 10.81%, respectively. Sketch the mass spectrometer reading, determine the average atomic mass of the element, estimate its atomic number, and determine its approximate location on the periodic table.

Answers to Practice Exercises

5.1 5.98×10^{24} atoms

5.2 $^{16}_8O$ 8p, 8e$^-$, 8n
$^{17}_8O$ 8p, 8e$^-$, 9n
$^{18}_8O$ 8p, 8e$^-$, 10n

5.3

	protons	neutrons	electrons
(a)	8	8	8
(b)	35	45	35
(c)	92	143	92
(d)	29	35	29

5.4 (a) atomic number   9
        mass number   19
        name   potassium
     (b) atomic number   24
        mass number   52
        name   chromium
     (c) atomic number   79
        mass number   179
        name   gold

5.5 107.9 amu

Nomenclature of Inorganic Compounds

This exquisite chambered nautilus shell is formed from calcium carbonate, commonly called limestone.

Chapter Outline

6.1 Common and Systematic Names

6.2 Elements and Ions

6.3 Writing Formulas from Names of Ionic Compounds

6.4 Naming Binary Compounds

6.5 Naming Compounds Containing Polyatomic Ions

6.6 Acids

As children, we begin to communicate with other people in our lives by learning the names of objects around us. As we continue to develop, we learn to speak and use language to complete a wide variety of tasks. As we enter school, we begin to learn of other languages—the languages of mathematics, of other cultures, of computers. In each case, we begin by learning the names of the building blocks and then proceed to more abstract concepts. Chemistry has a language all its own—a whole new way of describing the objects so familiar to us in our daily lives. Once we learn the language, we can begin to understand the macroscopic and microscopic world of chemistry.

6.1   Common and Systematic Names

Chemical nomenclature is the system of names that chemists use to identify compounds. When a new substance is formulated, it must be named in order to distinguish it from all other substances (see Figure 6.1). In this chapter, we will restrict our discussion to the nomenclature of inorganic compounds—compounds that do not generally contain carbon.

Common names are arbitrary names that are not based on the chemical composition of compounds. Before chemistry was systematized, a substance was given a name that generally associated it with one of its outstanding physical or chemical properties. For example, *quicksilver* is a common name for mercury, and nitrous oxide (N_2O), used as an anesthetic in dentistry, has been called *laughing gas* because it induces laughter when inhaled. Water and ammonia are also common names because neither provides any information about the chemical composition of the compounds. If every substance were assigned a common name, the amount of memorization required to learn over 12 million names would be astronomical.

Common names have distinct limitations, but they remain in frequent use. Common names continue to be used because the systematic name is too long or too technical for everyday use. For example, calcium oxide (CaO) is called *lime* by plasterers; photographers refer to *hypo* rather than sodium thiosulfate ($Na_2S_2O_3$); and nutritionists use the name *vitamin D_3*, instead of 9,10-secocholesta-5,7,10(19)-trien-3-β-ol ($C_{27}H_{44}O$). Table 6.1 lists the common names, formulas, and systematic names of some familiar substances.

Water

Ammonia

Water (H_2O) and ammonia (NH_3) are almost always referred to by their common names.

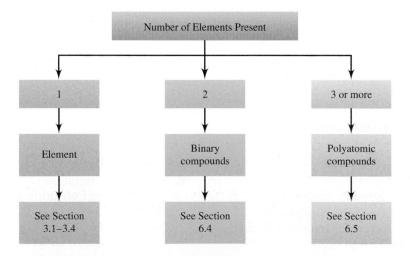

Figure 6.1
Where to find rules for naming inorganic substances in this book.

Table 6.1 Common Names, Formulas, and Chemical Names of Familiar Substances

Common names	Formula	Chemical name
Acetylene	C_2H_2	ethyne
Lime	CaO	calcium oxide
Slaked lime	$Ca(OH)_2$	calcium hydroxide
Water	H_2O	water
Galena	PbS	lead(II) sulfide
Alumina	Al_2O_3	aluminum oxide
Baking soda	$NaHCO_3$	sodium hydrogen carbonate
Cane or beet sugar	$C_{12}H_{22}O_{11}$	sucrose
Borax	$Na_2B_4O_7 \cdot 10\,H_2O$	sodium tetraborate decahydrate
Brimstone	S	sulfur
Calcite, marble, limestone	$CaCO_3$	calcium carbonate
Cream of tartar	$KHC_4H_4O_6$	potassium hydrogen tartrate
Epsom salts	$MgSO_4 \cdot 7\,H_2O$	magnesium sulfate heptahydrate
Gypsum	$CaSO_4 \cdot 2\,H_2O$	calcium sulfate dihydrate
Grain alcohol	C_2H_5OH	ethanol, ethyl alcohol
Hypo	$Na_2S_2O_3$	sodium thiosulfate
Laughing gas	N_2O	dinitrogen monoxide
Lye, caustic soda	NaOH	sodium hydroxide
Milk of magnesia	$Mg(OH)_2$	magnesium hydroxide
Muriatic acid	HCl	hydrochloric acid
Plaster of paris	$CaSO_4 \cdot \frac{1}{2}\,H_2O$	calcium sulfate hemihydrate
Potash	K_2CO_3	potassium carbonate
Pyrite (fool's gold)	FeS_2	iron disulfide
Quicksilver	Hg	mercury
Saltpeter (chile)	$NaNO_3$	sodium nitrate
Table salt	NaCl	sodium chloride
Vinegar	$HC_2H_3O_2$	acetic acid
Washing soda	$Na_2CO_3 \cdot 10\,H_2O$	sodium carbonate decahydrate
Wood alcohol	CH_3OH	methanol, methyl alcohol

Chemists prefer systematic names that precisely identify the chemical composition of chemical compounds. The system for inorganic nomenclature was devised by the International Union of Pure and Applied Chemistry (IUPAC), which was founded in 1921. The IUPAC meets regularly and constantly reviews and updates the system.

6.2 Elements and Ions

In Chapter 3, we studied the names and symbols for the elements as well as their location on the periodic table. In Chapter 5, we investigated the composition of the atom and learned that all atoms are composed of protons, electrons, and neutrons; that a particular element is defined by the number of protons it contains; and that atoms are uncharged because they contain equal numbers of protons and electrons.

When a scientist discovered a new element in the early days of chemistry, he or she had the honor of naming it. Now researchers must submit their choices for a name to an international committee called the International Union of Pure and Applied Chemistry before they can be placed on the periodic table. In 1997, the IUPAC decided on names for the elements from 104 through 111. These eight elements are now called rutherfordium (Rf), dubnium (Db), sea-

borgium (Sg), bohrium (Bh), hassium (Hs), meitnerium (Mt), darmstadtium (Ds), and roentgenium (Rg).

The new names are a compromise among choices presented by different research teams. The Russians gained recognition for work done at a laboratory in Dubna. Americans gained recognition for Glenn Seaborg, the first living scientist to have an element named after him. The British recognized Ernest Rutherford, who discovered the atomic nucleus. The Germans won recognition

both for Lise Meitner, who co-discovered atomic fission, and for one of their labs in the German state of Hesse. Both the Germans and the Russians won recognition for Niels Bohr, whose model of the atom led the way toward modern ideas about atomic structure. The German group that discovered elements 107–111 is recognized in the name darmstadtium, for their lab. Element 111 was named in honor of Wilhelm Roentgen, who discovered X-rays and won the first Nobel Prize in physics.

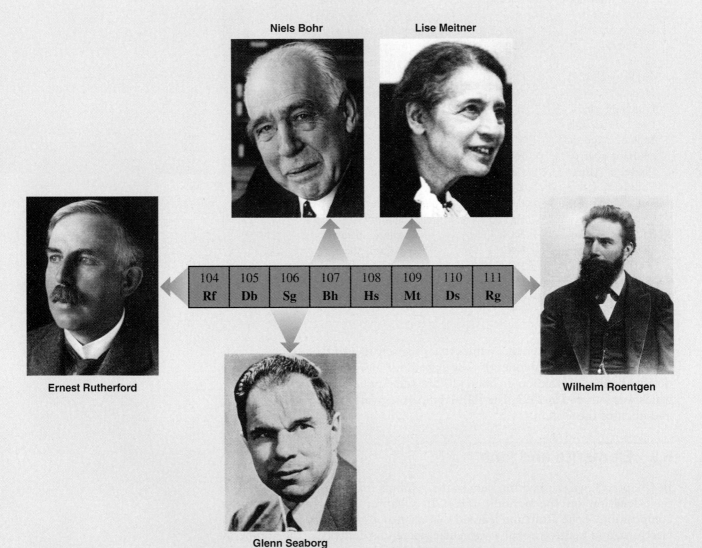

Niels Bohr

Lise Meitner

Ernest Rutherford

Wilhelm Roentgen

104	105	106	107	108	109	110	111
Rf	Db	Sg	Bh	Hs	Mt	Ds	Rg

Glenn Seaborg

The formula for most elements is simply the symbol of the element. In chemical reactions or mixtures, an element behaves as though it were a collection of individual particles. A small number of elements have formulas that are not single atoms at normal temperatures. Seven of the elements are *diatomic* molecules— that is, two atoms bonded together to form a molecule. These diatomic elements are hydrogen, H_2; oxygen, O_2; nitrogen, N_2; fluorine, F_2; chlorine, Cl_2; bromine, Br_2; and iodine, I_2. Two other elements that are commonly polyatomic are S_8, sulfur; and P_4, phosphorus.

Elements Occurring as Polyatomic Molecules

Hydrogen	H_2	Chlorine	Cl_2	Sulfur	S_8
Oxygen	O_2	Fluorine	F_2	Phosphorus	P_4
Nitrogen	N_2	Bromine	Br_2		
		Iodine	I_2		

We have learned that a charged particle, known as an *ion*, can be produced by adding or removing one or more electrons from a neutral atom. For example, potassium atoms contain 19 protons and 19 electrons. To make a potassium ion, we remove one electron, leaving 19 protons and only 18 electrons. This gives an ion with a positive one ($+1$) charge:

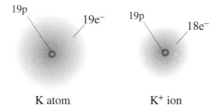

K atom K^+ ion

Written in the form of an equation, $K \rightarrow K^+ + e^-$. A positive ion is called a *cation*. Any neutral atom that *loses* an electron will form a cation. Sometimes an atom may lose one electron, as in the potassium example. Other atoms may lose more than one electron:

$$Mg \rightarrow Mg^{2+} + 2e^-$$

or

$$Al \rightarrow Al^{3+} + 3e^-$$

Cations are named the same as their parent atoms, as shown here:

Atom		Ion	
K	potassium	K^+	potassium ion
Mg	magnesium	Mg^{2+}	magnesium ion
Al	aluminum	Al^{3+}	aluminum ion

Ions can also be formed by adding electrons to a neutral atom. For example, the chlorine atom contains 17 protons and 17 electrons. The equal number of positive charges and negative charges results in a net charge of zero for the atom.

If one electron is added to the chlorine atom, it now contains 17 protons and 18 electrons, resulting in a net charge of negative one (-1) on the ion:

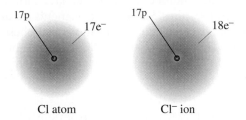

Cl atom   Cl⁻ ion

In a chemical equation, this process is summarized as $Cl + e^- \rightarrow Cl^-$. A negative ion is called an *anion*. Any neutral atom that *gains* an electron will form an anion. Atoms may gain more than one electron to form anions with different charges:

$$O + 2e^- \rightarrow O^{2-}$$
$$N + 3e^- \rightarrow N^{3-}$$

Anions are named differently from cations. To name an anion consisting of only one element, use the stem of the parent element name and change the ending to *-ide*. For example, the Cl^- ion is named by using the stem *chlor-* from chlorine and adding *-ide* to form chloride ion. Here are some examples:

Symbol	Name of atom	Ion	Name of ion
F	fluorine	F^-	fluoride ion
Br	bromine	Br^-	bromide ion
Cl	chlorine	Cl^-	chloride ion
I	iodine	I^-	iodide ion
O	oxygen	O^{2-}	oxide ion
N	nitrogen	N^{3-}	nitride ion

Ions are always formed by adding or removing electrons from an atom. Atoms do not form ions on their own. Most often ions are formed when metals combine with nonmetals.

The charge on an ion can often be predicted from the position of the element on the periodic table. Figure 6.2 shows the charges of selected ions from several groups on the periodic table. Notice that all the metals and hydrogen in the far left column (Group 1A) are ($1+$), all those in the next column (Group 2A) are ($2+$), and the metals in the next tall column (Group 3A) form ($3+$) ions. The elements in the lower center part of the table are called *transition metals*. These elements tend to form cations with various positive charges. There is no easy way to predict the charges on these cations. All metals lose electrons to form positive ions.

In contrast, the nonmetals form anions by gaining electrons. On the right side of the periodic table in Figure 6.2, you can see that the atoms in Group 7A form ($1-$) ions. The nonmetals in Group 6A form ($2-$) ions. It's important to learn the charges on the ions shown in Figure 6.2 and their relationship to the group number at the top of the column. For nontransition metals the charge is equal to the group number. For nonmetals, the charge is equal to the group number minus 8. We will learn more about why these ions carry their particular charges later in the course.

1A																	
H^+	2A											3A	4A	5A	6A	7A	
Li^+	Be^{2+}													N^{3-}	O^{2-}	F^-	
Na^+	Mg^{2+}											Al^{3+}		P^{3-}	S^{2-}	Cl^-	
K^+	Ca^{2+}			Cr^{2+} Cr^{3+}		Fe^{2+} Fe^{3+}			Cu^+ Cu^{2+}	Zn^{2+}						Br^-	
Rb^+	Sr^{2+}			Transition metals					Ag^+	Cd^{2+}						I^-	
Cs^+	Ba^{2+}																

Figure 6.2
Charges of selected ions in the periodic table.

6.3 Writing Formulas from Names of Ionic Compounds

In Chapters 3 and 5, we learned that compounds can be composed of ions. These substances are called *ionic compounds* and will conduct electricity when dissolved in water. An excellent example of an ionic compound is ordinary table salt. It's composed of crystals of sodium chloride. When dissolved in water, sodium chloride conducts electricity very well, as shown in Figure 6.3.

A chemical compound must have a net charge of zero. If it contains ions, the charges on the ions must add up to zero in the formula for the compound. This is relatively easy in the case of sodium chloride. The sodium ion (1+) and the chloride ion (1−) add to zero, resulting in the formula NaCl. Now consider an ionic compound containing calcium (Ca^{2+}) and fluoride (F^-) ions. How can we write a formula with a net charge of zero? To do this we need one Ca^{2+} and two F^- ions. The correct formula is CaF_2. The subscript 2 indicates that two fluoride ions are needed for each calcium ion. Aluminum oxide is a bit more complicated because it consists of Al^{3+} and O^{2-} ions. Since 6 is the least common multiple of 3 and 2, we have $2(3+) + 3(2-) = 0$, or a formula containing 2 Al^{3+} ions and 3 O^{2-} ions for Al_2O_3. Here are a few more examples of formula writing for ionic compounds:

Name of compound	Ions	Least common multiple	Sum of charges on ions	Formula
Sodium bromide	Na^+, Br^-	1	$(+1) + (-1) = 0$	NaBr
Potassium sulfide	K^+, S^{2-}	2	$2(+1) + (-2) = 0$	K_2S
Zinc sulfate	Zn^{2+}, SO_4^{2-}	2	$(+2) + (-2) = 0$	$ZnSO_4$
Ammonium phosphate	NH_4^+, PO_4^{3-}	3	$3(+1) + (-3) = 0$	$(NH_4)_3PO_4$
Aluminum chromate	Al^{3+}, CrO_4^{2-}	6	$2(+3) + 3(-2) = 0$	$Al_2(CrO_4)_3$

Na^+

Cl^-

H_2O

Figure 6.3
A solution of salt water contains Na^+ and Cl^- ions in addition to water molecules. The ions cause the solution to conduct electricity, lighting the bulb.

Example 6.1   Write formulas for (a) calcium chloride, (b) magnesium oxide, and (c) barium phosphide.

SOLUTION   (a) Use the following steps for calcium chloride.

Step 1   From the name, we know that calcium chloride is composed of calcium and chloride ions. First write down the formulas of these ions:

Ca^{2+} Cl^-

Step 2   To write the formula of the compound, combine the smallest numbers of Ca^{2+} and Cl^- ions to give the charge sum equal to zero. In this case the lowest common multiple of the charges is 2:

$(Ca^{2+}) + 2(Cl^-) = 0$

$(2+) \ \ + 2(1-) = 0$

Therefore the formula is $CaCl_2$.

(b) Use the same procedure for magnesium oxide:

Step 1   From the name, we know that magnesium oxide is composed of magnesium and oxide ions. First write down the formulas of these ions:

Mg^{2+} O^{2-}

Step 2   To write the formula of the compound, combine the smallest numbers of Mg^{2+} and O^{2-} ions to give the charge sum equal to zero:

$(Mg^{2+}) + (O^{2-}) = 0$

$(2+) \ \ + (2-) = 0$

The formula is MgO.

(c) Use the same procedure for barium phosphide:

Step 1   From the name, we know that barium phosphide is composed of barium and phosphide ions. First write down the formulas of these ions:

Ba^{2+} P^{3-}

Step 2   To write the formula of this compound, combine the smallest numbers of Ba^{2+} and P^{3-} ions to give the charge sum equal to zero. In this case the lowest common multiple of the charges is 6:

$3(Ba^{2+}) + 2(P^{3-}) = 0$

$3(2+) \ \ + 2(3-) = 0$

The formula is Ba_3P_2.

Practice 6.1 _____

Write formulas for compounds containing the following ions:

(a) K^+ and F^- (d) Na^+ and S^{2-}

(b) Ca^{2+} and Br^- (e) Ba^{2+} and O^{2-}

(c) Mg^{2+} and N^{3-} (f) As^{5+} and CO_3^{2-}

6.4 Naming Binary Compounds

Binary compounds contain only two different elements. Many binary compounds are formed when a metal combines with a nonmetal to form a *binary ionic compound*. The metal loses one or more electrons to become a cation while the nonmetal gains one or more electrons to become an anion. The cation is written first in the formula, followed by the anion.

Binary Ionic Compounds Containing a Metal Forming Only One Type of Cation

The chemical name is composed of the name of the metal followed by the name of the nonmetal, which has been modified to an identifying stem plus the suffix *-ide*.

For example, sodium chloride, NaCl, is composed of one atom each of sodium and chlorine. The name of the metal, sodium, is written first and is not modified. The second part of the name is derived from the nonmetal, chlorine, by using the stem *chlor-* and adding the ending *-ide*; it is named *chloride*. The compound name is sodium chloride.

NaCl

Elements:	Sodium (metal)
	Chlorine (nonmetal)
	name modified to the stem *chlor-* + *-ide*
Name of compound:	Sodium chloride

To name these compounds:
1. **Write the name of the cation.**
2. **Write the stem for the anion with the suffix *–ide*.**

Stems of the more common negative-ion-forming elements are shown in Table 6.2.

Compounds may contain more than one atom of the same element, but as long as they contain only two different elements and only one compound of these two elements exists, the name follows the rules for binary compounds:

Examples:

$CaBr_2$	Mg_3N_2	Li_2O
calcium bromide	magnesium nitride	lithium oxide

Table 6.2 Examples of Elements Forming Anions

Symbol	Element	Stem	Anion name
Br	bromine	brom	bromide
Cl	chlorine	chlor	chloride
F	fluorine	fluor	fluoride
H	hydrogen	hydr	hydride
I	iodine	iod	iodide
N	nitrogen	nitr	nitride
O	oxygen	ox	oxide
P	phosphorus	phosph	phosphide
S	sulfur	sulf	sulfide

Table 6.3 lists some compounds with names ending in *-ide*.

Table 6.3 Examples of Compounds with Names Ending in *–ide*

Formula	Name	Formula	Name
$AlCl_3$	aluminum chloride	BaS	barium sulfide
Al_2O_3	aluminum oxide	LiI	lithium iodide
CaC_2	calcium carbide	$MgBr_2$	magnesium bromide
HCl	hydrogen chloride	NaH	sodium hydride
HI	hydrogen iodide	Na_2O	sodium oxide

Example 6.2   Name the compound CaF_2.

SOLUTION   From the formula it is a two-element compound and follows the rules for binary compounds.

Step 1   The compound is composed of Ca, a metal, and F, a nonmetal. Elements in the 2A column of the periodic table form only one type of cation. Thus, we name the positive part of the compound *calcium*.

Step 2   Modify the name of the second element to the stem *fluor-* and add the binary ending *-ide* to form the name of the negative part, *fluoride*.

The name of the compound is therefore *calcium fluoride*.

Practice 6.2 _____

Write formulas for these compounds:
(a) strontium chloride   (d) calcium sulfide
(b) potassium iodide   (e) sodium oxide
(c) aluminum nitride

Binary Ionic Compounds Containing a Metal That Can Form Two or More Types of Cations

The metals in the center of the periodic table (including the transition metals) often form more than one type of cation. For example, iron can form Fe^{2+} and Fe^{3+} ions, and copper can form Cu^+ and Cu^{2+} ions. This can be confusing when you are naming compounds. For example, copper chloride could be $CuCl_2$ or CuCl. To resolve this difficulty the IUPAC devised a system, known as the **Stock System**, to name these compounds. This system is currently recognized as the official system to name these compounds, although another older system is sometimes used. In the Stock System, when a compound contains a metal that can form more than one type of cation, the charge on the cation of the metal is designated by a Roman numeral placed in parentheses immediately following the name of the metal. The negative element is treated in the usual manner for binary compounds:

Stock System

Cation charge	+1	+2	+3	+4	+5
Roman numeral	(I)	(II)	(III)	(IV)	(V)

Examples:	FeCl$_2$	iron(II) chloride	Fe^{2+}
	FeCl$_3$	iron(III) chloride	Fe^{3+}
	CuCl	copper(I) chloride	Cu$^+$
	CuCl$_2$	copper(II) chloride	Cu^{2+}

The fact that FeCl$_2$ has two chloride ions, each with a -1 charge, establishes that the charge on Fe is $+2$. To distinguish between the two iron chlorides, FeCl$_2$ is named iron(II) chloride and FeCl$_3$ is named iron(III) chloride:

charge:	$+2$	-1	$+3$	-1
name:	iron(II)	chloride	iron(III)	chloride

When a metal forms only one possible cation, we need not distinguish one cation from another, so Roman numerals are not needed. Thus we do not say - calcium(II) chloride for CaCl$_2$, but rather calcium chloride, since the charge of calcium is understood to be $+2$.

To name compounds using the Stock System, do the following:

1. Write the name of the cation.
2. Write the charge on the cation as a Roman numeral in parentheses.
3. Write the stem of the anion with the suffix *-ide*.

In classical nomenclature, when the metallic ion has only two cation types, the name of the metal (usually the Latin name) is modified with the suffixes *-ous* and *-ic* to distinguish between the two. The lower-charge cation is given the *-ous* ending, and the higher one, the *-ic* ending.

Examples:	FeCl$_2$	ferrous chloride	Fe^{2+}	(lower charge cation)
	FeCl$_3$	ferric chloride	Fe^{3+}	(higher charge cation)
	CuCl	cuprous chloride	Cu$^+$	(lower charge cation)
	CuCl$_2$	cupric chloride	Cu^{2+}	(higher charge cation)

Table 6.4 lists some common metals that have more than one type of cation.

Notice that the *ous–ic* naming system does not give the charge of the cation of an element but merely indicates that at least two types of cations exist. The Stock System avoids any possible uncertainty by clearly stating the charge on the cation.

To name compounds using the Stock System:
1. Write the name of the cation.
2. Write the charge on the cation as a Roman numeral in parentheses.
3. Write the stem of the anion with the suffix *-ide*.

In this book we will use mainly the Stock System.

Table 6.4 Names and Charges of Some Common Metal Ions That Have More Than One Type of Cation

Formula	Stock System name	Classical name	Formula	Stock System name	Classical name
Cu^{1+}	copper(I)	cuprous	Sn^{4+}	tin(IV)	stannic
Cu^{2+}	copper(II)	cupric	Pb^{2+}	lead(II)	plumbous
Hg^{1+}(Hg$_2$)$^{2+}$	mercury(I)	mercurous	Pb^{4+}	lead(IV)	plumbic
Hg^{2+}	mercury(II)	mercuric	As^{3+}	arsenic(III)	arsenous
Fe^{2+}	iron(II)	ferrous	As^{5+}	arsenic(V)	arsenic
Fe^{3+}	iron(III)	ferric	Ti^{3+}	titanium(III)	titanous
Sn^{2+}	tin(II)	stannous	Ti^{4+}	titanium(IV)	titanic

Example 6.3   Name the compound FeS.

SOLUTION   This compound follows the rules for a binary compound.

Steps 1,2 It is a compound of Fe, a metal, and S, a nonmetal, and Fe is a transition metal that has more than one type of cation. In sulfides, the charge on the S is -2. Therefore, the charge on Fe must be $+2$, and the name of the positive part of the compound is *iron(II)*.

Step 3   We have already determined that the name of the negative part of the compound will be *sulfide*.

The name of FeS is *iron(II) sulfide*.

Practice 6.3
Write the name for each of the following compounds using the Stock System:
(a) PbI_2 (b) SnF_4 (c) Fe_2O_3 (d) CuO

Practice 6.4
Write formulas for the following compounds:
(a) tin(IV) chromate   (c) tin(II) fluoride
(b) chromium(III) bromide   (d) copper(I) oxide

Binary Compounds Containing Two Nonmetals

Prefix	Number
mono	1
di	2
tri	3
tetra	4
penta	5
hexa	6
hepta	7
octa	8
nona	9
deca	10

Compounds between nonmetals are molecular, not ionic. Therefore, a different system for naming them is used. In a compound formed between two nonmetals, the element that occurs first in this series is written and named first:

Si, B, P, H, C, S, I, Br, N, Cl, O, F

The name of the second element retains the *-ide* ending as though it were an anion. A Latin or Greek prefix (*mono-*, *di-*, *tri-*, and so on) is attached to the name of each element to indicate the number of atoms of that element in the molecule. The prefix *mono-* is rarely used for naming the first element. Some common prefixes and their numerical equivalences are as shown in the margin table.

Here are some examples of compounds that illustrate this system:

$$N_2O_3$$

(di) nitrogen   (tri) oxide
Indicates two   Indicates three
nitrogen atoms   oxygen atoms

To name these compounds:
1. **Write the name for the first element using a prefix if there is more than one atom of this element.**
2. **Write the stem for the second element followed by the suffix *–ide*. Use a prefix to indicate the number of atoms for the second element.**

CO	carbon monoxide
CO_2	carbon dioxide
PCl_3	phosphorus trichloride
SO_2	sulfur dioxide
P_2O_5	diphosphorus pentoxide
CCl_4	carbon tetrachloride

N_2O	dinitrogen monoxide
N_2O_4	dinitrogen tetroxide
NO	nitrogen monoxide
N_2O_3	dinitrogen trioxide
S_2Cl_2	disulfur dichloride
S_2F_{10}	disulfur decafluoride

These examples illustrate that we sometimes drop the final *o* (mono) or *a* (penta) of the prefix when the second element is oxygen. This avoids creating a name that is awkward to pronounce. For example, CO is carbon monoxide instead of carbon monooxide.

Example 6.4

SOLUTION

Phosphorus pentachloride

Name the compound PCl_5.

Step 1 Phosphorus and chlorine are nonmetals, so the rules for naming binary compounds containing two nonmetals apply. Phosphorus is named first. Therefore the compound is a chloride.

Step 2 No prefix is needed for phosphorus because each molecule has only one atom of phosphorus. The prefix *penta-* is used with chloride to indicate the five chlorine atoms. (PCl_3 is also a known compound.)

Step 3 The name for PCl_5 is *phosphorus pentachloride*.

Practice 6.5 _____

Name these compounds:
(a) Cl_2O (b) SO_2 (c) CBr_4 (d) N_2O_5 (e) NH_3 (f) ICl_3

Acids Derived from Binary Compounds

Certain binary hydrogen compounds, when dissolved in water, form solutions that have *acid* properties. Because of this property, these compounds are given acid names in addition to their regular *-ide* names. For example, HCl is a gas and is called *hydrogen chloride*, but its water solution is known as *hydrochloric acid*. Binary acids are composed of hydrogen and one other nonmetallic element. However, not all binary hydrogen compounds are acids. To express the formula of a binary acid, it's customary to write the symbol of hydrogen first, followed by the symbol of the second element (e.g., HCl, HBr, or H_2S). When we see formulas such as CH_4 or NH_3, we understand that these compounds are not normally considered to be acids.

To name a binary acid, place the prefix *hydro-* in front of, and the suffix *-ic* after, the stem of the nonmetal name. Then add the word *acid*:

To name binary acids:
1. **Write the prefix** *hydro–* **followed by the stem of the second element and add the suffix** *–ic.*
2. **Write the word** *acid.*

	HCl	H_2S
Examples:	*Hydro-chlor-ic acid* (hydrochloric acid)	*Hydro-sulfur-ic acid* (hydrosulfuric acid)

Acids are hydrogen-containing substances that liberate hydrogen ions when dissolved in water. The same formula is often used to express binary hydrogen compounds, such as HCl, regardless of whether or not they are dissolved in water. Table 6.5 shows several examples of binary acids.

Table 6.5 Names and Formulas of Selected Binary Acids

Formula	Acid name	Formula	Acid name
HF	Hydrofluoric acid	HI	Hydroiodic acid
HCl	Hydrochloric acid	H_2S	Hydrosulfuric acid
HBr	Hydrobromic acid	H_2Se	Hydroselenic acid

Naming binary compounds is summarized in Figure 6.4.

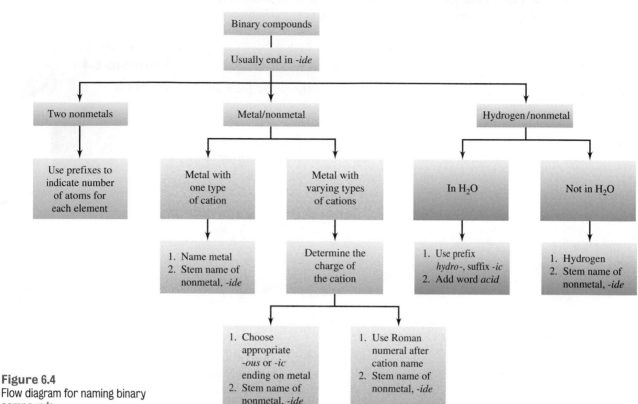

Figure 6.4
Flow diagram for naming binary compounds.

Practice 6.6 _____

Name these binary compounds:
(a) KBr   (b) Ca_3N_2 (c) SO_3 (d) SnF_2 (e) $CuCl_2$ (f) N_2O_4

6.5   Naming Compounds Containing Polyatomic Ions

polyatomic ion

Potassium permanganate crystals are dark purple.

A **polyatomic ion** is an ion that contains two or more elements. Compounds containing polyatomic ions are composed of three or more elements and usually consist of one or more cations combined with a negative polyatomic ion. In general, naming compounds containing polyatomic ions is similar to naming binary compounds. The cation is named first, followed by the name for the negative polyatomic ion.

To name these compounds, you must learn to recognize the common polyatomic ions (Table 6.6) and know their charges. Consider the formula $KMnO_4$. You must be able to recognize that it consists of two parts $KMnO_4$. These parts are composed of a K^+ ion and a MnO_4^- ion. The correct name for this compound is potassium permanganate. Many polyatomic ions that contain oxygen are called *oxy-anions*, and generally have the suffix *-ate* or *-ite*. Unfortunately, the suffix doesn't indicate the number of oxygen atoms present. The *-ate* form

Table 6.6 Names, Formulas, and Charges of Some Common Polyatomic Ions

Name	Formula	Charge	Name	Formula	Charge
Acetate	$C_2H_3O_2^-$	-1	Cyanide	CN^-	-1
Ammonium	NH_4^+	$+1$	Dichromate	$Cr_2O_7^{2-}$	-2
Arsenate	AsO_4^{3-}	-3	Hydroxide	OH^-	-1
Hydrogen carbonate	HCO_3^-	-1	Nitrate	NO_3^-	-1
Hydrogen sulfate	HSO_4^-	-1	Nitrite	NO_2^-	-1
Bromate	BrO_3^-	-1	Permanganate	MnO_4^-	-1
Carbonate	CO_3^{2-}	-2	Phosphate	PO_4^{3-}	-3
Chlorate	ClO_3^-	-1	Sulfate	SO_4^{2-}	-2
Chromate	CrO_4^{2-}	-2	Sulfite	SO_3^{2-}	-2

More ions are listed on the back endpapers.

contains more oxygen atoms than the *-ite* form. Examples include sulfate (SO_4^{2-}), sulfite (SO_3^{2-}), nitrate (NO_3^-), and nitrite (NO_2^-).

Some elements form more than two different polyatomic ions containing oxygen. To name these ions, prefixes are used in addition to the suffix. To indicate more oxygen than in the *-ate* form, we add the prefix *per-*, which is a short form of *hyper-*, meaning "more." The prefix *hypo-*, meaning "less" (oxygen in this case), is used for the ion containing less oxygen than the *-ite* form. An example of this system is shown for the polyatomic ions containing chlorine and oxygen in Table 6.7. The prefixes are also used with other similar ions, such as iodate (IO_3^-), bromate (BrO_3^-), and phosphate (PO_4^{3-}).

To name these compounds:
1. **Write the name of the cation.**
2. **Write the name of the anion.**

Table 6.7 Oxy-Anions and Oxy-Acids of Chlorine

Anion formula	Anion name	Acid formula	Acid name
ClO^-	*hypo*chlor*ite*	$HClO$	*hypo*chlor*ous* acid
ClO_2^-	chlor*ite*	$HClO_2$	chlor*ous* acid
ClO_3^-	chlor*ate*	$HClO_3$	chlor*ic* acid
ClO_4^-	*per*chlor*ate*	$HClO_4$	*per*chlor*ic* acid

Only four of the common negatively charged polyatomic ions do not use the *ate–ite* system. These exceptions are hydroxide (OH^-), hydrogen sulfide (HS^-), peroxide (O_2^{2-}), and cyanide (CN^-). Care must be taken with these, as their endings can easily be confused with the *-ide* ending for binary compounds (Section 6.4).

There are three common positively charged polyatomic ions as well—the ammonium, the mercury(I) (Hg_2^{2+}), and the hydronium(H_3O^+) ions. The ammonium ion (NH_4^+) is frequently found in polyatomic compounds (Section 6.5), whereas the hydronium ion (H_3O^+) is usually associated with aqueous solutions of acids (Chapter 15).

Practice 6.7

Name these compounds:

(a) $NaNO_3$ (b) $Ca_3(PO_4)_2$ (c) KOH (d) Li_2CO_3 (e) $NaClO_3$

Inorganic compounds are also formed from more than three elements (see Table 6.8). In these cases, one or more of the ions is often a polyatomic ion.

Table 6.8 Names of Selected Compounds That Contain More Than One Kind of Positive Ion

Formula	Name of compound
$KHSO_4$	potassium hydrogen sulfate
$Ca(HSO_3)_2$	calcium hydrogen sulfite
NH_4HS	ammonium hydrogen sulfide
$MgNH_4PO_4$	magnesium ammonium phosphate
NaH_2PO_4	sodium dihydrogen phosphate
Na_2HPO_4	sodium hydrogen phosphate
KHC_2O_4	potassium hydrogen oxalate
$KAl(SO_4)_2$	potassium aluminum sulfate
$Al(HCO_3)_3$	aluminum hydrogen carbonate

Once you have learned to recognize the polyatomic ions, naming these compounds follows the patterns we have already learned. First identify the ions. Name the cations in the order given, and follow them with the names of the anions. Study the following examples:

Compound	Ions	Name
$NaHCO_3$	Na^+; HCO_3^-	sodium hydrogen carbonate
$NaHS$	Na^+; HS^-	sodium hydrogen sulfide
$MgNH_4PO_4$	Mg^{2+}; NH_4^+; PO_4^{3-}	magnesium ammonium phosphate
$NaKSO_4$	Na^+; K^+; SO_4^{2-}	sodium potassium sulfate

6.6 Acids

While we will learn much more about acids later (see Chapter 15), it is helpful to be able to recognize and name common acids both in the laboratory and in class. The simplest way to recognize many acids is to know that acid formulas often begin with hydrogen. Naming binary acids was covered in Section 6.4. Inorganic compounds containing hydrogen, oxygen, and one other element are called *oxy-acids*.

The first step in naming these acids is to determine that the compound in question is really an oxy-acid. The keys to identification are (1) hydrogen is the first element in the compound's formula and (2) the second part of the formula consists of a polyatomic ion containing oxygen.

Hydrogen in an oxy-acid is not specifically designated in the acid name. The presence of hydrogen in the compound is indicated by the use of the word *acid* in the name of the substance. To determine the particular type of acid, the polyatomic ion following hydrogen must be examined. The name of the polyatomic ion is modified in the following manner: (1) *-ate* changes to an *-ic* ending; (2) *-ite* changes to an *-ous* ending. (See Table 6.9.) The compound with the *-ic* ending contains more oxygen than the one with the *-ous* ending.

Table 6.9   Comparison of Acid and Anion Names for Selected Oxy-Acids

Acid	Anion	Acid	Anion	Acid	Anion
H_2SO_4 Sulfuric acid	SO_4^{2-} Sulfate ion	H_2CO_3 Carbonic acid	CO_3^{2-} Carbonate ion	HIO_3 Iodic acid	IO_3^- Iodate ion
H_2SO_3 Sulfurous acid	SO_3^{2-} Sulfite ion	H_3BO_3 Boric acid	BO_3^{3-} Borate ion	$HC_2H_3O_2$ Acetic acid	$C_2H_3O_2^-$ Acetate ion
HNO_3 Nitric acid	NO_3^- Nitrate ion	H_3PO_4 Phosphoric acid	PO_4^{3-} Phosphate ion	$H_2C_2O_4$ Oxalic acid	$C_2O_4^{2-}$ Oxalate ion
HNO_2 Nitrous acid	NO_2^- Nitrite ion	H_3PO_3 Phosphorous acid	PO_3^{3-} Phosphite ion	$HBrO_3$ Bromic acid	BrO_3^- Bromate ion

Consider these examples:

SO_4^{2-} sulf*ate* $\longrightarrow$ H_2SO_4 sulfur*ic* acid (contains four oxygens)

SO_3^{2-} sulf*ite* $\longrightarrow$ H_2SO_3 sulfur*ous* acid (three oxygens)

NO_3^- nitr*ate* $\longrightarrow$ HNO_3 nitr*ic* acid (three oxygens)

NO_2^- nitr*ite* $\longrightarrow$ HNO_2 nitr*ous* acid (two oxygens)

The complete system for naming oxy-acids is shown in Table 6.7 for the various oxy-acids containing chlorine.

Naming polyatomic compounds is summarized in Figure 6.5. We have now looked at ways of naming a variety of inorganic compounds—binary compounds consisting of a metal and a nonmetal and of two nonmetals, binary acids, and polyatomic compounds. These compounds are just a small part of the classified chemical compounds. Most of the remaining classes are in the broad field of organic chemistry under such categories as hydrocarbons, alcohols, ethers, aldehydes, ketones, phenols, and carboxylic acids.

Figure 6.5
Flow diagram for naming polyatomic compounds.

Practice 6.8 _____

Name these compounds:
(a) Cu_2CO_3 (b) $Fe(ClO)_3$ (c) $Sn(C_2H_3O_2)_2$ (d) $HBrO_3$ (e) HBr

Practice 6.9 _____

Write formulas for
(a) lead(II) nitrate   (c) mercury(II) cyanide
(b) potassium phosphate   (d) ammonium chromate

Chapter 6 Review

6.1 Common and Systematic Names

- Common names are arbitrary and do not describe the chemical composition of compounds.
  - Examples—ammonia, water
- A standard system of nomenclature was devised and is maintained by the IUPAC.

6.2 Elements and Ions

- Diatomic elements—H_2, O_2, N_2, F_2, Cl_2, Br_2, I_2.
- Polyatomic elements—P_4, S_8.
- Ions form through the addition or loss of electrons from neutral atoms.
- The charge on ions can often be predicted from the periodic table.

6.3 Writing Formulas from the Names of Ionic Compounds

- Compounds must have a net charge of zero.
- To write a formula from a name:
  - Identify and write the symbols for the elements in the compound.
  - Combine the smallest number of ions required to produce a net charge of zero.

6.4 Naming Binary Compounds

KEY TERM

Stock System

- The rules for naming binary compounds can be summarized in the chart below:

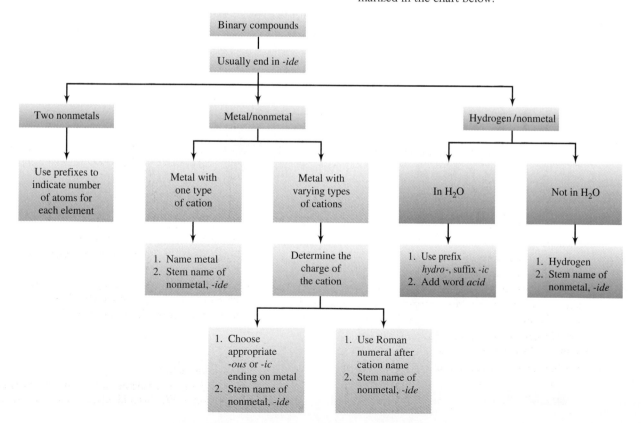

6.5 Naming Compounds Containing Polyatomic Ions

KEY TERM

Polyatomic ion

- The rules for naming these compounds are included in the chart for naming acids below.

6.6 Acids

- The rules for naming acids can be summarized in the chart below:

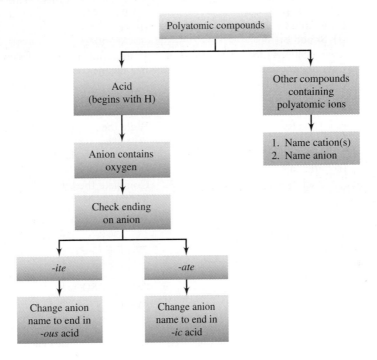

Review Questions

All questions with blue numbers have answers in the appendix of the text.

1. Use the common ion table on the back endpapers of your text to determine the formulas for compounds composed of the following ions:
   (a) sodium and chlorate
   (b) hydrogen and sulfate
   (c) tin(II) and acetate
   (d) copper(I) and oxide
   (e) zinc and hydrogen carbonate
   (f) iron(III) and carbonate

2. Does the fact that two elements combine in a one-to-one atomic ratio mean that the charges on their ions are both 1? Explain.

3. Write the names and formulas for the four oxy-acids containing (a) bromine, (b) iodine. (Table 6.7)

4. Explain why N_2O_5 is called dinitrogen pentoxide.

5. Write the formulas for the compounds formed when a chromium(III) ion is combined with the following anions:
   (a) hydroxide   (f) dichromate
   (b) nitrate   (g) phosphate
   (c) nitrite   (h) oxalate
   (d) hydrogen carbonate   (i) oxide
   (e) carbonate   (j) fluoride
   (See Table of Names, Formulas, and Charges of Common Ions on the back endpapers.)

6. Explain why the name for $MgCl_2$ is magnesium chloride but the name for $CuCl_2$ is copper(II) chloride.

7. Where are each of the following located in the periodic table?
   (a) metals
   (b) nonmetals
   (c) transition metals

8. Name two compounds almost always referred to by their common names. What would be their systematic names?

Paired Exercises

All exercises with blue numbers have answers in the appendix of the text.

1. Write the formula of the compound that would be formed between these elements:
   (a) Na and I   (d) K and S
   (b) Ba and F   (e) Cs and Cl
   (c) Al and O   (f) Sr and Br

2. Write the formula of the compound that would be formed between these elements:
   (a) Ba and O   (d) Be and Br
   (b) H and S   (e) Li and Se
   (c) Al and Cl   (f) Mg and P

3. Write formulas for the following cations (don't forget to include the charges): sodium, magnesium, aluminum, copper(II), iron(II), iron(III), lead(II), silver, cobalt(II), barium, hydrogen, mercury(II), tin(II), chromium(III), tin(IV), manganese(II), bismuth(III).

4. Write formulas for the following anions (don't forget to include the charges): chloride, bromide, fluoride, iodide, cyanide, oxide, hydroxide, sulfide, sulfate, hydrogen sulfate, hydrogen sulfite, chromate, carbonate, hydrogen carbonate, acetate, chlorate, permanganate, oxalate.

5. Write the systematic names for the following:
   (a) laughing gas (N_2O)   (d) galena (PbS)
   (b) calcite ($CaCO_3$)   (e) muriatic acid (HCl)
   (c) alumina (Al_2O_3)   (f) table salt (NaCl)

6. Write the systematic names for the following:
   (a) slaked lime ($Ca(OH)_2$)   (d) baking soda ($NaHCO_3$)
   (b) saltpeter ($NaNO_3$)   (e) pyrite (FeS_2)
   (c) brimstone (S)   (f) potash (K_2CO_3)

7. Complete the table, filling in each box with the proper formula.

Cations	Anions				
	Br^-	O^{2-}	NO_3^-	PO_4^{3-}	CO_3^{2-}
K^+	KBr				
Mg^{2+}					
Al^{3+}					
Zn^{2+}				$Zn_3(PO_4)_2$	
H^+					

8. Complete the table, filling in each box with the proper formula.

Cations	Anions				
	SO_4^{2-}	OH^-	AsO_4^{3-}	$C_2H_3O_2^-$	CrO_4^{2-}
NH_4^+			$(NH_4)_3AsO_4$		
Ca^{2+}					
Fe^{3+}	$Fe_2(SO_4)_3$				
Ag^+					
Cu^{2+}					

9. Write the names of each of the compounds formed in Question 7.

10. Write the names of each of the compounds formed in Question 8.

11. Write formulas for these binary compounds, all of which are composed of nonmetals:
    (a) carbon monoxide   (f) dinitrogen pentoxide
    (b) sulfur trioxide   (g) iodine monobromide
    (c) carbon tetrabromide   (h) silicon tetrachloride
    (d) phosphorus trichloride   (i) phosphorus pentaiodide
    (e) nitrogen dioxide   (j) diboron trioxide

12. Write formulas for these compounds:
    (a) sodium nitrate   (e) silver carbonate
    (b) magnesium fluoride   (f) calcium phosphate
    (c) barium hydroxide   (g) potassium nitrite
    (d) ammonium sulfate   (h) strontium oxide

13. Name these binary compounds, all of which are composed of nonmetals:
    (a) CO_2 (f) N_2O_4
    (b) N_2O (g) P_2O_5
    (c) PCl_5 (h) OF_2
    (d) CCl_4 (i) NF_3
    (e) SO_2 (j) CS_2

14. Name these compounds:
    (a) K_2O (e) Na_3PO_4
    (b) NH_4Br (f) Al_2O_3
    (c) CaI_2 (g) $Zn(NO_3)_2$
    (d) $BaCO_3$ (h) Ag_2SO_4

15. Name these compounds by the Stock (IUPAC) System:
    (a) $CuCl_2$ (d) $FeCl_3$
    (b) CuBr   (e) SnF_2
    (c) $Fe(NO_3)_2$ (f) $HgCO_3$

16. Write formulas for these compounds:
    (a) tin(IV) bromide   (d) mercury(II) nitrite
    (b) copper(I) sulfate   (e) titanium(IV) sulfide
    (c) iron(III) carbonate   (f) iron(II) acetate

17. Write formulas for these acids:
    (a) hydrochloric acid   (d) carbonic acid
    (b) chloric acid   (e) sulfurous acid
    (c) nitric acid   (f) phosphoric acid

18. Write formulas for these acids:
    (a) acetic acid   (d) boric acid
    (b) hydrofluoric acid   (e) nitrous acid
    (c) hydrosulfuric acid   (f) hypochlorous acid

19. Name these acids:
    (a) HNO_2 (d) HBr   (g) HF
    (b) H_2SO_4 (e) H_3PO_3 (h) $HBrO_3$
    (c) $H_2C_2O_4$ (f) $HC_2H_3O_2$

20. Name these acids:
    (a) H_3PO_4 (d) HCl   (g) HI
    (b) H_2CO_3 (e) HClO   (h) $HClO_4$
    (c) HIO_3 (f) HNO_3

21. Write formulas for these compounds:
(a) silver sulfite
(b) cobalt(II) bromide
(c) tin(II) hydroxide
(d) aluminum sulfate
(e) manganese(II) fluoride
(f) ammonium carbonate
(g) chromium(III) oxide
(h) copper(II) chloride
(i) potassium permanganate
(j) barium nitrite
(k) sodium peroxide
(l) iron(II) sulfate
(m) potassium dichromate
(n) bismuth(III) chromate

22. Write formulas for these compounds:
(a) sodium chromate
(b) magnesium hydride
(c) nickel(II) acetate
(d) calcium chlorate
(e) lead(II) nitrate
(f) potassium dihydrogen phosphate
(g) manganese(II) hydroxide
(h) cobalt(II) hydrogen carbonate
(i) sodium hypochlorite
(j) arsenic(V) carbonate
(k) chromium(III) sulfite
(l) antimony(III) sulfate
(m) sodium oxalate
(n) potassium thiocyanate

23. Write the name of each compound:
(a) $ZnSO_4$
(b) $HgCl_2$
(c) $CuCO_3$
(d) $Cd(NO_3)_2$
(e) $Al(C_2H_3O_2)_3$
(f) CoF_2
(g) $Cr(ClO_3)_3$
(h) Ag_3PO_4
(i) NiS
(j) $BaCrO_4$

24. Write the name of each compound:
(a) $Ca(HSO_4)_2$
(b) $As_2(SO_3)_3$
(c) $Sn(NO_2)_2$
(d) $FeBr_3$
(e) $KHCO_3$
(f) $BiAsO_4$
(g) $Fe(BrO_3)_2$
(h) $(NH_4)_2HPO_4$
(i) $NaClO$
(j) $KMnO_4$

25. Write the chemical formula for these substances:
(a) baking soda
(b) lime
(c) Epsom salts
(d) muriatic acid
(e) vinegar
(f) potash
(g) lye

26. Write the chemical formula for these substances:
(a) fool's gold
(b) saltpeter
(c) limestone
(d) cane sugar
(e) milk of magnesia
(f) washing soda
(g) grain alcohol

Additional Exercises

All exercises with blue numbers have answers in the appendix of the text.

27. Write equations similar to those found in Section 6.2 for the formation of:
(a) potassium ion
(b) iodide ion
(c) bromide ion
(d) iron(II) ion
(e) calcium ion
(f) oxide ion

28. Name each of the following polyatomic ions:

29. Write formulas for all possible compounds formed between the calcium ion and the anions shown in Question 28.

30. Write formulas for all possible compounds formed between the potassium ion and the anions shown in Question 28.

31. Write the formula and name for each of the following compounds:

(a)

(b)

(c)

32. Write the formula and name for the following compound:

K⁺   Cl⁻

33. State how each of the following is used in naming inorganic compounds: *ide, ous, ic, hypo, per, ite, ate,* Roman numerals.

34. Translate the following sentences into unbalanced formula chemical equations:
(a) Silver nitrate and sodium chloride react to form silver chloride and sodium nitrate.
(b) Iron(III) sulfate and calcium hydroxide react to form iron(III) hydroxide and calcium sulfate.
(c) Potassium hydroxide and sulfuric acid react to form potassium sulfate and water.

35. How many of each type of subatomic particle (protons and electrons) is in
(a) an atom of tin?   (c) a Sn^{4+} ion?
(b) a Sn^{2+} ion?

36. The compound X_2Y_3 is a stable solid. What ionic charge do you expect for X and Y? Explain.

37. The ferricyanide ion has the formula $Fe(CN)_6^{3-}$. Write the formula for the compounds that ferricyanide would form with the cations of elements 3, 13, and 30.

38. Compare and contrast the formulas of
(a) nitride with nitrite
(b) nitrite with nitrate
(c) nitrous acid with nitric acid

Challenge Exercise

All exercises with blue numbers have answers in the appendix of the text.

*39. The compound X_2Y has a mass of 110.27 g. The ratio of $X:Y$ (by mass) in the compound is 2.44 : 1. What is the compound?

Answers to Practice Exercises

6.1 (a) KF; (b) $CaBr_2$; (c) Mg_3N_2; (d) Na_2S; (e) BaO; (f) $As_2(CO_3)_5$

6.2 (a) $SrCl_2$; (b) KI; (c) AlN; (d) CaS; (e) Na_2O

6.3 (a) lead(II) iodide; (b) tin(IV) fluoride; (c) iron(III) oxide; (d) copper(II) oxide

6.4 (a) $Sn(CrO_4)_2$; (b) $CrBr_3$; (c) SnF_2; (d) Cu_2O

6.5 (a) dichlorine monoxide; (b) sulfur dioxide; (c) carbon tetrabromide; (d) dinitrogen pentoxide; (e) ammonia; (f) iodine trichloride

6.6 (a) potassium bromide; (b) calcium nitride; (c) sulfur trioxide; (d) tin(II) fluoride; (e) copper(II) chloride; (f) dinitrogen tetroxide

6.7 (a) sodium nitrate; (b) calcium phosphate; (c) potassium hydroxide; (d) lithium carbonate; (e) sodium chlorate

6.8 (a) copper(I) carbonate; (b) iron(III) hypochlorite; (c) tin(II) acetate; (d) bromic acid; (e) hydrobromic acid

6.9 (a) $Pb(NO_3)_2$; (b) K_3PO_4; (c) $Hg(CN)_2$; (d) $(NH_4)_2CrO_4$

PUTTING IT TOGETHER

Multiple Choice:

Choose the correct answer to each of the following.

1. The concept of positive charge and a small, "heavy" nucleus surrounded by electrons was the contribution of
   (a) Dalton   (c) Thomson
   (b) Rutherford   (d) Chadwick

2. The neutron was discovered in 1932 by
   (a) Dalton   (c) Thomson
   (b) Rutherford   (d) Chadwick

3. An atom of atomic number 53 and mass number 127 contains how many neutrons?
   (a) 53   (c) 127
   (b) 74   (d) 180

4. How many electrons are in an atom of $^{40}_{18}Ar$?
   (a) 20   (c) 40
   (b) 22   (d) no correct answer given

5. The number of neutrons in an atom of $^{139}_{56}Ba$ is
   (a) 56   (c) 139
   (b) 83   (d) no correct answer given

6. The name of the isotope containing one proton and two neutrons is
   (a) protium   (c) deuterium
   (b) tritium   (d) helium

7. Each atom of a specific element has the same
   (a) number of protons   (c) number of neutrons
   (b) atomic mass   (d) no correct answer given

8. Which pair of symbols represents isotopes?
   (a) $^{23}_{11}Na$ and $^{23}_{12}Na$ (c) $^{63}_{29}Cu$ and $^{29}_{64}Cu$
   (b) $^{7}_{3}Li$ and $^{6}_{3}Li$ (d) $^{12}_{24}Mg$ and $^{12}_{26}Mg$

9. Two naturally occurring isotopes of an element have masses and abundance as follows: 54.00 amu (20.00%) and 56.00 amu (80.00%). What is the relative atomic mass of the element?
   (a) 54.20   (c) 54.80
   (b) 54.40   (d) 55.60

10. Substance X has 13 protons, 14 neutrons, and 10 electrons. Determine its identity.
    (a) ^{27}Mg (c) $^{27}Al^{3+}$
    (b) ^{27}Ne (d) ^{27}Al

11. The mass of a chlorine atom is 5.90×10^{-23} g. How many atoms are in a 42.0-g sample of chlorine?
    (a) 2.48×10^{-21} (c) 1.40×10^{-24}
    (b) 7.12×10^{23} (d) no correct answer given

12. The number of neutrons in an atom of $^{108}_{47}Ag$ is
    (a) 47   (c) 155
    (b) 108   (d) no correct answer given

13. The number of electrons in an atom of $^{27}_{13}Al$ is
    (a) 13   (c) 27
    (b) 14   (d) 40

Review for Chapters 5–6

14. The number of protons in an atom of $^{65}_{30}Zn$ is
    (a) 65   (c) 30
    (b) 35   (d) 95

15. The number of electrons in the nucleus of an atom of $^{24}_{12}Mg$ is
    (a) 12   (c) 36
    (b) 24   (d) no correct answer given

Names and Formulas:

In which of the following is the formula correct for the name given?

1. copper(II) sulfate, $CuSO_4$
2. ammonium hydroxide, NH_4OH
3. mercury(I) carbonate, $HgCO_3$
4. phosphorus triiodide, PI_3
5. calcium acetate, $Ca(C_2H_3O_2)_2$
6. hypochlorous acid, $HClO$
7. dichlorine heptoxide, Cl_2O_7
8. magnesium iodide, MgI
9. sulfurous acid, H_2SO_3
10. potassium manganate, $KMnO_4$
11. lead(II) chromate, $PbCrO_4$
12. ammonium hydrogen carbonate, NH_4HCO_3
13. iron(II) phosphate, $FePO_4$
14. calcium hydrogen sulfate, $CaHSO_4$
15. mercury(II) sulfate, $HgSO_4$
16. dinitrogen pentoxide, N_2O_5
17. sodium hypochlorite, $NaClO$
18. sodium dichromate, $Na_2Cr_2O_7$
19. cadmium cyanide, $Cd(CN)_2$
20. bismuth(III) oxide, Bi_3O_2
21. carbonic acid, H_2CO_3
22. silver oxide, Ag_2O
23. ferric iodide, FeI_2
24. tin(II) fluoride, TiF_2
25. carbon monoxide, CO
26. phosphoric acid, H_3PO_3
27. sodium bromate, Na_2BrO_3
28. hydrosulfuric acid, H_2S
29. potassium hydroxide, POH
30. sodium carbonate, Na_2CO_3
31. zinc sulfate, $ZnSO_3$
32. sulfur trioxide, SO_3

33. tin(IV) nitrate, $Sn(NO_3)_4$

34. ferrous sulfate, $FeSO_4$

35. chloric acid, HCl

36. aluminum sulfide, Al_2S_3

37. cobalt(II) chloride, $CoCl_2$

38. acetic acid, $HC_2H_3O_2$

39. zinc oxide, ZnO_2

40. stannous fluoride, SnF_2

Free Response Questions:

Answer each of the following. Be sure to include your work and explanations in a clear, logical form.

1. (a) What is an ion?
  (b) The average mass of a calcium atom is 40.08 amu. Why do we also use 40.08 amu as the average mass of a calcium ion (Ca^{2+})?

2. Congratulations! You discover a new element you name wyzzlebium (Wz). The average atomic mass of Wz was found to be 303.001 amu and its atomic number is 120.
  (a) If the masses of the two isotopes of wyzzlebium are 300.9326 amu and 303.9303 amu, what is the relative abundance of each isotope?
  (b) What are the isotopic notations of the two isotopes? (e.g., $_Z^A Wz$)
  (c) How many neutrons are in one atom of the more abundant isotope?

3. How many protons are in one molecule of dichlorine heptoxide? Is it possible to determine precisely how many electrons and neutrons are in a molecule of dichlorine heptoxide? Why or why not?

4. An unidentified metal forms an ionic compound with phosphate. The metal forms a 2+ cation. If the minimum ratio of protons in the metal to the phosphorus is 6:5, what metal is it? (*Hint:* First write the formula for the ionic compound formed with phosphate anion.)

5. For each of the following compounds, indicate what is wrong with the name and why. If possible, fix the name.
  (a) iron hydroxide
  (b) dipotassium dichromium heptoxide
  (c) sulfur oxide

6. Sulfur dioxide is a gas formed as a by-product of burning coal. Sulfur trioxide is a significant contributor to acid rain. Does the existence of these two substances violate the law of multiple proportions? Explain.

7. (a) Which subatomic particles are not in the nucleus?
  (b) What happens to the size of an atom when it becomes an anion?
  (c) What do an ion of Ca and an atom of Ar have in common?

8. An unidentified atom is found to have an atomic mass 7.18 times that of the carbon-12 isotope.
  (a) What is the mass of the unidentified atom?
  (b) What are the possible identities of this atom?
  (c) Why are you unable to positively identify the element based on the atomic mass and the periodic table?
  (d) If the element formed a compound M_2O, where M is the unidentified atom, identify M by writing the isotopic notation for the atom.

9. Scientists such as Dalton, Thomson, and Rutherford proposed important models, which were ultimately challenged by later technology. What do we know to be false in Dalton's atomic model? What was missing in Thomson's model of the atom? What was Rutherford's experiment that led to the current model of the atom?

Quantitative Composition of Compounds

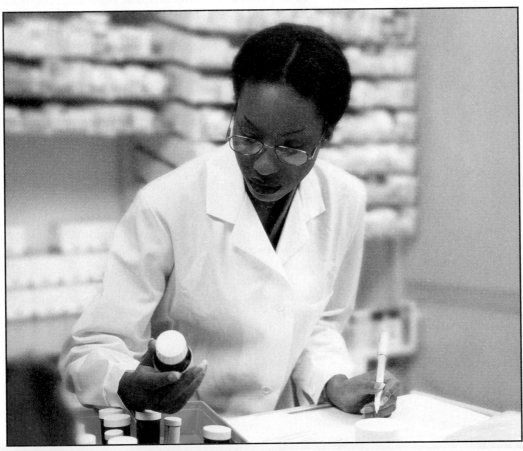

A compounding pharmacist adjusts the quantities of the components for a prescription to precisely meet the needs of his patients.

Chapter Outline

7.1 The Mole

7.2 Molar Mass of Compounds

7.3 Percent Composition of Compounds

7.4 Empirical Formula versus Molecular Formula

7.5 Calculating Empirical Formulas

7.6 Calculating the Molecular Formula from the Empirical Formula

Cereals, cleaning products, and pain remedies all list their ingredients on the package label. The ingredients are listed in order from most to least, but the amounts are rarely given. However, it is precisely these amounts that give products their desired properties and distinguish them from the competition. Understandably, manufacturers carefully regulate the amounts of ingredients to maintain quality and hopefully their customers' loyalty. In the medicines we purchase, these quantities are especially important for safety reasons—for example, they determine whether a medicine is given to children or is safe only for adults.

The composition of compounds is an important concept in chemistry. Determining numerical relationships among the elements in compounds and measuring exact quantities of particles are fundamental tasks that chemists routinely perform in their daily work.

7.1   The Mole

The atom is an incredibly tiny object. Its mass is far too small to measure on an ordinary balance. In Chapter 5 (Section 5.9), we learned to compare atoms using a table of atomic mass units. These units are valuable when we compare the masses of individual atoms (mentally), but they have no practical use in the laboratory. The mass in grams for an "average" carbon atom (atomic mass 12.01 amu) would be 2.00×10^{-23} g, which is much too tiny for the best laboratory balance.

So how can we confidently measure masses for these very tiny atoms? We increase the number of atoms in a sample until we have an amount large enough to measure on a laboratory balance. The problem then is how to count our sample of atoms.

Consider for a moment the produce in a supermarket. Frequently apples and oranges are sorted by size and then sold by weight, not by the piece of fruit. The grocer is counting by weighing. To do this, he needs to know the mass of an "average" apple (235 g) and the mass of an "average" orange (186 g). Now suppose he has an order from the local college for 275 apples and 350 oranges. It would take a long time to count and package this order. The grocer can quickly count fruit by weighing. For example,

$$(275\ \text{apples})\left(\frac{235\ \text{g}}{\text{apple}}\right) = 6.46 \times 10^4\ \text{g} = 64.6\ \text{kg}$$

$$(350\ \text{oranges})\left(\frac{186\ \text{g}}{\text{orange}}\right) = 6.51 \times 10^4\ \text{g} = 65.1\ \text{kg}$$

He can now weigh 64.6 kg of apples and 65.1 kg of oranges and pack them without actually counting them. Manufacturers and suppliers often count by weighing. Other examples of counting by weighing include nuts, bolts, and candy.

Chemists also count atoms by weighing. We know the "average" masses of atoms, so we can count atoms by defining a unit to represent a larger number of atoms. Chemists have chosen the mole (mol) as the unit for counting atoms. The mole is a unit for counting just as a dozen or a ream or a gross is used to count:

Oranges can be "counted" by weighing them in the store.

1 dozen = 12 objects

1 ream = 500 objects

1 gross = 144 objects

1 mole = 6.022×10^{23} objects

Note that we use a unit only when it is appropriate. A dozen eggs is practical in our kitchen, a gross might be practical for a restaurant, but a ream of eggs would not be very practical. Chemists can't use dozens, grosses, or reams because atoms are so tiny that a dozen, gross, or ream of atoms still couldn't be measured in the laboratory.

The number represented by 1 mol, 6.022×10^{23}, is called **Avogadro's number**, in honor of Amadeo Avogadro (1776–1856), who investigated several quantitative aspects in chemistry. It's difficult to imagine how large Avogadro's number really is, but this example may help: If 10,000 people started to count Avogadro's number, and each counted at the rate of 100 numbers per minute each minute of the day, it would take them over 1 trillion (10^{12}) years to count the total number. So even the tiniest amount of matter contains extremely large numbers of atoms.

Avogadro's number

Avogadro's number has been experimentally determined by several methods. How does it relate to atomic mass units? Remember that the atomic mass for an element is the average relative mass of all the isotopes for the element. The atomic mass (expressed in grams) of 1 mole of any element contains the same number of particles (Avogadro's number) as there are in exactly 12 g of ^{12}C. Thus, 1 **mole** of anything is the amount of the substance that contains the same number of items as there are atoms in exactly 12 g of ^{12}C.

Remember that ^{12}C is the reference isotope for atomic masses.

mole

1 mole = 6.022×10^{23} items

From the definition of mole, we can see that the atomic mass in grams of any element contains 1 mol of atoms. The term *mole* is so commonplace in chemistry that chemists use it as freely as the words *atom* or *molecule*. A mole of atoms, molecules, ions, or electrons represents Avogadro's number of these particles. If we can speak of a mole of atoms, we can also speak of a mole of

(a)

(b)

(c)

Units of measurement need to be appropriate for the object being measured. (a) Eggs are measured by the dozen, (b) paper is measured by the ream (500 sheets), and (c) pencils are measured by the gross (144).

molecules, a mole of electrons, or a mole of ions, understanding that in each case we mean 6.022×10^{23} particles:

> 1 mol of atoms $= 6.022 \times 10^{23}$ atoms
> 1 mol of molecules $= 6.022 \times 10^{23}$ molecules
> 1 mol of ions $= 6.022 \times 10^{23}$ ions

Practice 7.1

How many atoms are in 1.000 mole of
(a) Fe   (b) H_2 (c) H_2SO_4

molar mass

The atomic mass of an element in grams contains Avogadro's number of atoms and is defined as the **molar mass** of that element. To determine molar mass, we change the units of the atomic mass (found in the periodic table) from atomic mass units to grams. For example, sulfur has an atomic mass of 32.07 amu, so 1 mol of sulfur has a molar mass of 32.07 g and contains 6.022×10^{23} atoms of sulfur. Here are some other examples:

Element	Atomic mass	Molar mass	Number of atoms
H	1.008 amu	1.008 g	6.022×10^{23}
Mg	24.31 amu	24.31 g	6.022×10^{23}
Na	22.99 amu	22.99 g	6.022×10^{23}

To summarize:

In this text molar masses of elements are given to 4 significant figures.

1. The atomic mass expressed in grams is the *molar mass* of an element. It is different for each element. In this text, molar masses are expressed to four significant figures.

   1 molar mass = atomic mass of an element in grams

2. One mole of any element contains Avogadro's number of atoms.

   1 mol of atoms $= 6.022 \times 10^{23}$ atoms

We can use these relationships to make conversions between number of atoms, mass, and moles, as shown in the following examples.

Example 7.1 How many moles of iron does 25.0 g of iron (Fe) represent?

SOLUTION We need to change grams of Fe to moles of Fe. The atomic mass of Fe (from either the periodic table or the table of atomic masses) is 55.85. Use the proper conversion factor to obtain moles:

$$\text{grams Fe} \longrightarrow \text{moles Fe} \qquad (\text{grams Fe})\left(\frac{1 \text{ mol Fe}}{55.85 \text{ g Fe}}\right)$$

$$(25.0 \ \cancel{\text{g Fe}})\left(\frac{1 \text{ mol Fe}}{55.85 \ \cancel{\text{g Fe}}}\right) = 0.448 \text{ mol Fe}$$

How many magnesium atoms are contained in 5.00 g of Mg?

We need to change grams of Mg to atoms of Mg:

grams Mg $\longrightarrow$ atoms Mg

We find the atomic mass of magnesium to be 24.31 and set up the calculation using a conversion factor between atoms and grams:

$$(\text{grams Mg})\left(\frac{6.022 \times 10^{23} \text{ atoms Mg}}{24.31 \text{ g Mg}}\right)$$

$$(5.00 \text{ g Mg})\left(\frac{6.022 \times 10^{23} \text{ atoms Mg}}{24.31 \text{ g Mg}}\right) = 1.24 \times 10^{23} \text{ atoms Mg}$$

Alternatively, we could first convert grams of Mg to moles of Mg, which are then changed to atoms of Mg:

grams Mg $\longrightarrow$ moles Mg $\longrightarrow$ atoms Mg

Use conversion factors for each step. The calculation setup is

$$(5.00 \text{ g Mg})\left(\frac{1 \text{ mol Mg}}{24.31 \text{ g Mg}}\right)\left(\frac{6.022 \times 10^{23} \text{ atoms Mg}}{1 \text{ mol Mg}}\right) = 1.24 \times 10^{23} \text{ atoms Mg}$$

Thus, 1.24×10^{23} atoms of Mg are contained in 5.00 g of Mg.

Example 7.2

SOLUTION

What is the mass, in grams, of one atom of carbon (C)?

The molar mass of C is 12.01 g. We need to change atoms of C to grams of C:

atoms C $\longrightarrow$ grams C

Our conversion will be

$$(\text{atoms C})\left(\frac{12.01 \text{ g C}}{6.022 \times 10^{23} \text{ atoms C}}\right)$$

$$(1 \text{ atom C})\left(\frac{12.01 \text{ g C}}{6.022 \times 10^{23} \text{ atoms C}}\right) = 1.994 \times 10^{-23} \text{ g C}$$

Example 7.3

SOLUTION

What is the mass of 3.01×10^{23} atoms of sodium (Na)?

We need the molar mass of Na (22.99 g) and a conversion factor between molar mass and atoms:

atoms Na $\longrightarrow$ grams Na

$$(\text{atoms Na})\left(\frac{22.99 \text{ g Na}}{6.022 \times 10^{23} \text{ atoms Na}}\right)$$

$$(3.01 \times 10^{23} \text{ atoms Na})\left(\frac{22.99 \text{ g Na}}{6.022 \times 10^{23} \text{ atoms Na}}\right) = 11.5 \text{ g Na}$$

Example 7.4

SOLUTION

Example 7.5   What is the mass of 0.252 mol of copper (Cu)?

SOLUTION   We need the molar mass of Cu (63.55 g) and a conversion factor between molar mass and moles:

moles Cu $\longrightarrow$ grams Cu

$$(\text{moles Cu})\left(\frac{1 \text{ molar mass Cu}}{1 \text{ mol Cu}}\right)$$

$$(0.252 \text{ mol Cu})\left(\frac{63.55 \text{ g Cu}}{1 \text{ mol Cu}}\right) = 16.0 \text{ g Cu}$$

Example 7.6   How many oxygen atoms are present in 1.00 mol of oxygen molecules?

SOLUTION   Oxygen is a diatomic molecule with the formula O_2. Therefore, a molecule of oxygen contains two oxygen atoms:

$$\frac{2 \text{ atoms O}}{1 \text{ molecule O}_2}$$

The sequence of conversions is

moles O_2 $\longrightarrow$ molecules O_2 $\longrightarrow$ atoms O

The two conversion factors needed are

$$\frac{6.022 \times 10^{23} \text{ molecules O}_2}{1 \text{ mol O}_2} \quad \text{and} \quad \frac{2 \text{ atoms O}}{1 \text{ molecule O}_2}$$

The calculation is

$$(1.00 \text{ mol O}_2)\left(\frac{6.022 \times 10^{23} \text{ molecules O}_2}{1 \text{ mol O}_2}\right)\left(\frac{2 \text{ atoms O}}{1 \text{ molecules O}_2}\right)$$
$$= 1.20 \times 10^{24} \text{ atoms O}$$

Practice 7.2 _____
What is the mass of 2.50 mol of helium (He)?

Practice 7.3 _____
How many atoms are present in 0.025 mol of iron?

7.2   Molar Mass of Compounds

A formula unit is indicated by the formula, e.g., Mg, MgS, H_2O, NaCl.

One mole of a compound contains Avogadro's number of *formula units* of that compound. The terms *molecular weight*, *molecular mass*, *formula weight*, and *formula mass* have been used in the past to refer to the mass of 1 mol of a compound. However, the term *molar mass* is more inclusive, because it can be used for all types of compounds.

If the formula of a compound is known, its molar mass can be determined by adding the molar masses of all the atoms in the formula. If more than one atom of any element is present, its mass must be added as many times as it appears in the compound.

Example 7.7

The formula for water is H_2O. What is its molar mass?

SOLUTION

First we look up the molar masses of H (1.008 g) and O (16.00 g); then we add the masses of all the atoms in the formula unit. Water contains two H atoms and one O atom. Thus

$$
\begin{aligned}
2\,H &= 2(1.008\,g) = 2.016\,g \\
1\,O &= 1(16.00\,g) = \underline{16.00\ \ g} \\
&\qquad\qquad\qquad\ \ 18.02\ \ g = \text{molar mass of } H_2O
\end{aligned}
$$

Example 7.8

Calculate the molar mass of calcium hydroxide ($Ca(OH)_2$).

SOLUTION

The formula of this substance contains one atom of Ca and two atoms each of O and H. We proceed as in Example 7.7:

$$
\begin{aligned}
1\,Ca &= 1(40.08\,g) = 40.08\ \ g \\
2\,O &= 2(16.00\,g) = 32.00\ \ g \\
2\,H &= 2(1.008\,g) = \underline{2.016\,g} \\
&\qquad\qquad\qquad\ \ 74.10\ \ g = \text{molar mass of } Ca(OH)_2
\end{aligned}
$$

Practice 7.4

Calculate the molar mass of KNO_3.

In this text we round all molar masses to four significant figures, although you may need to use a different number of significant figures for other work (in the lab).

The mass of 1 mol of a compound contains Avogadro's number of formula units. Consider the compound hydrogen chloride (HCl). One atom of H combines with one atom of Cl to form HCl. When 1 mol of H (1.008 g of H or 6.022×10^{23} H atoms) combines with 1 mol of Cl (35.45 g of Cl or 6.022×10^{23} Cl atoms), 1 mol of HCl (36.46 g of HCl or 6.022×10^{23} HCl

A mole of table salt (in front of a salt shaker) and a mole of water (in the film container) have different sizes but both contain Avogadro's number of formula units.

molecules) is produced. These relationships are summarized in the following table:

H	Cl	HCl
6.022×10^{23} H *atoms*	6.022×10^{23} Cl *atoms*	6.022×10^{23} HCl *molecules*
1 mol H *atoms*	1 mol Cl *atoms*	1 mol HCl *molecules*
1.008 g H	35.45 g Cl	36.46 g HCl
1 molar mass H *atoms*	1 molar mass Cl *atoms*	1 molar mass HCl

In dealing with diatomic elements (H_2, O_2, N_2, F_2, Cl_2, Br_2, and I_2), we must take special care to distinguish between a mole of atoms and a mole of molecules. For example, consider 1 mol of oxygen molecules, which has a mass of 32.00 g. This quantity is equal to 2 mol of oxygen atoms. Remember that 1 mol represents Avogadro's number of the particular chemical entity that is under consideration:

$$1 \text{ mol } H_2O = 18.02 \text{ g } H_2O = 6.022 \times 10^{23} \text{ molecules}$$
$$1 \text{ mol NaCl} = 58.44 \text{ g NaCl} = 6.022 \times 10^{23} \text{ formula units}$$
$$1 \text{ mol } H_2 = 2.016 \text{ g } H_2 = 6.022 \times 10^{23} \text{ molecules}$$
$$1 \text{ mol } HNO_3 = 63.02 \text{ g } HNO_3 = 6.022 \times 10^{23} \text{ molecules}$$
$$1 \text{ mol } K_2SO_4 = 174.3 \text{ g } K_2SO_4 = 6.022 \times 10^{23} \text{ formula units}$$

Formula units are often used in place of molecules for substances that contain ions.

Create the appropriate conversion factor by placing the unit desired in the numerator and the unit to be eliminated in the denominator.

$1 \text{ mol} = 6.022 \times 10^{23}$ formula units or molecules
$= 1$ molar mass of a compound

Example 7.9   What is the mass of 1 mol of sulfuric acid (H_2SO_4)?

SOLUTION   We look up the masses of hydrogen, sulfur, and oxygen and solve in a manner similar to Examples 7.7 and 7.8:

$$2 \text{ H} = 2(\ 1.008 \text{ g}) = \ 2.016 \text{ g}$$
$$1 \text{ S} = 1(32.07 \ \text{ g}) = 32.07 \ \text{ g}$$
$$4 \text{ O} = 4(16.00 \ \text{ g}) = \underline{64.00 \ \text{ g}}$$
$$98.09 \ \text{ g} = \text{mass of 1 mol of } H_2SO_4$$

Example 7.10   How many moles of sodium hydroxide (NaOH) are there in 1.00 kg of sodium hydroxide?

SOLUTION   First we know that

$$\overset{\text{Na}}{} \overset{\text{O}}{} \overset{\text{H}}{}$$
$$\text{molar mass} = (22.99 \text{ g} + 16.00 \text{ g} + 1.008 \text{ g}) \text{ or } 40.00 \text{ g NaOH}$$

To convert grams to moles we use the conversion factor $\dfrac{1 \text{ mol NaOH}}{40.00 \text{ g NaOH}}$ and this conversion sequence:

$$\text{kg NaOH} \longrightarrow \text{g NaOH} \longrightarrow \text{mol NaOH}$$

The calculation is

$$(1.00 \text{ kg NaOH})\left(\frac{1000 \text{ g NaOH}}{\text{kg NaOH}}\right)\left(\frac{1 \text{ mol NaOH}}{40.00 \text{ g NaOH}}\right) = 25.0 \text{ mol NaOH}$$

1.00 kg NaOH = 25.0 mol NaOH

What is the mass of 5.00 mol of water?

First we know that

1 mol H_2O = 18.02 g (Example 7.7)

The conversion is

mol H_2O ⟶ g H_2O

To convert moles to grams, we use the conversion factor $\dfrac{18.02 \text{ g } H_2O}{1 \text{ mol } H_2O}$

The calculation is

$$(5.00 \text{ mol } H_2O)\left(\frac{18.02 \text{ g } H_2O}{1 \text{ mol } H_2O}\right) = 90.1 \text{ g } H_2O$$

Example 7.11

SOLUTION

How many molecules of hydrogen chloride (HCl) are there in 25.0 g of hydrogen chloride?

From the formula, we find that the molar mass of HCl is

$$\overset{H}{} \quad \overset{Cl}{}$$
36.46 g (1.008 g + 35.45 g). The sequence of conversions is

g HCl ⟶ mol HCl ⟶ molecules HCl

Using the conversion factors

$$\frac{1 \text{ mol HCl}}{36.46 \text{ g HCl}} \quad \text{and} \quad \frac{6.022 \times 10^{23} \text{ molecules HCl}}{1 \text{ mol HCl}}$$

we get

$$(25.0 \text{ g HCl})\left(\frac{1 \text{ mol HCl}}{36.46 \text{ g HCl}}\right)\left(\frac{6.022 \times 10^{23} \text{ molecules HCl}}{1 \text{ mol HCl}}\right)$$
$$= 4.13 \times 10^{23} \text{ molecules HCl}$$

Example 7.12

SOLUTION

Practice 7.5

What is the mass of 0.150 mol of Na_2SO_4?

Practice 7.6

How many moles and molecules are there in 500.0 g of $HC_2H_3O_2$?

7.3   Percent Composition of Compounds

percent composition of a compound

Percent means parts per 100 parts. Just as each piece of pie is a percent of the whole pie, each element in a compound is a percent of the whole compound. The **percent composition of a compound** is the *mass percent* of each element in the compound. The molar mass represents the total mass, or 100%, of the compound. Thus, the percent composition of water, H_2O, is 11.19% H and 88.79% O by mass. According to the law of definite composition, the percent composition must be the same no matter what size sample is taken.

The percent composition of a compound can be determined (1) from knowing its formula or (2) from experimental data.

H
11.19%

O
88.79%

Percent Composition from Formula

If the formula of a compound is known, a two-step process is needed to determine the percent composition:

Step 1 Calculate the molar mass (Section 7.2).
Step 2 Divide the total mass of each element in the formula by the molar mass and multiply by 100. This gives the percent composition:

$$\frac{\text{total mass of the element}}{\text{molar mass}} \times 100 = \text{percent of the element}$$

Example 7.13   Calculate the percent composition of sodium chloride (NaCl).

SOLUTION   **Step 1**   Calculate the molar mass of NaCl:

$$1\,Na = 1(22.99\,g) = 22.99\,g$$
$$1\,Cl = 1(35.45\,g) = \underline{35.45\,g}$$
$$58.44\,g\,(\text{molar mass})$$

Na
39.34%

Cl
60.66%

Step 2   Now calculate the percent composition. We know there are 22.99 g Na and 35.45 g Cl in 58.44 g NaCl:

$$Na\!: \left(\frac{22.99\,g\,Na}{58.44\,g}\right)(100) = 39.34\%\,Na$$

$$Cl\!: \left(\frac{35.45\,g\,Cl}{58.44\,g}\right)(100) = 60.66\%\,Cl$$

100.00% total

In any two-component system, if the percent of one component is known, the other is automatically defined by the difference; that is, if Na is 39.34%, then Cl is 100% − 39.34% = 60.66%. However, the calculation of the percent of each component should be carried out, since this provides a check against possible error. The percent composition data should add up to 100 ± 0.2%.

Calculate the percent composition of potassium chloride (KCl).

Example 7.14

SOLUTION

Step 1 Calculate the molar mass of KCl:

$$1\,K = 1(39.10\,g) = 39.10\,g$$
$$1\,Cl = 1(35.45\,g) = \underline{35.45\,g}$$
$$74.55\,g\,(molar\,mass)$$

Step 2 Now calculate the percent composition. We know there are 39.10 g K and 35.45 g Cl in 74.55 g KCl:

$$K: \left(\frac{39.10\,g\,K}{74.55\,g}\right)(100) = 52.45\%\,K$$

$$Cl: \left(\frac{35.45\,g\,Cl}{74.55\,g}\right)(100) = \underline{47.55\%\,Cl}$$
$$100.00\%\,total$$

Comparing the results calculated for NaCl and for KCl, we see that NaCl contains a higher percentage of Cl by mass, although each compound has a one-to-one atom ratio of Cl to Na and Cl to K. The reason for this mass percent difference is that Na and K do not have the same atomic masses.

It is important to realize that when we compare *1 mol of NaCl with 1 mol of KCl,* each quantity contains the same number of Cl atoms—namely, 1 mol of Cl atoms. However, if we compare *equal masses* of NaCl and KCl, there will be more Cl atoms in the mass of NaCl, since NaCl has a higher mass percent of Cl.

1 mol NaCl contains	100.00 g NaCl contains	1 mol KCl contains	100.00 g KCl contains
1 mol Na	39.34 g Na	1 mol K	52.45 g K
1 mol Cl	60.66 g Cl	1 mol Cl	47.55 g Cl
	60.66% Cl		47.55% Cl

Calculate the percent composition of potassium sulfate (K_2SO_4):

Example 7.15

SOLUTION

Step 1 Calculate the molar mass of K_2SO_4:

$$2\,K = 2(39.10\,g) = 78.20\,g$$
$$1\,S = 1(32.07\,g) = 32.07\,g$$
$$4\,O = 4(16.00\,g) = \underline{64.00\,g}$$
$$174.3\ \ g\,(molar\,mass)$$

Step 2 Now calculate the percent composition. We know there are 78.20 g of K, 32.07 g of S, and 64.00 g of O in 174.3 g of K_2SO_4:

$$K: \left(\frac{78.20 \text{ g K}}{174.3 \text{ g}}\right)(100) = 44.87\% \text{ K}$$

$$S: \left(\frac{32.07 \text{ g S}}{174.3 \text{ g}}\right)(100) = 18.40\% \text{ S}$$

$$O: \left(\frac{64.00 \text{ g O}}{174.3 \text{ g}}\right)(100) = \underline{36.72\% \text{ O}}$$

$$99.99\% \text{ total}$$

Practice 7.7
Calculate the percent composition of $Ca(NO_3)_2$.

Practice 7.8
Calculate the percent composition of K_2CrO_4.

Percent Composition from Experimental Data

The percent composition can be determined from experimental data without knowing the formula of a compound:

Step 1 Calculate the mass of the compound formed.
Step 2 Divide the mass of each element by the total mass of the compound and multiply by 100.

Example 7.16 Zinc oxide is a compound with many uses from preventing sunburn to a pigment in white paint. When heated in the air, 1.63 g of zinc (Zn) combines with 0.40 g of oxygen (O_2) to form zinc oxide. Calculate the percent composition of the compound formed:

SOLUTION   **Step 1** First calculate the total mass of the compound formed:

1.63 g Zn
0.40 g O_2
―――――――
2.03 g   = total mass of product

Step 2 Divide the mass of each element by the total mass (2.03 g) and multiply by 100:

$$\left(\frac{1.63 \text{ g Zn}}{2.03 \text{ g}}\right)(100) = 80.3\% \text{ Zn} \qquad \left(\frac{0.40 \text{ g O}}{2.03 \text{ g}}\right)(100) = 20.\% \text{ O}$$

The compound formed contains 80.3% Zn and 20.% O.

Modern technology is changing our coins in some pretty interesting ways. The first U.S. coins were produced during the late 1700s from "coin silver," an alloy of 90% silver and 10% copper. Coins were made following a very simple rule: The mass of the coin reflects is relative value. Therefore a half dollar weighed half as much as a dollar coin and a quarter weighed $\frac{1}{4}$ as much as a dollar coin.

In the twentieth century our society began using machines to collect coins. We use vending machines, parking meters, toll baskets, Laundromats, slot machines, and video games to make our lives more convenient and fun. This change produced a huge demand for coins. At the same time, the price of silver rose rapidly.

The solution? Make coins from a different alloy that would still work in all of our machines. This was a tricky process since the machines require a specific mass and electric "resistivity." In the mid-1960s new coins which sandwiched a layer of copper between layers of an alloy of copper and nickel began to replace quarters and dimes. And a new Susan B. Anthony dollar appeared briefly in 1979. Finally, in 1997 Congress decided to revive the dollar coin once more.

Congress specified that the new dollar coin had to look and feel different to consumers, be golden in color, and have a distinctive edge. At the same time, it needed to fool vending machines so that both the Susan B. Anthony dollar and the new coin would work. Designers decided on the portrait of Sacagawea, a Shoshone woman and her child. She was a guide on the Lewis and Clark expedition. On the back is an eagle. The chemists struggled to find a golden alloy for the coin. It is slightly larger than a quarter and weighs 8.1 grams. Vending machines determine a coin's value by its size, weight and its electromagnetic signature. This electromagnetic signature is hard to duplicate. All the golden coins tested had three times the conductivity of the silver Anthony coin. Metallurgists finally tried adding manganese and zinc to the copper core and found a "golden" alloy that fooled the vending machines! Sacagawea dollars are made of an alloy which is 77% copper, 12% zinc, 7% manganese and 4% nickel. And on top of it all, the new coins only cost 12 cents each to make, leaving an 88 cent profit for the mint on each coin!

The future of coins is becoming less certain, however, as our machines are converted to electronic devices which use magnetic strips and a swipe of a card in place of "coins" and scan our cars for toll pass information. Some day soon, coins may vanish altogether from our pockets into collector's albums.

Sacagawea coin

Practice 7.9

Aluminum chloride is formed by reacting 13.43 g aluminum with 53.18 g chlorine. What is the percent composition of the compound?

7.4 Empirical Formula versus Molecular Formula

The **empirical formula,** or *simplest formula,* gives the smallest whole-number ratio of atoms present in a compound. This formula gives the relative number of atoms of each element in the compound.

The **molecular formula** is the true formula, representing the total number of atoms of each element present in one molecule of a compound. It is entirely possible that two or more substances will have the same percent composition yet be distinctly different compounds. For example, acetylene (C_2H_2) is a common gas used in welding; benzene (C_6H_6) is an important solvent obtained from

empirical formula

molecular formula

coal tar and is used in the synthesis of styrene and nylon. Both acetylene and benzene contain 92.3% C and 7.7% H. The smallest ratio of C and H corresponding to these percentages is CH (1:1). Therefore, the *empirical* formula for both acetylene and benzene is CH, even though the *molecular* formulas are C_2H_2 and C_6H_6, respectively. Often the molecular formula is the same as the empirical formula. If the molecular formula is not the same, it will be an integral (whole number) multiple of the empirical formula. For example,

CH = empirical formula

$(CH)_2 = C_2H_2$ = acetylene   (molecular formula)

$(CH)_6 = C_6H_6$ = benzene   (molecular formula)

Table 7.1 compares the formulas of these substances. Table 7.2 shows empirical and molecular formula relationships of other compounds.

Table 7.1 Molecular Formulas of Two Compounds Having an Empirical Formula with a 1:1 Ratio of Carbon and Hydrogen Atoms

Formula	Composition		Molar mass
	% C	% H	
CH (empirical)	92.3	7.7	13.02 (empirical)
C_2H_2 (acetylene)	92.3	7.7	26.04 (2×13.02)
C_6H_6 (benzene)	92.3	7.7	78.12 (6×13.02)

Table 7.2 Some Empirical and Molecular Formulas

Substance	Empirical formula	Molecular formula	Substance	Empirical formula	Molecular formula
Acetylene	CH	C_2H_2	Diborane	BH_3	B_2H_6
Benzene	CH	C_6H_6	Hydrazine	NH_2	N_2H_4
Ethylene	CH_2	C_2H_4	Hydrogen	H	H_2
Formaldehyde	CH_2O	CH_2O	Chlorine	Cl	Cl_2
Acetic acid	CH_2O	$C_2H_4O_2$	Bromine	Br	Br_2
Glucose	CH_2O	$C_6H_{12}O_6$	Oxygen	O	O_2
Hydrogen chloride	HCl	HCl	Nitrogen	N	N_2
Carbon dioxide	CO_2	CO_2	Iodine	I	I_2

7.5 Calculating Empirical Formulas

We can establish empirical formulas because (1) individual atoms in a compound are combined in whole-number ratios, and (2) each element has a specific atomic mass.

To calculate an empirical formula, we need to know (1) the elements that are combined, (2) their atomic masses, and (3) the ratio by mass or percentage in which they are combined. If elements A and B form a compound, we may represent the empirical formula as A_xB_y, where x and y are small whole numbers that represent the atoms of A and B. To write the empirical formula, we must determine x and y:

Step 1	Assume a definite starting quantity (usually 100.0 g) of the compound, if not given, and express the mass of each element in grams.
Step 2	Convert the grams of each element into moles using each element's molar mass. This conversion gives the number of moles of atoms of each element in the quantity assumed in Step 1. At this point, these numbers will usually not be whole numbers.
Step 3	Divide each value obtained in Step 2 by the smallest of these values. If the numbers obtained are whole numbers, use them as subscripts and write the empirical formula. If the numbers obtained are not whole numbers, go on to Step 4.
Step 4	Multiply the values obtained in Step 3 by the smallest number that will convert them to whole numbers. Use these whole numbers as the subscripts in the empirical formula. For example, if the ratio of A to B is 1.0 : 1.5, multiply both numbers by 2 to obtain a ratio of 2 : 3. The empirical formula then is A_2B_3.

In many of these calculations, results will vary somewhat from an exact whole number; this can be due to experimental errors in obtaining the data or from rounding off numbers. Calculations that vary by no more than ±0.1 from a whole number usually are rounded off to the nearest whole number. Deviations greater than about 0.1 unit usually mean that the calculated ratios need to be multiplied by a factor to make them all whole numbers. For example, an atom ratio of 1 : 1.33 should be multiplied by 3 to make the ratio 3 : 4.

Some common fractions and their decimal equivalents are

$$\tfrac{1}{4} = 0.25$$
$$\tfrac{1}{3} = 0.333\ldots$$
$$\tfrac{2}{3} = 0.666\ldots$$
$$\tfrac{1}{2} = 0.5$$
$$\tfrac{3}{4} = 0.75$$

Multiply the decimal equivalent by the number in the denominator of the fraction to get a whole number: 4(0.75) = 3.

Calculate the empirical formula of a compound containing 11.19% hydrogen (H) and 88.79% oxygen (O):

Example 7.17

Step 1 Express each element in grams. Assuming 100.00 g of material, we know that the percent of each element equals the grams of each element and that the percent sign can be omitted:

SOLUTION

H = 11.19 g

O = 88.79 g

Step 2 Convert the grams of each element to moles:

$$\text{H: } (11.19 \text{ g H})\left(\frac{1 \text{ mol H atoms}}{1.008 \text{ g H}}\right) = 11.10 \text{ mol H atoms}$$

$$\text{O: } (88.79 \text{ g O})\left(\frac{1 \text{ mol O atoms}}{16.00 \text{ g O}}\right) = 5.549 \text{ mol O atoms}$$

The formula could be expressed as $H_{11.10}O_{5.549}$. However, it's customary to use the smallest whole-number ratio of atoms. This ratio is calculated in Step 3.

Step 3 Change these numbers to whole numbers by dividing them by the smaller number:

$$H = \frac{11.10\,\text{mol}}{5.549\,\text{mol}} = 2.000 \qquad O = \frac{5.549\,\text{mol}}{5.549\,\text{mol}} = 1.000$$

In this step, the ratio of atoms has not changed, because we divided the number of moles of each element by the same number:

The simplest ratio of H to O is $2:1$.

Empirical formula $= H_2O$

Example 7.18   The analysis of a salt shows that it contains 56.58% potassium (K); 8.68% carbon (C); and 34.73% oxygen (O). Calculate the empirical formula for this substance.

SOLUTION   **Steps 1 and 2** After changing the percentage of each element to grams, we find the relative number of moles of each element by multiplying by the proper conversion factor:

$$K:\ (56.58\,\cancel{g\,K})\left(\frac{1\,\text{mol K atoms}}{39.10\,\cancel{g\,K}}\right) = 1.447\,\text{mol K atoms}$$

$$C:\ (8.68\,\cancel{g\,C})\left(\frac{1\,\text{mol C atoms}}{12.01\,\cancel{g\,C}}\right) = 0.723\,\text{mol C atoms}$$

$$O:\ (34.73\,\cancel{g\,O})\left(\frac{1\,\text{mol O atoms}}{16.00\,\cancel{g\,O}}\right) = 2.171\,\text{mol O atoms}$$

Step 3 Divide each number of moles by the smallest value:

$$K = \frac{1.447\,\text{mol}}{0.723\,\text{mol}} = 2.00$$

$$C = \frac{0.723\,\text{mol}}{0.723\,\text{mol}} = 1.00$$

$$O = \frac{2.171\,\text{mol}}{0.723\,\text{mol}} = 3.00$$

The simplest ratio of $K:C:O$ is $2:1:3$.

Empirical formula $= K_2CO_3$

Example 7.19   A sulfide of iron was formed by combining 2.233 g of iron (Fe) with 1.926 g of sulfur (S). What is the empirical formula of the compound?

SOLUTION   **Steps 1 and 2** The grams of each element are given, so we use them directly in our calculations. Calculate the relative number of moles of each element by multiplying the grams of each element by the proper conversion factor:

$$\text{Fe: } (2.233 \text{ gFe})\left(\frac{1 \text{ mol Fe atoms}}{55.85 \text{ gFe}}\right) = 0.03998 \text{ mol Fe atoms}$$

$$\text{S: } (1.926 \text{ gS})\left(\frac{1 \text{ mol S atoms}}{32.07 \text{ gS}}\right) = 0.06006 \text{ mol S atoms}$$

Step 3 Divide each number of moles by the smaller of the two numbers:

$$\text{Fe} = \frac{0.03998 \text{ mol}}{0.03998 \text{ mol}} = 1.000 \qquad \text{S} = \frac{0.06006 \text{ mol}}{0.03998 \text{ mol}} = 1.502$$

Step 4 We still have not reached a ratio that gives whole numbers in the formula, so we double each value to obtain a ratio of 2.000 Fe atoms to 3.000 S atoms. Doubling both values does not change the ratio of Fe and S atoms:

$$\text{Fe: } (1.000)2 = 2.000$$
$$\text{S: } (1.502)2 = 3.004$$

Empirical formula = Fe_2S_3

Practice 7.10

Calculate the empirical formula of a compound containing 52.14% C, 13.12% H, and 34.73% O.

Practice 7.11

Calculate the empirical formula of a compound that contains 43.7% phosphorus and 56.3% O by mass.

7.6 Calculating the Molecular Formula from the Empirical Formula

The molecular formula can be calculated from the empirical formula if the molar mass is known. The molecular formula, as stated in Section 7.4, will be equal either to the empirical formula or some multiple of it. For example, if the empirical formula of a compound of hydrogen and fluorine is HF, the molecular formula can be expressed as $(HF)_n$, where $n = 1, 2, 3, 4, \ldots$. This n means that the molecular formula could be HF, H_2F_2, H_3F_3, H_4F_4, and so on. To determine the molecular formula, we must evaluate n:

$$n = \frac{\text{molar mass}}{\text{mass of empirical formula}} = \text{number of empirical formula units}$$

What we actually calculate is the number of units of the empirical formula contained in the molecular formula.

Example 7.20   A compound of nitrogen and oxygen with a molar mass of 92.00 g was found to have an empirical formula of NO_2. What is its molecular formula?

SOLUTION   Let n be the number of NO_2 units in a molecule; then the molecular formula is $(NO_2)_n$. Each NO_2 unit has a mass of $[14.01\text{ g} + 2(16.00\text{ g})]$ or 46.01 g. The molar mass of $(NO_2)_n$ is 92.00 g, and the number of 46.01 units in 92.00 is 2:

$$n = \frac{92.00\text{ g}}{46.01\text{ g}} = 2 \quad \text{(empirical formula units)}$$

N_2O_4

The molecular formula is $(NO_2)_2$, or N_2O_4.

Example 7.21   Propylene is a compound frequently polymerized to make polypropylene which is used for rope, laundry bags, blankets, carpets and textile fibers. Propylene has a molar mass of 42.08 g and contains 14.3% H and 85.7% C. What is its molecular formula?

SOLUTION   First find the empirical formula:

C_3H_6

$$\text{C:}\quad (85.7\text{ g C})\left(\frac{1\text{ mol C atoms}}{12.01\text{ g C}}\right) = 7.14\text{ mol C atoms}$$

$$\text{H:}\quad (14.3\text{ g H})\left(\frac{1\text{ mol H atoms}}{1.008\text{ g H}}\right) = 14.2\text{ mol H atoms}$$

Divide each value by the smaller number of moles:

$$C = \frac{7.14\text{ mol}}{7.14\text{ mol}} = 1.00$$

$$H = \frac{14.2\text{ mol}}{7.14\text{ mol}} = 1.99$$

Empirical formula $= CH_2$

We determine the molecular formula from the empirical formula and the molar mass:

Molecular formula $= (CH_2)_n$
Molar mass $= 42.08$ g

Each CH_2 unit has a mass of $(12.01\text{ g} + 2.016\text{ g})$ or 14.03 g. The number of CH_2 units in 42.08 g is 3:

$$n = \frac{42.08\text{ g}}{14.03\text{ g}} = 3 \quad \text{(empirical formula units)}$$

The molecular formula is $(CH_2)_3$, or C_3H_6.

When propylene is polymerized into polypropylene it can be used for rope and other recreational products.

Practice 7.12

Calculate the empirical and molecular formulas of a compound that contains 80.0% C, 20.0% H, and has a molar mass of 30.00 g.

Chapter 7 Review

7.1 The Mole

KEY TERMS

Avogadro's number
Mole

- We count atoms by weighing them since they are so tiny.
- 1 mole = 6.022×10^{23} items.
- Avogadro's number is 6.022×10^{23}.

7.2 Molar Mass of Compounds

- One mole of a compound contains Avogadro's number of formula units of that compound.
- The mass (grams) of one mole of a compound is the molar mass
- Molar mass is determined by adding the molar masses of all the atoms in a formula.
- Molar masses are given to four significant figures in this text.

7.3 Percent Composition of Compounds

KEY TERM

Percent composition of a compound

- To determine the percent composition from a formula:
  - Calculate the molar mass
  - For each element in the formula

  $$\frac{\text{total mass of the element}}{\text{molar mass of the compound}} \times 100$$

  $$= \text{percent of the element}$$

- To determine percent composition from experimental data:
  - Calculate the mass of the compound formed
  - For each element in the formula

  $$\frac{\text{mass of the element}}{\text{mass of the compound formed}} \times 100$$

  $$= \text{percent of the element}$$

7.4 Empirical Formula versus Molecular Formula

KEY TERMS

Empirical formula
Molecular formula

- The empirical formula is the simplest formula giving the smallest whole number ratio of atoms present in a compound
- The molecular formula is the true formula representing the total number of atoms of each element present in one molecule of the compound
- Two or more substances may have the same empirical formulas but different molecular formulas

7.5 Calculating Empirical Formulas

- To determine the empirical formula for a compound you need to know:
  - The elements that are combined
  - Their atomic masses
  - The ratio of masses or percentage in which they are combined
- Empirical formulas are represented in the form A_xB_y. To determine the empirical formula of this compound:
  - Assume a starting quantity (100.0 g is a good choice)
  - Convert mass (g) to moles for each element
  - Divide each element's moles by the smallest number of moles.
    - If the ratios are whole numbers, use them as subscripts and write the empirical formula.
    - If the ratios are not whole numbers, multiply them all by the smallest number which will convert them to whole number.
    - Use the whole numbers to write the empirical formula

7.6 Calculating the Molecular Formula from the Empirical Formula

- The molecular formula is calculated from the empirical formula when the molar mass is known:

  $$n = \frac{\text{molar mass}}{\text{mass of empirical formula}}$$

  $$= \text{number of empirical formula units}$$

  - The molecular formula is $(A_xB_y)_n$

Review Questions

All questions with blue numbers have answers in the appendix of the text.

1. What is a mole?

2. Which would have a higher mass: a mole of K atoms or a mole of Au atoms?

3. Which would contain more atoms: a mole of K atoms or a mole of Au atoms?

4. Which would contain more electrons: a mole of K atoms or a mole of Au atoms?

5. What is molar mass?

6. If the atomic mass scale had been defined differently, with an atom of $^{12}_{6}C$ defined as a mass of 50 amu, would this have any effect on the value of Avogadro's number? Explain.

7. What is the numerical value of Avogadro's number?

8. What is the relationship between Avogadro's number and the mole?

9. Complete these statements, supplying the proper quantity.
   (a) A mole of O atoms contains _____ atoms.
   (b) A mole of O_2 molecules contains _____ molecules.
   (c) A mole of O_2 molecules contains _____ atoms.
   (d) A mole of O atoms has a mass of _____ grams.
   (e) A mole of O_2 molecules has a mass of _____ grams.

10. How many molecules are present in 1 molar mass of sulfuric acid (H_2SO_4)? How many atoms are present?

11. What is the difference between an empirical formula and a molecular formula?

12. In calculating the empirical formula of a compound from its percent composition, why do we choose to start with 100.0 g of the compound?

Paired Exercises

All exercises with blue numbers have answers in the appendix of the text.

Molar Masses

1. Determine the molar masses of these compounds:
   (a) KBr
   (b) Na_2SO_4
   (c) $Pb(NO_3)_2$
   (d) C_2H_5OH
   (e) $HC_2H_3O_2$
   (f) Fe_3O_4
   (g) $C_{12}H_{22}O_{11}$
   (h) $Al_2(SO_4)_3$
   (i) $(NH_4)_2HPO_4$

2. Determine the molar masses of these compounds:
   (a) NaOH
   (b) Ag_2CO_3
   (c) Cr_2O_3
   (d) $(NH_4)_2CO_3$
   (e) $Mg(HCO_3)_2$
   (f) C_6H_5COOH
   (g) $C_6H_{12}O_6$
   (h) $K_4Fe(CN)_6$
   (i) $BaCl_2 \cdot 2\,H_2O$

Moles and Avogadro's Number

3. How many moles of atoms are contained in the following?
   (a) 22.5 g Zn
   (b) 0.688 g Mg
   (c) 4.5×10^{22} atoms Cu
   (d) 382 g Co
   (e) 0.055 g Sn
   (f) 8.5×10^{24} molecules N_2

4. How many moles are contained in the following?
   (a) 25.0 g NaOH
   (b) 44.0 g Br_2
   (c) 0.684 g $MgCl_2$
   (d) 14.8 g CH_3OH
   (e) 2.88 g Na_2SO_4
   (f) 4.20 lb ZnI_2

5. Calculate the number of grams in each of the following:
   (a) 0.550 mol Au
   (b) 15.8 mol H_2O
   (c) 12.5 mol Cl_2
   (d) 3.15 mol NH_4NO_3

6. Calculate the number of grams in each of the following:
   (a) 4.25×10^{-4} mol H_2SO_4
   (b) 4.5×10^{22} molecules CCl_4
   (c) 0.00255 mol Ti
   (d) 1.5×10^{16} atoms S

7. How many molecules are contained in each of the following:
   (a) 1.26 mol O_2
   (b) 0.56 mol C_6H_6
   (c) 16.0 g CH_4
   (d) 1000. g HCl

8. How many molecules are contained in each of the following:
   (a) 1.75 mol Cl_2
   (b) 0.27 mol C_2H_6O
   (c) 12.0 g CO_2
   (d) 1000. g CH_4

9. How many atoms are contained in each of the following:
   (a) 11 molecules C_2H_5OH
   (b) 25.0 atom Ag
   (c) 0.0986 g Xe
   (d) 72.5 g $CHCl_3$

10. How many atoms are contained in each of the following:
    (a) 18 molecules N_2O_5
    (b) 10.0 mol Au
    (c) 75.2 g BF_3
    (d) 15.2 g U

11. Calculate the mass in grams of each of the following:
    (a) 1 atom Pb
    (b) 1 atom Ag
    (c) 1 molecule H_2O
    (d) 1 molecule $C_3H_5(NO_3)_3$

12. Calculate the mass in grams of each of the following:
    (a) 1 atom Au
    (b) 1 atom U
    (c) 1 molecule NH_3
    (d) 1 molecule $C_6H_4(NH_2)_2$

13. Make the following conversions:
    (a) 8.66 mol Cu to grams Cu
    (b) 125 mol Au to kilograms Au
    (c) 10. atoms C to moles C
    (d) 5000 molecules CO_2 to moles CO_2

14. Make the following conversions:
    (a) 28.4 g S to moles S
    (b) 2.50 kg NaCl to moles NaCl
    (c) 42.4 g Mg to atoms Mg
    (d) 485 mL Br_2 ($d = 3.12$ g/mL) to moles Br_2

15. Exactly 1 mol of carbon disulfide contains
    (a) how many carbon disulfide molecules?
    (b) how many carbon atoms?
    (c) how many sulfur atoms?
    (d) how many total atoms of all kinds?

16. One (1.000) mole of ammonia contains
    (a) how many ammonia molecules?
    (b) how many nitrogen atoms?
    (c) how many hydrogen atoms?
    (d) how many total atoms of all kinds?

17. How many atoms of oxygen are contained in each of the following?
    (a) 16.0 g O_2
    (b) 0.622 mol MgO
    (c) 6.00×10^{22} molecules $C_6H_{12}O_6$

18. How many atoms of oxygen are contained in each of the following?
    (a) 5.0 mol MnO_2
    (b) 255 g $MgCO_3$
    (c) 5.0×10^{18} molecules H_2O

19. Calculate the number of
    (a) grams of silver in 25.0 g AgBr
    (b) grams of nitrogen in 6.34 mol $(NH_4)_3PO_4$
    (c) grams of oxygen in 8.45×10^{22} molecules SO_3

20. Calculate the number of
    (a) grams of chlorine in 5.0 g $PbCl_2$
    (b) grams of hydrogen in 4.50 mol H_2SO_4
    (c) grams of hydrogen in 5.45×10^{22} molecules NH_3

Percent Composition

21. Calculate the percent composition by mass of these compounds:
    (a) NaBr   (d) $SiCl_4$
    (b) $KHCO_3$ (e) $Al_2(SO_4)_3$
    (c) $FeCl_3$ (f) $AgNO_3$

22. Calculate the percent composition by mass of these compounds:
    (a) $ZnCl_2$ (d) $(NH_4)_2SO_4$
    (b) $NH_4C_2H_3O_2$ (e) $Fe(NO_3)_3$
    (c) MgP_2O_7 (f) ICl_3

23. Calculate the percent of iron in the following compounds:
    (a) FeO
    (b) Fe_2O_3
    (c) Fe_3O_4
    (d) $K_4Fe(CN)_6$

24. Which of the following chlorides has the highest and which has the lowest percentage of chlorine, by mass, in its formula?
    (a) KCl   (c) $SiCl_4$
    (b) $BaCl_2$ (d) LiCl

25. A 6.20-g sample of phosphorus was reacted with oxygen to form an oxide with a mass of 14.20 g. Calculate the percent composition of the compound.

26. A sample of ethylene chloride was analyzed to contain 6.00 g of C, 1.00 g of H, and 17.75 g of Cl. Calculate the percent composition of ethylene chloride.

27. Examine the following formulas. Which compound has the
    (a) higher percent by mass of hydrogen: H_2O or H_2O_2?
    (b) lower percent by mass of nitrogen: NO or N_2O_3?
    (c) higher percent by mass of oxygen: NO_2 or N_2O_4?

    Check your answers by calculation if you wish.

28. Examine the following formulas. Which compound has the
    (a) lower percent by mass of chlorine: $NaClO_3$ or $KClO_3$?
    (b) higher percent by mass of sulfur: $KHSO_4$ or K_2SO_4?
    (c) lower percent by mass of chromium: Na_2CrO_4 or $Na_2Cr_2O_7$?

    Check your answers by calculation if you wish.

Empirical and Molecular Formulas

29. Determine the empirical formula for the following:
    (a) a compound made from 5.94% hydrogen and 94.06% oxygen with a total molar mass of 34.02 g
    (b) a compound made from 80.34% zinc and 19.66% oxygen with a total molar mass of 81.39 g
    (c) a compound made from 35.18% iron, 44.66% chlorine and 20.16% oxygen with a total molar mass of 158.75 g
    (d) a compound made from 26.19% nitrogen, 7.55% hydrogen and 66.26% chlorine with a total molar mass of 53.50 g

30. Determine the empirical formula for the following:
    (a) a compound made from 32.86% potassium and 67.14% bromine with a total molar mass of 119.00 g
    (b) a compound made from 63.50% silver, 8.25% nitrogen and 28.25% oxygen with a total molar mass of 169.91 g
    (c) a compound made from 54.09% calcium, 2.72% hydrogen and 43.18% oxygen with a total molar mass of 74.10 g
    (d) a compound made from 2.06% hydrogen, 32.69% sulfur and 65.25% oxygen with a total molar mass of 98.09 g

31. Calculate the empirical formula of each compound from the percent compositions given:
    (a) 63.6% N, 36.4% O
    (b) 46.7% N, 53.3% O
    (c) 25.9% N, 74.1% O
    (d) 43.4% Na, 11.3% C, 45.3% O
    (e) 18.8% Na, 29.0% Cl, 52.3% O
    (f) 72.02% Mn, 27.98% O

32. Calculate the empirical formula of each compound from the percent compositions given:
    (a) 64.1% Cu, 35.9% Cl
    (b) 47.2% Cu, 52.8% Cl
    (c) 51.9% Cr, 48.1% S
    (d) 55.3% K, 14.6% P, 30.1% O
    (e) 38.9% Ba, 29.4% Cr, 31.7% O
    (f) 3.99% P, 82.3% Br, 13.7% Cl

33. A sample of tin having a mass of 3.996 g was oxidized and found to have combined with 1.077 g of oxygen. Calculate the empirical formula of this oxide of tin.

34. A 3.054-g sample of vanadium (V) combined with oxygen to form 5.454 g of product. Calculate the empirical formula for this compound.

35. In a laboratory experiment a student burned a 2.465-g sample of copper, which reacted with oxygen from the air, to form an oxide of copper. The oxide had a mass of 2.775 g. Calculate the empirical formula for this compound.

36. A 5.276-g sample of an unknown compound was found to contain 3.989 g of mercury and the rest, chlorine. Calculate the empirical formula of this chloride of mercury.

37. Hydroquinone is an organic compound commonly used as a photographic developer. It has a molar mass of 110.1 g/mol and a composition of 65.45% C, 5.45% H, and 29.09% O. Calculate the molecular formula of hydroquinone.

38. Fructose is a very sweet natural sugar that is present in honey, fruits, and fruit juices. It has a molar mass of 180.1 g/mol and a composition of 40.0% C, 6.7% H, and 53.3% O. Calculate the molecular formula of fructose.

39. Ethanedioic acid, a compound that is present in many vegetables, has a molar mass of 90.04 g/mol and a composition of 26.7% C, 2.2% H, and 71.1% O. What is the molecular formula for this substance?

40. Butyric acid is a compound that is present in butter. It has a molar mass of 88.11 g/mol and is composed of 54.5% C, 9.2% H, and 36.3% O. What is the molecular formula for this subtance?

41. Calculate the percent composition and determine the molecular formula and the empirical formula for the nitrogen-oxygen compound that results when 12.04 g of nitrogen is reacted with enough oxygen to produce 39.54 g of product. The molar mass of the product is 92.02 g.

42. Calculate the percent composition and determine the molecular formula and the empirical formula for the carbon-hydrogen-oxygen compound that results when 30.21 g of carbon, 40.24 g of oxygen, and 5.08 g of hydrogen are reacted to produce a product with a molar mass of 180.18 g.

43. The compound XYZ_3 has a molar mass of 100.09 g and a percent composition (by mass) of 40.04% X, 12.00% Y, and 47.96% Z. What is the compound?

44. The compound $X_2(YZ_3)_3$ has a molar mass of 282.23 g and a percent composition (by mass) of 19.12% X, 29.86% Y, and 51.02% Z. What is the compound?

Additional Exercises

All exercises with blue numbers have answers in the appendix of the text.

45. White phosphorus is one of several forms of phosphorus and exists as a waxy solid consisting of P_4 molecules. How many atoms are present in 0.350 mol of P_4?

46. How many grams of sodium contain the same number of atoms as 10.0 g of potassium?

47. One atom of an unknown element is found to have a mass of 1.79×10^{-23} g. What is the molar mass of this element?

48. How many molecules of sugar are in a 5-lb bag of sugar? The formula for table sugar or sucrose is $C_{12}H_{22}O_{11}$.

49. If a stack of 500 sheets of paper is 4.60 cm high, what will be the height, in meters, of a stack of Avogadro's number of sheets of paper?

50. There are about 6.1 billion (6.1×10^9) people on earth. If exactly 1 mol of dollars were distributed equally among these people, how many dollars would each person receive?

51. If 20. drops of water equal 1.0 mL (1.0 cm^3),
(a) how many drops of water are there in a cubic mile of water?
(b) what would be the volume in cubic miles of a mole of drops of water?

52. Silver has a density of 10.5 g/cm^3. If 1.00 mol of silver were shaped into a cube,
(a) what would be the volume of the cube?
(b) what would be the length of one side of the cube?

53. Given 1.00-g samples of each of the compounds, CO_2, O_2, H_2O, and CH_3OH,
(a) which sample will contain the largest number of molecules?
(b) which sample will contain the largest number of atoms? Show proof for your answers.

54. How many grams of Fe_2S_3 will contain a total number of atoms equal to Avogadro's number?

55. How many grams of calcium must be combined with phosphorus to form the compound Ca_3P_2?

56. An iron ore contains 5% $FeSO_4$ by mass. How many grams of iron could be obtained from 1.0 ton of this ore?

57. How many grams of lithium will combine with 20.0 g of sulfur to form the compound Li_2S?

58. Calculate the percentage of
(a) mercury in $HgCO_3$
(b) oxygen in $Ca(ClO_3)_2$
(c) nitrogen in $C_{10}H_{14}N_2$ (nicotine)
(d) Mg in $C_{55}H_{72}MgN_4O_5$ (chlorophyll)

59. Zinc and sulfur react to form zinc sulfide, ZnS. If we mix 19.5 g of zinc and 9.40 g of sulfur, have we added sufficient sulfur to fully react all the zinc? Show evidence for your answer.

60. Pyrethrins are a group of compounds that are isolated from flowers and used as insecticidal agents. One form has the formula $C_{21}H_{28}O_3$. What is the percent composition of this compound?

61. Diphenhydramine hydrochloride, a drug used commonly as an antihistamine, has the formula $C_{17}H_{21}NO \cdot HCl$. What is the percent composition of each element in this compound?

62. What is the percent composition of each element in sucrose, $C_{12}H_{22}O_{11}$?

63. Aspirin is well known as a pain reliever (analgesic) and as a fever reducer (antipyretic). It has a molar mass of 180.2 g/mol and a composition of 60.0% C, 4.48% H, and 35.5% O. Calculate the molecular formula of aspirin.

64. How many grams of oxygen are contained in 8.50 g of $Al_2(SO_4)_3$?

65. Gallium arsenide is one of the newer materials used to make semiconductor chips for use in supercomputers. Its composition is 48.2% Ga and 51.8% As. What is the empirical formula?

66. Calcium tartrate is used as a preservative for certain foods and as an antacid. It contains 25.5% C, 2.1% H, 21.3% Ca, and 51.0% O. What is the empirical formula for calcium tartrate?

67. The compositions of four different compounds of carbon and chlorine follow. Determine the empirical formula and the molecular formula for each compound.

	Percent C	Percent Cl	Molar mass (g)
(a)	7.79	92.21	153.8
(b)	10.13	89.87	236.7
(c)	25.26	74.74	284.8
(d)	11.25	88.75	319.6

68. How many years is a mole of seconds?

69. A normal penny has a mass of about 2.5 g. If we assume the penny to be pure copper (which means the penny is very old since newer pennies are a mixture of copper and zinc), how many atoms of copper does it contain?

70. What would be the mass (in grams) of one thousand trillion molecules of glycerin ($C_3H_8O_3$)?

71. If we assume that there are 6.1 billion people on the Earth, how many moles of people is this?

72. An experimental catalyst used to make polymers has the following composition: Co, 23.3%; Mo, 25.3%; and Cl, 51.4%. What is the empirical formula for this compound?

73. If a student weighs 18 g of aluminum and needs twice as many atoms of magnesium as she has of aluminum, how many grams of Mg does she need?

74. If 10.0 g of an unknown compound composed of carbon, hydrogen, and nitrogen contains 17.7% N and 3.8×10^{23} atoms of hydrogen, what is its empirical formula?

75. A substance whose formula is A_2O (A is a mystery element) is 60.0% A and 40.0% O. Identify the element A.

76. For the following compounds whose molecular formulas are given, indicate the empirical formula:
(a) $C_6H_{12}O_6$ glucose
(b) C_8H_{18} octane
(c) $C_3H_6O_3$ lactic acid
(d) $C_{25}H_{52}$ paraffin
(e) $C_{12}H_4Cl_4O_2$ dioxin (a powerful poison)

Challenge Exercises

All exercises with blue numbers have answers in the appendix of the text.

77. The compound $A(BC)_3$ has a molar mass of 78.01 grams and a percent composition of 34.59% A, 61.53% B, and 3.88% C. Determine the identity of the elements A, B, and C. What is the percent composition by mass of the compound A_2B_3?

78. A 2.500 g sample of an unknown compound containing only C, H, and O was burned in O_2. The products were 4.776 g of CO_2 and 2.934 g of H_2O.
(a) What is the percent composition of the original compound?
(b) What is the empirical formula of the compound?

Answers to Practice Exercises

7.1 (a) 6.022×10^{23} atoms Fe
(b) 1.204×10^{24} atoms H
(c) 4.215×10^{24} atoms

7.2 10.0 g helium

7.3 1.5×10^{22} atoms

7.4 101.1 g KNO_3

7.5 21.3 g Na_2SO_4

7.6 8.326 mol and 5.014×10^{24} molecules $HC_2H_3O_2$

7.7 24.42% Ca; 17.07% N; 58.50% O

7.8 40.27% K; 26.78% Cr; 32.96% O

7.9 20.16% Al; 79.84% Cl

7.10 C_2H_6O

7.11 P_2O_5

7.12 The empirical formula is CH_3; the molecular formula is C_2H_6.

CHAPTER 8

Chemical Equations

Chapter Outline

8.1 The Chemical Equation

8.2 Writing and Balancing Chemical Equations

8.3 Information in a Chemical Equation

8.4 Types of Chemical Equations

8.5 Heat in Chemical Reactions

8.6 Global Warming: The Greenhouse Effect

The thermite reaction is a reaction between elemental aluminum and iron oxide. This reaction produces so much energy the iron becomes molten. One use of the thermite reaction is to weld railroad rails.

In the world today, we continually strive to express information in a concise, useful manner. From early childhood, we are taught to translate our ideas and desires into sentences. In mathematics, we learn to describe numerical relationships and situations through mathematical expressions and equations. Historians describe thousands of years of history in 500-page textbooks. Filmmakers translate entire events, such as the Olympics, into a few hours of entertainment.

Chemists use chemical equations to describe reactions they observe in the laboratory or in nature. Chemical equations provide us with the means to (1) summarize the reaction, (2) display the substances that are reacting, (3) show the products, and (4) indicate the amounts of all component substances in a reaction.

8.1 The Chemical Equation

Chemical reactions always involve change. Atoms, molecules, or ions rearrange to form different substances, sometimes in a spectacular manner. For example, the thermite reaction (shown in the chapter opening photo) is a reaction between aluminum metal and iron(III) oxide, which produces molten iron and aluminum oxide. The substances entering the reaction are called the **reactants** and the substances formed are called the **products**. In our example,

reactants
products

reactants	aluminum
	iron(III) oxide
products	iron
	aluminum oxide

During reactions chemical bonds are broken and new bonds are formed. The reactants and products may be present as solids, liquids, gases, or in solution.

In a chemical reaction atoms are neither created nor destroyed. All atoms present in the reactants must also be present in the products.

A **chemical equation** is a shorthand expression for a chemical change or reaction. A chemical equation uses the chemical symbols and formulas of the reactants and products and other symbolic terms to represent a chemical reaction. The equations are written according to this general format:

chemical equation

1. Reactants are separated from products by an arrow ($\longrightarrow$) that indicates the direction of the reaction. The reactants are placed to the left and the products to the right of the arrow. A plus sign ($+$) is placed between reactants and between products when needed.

$$Al + Fe_2O_3 \longrightarrow Fe + Al_2O_3$$
$$\text{reactants} \qquad\qquad \text{products}$$

2. Coefficients (whole numbers) are placed in front of substances to balance the equation and to indicate the number of units (atoms, molecules, moles, ions) of each substance reacting or being produced. When no number is shown, it is understood that one unit of the substance is indicated.

$$2\,Al + Fe_2O_3 \longrightarrow 2\,Fe + Al_2O_3$$

3. Conditions required to carry out the reaction may, if desired, be placed above or below the arrow or equality sign. For example, a delta sign placed over the arrow ($\xrightarrow{\Delta}$) indicates that heat is supplied to the reaction.

$$2\,Al + Fe_2O_3 \xrightarrow{\Delta} 2\,Fe + Al_2O_3$$

4. The physical state of a substance is indicated by the following symbols: (s) for solid state; (l) for liquid state; (g) for gaseous state; and (aq) for substances in aqueous solution. States are not always given in chemical equations.

$$2\,Al(s) + Fe_2O_3(s) \xrightarrow{\Delta} Fe(l) + Al_2O_3(s)$$

Commonly used symbols are given in Table 8.1.

Table 8.1 Symbols Commonly Used in Chemical Equations

Symbol	Meaning
+	Plus or added to (placed between substances)
$\longrightarrow$	Yields; produces (points to products)
(s)	Solid state (written after a substance)
(l)	Liquid state (written after a substance)
(g)	Gaseous state (written after a substance)
(aq)	Aqueous solution (substance dissolved in water)
Δ	Heat is added (when written above or below arrow)

8.2   Writing and Balancing Chemical Equations

balanced equation

To represent the quantitative relationships of a reaction, the chemical equation must be balanced. A **balanced equation** contains the same number of each kind of atom on each side of the equation. The balanced equation therefore obeys the law of conservation of mass.

Every chemistry student must learn to *balance* equations. Many equations are balanced by trial and error, but care and attention to detail are still required. The way to balance an equation is to adjust the number of atoms of each element so that they are the same on each side of the equation, but a correct formula is never changed in order to balance an equation. The general procedure for balancing equations is as follows:

Correct formulas are not changed to balance an equation.

Study this procedure carefully and refer to it when you work examples.

Step 1   **Identify the reaction.** Write a description or word equation for the reaction. For example, let's consider mercury(II) oxide decomposing into mercury and oxygen.

mercury(II) oxide $\xrightarrow{\Delta}$ mercury + oxygen

Step 2   **Write the unbalanced (skeleton) equation.** Make sure that the formula for each substance is correct and that reactants are written to the left and products to the right of the arrow. For our example,

$HgO \xrightarrow{\Delta} Hg + O_2$

The correct formulas must be known or determined from the periodic table, lists of ions, or experimental data.

Step 3 **Balance the equation.** Use the following process as necessary:

(a) Count and compare the number of atoms of each element on each side of the equation and determine those that must be balanced:

Hg is balanced (1 on each side)

O needs to be balanced

(1 on reactant side, 2 on product side)

(b) Balance each element, one at a time, by placing whole numbers (coefficients) in front of the formulas containing the unbalanced element. It is usually best to balance metals first, then nonmetals, then hydrogen and oxygen. Select the smallest coefficients that will give the same number of atoms of the element on each side. A coefficient placed before a formula multiplies every atom in the formula by that number (e.g., $2\,H_2SO_4$ means two molecules of sulfuric acid and also means four H atoms, two S atoms, and eight O atoms). Place a 2 in front of HgO to balance O:

$$2\,HgO \xrightarrow{\Delta} Hg + O_2$$

(c) Check all other elements after each individual element is balanced to see whether, in balancing one element, other elements have become unbalanced. Make adjustments as needed. Now Hg is not balanced. To adjust this, we write a 2 in front of Hg:

$$2\,HgO \xrightarrow{\Delta} 2\,Hg + O_2 \quad \text{(balanced)}$$

(d) Do a final check, making sure that each element and/or polyatomic ion is balanced and that the smallest possible set of whole-number coefficients has been used:

$$2\,HgO \xrightarrow{\Delta} 2\,Hg + O_2 \quad \text{(correct form)}$$

$$4\,HgO \xrightarrow{\Delta} 4\,Hg + 2\,O_2 \quad \text{(incorrect form)}$$

Leave elements that are in two or more formulas (on the same side of the equation) unbalanced until just before balancing hydrogen and oxygen.

Not all chemical equations can be balanced by the simple method of inspection just described. The following examples show *stepwise* sequences leading to balanced equations. Study each one carefully.

Write the balanced equation for the reaction that takes place when magnesium metal is burned in air to produce magnesium oxide.

Example 8.1

Step 1 *Word equation:*

magnesium + oxygen ⟶ magnesium oxide

reactants (R) product (P)

SOLUTION

Step 2 *Skeleton equation:*

$$Mg + O_2 \longrightarrow MgO \quad \text{(unbalanced)}$$

Step 3 *Balance:*

(a) Mg is balanced.

Oxygen is not balanced. Two O atoms appear on the left side and one on the right side.

R = reactant

P = product

R	1 Mg	2 O
P	1 Mg	1 O

(b) Place the coefficient 2 in front of MgO:

R	2 Mg	2 O
P	2 Mg	2 O

$$Mg + O_2 \longrightarrow 2\,MgO \quad \text{(unbalanced)}$$

(c) Now Mg is not balanced. One Mg atom appears on the left side and two on the right side. Place a 2 in front of Mg:

R	2 Mg	2 O
P	2 Mg	2 O

$$2\,Mg + O_2 \longrightarrow 2\,MgO \quad \text{(balanced)}$$

(d) *Check:* Each side has two Mg and two O atoms.

Example 8.2   When methane, CH_4, undergoes complete combustion, it reacts with oxygen to produce carbon dioxide and water. Write the balanced equation for this reaction.

SOLUTION   **Step 1**   *Word equation:*

$$\text{methane} + \text{oxygen} \longrightarrow \text{carbon dioxide} + \text{water}$$

Step 2   *Skeleton equation:*

R	1 C	4 H	2 O
P	1 C	2 H	3 O

$$CH_4 + O_2 \longrightarrow CO_2 + H_2O \quad \text{(unbalanced)}$$

Step 3   *Balance:*
(a) Carbon is balanced.
    Hydrogen and oxygen are not balanced.
(b) Balance H atoms by placing a 2 in front of H_2O:

R	1 C	4 H	2 O
P	1 C	4 H	4 O

$$CH_4 + O_2 \longrightarrow CO_2 + 2\,H_2O \quad \text{(unbalanced)}$$

Each side of the equation has four H atoms; oxygen is still not balanced. Place a 2 in front of O_2 to balance the oxygen atoms:

$$CH_4 + 2\,O_2 \longrightarrow CO_2 + 2\,H_2O \quad \text{(balanced)}$$

R	1 C	4 H	4 O
P	1 C	4 H	4 O

(c) The other atoms remain balanced.
(d) *Check:* The equation is correctly balanced; it has one C, four O, and four H atoms on each side.

Example 8.3   Oxygen and potassium chloride are formed by heating potassium chlorate. Write a balanced equation for this reaction:

SOLUTION   **Step 1**   *Word equation:*

$$\text{potassium chlorate} \xrightarrow{\Delta} \text{potassium chloride} + \text{oxygen}$$

Step 2   *Skeleton equation:*

$$KClO_3 \xrightarrow{\Delta} KCl + O_2 \quad \text{(unbalanced)}$$

Step 3   *Balance:*

R	1 K	1 Cl	3 O
P	1 K	1 Cl	2 O

(a) Potassium and chlorine are balanced.
    Oxygen is unbalanced (three O atoms on the left and two on the right side).
(b) How many oxygen atoms are needed? The subscripts of oxygen (3 and 2) in $KClO_3$ and O_2 have a least common multiple of 6. Therefore, coefficients for $KClO_3$ and O_2 are needed to get six O atoms on each side. Place a 2 in front of $KClO_3$ and a 3 in front of O_2:

R	2 K	2 Cl	6 O
P	1 K	1 Cl	6 O

$$2\,KClO_3 \xrightarrow{\Delta} KCl + 3\,O_2 \quad \text{(unbalanced)}$$

(c) Now K and Cl are not balanced. Place a 2 in front of KCl, which balances both K and Cl at the same time:

$$2\,KClO_3 \xrightarrow{\Delta} 2\,KCl + 3\,O_2 \qquad \text{(balanced)}$$

R	2 K	2 Cl	6 O
P	2 K	2 Cl	6 O

(d) *Check:* Each side now contains two K, two Cl, and six O atoms.

Silver nitrate reacts with hydrogen sulfide to produce silver sulfide and nitric acid. Write a balanced equation for this reaction.

Example 8.4

Step 1 *Word equation:*

silver nitrate + hydrogen sulfide $\longrightarrow$ silver sulfide + nitric acid

SOLUTION

Step 2 *Skeleton equation:*

$$AgNO_3 + H_2S \longrightarrow Ag_2S + HNO_3 \qquad \text{(unbalanced)}$$

R	1 Ag	2 H	1 S	1 NO$_3$
P	2 Ag	1 H	1 S	1 NO$_3$

Step 3 *Balance:*

(a) Ag and H are unbalanced.

(b) Place a 2 in front of AgNO$_3$ to balance Ag:

$$2\,AgNO_3 + H_2S \longrightarrow Ag_2S + HNO_3 \qquad \text{(unbalanced)}$$

R	2 Ag	2 H	1 S	2 NO$_3$
P	2 Ag	1 H	1 S	1 NO$_3$

(c) H and NO$_3^-$ are still unbalanced. Balance by placing a 2 in front of HNO$_3$:

$$2\,AgNO_3 + H_2S \longrightarrow Ag_2S + 2\,HNO_3 \qquad \text{(balanced)}$$

R	2 Ag	2 H	1 S	2 NO$_3$
P	2 Ag	2 H	1 S	2 NO$_3$

In this example, N and O atoms are balanced by balancing the NO$_3^-$ ion as a unit.

(d) The other atoms remain balanced.

(e) *Check:* Each side has two Ag, two H, and one S atom. Also, each side has two NO$_3^-$ ions.

When aluminum hydroxide is mixed with sulfuric acid, the products are aluminum sulfate and water. Write a balanced equation for this reaction.

Example 8.5

Step 1 *Word equation:*

aluminum hydroxide + sulfuric acid $\longrightarrow$ aluminum sulfate + water

SOLUTION

Step 2 *Skeleton equation:*

$$Al(OH)_3 + H_2SO_4 \longrightarrow Al_2(SO_4)_3 + H_2O \qquad \text{(unbalanced)}$$

R	1 Al	1 SO$_4$	3 O	5 H
P	2 Al	3 SO$_4$	1 O	2 H

Step 3 *Balance:*

(a) All elements are unbalanced.

(b) Balance Al by placing a 2 in front of Al(OH)$_3$. Treat the unbalanced SO$_4^{2-}$ ion as a unit and balance by placing a 3 in front of H$_2$SO$_4$:

$$2\,Al(OH)_3 + 3\,H_2SO_4 \longrightarrow Al_2(SO_4)_3 + H_2O \qquad \text{(unbalanced)}$$

R	2 Al	3 SO$_4$	6 O	12 H
P	2 Al	3 SO$_4$	1 O	2 H

Balance the unbalanced H and O by placing a 6 in front of H$_2$O:

$$2\,Al(OH)_3 + 3\,H_2SO_4 \longrightarrow Al_2(SO_4)_3 + 6\,H_2O \qquad \text{(balanced)}$$

(c) The other atoms remain balanced.

(d) *Check:* Each side has 2 Al, 12 H, 3 S, and 18 O atoms.

R	2 Al	3 SO$_4$	6 O	12 H
P	2 Al	3 SO$_4$	6 O	12 H

Example 8.6　When the fuel in a butane gas stove undergoes complete combustion, it reacts with oxygen to form carbon dioxide and water. Write the balanced equation for this reaction.

SOLUTION　**Step 1**　*Word equation:*

butane + oxygen $\longrightarrow$ carbon dioxide + water

Step 2　*Skeleton equation:*

$C_4H_{10} + O_2 \longrightarrow CO_2 + H_2O$　　　(unbalanced)

Step 3　*Balance:*

(a) All elements are unbalanced.

(b) Balance C by placing a 4 in front of CO_2:

$C_4H_{10} + O_2 \longrightarrow 4\,CO_2 + H_2O$　　　(unbalanced)

Balance H by placing a 5 in front of H_2O:

$C_4H_{10} + O_2 \longrightarrow 4\,CO_2 + 5\,H_2O$　　　(unbalanced)

Oxygen remains unbalanced. The oxygen atoms on the right side are fixed because $4\,CO_2$ and $5\,H_2O$ are derived from the single C_4H_{10} molecule on the left. When we try to balance the O atoms, we find that there is no whole number that can be placed in front of O_2 to bring about a balance, so we double the coefficients of each substance, and then balance the oxygen:

$2\,C_4H_{10} + 13\,O_2 \longrightarrow 8\,CO_2 + 10\,H_2O$　　　(balanced)

(c) The other atoms remain balanced.

(d) *Check:* Each side now has 8 C, 20 H, and 26 O atoms.

Butane

R	4 C	10 H	2 O
P	1 C	2 H	3 O

R	4 C	10 H	2 O
P	4 C	10 H	13 O

R	8 C	20 H	26 O
P	8 C	20 H	26 O

Practice 8.1 _____

Write a balanced formula equation:

aluminum + oxygen $\longrightarrow$ aluminum oxide

Practice 8.2 _____

Write a balanced formula equation:

magnesium hydroxide + phosphoric acid $\longrightarrow$

magnesium phosphate + water

8.3　Information in a Chemical Equation

Depending on the particular context in which it is used, a formula can have different meanings. A formula can refer to an individual chemical entity (atom, ion, molecule, or formula unit) or to a mole of that chemical entity. For example, the formula H_2O can mean any of the following:

1. 2 H atoms and 1 O atom
2. 1 molecule of water
3. 1 mol of water
4. 6.022×10^{23} molecules of water
5. 18.02 g of water

Carbon monoxide is one possible product when carbon-containing fuels such as natural gas, propane, heating oil, or gasoline are burned. For example, when methane, the main component of natural gas, burns, the reaction is

$$CH_4(g) + 2\,O_2(g) \longrightarrow$$
$$2\,H_2O(g) + CO_2(g)$$

However, if not enough oxygen is available, carbon monoxide will be produced instead of carbon dioxide:

$$2\,CH_4(g) + 3\,O_2(g) \longrightarrow$$
$$4\,H_2O(g) + 2\,CO(g)$$

Carbon monoxide is colorless and has no flavor or smell. It often can poison several people at once, since it can build up undetected in enclosed spaces. The symptoms of carbon monoxide poisoning are so easily missed that people often don't realize that they have been poisoned. Common symptoms are headache, ringing in ears, nausea, dizziness, weakness, confusion, and drowsiness. A low concentration of CO in the air may go completely unnoticed while the toxicity level increases in the blood.

Carbon monoxide in the blood is poisonous because of its ability to stick to a hemoglobin molecule. Both CO and O_2 bind to the same place on the hemoglobin molecule. The CO sticks more tightly than O_2. We use hemoglobin to transport oxygen to the various tissues from the lungs. Unfortunately, CO also binds to hemoglobin, forming a molecule called carboxyhemoglobin. In fact, hemoglobin prefers CO to O_2. Once a CO is bound to a hemoglobin molecule, it cannot carry an oxygen molecule. So, until the CO is released, that molecule of hemoglobin is lost as an O_2 carrier. If you breathe normal air (about 20% oxygen) containing 0.1%

CO, in just one hour the CO binds to 50% of the hemoglobin molecules. To keep CO from being a silent killer, there are three possible actions:

1. Immediate treatment of victims to restore the hemoglobin
2. Detection of CO levels before poisoning occurs
3. Conversion of CO to CO_2 to eliminate the threat.

If a person already has CO poisoning, the treatment must focus on getting CO released from the hemoglobin. This can be done by giving the victim oxygen or by placing the victim in a hyperbaric chamber (at about 3 atm pressure). These treatments reduce the time required for carboxyhemoglobin to release CO from between 4 and 6 h to 30 to 90 min. This treatment also supplies more oxygen to the system to help keep the victim's brain functioning.

A CO detector can sound an alarm before the CO level becomes toxic in a

home. Current CO detectors sound an alarm when levels increase rapidly or when lower concentrations are present over a long time. Inside a CO detector is a beam of infrared light that shines on a chromophore (a substance that turns darker with increasing CO levels). If the infrared light passing through the chromophore is too low, the alarm sounds. Typically, these alarms ring when CO levels reach 70 ppm, or greater than 30 ppm for 30 days.

The final option is to convert CO to CO_2 before toxicity occurs. David Schreyer from NASA's Langley Research Center has devised a catalyst (tin hydroxide with small amounts of Pd) that speeds up the conversion of CO to CO_2 at room temperature. He and his colleagues have adapted their catalyst to home ventilation systems to eliminate even low levels of CO before it has a chance to build up. Using all of these approaches can help to eliminate 500–10,000 cases of CO poisoning each year.

People in a hyperbaric chamber.

Formulas used in equations can represent units of individual chemical entities or moles, the latter being more commonly used. For example, in the reaction of hydrogen and oxygen to form water,

$2\,H_2$	+	O_2	$\longrightarrow$	$2\,H_2O$
2 molecules hydrogen		1 molecule oxygen		2 molecules water
2 mol hydrogen		1 mol oxygen		2 mol water

We generally use moles in equations because molecules are so small.

As indicated earlier, a chemical equation is a shorthand description of a chemical reaction. Interpretation of a balanced equation gives us the following information:

1. What the reactants are and what the products are
2. The formulas of the reactants and products
3. The number of molecules or formula units of reactants and products in the reaction
4. The number of atoms of each element involved in the reaction
5. The number of moles of each substance

Consider the reaction that occurs when propane gas (C_3H_8) is burned in air; the products are carbon dioxide (CO_2) and water (H_2O). The balanced equation and its interpretation are as follows:

Propane	Oxygen	Carbon dioxide	Water
$C_3H_8(g)$ +	$5\,O_2(g)$ →	$3\,CO_2(g)$ +	$4\,H_2O(g)$
1 molecule	5 molecules	3 molecules	4 molecules
3 atoms C / 8 atoms H	10 atoms O	3 atoms C / 6 atoms O	8 atoms H / 4 atoms O
1 mol	5 mol	3 mol	4 mol
44.09 g	5(32.00 g) = (160.0 g)	3(44.01 g) = (132.0 g)	4(18.02 g) = (72.08 g)

Practice 8.3

Consider the reaction that occurs when hydrogen gas reacts with chlorine gas to produce gaseous hydrogen chloride:

(a) Write a word equation for this reaction.
(b) Write a balanced formula equation including the state for each substance.
(c) Label each reactant and product to show the relative amounts of each substance.
   (1) number of molecules
   (2) number of atoms
   (3) number of moles
   (4) mass
(d) What mass of HCl would be produced if you reacted 2 mol hydrogen gas with 2 mol chlorine gas?

The quantities involved in chemical reactions are important when working in industry or the laboratory. We will study the relationship among quantities of reactants and products in the next chapter.

8.4   Types of Chemical Equations

Chemical equations represent chemical changes or reactions. Reactions are classified into types to assist in writing equations and in predicting other reactions. Many chemical reactions fit one or another of the four principal reaction types that we discuss in the following paragraphs. Reactions are also classified as oxidation–reduction. Special methods are used to balance complex oxidation–reduction equations. (See Chapter 17.)

Combination Reaction

In a **combination reaction**, two reactants combine to give one product. The general form of the equation is

$$A + B \longrightarrow AB$$

in which A and B are either elements or compounds and AB is a compound. The formula of the compound in many cases can be determined from a knowledge of the ionic charges of the reactants in their combined states. Some reactions that fall into this category are given here:

combination reaction

1. metal + oxygen $\longrightarrow$ metal oxide:

$$2\,Mg(s) + O_2(g) \xrightarrow{\Delta} 2\,MgO(s)$$
$$4\,Al(s) + 3\,O_2(g) \xrightarrow{\Delta} 2\,Al_2O_3(s)$$

2. nonmetal + oxygen $\longrightarrow$ nonmetal oxide:

$$S(s) + O_2(g) \xrightarrow{\Delta} SO_2(g)$$
$$N_2(g) + O_2(g) \xrightarrow{\Delta} 2\,NO(g)$$

3. metal + nonmetal $\longrightarrow$ salt:

$$2\,Na(s) + Cl_2(g) \longrightarrow 2\,NaCl(s)$$
$$2\,Al(s) + 3\,Br_2(l) \longrightarrow 2\,AlBr_3(s)$$

4. metal oxide + water $\longrightarrow$ metal hydroxide:

$$Na_2O(s) + H_2O(l) \longrightarrow 2\,NaOH(aq)$$
$$CaO(s) + H_2O(l) \longrightarrow Ca(OH)_2(aq)$$

Flames and sparks result when aluminum foil is dropped into liquid bromine.

5. nonmetal oxide + water $\longrightarrow$ oxy-acid:

$$SO_3(g) + H_2O(l) \longrightarrow H_2SO_4(aq)$$
$$N_2O_5(s) + H_2O(l) \longrightarrow 2\,HNO_3(aq)$$

Decomposition Reaction

In a **decomposition reaction**, a single substance is decomposed, or broken down, to give two or more different substances. This reaction may be considered the reverse of combination. The starting material must be a compound, and the products may be elements or compounds. The general form of the equation is

$$AB \longrightarrow A + B$$

decomposition reaction

Hydrogen peroxide decomposes to steam ($H_2O(g)$) and oxygen.

Predicting the products of a decomposition reaction can be difficult and requires an understanding of each individual reaction. Heating oxygen-containing compounds often results in decomposition. Some reactions that fall into this category are

1. Metal oxides. Some metal oxides decompose to yield the free metal plus oxygen; others give another oxide, and some are very stable, resisting decomposition by heating:

$$2\,HgO(s) \xrightarrow{\Delta} 2\,Hg(l) + O_2(g)$$

$$2\,PbO_2(s) \xrightarrow{\Delta} 2\,PbO(s) + O_2(g)$$

2. Carbonates and hydrogen carbonates decompose to yield CO_2 when heated:

$$CaCO_3(s) \xrightarrow{\Delta} CaO(s) + CO_2(g)$$

$$2\,NaHCO_3(s) \xrightarrow{\Delta} Na_2CO_3(s) + H_2O(g) + CO_2(g)$$

3. Miscellaneous reactions in this category:

$$2\,KClO_3(s) \xrightarrow{\Delta} 2\,KCl(s) + 3\,O_2(g)$$

$$2\,NaNO_3(s) \xrightarrow{\Delta} 2\,NaNO_2(g) + O_2(g)$$

$$2\,H_2O_2(l) \xrightarrow{\Delta} 2\,H_2O(l) + O_2(g)$$

Single-Displacement Reaction

single-displacement reaction

In a **single-displacement reaction**, one element reacts with a compound to replace one of the elements of that compound, yielding a different element and a different compound. The general forms of the equation follow.

If A is a *metal*, A will replace B to form AC, provided that A is a more reactive metal than B.

$$A + BC \longrightarrow B + AC$$

If A is a *halogen*, it will replace C to form BA, provided that A is a more reactive halogen than C.

$$A + BC \longrightarrow C + BA$$

A brief activity series of selected metals (and hydrogen) and halogens is shown in Table 8.2. This series is listed in descending order of chemical activity, with the most active metals and halogens at the top. Many chemical reactions can be predicted from an activity series because the atoms of any element in the series will replace the atoms of those elements below it. For example, zinc metal will replace hydrogen from a hydrochloric acid solution. But copper metal, which is below hydrogen on the list and thus less reactive than hydrogen, will not replace hydrogen from a hydrochloric acid solution. Here are some reactions that fall into this category:

When pieces of zinc metal are placed in hydrochloric acid, hydrogen bubbles form immediately.

1. metal + acid $\longrightarrow$ hydrogen + salt

$$Zn(s) + 2\,HCl(aq) \longrightarrow H_2(g) + ZnCl_2(aq)$$

$$2\,Al(s) + 3\,H_2SO_4(aq) \longrightarrow 3\,H_2(g) + Al_2(SO_4)_3(aq)$$

2. metal + water ⟶ hydrogen + metal hydroxide or metal oxide

$$2\,Na(s) + 2\,H_2O \longrightarrow H_2(g) + 2\,NaOH(aq)$$

$$Ca(s) + 2\,H_2O \longrightarrow H_2(g) + Ca(OH)_2(aq)$$

$$3\,Fe(s) + 4\,H_2O(g) \longrightarrow 4\,H_2(g) + Fe_3O_4(s)$$
        steam

3. metal + salt ⟶ metal + salt

$$Fe(s) + CuSO_4(aq) \longrightarrow Cu(s) + FeSO_4(aq)$$

$$Cu(s) + 2\,AgNO_3(aq) \longrightarrow 2\,Ag(s) + Cu(NO_3)_2(aq)$$

4. halogen + halide salt ⟶ halogen + halide salt

$$Cl_2(g) + 2\,NaBr(aq) \longrightarrow Br_2(l) + 2\,NaCl(aq)$$

$$Cl_2(g) + 2\,KI(aq) \longrightarrow I_2(s) + 2\,KCl(aq)$$

A common chemical reaction is the displacement of hydrogen from water or acids (shown in 1 and 2 above). This reaction is a good illustration of the relative reactivity of metals and the use of the activity series. For example,

- K, Ca, and Na displace hydrogen from cold water, steam (H_2O), and acids.
- Mg, Al, Zn, and Fe displace hydrogen from steam and acids.
- Ni, Sn, and Pb displace hydrogen only from acids.
- Cu, Ag, Hg, and Au do not displace hydrogen.

Table 8.2 Activity Series

Metals	Halogens
K	F_2
Ca	Cl_2
Na	Br_2
Mg	I_2
Al	
Zn	
Fe	
Ni	
Sn	
Pb	
H	
Cu	
Ag	
Hg	
Au	

(increasing activity ↑)

Will a reaction occur between (a) nickel metal and hydrochloric acid and (b) tin metal and a solution of aluminum chloride? Write balanced equations for the reactions.

Example 8.7

(a) Nickel is more reactive than hydrogen, so it will displace hydrogen from hydrochloric acid. The products are hydrogen gas and a salt of Ni^{2+} and Cl^- ions:

$$Ni(s) + 2\,HCl(aq) \longrightarrow H_2(g) + NiCl_2(aq)$$

SOLUTION

(b) According to the activity series, tin is less reactive than aluminum, so no reaction will occur:

$$Sn(s) + AlCl_3(aq) \longrightarrow no\ reaction$$

Practice 8.4 _____

Write balanced equations for these reactions:
(a) iron metal and a solution of magnesium chloride
(b) zinc metal and a solution of lead(II) nitrate

Double-Displacement Reaction

In a **double-displacement reaction**, two compounds exchange partners with each other to produce two different compounds. The general form of the equation is

$$AB + CD \longrightarrow AD + CB$$

double-displacement reaction

A double-displacement reaction results from pouring a clear, colorless solution of $Pb(NO_3)_2$ into a clear, colorless solution of KI, forming a yellow precipitate of PbI_2.

This reaction can be thought of as an exchange of positive and negative groups, in which A combines with D and C combines with B. In writing formulas for the products, we must account for the charges of the combining groups.

It's also possible to write an equation in the form of a double-displacement reaction when a reaction has not occurred. For example, when solutions of sodium chloride and potassium nitrate are mixed, the following equation can be written:

$$NaCl(aq) + KNO_3(aq) \longrightarrow NaNO_3(aq) + KCl(aq)$$

When the procedure is carried out, no physical changes are observed, indicating that no chemical reaction has taken place.

A double-displacement reaction is accompanied by evidence of such reactions as

1. The evolution of heat
2. The formation of an insoluble precipitate
3. The production of gas bubbles

Let's look at some of these reactions more closely:

Neutralization of an Acid and a Base The production of a molecule of water from an H^+ and an OH^- ion is accompanied by a *release of heat*, which can be detected by touching the reaction container. For neutralization reactions, $H^+ + OH^- \longrightarrow H_2O$:

$$\text{acid} + \text{base} \longrightarrow \text{salt} + \text{water}$$
$$HCl(aq) + NaOH(aq) \longrightarrow NaCl(aq) + H_2O(l)$$
$$H_2SO_4(aq) + Ba(OH)_2(aq) \longrightarrow BaSO_4(s) + 2\,H_2O(l)$$

Metal Oxide + Acid *Heat is released* by the production of a molecule of water.

$$\text{metal oxide} + \text{acid} \longrightarrow \text{salt} + \text{water}$$
$$CuO(s) + 2\,HNO_3(aq) \longrightarrow Cu(NO_3)_2(aq) + H_2O(l)$$
$$CaO(s) + 2\,HCl(aq) \longrightarrow CaCl_2(aq) + H_2O(l)$$

Formation of an Insoluble Precipitate The solubilities of the products can be determined by consulting the solubility table in Appendix V to see whether one or both of the products are insoluble in water. An insoluble product (precipitate) is indicated by placing (s) after its formula in the equation.

$$BaCl_2(aq) + 2\,AgNO_3(aq) \longrightarrow 2\,AgCl(s) + Ba(NO_3)_2(aq)$$
$$Pb(NO_3)_2(aq) + 2\,KI(aq) \longrightarrow PbI_2(s) + 2\,KNO_3(aq)$$

Formation of a Gas A gas such as HCl or H_2S may be produced directly, as in these two examples:

$$H_2SO_4(l) + NaCl(s) \longrightarrow NaHSO_4(s) + HCl(g)$$
$$2\,HCl(aq) + ZnS(s) \longrightarrow ZnCl_2(aq) + H_2S(g)$$

A gas can also be produced indirectly. Some unstable compounds formed in a double-displacement reaction, such as H_2CO_3, H_2SO_3, and NH_4OH, will decompose to form water and a gas:

$$2\,HCl(aq) + Na_2CO_3(aq) \longrightarrow 2\,NaCl(aq) + H_2CO_3(aq) \longrightarrow 2\,NaCl(aq) + H_2O(l) + CO_2(g)$$
$$2\,HNO_3(aq) + K_2SO_3(aq) \longrightarrow 2\,KNO_3(aq) + H_2SO_3(aq) \longrightarrow 2\,KNO_3(aq) + H_2O(l) + SO_2(g)$$
$$NH_4Cl(aq) + NaOH(aq) \longrightarrow NaCl(aq) + NH_4OH(aq) \longrightarrow NaCl(aq) + H_2O(l) + NH_3(g)$$

Write the equation for the reaction between aqueous solutions of hydrobromic acid and potassium hydroxide.

Example 8.8

First we write the formulas for the reactants. They are HBr and KOH. Then we classify the type of reaction that would occur between them. Because the reactants are compounds, one an acid and the other a base, the reaction will be of the neutralization type:

SOLUTION

$$\text{acid} + \text{base} \longrightarrow \text{salt} + \text{water}$$

Now rewrite the equation using the formulas for the known substances:

$$\text{HBr}(aq) + \text{KOH}(aq) \longrightarrow \text{salt} + \text{H}_2\text{O}$$

In this reaction, which is a double-displacement type, the H^+ from the acid combines with the OH^- from the base to form water. The ionic compound must be composed of the other two ions, K^+ and Br^-. We determine the formula of the ionic compound to be KBr from the fact that K is a $+1$ cation and Br is a -1 anion. The final balanced equation is

$$\text{HBr}(aq) + \text{KOH}(aq) \longrightarrow \text{KBr}(aq) + \text{H}_2\text{O}(l)$$

Complete and balance the equation for the reaction between aqueous solutions of barium chloride and sodium sulfate.

Example 8.9

First determine the formulas for the reactants. They are BaCl_2 and Na_2SO_4. Then classify these substances as acids, bases, or ionic compounds. Both substances are ionic compounds. Since both substances are compounds, the reaction will be of the double-displacement type. Start writing the equation with the reactants:

SOLUTION

$$\text{BaCl}_2(aq) + \text{Na}_2\text{SO}_4(aq) \longrightarrow$$

If the reaction is double-displacement, Ba^{2+} will be written combined with SO_4^{2-}, and Na^+ with Cl^- as the products. The balanced equation is

$$\text{BaCl}_2(aq) + \text{Na}_2\text{SO}_4(aq) \longrightarrow \text{BaSO}_4 + 2\,\text{NaCl}$$

The final step is to determine the nature of the products, which controls whether or not the reaction will take place. If both products are soluble, we have a mixture of all the ions in solution. But if an insoluble precipitate is formed, the reaction will definitely occur. We know from experience that NaCl is fairly soluble in water, but what about BaSO_4? Consulting the solubility table in Appendix V, we see that BaSO_4 is insoluble in water, so it will be a precipitate in the reaction. Thus, the reaction will occur, forming a precipitate. The equation is

When barium chloride is poured into a solution of sodium sulfate, a white precipitate of barium sulfate forms.

$$\text{BaCl}_2(aq) + \text{Na}_2\text{SO}_4(aq) \longrightarrow \text{BaSO}_4(s) + 2\,\text{NaCl}(aq)$$

Practice 8.5

Complete and balance the equations for these reactions:
(a) potassium phosphate + barium chloride
(b) hydrochloric acid + nickel carbonate
(c) ammonium chloride + sodium nitrate

(a)

(b)

(c)

Figure 8.1
(a) A solution of silver nitrate contains Ag^+ ions and NO_3^- ions. (b) A copper wire is placed in a solution of silver nitrate. After 24 hours, crystals of silver are seen hanging on the copper wire and the solution has turned blue, indicating copper ions are present there. (c) The silver metal clings to the copper wire and the blue solution contains Cu^{2+} and NO_3^- ions.

NO_3^-

Ag^+

NO_3^-

Cu^{2+}

    Some of the reactions you attempt may fail because the substances are not reactive or because the proper conditions for reaction are not present. For example, mercury(II) oxide does not decompose until it is heated; magnesium does not burn in air or oxygen until the temperature reaches a certain point. When silver is placed in a solution of copper(II) sulfate, no reaction occurs. When copper wire is placed in a solution of silver nitrate, a single-displacement reaction takes place because copper is a more reactive metal than silver. (See Figure 8.1.)

    The successful prediction of the products of a reaction is not always easy. The ability to predict products correctly comes with knowledge and experience. Although you may not be able to predict many reactions at this point, as you continue to experiment you will find that reactions can be categorized and that prediction of the products becomes easier, if not always certain.

8.5   Heat in Chemical Reactions

Energy changes always accompany chemical reactions. One reason reactions occur is that the products attain a lower, more stable energy state than the reactants. When the reaction leads to a more stable state, energy is released to the surroundings as heat and/or work. When a solution of a base is neutralized by the addition of an acid, the liberation of heat energy is signaled by an immediate rise in the temperature of the solution. When an automobile engine burns gasoline, heat is certainly liberated; at the same time, part of the liberated energy does the work of moving the automobile.

Reactions are either exothermic or endothermic. **Exothermic reactions** liberate heat; **endothermic reactions** absorb heat. In an exothermic reaction, heat is a product and may be written on the right side of the equation for the reaction. In an endothermic reaction, heat can be regarded as a reactant and is written on the left side of the equation. Here are two examples:

exothermic reaction
endothermic reaction

$$H_2(g) + Cl_2(g) \longrightarrow 2\,HCl(g) + 185\,kJ \quad (exothermic)$$

$$N_2(g) + O_2(g) + 181\,kJ \longrightarrow 2\,NO(g) \quad (endothermic)$$

The quantity of heat produced by a reaction is known as the **heat of reaction**. The units used can be kilojoules or kilocalories. Consider the reaction represented by this equation:

heat of reaction

$$C(s) + O_2(g) \longrightarrow CO_2(g) + 393\,kJ$$

When the heat released is expressed as part of the equation, the substances are expressed in units of moles. Thus, when 1 mol (12.01 g) of C combines with 1 mol (32.00 g) of O_2, 1 mol (44.01 g) of CO_2 is formed and 393 kJ of heat are released. In this reaction, as in many others, the heat energy is more useful than the chemical products.

This cornfield is a good example of the endothermic reactions happening through photosynthesis in plants.

Glucose

hydrocarbon

Aside from relatively small amounts of energy from nuclear processes, the sun is the major provider of energy for life on Earth. The sun maintains the temperature necessary for life and also supplies light energy for the endothermic photosynthetic reactions of green plants. In photosynthesis, carbon dioxide and water are converted to free oxygen and glucose:

$$6\,CO_2 + 6\,H_2O + 2519\,kJ \longrightarrow C_6H_{12}O_6 + 6\,O_2$$
$$\text{glucose}$$

Nearly all of the chemical energy used by living organisms is obtained from glucose or compounds derived from glucose.

The major source of energy for modern technology is fossil fuel—coal, petroleum, and natural gas. The energy is obtained from the combustion (burning) of these fuels, which are converted to carbon dioxide and water. Fossil fuels are mixtures of **hydrocarbons**, compounds containing only hydrogen and carbon.

Natural gas is primarily methane, CH_4. Petroleum is a mixture of hydrocarbons (compounds of carbon and hydrogen). Liquefied petroleum gas (LPG) is a mixture of propane (C_3H_8) and butane (C_4H_{10}).

The combustion of these fuels releases a tremendous amount of energy, but reactions won't occur to a significant extent at ordinary temperatures. A spark or a flame must be present before methane will ignite. The amount of energy that must be supplied to start a chemical reaction is called the **activation energy**. In an exothermic reaction, once this activation energy is provided, enough energy is then generated to keep the reaction going.

activation energy

Here are some examples:

$$CH_4(g) + 2\,O_2(g) \longrightarrow CO_2(g) + 2\,H_2O(g) + 890\,kJ$$

$$C_3H_8(g) + 5\,O_2(g) \longrightarrow 3\,CO_2(g) + 4\,H_2O(g) + 2200\,kJ$$

$$2\,C_8H_{18}(l) + 25\,O_2(g) \longrightarrow 16\,CO_2(g) + 18\,H_2O(g) + 10{,}900\,kJ$$

Be careful not to confuse an exothermic reaction that requires heat (activation energy) to get it *started* with an endothermic process that requires energy to keep it going. The combustion of magnesium, for example, is highly exothermic, yet magnesium must be heated to a fairly high temperature in air before combustion begins. Once started, however, the combustion reaction goes very vigorously until either the magnesium or the available supply of oxygen is exhausted. The electrolytic decomposition of water to hydrogen and oxygen is highly endothermic. If the electric current is shut off when this process is going on, the reaction stops instantly. The relative energy levels of reactants and products in exothermic and in endothermic processes are presented graphically in Figures 8.2 and 8.3.

Examples of endothermic and exothermic processes can be easily demonstrated. In Figure 8.2, the products are at a higher potential energy than the reactants. Energy has therefore been absorbed, and the reaction is endothermic. An endothermic reaction takes place when you apply a cold pack to an injury. When a cold pack is activated, ammonium chloride (NH_4Cl) dissolves in water. For example, temperature changes from 24.5°C to 18.1°C result

Combustion of Mg inside a block of dry ice (CO_2) makes a glowing lantern.

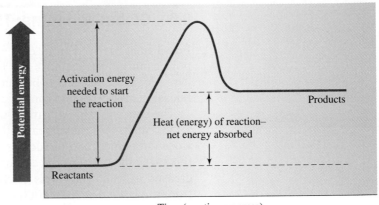

Endothermic reaction

Figure 8.2
An endothermic process occurs when a cold pack containing an ampule of solid NH_4Cl is broken, releasing the crystals into water in the pack. The dissolving of NH_4Cl in water is endothermic, producing a salt solution that is cooler than the surroundings. This process is represented graphically to show the energy changes between the reactants and the products.

when 10 g of NH_4Cl are added to 100 mL of water. Energy, in the form of heat, is taken from the immediate surroundings (water), causing the salt solution to become cooler.

In Figure 8.3, the products are at a lower potential energy than the reactants. Energy (heat) is given off, producing an exothermic reaction. Here, potassium chlorate ($KClO_3$) and sugar are mixed and placed into a pile. A drop of concentrated sulfuric acid is added, creating a spectacular exothermic reaction.

Exothermic reaction

Figure 8.3
Exothermic reaction between $KClO_3$ and sugar. A sample of $KClO_3$ and sugar are well mixed and placed on a fireproof pad. Several drops of concentrated H_2SO_4 are used to ignite the mixture.

8.6   Global Warming: The Greenhouse Effect

Fossil fuels, derived from coal and petroleum, provide the energy we use to power our industries, heat and light our homes and workplaces, and run our cars. As these fuels are burned, they produce carbon dioxide and water, releasing over 50 billion tons of carbon dioxide into our atmosphere each year.

The concentration of carbon dioxide has been monitored by scientists since 1958. Analysis of the air trapped in a core sample of snow from Antartica provides data on carbon dioxide levels for the past 160,000 years. The results of this study show that as the carbon dioxide increased, the global temperature increased as well. The levels of carbon dioxide remained reasonably constant from the last ice age, 100,000 years ago, until the Industrial Revolution. Since then, the concentration of carbon dioxide in our atmosphere has risen 15% to an all-time high. (See Figure 8.4.)

Carbon dioxide is a minor component in our atmosphere and is not usually considered to be a pollutant. The concern expressed by scientists arises from the dramatic increase occurring in the Earth's atmosphere. Without the influence of humans in the environment, the exchange of carbon dioxide between plants and animals would be relatively balanced. Our continued use of fossil fuels led to an increase of 7.4% in carbon dioxide between 1900 and 1970 and an additional 3.5% increase during the 1980s. Continued increases were observed in the 1990s.

Besides our growing consumption of fossil fuels, other factors contribute to increased carbon dioxide levels in the atmosphere: Rain forests are being destroyed by cutting and burning to make room for increased population and agricultural needs. Carbon dioxide is added to the atmosphere during the burning, and the loss of trees diminishes the uptake of carbon dioxide by plants.

About half of all the carbon dioxide released into our atmosphere each year remains there, thus increasing its concentration. The other half is absorbed by plants during photosynthesis or is dissolved in the ocean to form hydrogen carbonates and carbonates.

Methane is the second most important greenhouse gas. Its concentration in the atmosphere has also increased significantly since the 1850s, as shown in

Figure 8.4
Concentration of CO_2 in the atmosphere.

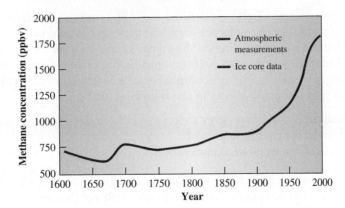

Figure 8.5
Concentration of methane in the atmosphere.

Figure 8.5. Methane is produced by cows, termites, agriculture, and anaerobic bacteria. Coal mining and oil wells also release methane into the atmosphere. The greenhouse effect of methane is 20 times more than that of CO_2, but there is less methane in the atmosphere. It presently contributes about 25% of the global warming.

Carbon dioxide and other greenhouse gases, such as methane and water, act to warm our atmosphere by trapping heat near the surface of the Earth. Solar radiation strikes the Earth and warms the surface. The warmed surface then reradiates this energy as heat. The greenhouse gases absorb some of this heat energy from the surface, which then warms our atmosphere. (See Figure 8.6.) A similar principle is illustrated in a greenhouse, where sunlight comes through the glass yet heat cannot escape. The air in the greenhouse warms, producing a climate considerably different than that in nature. In the atmosphere, these greenhouse gases are producing dramatic changes in our climate.

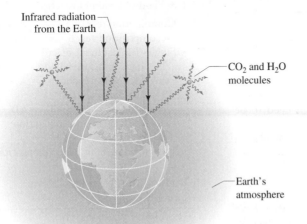

Figure 8.6
Visible light from the sun (red arrows) penetrates the atmosphere and strikes the Earth. Part of this light is changed to infrared radiation. Molecules of CO_2 and H_2O in the atmosphere absorb infrared radiation, acting like the glass of a greenhouse by trapping the energy.

Long-term effects of global warming are still a matter of debate. One consideration is whether the polar ice caps will melt; this would cause a rise in sea level and lead to major flooding on the coasts of our continents. Further effects could include shifts in rainfall patterns, producing droughts and extreme seasonal changes in such major agricultural regions as California. To reverse these trends will require major efforts in the following areas:

• The development of new energy sources to cut our dependence on fossil fuels
• An end to deforestation worldwide
• Intense efforts to improve conservation

On an individual basis each of us can play a significant role. Recycling, switching to more fuel-efficient cars, and energy-efficient appliances, heaters, and air conditioners all would result in decreased energy consumption and less carbon dioxide being released into our atmosphere.

Chapter 8 Review

8.1 The Chemical Equation

KEY TERMS

Reactants
Products
Chemical equation

• A chemical equation is shorthand for expressing a chemical change or reaction.
• In a chemical reaction atoms are neither created nor destroyed.
• All atoms in the reactants must be present in the products.

8.2 Writing and Balancing Chemical Equations

KEY TERM

Balanced equation

• To balance a chemical equation:
  · Identify the reaction.
  · Write the unbalanced (skeleton) equation.
  · Balance the equation:
    • Count the number of atoms of each element on each side and determine which need to be balanced.
    • Balance each element (one at a time) by placing whole numbers (coefficients) in front of the formulas containing the unbalanced element:
      • Begin with metals, then nonmetals, then H and O.
    • Check the other elements to see if they have become unbalanced in the process of balancing the chosen element. If so, rebalance as needed.
    • Do a final check to make sure all elements are balanced.

8.3 Information in a Chemical Equation

• The following information can be found in a chemical equation:
  · Identity of reactants and products
  · Formulas for reactants and products
  · Number of formula units for reactants and products
  · Number of atoms of each element in the reaction
  · Number of moles of each substance

8.4 Types of Chemical Equations

KEY TERMS

Combination reaction
Decomposition reaction
Single-displacement reaction
Double-displacement reaction

• Combination reactions $A + B \rightarrow AB$
• Decomposition reactions $AB \rightarrow A + B$
• Single-displacement reactions:
    In which A is a metal $A + BC \rightarrow B + AC$
    In which A is a halogen $A + BC \rightarrow C + BA$
  • Double-displacement
    reactions $AB + CD \rightarrow AD + CB$
• Evidence for a chemical reaction:
  · Evolution of heat
  · Formation of an insoluble precipitate
  · Production of a gas

8.5 Heat in Chemical Reactions

KEY TERMS

Exothermic reactions
Endothermic reactions
Heat of reaction
Hydrocarbons
Activation energy

- Exothermic reactions release heat.

Time (reaction progress)

- Endothermic reactions absorb heat.

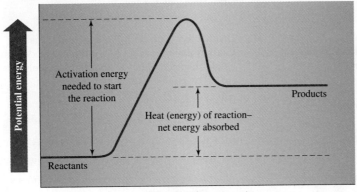

Time (reaction progress)

- The amount of heat released or absorbed in a chemical reaction is called the heat of reaction:
  - It can be written as a reactant or product in the chemical equation.
  - Units are joules (J) or kilojoules (kJ).
- The major source of energy for modern technology is hydrocarbon combustion.
- To initiate a chemical reaction, activation energy is required:
  - In exothermic reactions this energy is returned and more is released, which allows the reaction to continue on its own.
  - In endothermic reactions energy must be added to start and continue to be added to sustain the reaction.

8.6 Global Warming: The Greenhouse Effect

- Carbon dioxide levels are now increasing on our planet every year.
- Carbon dioxide and other greenhouse gases warm the atmosphere by trapping heat near the surface of the Earth.

Review Questions

All questions with blue numbers have answers in the appendix of the text.

1. What is the purpose of balancing equations?
2. What is represented by the numbers (coefficients) that are placed in front of the formulas in a balanced equation?

3. In a balanced chemical equation:
   (a) are atoms conserved?
   (b) are molecules conserved?
   (c) are moles conserved?
   Explain yours answers briefly.

4. Explain how endothermic reactions differ from exothermic reactions.

5. In Section 8.2, there are small charts in color along the margins. What information is represented in those charts?

6. What is meant by the physical state of a substance? What symbols are used to represent these physical states and what does each symbol mean?

7. What information does the activity series shown in Table 8.2 give?

8. What is a combustion reaction? How can you identify whether a reaction is a combustion reaction?

Paired Exercises

All exercises with blue numbers have answers in the appendix of the text.

1. Classify the following as an endothermic or exothermic reaction.
   (a) freezing water
   (b) the reaction inside an ice pack
   (c) burning wood
   (d) combustion of Mg in dry ice
   (e) melting ice

2. Classify the following as an endothermic or exothermic reaction.
   (a) making popcorn in a microwave
   (b) a burning match
   (c) boiling water
   (d) burning rocket fuel
   (e) the reaction inside a heat pack

3. Balance each of the following equations.
   Classify each reaction as combination, decomposition, single-displacement, or double-displacement.
   (a) $H_2 + O_2 \longrightarrow H_2O$
   (b) $C + Fe_2O_3 \longrightarrow Fe + CO$
   (c) $H_2SO_4 + NaOH \longrightarrow H_2O + Na_2SO_4$
   (d) $Al_2(CO_3)_3 \overset{\Delta}{\longrightarrow} Al_2O_3 + CO_2$
   (e) $NH_4I + Cl_2 \longrightarrow NH_4Cl + I_2$

4. Balance each of the following equations.
   Classify each reaction as combination, decomposition, single-displacement, or double-displacement.
   (a) $H_2 + Br_2 \longrightarrow HBr$
   (b) $Al + C \overset{\Delta}{\longrightarrow} Al_4C_3$
   (c) $Ba(ClO_3)_2 \overset{\Delta}{\longrightarrow} BaCl_2 + O_2$
   (d) $CrCl_3 + AgNO_3 \longrightarrow Cr(NO_3)_3 + AgCl$
   (e) $H_2O_2 \longrightarrow H_2O + O_2$

5. What reactant(s) is (are) required to form an oxide product?

6. What reactant(s) is (are) required to form a salt?

7. Balance the following equations:
   (a) $MnO_2 + CO \longrightarrow Mn_2O_3 + CO_2$
   (b) $Mg_3N_2 + H_2O \longrightarrow Mg(OH)_2 + NH_3$
   (c) $C_3H_5(NO_3)_3 \longrightarrow CO_2 + H_2O + N_2 + O_2$
   (d) $FeS + O_2 \longrightarrow Fe_2O_3 + SO_2$
   (e) $Cu(NO_3)_2 \longrightarrow CuO + NO_2 + O_2$
   (f) $NO_2 + H_2O \longrightarrow HNO_3 + NO$
   (g) $Al + H_2SO_4 \longrightarrow Al_2(SO_4)_3 + H_2$
   (h) $HCN + O_2 \longrightarrow N_2 + CO_2 + H_2O$
   (i) $B_5H_9 + O_2 \longrightarrow B_2O_3 + H_2O$

8. Balance the following equations:
   (a) $SO_2 + O_2 \longrightarrow SO_3$
   (b) $Al + MnO_2 \overset{\Delta}{\longrightarrow} Mn + Al_2O_3$
   (c) $Na + H_2O \longrightarrow NaOH + H_2$
   (d) $AgNO_3 + Ni \longrightarrow Ni(NO_3)_2 + Ag$
   (e) $Bi_2S_3 + HCl \longrightarrow BiCl_3 + H_2S$
   (f) $PbO_2 \overset{\Delta}{\longrightarrow} PbO + O_2$
   (g) $LiAlH_4 \overset{\Delta}{\longrightarrow} LiH + Al + H_2$
   (h) $KI + Br_2 \longrightarrow KBr + I_2$
   (i) $K_3PO_4 + BaCl_2 \longrightarrow KCl + Ba_3(PO_4)_2$

9. Change these word equations into formula equations and balance them:
   (a) water $\longrightarrow$ hydrogen + oxygen
   (b) acetic acid + potassium hydroxide $\longrightarrow$
       potassium acetate + water
   (c) phosphorus + iodine $\longrightarrow$ phosphorus triiodide
   (d) aluminum + copper(II) sulfate $\longrightarrow$
       copper + aluminum sulfate
   (e) ammonium sulfate + barium chloride $\longrightarrow$
       ammonium chloride + barium sulfate
   (f) sulfur tetrafluoride + water $\longrightarrow$
       sulfur dioxide + hydrogen fluoride
   (g) chromium(III) carbonate $\overset{\Delta}{\longrightarrow}$
       chromium(III) oxide + carbon dioxide

10. Change these word equations into formula equations and balance them:
    (a) copper + sulfur $\overset{\Delta}{\longrightarrow}$ copper(I) sulfide
    (b) phosphoric acid + calcium hydroxide $\longrightarrow$
        calcium phosphate + water
    (c) silver oxide $\overset{\Delta}{\longrightarrow}$ silver + oxygen
    (d) iron(III) chloride + sodium hydroxide $\longrightarrow$
        iron(III) hydroxide + sodium chloride
    (e) nickel(II) phosphate + sulfuric acid $\longrightarrow$
        nickel(II) sulfate + phosphoric acid
    (f) zinc carbonate + hydrochloric acid $\longrightarrow$
        zinc chloride + water + carbon dioxide
    (g) silver nitrate + aluminum chloride $\longrightarrow$
        silver chloride + aluminum nitrate

11. For each of the following reactions, predict the products, converting each to a balanced formula equation:
    (a) Aqueous solutions of sulfuric acid and sodium hydroxide are mixed together. (Heat is released during the reaction.)
    (b) Aqueous solutions of lead(II) nitrate and potassium bromide are mixed together. (The solution turns cloudy white during the reaction.)
    (c) Aqueous solutions of ammonium chloride and silver nitrate are mixed together. (The solution turns cloudy white during the reaction.)
    (d) Solid calcium carbonate is mixed with acetic acid. (Bubbles of gas are formed during the reaction.)

12. For each of the following reactions, predict the products, converting each to a balanced formula equation:
    (a) Aqueous solutions of copper(II) sulfate and potassium hydroxide are mixed together. (The solution turns cloudy and light blue during the reaction.)
    (b) Aqueous solutions of phosphoric acid and sodium hydroxide are mixed together. (Heat is produced during the reaction.)
    (c) Solid sodium bicarbonate is mixed with phosphoric acid. (Bubbles of gas are formed during the reaction.)
    (d) Aqueous solutions of aluminum chloride and lead(II) nitrate are mixed together. (The solution turns cloudy white during the reaction.)

13. Use the activity series to predict which of the following reactions will occur. Complete and balance the equations. Where no reaction will occur, write "no reaction" as the product.
    (a) $Ag(s) + H_2SO_4(aq) \longrightarrow$
    (b) $Cl_2(g) + NaBr(aq) \longrightarrow$
    (c) $Mg(s) + ZnCl_2(aq) \longrightarrow$
    (d) $Pb(s) + AgNO_3(aq) \longrightarrow$

14. Use the activity series to predict which of the following reactions will occur. Complete and balance the equations. Where no reaction will occur, write "no reaction" as the product.
    (a) $Cu(s) + FeCl_3(aq) \longrightarrow$
    (b) $H_2(g) + Al_2O_3(aq) \longrightarrow$
    (c) $Al(s) + HBr(aq) \longrightarrow$
    (d) $I_2(s) + HCl(aq)$

15. Complete and balance the equations for these reactions. All reactions yield products.
    (a) $H_2 + I_2 \longrightarrow$
    (b) $CaCO_3 \overset{\Delta}{\longrightarrow}$
    (c) $Mg + H_2SO_4 \longrightarrow$
    (d) $FeCl_2 + NaOH \longrightarrow$

16. Complete and balance the equations for these reactions. All reactions yield products.
    (a) $SO_2 + H_2O \longrightarrow$
    (b) $SO_3 + H_2O \longrightarrow$
    (c) $Ca + H_2O \longrightarrow$
    (d) $Bi(NO_3)_3 + H_2S \longrightarrow$

17. Complete and balance the equations for these reactions. All reactions yield products.
    (a) $Ba + O_2 \longrightarrow$
    (b) $NaHCO_3 \overset{\Delta}{\longrightarrow} Na_2CO_3 +$
    (c) $Ni + CuSO_4 \longrightarrow$
    (d) $MgO + HCl \longrightarrow$
    (e) $H_3PO_4 + KOH \longrightarrow$

18. Complete and balance the equations for these reactions. All reactions yield products.
    (a) $C + O_2 \longrightarrow$
    (b) $Al(ClO_3)_3 \overset{\Delta}{\longrightarrow} O_2 +$
    (c) $CuBr_2 + Cl_2 \longrightarrow$
    (d) $SbCl_3 + (NH_4)_2S \longrightarrow$
    (e) $NaNO_3 \overset{\Delta}{\longrightarrow} NaNO_2 +$

19. Interpret these chemical reactions in terms of the number of moles of each reactant and product:
    (a) $MgBr_2 + 2\,AgNO_3 \longrightarrow Mg(NO_3)_2 + 2\,AgBr$
    (b) $N_2 + 3\,H_2 \longrightarrow 2\,NH_3$
    (c) $2\,C_3H_7OH + 9\,O_2 \longrightarrow 6\,CO_2 + 8\,H_2O$

20. Interpret these equations in terms of the relative number of moles of each substance involved and indicate whether the reaction is exothermic or endothermic:
    (a) $2\,Na + Cl_2 \longrightarrow 2\,NaCl + 822\,kJ$
    (b) $PCl_5 + 92.9\,kJ \longrightarrow PCl_3 + Cl_2$

21. Write balanced equations for each of these reactions, including the heat term:
    (a) Lime, CaO, is converted to slaked lime $Ca(OH)_2$ by reaction with water. The reaction liberates 65.3 kJ of heat for each mole of lime reacted.
    (b) The industrial production of aluminum metal from aluminum oxide is an endothermic electrolytic process requiring 1630 kJ per mole of Al_2O_3. Oxygen is also a product.

22. Write a balanced equation for these reactions. Include a heat term on the appropriate side of the equation.
    (a) Powdered aluminum will react with crystals of iodine when moistened with dishwashing detergent. The reaction produces violet sparks and flaming aluminum. The major product is aluminum iodide (AlI_3). The detergent is not a reactant.
    (b) Copper(II) oxide (CuO), a black powder, can be decomposed to produce pure copper by heating the powder in the presence of methane gas (CH_4). The products are copper, carbon dioxide, and water vapor.

23. Determine what reactants would form the given products. Give a balanced equation for each and classify the reaction as a combination, decomposition, single-displacement, or double-displacement reaction.
(a) $AgCl(s) + O_2(g)$
(b) $H_2(g) + FeSO_4(aq)$
(c) $ZnCl_2(s)$
(d) $KBr(aq) + H_2O(l)$

24. Determine what reactants would form the given products. Give a balanced equation for each and classify the reaction as a combination, decomposition, single-displacement, or double-displacement reaction.
(a) $Pb(s) + Ni(NO_3)_2(aq)$
(b) $Mg(OH)_2(s)$
(c) $Hg(l) + O_2(g)$
(d) $PbCO_3(s) + NH_4Cl(aq)$

Additional Exercises

All exercises with blue numbers have answers in the appendix of the text.

25. Name one piece of evidence that a chemical reaction is actually taking place in each of these situations:
(a) making a piece of toast
(b) frying an egg
(c) striking a match

26. Balance this equation, using the smallest possible whole numbers. Then determine how many atoms of oxygen appear on each side of the equation:

$$P_4O_{10} + HClO_4 \longrightarrow Cl_2O_7 + H_3PO_4$$

27. Suppose that in a balanced equation the term $7\,Al_2(SO_4)_3$ appears.
(a) How many atoms of aluminum are represented?
(b) How many atoms of sulfur are represented?
(c) How many atoms of oxygen are represented?
(d) How many atoms of all kinds are represented?

28. Name two pieces of information that can be obtained from a balanced chemical equation. Name two pieces of information that the equation does not provide.

29. Make a drawing to show six molecules of ammonia gas decomposing to form hydrogen and nitrogen gases.

30. Explain briefly why this single-displacement reaction will not take place:

$$Zn + Mg(NO_3)_2 \longrightarrow \text{no reaction}$$

31. A student does an experiment to determine where titanium metal should be placed on the activity series chart. He places newly cleaned pieces of titanium into solutions of nickel(II) nitrate, lead(II) nitrate, and magnesium nitrate. He finds that the titanium reacts with the nickel(II) nitrate and lead(II) nitrate solutions, but not with the magnesium nitrate solution. From this information, place titanium in the activity series in a position relative to these ions.

32. Complete and balance the equations for these combination reactions:
(a) $K + O_2 \longrightarrow$
(b) $Al + Cl_2 \longrightarrow$
(c) $CO_2 + H_2O \longrightarrow$
(d) $CaO + H_2O \longrightarrow$

33. Complete and balance the equations for these decomposition reactions:
(a) $HgO \xrightarrow{\Delta}$
(b) $NaClO_3 \xrightarrow{\Delta}$
(c) $MgCO_3 \xrightarrow{\Delta}$
(d) $PbO_2 \xrightarrow{\Delta} PbO +$

34. Complete and balance the equations for these single-displacement reactions:
(a) $Zn + H_2SO_4 \longrightarrow$
(b) $AlI_3 + Cl_2 \longrightarrow$
(c) $Mg + AgNO_3 \longrightarrow$
(d) $Al + CoSO_4 \longrightarrow$

35. Complete and balance the equations for these double-displacement reactions:
(a) $ZnCl_2 + KOH \longrightarrow$
(b) $CuSO_4 + H_2S \longrightarrow$
(c) $Ca(OH)_2 + H_3PO_4 \longrightarrow$
(d) $(NH_4)_3PO_4 + Ni(NO_3)_2 \longrightarrow$
(e) $Ba(OH)_2 + HNO_3 \longrightarrow$
(f) $(NH_4)_2S + HCl \longrightarrow$

36. Predict which of the following double-displacement reactions will occur. Complete and balance the equations. Where no reaction will occur, write "no reaction" as the product.
(a) $AgNO_3(aq) + KCl(aq) \longrightarrow$
(b) $Ba(NO_3)_2(aq) + MgSO_4(aq) \longrightarrow$
(c) $H_2SO_4(aq) + Mg(OH)_2(aq) \longrightarrow$
(d) $MgO(s) + H_2SO_4(aq) \longrightarrow$
(e) $Na_2CO_3(aq) + NH_4Cl(aq) \longrightarrow$

37. Write balanced equations for each of the following combustion reactions:

(a)

(b)

38. Write balanced equations for the complete combustion of these hydrocarbons:
    (a) ethane, C_2H_6
    (b) benzene, C_6H_6
    (c) heptane, C_7H_{16}
39. List the factors that contribute to an increase in carbon dioxide in our atmosphere.

40. List three gases considered to be greenhouse gases. Explain why they are given this name.
41. How can the effects of global warming be reduced?
42. What happens to carbon dioxide released into our atmosphere?

Challenge Exercise

All exercises with blue numbers have answers in the appendix of the text.

*43. You are given a solution of the following cations in water: Ag^+, Co^{2+}, Ba^{2+}, Zn^{2+}, and Sn^{2+}. You want to separate the ions out *one at a time* by precipitation using the following reagents: NaF, NaI, Na_2SO_4, and NaCl.
    (a) In what order would you add the reagents to ensure that *only one cation* is precipitating out at a time?

Use the solubility table in Appendix V to help you. (*Note:* A precipitate that is slightly soluble in water is still considered to precipitate out of the solution.)
    (b) Why are the anionic reagents listed above all sodium salts?

Answers to Practice Exercises

8.1 $4\,Al + 3\,O_2 \longrightarrow 2\,Al_2O_3$

8.2 $3\,Mg(OH)_2 + 2\,H_3PO_4 \longrightarrow Mg_3(PO_4)_2 + 6\,H_2O$

8.3 (a) hydrogen gas + chlorine gas $\longrightarrow$ hydrogen chloride gas

(b) $H_2(g) + Cl_2(g) \longrightarrow 2\,HCl(g)$

(c)

$H_2(g)$	+	$Cl_2(g)$	$\longrightarrow$	$2\,HCl(g)$
1 molecule		1 molecule		2 molecules
2 atoms H		2 atoms Cl		2 atoms H 2 atoms Cl
1 mol 2.016 g		1 mol 70.90 g		2 mol 2 (36.46 g) (72.92 g)

(d) $2\,mol\,H_2 + 2\,mol\,Cl_2 \longrightarrow 4\,mol\,HCl(145.8\,g)$

8.4 (a) $Fe + MgCl_2 \longrightarrow$ no reaction

(b) $Zn(s) + Pb(NO_3)_2(aq) \longrightarrow Pb(s) + Zn(NO_3)_2(aq)$

8.5 (a) $2\,K_3PO_4(aq) + 3\,BaCl_2(aq) \longrightarrow Ba_3(PO_4)_2(s) + 6\,KCl(aq)$

(b) $2\,HCl(aq) + NiCO_3(aq) \longrightarrow NiCl_2(aq) + H_2O(l) + CO_2(g)$

(c) $NH_4Cl(aq) + NaNO_3(aq) \longrightarrow$ no reaction

CHAPTER 9

Calculations from Chemical Equations

We quickly learn as children that accurate measurement and calculation are required to bake a masterpiece. Although we don't see the chemical reactions occurring, we must scale up or down correctly to increase or decrease a recipe.

Chapter Outline

9.1 A Short Review

9.2 Introduction to Stoichiometry

9.3 Mole–Mole Calculations

9.4 Mole–Mass Calculations

9.5 Mass–Mass Calculations

9.6 Limiting Reactant and Yield Calculations

The old adage "waste not, want not" is appropriate in our daily life and in the laboratory. Determining correct amounts comes into play in almost all professions. A seamstress determines the amount of material, lining, and trim necessary to produce a gown for her client by relying on a pattern or her own experience to guide the selection. A carpet layer determines the correct amount of carpet and padding necessary to recarpet a customer's house by calculating the floor area. The IRS determines the correct deduction for federal income taxes from your paycheck based on your expected annual income.

The chemist also finds it necessary to calculate amounts of products or reactants by using a balanced chemical equation. With these calculations, the chemist can control the amount of product by scaling the reaction up or down to fit the needs of the laboratory, and can thereby minimize waste or excess materials formed during the reaction.

9.1   A Short Review

Molar Mass   The sum of the atomic masses of all the atoms in an element or compound is **molar mass**. The term *molar mass* also applies to the mass of a mole of any formula unit—atoms, molecules, or ions; it is the atomic mass of an atom or the sum of the atomic masses in a molecule or an ion (in grams).

molar mass

Relationship between Molecule and Mole   A molecule is the smallest unit of a molecular substance (e.g., Br_2), and a mole is Avogadro's number (6.022×10^{23}) of molecules of that substance. A mole of bromine (Br_2) has the same number of molecules as a mole of carbon dioxide, a mole of water, or a mole of any other molecular substance. When we relate molecules to molar mass, 1 molar mass is equivalent to 1 mol, or 6.022×10^{23} molecules.

The term *mole* also refers to any chemical species. It represents a quantity (6.022×10^{23} particles) and may be applied to atoms, ions, electrons, and formula units of nonmolecular substances. In other words,

$$1 \text{ mole} = \begin{cases} 6.022 \times 10^{23} \text{ molecules} \\ 6.022 \times 10^{23} \text{ formula units} \\ 6.022 \times 10^{23} \text{ atoms} \\ 6.022 \times 10^{23} \text{ ions} \end{cases}$$

Other useful mole relationships are

$$\text{molar mass} = \frac{\text{grams of a substance}}{\text{number of moles of the substance}}$$

$$\text{molar mass} = \frac{\text{grams of a monatomic element}}{\text{number of moles of the element}}$$

$$\text{number of moles} = \frac{\text{number of molecules}}{6.022 \times 10^{23} \text{ molecules/mole}}$$

A mole of water, salt, and any gas all have the same number of particles (6.022×10^{23}).

Balanced Equations   When using chemical equations for calculations of mole–mass–volume relationships between reactants and products, the equations must be balanced. *Remember:* The number in front of a formula in a balanced chemical equation represents the number of moles of that substance in the chemical reaction.

9.2   Introduction to Stoichiometry

stoichiometry

mole ratio

We often need to calculate the amount of a substance that is either produced from, or needed to react with, a given quantity of another substance. The area of chemistry that deals with quantitative relationships among reactants and products is known as **stoichiometry** (*stoy-key-ah-meh-tree*). Solving problems in stoichiometry requires the use of *moles* in the form of *mole ratios*.

A **mole ratio** is a ratio between the number of moles of any two species involved in a chemical reaction. For example, in the reaction

$$2\,H_2 + O_2 \longrightarrow 2\,H_2O$$

2 mol   1 mol   2 mol

six mole ratios can be written:

$$\frac{2\,\text{mol}\,H_2}{1\,\text{mol}\,O_2} \qquad \frac{2\,\text{mol}\,H_2}{2\,\text{mol}\,H_2O} \qquad \frac{1\,\text{mol}\,O_2}{2\,\text{mol}\,H_2}$$

$$\frac{1\,\text{mol}\,O_2}{2\,\text{mol}\,H_2O} \qquad \frac{2\,\text{mol}\,H_2O}{2\,\text{mol}\,H_2} \qquad \frac{2\,\text{mol}\,H_2O}{1\,\text{mol}\,O_2}$$

We use the mole ratio to convert the number of moles of one substance to the corresponding number of moles of another substance in a chemical reaction. For example, if we want to calculate the number of moles of H_2O that can be obtained from 4.0 mol of O_2, we use the mole ratio 2 mol H_2O/1 mol O_2:

$$(4.0 \,\cancel{\text{mol}\,O_2})\left(\frac{2\,\text{mol}\,H_2O}{1\,\cancel{\text{mol}\,O_2}}\right) = 8.0\,\text{mol}\,H_2O$$

There are three basic steps to use in solving stoichiometry problems:

Write and balance the equation before you begin the problem.

1. Convert the quantity of starting substance to moles (if it is not given in moles).

2. Convert the moles of starting substance to moles of desired substance.

3. Convert the moles of desired substance to the units specified in the problem.

Like balancing chemical equations, making stoichiometric calculations requires practice. Several worked examples follow. Study this material and practice on the problems at the end of this chapter.

Use a balanced equation.

Step 1 **Determine the number of moles of starting substance.**
Identify the starting substance from the data given in the problem statement. Convert the quantity of the starting substance to moles, if it is not already done:

$$\text{moles} = (\text{grams})\left(\frac{1\ \text{mole}}{\text{molar mass}}\right)$$

Step 2 **Determine the mole ratio of the desired substance to the starting substance.**
The number of moles of each substance in the balanced equation is indicated by the coefficient in front of each substance. Use these coefficients to set up the mole ratio:

$$\text{mole ratio} = \frac{\text{moles of}\ \boxed{\text{desired substance}}\ \text{in the equation}}{\text{moles of}\ \boxed{\text{starting substance}}\ \text{in the equation}}$$

> As in all problems with units, the desired quantity is in the numerator, and the quantity to be eliminated is in the denominator.

Multiply the number of moles of starting substance (from Step 1) by the mole ratio to obtain the number of moles of desired substance:

$$\begin{array}{c}\text{moles of desired}\\\text{substance}\end{array} = \overset{\text{From Step 1}}{\left(\begin{array}{c}\text{moles of starting}\\\text{substance}\end{array}\right)} \left(\dfrac{\begin{array}{c}\text{moles of desired substance}\\\text{in the balanced equation}\end{array}}{\begin{array}{c}\text{moles of starting substance}\\\text{in the balanced equation}\end{array}}\right)$$

> Units of moles of starting substance cancel in the numerator and denominator.

Step 3 **Calculate the desired substance in the units specified in the problem.**
If the answer is to be in moles, the calculation is complete. If units other than moles are wanted, multiply the moles of the desired substance (from Step 2) by the appropriate factor to convert moles to the units required. For example, if grams of the desired substance are wanted,

$$\text{grams} = \overset{\text{From Step 2}}{(\text{moles})}\left(\frac{\text{molar mass}}{1\ \text{mol}}\right)$$

If moles $\longrightarrow$ atoms, use $\dfrac{6.022 \times 10^{23}\ \text{atoms}}{1\ \text{mol}}$

If moles $\longrightarrow$ molecules, use $\dfrac{6.022 \times 10^{23}\ \text{molecules}}{1\ \text{mol}}$

The steps for converting the mass of a starting substance A to either the mass, atoms, or molecules of desired substance B are summarized in Figure 9.1.

Figure 9.1
Steps for converting starting substance A to mass, atoms, or molecules of desired substance B.

9.3   Mole–Mole Calculations

Let's solve stoichiometric problems for mole–mole calculations. The quantity of starting substance is given in moles and the quantity of desired substance is requested in moles.

Example 9.1   How many moles of carbon dioxide will be produced by the complete reaction of 2.0 mol of glucose ($C_6H_{12}O_6$), according to the following equation?

$$C_6H_{12}O_6 + 6\,O_2 \longrightarrow 6\,CO_2 + 6\,H_2O$$
$$\text{1 mol} \qquad \text{6 mol} \qquad \text{6 mol} \qquad \text{6 mol}$$

SOLUTION   The balanced equation states that 6 mol of CO_2 will be produced from 1 mol of $C_6H_{12}O_6$. Even though we can readily see that 12 mol of CO_2 will be formed from 2.0 mol of $C_6H_{12}O_6$, let's use the mole-ratio method to solve the problem.

Step 1   The number of moles of starting substance is 2.0 mol $C_6H_{12}O_6$.
Step 2   The conversion needed is

$$\text{moles } C_6H_{12}O_6 \longrightarrow \text{moles } CO_2$$

The mole ratio (shown in color) is exact and does not affect the number of significant figures in the answer.

Multiply 2.0 mol of glucose (given in the problem) by this mole ratio:

$$(2.0 \; \cancel{\text{mol } C_6H_{12}O_6})\left(\frac{6 \text{ mol } CO_2}{1 \; \cancel{\text{mol } C_6H_{12}O_6}}\right) = 12 \text{ mol } CO_2$$

Note how the units work; the moles of $C_6H_{12}O_6$ cancel, leaving the answer in units of moles of CO_2.

Example 9.2   How many moles of ammonia can be produced from 8.00 mol of hydrogen reacting with nitrogen? The balanced equation is

$$3\,H_2 + N_2 \longrightarrow 2\,NH_3$$

SOLUTION   **Step 1**   The starting substance is 8.00 mol of H_2.
Step 2   The conversion needed is

$$\text{moles } H_2 \longrightarrow \text{moles } NH_3$$

Ammonia

The balanced equation states that we get 2 mol of NH_3 for every 3 mol of H_2 that react. Set up the mole ratio of desired substance (NH_3) to starting substance (H_2):

$$\text{mole ratio} = \frac{2 \text{ mol } NH_3}{3 \text{ mol } H_2} \quad \text{(from balanced equation)}$$

Multiply the 8.00 mol H_2 by the mole ratio:

$$(8.00 \; \cancel{\text{mol } H_2})\left(\frac{2 \text{ mol } NH_3}{3 \; \cancel{\text{mol } H_2}}\right) = 5.33 \text{ mol } NH_3$$

Given the balanced equation

Example 9.3

$$K_2Cr_2O_7 + 6\,KI + 7\,H_2SO_4 \longrightarrow Cr_2(SO_4)_3 + 4\,K_2SO_4 + 3\,I_2 + 7\,H_2O$$

1 mol   6 mol   3 mol

calculate (a) the number of moles of potassium dichromate ($K_2Cr_2O_7$) that will react with 2.0 mol of potassium iodide (KI) and (b) the number of moles of iodine (I_2) that will be produced from 2.0 mol of potassium iodide.

Since the equation is balanced, we are concerned only with $K_2Cr_2O_7$, KI, and I_2, and we can ignore all the other substances. The equation states that 1 mol of $K_2Cr_2O_7$ will react with 6 mol of KI to produce 3 mol of I_2.

SOLUTION

(a)   Calculate the number of moles of $K_2Cr_2O_7$.

Step 1   The starting substance is 2.0 mol of KI.

Step 2   The conversion needed is

$$\text{moles KI} \longrightarrow \text{moles } K_2Cr_2O_7$$

Set up the mole ratio of desired substance to starting substance:

$$\text{mole ratio} = \frac{1\text{ mol } K_2Cr_2O_7}{6\text{ mol KI}} \quad \text{(from balanced equation)}$$

Multiply the moles of starting material by this ratio:

$$(2.0\ \cancel{\text{mol KI}})\left(\frac{1\text{ mol } K_2Cr_2O_7}{6\ \cancel{\text{mol KI}}}\right) = 0.33\text{ mol } K_2Cr_2O_7$$

(b)   Calculate the number of moles of I_2:

Step 1   The moles of starting substance are 2.0 mol KI as in part (a).

Step 2   The conversion needed is

$$\text{moles KI} \longrightarrow \text{moles } I_2$$

Set up the mole ratio of desired substance to starting substance:

$$\text{mole ratio} = \frac{3\text{ mol } I_2}{6\text{ mol KI}} \quad \text{(from balanced equation)}$$

Multiply the moles of starting material by this ratio:

$$(2.0\ \cancel{\text{mol KI}})\left(\frac{3\text{ mol } I_2}{6\ \cancel{\text{mol KI}}}\right) = 1.0\text{ mol } I_2$$

How many molecules of water can be produced by reacting 0.010 mol of oxygen with hydrogen?

Example 9.4

The balanced equation is $2\,H_2 + O_2 \longrightarrow 2\,H_2O$.

SOLUTION

The sequence of conversions needed in the calculation is

$$\text{moles } O_2 \longrightarrow \text{moles } H_2O \longrightarrow \text{molecules } H_2O$$

Step 1   The starting substance is 0.010 mol O_2.

Step 2   The conversion needed is moles $O_2 \longrightarrow$ moles H_2O. Set up the mole ratio of desired substance to starting substance:

$$\text{mole ratio} = \frac{2 \text{ mol } H_2O}{1 \text{ mol } O_2} \quad \text{(from balanced equation)}$$

Multiply 0.010 mol O_2 by the mole ratio:

$$(0.010 \cancel{\text{ mol } O_2})\left(\frac{2 \text{ mol } H_2O}{1 \cancel{\text{ mol } O_2}}\right) = 0.020 \text{ mol } H_2O$$

Step 3   Since the problem asks for molecules instead of moles of H_2O, we must convert moles to molecules. Use the conversion factor $(6.022 \times 10^{23} \text{ molecules})/\text{mole}$:

$$(0.020 \cancel{\text{ mol }} H_2O)\left(\frac{6.022 \times 10^{23} \text{ molecules}}{1 \cancel{\text{ mol}}}\right) = 1.2 \times 10^{22} \text{ molecules } H_2O$$

Note that 0.020 mol is still quite a large number of water molecules.

Practice 9.1 _____

How many moles of aluminum oxide will be produced from 0.50 mol of oxygen?

$$4 \text{ Al} + 3 \text{ O}_2 \longrightarrow 2 \text{ Al}_2O_3$$

Practice 9.2 _____

How many moles of aluminum hydroxide are required to produce 22.0 mol of water?

$$2 \text{ Al(OH)}_3 + 3 \text{ H}_2SO_4 \longrightarrow \text{Al}_2(SO_4)_3 + 6 \text{ H}_2O$$

9.4   Mole–Mass Calculations

The object of this type of problem is to calculate the mass of one substance that reacts with or is produced from a given number of moles of another substance in a chemical reaction. If the mass of the starting substance is given, we need to convert it to moles. We use the mole ratio to convert moles of starting substance to moles of desired substance. We can then change moles of desired substance to mass. Each example is solved in two ways.

> Method A. Step by step
> **Method B. Continuous calculation**, where the individual steps are combined in a single line.

Select the method that is easier for you and use it to practice solving problems.

Example 9.5   What mass of hydrogen can be produced by reacting 6.0 mol of aluminum with hydrochloric acid?

SOLUTION   The balanced equation is $2 \text{ Al}(s) + 6 \text{ HCl}(aq) \longrightarrow 2 \text{ AlCl}_3(aq) + 3 \text{ H}_2(g)$.

Method A. Step by Step

Step 1 The starting substance is 6.0 mol of aluminum.

Step 2 Calculate moles of H_2

moles Al $\longrightarrow$ moles H_2

$$(6.0 \; \cancel{\text{mol Al}})\left(\frac{3 \text{ mol } H_2}{2 \; \cancel{\text{mol Al}}}\right) = 9.0 \text{ mol } H_2$$

Step 3 Convert moles of H_2 to grams:

moles H_2 $\longrightarrow$ grams H_2

$$(9.0 \; \cancel{\text{mol } H_2})\left(\frac{2.016 \text{ g } H_2}{1 \; \cancel{\text{mol } H_2}}\right) = 18 \text{ g } H_2$$

We see that 18 g of H_2 can be produced by reacting 6.0 mol of Al with HCl.

Method B. Continuous Calculation

moles Al $\longrightarrow$ moles H_2 $\longrightarrow$ grams H_2

$$(6.0 \; \cancel{\text{mol Al}})\left(\frac{3 \; \cancel{\text{mol } H_2}}{2 \; \cancel{\text{mol Al}}}\right)\left(\frac{2.016 \text{ g } H_2}{1 \; \cancel{\text{mol } H_2}}\right) = 18 \text{ g } H_2$$

How many moles of water can be produced by burning 325 g of octane (C_8H_{18})? The balanced equation is $2\,C_8H_{18}(l) + 25\,O_2(g) \longrightarrow 16\,CO_2(g) + 18\,H_2O(g)$.

Example 9.6

Method A. Step by Step

Step 1 The starting substance is 325 g C_8H_{18}. Convert 325 g of C_8H_{18} to moles:

SOLUTION

grams C_8H_{18} $\longrightarrow$ moles C_8H_{18}

$$(325 \; \cancel{\text{g } C_8H_{18}})\left(\frac{1 \text{ mol } C_8H_{18}}{114.2 \; \cancel{\text{g } C_8H_{18}}}\right) = 2.85 \text{ mol } C_8H_{18}$$

Step 2 Calculate the moles of water:

moles C_8H_{18} $\longrightarrow$ moles H_2O

$$(2.85 \; \cancel{\text{mol } C_8H_{18}})\left(\frac{18 \text{ mol } H_2O}{2 \; \cancel{\text{mol } C_8H_{18}}}\right) = 25.7 \text{ mol } H_2O$$

Octane

Method B. Continuous Calculation

grams C_8H_{18} $\longrightarrow$ moles C_8H_{18} $\longrightarrow$ moles H_2O

$$(325 \; \cancel{\text{g } C_8H_{18}})\left(\frac{1 \; \cancel{\text{mol } C_8H_{18}}}{114.2 \; \cancel{\text{g } C_8H_{18}}}\right)\left(\frac{18 \text{ mol } H_2O}{2 \; \cancel{\text{mol } C_8H_{18}}}\right) = 25.6 \text{ mol } H_2O$$

The answers for the different methods vary in the last digit. This results from rounding off at different times in the calculation. Check with your instructor to find the appropriate rules for your course.

> **Practice 9.3**
> How many moles of potassium chloride and oxygen can be produced from 100.0 g of potassium chlorate?
>
> $$2\,KClO_3 \longrightarrow 2\,KCl + 3\,O_2$$
>
> **Practice 9.4**
> How many grams of silver nitrate are required to produce 0.25 mol of silver sulfide?
>
> $$2\,AgNO_3 + H_2S \longrightarrow Ag_2S + 2\,HNO_3$$

9.5 Mass–Mass Calculations

Use either the step-by-step method or the continuous-calculation method to solve these problems.

Solving mass–mass stoichiometry problems requires all the steps of the mole-ratio method. The mass of starting substance is converted to moles. The mole ratio is then used to determine moles of desired substance, which, in turn, is converted to mass.

Example 9.7   What mass of carbon dioxide is produced by the complete combustion of 100. g of the hydrocarbon pentane, C_5H_{12}? The balanced equation is

$$C_5H_{12} + 8\,O_2 \longrightarrow 5\,CO_2 + 6\,H_2O$$

SOLUTION   Method A. Step by Step

Pentane

Step 1 The starting substance is 100. g of C_5H_{12}. Convert 100. g of C_5H_{12} to moles:

grams $\longrightarrow$ moles

$$(100.\ g\,C_5H_{12})\left(\frac{1\ mol\ C_5H_{12}}{72.15\ g\,C_5H_{12}}\right) = 1.39\ mol\ C_5H_{12}$$

Step 2 Calculate the moles of CO_2:

moles $C_5H_{12} \longrightarrow$ moles CO_2

$$(1.39\ mol\,C_5H_{12})\left(\frac{5\ mol\ CO_2}{1\ mol\,C_5H_{12}}\right) = 6.95\ mol\ CO_2$$

Step 3 Convert moles of CO_2 to grams:

moles $CO_2 \longrightarrow$ grams CO_2

$$(6.95\ mol\,CO_2)\left(\frac{44.01\ g\ CO_2}{1\ mol\,CO_2}\right) = 306\ g\ CO_2$$

Remember: Round off as appropriate for your particular course.

Method B. Continuous Calculation

grams $C_5H_{12} \longrightarrow$ moles $C_5H_{12} \longrightarrow$ moles $CO_2 \longrightarrow$ grams CO_2

$$(100.\ g\,C_5H_{12})\left(\frac{1\ mol\,C_5H_{12}}{72.15\ g\,C_5H_{12}}\right)\left(\frac{5\ mol\,CO_2}{1\ mol\,C_5H_{12}}\right)\left(\frac{44.01\ g\ CO_2}{1\ mol\,CO_2}\right) = 305\ g\ CO_2$$

How many grams of nitric acid (HNO_3) are required to produce 8.75 g of dini- **Example 9.8**
trogen monoxide (N_2O) according to the following equation?

$$4\,Zn(s) + 10\,HNO_3(aq) \longrightarrow 4\,Zn(NO_3)_2(aq) + N_2O(g) + 5\,H_2O(l)$$ **SOLUTION**

10 mol   1 mol

Method A. Step by Step

Step 1   The starting substance for this calculation is 8.75 g of N_2O:

grams N_2O $\longrightarrow$ moles N_2O

$$(8.75\ \cancel{g\,N_2O})\left(\frac{1\ mol\ N_2O}{44.02\ \cancel{g\,N_2O}}\right) = 0.199\ mol\ N_2O$$

Step 2   Calculate the moles of HNO_3 by the mole-ratio method:

moles N_2O $\longrightarrow$ moles HNO_3

$$(0.199\ \cancel{mol\,N_2O})\left(\frac{10\ mol\ HNO_3}{1\ \cancel{mol\,N_2O}}\right) = 1.99\ mol\ HNO_3$$

Step 3   Convert moles of HNO_3 to grams:

moles HNO_3 $\longrightarrow$ grams HNO_3

$$(1.99\ \cancel{mol\,HNO_3})\left(\frac{63.02\ g\ HNO_3}{1\ \cancel{mol\,HNO_3}}\right) = 125\ g\ HNO_3$$

Method B. Continuous Calculation

grams N_2O $\longrightarrow$ moles N_2O $\longrightarrow$ moles HNO_3 $\longrightarrow$ grams HNO_3

$$(8.75\ \cancel{g\,N_2O})\left(\frac{1\ \cancel{mol\,N_2O}}{44.02\ \cancel{g\,N_2O}}\right)\left(\frac{10\ \cancel{mol\,HNO_3}}{1\ \cancel{mol\,N_2O}}\right)\left(\frac{63.02\,g\ HNO_3}{1\ \cancel{mol\,HNO_3}}\right) = 125\,g\ HNO_3$$

Practice 9.5 _____

How many grams of chromium(III) chloride are required to produce 75.0 g
of silver chloride?

$$CrCl_3 + 3\,AgNO_3 \longrightarrow Cr(NO_3)_3 + 3\,AgCl$$

Practice 9.6 _____

What mass of water is produced by the complete combustion of 225.0 g of
butane (C_4H_{10})?

$$2\,C_4H_{10} + 13\,O_2 \longrightarrow 8\,CO_2 + 10\,H_2O$$

The microchip has revolutionized the field of electronics. Engineers at Bell Laboratories, Massachusetts Institute of Technology, the University of California, and Stanford University are racing to produce parts for tiny machines and robots. New techniques now produce gears smaller than a grain of sand and motors lighter than a speck of dust.

To produce ever smaller computers, calculators, and even microbots (micro-sized robots), precise quantities of chemicals in exact proportions are required. The secret to producing minute circuits is to print the entire circuit or blueprint at one time. Computers are used to draw a chip. This image is then transferred onto a pattern, or mask, with details finer than a human hair. In a process similar to photography, light is then shined through the mask onto a silicon-coated surface. The areas created on the silicon exhibit high or low resistance to chemical etching. Chemicals then etch away the silicon.

Micromachinery is produced in the same way. First a thin layer of silicon dioxide is applied (sacrificial material),

then a layer of polysilicon is carefully applied (structural material). A mask is then applied and the whole structure is covered with plasma (excited gas). The plasma acts as a tiny sandblaster, removing everything the mask doesn't protect. This process is repeated as the entire machine is constructed. When the entire assembly is complete, the whole machine is placed in hydrofluoric acid, which dissolves all the sacrificial material and permits the various parts of the machine to move.

To turn these micromachines into true microbots, current research is focusing on locomotion and sensing imaging systems. Possible uses for these microbots include "smart" pills, which could contain sensors or drug reservoirs (currently used in birth control). Tiny pumps, once inside the body, will dispense the proper amount of medication at precisely the correct site. These microbots are currently in production for the treatment of diabetes (to release insulin).

These nanorobots attack themselves to sed blood cells.

9.6 Limiting Reactant and Yield Calculations

In many chemical processes, the quantities of the reactants used are such that one reactant is in excess. The amount of the product(s) formed in such a case depends on the reactant that is not in excess. This reactant is called the **limiting reactant**—it limits the amount of product that can be formed.

limiting reactant

Consider the case illustrated in Figure 9.2. How many bicycles can be assembled from the parts shown? The limiting part in this case is the number of pedal assemblies; only three bicycles can be built because there are only three pedal assemblies. The wheels and frames are parts in excess.

Let's consider a chemical example at the molecular level in which seven molecules of H_2 are combined with four molecules of Cl_2 (Figure 9.3a). How many molecules of HCl can be produced according to this reaction?

$$H_2 + Cl_2 \longrightarrow 2\,HCl$$

If the molecules of H_2 and Cl_2 are taken apart and recombined as HCl (Figure 9.3b), we see that eight molecules of HCl can be formed before we run out of Cl_2. Therefore the Cl_2 is the limiting reactant and H_2 is in excess—three molecules of H_2 remain unreacted.

8 wheels

4 frames

3 pedal assemblies

3 bicycles

Figure 9.2
The number of bicycles that can be built from these parts is determined by the "limiting reactant" (the pedal assemblies).

When problem statements give the amounts of two reactants, one of them is usually a limiting reactant. We can identify the limiting reactant using the following method:

> **1.** Calculate the amount of product (moles or grams, as needed) formed from each reactant.
>
> **2.** Determine which reactant is limiting. (The reactant that gives the least amount of product is the limiting reactant; the other reactant is in excess.)

H

Cl

(a) Before reaction

(b) After reaction

Figure 9.3
Reaction of H_2 with Cl_2. Eight molecules of HCl are formed and three molecules of H_2 remain. The limiting reactant is Cl_2.

Once we know the limiting reactant, the amount of product formed can be determined.

Sometimes we must find the amount of the other reactant remaining after the reaction, so we add a third step:

> 3. Calculate the amount of the other reactant required to react with the limiting reactant, then subtract this amount from the starting quantity of the reactant. This gives the amount of that substance that remains unreacted.

Example 9.9

How many moles of HCl can be produced by reacting 4.0 mol H_2 and 3.5 mol Cl_2? Which compound is the limiting reactant?

SOLUTION

$$H_2(g) + Cl_2(g) \longrightarrow 2\,HCl(g)$$

Since we have the amounts of both reactants, we need to determine which reactant limits the amount of product formed.

Step 1 Calculate the moles of HCl that can be formed from each reactant:

Moles of H_2 or $Cl_2 \longrightarrow$ moles HCl

$$(4.0\,\text{mol}\,H_2)\frac{(2\,\text{mol}\,HCl)}{(1\,\text{mol}\,H_2)} = 8.0\,\text{mol}\,HCl$$

$$(3.5\,\text{mol}\,Cl_2)\frac{(2\,\text{mol}\,HCl)}{(1\,\text{mol}\,Cl_2)} = 7.0\,\text{mol}\,HCl$$

Step 2 Determine the limiting reactant. The limiting reactant is Cl_2 because it produces less HCl than H_2. The H_2 is in excess. The yield of product is 7.0 mol HCl. (*Note:* In this reaction, the limiting reactant can be determined by inspection. From the equation, you can see that 1 mol H_2 reacts with 1 mol Cl_2. Therefore, when reacting 4.0 mol H_2 and 3.5 mol Cl_2, Cl_2 is the reactant that limits the amount of HCl produced, since it is present in smaller amount.)

Example 9.10

How many moles of Fe_3O_4 can be obtained by reacting 16.8 g Fe with 10.0 g H_2O? Which substance is the limiting reactant? Which substance is in excess?

SOLUTION

$$3\,Fe(s) + 4\,H_2O(g) \xrightarrow{\Delta} Fe_3O_4(s) + 4\,H_2(g)$$

Step 1 Calculate the moles of Fe_3O_4 that can be formed from each reactant:

g reactant $\longrightarrow$ mol reactant $\longrightarrow$ mol Fe_3O_4

The continuous-calculation method is shown here. You can also use the step-by-step method to determine the requested substance.

$$(16.8\,\text{g Fe})\left(\frac{1\,\text{mol Fe}}{55.85\,\text{g Fe}}\right)\left(\frac{1\,\text{mol}\,Fe_3O_4}{3\,\text{mol Fe}}\right) = 0.100\,\text{mol}\,Fe_3O_4$$

$$(10.0\,\text{g}\,H_2O)\left(\frac{1\,\text{mol}\,H_2O}{18.02\,\text{g}\,H_2O}\right)\left(\frac{1\,\text{mol}\,Fe_3O_4}{4\,\text{mol}\,H_2O}\right) = 0.139\,\text{mol}\,Fe_3O_4$$

Step 2 Determine the limiting reactant. The limiting reactant is Fe because it produces less Fe_3O_4; the H_2O is in excess. The yield of product is 0.100 mol of Fe_3O_4.

How many grams of silver bromide (AgBr) can be formed when solutions con- **Example 9.11**
taining 50.0 g of $MgBr_2$ and 100.0 g of $AgNO_3$ are mixed together? How many
grams of the excess reactant remain unreacted?

$$MgBr_2(aq) + 2\,AgNO_3(aq) \longrightarrow 2\,AgBr(s) + Mg(NO_3)_2(aq)$$ **SOLUTION**

Step 1   Calculate the grams of AgBr that can be formed from each reactant.

$$g\,reactant \longrightarrow mol\,reactant \longrightarrow mol\,AgBr \longrightarrow g\,AgBr$$

$$(50.0\;\cancel{g\,MgBr_2})\left(\frac{1\;\cancel{mol\,MgBr_2}}{184.1\;\cancel{g\,MgBr_2}}\right)\left(\frac{2\;\cancel{mol\,AgBr}}{1\;\cancel{mol\,MgBr_2}}\right)\left(\frac{187.8\;g\,AgBr}{1\;\cancel{mol\,AgBr}}\right) = 102\;g\,AgBr$$

$$(100.0\;\cancel{g\,AgNO_3})\left(\frac{1\;\cancel{mol\,AgNO_3}}{169.9\;\cancel{g\,AgNO_3}}\right)\left(\frac{2\;\cancel{mol\,AgBr}}{2\;\cancel{mol\,AgNO_3}}\right)\left(\frac{187.8\;g\,AgBr}{1\;\cancel{mol\,AgBr}}\right) = 110.5\;g\,AgBr$$

Step 2   Determine the limiting reactant. The limiting reactant is $MgBr_2$ because it gives less AgBr; $AgNO_3$ is in excess. The yield is 102 g AgBr.

Step 3   Calculate the grams of unreacted $AgNO_3$. Calculate the grams of
$AgNO_3$ that will react with 50.0 g of $MgBr_2$:

$$g\,MgBr_2 \longrightarrow mol\,MgBr_2 \longrightarrow mol\,AgNO_3 \longrightarrow g\,AgNO_3$$

$$(50.0\;\cancel{g\,MgBr_2})\left(\frac{1\;\cancel{mol\,MgBr_2}}{184.1\;\cancel{g\,MgBr_2}}\right)\left(\frac{2\;\cancel{mol\,AgNO_3}}{1\;\cancel{mol\,MgBr_2}}\right)\left(\frac{169.9\;g\,AgNO_3}{1\;\cancel{mol\,AgNO_3}}\right) = 92.3\;g\,AgNO_3$$

Thus 92.3 g of $AgNO_3$ reacts with 50.0 g of $MgBr_2$. The amount of
$AgNO_3$ that remains unreacted is

$$100.0\;g\,AgNO_3 - 92.3\;g\,AgNO_3 = 7.7\;g\,AgNO_3\;unreacted$$

The final mixture will contain 102 g AgBr(s), 7.7 g $AgNO_3$, and an
undetermined amount of $Mg(NO_3)_2$ in solution.

Practice 9.7 _____

How many grams of hydrogen chloride can be produced from 0.490 g of
hydrogen and 50.0 g of chlorine?

$$H_2(g) + Cl_2(g) \longrightarrow 2\,HCl(g)$$

Practice 9.8 _____

How many grams of barium sulfate will be formed from 200.0 g of barium
nitrate and 100.0 g of sodium sulfate?

$$Ba(NO_3)_2(aq) + Na_2SO_4(aq) \longrightarrow BaSO_4(s) + 2\,NaNO_3(aq)$$

The quantities of the products we have been calculating from chemical equations represent the maximum yield (100%) of product according to the reaction
represented by the equation. Many reactions, especially those involving organic substances, fail to give a 100% yield of product. The main reasons for this
failure are the side reactions that give products other than the main product and

theoretical yield

actual yield
percent yield

Round off as appropriate for
your particular course.

the fact that many reactions are reversible. In addition, some product may be lost in handling and transferring from one vessel to another. The **theoretical yield** of a reaction is the calculated amount of product that can be obtained from a given amount of reactant, according to the chemical equation. The **actual yield** is the amount of product that we finally obtain.

The **percent yield** is the ratio of the actual yield to the theoretical yield multiplied by 100. Both the theoretical and the actual yields must have the same units to obtain a percent:

$$\frac{\text{actual yield}}{\text{theoretical yield}} \times 100 = \text{percent yield}$$

For example, if the theoretical yield calculated for a reaction is 14.8 g, and the amount of product obtained is 9.25 g, the percent yield is

$$\text{percent yield} = \left(\frac{9.25\ \text{g}}{14.8\ \text{g}}\right)(100) = 62.5\%$$

Example 9.12   Carbon tetrachloride (CCl_4) was prepared by reacting 100. g of carbon disulfide and 100. g of chlorine. Calculate the percent yield if 65.0 g of CCl_4 was obtained from the reaction:

$$CS_2 + 3\,Cl_2 \longrightarrow CCl_4 + S_2Cl_2$$

SOLUTION   We need to determine the limiting reactant to calculate the quantity of CCl_4 (theoretical yield) that can be formed. Then we can compare that amount with the 65.0 g CCl_4 (actual yield) to calculate the percent yield.

Step 1   Determine the theoretical yield. Calculate the grams of CCl_4 that can be formed from each reactant:

$$\text{g reactant} \longrightarrow \text{mol reactant} \longrightarrow \text{mol}\,CCl_4 \longrightarrow \text{g}\,CCl_4$$

$$(100.\ \text{g}\,CS_2)\left(\frac{1\ \text{mol}\,CS_2}{76.15\ \text{g}\,CS_2}\right)\left(\frac{1\ \text{mol}\,CCl_4}{1\ \text{mol}\,CS_2}\right)\left(\frac{153.8\ \text{g}\,CCl_4}{1\ \text{mol}\,CCl_4}\right) = 202\ \text{g}\,CCl_4$$

$$(100.\ \text{g}\,Cl_2)\left(\frac{1\ \text{mol}\,Cl_2}{70.90\ \text{g}\,Cl_2}\right)\left(\frac{1\ \text{mol}\,CCl_4}{3\ \text{mol}\,Cl_2}\right)\left(\frac{153.8\ \text{g}\,CCl_4}{1\ \text{mol}\,CCl_4}\right) = 72.3\ \text{g}\,CCl_4$$

Step 2   Determine the limiting reactant. The limiting reactant is Cl_2 because it gives less CCl_4. The CS_2 is in excess. The theoretical yield is 72.3 g CCl_4.

Step 3   Calculate the percent yield. According to the equation, 72.3 g of CCl_4 are the maximum amount or theoretical yield of CCl_4 possible from 100. g of Cl_2. Actual yield is 65.0 g of CCl_4:

$$\text{percent yield} = \left(\frac{65.0\ \text{g}}{72.3\ \text{g}}\right)(100) = 89.9\%$$

Silver bromide was prepared by reacting 200.0 g of magnesium bromide and an adequate amount of silver nitrate. Calculate the percent yield if 375.0 g of silver bromide were obtained from the reaction:

Example 9.13

$$MgBr_2 + 2\,AgNO_3 \longrightarrow Mg(NO_3)_2 + 2\,AgBr$$

Step 1 Determine the theoretical yield. Calculate the grams of AgBr that can be formed:

SOLUTION

$$g\,MgBr_2 \longrightarrow mol\,MgBr_2 \longrightarrow mol\,AgBr \longrightarrow g\,AgBr$$

$$(200.0\text{ g }MgBr_2)\left(\frac{1\text{ mol }MgBr_2}{184.1\text{ g }MgBr_2}\right)\left(\frac{2\text{ mol }AgBr}{1\text{ mol }MgBr_2}\right)\left(\frac{187.8\text{ g }AgBr}{1\text{ mol }AgBr}\right) = 408.0\text{ g }AgBr$$

The theoretical yield is 408.0 g AgBr.

Step 2 Calculate the percent yield. According to the equation, 408.0 g AgBr are the maximum amount of AgBr possible from 200.0 g MgBr_2. Actual yield is 375.0 g AgBr:

$$\text{percent yield} = \frac{375.0\text{ g }AgBr}{408.0\text{ g }AgBr} \times 100 = 91.91\%$$

Practice 9.9

Aluminum oxide was prepared by heating 225 g of chromium(II) oxide with 125 g of aluminum. Calculate the percent yield if 100.0 g of aluminum oxide were obtained:

$$2\,Al + 3\,CrO \longrightarrow Al_2O_3 + 3\,Cr$$

When solving problems, you will achieve better results if you work in an organized manner.

1. Write data and numbers in a logical, orderly manner.

2. Make certain that the equations are balanced and that the computations are accurate and expressed to the correct number of significant figures.

3. Check the units. Units are very important; a number without units has little or no meaning.

Chapter 9 Review

9.1 A Short Review

KEY TERM
Molar mass

- The molar mass is the sum of the atomic masses of all the atoms in an element or compound.

- $1\ \text{mole} = \begin{cases} 6.022 \times 10^{23}\ \text{molecules} \\ 6.022 \times 10^{23}\ \text{formula units} \\ 6.022 \times 10^{23}\ \text{atoms} \\ 6.022 \times 10^{23}\ \text{ions} \end{cases}$

- When you use chemical equations for calculations, the chemical equations must be balanced first.

9.2 Introduction to Stoichiometry

KEY TERMS
Stoichiometry
Mole ratio

- Solving stoichiometry problems requires the use of moles and mole ratios.
- To solve a stoichiometry problem:
  - Convert the quantity of starting substance to moles.
  - Convert the moles of starting substance to moles of desired substance.
  - Convert the moles of desired substance to the appropriate unit.

9.3 Mole–Mole Calculations
- If the starting substance is given in moles and the unit specified for the answer is moles:
  - Convert the moles of starting substance to moles of desired substance.

9.4 Mole–Mass Calculations
- If the starting substance is given in moles and the unit specified for the answer is grams:
  - Convert the moles of starting substance to moles of desired substance.
  - Convert the moles of desired substance to grams.

9.5 Mass–Mass Calculations
- If the starting substance is given in grams and the unit specified for the answer is grams:
  - Convert the quantity of starting substance to moles.
  - Convert the moles of starting substance to moles of desired substance.
  - Convert the moles of desired substance to grams.

9.6 Limiting Reactant and Yield Calculations

KEY TERMS
Limiting reactant
Theoretical yield
Actual yield
Percent yield

- To identify the limiting reactant in a reaction:
  - Calculate the amount of product formed from each reactant.
  - Determine the limiting reactant by selecting the one that gives the least amount of product.
- To determine the actual amount of product formed in a limiting reactant situation:
  - Use the result calculated that is the least amount of product.
- To determine the amount of the other reactants required to react with the limiting reactant:
  - Calculate the amount of the other reactant needed to react with the limiting reactant.
  - Subtract this amount from the original amount of the other reactant to find the amount of excess reactant (unreacted).
- The theoretical yield of a chemical reaction is the calculated amount of product from a given amount of reactant (the limiting reactant).
- The actual yield is the amount of product actually obtained experimentally.
- The percent yield is

$$\frac{\text{actual yield}}{\text{theoretical yield}} \times 100 = \text{percent yield}$$

Review Questions

All questions with blue numbers have answers in the appendix of the text.

1. What is a mole ratio?
2. What piece of information is needed to convert grams of a compound to moles of the same compound?
3. Phosphine (PH_3) can be prepared by the hydrolysis of calcium phosphide, Ca_3P_2:

$$Ca_3P_2 + 6\,H_2O \longrightarrow 3\,Ca(OH)_2 + 2\,PH_3$$

Based on this equation, which of the following statements are correct? Show evidence to support your answer.
   (a) One mole of Ca_3P_2 produces 2 mol of PH_3.
   (b) One gram of Ca_3P_2 produces 2 g of PH_3.
   (c) Three moles of $Ca(OH)_2$ are produced for each 2 mol of PH_3 produced.
   (d) The mole ratio between phosphine and calcium phosphide is
   $$\frac{2\text{ mol PH}_3}{1\text{ mol Ca}_3\text{P}_2}.$$
   (e) When 2.0 mol of Ca_3P_2 and 3.0 mol of H_2O react, 4.0 mol of PH_3 can be formed.
   (f) When 2.0 mol of Ca_3P_2 and 15.0 mol of H_2O react, 6.0 mol of $Ca(OH)_2$ can be formed.
   (g) When 200. g of Ca_3P_2 and 100. g of H_2O react, Ca_3P_2 is the limiting reactant.
   (h) When 200. g of Ca_3P_2 and 100. g of H_2O react, the theoretical yield of PH_3 is 57.4 g.

4. The equation representing the reaction used for the commercial preparation of hydrogen cyanide is

$$2\,CH_4 + 3\,O_2 + 2\,NH_3 \longrightarrow 2\,HCN + 6\,H_2O$$

Based on this equation, which of the following statements are correct? Rewrite incorrect statements to make them correct.
   (a) Three moles of O_2 are required for 2 mol of NH_3.
   (b) Twelve moles of HCN are produced for every 16 mol of O_2 that react.
   (c) The mole ratio between H_2O and CH_4 is
   $$\frac{6\text{ mol H}_2\text{O}}{2\text{ mol CH}_4}$$
   (d) When 12 mol of HCN are produced, 4 mol of H_2O will be formed.
   (e) When 10 mol of CH_4, 10 mol of O_2, and 10 mol of NH_3 are mixed and reacted, O_2 is the limiting reactant.
   (f) When 3 mol each of CH_4, O_2, and NH_3 are mixed and reacted, 3 mol of HCN will be produced.

5. What is the difference between the theoretical and the actual yield of a chemical reaction?
6. How is the percent yield of a reaction calculated?

Paired Exercises

All exercises with blue numbers have answers in the appendix of the text.

Mole Review

1. Calculate the number of moles in these quantities:
   (a) 25.0 g KNO_3
   (b) 56 millimol NaOH
   (c) 5.4×10^2 g $(NH_4)_2C_2O_4$
   (d) 16.8 mL H_2SO_4 solution ($d = 1.727$ g/mL, 80.0% H_2SO_4 by mass)

2. Calculate the number of moles in these quantities:
   (a) 2.10 kg $NaHCO_3$
   (b) 525 mg $ZnCl_2$
   (c) 9.8×10^{24} molecules CO_2
   (d) 250 mL ethyl alcohol, C_2H_5OH ($d = 0.789$ g/mL)

3. Calculate the number of grams in these quantities:
   (a) 2.55 mol $Fe(OH)_3$
   (b) 125 kg $CaCO_3$
   (c) 10.5 mol NH_3
   (d) 72 millimol HCl
   (e) 500.0 mL of liquid Br_2 ($d = 3.119$ g/mL)

4. Calculate the number of grams in these quantities:
   (a) 0.00844 mol $NiSO_4$
   (b) 0.0600 mol $HC_2H_3O_2$
   (c) 0.725 mol Bi_2S_3
   (d) 4.50×10^{21} molecules glucose, $C_6H_{12}O_6$
   (e) 75 mL K_2CrO_4 solution ($d = 1.175$ g/mL, 20.0% K_2CrO_4 by mass)

5. Which contains the larger number of molecules, 10.0 g H_2O or 10.0 g H_2O_2? Show evidence for your answer.

6. Which contains the larger numbers of molecules, 25.0 g HCl or 85.0 g $C_6H_{12}O_6$? Show evidence for your answer.

Mole Ratio

7. Given the equation for the combustion of isopropyl alcohol

$$2\,C_3H_7OH + 9\,O_2 \longrightarrow 6\,CO_2 + 8\,H_2O$$

set up the mole ratio of

(a) CO_2 to C_3H_7OH (d) H_2O to C_3H_7OH

(b) C_3H_7OH to O_2 (e) CO_2 to H_2O

(c) O_2 to CO_2 (f) H_2O to O_2

8. For the reaction

$$3\,CaCl_2 + 2\,H_3PO_4 \longrightarrow Ca_3(PO_4)_2 + 6\,HCl$$

set up the mole ratio of

(a) $CaCl_2$ to $Ca_3(PO_4)_2$ (d) $Ca_3(PO_4)_2$ to H_3PO_4

(b) HCl to H_3PO_4 (e) HCl to $Ca_3(PO_4)_2$

(c) $CaCl_2$ to H_3PO_4 (f) H_3PO_4 to HCl

9. How many moles of CO_2 can be produced from 7.75 mol C_2H_5OH? (Balance the equation first.)

$$C_2H_5OH + O_2 \longrightarrow CO_2 + H_2O$$

10. How many moles of Cl_2 can be produced from 5.60 mol HCl?

$$HCl + O_2 \longrightarrow Cl_2 + H_2O$$

11. Given the equation

$$MnO_2(s) + HCl(aq) \longrightarrow$$
$$Cl_2(g) + MnCl_2(aq) + H_2O(l) \text{ (unbalanced)}$$

(a) How many moles of HCl will react with 1.05 mol of MnO_2?

(b) How many moles of $MnCl_2$ will be produced when 1.25 mol of H_2O are formed?

12. Given the equation

$$Al_4C_3 + 12\,H_2O \longrightarrow 4\,Al(OH)_3 + 3\,CH_4$$

(a) How many moles of water are needed to react with 100. g of Al_4C_3?

(b) How many moles of $Al(OH)_3$ will be produced when 0.600 mol of CH_4 is formed?

13. How many grams of sodium hydroxide can be produced from 500 g of calcium hydroxide according to this equation?

$$Ca(OH)_2 + Na_2CO_3 \longrightarrow 2\,NaOH + CaCO_3$$

14. How many grams of zinc phosphate, $Zn_3(PO_4)_2$, are formed when 10.0 g of Zn are reacted with phosphoric acid?

$$3\,Zn + 2\,H_3PO_4 \longrightarrow Zn_3(PO_4)_2 + 3\,H_2$$

15. In a blast furnace, iron(III) oxide reacts with coke (carbon) to produce molten iron and carbon monoxide:

$$Fe_2O_3 + 3\,C \xrightarrow{\Delta} 2\,Fe + 3\,CO$$

How many kilograms of iron would be formed from 125 kg of Fe_2O_3?

16. How many grams of steam and iron must react to produce 375 g of magnetic iron oxide, Fe_3O_4?

$$3\,Fe(s) + 4\,H_2O(g) \xrightarrow{\Delta} Fe_3O_4(s) + 4\,H_2(g)$$

17. Ethane gas, C_2H_6, burns in air (i.e., reacts with the oxygen in air) to form carbon dioxide and water:

$$2\,C_2H_6 + 7\,O_2 \longrightarrow 4\,CO_2 + 6\,H_2O$$

(a) How many moles of O_2 are needed for the complete combustion of 15.0 mol of ethane?

(b) How many grams of CO_2 are produced for each 8.00 g of H_2O produced?

(c) How many grams of CO_2 will be produced by the combustion of 75.0 g of C_2H_6?

(d) If the reaction produces 2.75 mol of H_2O, how many grams of CO_2 are also produced?

(e) How many grams of O_2 are need to completely react with 25.0 mol C_2H_6?

(f) How many grams of C_2H_6 will be needed to produce 125 g of water?

18. Given the equation

$$4\,FeS_2 + 11\,O_2 \longrightarrow 2\,Fe_2O_3 + 8\,SO_2$$

(a) How many moles of Fe_2O_3 can be made from 1.00 mol of FeS_2?

(b) How many moles of O_2 are required to react with 4.50 mol of FeS_2?

(c) If the reaction produces 1.55 mol of Fe_2O_3, how many moles of SO_2 are produced?

(d) How many grams of SO_2 can be formed from 0.512 mol of FeS_2?

(e) If the reaction produces 40.6 g of SO_2, how many moles of O_2 were reacted?

(f) How many grams of FeS_2 are needed to produce 221 g of Fe_2O_3?

Limiting Reactant and Percent Yield

19. Draw the molecules of product(s) formed for each of the following reactions; then determine which substance is the limiting reactant:

(a)

H₂
O₂

(b)

H₂
Br₂

20. Draw the molecules of product(s) formed for the following reactions; then determine which substance is the limiting reactant:

(a)

Li
I₂

(b)

Ag
Cl₂

21. Draw pictures similar to those in Question 19 and determine which is the limiting reactant in the following reactions:
   (a) Eight atoms of potassium react with five molecules of chlorine to produce potassium chloride.
   (b) Ten atoms of aluminum react with three molecules of oxygen to produce aluminum oxide.

22. Draw pictures similar to those in Question 20 and determine which is the limiting reactant in the following reactions:
   (a) Eight molecules of nitrogen react with six molecules of oxygen to produce nitrogen dioxide.
   (b) Fifteen atoms of iron react with twelve molecules of water to produce magnetic iron oxide (Fe_3O_4) and hydrogen.

23. In the following equations, determine which reactant is the limiting reactant and which reactant is in excess. The amounts used are given below each reactant. Show evidence for your answers.
   (a) $KOH + HNO_3 \longrightarrow KNO_3 + H_2O$
      16.0 g   12.0 g
   (b) $2\,NaOH + H_2SO_4 \longrightarrow Na_2SO_4 + H_2O$
      10.0 g   10.0 g

24. In the following equations, determine which reactant is the limiting reactant and which reactant is in excess. The amounts used are given below each reactant. Show evidence for your answers.
   (a) $2\,Bi(NO_3)_3 + 3\,H_2S \longrightarrow Bi_2S_3 + 6\,HNO_3$
      50.0 g   6.00 g
   (b) $3\,Fe + 4\,H_2O \longrightarrow Fe_3O_4 + 4\,H_2$
      40.0 g   16.0 g

25. The reaction for the combustion of propane is
   $$C_3H_8 + 5\,O_2 \longrightarrow 3\,CO_2 + 4\,H_2O$$
   (a) If 20.0 g of C_3H_8 and 20.0 g of O_2 are reacted, how many moles of CO_2 can be produced?
   (b) If 20.0 g of C_3H_8 and 80.0 g of O_2 are reacted, how many moles of CO_2 can be produced?
   (c) If 2.0 mol of C_3H_8 and 14.0 mol of O_2 are placed in a closed container and they react to completion (until one reactant is completely used up), what compounds will be present in the container after the reaction, and how many moles of each compound are present?

26. The reaction for the combustion of propene is
   $$2\,C_3H_6 + 9\,O_2 \longrightarrow 6\,CO_2 + 6\,H_2O$$
   (a) If 15.0 g of C_3H_6 and 15 g of O_2 are reacted, how many moles of H_2O can be produced?
   (b) If 12.0 g of C_3H_6 and 25.0 g of O_2 are reacted, how many moles of H_2O can be produced?
   (c) If 5.0 mol of C_3H_6 and 15.0 mol of O_2 are reacted to completion (until one reactant is completely used up), how many moles of CO_2 can be produced? Which reactant is left over when the reaction has gone to completion in a closed container?

*27. When a certain nonmetal whose formula is X_8 burns in air, XO_3 forms. Write a balanced equation for this reaction. If 120.0 g of oxygen gas are consumed completely, along with 80.0 g of X_8, identify element X.

29. Aluminum reacts with bromine to form aluminum bromide:

$$2\,Al + 3\,Br_2 \longrightarrow 2\,AlBr_3$$

If 25.0 g of Al and 100. g of Br_2 are reacted, and 64.2 g of $AlBr_3$ product are recovered, what is the percent yield for the reaction?

*31. Carbon disulfide, CS_2, can be made from coke (C) and sulfur dioxide (SO_2):

$$3\,C + 2\,SO_2 \longrightarrow CS_2 + CO_2$$

If the actual yield of CS_2 is 86.0% of the theoretical yield, what mass of coke is needed to produce 950 g of CS_2?

*28. When a particular metal X reacts with HCl, the resulting products are XCl_2 and H_2. Write and balance the equation. When 78.5 g of the metal react completely, 2.42 g of hydrogen gas result. Identify the element X.

30. Iron was reacted with a solution containing 400. g of copper(II) sulfate. The reaction was stopped after 1 hour, and 151 g of copper were obtained. Calculate the percent yield of copper obtained:

$$Fe(s) + CuSO_4(aq) \longrightarrow Cu(s) + FeSO_4(aq)$$

*32. Acetylene (C_2H_2) can be manufactured by the reaction of water and calcium carbide, CaC_2:

$$CaC_2(s) + 2\,H_2O(l) \longrightarrow C_2H_2(g) + Ca(OH)_2(aq)$$

When 44.5 g of commercial-grade (impure) calcium carbide are reacted, 0.540 mol of C_2H_2 is produced. Assuming that all of the CaC_2 was reacted to C_2H_2, what is the percent of CaC_2 in the commercial-grade material?

Additional Exercises

All exercises with blue *numbers have answers in the appendix of the text.*

33. A tool set contains 6 wrenches, 4 screwdrivers, and 2 pliers. The manufacturer has 1000 pliers, 2000 screwdrivers, and 3000 wrenches in stock. Can an order for 600 tool sets be filled? Explain briefly.

34. What is the difference between using a number as a subscript and using a number as a coefficient in a chemical equation?

35. For mass–mass calculations, why is it necessary to convert the grams of starting material to moles of starting material, then determine the moles of product from the moles of starting material, and convert the moles of product to grams of product? Why can't you calculate the grams of product directly from the grams of starting material?

*36. Oxygen masks for producing O_2 in emergency situations contain potassium superoxide (KO_2). It reacts according to this equation:

$$4\,KO_2 + 2\,H_2O + 4\,CO_2 \longrightarrow 4\,KHCO_3 + 3\,O_2$$

(a) If a person wearing such a mask exhales 0.85 g of CO_2 every minute, how many moles of KO_2 are consumed in 10.0 minutes?

(b) How many grams of oxygen are produced in 1.0 hour?

37. Ethyl alcohol is the result of fermentation of sugar, $C_6H_{12}O_6$:

$$C_6H_{12}O_6 \longrightarrow 2\,C_2H_5OH + 2\,CO_2$$

(a) How many grams of ethyl alcohol and how many grams of carbon dioxide can be produced from 750 g of sugar?

(b) How many milliliters of alcohol ($d = 0.79$ g/mL) can be produced from 750 g of sugar?

38. The methyl alcohol (CH_3OH) used in alcohol burners combines with oxygen gas to form carbon dioxide and water. How many grams of oxygen are required to burn 60.0 mL of methyl alcohol ($d = 0.72$ g/mL)?

39. Hydrazine (N_2H_4) and hydrogen peroxide (H_2O_2) have been used as rocket propellants. They react according to the following equation:

$$7\,H_2O_2 + N_2H_4 \longrightarrow 2\,HNO_3 + 8\,H_2O$$

(a) When 75 kg of hydrazine react, how many grams of nitric acid can be formed?

(b) When 250 L of hydrogen peroxide ($d = 1.41$ g/mL) react, how many grams of water can be formed?

(c) How many grams of hydrazine will be required to react with 725 g hydrogen peroxide?

(d) How many grams of water can be produced when 750 g hydrazine combine with 125 g hydrogen peroxide?

(e) How many grams of the excess reactant in part (d) are left unreacted?

40. Chlorine gas can be prepared according to the reaction

$$16\,HCl + 2\,KMnO_4 \longrightarrow$$
$$5\,Cl_2 + 2\,KCl + 2\,MnCl_2 + 8\,H_2O$$

(a) How many moles of $MnCl_2$ can be produced when 25 g $KMnO_4$ is mixed with 85 g HCl?

(b) How many grams of water will be produced when 75 g KCl are produced?

(c) What is the percent yield of Cl_2 if 150 g HCl is reacted, producing 75 g of Cl_2?

(d) When 25 g HCl react with 25 g $KMnO_4$, how many grams of Cl_2 can be produced?

(e) How many grams of the excess reactant in part (d) are left unreacted?

41. Silver tarnishes in the presence of hydrogen sulfide (which smells like rotten eggs) and oxygen because of the reaction

$$4\,Ag + 2\,H_2S + O_2 \longrightarrow 2\,Ag_2S + 2\,H_2O$$

(a) How many grams of silver sulfide can be formed from a mixture of 1.1 g Ag, 0.14 g H_2S, and 0.080 g O_2?
(b) How many more grams of H_2S would be needed to completely react all of the Ag?

42. Solid CaO picks up water from the atmosphere to form $Ca(OH)_2$.

$$CaO(s) + H_2O(g) \longrightarrow Ca(OH)_2(s)$$

A beaker containing CaO weighs 26.095 g. When allowed to sit open to the atmosphere, it absorbs water to a final mass of 26.500 g. What is the mass of the beaker?

43. When a solution containing 15 g of lead(II) nitrate and a solution containing 15 g of potassium iodide are mixed, a yellow precipitate is formed.
(a) What is the name and formula for the solid product formed?
(b) What type of reaction occurred?
(c) After filtration and drying, the solid product has a mass of 6.68 g. What is the percent yield for the reaction?

44. A 10.00-g mixture of KNO_3 and KCl is reacted with $AgNO_3$ to give 4.33 g AgCl. What is the percent composition for the mixture?

45. After 180.0 g of zinc were dropped into a beaker of hydrochloric acid and the reaction ceased, 35 g of unreacted zinc remained in the beaker:

$$Zn + HCl \longrightarrow ZnCl_2 + H_2$$

(a) How many grams of hydrogen gas were produced?
(b) How many grams of HCl were reacted?
(c) How many more grams of HCl would be required to completely react with the original sample of zinc?

46. Use this equation to answer (a) and (b):

$$Fe(s) + CuSO_4(aq) \longrightarrow Cu(s) + FeSO_4(aq)$$

(a) When 2.0 mol of Fe and 3.0 mol of $CuSO_4$ are reacted, what substances will be present when the reaction is over? How many moles of each substance are present?
(b) When 20.0 g of Fe and 40.0 g of $CuSO_4$ are reacted, what substances will be present when the reaction is over? How many grams of each substance are present?

47. Methyl alcohol (CH_3OH) is made by reacting carbon monoxide and hydrogen in the presence of certain metal oxide catalysts. How much alcohol can be obtained by reacting 40.0 g of CO and 10.0 g of H_2? How many grams of excess reactant remain unreacted?

$$CO(g) + 2\,H_2(g) \longrightarrow CH_3OH(l)$$

*48. Ethyl alcohol (C_2H_5OH), also called grain alcohol, can be made by the fermentation of sugar:

$$\underset{\text{glucose}}{C_6H_{12}O_6} \longrightarrow \underset{\text{ethyl alcohol}}{2\,C_2H_5OH} + 2\,CO_2$$

If an 84.6% yield of ethyl alcohol is obtained,
(a) what mass of ethyl alcohol will be produced from 750 g of glucose?
(b) what mass of glucose should be used to produce 475 g of C_2H_5OH?

*49. Both $CaCl_2$ and $MgCl_2$ react with $AgNO_3$ to precipitate AgCl. When solutions containing equal masses of $CaCl_2$ and $MgCl_2$ are reacted with $AgNO_3$, which salt solution will produce the larger amount of AgCl? Show proof.

*50. An astronaut excretes about 2500 g of water a day. If lithium oxide (Li_2O) is used in the spaceship to absorb this water, how many kilograms of Li_2O must be carried for a 30-day space trip for three astronauts?

$$Li_2O + H_2O \longrightarrow 2\,LiOH$$

*51. Much commercial hydrochloric acid is prepared by the reaction of concentrated sulfuric acid with sodium chloride:

$$H_2SO_4 + 2\,NaCl \longrightarrow Na_2SO_4 + 2\,HCl$$

How many kilograms of concentrated H_2SO_4, 96% H_2SO_4 by mass, are required to produce 20.0 L of concentrated hydrochloric acid ($d = 1.20$ g/mL, 42.0% HCl by mass)?

52. Three chemical reactions that lead to the formation of sulfuric acid are

$$S + O_2 \longrightarrow SO_2$$
$$2\,SO_2 + O_2 \longrightarrow 2\,SO_3$$
$$SO_3 + H_2O \longrightarrow H_2SO_4$$

Starting with 100.0 g of sulfur, how many grams of sulfuric acid will be formed, assuming a 10% loss in each step? What is the percent yield of H_2SO_4?

53. A 10.00-g mixture of $NaHCO_3$ and Na_2CO_3 was heated and yielded 0.0357 mol H_2O and 0.1091 mol CO_2. Calculate the percent composition of the mixture:

$$NaHCO_3 \longrightarrow Na_2O + CO_2 + H_2O$$
$$Na_2CO_3 \longrightarrow Na_2O + CO_2$$

Challenge Exercises

All exercises with blue numbers have answers in the appendix of the text.

*54. When 12.82 g of a mixture of $KClO_3$ and NaCl are heated strongly, the $KClO_3$ reacts according to this equation:

$$2\,KClO_3(s) \longrightarrow 2\,KCl(s) + 3\,O_2(g)$$

The NaCl does not undergo any reaction. After the heating, the mass of the residue (KCl and NaCl) is 9.45 g. Assuming that all the loss of mass represents loss of oxygen gas, calculate the percent of $KClO_3$ in the original mixture.

*55. Gastric juice contains about 3.0 g HCl per liter. If a person produces about 2.5 L of gastric juice per day, how many antacid tablets, each containing 400. mg of $Al(OH)_3$, are needed to neutralize all the HCl produced in 1 day?

$$Al(OH)_3(s) + 3\,HCl(aq) \longrightarrow AlCl_3(aq) + 3\,H_2O(l)$$

*56. Phosphoric acid, H_3PO_4, can be synthesized from phosphorus, oxygen, and water according to these two reactions:

$$4\,P + 5\,O_2 \longrightarrow P_4O_{10}$$
$$P_4O_{10} + 6\,H_2O \longrightarrow 4\,H_3PO_4$$

Starting with 20.0 g P, 30.0 g O_2, and 15.0 g H_2O, what is the mass of phosphoric acid that can be formed?

Answers to Practice Exercises

9.1   0.33 mol Al_2O_3

9.2   7.33 mol $Al(OH)_3$

9.3   0.8157 mol KCl

9.4   85 g $AgNO_3$

9.5   27.6 g $CrCl_3$

9.6   348.8 g H_2O

9.7   17.7 g HCl

9.8   164.3 g $BaSO_4$

9.9   88.5% yield

PUTTING IT TOGETHER: Review for Chapters 7–9

Multiple Choice:

Choose the correct answer to each of the following.

1. 4.0 g of oxygen contain
   (a) 1.5×10^{23} atoms of oxygen
   (b) 4.0 molar masses of oxygen
   (c) 0.50 mol of oxygen
   (d) 6.022×10^{23} atoms of oxygen

2. One mole of hydrogen atoms contains
   (a) 2.0 g of hydrogen
   (b) 6.022×10^{23} atoms of hydrogen
   (c) 1 atom of hydrogen
   (d) 12 g of carbon-12

3. The mass of one atom of magnesium is
   (a) 24.31 g   (c) 12.00 g
   (b) 54.94 g   (d) 4.037×10^{-23} g

4. Avogadro's number of magnesium atoms
   (a) has a mass of 1.0 g
   (b) has the same mass as Avogadro's number of sulfur atoms
   (c) has a mass of 12.0 g
   (d) is 1 mol of magnesium atoms

5. Which of the following contains the largest number of moles?
   (a) 1.0 g Li   (c) 1.0 g Al
   (b) 1.0 g Na   (d) 1.0 g Ag

6. The number of moles in 112 g of acetylsalicylic acid (aspirin), $C_9H_8O_4$, is
   (a) 1.61   (c) 112
   (b) 0.622   (d) 0.161

7. How many moles of aluminum hydroxide are in one antacid tablet containing 400 mg of $Al(OH)_3$?
   (a) 5.13×10^{-3} (c) 5.13
   (b) 0.400   (d) 9.09×10^{-3}

8. How many grams of Au_2S can be obtained from 1.17 mol of Au?
   (a) 182 g   (c) 364 g
   (b) 249 g   (d) 499 g

9. The molar mass of $Ba(NO_3)_2$ is
   (a) 199.3   (c) 247.3
   (b) 261.3   (d) 167.3

10. A 16-g sample of oxygen
    (a) is 1 mol of O_2
    (b) contains 6.022×10^{23} molecules of O_2
    (c) is 0.50 molecule of O_2
    (d) is 0.50 molar mass of O_2

11. What is the percent composition for a compound formed from 8.15 g of zinc and 2.00 g of oxygen?
    (a) 80.3% Zn, 19.7% O   (c) 70.3% Zn, 29.7% O
    (b) 80.3% O, 19.7% Zn   (d) 65.3% Zn, 34.7% O

12. Which of these compounds contains the largest percentage of oxygen?
    (a) SO_2 (c) N_2O_3
    (b) SO_3 (d) N_2O_5

13. 2.00 mol of CO_2
    (a) have a mass of 56.0 g
    (b) contain 1.20×10^{24} molecules
    (c) have a mass of 44.0 g
    (d) contain 6.00 molar masses of CO_2

14. In Ag_2CO_3, the percent by mass of
    (a) carbon is 43.5%   (c) oxygen is 17.4%
    (b) silver is 64.2%   (d) oxygen is 21.9%

15. The empirical formula of the compound containing 31.0% Ti and 69.0% Cl is
    (a) TiCl   (c) $TiCl_3$
    (b) $TiCl_2$ (d) $TiCl_4$

16. A compound contains 54.3% C, 5.6% H, and 40.1% Cl. The empirical formula is
    (a) CH_3Cl (c) $C_2H_4Cl_2$
    (b) C_2H_5Cl (d) C_4H_5Cl

17. A compound contains 40.0% C, 6.7% H, and 53.3% O. The molar mass is 60.0 g/mol. The molecular formula is
    (a) $C_2H_3O_2$ (c) C_2HO
    (b) C_3H_8O (d) $C_2H_4O_2$

18. How many chlorine atoms are in 4.0 mol of PCl_3?
    (a) 3   (c) 12
    (b) 7.2×10^{24} (d) 2.4×10^{24}

19. What is the mass of 4.53 mol of Na_2SO_4?
    (a) 142.1 g   (c) 31.4 g
    (b) 644 g   (d) 3.19×10^{-2} g

20. The percent composition of Mg_3N_2 is
    (a) 72.2% Mg, 27.8% N   (c) 83.9% Mg, 16.1% N
    (b) 63.4% Mg, 36.6% N   (d) no correct answer given

21. How many grams of oxygen are contained in 0.500 mol of Na_2SO_4?
    (a) 16.0 g   (c) 64.0 g
    (b) 32.0 g   (d) no correct answer given

22. The empirical formula of a compound is CH. If the molar mass of this compound is 78.11, then the molecular formula is
    (a) C_2H_2 (c) C_6H_6
    (b) C_5H_{18} (d) no correct answer given

23. The reaction

    $$BaCl_2 + (NH_4)_2CO_3 \longrightarrow BaCO_3 + 2\,NH_4Cl$$

    is an example of
    (a) combination   (c) single displacement
    (b) decomposition (d) double displacement

24. When the equation

$$Al + O_2 \longrightarrow Al_2O_3$$

is properly balanced, which of the following terms appears?
(a) 2 Al   (c) 3 Al
(b) 2 Al$_2$O$_3$ (d) 2 O$_2$

25. Which equation is *incorrectly* balanced?

(a) $2 KNO_3 \xrightarrow{\Delta} 2 KNO_2 + O_2$

(b) $H_2O_2 \longrightarrow H_2O + O_2$

(c) $2 Na_2O_2 + 2 H_2O \longrightarrow 4 NaOH + O_2$

(d) $2 H_2O \xrightarrow[H_2SO_4]{\text{electrical energy}} 2 H_2 + O_2$

26. The reaction

$$2 Al + 3 Br_2 \longrightarrow 2 AlBr_3$$

is an example of
(a) combination   (c) decomposition
(b) single displacement   (d) double displacement

27. When the equation

$$PbO_2 \xrightarrow{\Delta} PbO + O_2$$

is balanced, one term in the balanced equation is
(a) PbO$_2$ (c) 3 PbO
(b) 3 O$_2$ (d) O$_2$

28. When the equation

$$Cr_2S_3 + HCl \longrightarrow CrCl_3 + H_2S$$

is balanced, one term in the balanced equation is
(a) 3 HCl   (c) 3 H$_2$S
(b) CrCl$_3$ (d) 2 Cr$_2$S$_3$

29. When the equation

$$F_2 + H_2O \longrightarrow HF + O_2$$

is balanced, one term in the balanced equation is
(a) 2 HF   (c) 4 HF
(b) 3 O$_2$ (d) 4 H$_2$O

30. When the equation

$$NH_4OH + H_2SO_4 \longrightarrow$$

is completed and balanced, one term in the balanced equation is
(a) NH$_4$SO$_4$ (c) H$_2$OH
(b) 2 H$_2$O   (d) 2 (NH$_4$)$_2$SO$_4$

31. When the equation

$$H_2 + V_2O_5 \longrightarrow V +$$

is completed and balanced, one term in the balanced equation is
(a) 2 V$_2$O$_5$ (c) 2 V
(b) 3 H$_2$O   (d) 8 H$_2$

32. When the equation

$$Al(OH)_3 + H_2SO_4 \longrightarrow Al_2(SO_4)_3 + H_2O$$

is balanced, the sum of the coefficients will be
(a) 9   (c) 12
(b) 11   (d) 15

33. When the equation

$$H_3PO_4 + Ca(OH)_2 \longrightarrow H_2O + Ca_3(PO_4)_2$$

is balanced, the proper sequence of coefficients is
(a) 3, 2, 1, 6   (c) 2, 3, 1, 6
(b) 2, 3, 6, 1   (d) 2, 3, 3, 1

34. When the equation

$$Fe_2(SO_4)_3 + Ba(OH)_2 \longrightarrow$$

is completed and balanced, one term in the balanced equation is
(a) Ba$_2$(SO$_4$)$_3$ (c) 2 Fe$_2$(SO$_4$)$_3$
(b) 2 Fe(OH)$_2$ (d) 2 Fe(OH)$_3$

35. For the reaction

$$2 H_2 + O_2 \longrightarrow 2 H_2O + 572.4 \text{ kJ}$$

which of the following is not true?
(a) The reaction is exothermic.
(b) 572.4 kJ of heat are liberated for each mole of water formed.
(c) 2 mol of hydrogen react with 1 mol of oxygen.
(d) 572.4 kJ of heat are liberated for each 2 mol of hydrogen reacted.

36. How many moles is 20.0 g of Na$_2$CO$_3$?
(a) 1.89 mol   (c) 212 mol
(b) 2.12 × 10^3 mol   (d) 0.189 mol

37. What is the mass in grams of 0.30 mol of BaSO$_4$?
(a) 7.0 × 10^3 g   (c) 70. g
(b) 0.13 g   (d) 700.2 g

38. How many molecules are in 5.8 g of acetone, C$_3$H$_6$O?
(a) 0.10 molecule
(b) 6.0 × 10^{22} molecules
(c) 3.5 × 10^{24} molecules
(d) 6.0 × 10^{23} molecules

Problems 39–45 refer to the reaction

$$2 C_2H_4 + 6 O_2 \longrightarrow 4 CO_2 + 4 H_2O$$

39. If 6.0 mol of CO$_2$ are produced, how many moles of O$_2$ were reacted?
(a) 4.0 mol   (c) 9.0 mol
(b) 7.5 mol   (d) 15.0 mol

40. How many moles of O$_2$ are required for the complete reaction of 45 g of C$_2$H$_4$?
(a) 1.3 × 10^2 mol   (c) 112.5 mol
(b) 0.64 mol   (d) 4.8 mol

41. If 18.0 g of CO$_2$ are produced, how many grams of H$_2$O are produced?
(a) 7.37 g   (c) 9.00 g
(b) 3.68 g   (d) 14.7 g

42. How many moles of CO$_2$ can be produced by the reaction of 5.0 mol of C$_2$H$_4$ and 12.0 mol of O$_2$?
(a) 4.0 mol   (c) 8.0 mol
(b) 5.0 mol   (d) 10. mol

43. How many moles of CO_2 can be produced by the reaction of 0.480 mol of C_2H_4 and 1.08 mol of O_2?
(a) 0.240 mol   (c) 0.720 mol
(b) 0.960 mol   (d) 0.864 mol

44. How many grams of CO_2 can be produced from 2.0 g of C_2H_4 and 5.0 g of O_2?
(a) 5.5 g   (c) 7.6 g
(b) 4.6 g   (d) 6.3 g

45. If 14.0 g of C_2H_4 are reacted and the actual yield of H_2O is 7.84 g, the percent yield in the reaction is
(a) 0.56%   (c) 87.1%
(b) 43.6%   (d) 56.0%

Problems 46–48 refer to the equation

$$H_3PO_4 + MgCO_3 \longrightarrow Mg_3(PO_4)_2 + CO_2 + H_2O$$

46. The sequence of coefficients for the balanced equation is
(a) 2, 3, 1, 3, 3   (c) 2, 2, 1, 2, 3
(b) 3, 1, 3, 2, 3   (d) 2, 3, 1, 3, 2

47. If 20.0 g of carbon dioxide are produced, the number of moles of magnesium carbonate used is
(a) 0.228 mol   (c) 0.910 mol
(b) 1.37 mol   (d) 0.454 mol

48. If 50.0 g of magnesium carbonate react completely with H_3PO_4, the number of grams of carbon dioxide produced is
(a) 52.2 g   (c) 13.1 g
(b) 26.1 g   (d) 50.0 g

49. When 10.0 g of $MgCl_2$ and 10.0 g of Na_2CO_3 are reacted in

$$MgCl_2 + Na_2CO_3 \longrightarrow MgCO_3 + 2\,NaCl$$

the limiting reactant is
(a) $MgCl_2$ (c) $MgCO_3$
(b) Na_2CO_3 (d) $NaCl$

50. When 50.0 g of copper are reacted with silver nitrate solution in

$$Cu + 2\,AgNO_3 \longrightarrow Cu(NO_3)_2 + 2\,Ag$$

148 g of silver are obtained. What is the percent yield of silver obtained?
(a) 87.1%   (c) 55.2%
(b) 84.9%   (d) no correct answer given

Free Response Questions:

Answer each of the following. Be sure to include your work and explanations in a clear, logical form.

1. Compound X requires 104 g of O_2 to produce 2 moles of CO_2 and 2.5 moles of H_2O.
(a) What is the empirical formula for X?
(b) What additional information would you need to determine the molecular formula for X?

2. Consider the reaction of sulfur dioxide with oxygen to form sulfur trioxide taking place in a closed container.

(a) Draw what you would expect to see in the box at the completion of the reaction.
(b) If you begin with 25 g of SO_2 and 5 g of oxygen gas, which is the limiting reagent?
(c) Is the following statement true or false? "When SO_2 is converted into SO_3, the percent composition of S in the compounds changes from 33% to 25%." Explain.

3. The percent composition of compound Z is 63.16% C and 8.77% H. When compound Z burns in air, the only products are carbon dioxide and water. The molar mass for Z is 114.
(a) What is the molecular formula for compound Z?
(b) What is the balanced reaction for Z burning in air?

4. Compound A decomposes at room temperature in an exothermic reaction, while compound B requires heating before it will decompose in an endothermic reaction.
(a) Draw reaction profiles (potential energy vs. reaction progress) for both reactions.
(b) The decomposition of 0.500 mol of $NaHCO_3$ to form sodium carbonate, water, and carbon dioxide requires 85.5 kJ of heat. How many grams of water could be collected and how many kJ of heat were absorbed when 24.0 g of CO_2 are produced?
(c) Could $NaHCO_3$ be compound A or compound B? Explain your reasoning.

5. Aqueous ammonium hydroxide reacts with aqueous cobalt(II) sulfate to produce aqueous ammonium sulfate and solid cobalt(II) hydroxide. When 38.0 g of one of the reactants were fully reacted with enough of the other reactant, 8.09 g of ammonium sulfate were obtained, which corresponded to a 25.0% yield.
(a) What type of reaction took place?
(b) Write the balanced chemical equation.
(c) What is the theoretical yield of ammonium sulfate?
(d) Which reactant was the limiting reagent?

6. $C_6H_{12}O_6 \longrightarrow 2\,C_2H_5OH + 2\,CO_2(g)$
(a) If 25.0 g of $C_6H_{12}O_6$ were used, and only 11.2 g of C_2H_5OH were produced, how much reactant was left over and what volume of gas was produced? (*Note*: One mole of gas occupies 24.0 L of space at room temperature.)
(b) What was the yield of the reaction?
(c) What type of reaction took place?

7. When solutions containing 25 g each of lead(II) nitrate and potassium iodide are mixed, a yellow precipitate results.
   (a) What type of reaction occurred?
   (b) What is the name and formula for the solid product?
   (c) If after filtration and drying, the solid product weighed 7.66 g, what was the percent yield for the reaction?

8. Consider the following unbalanced reaction:

   $$XNO_3 + CaCl_2 \longrightarrow XCl + Ca(NO_3)_2$$

   (a) If 30.8 g of $CaCl_2$ produced 79.6 g of XCl, what is X?
   (b) Would X be able to displace hydrogen from an acid?

9. Consider the following reaction: $H_2O_2 \longrightarrow H_2O + O_2$
   (a) If at the end of the reaction there are eight water molecules and eight oxygen molecules, what was in the flask at the start of the reaction?
   (b) Does the following reaction profile indicate the reaction is exothermic or endothermic?

Reaction progress

   (c) What type of reaction is given above?
   (d) What is the empirical formula for hydrogen peroxide?

CHAPTER 10

Modern Atomic Theory and the Periodic Table

Electrons moving between energy levels in neon produce the colors in this neon
walkway at Epcot Center.

Chapter Outline

10.1 A Brief History

10.2 Electromagnetic Radiation

10.3 The Bohr Atom

10.4 Energy Levels of Electrons

10.5 Atomic Structures of the First
18 Elements

10.6 Electron Structures and the
Periodic Table

How do we go about studying an object that is too small to see? Think back to that birthday present you could look at, but not yet open. Judging from the wrapping and size of the box was not very useful, but shaking, turning, and lifting the package all gave indirect clues to its contents. After all your experiments were done, you could make a fairly good guess about the contents. But was your guess correct? The only way to know for sure would be to open the package.

Chemists have the same dilemma when they study the atom. Atoms are so very small that it isn't possible to use the normal senses to describe them. We are essentially working in the dark with this package we call the atom. However, our improvements in instruments (X-ray machines and scanning tunneling microscopes) and measuring devices (spectrophotometers and magnetic resonance imaging, MRI) as well as in our mathematical skills are bringing us closer to revealing the secrets of the atom.

10.1 A Brief History

In the last 200 years, vast amounts of data have been accumulated to support atomic theory. When atoms were originally suggested by the early Greeks, no physical evidence existed to support their ideas. Early chemists did a variety of experiments, which culminated in Dalton's model of the atom. Because of the limitations of Dalton's model, modifications were proposed first by Thomson and then by Rutherford, which eventually led to our modern concept of the nuclear atom. These early models of the atom work reasonably well—in fact, we continue to use them to visualize a variety of chemical concepts. There remain questions that these models cannot answer, including an explanation of how atomic structure relates to the periodic table. In this chapter, we will present our modern model of the atom; we will see how it varies from and improves upon the earlier atomic models.

10.2 Electromagnetic Radiation

Scientists have studied energy and light for centuries, and several models have been proposed to explain how energy is transferred from place to place. One way energy travels through space is by *electromagnetic radiation*. Examples of electromagnetic radiation include light from the sun, X-rays in your dentist's office, microwaves from your microwave oven, radio and television waves, and radiant heat from your fireplace. While these examples seem quite different, they are all similar in some important ways. Each shows wavelike behavior, and all travel at the same speed in a vacuum (3.00×10^8 m/s).

Surfers judge the wavelength, frequency, and speed of waves to get the best ride.

wavelength

frequency
speed

The study of wave behavior is a topic for another course, but we need some basic terminology to understand atoms. Waves have three basic characteristics: wavelength, frequency, and speed. **Wavelength** (lambda, λ) is the distance between consecutive peaks (or troughs) in a wave, as shown in Figure 10.1. **Frequency** (nu, ν) tells how many waves pass a particular point per second. **Speed** (v) tells how fast a wave moves through space.

Light is one form of electromagnetic radiation and is usually classified by its wavelength, as shown in Figure 10.2. Visible light, as you can see, is only a tiny part of the electromagnetic spectrum. Some examples of electromagnetic radiation involved in energy transfer outside the visible region are hot coals in

Birds in the parrot family have an unusual way to attract their mates—their feathers glow in the dark! This phenomenon is called fluorescence. It results from the absorption of ultraviolet (UV) light, which is then reemitted at longer wavelengths that both birds and people can see. In everyday life this happens in a fluorescent bulb or in the many glow-in-the-dark products such as light sticks.

Kathleen Arnold from the University of Glasgow, Scotland, discovered that the feathers of parrots that fluoresced were only those used in display or those shown off during courtship.

She decided to experiment using budgerigars, with their natural colors. The researchers offered birds a choice of two companion birds, which were smeared with petroleum jelly. One of the potential companions also had a UV blocker in the petroleum jelly. The birds clearly preferred companions without the UV blocker, leading the researchers to conclude that the parrots prefer to court radiant partners. The researchers also tested same-sex companions and discovered they did not prefer radiant companions.

Perhaps we may find that the glow of candlelight really does add to the radiance of romance. Ultraviolet light certainly does for parrots!

Budgerigars under normal light and under UV light showing the glow used to attract mates.

your backyard grill, which transfer infrared radiation to cook your food, and microwaves, which transfer energy to water molecules in the food, causing them to move more quickly and thus raise the temperature of your food.

We have evidence for the wavelike nature of light. We also know that a beam of light behaves like a stream of tiny packets of energy called **photons**. So what is light exactly? Is it a particle? Is it a wave? Scientists have agreed to explain the properties of electromagnetic radiation by using both wave and particle properties. Neither explanation is ideal, but currently these are our best models.

photons

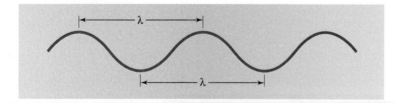

Figure 10.1
The wavelength of this wave is shown by λ. It can be measured from peak to peak or trough to trough.

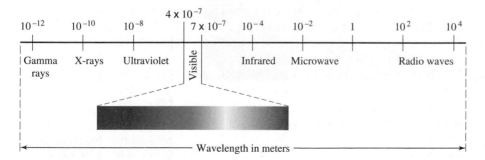

Figure 10.2
The electromagnetic spectrum.

10.3 The Bohr Atom

As scientists struggled to understand the properties of electromagnetic radiation, evidence began to accumulate that atoms could radiate light. At high temperatures, or when subjected to high voltages, elements in the gaseous state give off colored light. Brightly colored neon signs illustrate this property of matter very well. When the light emitted by a gas is passed through a prism or diffraction grating, a set of brightly colored lines called a **line spectrum** results (Figure 10.3). These colored lines indicate that the light is being emitted only at certain wavelengths, or frequencies, that correspond to specific colors. Each element possesses a unique set of these spectral lines that is different from the sets of all the other elements.

line spectrum

In 1912–1913, while studying the line spectrum of hydrogen, Niels Bohr (1885–1962), a Danish physicist, made a significant contribution to the rapidly growing knowledge of atomic structure. His research led him to believe that electrons exist in specific regions at various distances from the nucleus. He also visualized the electrons as revolving in orbits around the nucleus, like planets rotating around the sun, as shown in Figure 10.4.

Bohr's first paper in this field dealt with the hydrogen atom, which he described as a single electron revolving in an orbit about a relatively heavy nucleus. He applied the concept of energy quanta, proposed in 1900 by the German physicist Max Planck (1858–1947), to the observed line spectrum of hydrogen. Planck stated that energy is never emitted in a continuous stream but only in small, discrete packets called **quanta** (Latin, *quantus*, "how much"). From this, Bohr theorized that electrons have several possible energies corresponding to several possible orbits at different distances from the nucleus. Therefore an electron has to be in one specific energy level; it cannot exist between energy levels. In other words, the energy of the electron is said to be quantized. Bohr also stated that when a hydrogen atom absorbed one or more quanta of energy, its electron would "jump" to a higher energy level.

quanta

Bohr was able to account for spectral lines of hydrogen this way. A number of energy levels are available, the lowest of which is called the **ground state**.

ground state

Figure 10.3
Line spectrum of hydrogen. Each line corresponds to the wavelength of the energy emitted when the electron of a hydrogen atom, which has absorbed energy, falls back to a lower principal energy level.

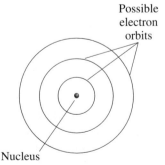

Figure 10.4
The Bohr model of the hydrogen atom described the electron revolving in certain allowed circular orbits around the nucleus.

When an electron falls from a high energy level to a lower one (say, from the fourth to the second), a quantum of energy is emitted as light at a specific frequency, or wavelength (Figure 10.5). This light corresponds to one of the lines visible in the hydrogen spectrum (Figure 10.3). Several lines are visible in this spectrum, each one corresponding to a specific electron energy-level shift within the hydrogen atom.

The chemical properties of an element and its position in the periodic table depend on electron behavior within the atoms. In turn, much of our knowledge of the behavior of electrons within atoms is based on spectroscopy. Niels Bohr contributed a great deal to our knowledge of atomic structure by (1) suggesting quantized energy levels for electrons and (2) showing that spectral lines result from the radiation of small increments of energy (Planck's quanta) when electrons shift from one energy level to another. Bohr's calculations succeeded very well in correlating the experimentally observed spectral lines with electron energy levels for the hydrogen atom. However, Bohr's methods of calculation did not succeed for heavier atoms. More theoretical work on atomic structure was needed.

In 1924, the French physicist Louis de Broglie suggested a surprising hypothesis: All objects have wave properties. De Broglie used sophisticated mathematics to show that the wave properties for an object of ordinary size, such as a baseball, are too small to be observed. But for smaller objects, such as an electron, the wave properties become significant. Other scientists confirmed de Broglie's hypothesis, showing that electrons do exhibit wave properties. In 1926, Erwin Schrödinger, an Austrian physicist, created a mathematical model that described electrons as waves. Using Schrödinger's wave mechanics, we can determine the *probability* of finding an electron in a certain region around the nucleus the atom.

This treatment of the atom led to a new branch of physics called *wave mechanics* or *quantum mechanics*, which forms the basis for our modern understanding of atomic structure. Although the wave-mechanical description of the atom is mathematical, it can be translated, at least in part, into a visual model. It's important to recognize that we cannot locate an electron precisely within an atom; however, it is clear that electrons are not revolving around the nucleus in orbits as Bohr postulated. The electrons are instead found in *orbitals*. An **orbital** is pictured in Figure 10.6 as a region in space around the nucleus where there is a high probability of finding a given electron.

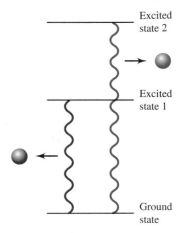

Figure 10.5
When an excited electron returns to the ground state, energy is emitted as a photon is released. The color (wavelength) of the light is determined by the difference in energy between the two states (excited and ground).

orbital

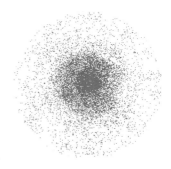

Figure 10.6
An orbital for a hydrogen atom. The intensity of the dots shows that the electron spends more time closer to the nucleus.

principal energy levels

sublevel

10.4 Energy Levels of Electrons

One of the ideas Bohr contributed to the modern concept of the atom was that the energy of the electron is quantized—that is, the electron is restricted to only certain allowed energies. The wave-mechanical model of the atom also predicts discrete **principal energy levels** within the atom. These energy levels are designated by the letter n, where n is a positive integer (Figure 10.7). The lowest principal energy level corresponds to $n = 1$, the next to $n = 2$, and so on. As n increases, the energy of the electron increases, and the electron is found on average farther from the nucleus.

Each principal energy level is divided into **sublevels**, which are illustrated in Figure 10.8. The first principal energy level has one sublevel. The second principal energy level has two sublevels, the third energy level has three sublevels, and so on. Each of these sublevels contains spaces for electrons called orbitals.

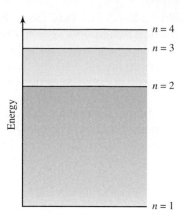

Figure 10.7
The first four principal energy levels in
the hydrogen atom. Each level is assigned
a principal quantum number n.

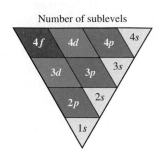

Figure 10.8
The types of orbitals on each
of the first four principal
energy levels.

In each sublevel the electrons are found within specified orbitals (s, p, d, f).
Let's consider each principal energy level in turn. The first principal energy
level ($n = 1$) has one sublevel or type of orbital. It is spherical in shape and is
designated as $1s$. It is important to understand what the spherical shape of the
$1s$ orbital means. The electron does *not* move around on the surface of the
sphere, but rather the surface encloses a space where there is a 90% probability that the electron may be found. It might help to consider these orbital
shapes in the same way we consider the atmosphere. There is no distinct dividing line between the atmosphere and "space." The boundary is quite fuzzy.
The same is true for atomic orbitals. Each has a region of highest density rough-
ly corresponding to its shape. The probability of finding the electron outside
this region drops rapidly but never quite reaches zero. Scientists often speak of
orbitals as electron "clouds" to emphasize the fuzzy nature of their boundaries.

spin   How many electrons can fit into a $1s$ orbital? To answer this question, we
need to consider one more property of electrons. This property is called **spin**.
Each electron appears to be spinning on an axis, like a globe. It can only spin
in two directions. We represent this spin with an arrow: ↑ or ↓. In order to occupy the same orbital, electrons must have *opposite* spins. That is, two electrons with the same spin cannot occupy the same orbital. This gives us the
answer to our question: An atomic orbital can hold a maximum of two elec-

Pauli exclusion principle   trons, which must have opposite spins. This rule is called the **Pauli exclusion
principle**. The first principal energy level contains one type of orbital ($1s$) that
holds a maximum of two electrons.

What happens with the second principal energy level ($n = 2$)? Here we find
two sublevels, $2s$ and $2p$. Like $1s$ in the first principal energy level, the $2s$ orbital
is spherical in shape but is larger in size and higher in energy. It also holds a
maximum of two electrons. The second type of orbital is designated by $2p$. The
$2p$ sublevel consists of three orbitals: $2p_x$, $2p_y$, and $2p_z$. The shape of p orbitals
is quite different from the s orbitals, as shown in Figure 10.9.

Each p orbital has two "lobes." Remember, the space enclosed by these surfaces represents the regions of probability for finding the electrons 90% of the
time. There are three separate p orbitals, each oriented in a different direction,

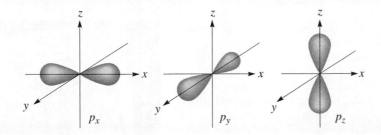

Figure 10.9
Perspective representation of the p_x, p_y, and p_z atomic orbitals.

and each p orbital can hold a maximum of two electrons. Thus the total number of electrons that can reside in all three p orbitals is six. To summarize our model, the first principal energy level of an atom has a $1s$ orbital. The second principal energy level has a $2s$ and three $2p$ orbitals labeled $2p_x$, $2p_y$, and $2p_z$, as shown in Figure 10.10.

Notice that there is a correspondence between the energy level and number of sublevels.

The third principal energy level has three sublevels labeled $3s$, $3p$, and $3d$. The $3s$ orbital is spherical and larger than the $1s$ and $2s$ orbitals. The $3p_x$, $3p_y$, $3p_z$ orbitals are shaped like those of the second level, only larger. The five $3d$ orbitals have the shapes shown in Figure 10.11. You don't need to memorize these shapes, but notice that they look different from the s or p orbitals.

Each time a new principal energy level is added, we also add a new sublevel. This makes sense because each energy level corresponds to a larger average distance from the nucleus, which provides more room on each level for new sublevels containing more orbitals.

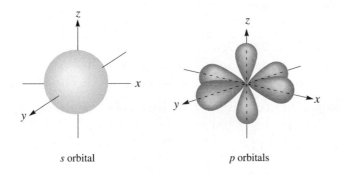

s orbital   p orbitals

Figure 10.10
Orbitals on the second principal energy level are one $2s$ and three $2p$ orbitals.

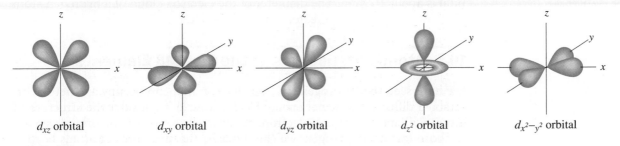

d_{xz} orbital d_{xy} orbital d_{yz} orbital d_{z^2} orbital $d_{x^2-y^2}$ orbital

Figure 10.11
The five d orbitals are found in the third principal energy level along with one $3s$ orbital and three $3p$ orbitals.

Imagine a clock that keeps time to within 1 second over a million years. The National Institute of Standards and Technology in Boulder, Colorado, has an atomic clock that does just that—a little better than your average alarm, grandfather, or cuckoo clock! This atomic clock serves as the international standard for time and frequency. How does it work?

Within the glistening case are several layers of magnetic shielding. In the heart of the clock is a small oven that heats cesium metal to release cesium atoms, which are collected into a narrow beam (1 mm wide). The beam of atoms passes down a long evacuated tube while being excited by a laser until all the cesium atoms are in the same electron state.

The atoms then enter another chamber filled with reflecting microwaves. The frequency of the microwaves (9,192,631,770 cycles per second) is exactly the same frequency required to excite a cesium atom from its ground state to the next higher energy level. These excited cesium atoms then release electromagnetic radiation in a process known as fluorescence. Electronic circuits maintain the microwave frequency at precisely the right level to keep the cesium atoms moving from one level to the next. One second is equal to 9,192,631,770 of these vibrations. The clock is set to this frequency and can keep accurate time for over a million years.

This clock automatically updates itself by comparing time with an atomic clock by radio signal.

The pattern continues with the fourth principal energy level. It has 4s, 4p, 4d, and 4f orbitals. There are one 4s, three 4p, five 4d, and seven 4f orbitals. The shapes of the s, p, and d orbitals are the same as those for lower levels, only larger. We will not consider the shapes of the f orbitals. Remember that for all s, p, d, and f orbitals, the maximum number of electrons per orbital is two. We summarize each principal energy level:

$n = 1$	1s			
$n = 2$	2s	2p 2p 2p		
$n = 3$	3s	3p 3p 3p	3d 3d 3d 3d 3d	
$n = 4$	4s	4p 4p 4p	4d 4d 4d 4d 4d	4f 4f 4f 4f 4f 4f 4f

The hydrogen atom consists of a nucleus (containing one proton) and one electron occupying a region outside of the nucleus. In its ground state, the electron occupies a 1s orbital, but by absorbing energy the electron can become *excited* and move to a higher energy level.

The hydrogen atom can be represented as shown in Figure 10.12. The diameter of the nucleus is about 10^{-13} cm, and the diameter of the electron orbital is about 10^{-8} cm. The diameter of the electron cloud of a hydrogen atom is about 100,000 times greater than the diameter of the nucleus.

Figure 10.12
The modern concept of a hydrogen atom consists of a proton and an electron in an s orbital. The shaded area represents a region where the electron may be found with 90% probability.

10.5 Atomic Structures of the First 18 Elements

We have seen that hydrogen has one electron that can occupy a variety of orbitals in different principal energy levels. Now let's consider the structure of atoms with more than one electron. Because all atoms contain orbitals similar to those found in hydrogen, we can describe the structures of atoms beyond hydrogen by systematically placing electrons in these hydrogenlike orbitals. We use the following guidelines:

1. No more than two electrons can occupy one orbital.
2. Electrons occupy the lowest energy orbitals available. They enter a higher energy orbital only when the lower orbitals are filled. For the atoms beyond hydrogen, orbital energies vary as $s < p < d < f$ for a given value of n.
3. Each orbital in a sublevel is occupied by a single electron before a second electron enters. For example, all three p orbitals must contain one electron before a second electron enters a p orbital.

We can use several methods to represent the atomic structures of atoms, depending on what we are trying to illustrate. When we want to show both the nuclear makeup and the electron structure of each principal energy level (without orbital detail), we can use a diagram such as Figure 10.13.

Figure 10.13
Atomic structure diagrams of fluorine, sodium, and magnesium atoms. The number of protons and neutrons is shown in the nucleus. The number of electrons is shown in each principal energy level outside the nucleus.

Often we are interested in showing the arrangement of the electrons in an atom in their orbitals. There are two ways to do this. The first method is called the **electron configuration**. In this method, we list each type of orbital, showing the number of electrons in it as an exponent. An electron configuration is read as follows:

electron configuration

We can also represent this configuration with an **orbital diagram** in which boxes represent the orbitals (containing small arrows indicating the electrons). When the orbital contains one electron, an arrow, pointing upward ($\uparrow$), is placed in the box. A second arrow, pointing downward ($\downarrow$), indicates the second electron in that orbital.

orbital diagram

Let's consider each of the first 18 elements on the periodic table in turn. The order of filling for the orbitals in these elements is $1s$, $2s$, $2p$, $3s$, $3p$, and $4s$. Hydrogen, the first element, has only one electron. The electron will be in the $1s$ orbital because this is the most favorable position (where it will have the greatest attraction for the nucleus). Both representations are shown here:

H $\boxed{\uparrow}$ $1s^1$

     Orbital   Electron
     diagram   configuration

Helium, with two electrons, can be shown as

He $\boxed{\uparrow\downarrow}$ $1s^2$

     Orbital   Electron
     diagram   configuration

The first energy level, which can hold a maximum of two electrons, is now full. An atom with three electrons will have its third electron in the second energy level. Thus, in lithium (atomic number 3), the first two electrons are in the $1s$ orbital, and the third electron is in the $2s$ orbital of the second energy level.

Lithium has the following structure:

Li $\boxed{\uparrow\downarrow}$ $\boxed{\uparrow}$ $1s^2 2s^1$
      1s   2s

All four electrons of beryllium are *s* electrons:

Be $\boxed{\uparrow\downarrow}$ $\boxed{\uparrow\downarrow}$ $1s^2 2s^2$
      1s   2s

The next six elements illustrate the filling of the *p* orbitals. Boron has the first *p* electron. Because *p* orbitals all have the same energy, it doesn't matter which of these orbitals fills first:

B $\boxed{\uparrow\downarrow}$ $\boxed{\uparrow\downarrow}$ $\boxed{\uparrow}\boxed{}\boxed{}$ $1s^2 2s^2 2p^1$
    1s   2s   2p

Carbon is the sixth element. It has two electrons in the 1*s* orbital, two electrons in the 2*s* orbital, and two electrons to place in the 2*p* orbitals. Because it is more difficult for the *p* electrons to pair up than to occupy a second *p* orbital, the second *p* electron is located in a different *p* orbital. We could show this by writing $2p_x^1 2p_y^1$, but we usually write it as $2p^2$; it is *understood* that the electrons are in different *p* orbitals. The spins on these electrons are alike, for reasons we will not explain here.

C $\boxed{\uparrow\downarrow}$ $\boxed{\uparrow\downarrow}$ $\boxed{\uparrow}\boxed{\uparrow}\boxed{}$ $1s^2 2s^2 2p^2$
    1s   2s   2p

Nitrogen has seven electrons. They occupy the 1*s*, 2*s*, and 2*p* orbitals. The third *p* electron in nitrogen is still unpaired and is found in the $2p_z$ orbital:

N $\boxed{\uparrow\downarrow}$ $\boxed{\uparrow\downarrow}$ $\boxed{\uparrow}\boxed{\uparrow}\boxed{\uparrow}$ $1s^2 2s^2 2p^3$
    1s   2s   2p

Oxygen is the eighth element. It has two electrons in both the 1*s* and 2*s* orbitals and four electrons in the 2*p* orbitals. One of the 2*p* orbitals is now occupied by a second electron, which has a spin opposite the electron already in that orbital:

O $\boxed{\uparrow\downarrow}$ $\boxed{\uparrow\downarrow}$ $\boxed{\uparrow\downarrow}\boxed{\uparrow}\boxed{\uparrow}$ $1s^2 2s^2 2p^4$
    1s   2s   2p

The next two elements are fluorine with 9 electrons and neon with 10 electrons:

F $\boxed{\uparrow\downarrow}$ $\boxed{\uparrow\downarrow}$ $\boxed{\uparrow\downarrow}\boxed{\uparrow\downarrow}\boxed{\uparrow}$ $1s^2 2s^2 2p^5$
    1s   2s   2p

Ne $\boxed{\uparrow\downarrow}$ $\boxed{\uparrow\downarrow}$ $\boxed{\uparrow\downarrow}\boxed{\uparrow\downarrow}\boxed{\uparrow\downarrow}$ $1s^2 2s^2 2p^6$
    1s   2s   2p

With neon, the first and second energy levels are filled as shown in Table 10.1. The second energy level can hold a maximum of eight electrons, $2s^2 2p^6$.

Sodium, element 11, has two electrons in the first energy level and eight electrons in the second energy level, with the remaining electron occupying the 3*s* orbital in the third energy level:

Na $\boxed{\uparrow\downarrow}$ $\boxed{\uparrow\downarrow}$ $\boxed{\uparrow\downarrow}\boxed{\uparrow\downarrow}\boxed{\uparrow\downarrow}$ $\boxed{\uparrow}$ $1s^2 2s^2 2p^6 3s^1$
    1s   2s   2p   3s

Table 10.1 Orbital Filling for the First Ten Elements*

Number	Element	Orbitals			Electron configuration
		$1s$	$2s$	$2p$	
1	H	↑			$1s^1$
2	He	↑↓			$1s^2$
3	Li	↑↓	↑		$1s^2 2s^1$
4	Be	↑↓	↑↓		$1s^2 2s^2$
5	B	↑↓	↑↓	↑	$1s^2 2s^2 2p^1$
6	C	↑↓	↑↓	↑ ↑	$1s^2 2s^2 2p^2$
7	N	↑↓	↑↓	↑ ↑ ↑	$1s^2 2s^2 2p^3$
8	O	↑↓	↑↓	↑↓ ↑ ↑	$1s^2 2s^2 2p^4$
9	F	↑↓	↑↓	↑↓ ↑↓ ↑	$1s^2 2s^2 2p^5$
10	Ne	↑↓	↑↓	↑↓ ↑↓ ↑↓	$1s^2 2s^2 2p^6$

*Boxes represent the orbitals grouped by sublevel. Electrons are shown by arrows.

Magnesium (12), aluminum (13), silicon (14), phosphorus (15), sulfur (16), chlorine (17), and argon (18) follow in order. Table 10.2 summarizes the filling of the orbitals for elements 11–18.

The electrons in the outermost (highest) energy level of an atom are called the **valence electrons**. For example, oxygen, which has the electron configuration of $1s^2 2s^2 2p^4$, has electrons in the first and second energy levels. Therefore the second principal energy level is the valence level for oxygen. The $2s$ and $2p$ electrons are the valence electrons. In the case of magnesium ($1s^2 2s^2 2p^6 3s^2$), the valence electrons are in the $3s$ orbital, since these are outermost electrons. Valence electrons are involved in bonding atoms together to form compounds and are of particular interest to chemists, as we will see in Chapter 11.

valence electrons

Table 10.2 Orbital Diagrams and Electron Configurations for Elements 11–18

Number	Element	Orbitals					Electron configuration
		$1s$	$2s$	$2p$	$3s$	$3p$	
11	Na	↑↓	↑↓	↑↓ ↑↓ ↑↓	↑		$1s^2 2s^2 2p^6 3s^1$
12	Mg	↑↓	↑↓	↑↓ ↑↓ ↑↓	↑↓		$1s^2 2s^2 2p^6 3s^2$
13	Al	↑↓	↑↓	↑↓ ↑↓ ↑↓	↑↓	↑	$1s^2 2s^2 2p^6 3s^2 3p^1$
14	Si	↑↓	↑↓	↑↓ ↑↓ ↑↓	↑↓	↑ ↑	$1s^2 2s^2 2p^6 3s^2 3p^2$
15	P	↑↓	↑↓	↑↓ ↑↓ ↑↓	↑↓	↑ ↑ ↑	$1s^2 2s^2 2p^6 3s^2 3p^3$
16	S	↑↓	↑↓	↑↓ ↑↓ ↑↓	↑↓	↑↓ ↑ ↑	$1s^2 2s^2 2p^6 3s^2 3p^4$
17	Cl	↑↓	↑↓	↑↓ ↑↓ ↑↓	↑↓	↑↓ ↑↓ ↑	$1s^2 2s^2 2p^6 3s^2 3p^5$
18	Ar	↑↓	↑↓	↑↓ ↑↓ ↑↓	↑↓	↑↓ ↑↓ ↑↓	$1s^2 2s^2 2p^6 3s^2 3p^6$

Practice 10.1 _____

Give the electron configuration for the valence electrons in these elements.
(a) B   (b) N   (c) Na   (d) Cl

10.6 Electron Structures and the Periodic Table

We have seen how the electrons are assigned for the atoms of elements 1–18. How do the electron structures of these atoms relate to their position on the periodic table? To answer this question, we need to look at the periodic table more closely.

The periodic table represents the efforts of chemists to organize the elements logically. Chemists of the early 19th century had sufficient knowledge of the properties of elements to recognize similarities among groups of elements. In 1869, Dimitri Mendeleev (1834–1907) of Russia and Lothar Meyer (1830–1895) of Germany independently published periodic arrangements of the elements based on increasing atomic masses. Mendeleev's arrangement is the precursor to the modern periodic table and his name is associated with it. The modern periodic table is shown in Figure 10.14 and on the inside front cover of the book.

period   Each horizontal row in the periodic table is called a **period**, as shown in Figure 10.14. There are seven periods of elements. The number of each period corresponds to the outermost energy level that contains electrons for elements in that period. Those in Period 1 contain electrons only in energy level 1 while those in Period 2 contain electrons in levels 1 and 2. In Period 3, electrons are found in levels 1, 2, and 3, and so on.

Figure 10.14
The periodic table of the elements.

Elements that behave in a similar manner are found in **groups** or **families**. These form the vertical columns on the periodic table. Several systems exist for numbering the groups. In one system, the columns are numbered from left to right using the numbers 1–18. However, we use a system that numbers the columns with numbers and the letters A and B, as shown in Figure 10.14. The A groups are known as the **representative elements**. The B groups and Group 8 are called the **transition elements**. In this book we will focus on the representative elements. The groups (columns) of the periodic table often have family names. For example, the group on the far right side of the periodic table (He, Ne, Ar, Kr, Xe, and Rn) is called the *noble gases*. Group 1A is called the *alkali metals*, Group 2A the *alkaline earth metals*, and Group 7A the *halogens*.

How is the structure of the periodic table related to the atomic structures of the elements? We've just seen that the periods of the periodic table are associated with the energy level of the outermost electrons of the atoms in that period. Look at the valence electron configurations of the elements we have just examined (Figure 10.15). Do you see a pattern? The valence electron configuration for the elements in each column is the same. The chemical behavior and properties of elements in a particular family must therefore be associated with the electron configuration of the elements. The number for the principal energy level is different. This is expected since each new period is associated with a different energy level for the valence electrons.

The electron configurations for elements beyond these first 18 become long and tedious to write. We often abbreviate the electron configuration using the following notation:

Na $[Ne]3s^1$

Look carefully at Figure 10.15 and you will see that the p orbitals are full at the noble gases. By placing the symbol for the noble gas in square brackets, we can abbreviate the complete electron configuration and focus our attention on the valence electrons (the electrons we will be interested in when we discuss bonding in Chapter 11). To write the abbreviated electron configuration for any element, go back to the previous noble gas and place its symbol in square brackets. Then list the valence electrons. Here are some examples:

B $1s^2 2s^2 2p^1$ $[He]2s^2 2p^1$

Cl $1s^2 2s^2 2p^6 3s^2 3p^5$ $[Ne]3s^2 3p^5$

Na $1s^2 2s^2 2p^6 3s^1$ $[Ne]3s^1$

groups, families

representative elements
transition elements

Both numbering systems
are shown on the inside
front cover.

							Noble gases
1A							
1 **H** $1s^1$							2 **He** $1s^2$
	2A	3A	4A	5A	6A	7A	
3 **Li** $2s^1$	4 **Be** $2s^2$	5 **B** $2s^2 2p^1$	6 **C** $2s^2 2p^2$	7 **N** $2s^2 2p^3$	8 **O** $2s^2 2p^4$	9 **F** $2s^2 2p^5$	10 **Ne** $2s^2 2p^6$
11 **Na** $3s^1$	12 **Mg** $3s^2$	13 **Al** $3s^2 3p^1$	14 **Si** $3s^2 3p^2$	15 **P** $3s^2 3p^3$	16 **S** $3s^2 3p^4$	17 **Cl** $3s^2 3p^5$	18 **Ar** $3s^2 3p^6$

Figure 10.15
Valence electron configurations for the first 18 elements.

The sequence for filling the orbitals is exactly as we would expect up through the 3p orbitals. The third energy level might be expected to fill with 3d electrons before electrons enter the 4s orbital, but this is not the case. The behavior and properties of the next two elements, potassium (19) and calcium (20), are very similar to the elements in Groups 1A and 2A respectively. They clearly belong in these groups. The other elements in Group 1A and Group 2A have electron configurations that indicate valence electrons in the s orbitals. For example, since the electron configuration is connected to the element's properties, we should place the last electrons for potassium and calcium in the 4s orbital. Their electron configurations are

| K | $1s^2 2s^2 2p^6 3s^2 3p^6 4s^1$ | or | $[Ar]4s^1$ |
| Ca | $1s^2 2s^2 2p^6 3s^2 3p^6 4s^2$ | or | $[Ar]4s^2$ |

Practice 10.2 _____

Write the abbreviated electron configuration for the following elements:
(a) Br   (b) Sr   (c) Ba   (d) Te

Elements 21–30 belong to the elements known as *transition elements*. Electrons are placed in the 3d orbitals for each of these elements. When the 3d orbitals are full, the electrons fill the 4p orbitals to complete the fourth period. Let's consider the overall relationship between orbital filling and the periodic table. Figure 10.16 illustrates the type of orbital filling and its location on the periodic table. The tall columns on the table (labeled 1A–7A, and noble gases) are often called the *representative elements*. Valence electrons in these elements

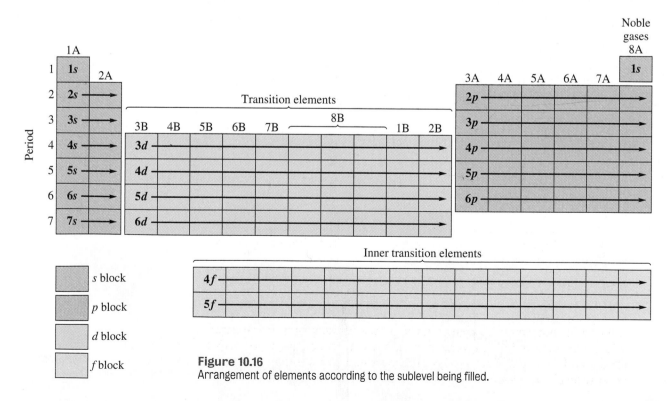

Figure 10.16
Arrangement of elements according to the sublevel being filled.

occupy s and p orbitals. The period number corresponds to the energy level of the valence electrons. The elements in the center of the periodic table (shown in ▣) are the transition elements where the d orbitals are being filled. Notice that the number for the d orbitals is one less than the period number. The two rows shown at the bottom of the table in Figure 10.16 are called the *inner transition elements* or the *lanthanide* and *actinide series*. The last electrons in these elements are placed in the f orbitals. The number for the f orbitals is always two less than that of the s and p orbitals. A periodic table is almost always available to you, so if you understand the relationship between the orbitals and the periodic table, you can write the electron configuration for any element. There are several minor variations to these rules, but we won't concern ourselves with them in this course.

Use the periodic table to write the electron configuration for phosphorus and tin.

Example 10.1

Phosphorus is element 15 and is located in Period 3, Group 5A. The electron configuration must have a full first and second energy level:

SOLUTION

   P $1s^2 2s^2 2p^6 3s^2 3p^3$ or $[Ne]3s^2 3p^3$

You can determine the electron configuration by looking across the period and counting the element blocks.

Tin is element 50 in Period 5, Group 4A, two places after the transition metals. It must have two electrons in the $5p$ series. Its electron configuration is

   Sn $1s^2 2s^2 2p^6 3s^2 3p^6 4s^2 3d^{10} 4p^6 5s^2 4d^{10} 5p^2$ or $[Kr]5s^2 4d^{10} 5p^2$

Notice that the d series of electrons always has a principal energy level number one less than its period.

Practice 10.3 _____

Use the periodic table to write the complete electron configuration for (a) O, (b) Ca, and (c) Ti.

The early chemists classified the elements based only on their observed properties, but modern atomic theory gives us an explanation for why the properties of elements vary periodically. For example, as we "build" atoms by filling orbitals with electrons, the same orbitals occur on each energy level. This means that the same electron configuration reappears regularly for each level. Groups of elements show similar chemical properties because of the similarity of these outermost electron configurations.

In Figure 10.17, only the electron configuration of the outermost electrons is given. This periodic table illustrates these important points:

1. The number of the period corresponds with the highest energy level occupied by electrons in that period.
2. The group numbers for the representative elements are equal to the total number of outermost electrons in the atoms of the group. For example, elements in Group 7A always have the electron configuration $ns^2 np^5$. The d and f electrons are always in a lower energy level than the highest energy level and so are not considered as outermost electrons.

Group number

Period	1A	2A												3A	4A	5A	6A	7A	Noble gases 8A
1	1 **H** $1s^1$																		2 **He** $1s^2$
2	3 **Li** $2s^1$	4 **Be** $2s^2$												5 **B** $2s^22p^1$	6 **C** $2s^22p^2$	7 **N** $2s^22p^3$	8 **O** $2s^22p^4$	9 **F** $2s^22p^5$	10 **Ne** $2s^22p^6$
3	11 **Na** $3s^1$	12 **Mg** $3s^2$	3B	4B	5B	6B	7B		8B		1B	2B		13 **Al** $3s^23p^1$	14 **Si** $3s^23p^2$	15 **P** $3s^23p^3$	16 **S** $3s^23p^4$	17 **Cl** $3s^23p^5$	18 **Ar** $3s^23p^6$
4	19 **K** $4s^1$	20 **Ca** $4s^2$	21 **Sc** $4s^23d^1$	22 **Ti** $4s^23d^2$	23 **V** $4s^23d^3$	24 **Cr** $4s^13d^5$	25 **Mn** $4s^23d^5$	26 **Fe** $4s^23d^6$	27 **Co** $4s^23d^7$	28 **Ni** $4s^23d^8$	29 **Cu** $4s^13d^{10}$	30 **Zn** $4s^23d^{10}$	31 **Ga** $4s^24p^1$	32 **Ge** $4s^24p^2$	33 **As** $4s^24p^3$	34 **Se** $4s^24p^4$	35 **Br** $4s^24p^5$	36 **Kr** $4s^24p^6$
5	37 **Rb** $5s^1$	38 **Sr** $5s^2$	39 **Y** $5s^24d^1$	40 **Zr** $5s^24d^2$	41 **Nb** $5s^14d^4$	42 **Mo** $5s^14d^5$	43 **Tc** $5s^14d^6$	44 **Ru** $5s^14d^7$	45 **Rh** $5s^14d^8$	46 **Pd** $5s^04d^{10}$	47 **Ag** $5s^14d^{10}$	48 **Cd** $5s^24d^{10}$	49 **In** $5s^25p^1$	50 **Sn** $5s^25p^2$	51 **Sb** $5s^25p^3$	52 **Te** $5s^25p^4$	53 **I** $5s^25p^5$	54 **Xe** $5s^25p^6$
6	55 **Cs** $6s^1$	56 **Ba** $6s^2$	57 **La** $6s^25d^1$	72 **Hf** $6s^25d^2$	73 **Ta** $6s^25d^3$	74 **W** $6s^25d^4$	75 **Re** $6s^25d^5$	76 **Os** $6s^25d^6$	77 **Ir** $6s^25d^7$	78 **Pt** $6s^15d^9$	79 **Au** $6s^15d^{10}$	80 **Hg** $6s^25d^{10}$	81 **Tl** $6s^26p^1$	82 **Pb** $6s^26p^2$	83 **Bi** $6s^26p^3$	84 **Po** $6s^26p^4$	85 **At** $6s^26p^5$	86 **Rn** $6s^26p^6$
7	87 **Fr** $7s^1$	88 **Ra** $7s^2$	89 **Ac** $7s^26d^1$	104 **Rf** $7s^26d^2$	105 **Db** $7s^26d^3$	106 **Sg** $7s^26d^4$	107 **Bh** $7s^26d^5$	108 **Hs** $7s^26d^6$	109 **Mt** $7s^26d^7$	110 **Ds** $7s^16d^9$	111 **Rg** $7s^16d^{10}$							

Figure 10.17
Outermost electron configurations.

3. The elements of a family have the same outermost electron configuration except that the electrons are in different energy levels.
4. The elements within each of the s, p, d, f blocks are filling the s, p, d, f orbitals, as shown in Figure 10.16.
5. Within the transition elements, some discrepancies in the order of filling occur. (Explanation of these discrepancies and similar ones in the inner transition elements are beyond the scope of this book.)

Example 10.2   Write the complete electron configuration of a zinc atom and a rubidium atom.

SOLUTION   The atomic number of zinc is 30; it therefore has 30 protons and 30 electrons in a neutral atom. Using Figure 10.14, we see that the electron configuration of a zinc atom is $1s^22s^22p^63s^23p^64s^23d^{10}$. Check by adding the superscripts, which should equal 30.

The atomic number of rubidium is 37; therefore it has 37 protons and 37 electrons in a neutral atom. With a little practice using a periodic table, you can write the electron configuration directly. The electron configuration of a rubidium atom is $1s^22s^22p^63s^23p^64s^23d^{10}4p^65s^1$. Check by adding the superscripts, which should equal 37.

Practice 10.4

Write the complete electron configuration for a gallium atom and a lead atom.

Chapter 10 Review

10.1 A Brief History

- Atomic theory has changed over the past 200 years:
  - Early Greek ideas
  - Dalton's model
  - Thomson's model
  - Rutherford's model
  - Modern atomic theory

10.2 Electromagnetic Radiation

KEY TERMS

Wavelength
Frequency
Speed
Photons

- Basic wave characteristics include:
  - Wavelength (λ)
  - Frequency (ν)
  - Speed (v)
- Light is a form of electromagnetic radiation.
- Light can also be considered to be composed of energy packets called photons, which behave like particles.

10.3 The Bohr Atom

KEY TERMS

Line spectrum
Quanta
Ground state
Orbital
Bohr atom

- The study of atomic spectra led Bohr to propose that:
  - Electrons are found in quantized energy levels in an atom.
  - Spectral lines result from the radiation of quanta of energy when the electron moves from a higher level to a lower level.
- The chemical properties of an element and its position on the periodic table depend on electrons.
- De Broglie suggested that all objects have wave properties.
- Schrödinger created a mathematical model to describe electrons as waves.
  - Electrons are located in orbitals, or regions of probability, around the nucleus of an atom.

10.4 Energy Levels of Electrons

KEY TERMS

Principal energy levels
Sublevels
Spin
Pauli exclusion principle

- Modern atomic theory predicts that
  - Electrons are found in discrete principal energy levels ($n = 1, 2, 3 \ldots$):
    - Energy levels contain sublevels.

Number of sublevels

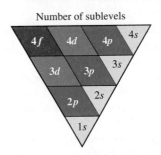

- Two electrons fit into each orbital but must have opposite spin to do so.

10.5 Atomic Structures of the First 18 Elements

KEY TERMS

Electron configuration
Orbital diagram
Valence electrons

- Guidelines for writing electron configurations:
  - Not more than two electrons per orbital
  - Electrons fill lowest-energy levels first:
    - $s < p < d < f$ for a given value of n
  - Orbitals on a given sublevel are each filled with a single electron before pairing of electrons begins to occur.
- For the representative elements, only electrons in the outermost energy level (valence electrons) are involved in bonding.

10.6 Electron Structures and the Periodic Table

KEY TERMS

Period
Group (family)
Representative element
Transition element

- Elements in horizontal rows on the periodic table contain elements whose valence electrons (s and p) are generally on the same energy level as the number of the row.
- Elements that are chemically similar are arranged in columns (groups) on the periodic table.
- The valence electron configuration of elements in a group or family are the same but they are located in different principal energy levels.

Review Questions

All questions with blue numbers have answers in the appendix of the text.

1. What is an orbital?
2. Under what conditions can a second electron enter an orbital already containing one electron?
3. What is a valence shell?
4. What are valence electrons and why are they important?
5. What is meant when we say the electron structure of an atom is in its ground state?
6. How do $1s$ and $2s$ orbitals differ? How are they alike?
7. What letters are used to designate the types of orbitals?
8. List the following orbitals in order of increasing energy: $2s, 2p, 4s, 1s, 3d, 3p, 4p, 3s$.
9. How many s electrons, p electrons, and d electrons are possible in any energy level?
10. What is the major difference between an orbital and a Bohr orbit?
11. Explain how and why Bohr's model of the atom was modified to include the cloud model of the atom.

12. Sketch the s, p_x, p_y, and p_z orbitals.
13. In the designation $3d^7$, give the significance of 3, d, and 7.
14. Describe the difference between transition and representative elements.
15. From the standpoint of electron structure, what do the elements in the s block have in common?
16. Write symbols for elements with atomic numbers 8, 16, 34, 52, and 84. What do these elements have in common?
17. Write the symbols of the family of elements that have seven electrons in their outermost energy level.
18. What is the greatest number of elements to be found in any period? Which periods have this number?
19. From the standpoint of energy level, how does the placement of the last electron in the Group A elements differ from that of the Group B elements?
20. Find the places in the periodic table where elements are not in proper sequence according to atomic mass. (See inside of front cover for periodic table.)

Paired Exercises

All exercises with blue numbers have answers in the appendix of the text.

1. How many protons are in the nucleus of an atom of these elements?
   (a) H   (c) Sc
   (b) B   (d) U

2. How many protons are in the nucleus of an atom of these elements?
   (a) F   (c) Br
   (b) Ag   (d) Sb

3. Give the electron configuration for these elements:
   (a) B   (d) Br
   (b) Ti   (e) Sr
   (c) Zn

4. Give the electron configuration for these elements:
   (a) chlorine   (d) iron
   (b) silver   (e) iodine
   (c) lithium

5. Explain how the spectral lines of hydrogen occur.

6. Explain how Bohr used the data from the hydrogen spectrum to support his model of the atom.

7. How many orbitals exist in the third principal energy level? What are they?

8. How many electrons can be present in the fourth principal energy level?

9. Write orbital diagrams for these elements:
   (a) N   (d) Zr
   (b) Cl   (e) I
   (c) Zn

10. Write orbital diagrams for these elements:
    (a) Si   (d) V
    (b) S   (e) P
    (c) Ar

11. For each of the orbital diagrams given, write out the corresponding electron configurations.

   (a) O $\uparrow\downarrow$ | $\uparrow\downarrow$ | $\uparrow\downarrow$ | $\uparrow$ | $\uparrow$

   (b) Ca $\uparrow\downarrow$ | $\uparrow\downarrow$ | $\uparrow\downarrow$ $\uparrow\downarrow$ $\uparrow\downarrow$ | $\uparrow\downarrow$ | $\uparrow\downarrow$ $\uparrow\downarrow$ $\uparrow\downarrow$ | $\uparrow\downarrow$

   (c) Ar $\uparrow\downarrow$ | $\uparrow\downarrow$ | $\uparrow\downarrow$ $\uparrow\downarrow$ $\uparrow\downarrow$ | $\uparrow\downarrow$ | $\uparrow\downarrow$ $\uparrow\downarrow$ $\uparrow\downarrow$

   (d) Br $\uparrow\downarrow$ | $\uparrow\downarrow$ | $\uparrow\downarrow$ $\uparrow\downarrow$ $\uparrow\downarrow$ | $\uparrow\downarrow$ | $\uparrow\downarrow$ $\uparrow\downarrow$ $\uparrow\downarrow$ | $\uparrow\downarrow$ | $\uparrow\downarrow$ $\uparrow\downarrow$ $\uparrow\downarrow$ $\uparrow\downarrow$ $\uparrow\downarrow$ | $\uparrow\downarrow$ $\uparrow\downarrow$ $\uparrow$

   (e) Fe $\uparrow\downarrow$ | $\uparrow\downarrow$ | $\uparrow\downarrow$ $\uparrow\downarrow$ $\uparrow\downarrow$ | $\uparrow\downarrow$ | $\uparrow\downarrow$ $\uparrow\downarrow$ $\uparrow\downarrow$ | $\uparrow\downarrow$ | $\uparrow\downarrow$ $\uparrow$ $\uparrow$ $\uparrow$ $\uparrow$

12. For each of the orbital diagrams given, write out the corresponding electron configurations.

(a) Li   [↑↓] [↑]

(b) P   [↑↓] [↑↓] [↑↓][↑↓][↑↓] [↑↓] [↑][↑][↑]

(c) Zn   [↑↓] [↑↓] [↑↓][↑↓][↑↓] [↑↓] [↑↓][↑↓][↑↓] [↑↓] [↑↓][↑↓][↑↓][↑↓][↑↓]

(d) Na   [↑↓] [↑↓] [↑↓][↑↓][↑↓] [↑]

(e) K   [↑↓] [↑↓] [↑↓][↑↓][↑↓] [↑↓] [↑↓][↑↓][↑↓] [↑]

13. Which elements have these electron configurations?
(a) $1s^2 2s^2 2p^6 3s^2$
(b) $1s^2 2s^2 2p^6 3s^2 3p^1$
(c) $1s^2 2s^2 2p^6 3s^2 3p^6 4s^2 3d^8$
(d) $1s^2 2s^2 2p^6 3s^2 3p^6 4s^2 3d^5$

14. Which elements have these electron configurations?
(a) $[Ar]4s^2 3d^1$
(b) $[Ar]4s^2 3d^{10}$
(c) $[Kr]5s^2 4d^{10} 5p^2$
(d) $[Xe]6s^1$

15. Show the electron configurations for elements with these atomic numbers:
(a) 8   (d) 23
(b) 11   (e) 28
(c) 17   (f) 34

16. Show the electron configurations for elements with these atomic numbers:
(a) 9   (d) 39
(b) 26   (e) 52
(c) 31   (f) 10

17. Identify the element and draw orbital diagrams for elements with these atomic numbers:
(a) 22   (d) 35
(b) 18   (e) 25
(c) 33

18. Identify the element and draw orbital diagrams for elements with these atomic numbers:
(a) 15   (d) 34
(b) 30   (e) 19
(c) 20

19. For each of the electron configurations given, write the corresponding orbital diagrams.
(a) F $1s^2 2s^2 2p^5$
(b) S $1s^2 2s^2 2p^6 3s^2 3p^4$
(c) Co $1s^2 2s^2 2p^6 3s^2 3p^6 4s^2 3d^7$
(d) Kr $1s^2 2s^2 2p^6 3s^2 3p^6 4s^2 3d^{10} 4p^6$
(e) Ru $1s^2 2s^2 2p^6 3s^2 3p^6 4s^2 3d^{10} 4p^6 5s^2 4d^6$

20. For each of the electron configurations given, write the corresponding orbital diagrams.
(a) Cl $1s^2 2s^2 2p^6 3s^2 3p^5$
(b) Mg $1s^2 2s^2 2p^6 3s^2$
(c) Ni $1s^2 2s^2 2p^6 3s^2 3p^6 4s^2 3d^8$
(d) Cu $1s^2 2s^2 2p^6 3s^2 3p^6 4s^2 3d^9$
(e) Ba $1s^2 2s^2 2p^6 3s^2 3p^6 4s^2 3d^{10} 4p^6 5s^2 4d^{10} 5p^6 6s^2$

21. Identify these elements from their atomic structure diagrams:

(a) (16p / 16n) $2e^-$ $8e^-$ $6e^-$

(b) (28p / 32n) $2e^-$ $8e^-$ $16e^-$ $2e^-$

22. Diagram the atomic structures (as you see in Exercise 21) for these elements:
(a) $^{27}_{13}Al$ (b) $^{48}_{22}Ti$

23. Why is the 11th electron of the sodium atom located in the third energy level rather than in the second energy level?

24. Why is the last electron in potassium located in the fourth energy level rather than in the third energy level?

25. What electron structure do the noble gases have in common?

26. What is unique about the noble gases, from an electron point of view?

27. How are elements in a period related to one another?

28. How are elements in a group related to one another?

29. How many valence electrons do each of the following elements have?
(a) C   (d) I
(b) S   (e) B
(c) K

30. How many valence electrons do each of the following elements have?
(a) N   (d) Ba
(b) P   (e) Al
(c) O

31. What do the electron structures of the alkali metals have in common?

32. Why would you expect the elements zinc, cadmium, and mercury to be in the same chemical family?

33. Pick the electron structures that represent elements in the same chemical family:
(a) $1s^2 2s^1$
(b) $1s^2 2s^2 2p^4$
(c) $1s^2 2s^2 2p^2$
(d) $1s^2 2s^2 2p^6 3s^2 3p^4$
(e) $1s^2 2s^2 2p^6 3s^2 3p^6$
(f) $1s^2 2s^2 2p^6 3s^2 3p^6 4s^2$
(g) $1s^2 2s^2 2p^6 3s^2 3p^6 4s^1$
(h) $1s^2 2s^2 2p^6 3s^2 3p^6 4s^2 3d^1$

34. Pick the electron structures that represent elements in the same chemical family:
(a) $[He]2s^2 2p^6$
(b) $[Ne]3s^1$
(c) $[Ne]3s^2$
(d) $[Ne]3s^2 3p^3$
(e) $[Ar]4s^2 3d^{10}$
(f) $[Ar]4s^2 3d^{10} 4p^6$
(g) $[Ar]4s^2 3d^5$
(h) $[Kr]5s^2 4d^{10}$

35. In the periodic table, calcium, element 20, is surrounded by elements 12, 19, 21, and 38. Which of these have physical and chemical properties most resembling calcium?

36. In the periodic table, phosphorus, element 15, is surrounded by elements 14, 7, 16, and 33. Which of these have physical and chemical properties most resembling phosphorus?

37. Classify the following elements as metals, nonmetals, or metalloids (review Chapter 3 if you need help):
(a) potassium
(b) plutonium
(c) sulfur
(d) antimony

38. Classify the following elements as metals, nonmetals, or metalloids (review Chapter 3 if you need help):
(a) iodine
(b) tungsten
(c) molybdenum
(d) germanium

39. In which period and group does an electron first appear in an f orbital?

40. In which period and group does an electron first appear in a d orbital?

41. How many electrons occur in the valence level of Group 7A and 7B elements? Why are they different?

42. How many electrons occur in the valence level of Group 3A and 3B elements? Why are they different?

Additional Exercises

All exercises with blue numbers have answers in the appendix of the text.

43. Using only the periodic table, explain how the valence energy level and the number of valence electrons could be determined.

44. Using the periodic table only, identify the valence energy level and the number of valence electrons for each of the following elements:
(a) Li   (d) S
(b) Cl   (e) Be
(c) Si

45. Which of the following would have the same number of valence electrons?
(a) Na^+
(b) O
(c) Li
(d) F^-
(e) Ne

46. Name the group in which each of the following elements appear:
(a) $1s^2 2s^2 2p^5$
(b) $1s^2 2s^2 2p^6 3s^2 3p^6 4s^2$
(c) $1s^2 2s^2 2p^6 3s^1$
(d) $1s^2 2s^2 2p^6 3s^2 3p^6 4s^2 3d^{10} 4p^6$
(e) $1s^2$
(f) $1s^2 2s^2 2p^6 3s^2 3p^6 4s^2 3d^{10} 4p^6 5s^1$

47. Write the electron configuration of each of the following neutral atoms. (You may need to refer to Figure 3.2.)
(a) the four most abundant elements in the Earth's crust, seawater, and air
(b) the five most abundant elements in the human body

48. What is the maximum number of electrons that can reside in the following:
(a) a p orbital
(b) a d sublevel
(c) the third principal energy level
(d) an s orbital
(e) an f sublevel

49. Give the names of each of the following elements based on the information given:
  (a) the second element in Period 3
  (b) [Ne]$3s^2 3p^3$
  (c)
       1s   2s   2p   3s   3p

50. Why does the emission spectrum for nitrogen reveal many more spectral lines than that for hydrogen?

51. Suppose we use a foam ball to represent a typical atom. If the radius of the ball is 1.5 cm and the radius of a typical atom is 1.0×10^{-8} cm, how much of an enlargement is this? Use a ratio to express your answer.

52. List the first element on the periodic table that satisfies each of these conditions:
  (a) a completed set of p orbitals
  (b) two $4p$ electrons
  (c) seven valence electrons
  (d) three unpaired electrons

53. Oxygen is a gas. Sulfur is a solid. What is it about their electron structures that causes them to be grouped in the same chemical family?

54. In which groups are the transition elements located?

55. How do the electron structures of the transition elements differ from those of the representative elements?

56. The atomic numbers of the noble gases are 2, 10, 18, 36, 54, and 86. What are the atomic numbers for the elements with six electrons in their outermost electron configuration?

57. What is the family name for
  (a) Group 1A?
  (b) Group 2A?
  (c) Group 7A?

58. What sublevel is being filled in
  (a) Period 3, Group 3A to 7A?
  (b) Period 5, transition elements?
  (c) the lanthanide series?

59. Classify each of the following as a noble gas, a representative element, or a transition metal. Also indicate whether the element is a metal, nonmetal, or metalloid.
  (a) Na   (d) Ra
  (b) N   (e) As
  (c) Mo   (f) Ne

60. If element 36 is a noble gas, in which groups would you expect elements 35 and 37 to occur?

61. Write a paragraph describing the general features of the periodic table.

62. Some scientists have proposed the existence of element 117. If it were to exist,
  (a) what would its electron configuration be?
  (b) how many valence electrons would it have?
  (c) what element would it likely resemble?
  (d) to what family and period would it belong?

63. What is the relationship between two elements if
  (a) one of them has 10 electrons, 10 protons, and 10 neutrons and the other has 10 electrons, 10 protons, and 12 neutrons?
  (b) one of them has 23 electrons, 23 protons, and 27 neutrons and the other has 24 electrons, 24 protons, and 26 neutrons?

64. Is there any pattern for the location of gases on the periodic table? for the location of liquids? for the location of solids?

Challenge Exercises

All exercises with blue numbers have answers in the appendix of the text.

*65. A valence electron in an atom of sulfur is excited by heating a sample. The electron jumps from the s orbital to the p orbital. What is the electron configuration of the excited sulfur atom, and what would the orbital diagram look like?

*66. Element 87 is in Group 1A, Period 7. In how many principal energy levels are electrons located? Describe its outermost energy level.

*67. Use the periodic table to explain why metals tend to lose electrons and nonmetals tend to gain electrons.

*68. Show how the periodic table helps determine the expected electron configuration for any element.

Answers to Practice Exercises

10.1 (a) $2s^2 2p^1$ (c) $3s^1$
       (b) $2s^2 2p^3$ (d) $3s^2 3p^5$

10.2 (a) [Ar]$4s^2 4p^5$ (c) [Xe]$6s^2$
       (b) [Kr]$5s^2$ (d) [Kr]$5s^2 5p^4$

10.3 (a) O $1s^2 2s^2 2p^4$
       (b) Ca $1s^2 2s^2 2p^6 3s^2 3p^6 4s^2$
       (c) Ti $1s^2 2s^2 2p^6 3s^2 3p^6 4s^2 3d^2$

10.4 Ga, $1s^2 2s^2 2p^6 3s^2 3p^6 4s^2 3d^{10} 4p^1$
       Pb, $1s^2 2s^2 2p^6 3s^2 3p^6 4s^2 3d^{10} 4p^6 5s^2 4d^{10} 5p^6 6s^2 4f^{14} 5d^{10} 6p^2$

Chemical Bonds:
The Formation of Compounds from Atoms

This colorfully lighted limestone cave reveals dazzling stalactites and stalagmites, formed from calcium carbonate.

Chapter Outline

11.1 Periodic Trends in Atomic Properties

11.2 Lewis Structures of Atoms

11.3 The Ionic Bond: Transfer of Electrons from One Atom to Another

11.4 Predicting Formulas of Ionic Compounds

11.5 The Covalent Bond: Sharing Electrons

11.6 Electronegativity

11.7 Lewis Structures of Compounds

11.8 Complex Lewis Structures

11.9 Compounds Containing Polyatomic Ions

11.10 Molecular Shape

11.11 The Valence Shell Electron Pair Repulsion (VSEPR) Model

For centuries we've been aware that certain metals cling to a magnet. We've seen balloons sticking to walls. Why? High-speed levitation trains are heralded to be the wave of the future. How do they function? In each case, forces of attraction and repulsion are at work.

Human interactions also suggest that "opposites attract" and "likes repel." Attractions draw us into friendships and significant relationships, whereas repulsive forces may produce debate and antagonism. We form and break apart interpersonal bonds throughout our lives.

In chemistry, we also see this phenomenon. Substances form chemical bonds as a result of electrical attractions. These bonds provide the tremendous diversity of compounds found in nature.

11.1 Periodic Trends in Atomic Properties

Although atomic theory and electron configuration help us understand the arrangement and behavior of the elements, it's important to remember that the design of the periodic table is based on observing properties of the elements. Before we use the concept of atomic structure to explain how and why atoms combine to form compounds, we need to understand the characteristic properties of the elements and the trends that occur in these properties on the periodic table. These trends allow us to use the periodic table to accurately predict properties and reactions of a wide variety of substances.

Metals and Nonmetals

In Section 3.5, we classified elements as metals, nonmetals, or metalloids. The heavy stair-step line beginning at boron and running diagonally down the periodic table separates the elements into metals and nonmetals. Metals are usually lustrous, malleable, and good conductors of heat and electricity. Nonmetals are just the opposite—nonlustrous, brittle, and poor conductors. Metalloids are found bordering the heavy diagonal line and may have properties of both metals and nonmetals.

Most elements are classified as metals (see Figure 11.1). Metals are found on the left side of the stair-step line, while the nonmetals are located toward the upper right of the table. Note that hydrogen does not fit into the division of metals and nonmetals. It displays nonmetallic properties under normal conditions, even though it has only one outermost electron like the alkali metals. Hydrogen is considered to be a unique element.

It is the chemical properties of metals and nonmetals that interest us most. Metals tend to lose electrons and form positive ions, while nonmetals tend to gain electrons and form negative ions. When a metal reacts with a nonmetal, electrons are often transferred from the metal to the nonmetal.

Atomic Radius

The relative radii of the representative elements are shown in Figure 11.2. Notice that the radii of the atoms tend to increase down each group and that they tend to decrease from left to right across a period.

The increase in radius down a group can be understood if we consider the electron structure of the atoms. For each step down a group, an additional

Figure 11.1
The elements are classified as metals, nonmetals, and metalloids.

Figure 11.2
Relative atomic radii for the representative elements. Atomic radius decreases across a period and increases down a group in the periodic table.

energy level is added to the atom. The average distance from the nucleus to the outside edge of the atom must increase as each new energy level is added. The atoms get bigger as electrons are placed in these new higher-energy levels.

Understanding the decrease in atomic radius across a period requires more thought, however. As we move from left to right across a period, electrons within the same block are being added to the same principal energy level. Within a given energy level, we expect the orbitals to have about the same size. We would then expect the atoms to be about the same size across the period. But each time an electron is added, a proton is added to the nucleus as well. The increase in positive charge (in the nucleus) pulls the electrons closer to the nucleus, which results in a gradual decrease in atomic radius across a period.

Ionization Energy

The **ionization energy** of an atom is the energy required to remove an electron from the atom. For example,

$$Na + ionization\ energy \longrightarrow Na^+ + e^-$$

ionization energy

The first ionization energy is the amount of energy required to remove the first electron from an atom, the second is the amount required to remove the second electron from that atom, and so on.

Table 11.1 gives the ionization energies for the removal of one to five electrons from several elements. The table shows that even higher amounts of energy are needed to remove the second, third, fourth, and fifth electrons. This makes sense because removing electrons leaves fewer electrons attracted to the same positive charge in the nucleus. The data in Table 11.1 also show that an extra-large ionization energy (blue) is needed when an electron is removed from a noble gas–like structure, clearly showing the stability of the electron structure of the noble gases.

First ionization energies have been experimentally determined for most elements. Figure 11.3 plots these energies for representative elements in the first four periods. Note these important points:

1. Ionization energy in Group A elements decreases from top to bottom in a group. For example, in Group 1A the ionization energy changes from 520 kJ/mol for Li to 419 kJ/mol for K.
2. Ionization energy gradually increases from left to right across a period. Noble gases have a relatively high value, confirming the nonreactive nature of these elements.

Table 11.1 Ionization Energies for Selected Elements*

Element	Required amounts of energy (kJ/mol)				
	1st e^-	2nd e^-	3rd e^-	4th e^-	5th e^-
H	1,314				
He	2,372	5,247			
Li	520	7,297	11,810		
Be	900	1,757	14,845	21,000	
B	800	2,430	3,659	25,020	32,810
C	1,088	2,352	4,619	6,222	37,800
Ne	2,080	3,962	6,276	9,376	12,190
Na	496	4,565	6,912	9,540	13,355

*Values are expressed in kilojoules per mole, showing energies required to remove 1 to 5 electrons per atom. Blue type indicates the energy needed to remove an electron from a noble gas electron structure.

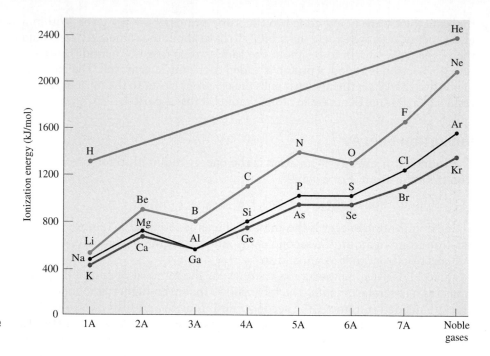

Figure 11.3
Periodic relationship of the first ionization energy for representative elements in the first four periods.

Metals don't behave in exactly the same manner. Some metals give up electrons much more easily than others. In the alkali metal family, cesium gives up its 6s electron much more easily than the metal lithium gives up its 2s electron. This makes sense when we consider that the size of the atoms increases down the group. The distance between the nucleus and the outer electrons increases and the ionization energy decreases. The most chemically active metals are located at the lower left of the periodic table.

Nonmetals have relatively large ionization energies compared to metals. Nonmetals tend to gain electrons and form anions. Since the nonmetals are located at the right side of the periodic table, it is not surprising that ionization energies tend to increase from left to right across a period. The most active nonmetals are found in the *upper* right corner of the periodic table (excluding the noble gases).

11.2 Lewis Structures of Atoms

Metals tend to form cations (positively charged ions) and nonmetals form anions (negatively charged ions) in order to attain a stable valence electron structure. For many elements this stable valence level contains eight electrons (two s and six p), identical to the valence electron configuration of the noble gases. Atoms undergo rearrangements of electron structure to lower their chemical potential energy (or to become more stable). These rearrangements are accomplished by losing, gaining, or sharing electrons with other atoms. For example, a hydrogen atom could accept a second electron and attain an electron structure the same as the noble gas helium. A fluorine atom could gain an electron and attain an electron structure like neon. A sodium atom could lose one electron to attain an electron structure like neon.

1A	2A	3A	4A	5A	6A	7A	Noble Gases
H·							He:
Li·	Be:	:Ḃ	:Ċ·	:Ṅ·	·Ö:	:F̈:	:N̈e:
Na·	Mg:	:Äl	:Si·	:Ṗ·	·S̈:	:C̈l:	:Är:
K·	Ca:						

Figure 11.4
Lewis structures of the first 20 elements. Dots represent electrons in the outermost s and p energy levels only.

The valence electrons in the outermost energy level of an atom are responsible for the electron activity that occurs to form chemical bonds. The **Lewis structure** of an atom is a representation that shows the valence electrons for that atom. American chemist Gilbert N. Lewis (1875–1946) proposed using the symbol for the element and dots for electrons. The number of dots placed around the symbol equals the number of s and p electrons in the outermost energy level of the atom. Paired dots represent paired electrons; unpaired dots represent unpaired electrons. For example, **H·** is the Lewis symbol for a hydrogen atom, $1s^1$; **:B** is the Lewis symbol for a boron atom, with valence electrons $2s^2 2p^1$. In the case of boron, the symbol B represents the boron nucleus and the $1s^2$ electrons; the dots represent only the $2s^2 2p^1$ electrons.

Lewis structure

Paired electrons ——→ :**B**← ——— Symbol of the element

·← ——— Unpaired electron

The Lewis method is used not only because of its simplicity of expression but also because much of the chemistry of the atom is directly associated with the electrons in the outermost energy level. Figure 11.4 shows Lewis structures for the elements hydrogen through calcium.

Write the Lewis structure for a phosphorus atom.

First establish the electron structure for a phosphorus atom, which is $1s^2 2s^2 2p^6 3s^2 3p^3$. Note that there are five electrons in the outermost energy level; they are $3s^2 3p^3$. Write the symbol for phosphorus and place the five electrons as dots around it.

$$:\overset{\textstyle\cdot}{\underset{\textstyle\cdot}{\mathbf{P}}}\cdot$$

The $3s^2$ electrons are paired and are represented by the paired dots. The $3p^3$ electrons, which are unpaired, are represented by the single dots.

Example 11.1
SOLUTION

A quick way to determine the correct number of dots (electrons) for a Lewis structure is to use the Group number. For the A groups on the periodic table, the Group number is the same as the number of electrons in the Lewis structure.

Practice 11.1 _____

Write the Lewis structure for the following elements:
(a) N   (b) Al   (c) Sr   (d) Br

11.3 The Ionic Bond: Transfer of Electrons from One Atom to Another

The chemistry of many elements, especially the representative ones, is to attain an outer electron structure like that of the chemically stable noble gases. With the exception of helium, this stable structure consists of eight electrons in the outermost energy level (see Table 11.2).

Table 11.2 Arrangement of Electrons in the Noble Gases*

Noble gas	Symbol	Electron structure					
		$n = 1$	2	3	4	5	6
Helium	He	$1s^2$					
Neon	Ne	$1s^2$	$2s^2 2p^6$				
Argon	Ar	$1s^2$	$2s^2 2p^6$	$3s^2 3p^6$			
Krypton	Kr	$1s^2$	$2s^2 2p^6$	$3s^2 3p^6 3d^{10}$	$4s^2 4p^6$		
Xenon	Xe	$1s^2$	$2s^2 2p^6$	$3s^2 3p^6 3d^{10}$	$4s^2 4p^6 4d^{10}$	$5s^2 5p^6$	
Radon	Rn	$1s^2$	$2s^2 2p^6$	$3s^2 3p^6 3d^{10}$	$4s^2 4p^6 4d^{10} 4f^{14}$	$5s^2 5p^6 5d^{10}$	$6s^2 6p^6$

*Each gas except helium has eight electrons in its outermost energy level.

Let's look at the electron structures of sodium and chlorine to see how each element can attain a structure of 8 electrons in its outermost energy level. A sodium atom has 11 electrons: 2 in the first energy level, 8 in the second energy level, and 1 in the third energy level. A chlorine atom has 17 electrons: 2 in the first energy level, 8 in the second energy level, and 7 in the third energy level. If a sodium atom transfers or loses its $3s$ electron, its third energy level becomes vacant, and it becomes a sodium ion with an electron configuration identical to that of the noble gas neon. This process requires energy:

Na atom
$(1s^2 2s^2 2p^6 3s^1)$ Na$^+$ ion
$(1s^2 2s^2 2p^6)$

An atom that has lost or gained electrons will have a positive or negative charge, depending on which particles (protons or electrons) are in excess. Remember that a charged particle or group of particles is called an *ion*.

By losing a negatively charged electron, the sodium atom becomes a positively charged particle known as a sodium ion. The charge, $+1$, results because the nucleus still contains 11 positively charged protons, and the electron orbitals contain only 10 negatively charged electrons. The charge is indicated by a plus sign $(+)$ and is written as a superscript after the symbol of the element (Na$^+$).

A chlorine atom with seven electrons in the third energy level needs one electron to pair up with its one unpaired $3p$ electron to attain the stable outer electron structure of argon. By gaining one electron, the chlorine atom becomes a chloride ion (Cl⁻), a negatively charged particle containing 17 protons and 18 electrons. This process releases energy:

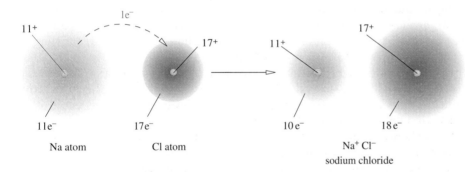

Cl atom
$(1s^22s^22p^63s^23p^5)$

Cl⁻ ion
$(1s^22s^22p^63s^23p^6)$

Consider sodium and chlorine atoms reacting with each other. The 3s electron from the sodium atom transfers to the half-filled 3p orbital in the chlorine atom to form a positive sodium ion and a negative chloride ion. The compound sodium chloride results because the Na⁺ and Cl⁻ ions are strongly attracted to each other by their opposite electrostatic charges. The force holding the oppositely charged ions together is called an ionic bond:

1e⁻

11+   17+   11+   17+

11e⁻   17e⁻   10 e⁻   18 e⁻

Na atom   Cl atom   Na⁺ Cl⁻
                          sodium chloride

The Lewis representation of sodium chloride formation is

$$\text{Na}\cdot + \cdot\ddot{\text{Cl}}: \longrightarrow [\text{Na}]^+ \left[:\ddot{\text{Cl}}:\right]^-$$

The chemical reaction between sodium and chlorine is a very vigorous one, producing considerable heat in addition to the salt formed. When energy is released in a chemical reaction, the products are more stable than the reactants. Note that in NaCl both atoms attain a noble gas electron structure.

Sodium chloride is made up of cubic crystals in which each sodium ion is surrounded by six chloride ions and each chloride ion by six sodium ions, except at the crystal surface. A visible crystal is a regularly arranged aggregate of millions of these ions, but the ratio of sodium to chloride ions is 1:1, hence the formula NaCl. The cubic crystalline lattice arrangement of sodium chloride is shown in Figure 11.5.

Figure 11.6 contrasts the relative sizes of sodium and chlorine atoms with those of their ions. The sodium ion is smaller than the atom due primarily to two factors: (1) The sodium atom has lost its outermost electron, thereby

These tiny NaCl crystals (on a penny) show the cubic structure illustrated in Figure 11.5.

Na⁺   Cl⁻   Na⁺   Cl⁻

Figure 11.5
Sodium chloride crystal. Diagram represents a small fragment of sodium chloride, which forms cubic crystals. Each sodium ion is surrounded by six chloride ions, and each chloride ion is surrounded by six sodium ions. The tiny NaCl crystals show the cubic crystal structure of salt.

Remember: A cation is always smaller than its parent atom whereas an anion is always larger than its parent atom.

reducing its size; and (2) the 10 remaining electrons are now attracted by 11 protons and are thus drawn closer to the nucleus. Conversely, the chloride ion is larger than the atom because (1) it has 18 electrons but only 17 protons and (2) the nuclear attraction on each electron is thereby decreased, allowing the chlorine atom to expand as it forms an ion.

We've seen that when sodium reacts with chlorine, each atom becomes an ion. Sodium chloride, like all ionic substances, is held together by the attrac-

ionic bond

tion existing between positive and negative charges. An **ionic bond** is the attraction between oppositely charged ions.

Ionic bonds are formed whenever one or more electrons are transferred from one atom to another. Metals, which have relatively little attraction for their valence electrons, tend to form ionic bonds when they combine with nonmetals.

It's important to recognize that substances with ionic bonds do not exist as molecules. In sodium chloride, for example, the bond does not exist solely between a single sodium ion and a single chloride ion. Each sodium ion in the crystal attracts six near-neighbor negative chloride ions; in turn, each negative chloride ion attracts six near-neighbor positive sodium ions (see Figure 11.5).

A metal will usually have one, two, or three electrons in its outer energy level. In reacting, metal atoms characteristically lose these electrons, attain

0.186 nm   0.095 nm   0.099 nm   0.181 nm

Figure 11.6
Relative radii of sodium and chlorine atoms and their ions.

Na atom   Na⁺ ion   Cl atom   Cl⁻ ion

Table 11.3 Change in Atomic Radii (nm) of Selected Metals and Nonmetals*

Atomic radius		Ionic radius		Atomic radius		Ionic radius	
Li	0.152	Li^+	0.060	F	0.071	F^-	0.136
Na	0.186	Na^+	0.095	Cl	0.099	Cl^-	0.181
K	0.227	K^+	0.133	Br	0.114	Br^-	0.195
Mg	0.160	Mg^{2+}	0.065	O	0.074	O^{2-}	0.140
Al	0.143	Al^{3+}	0.050	S	0.103	S^{2-}	0.184

*The metals shown lose electrons to become positive ions. The nonmetals gain electrons to become negative ions.

the electron structure of a noble gas, and become positive ions. A nonmetal, on the other hand, is only a few electrons short of having a noble gas electron structure in its outer energy level and thus has a tendency to gain electrons. In reacting with metals, nonmetal atoms characteristically gain one, two, or three electrons; attain the electron structure of a noble gas; and become negative ions. The ions formed by loss of electrons are much smaller than the corresponding metal atoms; the ions formed by gaining electrons are larger than the corresponding nonmetal atoms. The dimensions of the atomic and ionic radii of several metals and nonmetals are given in Table 11.3.

Practice 11.2

What noble gas structure is formed when an atom of each of these metals loses all its valence electrons? Write the formula for the metal ion formed.
(a) K   (b) Mg   (c) Al   (d) Ba

Study the following examples. Note the loss and gain of electrons between atoms; also note that the ions in each compound have a noble gas electron structure.

Explain how magnesium and chlorine combine to form magnesium chloride, $MgCl_2$.   **Example 11.2**

A magnesium atom of electron structure $1s^2 2s^2 2p^6 3s^2$ must lose two electrons   **SOLUTION**
or gain six to reach a stable electron structure. If magnesium reacts with chlorine and each chlorine atom can accept only one electron, two chlorine atoms will be needed for the two electrons from each magnesium atom. The compound formed will contain one magnesium ion and two chloride ions. The magnesium atom, having lost two electrons, becomes a magnesium ion with a +2 charge. Each chloride ion will have a −1 charge. The transfer of electrons from a magnesium atom to two chlorine atoms is shown in the following illustration:

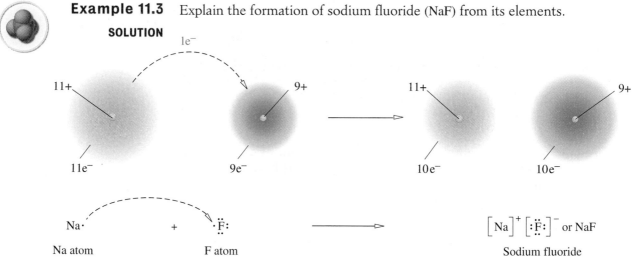

Example 11.3   Explain the formation of sodium fluoride (NaF) from its elements.

SOLUTION

The fluorine atom, with seven electrons in its outer energy level, behaves similarly to the chlorine atom.

Example 11.4   Explain the formation of aluminum fluoride (AlF_3) from its elements.

SOLUTION

$$\overset{..}{Al}\cdot\ +\ \overset{..}{F}:\ \longrightarrow\ [Al]^{3+}\left[:\overset{..}{\underset{..}{F}}:\right]^{-}\quad or\quad AlF_3$$

1 Al atom   3 F atoms   aluminum flouride

Each fluorine atom can accept only one electron. Therefore, three fluorine atoms are needed to combine with the three valence electrons of one aluminum atom. The aluminum atom has lost three electrons to become an aluminum ion (Al^{3+}) with a +3 charge.

Explain the formation of magnesium oxide (MgO) from its elements.

Example 11.5

SOLUTION

The magnesium atom, with two electrons in the outer energy level, exactly fills the need of one oxygen atom for two electrons. The resulting compound has a ratio of one magnesium atom to one oxygen atom. The oxygen (oxide) ion has a −2 charge, having gained two electrons. In combining with oxygen, magnesium behaves the same way as when it combines with chlorine—it loses two electrons.

Explain the formation of sodium sulfide (Na_2S) from its elements.

Example 11.6

SOLUTION

Two sodium atoms supply the electrons that one sulfur atom needs to make eight in its outer energy level.

Explain the formation of aluminum oxide (Al_2O_3) from its elements.

Example 11.7

SOLUTION

One oxygen atom, needing two electrons, cannot accommodate the three electrons from one aluminum atom. One aluminum atom falls one electron short of the four electrons needed by two oxygen atoms. A ratio of two atoms of aluminum to three atoms of oxygen, involving the transfer of six electrons (two to each oxygen atom), gives each atom a stable electron configuration.

Note that in each of these examples, outer energy levels containing eight electrons were formed in all the negative ions. This formation resulted from the pairing of all the s and p electrons in these outer energy levels.

11.4 Predicting Formulas of Ionic Compounds

In previous examples, we learned that when a metal and a nonmetal react to form an ionic compound, the metal loses one or more electrons to the nonmetal. In Chapter 6, where we learned to name compounds and write formulas, we saw that Group 1A metals always form +1 cations, whereas Group 2A form +2 cations. Group 7A elements form −1 anions and Group 6A form −2 anions.

It stands to reason, then, that this pattern is directly related to the stability of the noble gas configuration. Metals lose electrons to attain the electron configuration of a noble gas (the previous one on the periodic table). A nonmetal forms an ion by gaining enough electrons to achieve the electron configuration of the noble gas following it on the periodic table. These observations lead us to an important chemical principle:

> In almost all stable chemical compounds of representative elements, each atom attains a noble gas electron configuration. This concept forms the basis for our understanding of chemical bonding.

We can apply this principle in predicting the formulas of ionic compounds. To predict the formula of an ionic compound, we must recognize that chemical compounds are always electrically neutral. In addition, the metal will lose electrons to achieve noble gas configuration and the nonmetal will gain electrons to achieve noble gas configuration. Consider the compound formed between barium and sulfur. Barium has two valence electrons, whereas sulfur has six valence electrons:

Ba $[Xe]6s^2$ S $[Ne]3s^23p^4$

If barium loses two electrons, it will achieve the configuration of xenon. By gaining two electrons, sulfur achieves the configuration of argon. Consequently, a pair of electrons is transferred between atoms. Now we have Ba^{2+} and S^{2-}. Since compounds are electrically neutral, there must be a ratio of one Ba to one S, giving the formula BaS.

The same principle works for many other cases. Since the key lies in the electron configuration, the periodic table can be used to extend the prediction even further. Because of similar electron structures, the elements in a family

Table 11.4 Formulas of Compounds Formed by Alkali Metals

Lewis structure	Oxides	Chlorides	Bromides	Sulfates
Li·	Li_2O	LiCl	LiBr	Li_2SO_4
Na·	Na_2O	NaCl	NaBr	Na_2SO_4
K·	K_2O	KCl	KBr	K_2SO_4
Rb·	Rb_2O	RbCl	RbBr	Rb_2SO_4
Cs·	Cs_2O	CsCl	CsBr	Cs_2SO_4

generally form compounds with the same atomic ratios. In general, if we know the atomic ratio of a particular compound—say, NaCl—we can predict the atomic ratios and formulas of the other alkali metal chlorides. These formulas are LiCl, KCl, RbCl, CsCl, and FrCl (see Table 11.4).

Similarly, if we know that the formula of the oxide of hydrogen is H_2O, we can predict that the formula of the sulfide will be H_2S, because sulfur has the same valence electron structure as oxygen. Recognize, however, that these are only predictions; it doesn't necessarily follow that every element in a group will behave like the others or even that a predicted compound will actually exist. For example, knowing the formulas for potassium chlorate, bromate, and iodate to be $KClO_3$, $KBrO_3$, and KIO_3, we can correctly predict the corresponding sodium compounds to have the formulas $NaClO_3$, $NaBrO_3$, and $NaIO_3$. Fluorine belongs to the same family of elements (Group 7A) as chlorine, bromine, and iodine, so it would appear that fluorine should combine with potassium and sodium to give fluorates with the formulas KFO_3 and $NaFO_3$. However, potassium and sodium fluorates are not known to exist.

In the discussion in this section, we refer only to representative metals (Groups 1A, 2A, and 3A). The transition metals (Group B) show more complicated behavior (they form multiple ions), and their formulas are not as easily predicted.

The formula for calcium sulfide is CaS and that for lithium phosphide is Li_3P. Predict formulas for (a) magnesium sulfide, (b) potassium phosphide, and (c) magnesium selenide.

Example 11.8

(a) Look up calcium and magnesium in the periodic table; they are both in Group 2A. The formula for calcium sulfide is CaS, so it's reasonable to predict that the formula for magnesium sulfide is MgS.

SOLUTION

(b) Find lithium and potassium in the periodic table; they are in Group 1A. Since the formula for lithium phosphide is Li_3P, it's reasonable to predict that K_3P is the formula for potassium phosphide.

(c) Find selenium in the periodic table; it is in Group 6A just below sulfur. Therefore it's reasonable to assume that selenium forms selenide in the same way that sulfur forms sulfide. Since MgS was the predicted formula for magnesium sulfide in part (a), we can reasonably assume that the formula for magnesium selenide is MgSe.

O   C   Dry ice

Na⁺   Cl⁻

Sodium chloride

Figure 11.7
Solid carbon dioxide (dry ice) is composed of individual covalently bonded molecules of CO_2 closely packed together. Table salt is a large aggregate of Na^+ and Cl^- ions instead of molecules.

covalent bond

Practice 11.3
The formula for sodium oxide is Na_2O. Predict the formula for
(a) sodium sulfide
(b) rubidium oxide

Practice 11.4
The formula for barium phosphide is Ba_3P_2. Predict the formula for
(a) magnesium nitride
(b) barium arsenide

11.5 The Covalent Bond: Sharing Electrons

Some atoms do not transfer electrons from one atom to another to form ions. Instead they form a chemical bond by sharing pairs of electrons between them. A **covalent bond** consists of a pair of electrons shared between two atoms. This bonding concept was introduced in 1916 by G. N. Lewis. In the millions of known compounds, the covalent bond is the predominant chemical bond.

True molecules exist in substances in which the atoms are covalently bonded. It is proper to refer to molecules of such substances as hydrogen, chlorine, hydrogen chloride, carbon dioxide, water, or sugar (Figure 11.7). These substances contain only covalent bonds and exist as aggregates of molecules. We don't use the term *molecule* when talking about ionically bonded compounds such as sodium chloride, because such substances exist as large aggregates of positive and negative ions, not as molecules (Figure 11.7).

A study of the hydrogen molecule gives us an insight into the nature of the covalent bond and its formation. The formation of a hydrogen molecule (H_2) involves the overlapping and pairing of $1s$ electron orbitals from two hydrogen atoms, shown in Figure 11.8. Each atom contributes one electron of the pair that is shared jointly by two hydrogen nuclei. The orbital of the electrons now includes both hydrogen nuclei, but probability factors show that the most likely place to find the electrons (the point of highest electron density) is between the two nuclei. The two nuclei are shielded from each other by the pair of electrons, allowing the two nuclei to be drawn very close to each other.

The formula for chlorine gas is Cl_2. When the two atoms of chlorine combine to form this molecule, the electrons must interact in a manner similar to that shown in the hydrogen example. Each chlorine atom would be more stable with eight electrons in its outer energy level. But chlorine atoms are identical, and neither is able to pull an electron away from the other. What happens

Figure 11.8
The formation of a hydrogen molecule from two hydrogen atoms. The two $1s$ orbitals overlap, forming the H_2 molecule. In this molecule the two electrons are shared between the atoms, forming a covalent bond.

$1s$ orbitals

Overlapping $1s$ orbitals

H·   H·   H:H   Shared pair of electrons

hydrogen atoms   hydrogen molecule

p orbitals   Overlap of p orbitals   Paired p orbital

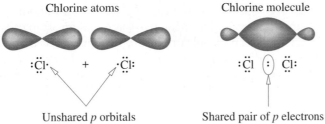

Chlorine atoms   Chlorine molecule

Unshared p orbitals   Shared pair of p electrons

Figure 11.9
Pairing of p electrons in the formation of a chlorine molecule.

is this: The unpaired 3p electron orbital of one chlorine atom overlaps the unpaired 3p electron orbital of the other atom, resulting in a pair of electrons that are mutually shared between the two atoms. Each atom furnishes one of the pair of shared electrons. Thus, each atom attains a stable structure of eight electrons by sharing an electron pair with the other atom. The pairing of the p electrons and the formation of a chlorine molecule are illustrated in Figure 11.9. Neither chlorine atom has a positive or negative charge, because both contain the same number of protons and have equal attraction for the pair of electrons being shared. Other examples of molecules in which electrons are equally shared between two atoms are hydrogen (H_2), oxygen (O_2), nitrogen (N_2), fluorine (F_2), bromine (Br_2), and iodine (I_2). Note that more than one pair of electrons may be shared between atoms:

H:H   :F̈:F̈:   :B̈r:B̈r: :Ï:Ï: :Ö::Ö: :N⋮⋮N:

hydrogen   fluorine   bromine   iodine   oxygen   nitrogen

The Lewis structure given for oxygen does not adequately account for all the properties of the oxygen molecule. Other theories explaining the bonding in oxygen molecules have been advanced, but they are complex and beyond the scope of this book.

In writing structures, we commonly replace the pair of dots used to represent a shared pair of electrons with a dash (—). One dash represents a single bond; two dashes, a double bond; and three dashes, a triple bond. The six structures just shown may be written thus:

H—H   :F̈—F̈:   :B̈r—B̈r: :Ï—Ï: :Ö=Ö: :N≡N:

The ionic bond and the covalent bond represent two extremes. In ionic bonding the atoms are so different that electrons are transferred between them, forming a charged pair of ions. In covalent bonding, two identical atoms share electrons equally. The bond is the mutual attraction of the two nuclei for the shared electrons. Between these extremes lie many cases in which the atoms are not different enough for a transfer of electrons but are different enough that the electron pair cannot be shared equally. This unequal sharing of electrons results in the formation of a **polar covalent bond**.

Molecular models for F_2 (green, single bond), O_2 (black, double bond), and N_2 (blue, triple bond).

Remember: A dash represents a shared pair of electrons.

polar covalent bond

11.6 Electronegativity

When two *different* kinds of atoms share a pair of electrons, a bond forms in which electrons are shared unequally. One atom assumes a partial positive charge and the other a partial negative charge with respect to each other. This difference in charge occurs because the two atoms exert unequal attraction for the pair of shared electrons. The attractive force that an atom of an element has for shared electrons in a molecule or polyatomic ion is known as its **electronegativity**. Elements differ in their electronegativities. For example, both hydrogen and chlorine need one electron to form stable electron configurations. They share a pair of electrons in hydrogen chloride (HCl). Chlorine is more electronegative and therefore has a greater attraction for the shared electrons than does hydrogen. As a result, the pair of electrons is displaced toward the chlorine atom, giving it a partial negative charge and leaving the hydrogen atom with a partial positive charge. Note that the electron is not transferred entirely to the chlorine atom (as in the case of sodium chloride) and that no ions are formed. The entire molecule, HCl, is electrically neutral. A partial charge is usually indicated by the Greek letter delta, δ. Thus, a partial positive charge is represented by $\delta+$ and a partial negative charge by $\delta-$.

electronegativity

$+\delta$ $-\delta$
H :Cl

hydrogen chloride

> The pair of shared electrons in HCl is closer to the more electronegative chlorine atom than to the hydrogen atom, giving chlorine a partial negative charge with respect to the hydrogen atom.

A scale of relative electronegativities, in which the most electronegative element, fluorine, is assigned a value of 4.0, was developed by the Nobel laureate (1954 and 1962) Linus Pauling (1901–1994). Table 11.5 shows that the relative electronegativity of the nonmetals is high and that of the metals is low. These electronegativities indicate that atoms of metals have a greater tendency to lose electrons than do atoms of nonmetals and that nonmetals have a greater tendency to gain electrons than do metals. The higher the electronegativity value, the greater the attraction for electrons. Note that electronegativity generally increases from left to right across a period and decreases down a group for the representative elements. The highest electronegativity is 4.0 for fluorine, and the lowest is 0.7 for francium and cesium. It's important to remember that the higher the electronegativity, the stronger an atom attracts electrons.

The polarity of a bond is determined by the difference in electronegativity values of the atoms forming the bond (see Figure 11.10). If the electronegativities are the same, the bond is **nonpolar covalent** and the electrons are shared equally. If the atoms have greatly different electronegativities, the bond is very *polar*. At the extreme, one or more electrons are actually transferred and an ionic bond results.

nonpolar covalent bond

dipole

A **dipole** is a molecule that is electrically asymmetrical, causing it to be oppositely charged at two points. A dipole is often written as ⊕⊖. A hydrogen chloride molecule is polar and behaves as a small dipole. The HCl dipole may be written as H⊢⟶Cl. The arrow points toward the negative end of the dipole. Molecules of H_2O, HBr, and ICl are polar:

H⊢⟶Cl   H⊢⟶Br   I⊢⟶Cl   H⟋O⟍H

Table 11.5 Three-Dimensional Representation of Electronegativity

How do we know whether a bond between two atoms is ionic or covalent? The difference in electronegativity between the two atoms determines the character of the bond formed between them. As the difference in electronegativity increases, the polarity of the bond (or percent ionic character) increases.

> If the electronegativity difference between two bonded atoms is greater than 1.7–1.9, the bond will be more ionic than covalent.

If the electronegativity difference is greater than 2.0, the bond is strongly ionic. If the electronegativity difference is less than 1.5, the bond is strongly covalent.

Practice 11.5

Which of these compounds would you predict to be ionic and which would be covalent?

(a) $SrCl_2$ (d) RbBr
(b) PCl_3 (e) LiCl
(c) NH_3 (f) CS_2

H_2 Cl_2

Nonpolar
molecules

HCl

Polar covalent
molecule

NaCl

Ionic
compound

Figure 11.10
Nonpolar, polar covalent, and ionic compounds.

Trans fats are virtually everywhere in the American diet. These fats remain when vegetable oils are converted into solid substances that are used in many processed foods. In fact, beginning January 1, 2006, manufacturers must label products to show their *trans* fat content. So just what is a *trans* fat?

Different categories of fat can be identified from the pattern of bonds and hydrogen atoms in the molecule. Fatty acids are one component of fats and contain long chains of carbon atoms with hydrogen atoms bonded to some or all of the carbon atoms. Unsaturated fats (such as corn or soybean oil) contain double bonds between some of the carbon atoms in the chains. A carbon that is doubly bonded to another carbon usually also bonds to a hydrogen atom. If the fatty acid has only one double bond, it is monounsaturated (such as olive oil). If the fatty acid contains more than one double bond, it is polyunsaturated. All fat with no double bonds is saturated—all the carbon atoms have the maximum possible hydrogen atoms bonded to them.

Trans fats contain a particular kind of unsaturated fatty acid. When there is a double bond between carbon atoms, the molecule bends in one of two ways: the *cis* or the *trans* direction. In *cis* configuration, the carbon chain on both sides of the double bond bends in to the same side of the double bond (see structure). In the *trans* configuration the chain on either side of the double bond bends toward opposite sides of the double bond (see structure). Most *trans* fats come from processing oils for prepared foods and from solid fats such as margarine.

When an oil is converted into a solid fat, some of the double bonds are converted to single bonds by adding hydrogen (hydrogenation). This process is easier at *cis* double bonds, and therefore the remaining double bonds are mainly in the *trans* configuration. These *trans* fatty acids tend to stack together, making a solid easier than the *cis* forms. Studies have linked diets high in *trans* fats to poor health, high cholesterol, heart disease, and diabetes. Food producers are working on ways to lower the *trans* fat content of foods. Gary List at USDA in Peoria, Illinois, has used high-pressure hydrogen gas on soybean oil at 140°–170°C to hydrogenate the oil, producing a soft margarine containing 5%–6% *trans* fat instead of the ~40% from the standard hydrogenation techniques. This could lead to a product that qualifies for a label of 0 g *trans* fat. Look for lots of new products and labels in your grocery stores as manufactures *trans*-form products into ones with less *trans* fat.

cis fatty acid

trans fatty acid

Care must be taken to distinguish between polar bonds and polar molecules. A covalent bond between different kinds of atoms is always polar. But a molecule containing different kinds of atoms may or may not be polar, depending on its shape or geometry. Molecules of HF, HCl, HBr, HI, and ICl are all polar because each contains a single polar bond. However, CO_2, CH_4, and CCl_4 are nonpolar molecules despite the fact that all three contain polar bonds. The carbon dioxide molecule O=C=O is nonpolar because the carbon–oxygen dipoles cancel each other by acting in opposite directions.

$$O=C=O$$

dipoles in opposite directions

Carbon tetrachloride (CCl_4) is nonpolar because the four C—Cl polar bonds are identical, and since these bonds emanate from the center to the corners of a tetrahedron in the molecule, their polarities cancel one another. Methane has the same molecular structure and is also nonpolar. We will discuss the shapes of molecules later in this chapter.

We have said that water is a polar molecule. If the atoms in water were linear like those in carbon dioxide, the two O—H dipoles would cancel each other, and the molecule would be nonpolar. However, water is definitely polar and has a nonlinear (bent) structure with an angle of 105° between the two O—H bonds.

Spacefilling molecular model CCl_4 (top) and methane (CH_4).

One of the most diverse elements in the periodic table is carbon. Graphite and diamond, two well-known forms of elemental carbon, both contain extended arrays of carbon atoms. In graphite the carbon atoms are arranged in sheets, and the bonding between the sheets is very weak. This property makes graphite useful as a lubricant and as a writing material. Diamond consists of transparent octahedral crystals in which each carbon atom is bonded to four other carbon atoms. This three-dimensional network of bonds gives diamond the property of hardness for which it is noted. In the 1980s, a new form of carbon was discovered in which the atoms are arranged in relatively small clusters.

When Harold Kroto of the University of Sussex, England, and Richard Smalley of Rice University, Texas, discovered a strange carbon molecule of formula C_{60}, they deduced that the most stable arrangement for the atoms would be in the shape of a soccer ball. In thinking about possible arrangements, the scientists considered the geodesic domes designed by R. Buckminster Fuller in the 1960s. This cluster form of carbon was thus named *buckminsterfullerene* and is commonly called buckyballs.

Buckyballs have captured the imagination of a variety of chemists. Research on buckyballs has led to a host of possible applications for these molecules. If metals are bound to the carbon atoms, the fullerenes become superconducting; that is, they conduct electrical current without resistance, at very low temperature. Scientists are now able to make buckyball compounds that superconduct at temperatures of 45 K. Other fullerenes are being used in lubricants and optical materials.

Chemists at Yale University have managed to trap helium and neon inside buckyballs. This is the first time chemists have ever observed helium or neon in a compound of any kind. They found that at temperatures from 1000°F to 1500°F one of the covalent bonds linking neighboring carbon atoms in the buckyball breaks. This opens a window in the fullerene molecule through which a helium or neon atom can enter the buckyball. When the fullerene is allowed to cool, the broken bond between carbon atoms re-forms, shutting the window and trapping the helium or neon atom inside the buckyball. Since the trapped helium or neon cannot react or share electrons with its host, the resulting compound has forced scientists to invent a new kind of chemical formula to describe the compound. The relationship between the "prisoner" helium or neon and the host buckyball is shown with an @ sign. A helium fullerene containing 60 carbon atoms would thus be He@C_{60}.

Buckyballs can be tailored to fit a particular size requirement. Raymond Schinazi of the Emory University School of Medicine, Georgia, made a buckyball to fit the active site of a key HIV enzyme that paralyzes the virus, making it noninfectious in human cells. The key to making this compound was preparing a water-soluble buckyball that would fit in the active site of the enzyme. Eventually, scientists created a water-soluble fullerene molecule that has two charged arms to grasp the binding site of the enzyme. It is toxic to the virus but doesn't appear to harm the host cells.

Scandium atoms trapped in a buckyball.

"Raspberries" of fullerene lubricant.

The relationships among types of bonds are summarized in Figure 11.11. It is important to realize that bonding is a continuum; that is, the difference between ionic and covalent is a gradual change.

Bond type

Covalent   Polar covalent   Ionic

0   Intermediate   3.3

Electronegativity difference

Figure 11.11
Relating bond type to electronegativity difference between atoms.

11.7 Lewis Structures of Compounds

As we have seen, Lewis structures are a convenient way of showing the covalent bonds in many molecules or ions of the representative elements. In writing Lewis structures, the most important consideration for forming a stable compound is that the atoms attain a noble gas configuration.

The most difficult part of writing Lewis structures is determining the arrangement of the atoms in a molecule or an ion. In simple molecules with more than two atoms, one atom will be the central atom surrounded by the other atoms. Thus, Cl_2O has two possible arrangements, Cl—Cl—O or Cl—O—Cl. Usually, but not always, the single atom in the formula (except H) will be the central atom.

Although Lewis structures for many molecules and ions can be written by inspection, the following procedure is helpful for learning to write them:

Remember: The number of valence electrons of Group A elements is the same as their group number in the periodic table.

Step 1   Obtain the total number of valence electrons to be used in the structure by adding the number of valence electrons in all the atoms in the molecule or ion. If you are writing the structure of an ion, add one electron for each negative charge or subtract one electron for each positive charge on the ion.

Step 2   Write the skeletal arrangement of the atoms and connect them with a single covalent bond (two dots or one dash). Hydrogen, which contains only one bonding electron, can form only one covalent bond. Oxygen atoms are not normally bonded to each other, except in compounds known to be peroxides. Oxygen atoms normally have a maximum of two covalent bonds (two single bonds or one double bond).

Step 3   Subtract two electrons for each single bond you used in Step 2 from the total number of electrons calculated in Step 1. This gives you the net number of electrons available for completing the structure.

Step 4   Distribute pairs of electrons (pairs of dots) around each atom (except hydrogen) to give each atom a noble gas structure.

Step 5   If there are not enough electrons to give these atoms eight electrons, change single bonds between atoms to double or triple bonds by shifting unbonded pairs of electrons as needed. Check to see that each atom has a noble gas electron structure (two electrons for hydrogen and eight for the others). A double bond counts as four electrons for each atom to which it is bonded.

Example 11.9   How many valence electrons are in each of these atoms: Cl, H, C, O, N, S, P, I?

SOLUTION   You can look at the periodic table to determine the electron structure, or, if the element is in Group A of the periodic table, the number of valence electrons is equal to the group number:

Atom	Group	Valence electrons
Cl	7A	7
H	1A	1
C	4A	4
O	6A	6
N	5A	5
S	6A	6
P	5A	5
I	7A	7

Write the Lewis structure for water (H_2O).

Example 11.10

SOLUTION

Step 1 The total number of valence electrons is eight, two from the two hydrogen atoms and six from the oxygen atom.

Step 2 The two hydrogen atoms are connected to the oxygen atom. Write the skeletal structure:

H O   or   H O H
   H

Place two dots between the hydrogen and oxygen atoms to form the covalent bonds:

H:O   or   H:O:H
   Ḧ

Step 3 Subtract the four electrons used in Step 2 from eight to obtain four electrons yet to be used.

Step 4 Distribute the four electrons in pairs around the oxygen atom. Hydrogen atoms cannot accommodate any more electrons:

H—Ö: or   H—Ö—H
   |
   H

These arrangements are Lewis structures because each atom has a noble gas electron structure. Note that the shape of the molecule is not shown by the Lewis structure.

Write Lewis structures for a molecule of methane (CH_4).

Example 11.11

SOLUTION

Step 1 The total number of valence electrons is eight, one from each hydrogen atom and four from the carbon atom.

Step 2 The skeletal structure contains four H atoms around a central C atom. Place two electrons between the C and each H.

     H   Ḧ
H C H   H:C:H
     H   Ḧ

Step 3 Subtract the eight electrons used in Step 2 from eight (obtained in Step 1) to obtain zero electrons yet to be placed. Therefore the Lewis structure must be as written in Step 2:

$$
\begin{array}{c}
\text{H} \\
\text{H} : \overset{..}{\underset{..}{\text{C}}} : \text{H} \\
\text{H}
\end{array}
\quad \text{or} \quad
\begin{array}{c}
\text{H} \\
| \\
\text{H} - \text{C} - \text{H} \\
| \\
\text{H}
\end{array}
$$

Example 11.12   Write the Lewis structure for a molecule of carbon tetrachloride (CCl_4).

SOLUTION   **Step 1** The total number of valence electrons to be used is 32, 4 from the carbon atom and 7 from each of the four chlorine atoms.

Step 2 The skeletal structure contains the four Cl atoms around a central C atom. Place 2 electrons between the C and each Cl:

$$
\begin{array}{c}
\text{Cl} \\
\text{Cl} \; \text{C} \; \text{Cl} \\
\text{Cl}
\end{array}
\qquad
\begin{array}{c}
\text{Cl} \\
\text{Cl} : \overset{..}{\underset{..}{\text{C}}} : \text{Cl} \\
\text{Cl}
\end{array}
$$

Step 3 Subtract the 8 electrons used in Step 2 from 32 (obtained in Step 1) to obtain 24 electrons yet to be placed.

Step 4 Distribute the 24 electrons (12 pairs) around the Cl atoms so that each Cl atom has 8 electrons around it:

$$
\begin{array}{c}
: \overset{..}{\underset{..}{\text{Cl}}} : \\
: \overset{..}{\underset{..}{\text{Cl}}} : \overset{..}{\underset{..}{\text{C}}} : \overset{..}{\underset{..}{\text{Cl}}} : \\
: \overset{..}{\underset{..}{\text{Cl}}} :
\end{array}
\quad \text{or} \quad
\begin{array}{c}
: \overset{..}{\underset{..}{\text{Cl}}} : \\
| \\
: \overset{..}{\underset{..}{\text{Cl}}} - \text{C} - \overset{..}{\underset{..}{\text{Cl}}} : \\
| \\
: \overset{..}{\underset{..}{\text{Cl}}} :
\end{array}
$$

This arrangement is the Lewis structure; CCl_4 contains four covalent bonds.

Example 11.13   Write a Lewis structure for CO_2.

SOLUTION   **Step 1** The total number of valence electrons is 16, 4 from the C atom and 6 from each O atom.

Step 2 The two O atoms are bonded to a central C atom. Write the skeletal structure and place 2 electrons between the C and each O atom.

O:C:O

Step 3 Subtract the 4 electrons used in Step 2 from 16 (found in Step 1) to obtain 12 electrons yet to be placed.

Step 4 Distribute the 12 electrons (six pairs) around the C and O atoms. Several possibilities exist:

$$
\underset{\text{I}}{: \overset{..}{\underset{..}{\text{O}}} : \text{C} : \overset{..}{\underset{..}{\text{O}}} :}
\qquad
\underset{\text{II}}{: \overset{..}{\text{O}} : \text{C} : \overset{..}{\text{O}} :}
\qquad
\underset{\text{III}}{: \overset{..}{\underset{..}{\text{O}}} : \overset{..}{\text{C}} : \text{O} :}
$$

Step 5   Not all the atoms have 8 electrons around them (noble gas structure). Remove one pair of unbonded electrons from each O atom in structure I and place one pair between each O and the C atom, forming two double bonds:

$$:\ddot{O}::C::\ddot{O}: \quad \text{or} \quad :\ddot{O}=C=\ddot{O}:$$

Each atom now has 8 electrons around it. The carbon is sharing four pairs of electrons, and each oxygen is sharing two pairs. These bonds are known as double bonds because each involves sharing two pairs of electrons.

Practice 11.6_____

Write the Lewis structures for the following:

(a) PBr_3 (b) $CHCl_3$ (c) HF   (d) H_2CO (e) N_2

Although many compounds attain a noble gas structure in covalent bonding, there are numerous exceptions. Sometimes it's impossible to write a structure in which each atom has 8 electrons around it. For example, in BF_3 the boron atom has only 6 electrons around it, and in SF_6 the sulfur atom has 12 electrons around it.

Although there are exceptions, many molecules can be described using Lewis structures where each atom has a noble gas electron configuration. This is a useful model for understanding chemistry.

11.8 Complex Lewis Structures

Most Lewis structures give bonding pictures that are consistent with experimental information on bond strength and length. There are some molecules and polyatomic ions for which no single Lewis structure consistent with all characteristics and bonding information can be written. For example, consider the nitrate ion, NO_3^-. To write a Lewis structure for this polyatomic ion, we use the following steps.

Step 1   The total number of valence electrons is 24, 5 from the nitrogen atom, 6 from each oxygen atom, and 1 from the -1 charge.

Step 2   The three O atoms are bonded to a central N atom. Write the skeletal structure and place two electrons between each pair of atoms. Since we have an extra electron in this ion, resulting in a -1 charge, we enclose the group of atoms in square brackets and add a $-$ charge as shown.

$$\left[\begin{matrix} O \\ O:\ddot{N}:O \end{matrix} \right]^-$$

Step 3   Subtract the 6 electrons used in Step 2 from 24 (found in Step 1) to obtain 18 electrons yet to be placed.

Keeping the shower area sparkling clean and free of mildew is a job none of us enjoy. Now thanks to chemist Bob Black of Jacksonville, Florida, there is a product that cleans the shower without any scrubbing! Black struggled with his home shower until he finally decided he really needed a new product that would solve the mildew and scrubbing problem.

His search was based on the following needs:

1. A molecule to lift deposits off the walls of the shower
2. A way to prevent hard-water deposits from forming
3. A wetting agent to wet the walls and rinse off deposits

In all cases, he limited his search to substances nontoxic and environmentally safe.

Black used a molecule called a glycol ether to lift deposits off the shower wall. This molecule is a long chain with a polar end and a nonpolar end. Substances that are nonpolar (such as grease, oils, and organic material) are attracted to the nonpolar end of the molecules. The molecules cluster together to form micelles (with the polar end pointed out). The polar sphere dissolves in the polar water from the shower, washing off the organic deposits.

A micelle

Preventing the hard-water deposits from forming on the shower walls required use of a molecule called EDTA. This molecule bonds to ions (like Ca^{2+}, Mg^{2+}, or Fe^{3+}) and prevents the formation of soap scum or hard-water deposits on the walls.

Lastly, Black added isopropyl alcohol to wet the shower wall and also to disturb the mildew fungus. He mixed all the ingredients in just the right proportions and found he no longer needed to work so hard to clean the shower.

Fortunately for all of us, Black shared his solution with friends, who also liked it. Black patented his new product and began mass production. You can now find Clean Shower in your local grocery store!

Simply spraying your shower regularly with Clean Shower will keep it free of deposits and mildew.

Step 4 Distribute the 18 electrons around the N and O atoms:

$$:\ddot{O} \longleftarrow \text{electron deficient}$$
$$:\ddot{O}:\ddot{N}:\ddot{O}:$$

Step 5 One pair of electrons is still needed to give all the N and O atoms a noble gas structure. Move the unbonded pair of electrons from the N atom and place it between the N and the electron-deficient O atom, making a double bond.

$$\left[\begin{array}{c} :\ddot{O} \\ \parallel \\ :\ddot{O} - N - \ddot{O}: \end{array} \right]^{-} \quad \text{or} \quad \left[\begin{array}{c} :\ddot{O}: \\ \mid \\ :\ddot{O} - N = \ddot{O}: \end{array} \right]^{-} \quad \text{or} \quad \left[\begin{array}{c} :\ddot{O}: \\ \mid \\ \ddot{O} = N - \ddot{O}: \end{array} \right]^{-}$$

Are these all valid Lewis structures? Yes, so there really are three possible Lewis structures for NO_3^-.

A molecule or ion that has multiple correct Lewis structures shows *resonance*.

resonance structure   Each of these Lewis structures is called a **resonance structure**. In this book, however, we will not be concerned with how to choose the correct resonance structure for a molecule or ion. Therefore any of the possible resonance structures may be used to represent the ion or molecule.

Write the Lewis structure for a carbonate ion (CO_3^{2-}).   **Example 11.14**

Step 1 These four atoms have 22 valence electrons plus 2 electrons from the   **SOLUTION**
−2 charge, which makes 24 electrons to be placed.

Step 2 In the carbonate ion, the carbon is the central atom surrounded by
the three oxygen atoms. Write the skeletal structure and place 2 electrons (or a single line) between each C and O:

$$
\begin{array}{c}
O \\
| \\
C\!-\!O \\
| \\
O
\end{array}
$$

Step 3 Subtract the 6 electrons used in Step 2 from 24 (from Step 1) to give
18 electrons yet to be placed.

Step 4 Distribute the 18 electrons around the three oxygen atoms and indicate that the carbonate ion has a −2 charge:

$$
\left[\begin{array}{c}
:\ddot{O}: \\
| \\
\ddot{O}\diagup^{C}\diagdown\ddot{O}:
\end{array} \right]^{2-}
$$

The difficulty with this structure is that the carbon atom has only six
electrons around it instead of a noble gas octet.

Step 5 Move one of the nonbonding pairs of electrons from one of the oxygens and place them between the carbon and the oxygen. Three
Lewis structures are possible:

$$
\left[\begin{array}{c}
:\ddot{O}: \\
| \\
\ddot{O}\!=\!\!C\diagdown\ddot{O}:
\end{array} \right]^{2-}
\quad \text{or} \quad
\left[\begin{array}{c}
:\ddot{O}: \\
| \\
\ddot{O}\diagup^{C}\!=\!\ddot{O}
\end{array} \right]^{2-}
\quad \text{or} \quad
\left[\begin{array}{c}
:O: \\
\| \\
\ddot{O}\diagup^{C}\diagdown\ddot{O}:
\end{array} \right]^{2-}
$$

Practice 11.7 _____

Write the Lewis structure for each of the following:

(a) NH_3 (b) H_3O^+ (c) NH_4^+ (d) HCO_3^-

11.9 Compounds Containing Polyatomic Ions

A polyatomic ion is a stable group of atoms that has either a positive or a negative charge and behaves as a single unit in many chemical reactions. Sodium
carbonate, Na_2CO_3, contains two sodium ions and a carbonate ion. The carbonate ion (CO_3^{2-}) is a polyatomic ion composed of one carbon atom and three
oxygen atoms and has a charge of −2. One carbon and three oxygen atoms have

a total of 22 electrons in their outer energy levels. The carbonate ion contains 24 outer electrons and therefore has a charge of −2. In this case, the 2 additional electrons come from the two sodium atoms, which are now sodium ions:

$[Na]^+$ sodium carbonate $[Na]^+$ carbonate ion

Sodium carbonate has both ionic and covalent bonds. Ionic bonds exist between each of the sodium ions and the carbonate ion. Covalent bonds are present between the carbon and oxygen atoms within the carbonate ion. One important difference between the ionic and covalent bonds in this compound can be demonstrated by dissolving sodium carbonate in water. It dissolves in water, forming three charged particles—two sodium ions and one carbonate ion—per formula unit of Na_2CO_3:

$$Na_2CO_3(s) \xrightarrow{water} 2\,Na^+(aq) + CO_3^{2-}(aq)$$

sodium carbonate   sodium ions   carbonate ion

The CO_3^{2-} ion remains as a unit, held together by covalent bonds; but where the bonds are ionic, dissociation of the ions takes place. Do not think, however, that polyatomic ions are so stable that they cannot be altered. Chemical reactions by which polyatomic ions can be changed to other substances do exist.

11.10 Molecular Shape

So far in our discussion of bonding we have used Lewis structures to represent valence electrons in molecules and ions, but they don't indicate anything regarding the molecular or geometric shape of a molecule. The three-dimensional arrangement of the atoms within a molecule is a significant feature in understanding molecular interactions. Let's consider several examples illustrated in Figure 11.12.

Figure 11.12
Geometric shapes of common molecules. Each molecule is shown as a ball-and-stick model (showing the bonds) and as a spacefilling model (showing the shape).

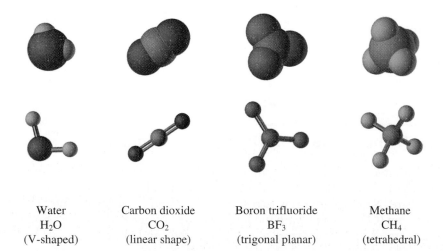

Water H_2O (V-shaped)   Carbon dioxide CO_2 (linear shape)   Boron trifluoride BF_3 (trigonal planar)   Methane CH_4 (tetrahedral)

What do color-changing pens, bullet-resistant vests, and calculators have in common? The chemicals that make each of them work are liquid crystals. These chemicals find numerous applications; you are probably most familiar with liquid crystal displays (LCDs) and color-changing products, but these chemicals are also used to make super-strong synthetic fibers.

Molecules in a normal crystal remain in an orderly arrangement, but in a liquid crystal the molecules can flow *and* maintain an orderly arrangement at the same time. Liquid crystal molecules are linear and polar. Since the atoms tend to lie in a relatively straight line, the molecules are generally much longer than they are wide. These polar molecules are attracted to each other and are able to line up in an orderly fashion, without solidifying.

Liquid crystals with twisted arrangements of molecules give us novelty color-changing products. In these liquid crystals the molecules lie side by side in a nearly flat layer. The next layer is similar, but at an angle to the one below. The closely packed flat layers have a special effect on light. As the light strikes the surface, some of it is reflected from the top layer and some from lower layers. When the same wavelength is reflected from many layers, we see a color. (This is similar to the rainbow of colors formed by oil in a puddle on the street or the film of a soap bubble.) As the temperature is increased, the molecules move faster, causing a change in the angle and the space between the layers. This results in a color change in the reflected light. Different compounds change color within different temperature ranges, allowing a variety of practical and amusing applications.

Liquid crystal (nematic) molecules that lie parallel to one another are used to manufacture very strong synthetic fibers. Perhaps the best example of these liquid crystals is Kevlar, a synthetic fiber used in bullet-resistant vests, canoes, and parts of the space shuttle. Kevlar is a synthetic polymer, like nylon or polyester, that gains strength by passing through a liquid crystal state during its manufacture.

In a typical polymer, the long molecular chains are jumbled together, somewhat like spaghetti. The strength of the material is limited by the disorderly arrangement. The trick is to get the molecules to line up parallel to each other. Once the giant molecules have been synthesized, they are dissolved in sulfuric acid. At the proper concentration the molecules align, and the solution is forced through tiny holes in a nozzle and further aligned. The sulfuric acid is removed in a water bath, thereby forming solid fibers in near-perfect alignment. One strand of Kevlar is stronger than an equal-sized strand of steel. It has a much lower density as well, making it a material of choice in bullet-resistant vests.

Kevlar is used to make protective vests for police.

Water is known to have the geometric shape known as "bent" or "V-shaped." Carbon dioxide exhibits a linear shape. BF_3 forms a third molecular shape called *trigonal planar* since all the atoms lie in one plane in a triangular arrangement. One of the more common molecular shapes is the tetrahedron, illustrated by the molecule methane (CH_4).

How do we predict the geometric shape of a molecule? We will now study a model developed to assist in making predictions from the Lewis structure.

11.11 The Valence Shell Electron Pair Repulsion (VSEPR) Model

The chemical properties of a substance are closely related to the structure of its molecules. A change in a single site on a large biomolecule can make a difference in whether or not a particular reaction occurs.

Instrumental analysis can be used to determine exact spatial arrangements of atoms. Quite often, though, we only need to be able to predict the approximate structure of a molecule. A relatively simple model has been developed to allow us to make predictions of shape from Lewis structures.

Nonbonding pairs of electrons are not shown here, so you can focus your attention on the shapes, not electron arrangement.

The VSEPR model is based on the idea that electron pairs will repel each other electrically and will seek to minimize this repulsion. To accomplish this minimization, the electron pairs will be arranged around a central atom as far apart as possible. Consider $BeCl_2$, a molecule with only two pairs of electrons surrounding the central atom. These electrons are arranged 180° apart for maximum separation:

$$Cl \overset{180°}{\frown} Be \frown Cl$$

linear structure

This molecular structure can now be labeled as a **linear structure**. When only two pairs of electrons surround a central atom, they should be placed 180° apart to give a linear structure.

What occurs when there are only three pairs of electrons around the central atom? Consider the BF_3 molecule. The greatest separation of electron pairs occurs when the angles between atoms are 120°:

$$\underset{120°}{F}\overset{120°}{\underset{F}{\diagdown B \diagup}}\underset{120°}{F}$$

trigonal planar structure

This arrangement of atoms is flat (planar) and, as noted earlier, is called **trigonal planar structure**. When three pairs of electrons surround an atom, they should be placed 120° apart to show the trigonal planar structure.

Now consider the most common situation (CH_4), with four pairs of electrons on the central carbon atom. In this case the central atom exhibits a noble gas electron structure. What arrangement best minimizes the electron pair repulsions? At first it seems that an obvious choice is a 90° angle with all the atoms in a single plane:

$$\overset{H}{\underset{H}{\underset{90° \ 90°}{H \mp C \pm H}}}$$

However, we must consider that molecules are three-dimensional. This concept results in a structure in which the electron pairs are actually 109.5° apart:

$$\overset{109.5°}{H \diagdown \diagup H}\\ H \blacktriangleleft \underset{C}{} \cdots H$$

tetrahedral structure

In this diagram the wedged line seems to protrude from the page whereas the dashed line recedes. Two representations of this arrangement, known as **tetrahedral structure**, are illustrated in Figure 11.13. When four pairs of electrons surround a central atom, they should be placed 109.5° apart to give them a tetrahedral structure.

The VSEPR model is based on the premise that we are counting electron pairs. It's quite possible that one or more of these electron pairs may be nonbonding (lone) pairs. What happens to the molecular structure in these cases? Consider the ammonia molecule. First we draw the Lewis structure to determine the number of electron pairs around the central atom:

$$H:\overset{..}{\underset{..}{N}}:H\\H$$

Methane, CH_4 CH_4 Carbon tetrachloride, CCl_4 CCl_4

Figure 11.13
Ball-and-stick models of methane and carbon tetrachloride. Methane and carbon tetrachloride are nonpolar molecules because their polar bonds cancel each other in the tetrahedral arrangement of their atoms. The carbon atoms are located in the centers of the tetrahedrons.

Since there are four pairs of electrons, the arrangement of electrons around the central atom will be tetrahedral (Figure 11.14a). However, only three of the pairs are bonded to another atom, so the molecule itself is pyramidal. It is important to understand that the placement of the electron pairs determines the shape but the name for the molecule is determined by the position of the atoms themselves. Therefore, ammonia is pyramidal. See Figure 11.14c.

Now consider the effect of two unbonded pairs of electrons in the water molecule. The Lewis structure for water is

H—Ö:
  |
  H

The four electron pairs indicate a tetrahedral electron arrangement is necessary (see Figure 11.15a). The molecule is not called tetrahedral because two of the electron pairs are unbonded pairs. The water molecule is "bent," as shown in Figure 11.15c.

The arrangement of electron pairs around an atom determines its shape, but we name the shape of molecules by the position of the atoms.

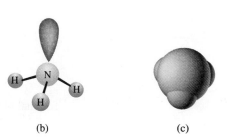

(a)   (b)   (c)

Figure 11.14
(a) The tetrahedral arrangement of electron pairs around the N atom in the NH_3 molecule.
(b) Three pairs are shared and one is unshared. (c) The NH_3 molecule is pyramidal.

(a)   (b)   (c)

Figure 11.15
(a) The tetrahedral arrangement of the four electron pairs around oxygen in the H_2O molecule.
(b) Two of the pairs are shared and two are unshared. (c) The H_2O molecule is bent.

Let's summarize the VSEPR model. To determine the molecular shape for a substance, follow these steps:

Step 1   Draw the Lewis structure for the molecule.

Step 2   Count the electron pairs and arrange them to minimize repulsions (as far apart as possible). This determines the electron pair arrangement.

Step 3   Determine the positions of the atoms.

Step 4   Name the molecular structure from the position of the atoms.

It is important to recognize that the placement of the electron pairs determines the structure but the name of the molecular structure is determined by the position of the atoms. Table 11.6 shows the results of this process. Note that when the number of electron pairs is the same as the number of atoms, the electron pair arrangement and the molecular structure are the same. But when the number of atoms and the number of electron pairs are not the same, the molecular structure is different from the electron pair arrangement. This is illustrated when the number of electron pairs is four (a tetrahedral arrangement) in Table 11.6.

Table 11.6 Arrangement of Electron Pairs and Molecular Structure

Number of electron pairs	Electron pair arrangement	Ball-and-stick model	Bonds	Molecular structure	Molecular structure model
2	Linear	180°	2	Linear	
3	Trigonal planar	120°	3	Trigonal planar	
4	Tetrahedral	109.5°	4	Tetrahedral	
4	Tetrahedral	109.5°	3	Trigonal pyramidal	
4	Tetrahedral	109.5°	2	Bent	

Practice 11.8

Predict the shape for CF_4, NF_3, and BeI_2.

Predict the molecular shape for these molecules: H_2S, CCl_4, $AlCl_3$.

1. Draw the Lewis structure.

2. Count the electron pairs around the central atom and determine the electron arrangement that will minimize repulsions.

3. Determine the positions of the atoms and name the shape of the molecule.

Example 11.15

SOLUTION

Molecule	Lewis structure	Number of electron pairs	Electron pair arrangement	Molecular shape
H_2S	H:S̈:H	4	tetrahedral	bent
CCl_4	:C̈l:C̈:C̈l: with :C̈l: above and :C̈l: below	4	tetrahedral	tetrahedral
$AlCl_3$	:C̈l:Äl:C̈l: with :C̈l: above	3	trigonal planar	trigonal planar

Chapter 11 Review

11.1 Periodic Trends in Atomic Properties

KEY TERM

Ionization energy

- Metals and nonmetals
- Atomic radius:
  - Increases down a group
  - Decreases across a row
- Ionization energy:
  - Energy required to remove an electron from an atom
  - Decreases down a group
  - Increases across a row

11.2 Lewis Structures of Atoms

KEY TERM

Lewis structure

- A Lewis structure is a representation of the atom where the symbol represents the element and dots around the symbol represent the valence electrons.

- To determine a Lewis structure for representative elements, use the Group number as the number of electrons to place around the symbol for the element.

11.3 The Ionic Bond: Transfer of Electrons from One Atoms to Another

KEY TERM

Ionic bond

- The goal of bonding is to achieve stability:
  - For representative elements, this stability can be achieved by attaining a valence electron structure of a noble gas.
- In an ionic bond stability is attained by transferring an electron from one atom to another:
  - The atom that loses an electron becomes a cation:
    - Positive ions are smaller than their parent atoms.
    - Metals tend to form cations.

- The atom gaining an electron becomes an anion:
  - Negative ions are larger than their parent atoms.
  - Nonmetals tend to form anions.
- Ionic compounds do not exist as molecules:
  - Ions are attracted by multiple ions of the opposite charge to form a crystalline structure.

11.4 Predicting Formulas of Ionic Compounds
- Chemical compounds are always electrically neutral.
- Metals lose electrons and nonmetals gain electrons to form compounds.
- Stability is achieved (for representative elements) by attaining a noble gas electron configuration.

11.5 The Covalent Bond: Sharing Electrons
KEY TERMS
Covalent bond

Polar covalent bond

- Covalent bonds are formed when two atoms share a pair of electrons between them:
  - This is the predominant type of bonding in compounds.
  - True molecules exist in covalent compounds.
  - Overlap of orbitals forms a covalent bond.
- Unequal sharing of electrons results in a polar covalent bond.

11.6 Electronegativity
KEY TERMS
Electronegativity

Nonpolar covalent bond

Dipole

- Electronegativity is the attractive force an atom has for a shared electrons in a molecule or polyatomic ion.
- Electrons spend more time closer to the more electronegative atom in a bond forming a polar bond.
- The polarity of a bond is determined by the electronegativity difference between the atoms involved in the bond:
  - The greater the difference, the more polar the bond is.
  - At the extremes:
    - Large differences result in ionic bonds.
    - Tiny differences (or no difference) result(s) in a nonpolar covalent bond.
  - A molecule that is electrically asymmetrical has a dipole, resulting in charged areas within the molecule.

hydrogen chloride

- If the electronegativity difference between two bonded atoms is greater than 1.7–1.9, the bond will be more ionic than covalent.
- Polar bonds do not always result in polar molecules.

11.7 Lewis Structures of Compounds
- To write a Lewis structure for a compound:
  - Add all the valence electrons in all the atoms in the compound. For ions, adjust the number accordingly.
  - Write the skeletal arrangement for the atoms and connect them with single bonds (2 electrons per bond).
  - Subtract the electrons used in the bonds from the total valence electrons.
  - Distribute pairs of electrons around each atom to give each atom a noble gas structure.
  - If there are not enough electrons, convert single bonds between atoms to multiple bonds to attain noble gas structures.

11.8 Complex Lewis Structures
KEY TERM
Resonance structure

- When a single unique Lewis structure cannot be drawn for a molecule, resonance structures (multiple Lewis structures) are used to represent the molecule

11.9 Compounds Containing Polyatomic Ions
- Polyatomic ions behave like a single unit in many chemical reactions.
- The bonds within a polyatomic ion are covalent.

11.10 Molecular Shape
- Lewis structures do not indicate the shape of a molecule.

11.11 The Valence Shell Electron Pair Repulsion (VSEPR) Model
KEY TERMS
Linear structure

Trigonal planar structure

Tetrahedral structure

Bent structure

- VSEPR model:
  - Electron pairs around an atom tend to orient themselves in space as far apart as possible to minimize repulsive forces.
  - The arrangement of electron pairs around an atom determines its structure, but the molecular shape is determined by the position of the atoms in the molecule

Review Questions

All questions with blue numbers have answers in the appendix of the text.

1. Rank these elements according to the radii of their atoms, from smallest to largest: Na, Mg, Cl, K, and Rb. (Figure 11.2)

2. Explain why much more ionization energy is required to remove the first electron from neon than from sodium. (Table 11.1)

3. Explain the large increase in ionization energy needed to remove the third electron from beryllium compared with that needed for the second electron. (Table 11.1)

4. Does the first ionization energy increase or decrease from top to bottom in the periodic table for the alkali metal family? Explain. (Figure 11.3)

5. Does the first ionization energy increase or decrease from top to bottom in the periodic table for the noble gas family? Explain. (Figure 11.3)

6. Why does barium (Ba) have a lower ionization energy than beryllium (Be)? (Figure 11.3)

7. Why is there such a large increase in the ionization energy required to remove the second electron from a sodium atom as opposed to the first? (Table 11.1)

8. Which element in the pair has the larger atomic radius? (Figure 11.2)

   (a) Na or K   (d) Br or I
   (b) Na or Mg   (e) Ti or Zr
   (c) O or F

9. In Groups 1A–7A, which element in each group has the smallest atomic radius? (Figure 11.2)

10. Why does the atomic size increase in going down any family of the period table?

11. Why are only valence electrons represented in a Lewis structure?

12. Why do metals tend to lose electrons and nonmetals tend to gain electrons when forming ionic bonds?

13. State whether the elements in each group gain or lose electrons in order to achieve a noble gas configuration. Explain.
    (a) Group 1A
    (b) Group 2A
    (c) Group 6A
    (d) Group 7A

14. What is the difference between a polar bond and a polar molecule?

15. What is the difference between electron pair arrangement and molecular shape?

16. What is the purpose of a Lewis structure?

17. In a polar covalent bond, how do you determine which atom has a partial negative charge ($\delta-$) and which has a partial positive charge ($\delta+$)?

18. In a Lewis structure, what do the dots represent and what do the lines represent?

19. All the atoms within each Group A family of elements can be represented by the same Lewis structure. Complete the following table, expressing the Lewis structure for each group. (Use E to represent the elements.) (Figure 11.4)

Group	1A	2A	3A	4A	5A	6A	7A
E·							

20. Draw the Lewis structure for Cs, Ba, Tl, Pb, Po, At, and Rn. How do these structures correlate with the group in which each element occurs?

21. In which general areas of the periodic table are the elements with (a) the highest and (b) the lowest electronegativities located?

22. What are valence electrons?

23. Explain why potassium usually forms a K^+ ion, but not a K^{2+} ion.

24. Why does an aluminum ion have a +3 charge?

Paired Exercises

All exercises with blue numbers have answers in the appendix of the text.

1. Which one in each pair has the larger radius? Explain.
   (a) a calcium atom or a calcium ion
   (b) a chlorine atom or a chloride ion
   (c) a magnesium ion or an aluminum ion
   (d) a sodium atom or a silicon atom
   (e) a potassium ion or a bromide ion

2. Which one in each pair has the larger radius? Explain.
   (a) Fe^{2+} or Fe^{3+}
   (b) a potassium atom or a potassium ion
   (c) a sodium ion or a chloride ion
   (d) a strontium atom or an iodine atom
   (e) a rubidium ion or a strontium ion

3. Using the table of electronegativity values (Table 11.5), indicate which element is more positive and which is more negative in these compounds:
(a) H_2O (d) PbS
(b) NaF   (e) NO
(c) NH_3 (f) CH_4

4. Using the table of electronegativity values (Table 11.5), indicate which element is more positive and which is more negative in these compounds:
(a) HCl   (d) IBr
(b) LiH   (e) MgH_2
(c) CCl_4 (f) OF_2

5. Classify the bond between these pairs of elements as principally ionic or principally covalent (use Table 11.5):
(a) sodium and chlorine
(b) carbon and hydrogen
(c) chlorine and carbon
(d) calcium and oxygen

6. Classify the bond between these pairs of elements as principally ionic or principally covalent (use Table 11.5):
(a) hydrogen and sulfur
(b) barium and oxygen
(c) fluorine and fluorine
(d) potassium and fluorine

7. Explain what happens to the electron structures of Mg and Cl atoms when they react to form $MgCl_2$.

8. Write an equation representing each of the following:
(a) the change of a fluorine atom to a fluoride ion
(b) the change of a calcium atom to a calcium ion

9. Use Lewis structures to show the electron transfer that enables these ionic compounds to be formed:
(a) MgF_2 (b) K_2O

10. Use Lewis structures to show the electron transfer that enables these ionic compounds to be formed:
(a) CaO   (b) NaBr

11. How many valence electrons are in each of these atoms? H, K, Mg, He, Al

12. How many valence electrons are in each of these atoms? Si, N, P, O, Cl

13. How many electrons must be gained or lost for the following to achieve a noble gas electron structure?
(a) a calcium atom
(b) a sulfur atom
(c) a helium atom

14. How many electrons must be gained or lost for the following to achieve a noble gas electron structure?
(a) a chloride ion
(b) a nitrogen atom
(c) a potassium atom

15. Determine whether the following atoms will form an ionic compound or a molecular compound and give the formula of the compound.
(a) sodium and chlorine
(b) carbon and 4 hydrogen
(c) magnesium and bromine
(d) 2 bromine
(e) carbon and 2 oxygen

16. Determine whether each of the following atoms will form a nonpolar covalent compound or a polar covalent compound, and give the formula of the compound.
(a) 2 oxygen
(b) hydrogen and bromine
(c) oxygen and 2 hydrogen
(d) 2 iodine

17. Let E be any representative element. Following the pattern in the table, write formulas for the hydrogen and oxygen compounds of the following:
(a) Na   (c) Al
(b) Ca   (d) Sn

18. Let E be any representative element. Following the pattern in the table, write formulas for the hydrogen and oxygen compounds of the following:
(a) Sb   (c) Cl
(b) Se   (d) C

Group						
1A	2A	3A	4A	5A	6A	7A
EH	EH_2	EH_3	EH_4	EH_3	H_2E	HE
E_2O	EO	E_2O_3	EO_2	E_2O_5	EO_3	E_2O_7

Group						
1A	2A	3A	4A	5A	6A	7A
EH	EH_2	EH_3	EH_4	EH_3	H_2E	HE
E_2O	EO	E_2O_3	EO_2	E_2O_5	EO_3	E_2O_7

19. The formula for sodium sulfate is Na_2SO_4. Write the names and formulas for the other alkali metal sulfates.

20. The formula for calcium bromide is $CaBr_2$. Write the names and formulas for the other alkaline earth metal bromides.

21. Write Lewis structures for the following:
(a) Na   (b) Br^- (c) O^{2-}

22. Write Lewis structures for the following:
(a) Ga   (b) Ga^{3+} (c) Ca^{2+}

23. Classify the bonding in each compound as ionic or covalent:
    (a) H_2O (c) MgO
    (b) NaCl   (d) Br_2

24. Classify the bonding in each compound as ionic or covalent:
    (a) HCl   (c) NH_3
    (b) $BaCl_2$ (d) SO_2

25. Predict the type of bond that would be formed between the following pairs of atoms:
    (a) Na and N
    (b) N and S
    (c) Br and I

26. Predict the type of bond that would be formed between the following pairs of atoms:
    (a) H and Si
    (b) O and F
    (c) Ca and I

27. Draw Lewis structures for the following:
    (a) H_2 (b) N_2 (c) Cl_2

28. Draw Lewis structures for the following:
    (a) O_2 (b) Br_2 (c) I_2

29. Draw Lewis structures for the following:
    (a) NCl_3 (c) C_2H_6
    (b) H_2CO_3 (d) $NaNO_3$

30. Draw Lewis structures for the following:
    (a) H_2S (c) NH_3
    (b) CS_2 (d) NH_4Cl

31. Draw Lewis structures for the following:
    (a) Ba^{2+} (d) CN^-
    (b) Al^{3+} (e) HCO_3^-
    (c) SO_3^{2-}

32. Draw Lewis structures for the following:
    (a) I^- (d) ClO_3^-
    (b) S^{2-} (e) NO_3^-
    (c) CO_3^{2-}

33. Classify these molecules as polar or nonpolar:
    (a) H_2O
    (b) HBr
    (c) CF_4

34. Classify these molecules as polar or nonpolar:
    (a) F_2
    (b) CO_2
    (c) NH_3

35. Give the number and arrangement of the electron pairs around the central atom:
    (a) C in CCl_4
    (b) S in H_2S
    (c) Al in AlH_3

36. Give the number and arrangement of the electron pairs around the central atom:
    (a) Ga in $GaCl_3$
    (b) N in NF_3
    (c) Cl in ClO_3^-

37. Use VSEPR theory to predict the structure of these polyatomic ions:
    (a) sulfate ion
    (b) chlorate ion
    (c) periodate ion

38. Use VSEPR theory to predict the structure of these polyatomic ions:
    (a) ammonium ion
    (b) sulfite ion
    (c) phosphate ion

39. Use VSEPR theory to predict the shape of these molecules:
    (a) SiH_4 (b) PH_3 (c) SeF_2

40. Use VSEPR theory to predict the shape of these molecules:
    (a) SiF_4 (b) OF_2 (c) Cl_2O

41. Element X reacts with sodium to form the compound Na_2X and is in the second period on the periodic table. Identify this element.

42. Element Y reacts with oxygen to form the compound Y_2O and has the lowest ionization energy of any Period 4 element on the periodic table. Identify this element.

Additional Exercises

All exercises with blue *numbers have answers in the appendix of the text.*

43. Write Lewis structures for hydrazine (N_2H_4) and hydrazoic acid (HN_3).

44. Draw Lewis structures and give the molecular or ionic shape of each of the following compounds:
    (a) NO_2^- (c) $SOCl_2$
    (b) SO_4^{2-} (d) Cl_2O

45. Draw Lewis structures for each of the following compounds:
    (a) ethane (C_2H_6)
    (b) ethylene (C_2H_4)
    (c) acetylene (C_2H_2)

46. Identify the element on the periodic table that satisfies each description:
    (a) transition metal with the largest atomic radius
    (b) alkaline earth metal with the greatest first ionization energy
    (c) least dense member of the nitrogen family
    (d) alkali metal with the greatest ratio of neutrons to protons
    (e) most electronegative transition metal

47. Choose the element that fits each description:
    (a) the lower electronegativity: As or Zn
    (b) the lower chemical reactivity: Ba or Be
    (c) the fewer valence electrons: N or Ne

48. Identify two reasons why fluorine has a much higher electronegativity than neon.

49. When one electron is removed from an atom of Li, it has two left. Helium atoms also have two electrons. Why is more energy required to remove the second electron from Li than to remove the first from He?

50. Group 1B elements (see the periodic table on the inside cover of your book) have one electron in their outer energy level, as do Group 1A elements. Would you expect them to form compounds such as CuCl, AgCl, and AuCl? Explain.

51. The formula for lead(II) bromide is $PbBr_2$: predict formulas for tin(II) and germanium(II) bromides.

52. Why is it not proper to speak of sodium chloride molecules?

53. What is a covalent bond? How does it differ from an ionic bond?

54. Briefly comment on the structure $Na \!:\! \overset{\cdot\cdot}{\underset{\cdot\cdot}{O}} \!:\! Na$ for the compound Na_2O.

55. What are the four most electronegative elements?

56. Rank these elements from highest electronegativity to lowest: Mg, S, F, H, O, Cs.

57. Is it possible for a molecule to be nonpolar even though it contains polar covalent bonds? Explain.

58. Why is CO_2 a nonpolar molecule, whereas CO is a polar molecule?

59. Estimate the bond angle between atoms in these molecules:
    (a) H_2S (c) NH_4^+
    (b) NH_3 (d) $SiCl_4$

60. Consider the two molecules BF_3 and NF_3. Compare and contrast them in terms of the following:
    (a) valence-level orbitals on the central atom that are used for bonding
    (b) shape of the molecule
    (c) number of lone electron pairs on the central atom
    (d) type and number of bonds found in the molecule

61. With respect to electronegativity, why is fluorine such an important atom? What combination of atoms on the periodic table results in the most ionic bond?

62. Why does the Lewis structure of each element in a given group of representative elements on the periodic table have the same number of dots?

63. A sample of an air pollutant composed of sulfur and oxygen was found to contain 1.40 g sulfur and 2.10 g oxygen. What is the empirical formula for this compound? Draw a Lewis structure to represent it.

64. A dry-cleaning fluid composed of carbon and chlorine was found to have the composition 14.5% carbon and 85.5% chlorine. Its known molar mass is 166 g/mol. Draw a Lewis structure to represent the compound.

Challenge Exercises

All exercises with blue numbers have answers in the appendix of the text.

*65. Determine whether the following Lewis structures are correct. If they are incorrect, state why and provide the correct Lewis structure.

(a) CO_2 $\overset{\cdot\cdot}{O} - \overset{\cdot\cdot}{C} = \overset{\cdot\cdot}{O}$

(b) ClO_2^- $[:\overset{\cdot\cdot}{\underset{\cdot\cdot}{O}} - \overset{\cdot\cdot}{\underset{\cdot\cdot}{Cl}} - \overset{\cdot\cdot}{\underset{\cdot\cdot}{O}}:]^-$

(c) SF_6 $\begin{matrix} :\overset{\cdot\cdot}{F}: & :\overset{\cdot\cdot}{F}: \\ :\overset{\cdot\cdot}{F} - S - \overset{\cdot\cdot}{F}: \\ :\underset{\cdot\cdot}{F}: & \underset{\cdot\cdot}{F}: \end{matrix}$

(d) NO_3^- $\left[\begin{matrix} \overset{\cdot\cdot}{O} = N = \overset{\cdot\cdot}{O} \\ | \\ :\underset{\cdot\cdot}{O}: \end{matrix} \right]^-$

(e) HCN $H - \overset{\cdot\cdot}{C} = N:$

(f) SO_4^{2-} $\left[\begin{matrix} :\overset{\cdot\cdot}{O}: \\ \| \\ :\overset{\cdot\cdot}{O} - S - \overset{\cdot\cdot}{O}: \\ \| \\ :\underset{\cdot\cdot}{O}: \end{matrix} \right]^{2-}$

*66. The first ionization energy for lithium is 520 kJ/mol. How much energy would be required to change 25 g of lithium atoms to lithium ions? (Refer to Table 11.1.)

*67. What is the total amount of energy required to remove the first two electrons from 15 moles of sodium atoms? (Refer to Table 11.1.)

Answers to Practice Exercises

11.1 (a) :$\dot{\ddot{N}}$·

(b) :$\dot{Al}$

(c) Sr:

(d) :$\ddot{\underset{..}{Br}}$·

11.2 (a) Ar; K^+
(b) Ar; Mg^{2+}
(c) Ne; Al^{3+}
(d) Xe; Ba^{2+}

11.3 (a) Na_2S
(b) Rb_2O

11.4 (a) Mg_3N_2
(b) Ba_3As_2

11.5 ionic: (a), (d), (e)
covalent: (b), (c), (f)

11.6 (a) :$\ddot{Br}$:
:P:$\ddot{Br}$:
:$\ddot{Br}$:

(b) :$\ddot{Cl}$:$\overset{H}{\underset{:\ddot{Cl}:}{\ddot{C}}}$:$\ddot{Cl}$:

(c) H:$\ddot{\underset{..}{F}}$:

(d) H:$\overset{\ddot{O}:}{C}$:H

(e) :N:::N:

11.7 (a) H:$\ddot{N}$:H
$\;\;\;\;\;$H

(b) $\left[\text{H} : \overset{..}{\underset{H}{O}} : \text{H} \right]^+$

(c) $\left[\text{H} : \overset{H}{\underset{H}{N}} : \text{H} \right]^+$

(d) $\left[\text{H} : \overset{:\ddot{O}:}{\ddot{O}} : \overset{}{C} :: \ddot{O} : \right]^-$

11.8 CF_4, tetrahedral; NF_3, pyramidal; BeI_2, linear

PUTTING IT TOGETHER: Review for Chapters 10–11

Multiple Choice:

Choose the correct answer to each of the following.

1. The concept of electrons existing in specific orbits around the nucleus was the contribution of
   - (a) Thomson
   - (b) Rutherford
   - (c) Bohr
   - (d) Schrödinger

2. The correct electron structure for a fluorine atom (F) is
   - (a) $1s^2 2s^2 2p^5$
   - (b) $1s^2 2s^2 2p^2 3s^2 3p^1$
   - (c) $1s^2 2s^2 2p^4 3s^1$
   - (d) $1s^2 2s^2 2p^3$

3. The correct electron structure for $_{48}$Cd is
   - (a) $1s^2 2s^2 2p^6 3s^2 3p^6 4s^2 3d^{10}$
   - (b) $1s^2 2s^2 2p^6 3s^2 3p^6 4s^2 3d^{10} 4p^6 5s^2 4d^{10}$
   - (c) $1s^2 2s^2 2p^6 3s^2 3p^6 4s^2 3d^{10} 4p^6 4d^4$
   - (d) $1s^2 2s^2 2p^6 3s^2 3p^6 4s^2 4p^6 4d^{10} 5s^2 5d^{10}$

4. The correct electron structure of $_{23}$V is
   - (a) $[Ar]4s^2 3d^3$
   - (b) $[Ar]4s^2 4p^3$
   - (c) $[Ar]4s^2 4d^3$
   - (d) $[Kr]4s^2 3d^3$

5. Which of the following is the correct atomic structure for $_{22}^{48}$Ti?

   (a)

   22+
   26n
   22e⁻

   (b)

   22+
   48n
   22e⁻

   (c)

   26+
   22n
   22e⁻

   (d)

   22+
   26n
   48e⁻

6. The number of orbitals in a *d* sublevel is
   - (a) 3
   - (b) 5
   - (c) 7
   - (d) no correct answer given

7. The number of electrons in the third principal energy level in an atom having the electron structure $1s^2 2s^2 2p^6 3s^2 3p^2$ is
   - (a) 2
   - (b) 4
   - (c) 6
   - (d) 8

8. The total number of orbitals that contain at least one electron in an atom having the structure $1s^2 2s^2 2p^6 3s^2 3p^2$ is
   - (a) 5
   - (b) 8
   - (c) 14
   - (d) no correct answer given

9. Which of these elements has two *s* and six *p* electrons in its outer energy level?
   - (a) He
   - (b) O
   - (c) Ar
   - (d) no correct answer given

10. Which element is not a noble gas?
    - (a) Ra
    - (b) Xe
    - (c) He
    - (d) Ar

11. Which element has the largest number of unpaired electrons?
    - (a) F
    - (b) S
    - (c) Cu
    - (d) N

12. How many unpaired electrons are in the electron structure of $_{24}$Cr, $[Ar]4s^1 3d^5$?
    - (a) 2
    - (b) 4
    - (c) 5
    - (d) 6

13. Groups 3A–7A plus the noble gases form the area of the periodic table where the electron sublevels being filled are
    - (a) *p* sublevels
    - (b) *s* and *p* sublevels
    - (c) *d* sublevels
    - (d) *f* sublevels

14. In moving down an A group on the periodic table, the number of electrons in the outermost energy level
    - (a) increases regularly
    - (b) remains constant
    - (c) decreases regularly
    - (d) changes in an unpredictable manner

15. Which of the following is an incorrect formula?
    - (a) NaCl
    - (b) K_2O
    - (c) AlO
    - (d) BaO

16. Elements of the noble gas family
    - (a) form no compounds at all
    - (b) have no valence electrons
    - (c) have an outer electron structure of $ns^2 np^6$ (helium excepted), where *n* is the period number
    - (d) no correct answer given

17. The lanthanide and actinide series of elements are
    - (a) representative elements
    - (b) transition elements
    - (c) filling in *d*-level electrons
    - (d) no correct answer given

18. The element having the structure $1s^2 2s^2 2p^6 3s^2 3p^2$ is in Group
    - (a) 2A
    - (b) 2B
    - (c) 4A
    - (d) 4B

19. In Group 5A, the element having the smallest atomic radius is

(a) Bi   (c) As
(b) P   (d) N

20. In Group 4A, the most metallic element is

(a) C   (c) Ge
(b) Si   (d) Sn

21. Which group in the periodic table contains the least reactive elements?

(a) 1A   (c) 3A
(b) 2A   (d) noble gases

22. Which group in the periodic table contains the alkali metals?

(a) 1A   (c) 3A
(b) 2A   (d) 4A

23. An atom of fluorine is smaller than an atom of oxygen. One possible explanation is that, compared to oxygen, fluorine has

(a) a larger mass number
(b) a smaller atomic number
(c) a greater nuclear charge
(d) more unpaired electrons

24. If the size of the fluorine atom is compared to the size of the fluoride ion,

(a) they would both be the same size.
(b) the atom is larger than the ion.
(c) the ion is larger than the atom.
(d) the size difference depends on the reaction.

25. Sodium is a very active metal because

(a) it has a low ionization energy.
(b) it has only one outermost electron.
(c) it has a relatively small atomic mass.
(d) all of the above

26. Which of the following formulas is not correct?

(a) Na^+ (c) Al^{3+}
(b) S^- (d) F^-

27. Which of the following molecules does not have a polar covalent bond?

(a) CH_4 (c) CH_3OH
(b) H_2O (d) Cl_2

28. Which of the following molecules is a dipole?

(a) HBr   (c) H_2
(b) CH_4 (d) CO_2

29. Which of the following has bonding that is ionic?

(a) H_2 (c) H_2O
(b) MgF_2 (d) CH_4

30. Which of the following is a correct Lewis structure?

(a) :Ö:C:Ö: (c) C̈l::C̈l

       :C̈l:
(b) :C̈l:C:C̈l: (d) :N̈:N̈:
       :C̈l:

31. Which of the following is an incorrect Lewis structure?

(a) H:N̈:H

                 H
(c) H:C̈:H
                 H

(b) :Ö:H
    H

(d) :N:::N:

32. The correct Lewis structure for SO_2 is

(a) :Ö:S̈:Ö: (c) :Ö::S::Ö:
(b) :Ö:S::Ö: (d) :Ö:S̈:Ö:

33. Carbon dioxide (CO_2) is a nonpolar molecule because

(a) oxygen is more electronegative than carbon
(b) the two oxygen atoms are bonded to the carbon atom
(c) the molecule has a linear structure with the carbon atom in the middle
(d) the carbon-oxygen bonds are polar covalent

34. When a magnesium atom participates in a chemical reaction, it is most likely to

(a) lose 1 electron
(b) gain 1 electron
(c) lose 2 electrons
(d) gain 2 electrons

35. If X represents an element of Group 3A, what is the general formula for its oxide?

(a) X_3O_4 (c) XO
(b) X_3O_2 (d) X_2O_3

36. Which of the following has the same electron structure as an argon atom?

(a) Ca^{2+} (c) Na^+
(b) Cl^0 (d) K^0

37. As the difference in electronegativity between two elements decreases, the tendency for the elements to form a covalent bond

(a) increases
(b) decreases
(c) remains the same
(d) sometimes increases and sometimes decreases

38. Which compound forms a tetrahedral molecule?

(a) NaCl   (c) CH_4
(b) CO_2 (d) $MgCl_2$

39. Which compound has a bent (V-shaped) molecular structure?

(a) NaCl   (c) CH_4
(b) CO_2 (d) H_2O

40. Which compound has double bonds within its molecular structure?

(a) NaCl   (c) CH_4
(b) CO_2 (d) H_2O

41. The total number of valence electrons in a nitrate ion, NO_3^- is

(a) 12   (c) 23

(b) 18   (d) 24

42. The number of electrons in a triple bond is

(a) 3   (c) 6

(b) 4   (d) 8

43. The number of unbonded pairs of electrons in H_2O is

(a) 0   (c) 2

(b) 1   (d) 4

44. Which of the following does not have a noble gas electron structure?

(a) Na   (c) Ar

(b) Sc^{3+} (d) O^{2-}

Free Response Questions:

Answer each of the following. Be sure to include your work and explanations in a clear, logical form.

1. An alkaline earth metal, M, combines with a halide, X. Will the resulting compound be ionic or covalent? Why? What is the Lewis structure for the compound?

2. "All electrons in atoms with even atomic numbers are paired." Is this statement true or false? Explain your answer using an example.

3. Discuss whether the following statement is true or false: "All nonmetals have two valence electrons in an *s* sublevel with the exception of the noble gases, which have at least one unpaired electron in a *p* sublevel."

4. The first ionization energy (IE) of potassium is lower than the first IE for calcium, but the second IE of calcium is lower than the second IE of potassium. Use an electron configuration or size argument to explain this trend in ionization energies.

5. Chlorine has a very large first ionization energy, yet it forms a chloride ion relatively easily. Explain.

6. Three particles have the same electron configuration. One is a cation of an alkali metal, one is an anion of the halide in the third period, and the third particle is an atom of a noble gas. What are the identities of the three particles (including charges)? Which particle should have the smallest atomic/ionic radius, which should have the largest, and why?

7. Why is the Lewis structure of $AlCl_3$ not written as

$$
\begin{array}{c}
\ddot{\text{F}}\text{:} \\
/ \\
\text{:}\ddot{\text{F}}\text{—Al} \\
\backslash \\
\text{:}\ddot{\text{F}}\text{:}
\end{array}
$$

What is the correct Lewis structure and which electrons are shown in a Lewis structure?

8. Why does carbon have a maximum of four covalent bonds?

9. Both NCl_3 and BF_3 have a central atom bonded to three other atoms, yet one is pyramidal and the other is trigonal planar. Explain.

10. Draw the Lewis structure of the atom whose electron configuration is $1s^2 2s^2 2p^6 3s^2 3p^6 4s^2 3d^{10} 4p^5$. Would you expect this atom to form an ionic, nonpolar covalent, or polar covalent bond with sulfur?

The Gaseous State of Matter

Properties of gases are transformed to art and sport at the hot-air balloon festival in Albuquerque, New Mexico.

Chapter Outline

12.1 General Properties

12.2 The Kinetic-Molecular Theory

12.3 Measurement of Pressure of Gases

12.4 Dependence of Pressure on Number of Molecules and Temperature

12.5 Boyle's Law

12.6 Charles' Law

12.7 Gay-Lussac's Law

12.8 Combined Gas Laws

12.9 Dalton's Law of Partial Pressures

12.10 Avogadro's Law

12.11 Mole–Mass–Volume Relationships of Gases

12.12 Density of Gases

12.13 Ideal Gas Law

12.14 Gas Stoichiometry

12.15 Real Gases

Our atmosphere is composed of a mixture of gases, including nitrogen, oxygen, argon, carbon dioxide, ozone, and trace amounts of others. These gases are essential to life, yet they can also create hazards to us. For example, carbon dioxide is valuable when it is taken in by plants and converted to carbohydrates, but it is also associated with the potentially hazardous greenhouse effect. Ozone surrounds the Earth at high altitudes and protects us from harmful ultraviolet rays, but it also destroys rubber and plastics. We require air to live, yet scuba divers must be concerned about oxygen poisoning and the "bends."

In chemistry, the study of the behavior of gases allows us to understand our atmosphere and the effects that gases have on our lives.

12.1 General Properties

A mole of water occupies 18 mL as a liquid but would fill this box (22.4 L) as a gas at the same temperature.

Gases are the least dense and most mobile of the three states of matter. A solid has a rigid structure, and its particles remain in essentially fixed positions. When a solid absorbs sufficient heat, it melts and changes into a liquid. Melting occurs because the molecules (or ions) have absorbed enough energy to break out of the rigid structure of the solid. The molecules or ions in the liquid are more energetic than they were in the solid, as indicated by their increased mobility. Molecules in the liquid state cling to one another. When the liquid absorbs additional heat, the more energetic molecules break away from the liquid surface and go into the gaseous state—the most mobile state of matter. Gas molecules move at very high velocities and have high kinetic energy. The average velocity of hydrogen molecules at 0°C is over 1600 meters (1 mile) per second. Mixtures of gases are uniformly distributed within the container in which they are confined.

The same quantity of a substance occupies a much greater volume as a gas than it does as a liquid or a solid. For example, 1 mol of water (18.02 g) has a volume of 18 mL at 4°C. This same amount of liquid water would occupy about 22,400 mL in the gaseous state—more than a 1200-fold increase in volume. We may assume from this difference in volume that (1) gas molecules are relatively far apart, (2) gases can be greatly compressed, and (3) the volume occupied by a gas is mostly empty space.

12.2 The Kinetic-Molecular Theory

kinetic-molecular theory
(KMT)

ideal gas

Careful scientific studies of the behavior and properties of gases were begun in the 17th century by Robert Boyle (1627–1691). His work was carried forward by many investigators, and the accumulated data were used in the second half of the 19th century to formulate a general theory to explain the behavior and properties of gases. This theory is called the **kinetic–molecular theory (KMT)**. The KMT has since been extended to cover, in part, the behavior of liquids and solids.

The KMT is based on the motion of particles, particularly gas molecules. A gas that behaves exactly as outlined by the theory is known as an **ideal gas**. No ideal gases exist, but under certain conditions of temperature and pressure, real gases approach ideal behavior or at least show only small deviations from it.

Under extreme conditions, such as very high pressure and low temperature, real gases deviate greatly from ideal behavior. For example, at low temperature and high pressure many gases become liquids.

The principal assumptions of the KMT are as follows:

1. Gases consist of tiny particles.
2. The distance between particles is large compared with the size of the particles themselves. The volume occupied by a gas consists mostly of empty space.
3. Gas particles have no attraction for one another.
4. Gas particles move in straight lines in all directions, colliding frequently with one another and with the walls of the container.
5. No energy is lost by the collision of a gas particle with another gas particle or with the walls of the container. All collisions are perfectly elastic.
6. The average kinetic energy for particles is the same for all gases at the same temperature, and its value is directly proportional to the Kelvin temperature.

The kinetic energy (KE) of a particle is expressed by the equation

$$KE = \frac{1}{2}mv^2$$

where m is the mass and v is the velocity of the particle.

All gases have the same kinetic energy at the same temperature. Therefore, from the kinetic energy equation we can see that, if we compare the velocities of the molecules of two gases, the lighter molecules will have a greater velocity than the heavier ones. For example, calculations show that the velocity of a hydrogen molecule is four times the velocity of an oxygen molecule.

Due to their molecular motion, gases have the property of **diffusion**, the ability of two or more gases to mix spontaneously until they form a uniform mixture. The diffusion of gases may be illustrated by the use of the apparatus shown in Figure 12.1. Two large flasks, one containing reddish-brown bromine vapors and the other dry air, are connected by a side tube. When the stopcock between the flasks is opened, the bromine and air will diffuse into each other. After standing awhile, both flasks will contain bromine and air.

If we put a pinhole in a balloon, the gas inside will effuse or flow out of the balloon. **Effusion** is a process by which gas molecules pass through a very small orifice (opening) from a container at higher pressure to one at lower pressure.

diffusion

effusion

Bromine   Air   Bromine and air   Bromine and air

Figure 12.1
Diffusion of gases. When the stopcock between the two flasks is opened, colored bromine molecules can be seen diffusing into the flask containing air.

Graham's law of effusion

Thomas Graham (1805–1869), a Scottish chemist, observed that the rate of effusion was dependent on the density of a gas. This observation led to **Graham's law of effusion**.

> The rates of effusion of two gases at the same temperature and pressure are inversely proportional to the square roots of their densities, or molar masses:
>
> $$\frac{\text{rate of effusion of gas A}}{\text{rate of effusion of gas B}} = \sqrt{\frac{d\text{B}}{d\text{A}}} = \sqrt{\frac{\text{molar mass B}}{\text{molar mass A}}}$$

12.3 Measurement of Pressure of Gases

pressure

Pressure is defined as force per unit area. When a rubber balloon is inflated with air, it stretches and maintains its larger size because the pressure on the inside is greater than that on the outside. Pressure results from the collisions of gas molecules with the walls of the balloon (see Figure 12.2). When the gas is released, the force or pressure of the air escaping from the small neck propels the balloon in a rapid, irregular flight. If the balloon is inflated until it bursts, the gas escaping all at once causes an explosive noise.

The effects of pressure are also observed in the mixture of gases in the atmosphere, which is composed of about 78% nitrogen, 21% oxygen, 1% argon, and other minor constituents by volume (see Table 12.1). The outer boundary of the atmosphere is not known precisely, but more than 99% of the atmosphere is below an altitude of 20 miles (32 km). Thus, the concentration of gas molecules in the atmosphere decreases with altitude, and at about 4 miles the amount of oxygen is insufficient to sustain human life. The gases in the

atmospheric pressure

atmosphere exert a pressure known as **atmospheric pressure**. The pressure exerted by a gas depends on the number of molecules of gas present, the temperature, and the volume in which the gas is confined. Gravitational forces hold the atmosphere relatively close to Earth and prevent air molecules from flying off into space. Thus, the atmospheric pressure at any point is due to the mass of the atmosphere pressing downward at that point.

barometer

The pressure of the gases in the atmosphere can be measured with a **barometer**. A mercury barometer may be prepared by completely filling a long tube with pure, dry mercury and inverting the open end into an open dish of mercury. If the tube is longer than 760 mm, the mercury level will drop to a point at which the column of mercury in the tube is just supported by the pressure of the atmosphere. If the tube is properly prepared, a vacuum will exist above

Figure 12.2
The pressure resulting from the collisions of gas molecules with the walls of the balloon keeps the balloon inflated.

Table 12.1 Average Composition of Dry Air

Gas	Percent by volume	Gas	Percent by volume
N_2	78.08	He	0.0005
O_2	20.95	CH_4	0.0002
Ar	0.93	Kr	0.0001
CO_2	0.033	Xe, H_2, and N_2O	Trace
Ne	0.0018		

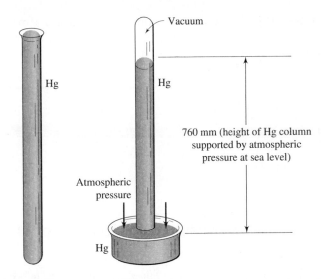

Figure 12.3
Preparation of a mercury barometer. The full tube of mercury at the left is inverted and placed in a dish of mercury.

the mercury column. The weight of mercury, per unit area, is equal to the pressure of the atmosphere. The column of mercury is supported by the pressure of the atmosphere, and the height of the column is a measure of this pressure (see Figure 12.3). The mercury barometer was invented in 1643 by the Italian physicist E. Torricelli (1608–1647), for whom the unit of pressure *torr* was named.

Air pressure is measured and expressed in many units. The standard atmospheric pressure, or simply **1 atmosphere** (atm), is the pressure exerted by a column of mercury 760 mm high at a temperature of 0°C. The normal pressure of the atmosphere at sea level is 1 atm, or 760 torr, or 760 mm Hg. The SI unit for pressure is the pascal (Pa), where 1 atm = 101,325 Pa, or 101.3 kPa. Other units for expressing pressure are inches of mercury, centimeters of mercury, the millibar (mbar), and pounds per square inch (lb/in.^{2} or psi). The values of these units equivalent to 1 atm are summarized in Table 12.2.

Atmospheric pressure varies with altitude. The average pressure at Denver, Colorado, 1.61 km (1 mile) above sea level, is 630 torr (0.83 atm). Atmospheric pressure is 0.5 atm at about 5.5 km (3.4 miles) altitude.

Pressure is often measured by reading the heights of mercury columns in millimeters on a barometer. Thus pressure may be recorded as mm Hg (torr). In problems dealing with gases, it is necessary to make interconversions among the various pressure units. Since atm, torr, and mm Hg are common pressure units, we give examples involving all three of these units:

1 atm = 760 torr = 760 mm Hg

1 atmosphere

Table 12.2 Pressure Units Equivalent to 1 Atmosphere

1 atm
760 torr
760 mm Hg
76 cm Hg
101.325 kPa
1013 mbar
29.9 in. Hg
14.7 lb/in.^{2}

The average atmospheric pressure at Walnut, California, is 740. mm Hg. Calculate this pressure in (a) torr and (b) atmospheres.

Let's use conversion factors that relate one unit of pressure to another.
(a) To convert mm Hg to torr, use the conversion factor 760 torr/760 mm Hg (1 torr/1 mm Hg):

$$(740. \text{ mm Hg})\left(\frac{1 \text{ torr}}{1 \text{ mm Hg}}\right) = 740. \text{ torr}$$

Example 12.1

SOLUTION

(b) To convert mm Hg to atm, use the conversion factor 1 atm/760. mm Hg:

$$(740. \text{ mm Hg})\left(\frac{1 \text{ atm}}{760. \text{ mm Hg}}\right) = 0.974 \text{ atm}$$

Practice 12.1

A barometer reads 1.12 atm. Calculate the corresponding pressure in (a) torr and (b) mm Hg.

12.4 Dependence of Pressure on Number of Molecules and Temperature

Pressure is produced by gas molecules colliding with the walls of a container. At a specific temperature and volume the number of collisions depends on the number of gas molecules present. The number of collisions can be increased by increasing the number of gas molecules present. If we double the number of molecules, the frequency of collisions and the pressure should double. We find, for an ideal gas, that this doubling is actually what happens. When the temperature and mass are kept constant, the pressure is directly proportional to the number of moles or molecules of gas present. Figure 12.4 illustrates this concept.

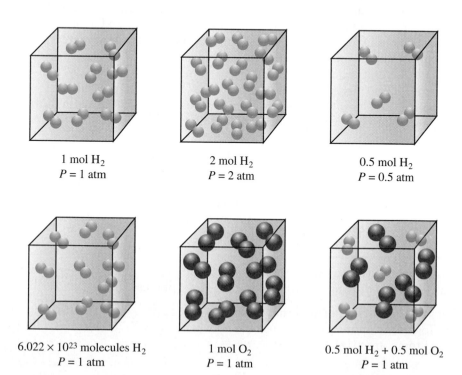

Figure 12.4
The pressure exerted by a gas is directly proportional to the number of molecules present. In each case shown, the volume is 22.4 L and the temperature is 0°C.

1 mol H$_2$
P = 1 atm

2 mol H$_2$
P = 2 atm

0.5 mol H$_2$
P = 0.5 atm

6.022×10^{23} molecules H$_2$
P = 1 atm

1 mol O$_2$
P = 1 atm

0.5 mol H$_2$ + 0.5 mol O$_2$
P = 1 atm

A good example of this molecule–pressure relationship may be observed in an ordinary cylinder of compressed gas equipped with a pressure gauge. When the valve is opened, gas escapes from the cylinder. The volume of the cylinder is constant, and the decrease in quantity (moles) of gas is registered by a drop in pressure indicated on the gauge.

The pressure of a gas in a fixed volume also varies with temperature. When the temperature is increased, the kinetic energy of the molecules increases, causing more frequent and more energetic collisions of the molecules with the walls of the container. This increase in collision frequency and energy results in a pressure increase (see Figure 12.5).

0°C
Volume = 1 liter
0.1 mole gas
$P = 2.24$ atm

100°C
Volume = 1 liter
0.1 mole gas
$P = 3.06$ atm

Figure 12.5
The pressure of a gas in a fixed volume increases with increasing temperature. The increased pressure is due to more frequent and more energetic collisions of the gas molecules with the walls of the container at the higher temperature.

12.5 Boyle's Law

Through a series of experiments, Robert Boyle (1627–1691) determined the relationship between the pressure (P) and volume (V) of a particular quantity of a gas. This relationship of P and V is known as **Boyle's law**:

Boyle's law

At constant temperature (T), the volume (V) of a fixed mass of a gas is inversely proportional to the pressure (P), which may be expressed as

$$V \propto \frac{1}{P} \quad \text{or} \quad P_1V_1 = P_2V_2$$

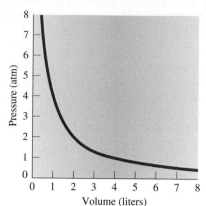

Figure 12.6
Graph of pressure versus volume showing the inverse PV relationship of an ideal gas.

This equation says that the volume varies ($\propto$) inversely with the pressure at constant mass and temperature. When the pressure on a gas is increased, its volume will decrease, and vice versa. The inverse relationship of pressure and volume is graphed in Figure 12.6.

When Boyle doubled the pressure on a specific quantity of a gas, keeping the temperature constant, the volume was reduced to one-half the original volume; when he tripled the pressure on the system, the new volume was one-third the original volume; and so on. His work showed that the product of volume and pressure is constant if the temperature is not changed:

$$PV = \text{constant} \quad \text{or} \quad PV = k \quad \text{(mass and temperature are constant)}$$

Let's demonstrate this law using a cylinder with a movable piston so that the volume of gas inside the cylinder may be varied by changing the external pressure (see Figure 12.7). Assume that the temperature and the number of gas molecules do not change. We start with a volume of 1000 mL and a pressure of 1 atm. When we change the pressure to 2 atm, the gas molecules are crowded closer together, and the volume is reduced to 500 mL. When we increase the pressure to 4 atm, the volume becomes 250 mL.

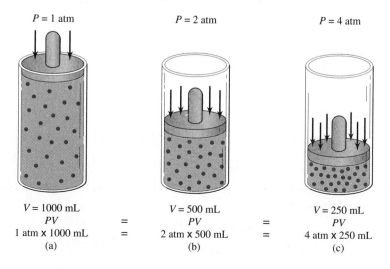

Figure 12.7
The effect of pressure on the volume of a gas.

Note that the product of the pressure times the volume is the same number in each case, substantiating Boyle's law. We may then say that

$$P_1V_1 = P_2V_2$$

where P_1V_1 is the pressure–volume product at one set of conditions, and P_2V_2 is the product at another set of conditions. In each case, the new volume may be calculated by multiplying the starting volume by a ratio of the two pressures involved. Of course, the ratio of pressures used must reflect the direction in which the volume should change. When the pressure is changed from 1 atm to 2 atm, the ratio to be used is 1 atm/2 atm. Now we can verify the results given in Figure 12.7:

1. Starting volume, 1000 mL; pressure change, 1 atm ⟶ 2 atm

   $$(1000 \text{ mL})\left(\frac{1 \text{ atm}}{2 \text{ atm}}\right) = 500 \text{ mL}$$

2. Starting volume, 1000 mL; pressure change, 1 atm ⟶ 4 atm

   $$(1000 \text{ mL})\left(\frac{1 \text{ atm}}{4 \text{ atm}}\right) = 250 \text{ mL}$$

3. Starting volume, 500 mL; pressure change, 2 atm ⟶ 4 atm

   $$(500 \text{ mL})\left(\frac{2 \text{ atm}}{4 \text{ atm}}\right) = 250 \text{ mL}$$

In summary, a change in the volume of a gas due to a change in pressure can be calculated by multiplying the original volume by a ratio of the two pressures. If the pressure is increased, the ratio should have the smaller pressure in the numerator and the larger pressure in the denominator. If the pressure is decreased, the larger pressure should be in the numerator and the smaller pressure in the denominator:

new volume = original volume × ratio of pressures

We use Boyle's law in the following examples. If no mention is made of temperature, assume that it remains constant.

What volume will 2.50 L of a gas occupy if the pressure is changed from 760. mm Hg to 630. mm Hg?   **Example 12.2**

Method A. Conversion Factors   **SOLUTION**

Step 1 Determine whether pressure is being increased or decreased:

pressure decreases ⟶ volume increases

Step 2 Multiply the original volume by a ratio of pressures that will result in an increase in volume:

$$V = (2.50 \text{ L})\left(\frac{760. \text{ mm Hg}}{630. \text{ mm Hg}}\right) = 3.02 \text{ L (new volume)}$$

Decide which method is the best for you and stick with it.

Method B. Algebraic Equation

Step 1   Organize the given information:

$$P_1 = 760. \text{ mm Hg} \qquad V_1 = 2.50 \text{ L}$$
$$P_2 = 630. \text{ mm Hg} \qquad V_2 = ?$$

Step 2   Write and solve this equation for the unknown:

$$P_1V_1 = P_2V_2 \qquad V_2 = \frac{P_1V_1}{P_2}$$

Step 3   Put the given information into this equation and calculate:

$$V_2 = \frac{(760. \text{ mm Hg})(2.50 \text{ L})}{630. \text{ mm Hg}} = 3.02 \text{ L}$$

Example 12.3   A given mass of hydrogen occupies 40.0 L at 700. torr. What volume will it occupy at 5.00 atm pressure?

SOLUTION   **Method A. Conversion Factors**

Step 1   Determine whether the pressure is being increased or decreased. Note that in order to compare the values, the units must be the same. We'll convert 700. torr to atm:

$$(700. \text{ torr})\left(\frac{1 \text{ atm}}{760 \text{ torr}}\right) = 0.921 \text{ atm}$$

The pressure is going from 0.921 atm to 5.00 atm:

pressure increases $\longrightarrow$ volume decreases

Step 2   Multiply the original volume by a ratio of pressures that will result in a decrease in volume:

$$V = (40.0 \text{ L})\left(\frac{0.921 \text{ atm}}{5.00 \text{ atm}}\right) = 7.37 \text{ L}$$

Method B. Algebraic Equation

Step 1   Organize the given information. Remember to make the pressure units the same:

$$P_1 = 700. \text{ torr} = 0.921 \text{ atm} \qquad V_1 = 40.0 \text{ L}$$
$$P_2 = 5.00 \text{ atm} \qquad\qquad\qquad\quad V_2 = ?$$

Step 2   Write and solve this equation for the unknown:

$$P_1V_1 = P_2V_2 \qquad V_2 = \frac{P_1V_1}{P_2}$$

Step 3   Put the given information into this equation and calculate:

$$V_2 = \frac{(0.921 \text{ atm})(40.0 \text{ L})}{5.00 \text{ atm}} = 7.37 \text{ L}$$

A gas occupies a volume of 200. mL at 400. torr pressure. To what pressure must the gas be subjected in order to change the volume to 75.0 mL? **Example 12.4**

SOLUTION

Method A. Conversion Factors

Step 1 Determine whether volume is being increased or decreased:

volume decreases $\longrightarrow$ pressure increases

Step 2 Multiply the original pressure by a ratio of volumes that will result in an increase in pressure:

new pressure = original pressure × ratio of volumes

$$P = (400.\text{ torr})\left(\frac{200.\text{ mL}}{75.0\text{ mL}}\right) = 1067\text{ torr} \quad \text{or} \quad 1.07 \times 10^3\text{ torr (new pressure)}$$

Method B. Algebraic Equation

Step 1 Organize the given information (remember to make units the same):

$P_1 = 400.\text{ torr} \qquad V_1 = 200.\text{ mL}$
$P_2 = ? \qquad\qquad V_2 = 75.0\text{ mL}$

Step 2 Write and solve this equation for the unknown:

$$P_1V_1 = P_2V_2 \qquad P_2 = \frac{P_1V_1}{V_2}$$

Step 3 Put the given information into the equation and calculate:

$$P_2 = \frac{P_1V_1}{V_2} = \frac{(400.\text{ torr})(200.\text{ mL})}{75.0\text{ mL}} = 1.07 \times 10^3\text{ torr}$$

Practice 12.2
A gas occupies a volume of 3.86 L at 0.750 atm. At what pressure will the volume be 4.86 L?

12.6 Charles' Law

The effect of temperature on the volume of a gas was observed in about 1787 by the French physicist J. A. C. Charles (1746–1823). Charles found that various gases expanded by the same fractional amount when they underwent the same change in temperature. Later it was found that if a given volume of any gas initially at 0°C was cooled by 1°C, the volume decreased by $\frac{1}{273}$; if cooled by 2°C, it decreased by $\frac{2}{273}$; if cooled by 20°C, by $\frac{20}{273}$; and so on. Since each degree of cooling reduced the volume by $\frac{1}{273}$, it was apparent that any quantity of any gas would have zero volume if it could be cooled to −273°C. Of course, no real gas can be cooled to −273°C for the simple reason that it would liquefy before that temperature is reached. However, −273°C (more precisely, −273.15°C) is referred to as **absolute zero**; this temperature is the zero point on the Kelvin (absolute) temperature scale—the temperature at which the volume of an ideal, or perfect, gas would become zero.

absolute zero

Figure 12.8
Volume–temperature relationship of methane (CH_4). Extrapolated portion of the graph is shown by the broken line.

The volume–temperature relationship for methane is shown graphically in Figure 12.8. Experimental data show the graph to be a straight line that, when extrapolated, crosses the temperature axis at $-273.15°C$, or absolute zero. This is characteristic for all gases.

Charles' law   In modern form, **Charles' law** is as follows:

> At *constant pressure* the volume of a fixed mass of any gas is directly proportional to the absolute temperature, which may be expressed as
>
> $$V \propto T \qquad \text{or} \qquad \frac{V_1}{T_1} = \frac{V_2}{T_2}$$

A capital T is usually used for absolute temperature (K) and a small t for °C.

Mathematically this states that the volume of a gas varies directly with the absolute temperature when the pressure remains constant. In equation form, Charles' law may be written as

$$V = kT \qquad \text{or} \qquad \frac{V}{T} = k \qquad \text{(at constant pressure)}$$

where k is a constant for a fixed mass of the gas. If the absolute temperature of a gas is doubled, the volume will double.

To illustrate, let's return to the gas cylinder with the movable or free-floating piston (see Figure 12.9). Assume that the cylinder labeled (a) contains a quantity of gas and the pressure on it is 1 atm. When the gas is heated, the molecules move faster, and their kinetic energy increases. This action should increase the number of collisions per unit of time and therefore increase the pressure. However, the increased internal pressure will cause the piston to rise to a level at which the internal and external pressures again equal 1 atm, as we see in cylinder (b). The net result is an increase in volume due to an increase in temperature.

Another equation relating the volume of a gas at two different temperatures is

$$\frac{V_1}{T_1} = \frac{V_2}{T_2} \qquad \text{(constant } P)$$

where V_1 and T_1 are one set of conditions and V_2 and T_2 are another set of conditions.

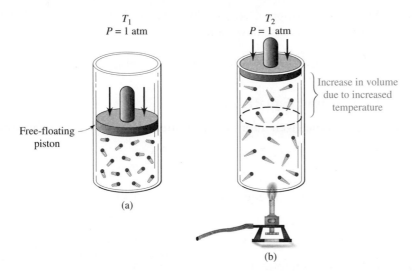

T_1
$P = 1$ atm

T_2
$P = 1$ atm

Increase in volume
due to increased
temperature

Free-floating
piston

(a)

(b)

Figure 12.9
The effect of temperature on the
volume of a gas. The gas in cylinder
(a) is heated from T_1 to T_2. With the
external pressure constant at 1 atm,
the free-floating piston rises,
resulting in an increased volume,
shown in cylinder (b).

A simple experiment showing the variation of the volume of a gas with temperature is illustrated in Figure 12.10. A balloon is placed in a beaker and liquid N_2 is poured over it. The volume is reduced, as shown by the collapse of the balloon; when the balloon is removed from the liquid N_2, the gas expands as it warms back to room temperature and the balloon increases in size.

(a)

(b)

(c)

Figure 12.10
The air-filled balloons in (a) are placed in liquid nitrogen (b). The volume of the air decreases tremendously at this temperature. In (c) the balloons are removed from the beaker and are beginning to return to their original volume as they warm back to room temperature.

Three liters of hydrogen at −20.°C are allowed to warm to a room temperature of 27°C. What is the volume at room temperature if the pressure remains constant?

Example 12.5

Method A. Conversion Factors

SOLUTION

Step 1 Determine whether temperature is being increased or decreased:

−20.°C + 273 = 253 K
27°C + 273 = 300. K

temperature increases ⟶ volume increases

Remember: Temperature must be changed to Kelvin in gas law problems. Note that we use 273 to convert instead of 273.15 since our original measurements are to the nearest degree.

Step 2   Multiply the original volume by a ratio of temperatures that will result in an increase in volume:

$$V = (3.00\ \text{L})\left(\frac{300.\ K}{253\ K}\right) = 3.56\ \text{L} \quad \text{(new volume)}$$

Method B. Algebraic Equation

Step 1   Organize the given information (remember to make units the same):

$$V_1 = 3.00\ \text{L} \qquad T_1 = -20.°\text{C} = 253\ \text{K}$$
$$V_2 = ? \qquad\qquad T_2 = 27°\text{C} = 300.\ \text{K}$$

Step 2   Write and solve the equation for the unknown:

$$\frac{V_1}{T_1} = \frac{V_2}{T_2} \qquad V_2 = \frac{V_1 T_2}{T_1}$$

Step 3   Put the given information into the equation and calculate:

$$V_2 = \frac{V_1 T_2}{T_1} = \frac{(3.00\ \text{L})(300.\ K)}{253\ K} = 3.56\ \text{L}$$

Example 12.6

SOLUTION

If 20.0 L of oxygen are cooled from 100.°C to 0.°C, what is the new volume?

Since no mention is made of pressure, assume that pressure does not change.

Method A. Conversion Factors

Step 1   Change °C to K:

$$100.°\text{C} + 273 = 373\ \text{K}$$
$$0.°\text{C} + 273 = 273\ \text{K}$$

Step 2   The ratio of temperature to be used is 273 K/373 K, because the final volume should be smaller than the original volume. The calculation is

$$V = (20.0\ \text{L})\left(\frac{273\ K}{373\ K}\right) = 14.6\ \text{L} \quad \text{(new volume)}$$

Method B. Algebraic Equation

Step 1   Organize the given information (remember to make units coincide):

$$V_1 = 20.0\ \text{L} \qquad T_1 = 100.°\text{C} = 373\ \text{K}$$
$$V_2 = ? \qquad\qquad T_2 = 0.°\text{C} = 273\ \text{K}$$

Step 2   Write and solve the equation for the unknown:

$$\frac{V_1}{T_1} = \frac{V_2}{T_2} \qquad V_2 = \frac{V_1 T_2}{T_1}$$

Step 3   Put the given information into the equation and calculate:

$$V_2 = \frac{V_1 T_2}{T_1} = \frac{(20.0\ \text{L})(273\ K)}{373\ K} = 14.6\ \text{L}$$

Practice 12.3 _____

A 4.50-L container of nitrogen gas at 28.0°C is heated to 56.0°C. Assuming that the volume of the container can vary, what is the new volume of the gas?

12.7 Gay-Lussac's Law

J. L. Gay-Lussac (1778–1850) was a French chemist involved in the study of volume relationships of gases. The three variables [pressure (P), volume (V), and temperature (T)] are needed to describe a fixed amount of a gas. Boyle's law ($PV = k$) relates pressure and volume at constant temperature; Charles' law ($V = kT$) relates volume and temperature at constant pressure. A third relationship involving pressure and temperature at constant volume is a modification of Charles' law and is sometimes called **Gay-Lussac's law** (Figure 12.11).

Gay-Lussac's law

The pressure of a fixed mass of a gas, at constant volume, is directly proportional to the Kelvin temperature:

$$P = kT \quad \text{or} \quad \frac{P_1}{T_1} = \frac{P_2}{T_2}$$

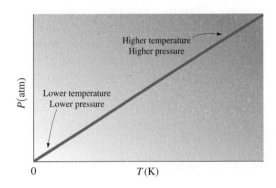

Figure 12.11
Gay-Lussac's law shows that pressure and temperature are directly proportional.

The pressure of a container of helium is 650. torr at 25°C. If the sealed container is cooled to 0.°C, what will the pressure be?

Example 12.7

Method A. Conversion Factors

SOLUTION

Step 1 Determine whether temperature is being increased or decreased:

temperature decreases ⟶ pressure decreases

Step 2 Multiply the original pressure by a ratio of Kelvin temperatures that will result in a decrease in pressure:

$$(650.\, \text{torr})\left(\frac{273\, K}{298\, K}\right) = 595\, \text{torr}$$

Method B. Algebraic Equation

Step 1 Organize the given information (remember to make units the same):

$$P_1 = 650.\,\text{torr} \qquad T_1 = 25°\text{C} = 298\,\text{K}$$
$$P_2 = ? \qquad\qquad T_2 = 0.°\text{C} = 273\,\text{K}$$

Step 2 Write and solve equation for the unknown:

$$\frac{P_1}{T_1} = \frac{P_2}{T_2} \qquad P_2 = \frac{P_1 T_2}{T_1}$$

Step 3 Put given information into equation and calculate:

$$P_2 = \frac{(650.\,\text{torr})(273\,\cancel{K})}{298\,\cancel{K}} = 595\,\text{torr}$$

Practice 12.4 _____

A gas cylinder contains 40.0 L of gas at 45.0°C and has a pressure of 650. torr. What will the pressure be if the temperature is changed to 100.°C?

12.8 Combined Gas Laws

standard conditions
standard temperature
and pressure (STP)

To compare volumes of gases, common reference points of temperature and pressure were selected and called **standard conditions** or **standard temperature and pressure** (abbreviated **STP**). Standard temperature is 273.15 K (0°C), and standard pressure is 1 atm or 760 torr or 760 mm Hg or 101.325 kPa. For purposes of comparison, volumes of gases are usually changed to STP conditions:

In this text we'll use 273 K for temperature conversions and calculations. Check with your instructor for rules in your class.

standard temperature = 273.15 K or 0.00°C
standard pressure = 1 atm or 760 torr or 760 mm Hg or 101.325 kPa

When temperature and pressure change at the same time, the new volume may be calculated by multiplying the initial volume by the correct ratios of both pressure and temperature as follows:

$$\text{final volume} = (\text{initial volume})\left(\begin{array}{c}\text{ratio of}\\\text{pressures}\end{array}\right)\left(\begin{array}{c}\text{ratio of}\\\text{temperatures}\end{array}\right)$$

This equation combines Boyle's and Charles' laws, and the same considerations for the pressure and temperature ratios should be used in the calculation. The four possible variations are as follows:

1. Both T and P cause an increase in volume.
2. Both T and P cause a decrease in volume.
3. T causes an increase and P causes a decrease in volume.
4. T causes a decrease and P causes an increase in volume.

The P, V, and T relationships for a given mass of any gas, in fact, may be expressed as a single equation, $PV/T = k$. For problem solving, this equation is usually written

$$\frac{P_1 V_1}{T_1} = \frac{P_2 V_2}{T_2}$$

where P_1, V_1, and T_1 are the initial conditions and P_2, V_2, and T_2 are the final conditions.

This equation can be solved for any one of the six variables and is useful in dealing with the pressure–volume–temperature relationships of gases. Note what happens to the combined gas law when one of the variables is constant:

- T constant $\rightarrow P_1 V_1 = P_2 V_2$ Boyle's law

- P constant $\rightarrow \dfrac{V_1}{T_1} = \dfrac{V_2}{T_2}$ Charles' law

- V constant $\rightarrow \dfrac{P_1}{T_1} = \dfrac{P_2}{T_2}$ Gay-Lussac's law

Note: In the examples below, the use of 273 K does not change the number of significant figures in the temperature. The converted temperature is expressed to the same precision as the original measurement.

Given 20.0 L of ammonia gas at 5°C and 730. torr, calculate the volume at 50.°C and 800. torr.

Example 12.8

SOLUTION

Step 1   Organize the given information, putting temperatures in kelvins:

$P_1 = 730.$ torr $P_2 = 800.$ torr
$V_1 = 20.0$ L $V_2 = ?$
$T_1 = 5°C = 278$ K $T_2 = 50.°C = 323$ K

Method A. Conversion Factors

Step 2   Set up ratios of T and P:

$$T \text{ ratio} = \frac{323 \text{ K}}{278 \text{ K}} \quad (\text{increase in } T \text{ should increase } V)$$

$$P \text{ ratio} = \frac{730. \text{ torr}}{800. \text{ torr}} \quad (\text{increase in } P \text{ should decrease } V)$$

Step 3   Multiply the original volume by the ratios:

$$V_2 = (20.0 \text{ L})\left(\frac{730. \text{ torr}}{800. \text{ torr}}\right)\left(\frac{323 \text{ K}}{278 \text{ K}}\right) = 21.2 \text{ L}$$

Method B. Algebraic Equation

Step 2   Write and solve the equation for the unknown. Solve

$$\frac{P_1 V_1}{T_1} = \frac{P_2 V_2}{T_2}$$

for V_2 by multiplying both sides of the equation by T_2/P_2 and rearranging to obtain

$$V_2 = \frac{V_1 P_1 T_2}{P_2 T_1}$$

Step 3   Put the given information into the equation and calculate:

$$V_2 = \frac{(20.0 \text{ L})(730. \text{ torr})(323 \text{ K})}{(800. \text{ torr})(278 \text{ K})} = 21.2 \text{ L}$$

Example 12.9 To what temperature (°C) must 10.0 L of nitrogen at 25°C and 700. torr be heated in order to have a volume of 15.0 L and a pressure of 760. torr?

SOLUTION **Step 1** Organize the given information, putting temperatures in kelvins:

$$P_1 = 700.\text{ torr} \qquad P_2 = 760.\text{ torr}$$
$$V_1 = 10.0\text{ L} \qquad V_2 = 15.0\text{ L}$$
$$T_1 = 25°\text{C} = 298\text{ K} \qquad T_2 = ?$$

Method A. Conversion Factors

Step 2 Set up ratios of V and P:

$$P\text{ ratio} = \frac{760.\text{ torr}}{700.\text{ torr}} \quad \text{(increase in } P \text{ should increase } T\text{)}$$

$$V\text{ ratio} = \frac{15.0\text{ L}}{10.0\text{ L}} \quad \text{(increase in } V \text{ should increase } T\text{)}$$

Step 3 Multiply the original temperature by the ratios:

$$T_2 = (298\text{ K})\left(\frac{760.\text{ torr}}{700.\text{ torr}}\right)\left(\frac{15.0\text{ L}}{10.0\text{ L}}\right) = 485\text{ K}$$

Method B. Algebraic Equation

Step 2 Write and solve the equation for the unknown:

$$\frac{P_1 V_1}{T_1} = \frac{P_2 V_2}{T_2} \qquad T_2 = \frac{T_1 P_2 V_2}{P_1 V_1}$$

Step 3 Put the given information into the equation and calculate:

$$T_2 = \frac{(298\text{ K})(760.\text{ torr})(15.0\text{ L})}{(700.\text{ torr})(10.0\text{ L})} = 485\text{ K}$$

In either method, since the problem asks for °C, we must subtract 273 from the Kelvin answer:

$$485\text{ K} - 273 = 212°\text{C}$$

Example 12.10 The volume of a gas-filled balloon is 50.0 L at 20.°C and 742 torr. What volume will it occupy at standard temperature and pressure (STP)?

SOLUTION **Step 1** Organize the given information, putting temperatures in kelvins:

$$P_1 = 742\text{ torr} \qquad P_2 = 760.\text{ torr (standard pressure)}$$
$$V_1 = 50.0\text{ L} \qquad V_2 = ?$$
$$T_1 = 20.°\text{C} = 293\text{ K} \qquad T_2 = 273\text{ K (standard temperature)}$$

Method A. Conversion Factors

Step 2 Set up ratios of T and P:

$$T\text{ ratio} = \frac{273\text{ K}}{293\text{ K}} \quad \text{(decrease in } T \text{ should decrease } V\text{)}$$

$$P\text{ ratio} = \frac{742\text{ torr}}{760.\text{ torr}} \quad \text{(increase in } P \text{ should decrease } V\text{)}$$

Step 3 Multiply the original volume by the ratios:

$$V_2 = (50.0 \text{ L})\left(\frac{273 \text{ K}}{293 \text{ K}}\right)\left(\frac{742 \text{ torr}}{760. \text{ torr}}\right) = 45.5 \text{ L}$$

Method B. Algebraic Equation

Step 2 Write and solve the equation for the unknown:

$$\frac{P_1V_1}{T_1} = \frac{P_2V_2}{T_2} \qquad V_2 = \frac{P_1V_1T_2}{P_2T_1}$$

Step 3 Put the given information into the equation and calculate:

$$V_2 = \frac{(742 \text{ torr})(50.0 \text{ L})(273 \text{ K})}{(760. \text{ torr})(293 \text{ K})} = 45.5 \text{ L}$$

Practice 12.5

15.00 L of gas at 45.0°C and 800. torr are heated to 400.°C and the pressure changed to 300. torr. What is the new volume?

Practice 12.6

To what temperature must 5.00 L of oxygen at 50.°C and 600. torr be heated in order to have a volume of 10.0 L and a pressure of 800. torr?

12.9 Dalton's Law of Partial Pressures

If gases behave according to the kinetic-molecular theory, there should be no difference in the pressure–volume–temperature relationships whether the gas molecules are all the same or different. This similarity in the behavior of gases is the basis for an understanding of **Dalton's law of partial pressures**:

Dalton's law of partial pressures

The total pressure of a mixture of gases is the sum of the partial pressures exerted by each of the gases in the mixture.

Each gas in the mixture exerts a pressure that is independent of the other gases present. These pressures are called **partial pressures**. Thus, if we have a mixture of three gases, A, B, and C, exerting partial pressures of 50. torr, 150. torr, and 400. torr, respectively, the total pressure will be 600. torr:

partial pressure

$$P_{\text{Total}} = P_A + P_B + P_C$$
$$P_{\text{Total}} = 50. \text{ torr} + 150. \text{ torr} + 400. \text{ torr} = 600. \text{ torr}$$

We can see an application of Dalton's law in the collection of insoluble gases over water. When prepared in the laboratory, oxygen is commonly collected by the downward displacement of water. Thus the oxygen is not pure but is mixed with water vapor (see Figure 12.12). When the water levels are adjusted to the same height inside and outside the bottle, the pressure of the oxygen plus water vapor inside the bottle is equal to the atmospheric pressure:

$$P_{\text{atm}} = P_{O_2} + P_{H_2O}$$

Traditional "messenger" molecules are amino acids (also known as the building blocks for proteins). You've probably heard of endorphins, which are the messenger molecules associated with "runner's high." Until recently, these molecules (called *neurotransmitters*) were thought to be specific; that is, each neurotransmitter fits into the target cell like a key in a lock. It was also thought that neurotransmitters were stored in tiny pouches where they are manufactured and then released when needed. But then two gases, nitrogen monoxide (NO) and carbon monoxide (CO), were found to act as neurotransmitters. These gases break all the "rules" for neurotransmission because, as gases, they are nonspecific. Gases freely diffuse into nearby cells, so NO and CO neurotransmitters must be made "on demand," since they cannot be stored. Gas neurotransmitters cannot use the lock-and-key model to act on cells, so they must use their chemical properties.

Nitrogen monoxide has been used for nearly a century to dilate blood vessels and increase blood flow, lowering blood pressure. In the late 1980s, scientists discovered that biologically produced NO is an important signaling molecule for nerve cells. Nitrogen monoxide mediates certain neurons that do not respond to traditional neurotransmitters. These NO-sensitive neurons are found in the cardiovascular, respiratory, digestive, and urogenital systems. Nitrogen monoxide also appears to play a role in regulating blood pressure, blood clotting, and neurotransmission. The 1998 Nobel Prize in medicine was awarded to three

Viagra, the medication for male impotence, is the latest application of our understanding of the role of NO in the body.

Americans for their work on nitrogen monoxide's role in the body.

Nitrogen monoxide is a key molecule in producing erections in men. When a man is sexually stimulated, NO is released into the penis, where it activates the release of an enzyme that increases the level of a molecule called cGMP and ultimately produces relaxation of smooth muscles, allowing blood to flow into the penis. Viagra, an impotence pill, enhances the effect of NO, inhibiting the enzyme that breaks down the cGMP molecule. This results in more cGMP molecules, and smooth muscle relaxation begins producing an erection.

After NO transmitters were discovered, researchers at Johns Hopkins Medical School reasoned that if one gas acted as a neurotransmitter, so might others. They proposed CO as a possible transmitter because the enzyme used to make it is localized in specific parts of the brain (those responsible for smell and long-term memory). The enzyme used to make messengers was found in exactly the same locations! Researchers then showed that nerve cells made the messenger molecule when stimulated with CO. An inhibitor for CO blocked the messenger molecule production. Researchers are exploring the possibility of still other roles for these gaseous messenger molecules.

To determine the amount of O_2 or any other gas collected over water, we subtract the pressure of the water vapor (vapor pressure) from the total pressure of the gases:

$$P_{O_2} = P_{atm} - P_{H_2O}$$

The vapor pressure of water at various temperatures is tabulated in Appendix IV.

Figure 12.12
Oxygen collected over water.

A 500.-mL sample of oxygen was collected over water at 23°C and 760. torr. What volume will the dry O_2 occupy at 23°C and 760. torr? The vapor pressure of water at 23°C is 21.2 torr.

Example 12.11

SOLUTION

To solve this problem, we must first determine the pressure of the oxygen alone, by subtracting the pressure of the water vapor (given in Appendix IV) present:

Step 1 Determine the pressure of dry O_2:

$$P_{Total} = 760. \text{torr} = P_{O_2} + P_{H_2O}$$
$$P_{O_2} = 760. \text{torr} - 21.2 \text{torr} = 739 \text{torr} \qquad (\text{dry } O_2)$$

Step 2 Organize the given information:

$$P_1 = 739 \text{torr} \qquad P_2 = 760. \text{torr}$$
$$V_1 = 500. \text{mL} \qquad V_2 = ?$$

T is constant

Step 3 Solve as a Boyle's law problem:

$$V = \frac{(500. \text{mL})(739 \text{torr})}{760. \text{torr}} = 486 \text{mL dry } O_2$$

Practice 12.7

Hydrogen gas was collected by downward displacement of water. A volume of 600.0 mL of gas was collected at 25.0°C and 740.0 torr. What volume will the dry hydrogen occupy at STP?

12.10 Avogadro's Law

Early in the 19th century, Gay-Lussac studied the volume relationships of reacting gases. His results, published in 1809, were summarized in a statement known as **Gay-Lussac's law of combining volumes**:

Gay-Lussac's law of combining volumes

When measured at the same temperature and pressure, the ratios of the volumes of reacting gases are small whole numbers.

Thus, H_2 and O_2 combine to form water vapor in a volume ratio of 2:1 (Figure 12.13); H_2 and Cl_2 react to form HCl in a volume ratio of 1:1; and H_2 and N_2 react to form NH_3 in a volume ratio of 3:1.

Two years later, in 1811, Amedeo Avogadro (1776–1856) used the law of combining volumes of gases to make a simple but significant and far-reaching generalization concerning gases. **Avogadro's law** states:

Avogadro's law

Equal volumes of different gases at the same temperature and pressure contain the same number of molecules.

Figure 12.13
Gay-Lussac's law of combining volumes of gases applied to the reaction of hydrogen and oxygen. When measured at the same temperature and pressure, hydrogen and oxygen react in a volume ratio of 2:1.

$$2\ H_2(g) \qquad + \qquad O_2(g) \longrightarrow \qquad 2\ H_2O(g)$$

H$_2$ H$_2$	+	O$_2$	→	H$_2$O H$_2$O
2 volumes		1 volume		2 volumes

This law was a real breakthrough in understanding the nature of gases.

1. It offered a rational explanation of Gay-Lussac's law of combining volumes of gases and indicated the diatomic nature of such elemental gases as hydrogen, chlorine, and oxygen.
2. It provided a method for determining the molar masses of gases and for comparing the densities of gases of known molar mass (see Sections 12.11 and 12.12).
3. It afforded a firm foundation for the development of the kinetic-molecular theory.

By Avogadro's law, equal volumes of hydrogen and chlorine at the same temperature and pressure contain the same number of molecules. On a volume basis, hydrogen and chlorine react thus (see Figure 12.14):

hydrogen + chlorine ⟶ hydrogen chloride
1 volume   1 volume   2 volumes

Therefore, hydrogen molecules react with chlorine molecules in a 1:1 ratio. Since two volumes of hydrogen chloride are produced, one molecule of hydrogen and one molecule of chlorine must produce two molecules of hydrogen chloride. Therefore, each hydrogen molecule and each chlorine molecule must consist of two atoms. The coefficients of the balanced equation for the reaction give the correct ratios for volumes, molecules, and moles of reactants and products:

$$H_2 \qquad + \qquad Cl_2 \qquad \longrightarrow \qquad 2\ HCl$$

1 volume   1 volume   2 volumes
1 molecule 1 molecule   2 molecules
1 mol   1 mol   2 mol

By like reasoning, oxygen molecules also must contain at least two atoms because one volume of oxygen reacts with two volumes of hydrogen to produce two volumes of water vapor.

The volume of a gas depends on the temperature, the pressure, and the number of gas molecules. Different gases at the same temperature have the same average kinetic energy. Hence, if two different gases are at the same temperature, occupy equal volumes, and exhibit equal pressures, each gas must contain the same number of molecules. This statement is true because systems with identical *PVT* properties can be produced only by equal numbers of molecules having the same average kinetic energy.

Figure 12.14
Avogadro's law proved the concept of diatomic molecules for hydrogen and chlorine.

H$_2$ + Cl$_2$ ⟶ 2 HCl

12.11 Mole–Mass–Volume Relationships of Gases

Because a mole contains 6.022×10^{23} molecules (Avogadro's number), a mole of any gas will have the same volume as a mole of any other gas at the same temperature and pressure. It has been experimentally determined that the volume occupied by a mole of any gas is 22.4 L at STP. This volume, 22.4 L, is known as the **molar volume** of a gas. The molar volume is a cube about 28.2 cm (11.1 in.) on a side. The molar masses of several gases, each occupying 22.4 L at STP, are shown in Figure 12.15:

> One mole of a gas occupies 22.4 L at STP.

The molar volume is useful for determining the molar mass of a gas or of substances that can be easily vaporized. If the mass and the volume of a gas at STP are known, we can calculate its molar mass. For example, 1.00 L of pure oxygen at STP has a mass of 1.429 g. The molar mass of oxygen may be calculated by multiplying the mass of 1.00 L by 22.4 L/mol:

$$\left(\frac{1.429 \text{ g}}{1.00 \text{ L}}\right)\left(\frac{22.4 \text{ L}}{1 \text{ mol}}\right) = 32.0 \text{ g/mol} \qquad \text{(molar mass)}$$

If the mass and volume are at other than standard conditions, we change the volume to STP and then calculate the molar mass.

The molar volume, 22.4 L/mol, is used as a conversion factor to convert grams per liter to grams per mole (molar mass) and also to convert liters to moles. The two conversion factors are

$$\frac{22.4 \text{ L}}{1 \text{ mol}} \quad \text{and} \quad \frac{1 \text{ mol}}{22.4 \text{ L}}$$

These conversions must be done at STP except under certain special circumstances.

molar volume

As with many constants, the molar volume is known more exactly to be 22.414 L. We use 22.4 L in our calculations, since the extra figures don't often affect the result, given the other measurements in the calculation.

Standard conditions apply only to pressure, temperature, and volume. Mass is not affected.

131.3 g Xe 71.90 g Cl_2 16.04 g CH_4
16.00 g O_2 36.46 g HCl
44.01 g CO_2 28.2 cm 64.07 g SO_2
22.4 L
28.02 g N_2 28.2 cm
17.03 g NH_3 28.2 cm 2.016 g H_2
39.95 g Ar 34.09 g H_2S

224 L

Figure 12.15
One mole of a gas occupies 22.4 L at STP. The mass given for each gas is the mass of 1 mol.

Example 12.12 If 2.00 L of a gas measured at STP have a mass of 3.23 g, what is the molar mass of the gas?

SOLUTION The unit of molar mass is g/mol; the conversion is from

$$\frac{g}{L} \longrightarrow \frac{g}{mol}$$

The starting amount is $\dfrac{3.23\ g}{2.00\ L}$. The conversion factor is $\dfrac{22.4\ L}{1\ mol}$. The calculation is

$$\left(\frac{3.23\ g}{2.00\ \cancel{L}}\right)\left(\frac{22.4\ \cancel{L}}{1\ mol}\right) = 36.2\ g/mol \qquad \text{(molar mass)}$$

Example 12.13 Measured at 40.°C and 630. torr, the mass of 691 mL of diethyl ether is 1.65 g. Calculate the molar mass of diethyl ether.

SOLUTION **Step 1** Organize the given information, converting temperatures to kelvin. Note that we must change to STP in order to determine molar mass.

Diethyl ether
$(C_2H_5)_2O$

$P_1 = 630.\ torr \qquad P_2 = 760.\ torr$
$V_1 = 691\ mL \qquad V_2 = ?$
$T_1 = 313\ K\,(40.°C) \qquad T_2 = 273\ K$

Step 2 Use either the conversion factor method or the algebraic method and the combined gas law to correct the volume (V_2) to STP:

$$V_2 = \frac{(691\ mL)(273\ \cancel{K})(630.\ \cancel{torr})}{(313\ \cancel{K})(760.\ \cancel{torr})} = 500.\ mL = 0.500\ L \quad \text{(at STP)}$$

Step 3 In the example, V_2 is the volume for 1.65 g of the gas, so we can now find the molar mass by converting g/L to g/mol:

$$\left(\frac{1.65\ g}{0.500\ \cancel{L}}\right)\left(\frac{22.4\ \cancel{L}}{mol}\right) = 73.9\ g/mol$$

Practice 12.8
A gas with a mass of 86 g occupies 5.00 L at 25°C and 3.00 atm pressure. What is the molar mass of the gas?

12.12 Density of Gases

The density, *d*, of a gas is its mass per unit volume, which is generally expressed in grams per liter as follows:

$$d = \frac{mass}{volume} = \frac{g}{L}$$

Because the volume of a gas depends on temperature and pressure, both should be given when stating the density of a gas. The volume of a solid or liquid is hardly affected by changes in pressure and is changed only slightly when the temperature is varied. Increasing the temperature from 0°C to 50°C will reduce the density of a gas by about 18% if the gas is allowed to expand, whereas a 50°C rise in the temperature of water (0°C $\longrightarrow$ 50°C) will change its density by less than 0.2%.

The density of a gas at any temperature and pressure can be determined by calculating the mass of gas present in 1 L. At STP, in particular, the density can be calculated by multiplying the molar mass of the gas by 1 mol/22.4 L:

$$d_{STP} = \text{molar mass} \left(\frac{1\ mol}{22.4\ L}\right)$$

$$\text{molar mass} = d_{STP} \left(\frac{22.4\ L}{1\ mol}\right)$$

Table 12.3 lists the densities of some common gases.

Table 12.3 Density of Common Gases at STP

Gas	Molar mass (g/mol)	Density (g/L at STP)	Gas	Molar mass (g/mol)	Density (g/L at STP)
H_2	2.016	0.0900	H_2S	34.09	1.52
CH_4	16.04	0.716	HCl	36.46	1.63
NH_3	17.03	0.760	F_2	38.00	1.70
C_2H_2	26.04	1.16	CO_2	44.01	1.96
HCN	27.03	1.21	C_3H_8	44.09	1.97
CO	28.01	1.25	O_3	48.00	2.14
N_2	28.02	1.25	SO_2	64.07	2.86
air	(28.9)	(1.29)	Cl_2	70.90	3.17
O_2	32.00	1.43			

Calculate the density of Cl_2 at STP.

Example 12.14

First calculate the molar mass of Cl_2. It is 70.90 g/mol. Since d = g/L, the conversion is

SOLUTION

$$\frac{g}{mol} \longrightarrow \frac{g}{L}$$

The conversion factor is $\frac{1\ mol}{22.4\ L}$:

$$d = \left(\frac{70.90\ g}{1\ mol}\right)\left(\frac{1\ mol}{22.4\ L}\right) = 3.17\ g/L$$

Practice 12.9

The molar mass of a gas is 20. g/mol. Calculate the density of the gas at STP.

12.13 Ideal Gas Law

We've used four variables in calculations involving gases: the volume (V), the pressure (P), the absolute temperature (T), and the number of molecules or moles (n). Combining these variables into a single expression, we obtain

$$V \propto \frac{nT}{P} \quad \text{or} \quad V = \frac{nRT}{P}$$

where R is a proportionality constant known as the *ideal gas constant*. The equation is commonly written as

$$PV = nRT$$

ideal gas law   and is known as the **ideal gas law**. This law summarizes in a single expression what we have considered in our earlier discussions. The value and units of R depend on the units of P, V, and T. We can calculate one value of R by taking 1 mol of a gas at STP conditions. Solve the equation for R:

$$R = \frac{PV}{nT} = \frac{(1\,\text{atm})(22.4\,\text{L})}{(1\,\text{mol})(273\,\text{K})} = 0.0821 \frac{\text{L-atm}}{\text{mol-K}}$$

The units of R in this case are liter-atmospheres (L-atm) per mole kelvin (mol-K). When the value of $R = 0.0821$ L-atm/mol-K, P is in atmospheres, n is in moles, V is in liters, and T is in kelvins.

The ideal gas equation can be used to calculate any one of the four variables when the other three are known.

Example 12.15   What pressure will be exerted by 0.400 mol of a gas in a 5.00-L container at 17°C?

SOLUTION   **Step 1**   Organize the given information, converting temperature to kelvins:

$P = ?$

$V = 5.00\,\text{L}$

$T = 290.\,\text{K}$

$n = 0.400\,\text{mol}$

Step 2   Write and solve the ideal gas equation for the unknown:

$$PV = nRT \quad \text{or} \quad P = \frac{nRT}{V}$$

Step 3   Substitute the given data into the equation and calculate:

$$P = \frac{(0.400\,\cancel{\text{mol}})(0.0821\,\cancel{\text{L}} \cdot \text{atm}/\cancel{\text{mol}} \cdot \cancel{\text{K}})(290.\,\cancel{\text{K}})}{5.00\,\cancel{\text{L}}} = 1.90\,\text{atm}$$

Example 12.16   How many moles of oxygen gas are in a 50.0-L tank at 22.0°C if the pressure gauge reads 2000. lb/in.^{2}?

Step 1   Organize the given information, converting temperature to kelvins and pressure to atmospheres:

$$P = \left(\frac{2000.\ \cancel{lb}}{\cancel{in.^2}}\right)\left(\frac{1\ atm}{14.7\ \cancel{lb/in.^2}}\right) = 136.1\ atm$$

$$V = 50.0\ L$$

$$T = 295\ K$$

$$n = ?$$

Step 2   Write and solve the ideal gas equation for the unknown:

$$PV = nRT \qquad or \qquad n = \frac{PV}{RT}$$

Step 3   Substitute the given information into the equation and calculate:

$$n = \frac{(136.1\ \cancel{atm})(50.0\ \cancel{L})}{(0.0821\ \cancel{L\cdot atm}/mol\cdot \cancel{K})(295\ \cancel{K})} = 281\ mol\ O_2$$

Practice 12.10 _____

A 23.8-L cylinder contains oxygen gas at 20.0°C and 732 torr. How many moles of oxygen are in the cylinder?

The molar mass of a gaseous substance can be determined using the ideal gas law. Since molar mass = g/mol, it follows that mol = g/molar mass. Using M for molar mass and g for grams, we can substitute g/M for n (moles) in the ideal gas law to get

$$PV = \frac{g}{M}RT \qquad or \qquad M = \frac{gRT}{PV} \qquad \text{(modified ideal gas law)}$$

This form of the ideal gas law is most useful in problems containing mass instead of moles.

which allows us to calculate the molar mass, M, for any substance in the gaseous state.

Calculate the molar mass of butane gas, if 3.69 g occupy 1.53 L at 20.0°C and 1.00 atm.   **Example 12.17**

Change 20.0°C to 293 K and substitute the data into the modified ideal gas law: **SOLUTION**

$$M = \frac{gRT}{PV} = \frac{(3.69\ g)(0.0821\ \cancel{L\cdot atm}/mol\cdot \cancel{K})(293\ \cancel{K})}{(1.00\ \cancel{atm})(1.53\ \cancel{L})} = 58.0\ g/mol$$

Practice 12.11 _____

A 0.286-g sample of a certain gas occupies 50.0 mL at standard temperature and 76.0 cm Hg. Determine the molar mass of the gas.

12.14 Gas Stoichiometry

Mole–Volume and Mass–Volume Calculations

Stoichiometric problems involving gas volumes can be solved by the general mole-ratio method outlined in Chapter 9. The factors 1 mol/22.4 L and 22.4 L/1 mol are used for converting volume to moles and moles to volume, respectively. (See Figure 12.16.) These conversion factors are used under the assumption that the gases are at STP and that they behave as ideal gases. In actual practice, gases are measured at other than STP conditions, and the volumes are converted to STP for stoichiometric calculations.

In a balanced equation, the number preceding the formula of a gaseous substance represents the number of moles or molar volumes (22.4 L at STP) of that substance.

Figure 12.16
Summary of the primary conversions involved in stoichiometry. The conversion for volumes of gases is included.

Volume A	→			→	Volume B
Grams A	→	Moles A	Moles B	→	Grams B
Atoms or molecules A	→			→	Atoms or molecules B

The following are examples of typical problems involving gases and chemical equations.

Example 12.18   What volume of oxygen (at STP) can be formed from 0.500 mol of potassium chlorate?

SOLUTION   **Step 1**   Write the balanced equation:

$$2\,KClO_3(s) \longrightarrow 2\,KCl(s) + 3\,O_2(g)$$

Step 2   The starting amount is 0.500 mol $KClO_3$. The conversion is from

$$mol\,KClO_3 \longrightarrow mol\,O_2 \longrightarrow L\,O_2$$

Step 3   Calculate the moles of O_2, using the mole-ratio method:

$$(0.500\ \text{mol KClO}_3)\left(\frac{3\,\text{mol}\,O_2}{2\ \text{mol KClO}_3}\right) = 0.750\,\text{mol}\,O_2$$

Step 4   Convert moles of O_2 to liters of O_2. The moles of a gas at STP are converted to liters by multiplying by the molar volume, 22.4 L/mol:

$$(0.750\ \text{mol}\,O_2)\left(\frac{22.4\,L}{1\ \text{mol}}\right) = 16.8\,L\,O_2$$

Alternatively, setting up a continuous calculation, we obtain

$$(0.500\ \text{mol KClO}_3)\left(\frac{3\,\text{mol}\,O_2}{2\ \text{mol KClO}_3}\right)\left(\frac{22.4\,L}{1\ \text{mol}}\right) = 16.8\,L\,O_2$$

How many grams of aluminum must react with sulfuric acid to produce 1.25 L of hydrogen gas at STP?

Example 12.19

SOLUTION

Step 1 The balanced equation is

$$2\,Al(s) + 3\,H_2SO_4(aq) \longrightarrow Al_2(SO_4)_3(aq) + 3\,H_2(g)$$

Step 2 We first convert liters of H_2 to moles of H_2. Then the familiar stoichiometric calculation from the equation is used. The conversion is

$$L\,H_2 \longrightarrow mol\,H_2 \longrightarrow mol\,Al \longrightarrow g\,Al$$

$$1.25\,\cancel{L}\,\cancel{H_2}\left(\frac{1\,\cancel{mol}}{22.4\,\cancel{L}}\right)\left(\frac{2\,\cancel{mol\,Al}}{3\,\cancel{mol\,H_2}}\right)\left(\frac{26.98\,g\,Al}{1\,\cancel{mol\,Al}}\right) = 1.00\,g\,Al$$

What volume of hydrogen, collected at 30.°C and 700. torr, will be formed by reacting 50.0 g of aluminum with hydrochloric acid?

Example 12.20

$$2\,Al(s) + 6\,HCl(aq) \longrightarrow 2\,AlCl_3(aq) + 3\,H_2(g)$$

SOLUTION

In this problem, the conditions are not at STP, so we cannot use the method shown in Example 12.18. Either we need to calculate the volume at STP from the equation and then convert this volume to the conditions given in the problem, or we can use the ideal gas law. Let's use the ideal gas law.

First calculate the moles of H_2 obtained from 50.0 g of Al. Then, using the ideal gas law, calculate the volume of H_2 at the conditions given in the problem:

Step 1 Moles of H_2: The conversion is

$$grams\,Al \longrightarrow moles\,Al \longrightarrow moles\,H_2$$

$$50.0\,\cancel{g\,Al}\left(\frac{1\,\cancel{mol\,Al}}{26.98\,\cancel{g\,Al}}\right)\left(\frac{3\,mol\,H_2}{2\,\cancel{mol\,Al}}\right) = 2.78\,mol\,H_2$$

Step 2 Liters of H_2: Solve $PV = nRT$ for V and substitute the data into the equation:

Convert °C to K: 30.°C + 273 = 303 K.

Convert torr to atm: (700. $\cancel{torr}$)(1 atm/760. $\cancel{torr}$) = 0.921 atm.

$$V = \frac{nRT}{P} = \frac{(2.78\,\cancel{mol}\,H_2)(0.0821\,L\text{-}\cancel{atm})(303\,\cancel{K})}{(0.921\,\cancel{atm})(\cancel{mol\text{-}K})} = 75.1\,L\,H_2$$

(*Note:* The volume at STP is 62.3 L H_2.)

Practice 12.12

If 10.0 g of sodium peroxide (Na_2O_2) react with water to produce sodium hydroxide and oxygen, how many liters of oxygen will be produced at 20.°C and 750. torr?

$$2\,Na_2O_2(s) + 2\,H_2O(l) \longrightarrow 4\,NaOH(aq) + O_2(g)$$

Volume–Volume Calculations

When all substances in a reaction are in the gaseous state, simplifications in the calculation can be made. These are based on Avogadro's law, which states that gases under identical conditions of temperature and pressure contain the same number of molecules and occupy the same volume. Under the standard conditions of temperature and pressure, the volumes of gases reacting are proportional to the numbers of moles of the gases in the balanced equation. Consider the reaction:

$$H_2(g) \quad + \quad Cl_2(g) \quad \longrightarrow \quad 2\,HCl(g)$$

1 mol	1 mol	2 mol
22.4 L	22.4 L	2×22.4 L
1 volume	1 volume	2 volumes
Y volume	Y volume	$2\,Y$ volumes

Remember: For gases at the same T and P, equal volumes contain equal numbers of particles.

   In this reaction, 22.4 L of hydrogen will react with 22.4 L of chlorine to give $2\,(22.4) = 44.8$ L of hydrogen chloride gas. This statement is true because these volumes are equivalent to the number of reacting moles in the equation. Therefore, Y volume of H_2 will combine with Y volume of Cl_2 to give $2\,Y$ volumes of HCl. For example, 100 L of H_2 react with 100 L of Cl_2 to give 200 L of HCl; if the 100 L of H_2 and of Cl_2 are at 50°C, they will give 200 L of HCl at 50°C. When the temperature and pressure before and after a reaction are the same, volumes can be calculated without changing the volumes to STP.

> For reacting gases at constant temperature and pressure, volume–volume relationships are the same as mole–mole relationships.

Example 12.21   What volume of oxygen will react with 150. L of hydrogen to form water vapor? What volume of water vapor will be formed?

SOLUTION   Assume that both reactants and products are measured at standard conditions. Calculate by using reacting volumes:

$$2\,H_2(g) \quad + \quad O_2(g) \quad \longrightarrow \quad 2\,H_2O(g)$$

2 mol	1 mol	2 mol
2×22.4 L	22.4 L	2×22.4 L
2 volumes	1 volume	2 volumes
150. L	75 L	150. L

For every two volumes of H_2 that react, one volume of O_2 reacts and two volumes of $H_2O(g)$ are produced:

$$(150.\ L\ H_2)\left(\frac{1\ \text{volume}\ O_2}{2\ \text{volumes}\ H_2}\right) = 75\ L\ O_2$$

$$(150.\ L\ H_2)\left(\frac{2\ \text{volumes}\ H_2O}{2\ \text{volumes}\ H_2}\right) = 150.\ L\ H_2O$$

Example 12.22   The equation for the preparation of ammonia is

$$3\,H_2(g) + N_2(g) \xrightarrow{\ 400°C\ } 2\,NH_3(g)$$

Assuming that the reaction goes to completion, determine the following:

(a) What volume of H_2 will react with 50.0 L of N_2?
(b) What volume of NH_3 will be formed from 50.0 L of N_2?
(c) What volume of N_2 will react with 100. mL of H_2?
(d) What volume of NH_3 will be produced from 100. mL of H_2?
(e) If 600. mL of H_2 and 400. mL of N_2 are sealed in a flask and allowed to react, what amounts of H_2, N_2, and NH_3 are in the flask at the end of the reaction?

The answers to parts (a)–(d) are shown in the boxes and can be determined from the equation by inspection, using the principle of reacting volumes: **SOLUTION**

$$3\,H_2(g) \quad + \quad N_2(g) \quad \longrightarrow \quad 2\,NH_3(g)$$

3 volumes   1 volume   2 volumes

(a) $\boxed{150.\,\text{L}}$ 50.0 L

(b)   50.0 L $\boxed{100.\,\text{L}}$

(c) 100. mL $\boxed{33.3\,\text{mL}}$

(d) 100. mL $\boxed{66.7\,\text{mL}}$

(e) Volume ratio from the equation $= \dfrac{3\ \text{volumes } H_2}{1\ \text{volume } N_2}$

Volume ratio used $= \dfrac{600.\ \text{mL } H_2}{400.\ \text{mL } N_2} = \dfrac{3\ \text{volumes } H_2}{2\ \text{volumes } N_2}$

Comparing these two ratios, we see that an excess of N_2 is present in the gas mixture. Therefore, the reactant limiting the amount of NH_3 that can be formed is H_2:

$$3\,H_2(g) + N_2(g) \longrightarrow 2\,NH_3(g)$$

600. mL   200. mL   400. mL

To have a $3:1$ ratio of volumes reacting, 600. mL of H_2 will react with 200. mL of N_2 to produce 400. mL of NH_3, leaving 200. mL of N_2 unreacted. At the end of the reaction the flask will contain 400. mL of NH_3 and 200. mL of N_2.

Practice 12.13 _____

What volume of oxygen will react with 15.0 L of propane (C_3H_8) to form carbon dioxide and water? What volume of carbon dioxide will be formed? What volume of water vapor will be formed?

$$C_3H_8(g) + 5\,O_2(g) \longrightarrow 3\,CO_2(g) + 4\,H_2O(g)$$

12.15 Real Gases

All the gas laws are based on the behavior of an ideal gas—that is, a gas with a behavior that is described exactly by the gas laws for all possible values of P, V, and T. Most real gases actually do behave very nearly as predicted by the gas laws over a fairly wide range of temperatures and pressures. However, when conditions are such that the gas molecules are crowded closely together (high pressure and/or low temperature), they show marked deviations from ideal behavior. Deviations occur because molecules have finite volumes and also have intermolecular attractions, which result in less compressibility at high pressures and greater compressibility at low temperatures than predicted by the gas laws. Many gases become liquids at high pressure and low temperature.

Air Pollution

Chemical reactions occur among the gases that are emitted into our atmosphere. In recent years, there has been growing concern over the effects these reactions have on our environment and our lives.

The outer portion (stratosphere) of the atmosphere plays a significant role in determining the conditions for life at the surface of the Earth. This stratosphere protects the surface from the intense radiation and particles bombarding our planet. Some of the high-energy radiation from the sun acts upon oxygen molecules, O_2, in the stratosphere, converting them into ozone, O_3.

allotrope

Different molecular forms of an element are called **allotropes** of that element. Thus, oxygen and ozone are allotropic forms of oxygen:

$$O_2 \xrightarrow{\text{sunlight}} O + O$$

oxygen atoms

$$O_2 + O \longrightarrow O_3$$

ozone

Ozone (O_3)

Ultraviolet radiation from the sun is highly damaging to living tissues of plants and animals. The ozone layer, however, shields the Earth by absorbing ultraviolet radiation and thus prevents most of this lethal radiation from reaching the Earth's surface. The reaction that occurs is the reverse of the preceding one:

$$O_3 \xrightarrow[\text{radiation}]{\text{ultraviolet}} O_2 + O + \text{heat}$$

Scientists have become concerned about a growing hazard to the ozone layer. Chlorofluorocarbon propellants, such as the Freons, CCl_3F and CCl_2F_2, which were used in aerosol spray cans and are used in refrigeration and air-conditioning units, are stable compounds and remain unchanged in the lower atmosphere. But when these chlorofluorocarbons are carried by convection currents to the stratosphere, they absorb ultraviolet radiation and produce chlorine atoms (chlorine free radicals), which in turn react with ozone. The following reaction sequence involving free radicals has been proposed to explain the partial destruction of the ozone layer by chlorofluorocarbons.

A free radical is a species containing an odd number of electrons. Free radicals are highly reactive.

$$CCl_3F \xrightarrow[\text{radiation}]{\text{ultraviolet}} \cdot CCl_2F \quad + \quad Cl\cdot \qquad (1)$$

fluorocarbon molecule   fluorocarbon free radical   chlorine free radical (atom)

$$Cl\cdot + O_3 \longrightarrow ClO\cdot + O_2 \qquad\qquad (2)$$

$$ClO\cdot + O \longrightarrow O_2 + Cl\cdot \qquad\qquad (3)$$

Because a chlorine atom is generated for each ozone molecule that is destroyed (reactions 2 and 3 can proceed repeatedly), a single chlorofluorocarbon molecule can be responsible for the destruction of many ozone molecules.

Volcanic eruptions send sulfur dioxide high into the atmosphere. The SO_2 gas forms tiny droplets that reflect sunlight back into space, cooling the ground temperatures for up to two years following an eruption. Eventually the sulfur dioxide combines with water vapor and oxygen to form sulfate compounds, which fall back to Earth as acid rain. Researchers in England suggest these sulfates may also cause climate change in the ground by suppressing the release of methane (a greenhouse gas) from wetland areas. The scientists spread powdered sodium sulfate on an area of peat bog in northern Scotland. They added an amount equal to their calculation of how much SO_2 was released during a volcanic eruption (and about 10 times the amount of SO_2 currently released by industrial sources.)

The scientists then measured atmospheric concentrations of methane over the bog and found a 42% reduction over the treated areas. Two years later (after halting the sulfate addition) the methane release was still 40% reduced. In fact, the data now suggest a five- to seven-year reduction in methane results from sulfate addition.

How does sulfate reduce methane emission? The methane comes primarily from soil bacteria. The additional sulfates increase the growth of a type of bacteria that uses sulfates to produce energy. This increase, called a bloom, causes the methane-producing bacteria to decrease. These complicated biological interactions can end up influencing atmospheric chemistry. Scientists think more evidence for these effects might be found in polar and glacial ice formed following volcanic eruptions.

Scientists have discovered an annual thinning in the ozone layer over Antarctica. This is what we call the "hole" in the ozone layer. If this hole were to occur over populated regions of the world, severe effects would result, including a rise in the cancer rate, increased climatic temperatures, and vision problems (which have been observed in the penguins of Antarctica). See Figure 12.17.

Sep 17 2001

Figure 12.17
Antarctica ozone hole.

Ozone can be prepared by passing air or oxygen through an electrical discharge:

$$3\,O_2(g) + 286\,kJ \xrightarrow[\text{discharge}]{\text{electrical}} 2\,O_3(g)$$

The characteristic pungent odor of ozone is noticeable in the vicinity of electrical machines and power transmission lines. Ozone is formed in the atmosphere during electrical storms and by the photochemical action of ultraviolet radiation on a mixture of nitrogen dioxide and oxygen. Areas with high air pollution are subject to high atmospheric ozone concentrations.

Ozone is not a desirable low-altitude constituent of the atmosphere because it is known to cause extensive plant damage, cracking of rubber, and the formation of eye-irritating substances. Concentrations of ozone greater than 0.1 part per million (ppm) of air cause coughing, choking, headache, fatigue, and reduced resistance to respiratory infection. Concentrations between 10 and 20 ppm are fatal to humans.

In addition to ozone, the air in urban areas contains nitrogen oxides, which are components of smog. The term *smog* refers to air pollution in urban environments. Often the chemical reactions occur as part of a *photochemical process*. Nitrogen monoxide (NO) is oxidized in the air or in automobile engines to produce nitrogen dioxide (NO_2). In the presence of light,

$$NO_2 \xrightarrow{\text{light}} NO + O$$

In addition to nitrogen oxides, combustion of fossil fuels releases CO_2, CO, and sulfur oxides. Incomplete combustion releases unburned and partially burned hydrocarbons.

Society is continually attempting to discover, understand, and control emissions that contribute to this sort of atmospheric chemistry. It is a problem that each one of us faces as we look to the future if we want to continue to support life as we know it on our planet.

Chapter 12 Review

12.1   General Properties

- Gases:
  - Particles are relatively far apart.
  - Particles are very mobile.
  - Gases take the shape and volume of the container.
  - Gases are easily compressible.

12.2   The Kinetic–Molecular Theory

KEY TERMS

Kinetic-molecular theory (KMT)
Ideal gas
Diffusion
Effusion
Graham's law of effusion

- Kinetic-molecular theory assumptions:
  - Gases are tiny particles with no attraction for each other.

  - The distance between particles is great compared to the size of the particles.
  - Gas particles move in straight lines.
  - No energy is lost in particle collisions (perfectly elastic collisions).
  - The average kinetic energy for particles is the same for all gases at the same temperature and pressure.
- A gas that follows the KMT is an ideal gas
- The kinetic energy of a particle is expressed as $KE = \frac{1}{2}mv^2$.
- Gases will diffuse to mix spontaneously.
- Gases can effuse through a small opening.
- Graham's law of effusion states

$$\frac{\text{rate of effusion for gas A}}{\text{rate of effusion for gas B}} = \sqrt{\frac{d\text{B}}{d\text{A}}} = \sqrt{\frac{\text{molar mass B}}{\text{molar mass A}}}$$

12.3 Measurement of Pressure of Gases

KEY TERMS

Pressure
Atmospheric pressure
Barometer
1 atmosphere

- Pressure is force per unit area.
- Pressure of the atmosphere is measured by using a barometer:
  - Units of pressure include:
    - Atmosphere (atm) = 760 mm Hg.
    - Pascal, 1 atm = 101,325 Pa = 101.3 kPa.
    - Torr, 1 atm = 760 torr.

12.4 Dependence of Pressure on Number of Molecules and Temperature

- Pressure is directly related to the number of molecules in the sample.
- Pressure is directly related to the Kelvin temperature of the sample.

12.5 Boyle's Law

KEY TERM

Boyle's law

- At constant temperature the volume of a gas is inversely proportional to the pressure of the gas:

$$V \propto \frac{1}{P} \quad \text{or} \quad P_1V_1 = P_2V_2$$

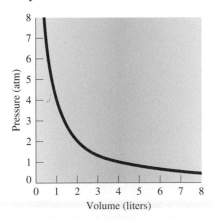

12.6 Charles' Law

KEY TERMS

Absolute zero
Charles' law

- At constant pressure the volume of a gas is directly proportional to the absolute temperature of the gas.

$$V \propto T \quad \text{or} \quad \frac{V_1}{T_1} = \frac{V_2}{T_2}$$

12.7 Gay-Lussac's Law

KEY TERM

Gay-Lussac's law

- At constant volume the pressure of a gas is directly proportional to the absolute temperature of the gas.

$$P \propto T \quad \text{or} \quad \frac{P_1}{T_1} = \frac{P_2}{T_2}$$

12.8 Combined Gas Law

KEY TERMS

Standard conditions
Standard temperature and pressure (STP)

- The $P\,V\,T$ relationship for gases can be expressed in a single equation known as the combined gas law:

$$\frac{P_1V_1}{T_1} = \frac{P_2V_2}{T_2}$$

12.9 Dalton's Law of Partial Pressures

KEY TERMS

Dalton's law of partial pressures
Partial pressure

- The total pressure of a mixture of gases is the sum of the partial pressures of the component gases in the mixture.
- When a gas is collected over water, the pressure of the collected gas is the difference between the atmospheric pressure and the vapor pressure of water at that temperature.

12.10 Avogadro's Law

KEY TERMS

Gay-Lussac's law of combining volumes
Avogadro's law

- Gay-Lussac's law of combining volumes states that when measured at constant T and P, the ratios of the volumes of reacting gases are small whole numbers.
- Avogadro's law states that equal volumes of different gases at the same T and P contain the same number of particles.

12.11 Mole–Mass–Volume Relationships of Gases

KEY TERM

Molar volume

- One mole of any gas occupies 22.4 L at STP.

12.12 Density of Gases

- The density of a gas is usually expressed in units of g/L.
- Temperature and pressure are given for a density of a gas since the volume of the gas depends on these conditions.

12.13 Ideal Gas Law

KEY TERM

Ideal gas law

- The ideal gas law combines all the variables involving gases into a single expression:

$$PV = nRT$$

- R is the ideal gas constant and can be expressed as

$$R = 0.0821\frac{\text{L-atm}}{\text{mol-K}}$$

12.14 Gas Stoichiometry

- Stoichiometry problems involving gases are solved the same way as other stoichiometry problems. See graphic below.

- For reacting gases (contant T and P) volume–volume relationships are the same as mole–mole relationships.

12.15 Real Gases

KEY TERM

Allotrope

- Real gases show deviation from the ideal gas law:
  - Deviations occur at:
    - High pressure
    - Low temperature
- These deviations occur because:
  - Molecules have finite volumes.
  - Molecules have intermolecular attractions.
- Ozone in our upper atmosphere protects us from harmful ultraviolet radiation.
- The ozone layer is being destroyed by pollutants such as Freons and other chlorinated hydrocarbons:
  - These molecules produce free radicals, which interact with ozone to convert it to oxygen.
- In the lower atmosphere ozone causes damage to plants, cracking of rubber, and irritations to humans.
- Other air pollutants such as nitrogen oxides and sulfur oxides are found in urban areas, creating smog, which can impact our daily lives.

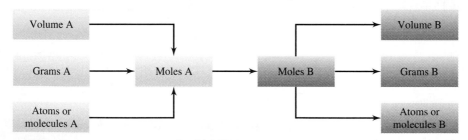

Review Questions

All questions with blue numbers have answers in the appendix of the text.

1. What evidence is used to show diffusion in Figure 12.1? If H_2 and O_2 were in the two flasks, how could you prove that diffusion had taken place?
2. What is meant by pressure (for a gas)?
3. How does the air pressure inside the balloon shown in Figure 12.2 compare with the air pressure outside the balloon? Explain.
4. According to Table 12.1, what two gases are the major constituents of dry air?

5. How does the pressure represented by 1 torr compare in magnitude to the pressure represented by 1 mm Hg? See Table 12.2.

6. In which container illustrated in Figure 12.5 are the molecules of gas moving faster? Assume both gases to be hydrogen.

7. In Figure 12.6, what gas pressure corresponds to a volume of 4 L?

8. How do the data illustrated in Figure 12.6 substantiate Boyle's law?

9. What effect would you observe in Figure 12.9 if T_2 were lower than T_1?

10. In the diagram shown in Figure 12.12, is the pressure of the oxygen plus water vapor inside the bottle equal to, greater than, or less than the atmospheric pressure outside the bottle? Explain.

11. List five gases in Table 12.3 that are more dense than air. Explain the basis for your selection.

12. What are the basic assumptions of the kinetic-molecular theory?

13. Arrange the following gases, all at standard temperature, in order of increasing relative molecular velocities: H_2, CH_4, Rn, N_2, F_2, and He. What is your basis for determining the order?

14. List, in descending order, the average kinetic energies of the molecules in Question 13.

15. What are the four parameters used to describe the behavior of a gas?

16. What are the characteristics of an ideal gas?

17. How is Boyle's law related to the ideal gas law?

18. How is Charles' law related to the ideal gas law?

19. Under what condition of temperature, high or low, is a gas least likely to exhibit ideal behavior? Explain.

20. Under what condition of pressure, high or low, is a gas least likely to exhibit ideal behavior? Explain.

21. Compare, at the same temperature and pressure, equal volumes of H_2 and O_2 as to the following:
(a) number of molecules
(b) mass
(c) number of moles
(d) average kinetic energy of the molecules
(e) rate of effusion
(f) density

22. How does the kinetic-molecular theory account for the behavior of gases as described by
(a) Boyle's law?
(b) Charles' law?
(c) Dalton's law of partial pressures?

23. Explain how the reaction

$$N_2(g) + O_2(g) \xrightarrow{\Delta} 2NO(g)$$

proves that nitrogen and oxygen are diatomic molecules.

24. What is the reason for comparing gases to STP?

25. Is the conversion of oxygen to ozone an exothermic or endothermic reaction? How do you know?

26. What compounds are responsible for damaging the ozone layer and why are they hazardous?

27. When constant pressure is maintained, what effect does heating a mole of N_2 gas have on
(a) its density?
(b) its mass?
(c) the average kinetic energy of its molecules?
(d) the average velocity of its molecules?
(e) the number of N_2 molecules in the sample?

28. Write formulas for an oxygen atom, an oxygen molecule, and an ozone molecule. How many electrons are in an oxygen molecule?

Paired Exercises

All exercises with blue numbers have answers in the appendix of the text.

Pressure Units

1. The barometer reads 715 mm Hg. Calculate the corresponding pressure in the following:
(a) atmospheres
(b) inches of Hg
(c) lb/in.^{2}

2. The barometer reads 715 mm Hg. Calculate the corresponding pressure in the following:
(a) torrs
(b) millibars
(c) kilopascals

3. Express the following pressures in atmospheres:
(a) 28 mm Hg
(b) 6000. cm Hg
(c) 795 torr
(d) 5.00 kPa

4. Express the following pressures in atmospheres:
(a) 62 mm Hg
(b) 4250. cm Hg
(c) 225 torr
(d) 0.67 kPa

Boyle's and Charles' Laws

5. A sample of a gas occupies a volume of 525 mL at 625 torr. At constant temperature, what will be the new volume when the pressure changes to the following:
(a) 1.5 atm
(b) 455 mm Hg

6. A sample of a gas occupies a volume of 635 mL at 722 torr. At constant temperature, what will be the new volume when the pressure changes to the following:
(a) 2.5 atm
(b) 795 mm Hg

7. A sample of a gas at 0.75 atm occupies a volume of 521 mL. If the temperature remains constant, what will be the new pressure if the volume increases to 776 mL?

8. A sample of a gas at 1.7 atm occupies a volume of 225 mL. If the temperature remains constant, what will be the new pressure if the volume decreases to 115 mL?

9. Given 6.00 L of N_2 gas at $-25°C$, what volume will the nitrogen occupy at (a) 0.0°C? (b) 100. K? (Assume constant pressure.)

10. Given 6.00 L of N_2 gas at $-25°C$, what volume will the nitrogen occupy at (a) 0.0°F? (b) 345. K? (Assume constant pressure.)

Combined Gas Laws

11. A gas occupies a volume of 410 mL at 27°C and 740 mm Hg pressure. Calculate the volume the gas would occupy at STP.

12. A gas occupies a volume of 410 mL at 27°C and 740 mm Hg pressure. Calculate the volume the gas would occupy at 250.°C and 680 mm Hg pressure.

13. An expandable balloon contains 1400. L of He at 0.950 atm pressure and 18°C. At an altitude of 22 miles (temperature 2.0°C and pressure 4.0 torr), what will be the volume of the balloon?

14. A gas occupies 22.4 L at 2.50 atm and 27°C. What will be its volume at 1.50 atm and $-5.00°C$?

15. A 775-mL sample of NO_2 gas is at STP. If the volume changes to 615 mL and the temperature changes to 25°C, what will be the new pressure?

16. A 2.5-L sample of SO_3 is at 19°C and 1.5 atm. What will be the new temperature in °C when the volume changes to 1.5 L and the pressure to 765 torr?

Dalton's Law of Partial Pressures

17. What would be the partial pressure of N_2 gas collected over water at 22°C and 721 torr pressure? (Check Appendix IV for the vapor pressure of water.)

18. What would be the partial pressure of N_2 gas collected over water at 25°C and 705 torr pressure? (Check Appendix IV for the vapor pressure of water.)

19. A mixture contains H_2 at 600. torr pressure, N_2 at 200. torr pressure, and O_2 at 300. torr pressure. What is the total pressure of the gases in the system?

20. A mixture contains H_2 at 325 torr pressure, N_2 at 475 torr pressure, and O_2 at 650. torr pressure. What is the total pressure of the gases in the system?

21. A sample of methane gas, CH_4, was collected over water at 25.0°C and 720. torr. The volume of the wet gas is 2.50 L. What will be the volume of the dry methane at standard pressure?

22. A sample of propane gas, C_3H_8, was collected over water at 22.5°C and 745 torr. The volume of the wet gas is 1.25 L. What will be the volume of the dry propane at standard pressure?

Mole–Mass–Volume Relationships

23. How many moles of O_2 will occupy a volume of 1.75 L at STP?

24. How many moles of N_2 will occupy a volume of 3.50 L at STP?

25. What volume will each of the following occupy at STP?
(a) 6.02×10^{23} molecules of CO_2
(b) 2.5 mol CH_4
(c) 12.5 g oxygen

26. What volume will each of the following occupy at STP?
(a) 1.80×10^{24} molecules of SO_3
(b) 7.5 mol C_2H_6
(c) 25.2 g chlorine

27. How many grams of NH_3 are present in 725 mL of the gas at STP?

28. How many grams of C_3H_6 are present in 945 mL of the gas at STP?

29. How many molecules of NH_3 gas are present in a 1.00-L flask of NH_3 gas at STP?

30. How many molecules of CH_4 gas are present in a 1.00-L flask of CH_4 gas at STP?

31. What volume would result if a balloon were filled with 10.0 grams of chlorine gas at STP?

32. How many grams of methane gas were used to fill a balloon to a volume of 3.0 L at STP?

Density of Gases

33. Calculate the density of the following gases at STP:
    (a) Xe
    (b) F_2
    (c) C_2H_6

34. Calculate the density of the following gases at STP:
    (a) Rn
    (b) CO_2
    (c) C_4H_8

35. Calculate the density of the following:
    (a) F_2 gas at STP
    (b) F_2 gas at 27°C and 1.00 atm pressure

36. Calculate the density of the following:
    (a) Cl_2 gas at STP
    (b) Cl_2 gas at 22°C and 0.500 atm pressure

Ideal Gas Law and Stoichiometry

37. At 27°C and 750 torr pressure, what will be the volume of 2.3 mol of Ne?

38. At 25°C and 725 torr pressure, what will be the volume of 0.75 mol of Kr?

39. What volume will a mixture of 5.00 mol of H_2 and 0.500 mol of CO_2 occupy at STP?

40. What volume will a mixture of 2.50 mol of N_2 and 0.750 mol of HCl occupy at STP?

41. What is the Kelvin temperature of a system in which 4.50 mol of a gas occupy 0.250 L at 4.15 atm?

42. How many moles of N_2 gas occupy 5.20 L at 250 K and 0.500 atm?

43. In the lab, students generated and collected hydrogen gas according to the following equation:

    $$Zn(s) + H_2SO_4(aq) \rightarrow H_2(g) + ZnSO_4(aq)$$

    (a) How many mL of hydrogen gas at STP were generated from 52.7 g of zinc metal?
    (b) If 525 mL of hydrogen gas at STP were needed, how many moles of H_2SO_4 would be required?

44. In the lab, students generated and collected oxygen gas according to the following equation:

    $$2 H_2O_2(aq) \rightarrow 2 H_2O(l) + O_2(g)$$

    (a) How many mL of oxygen gas at STP were generated from 50.0 g of H_2O_2?
    (b) If 225 mL of oxygen gas at STP were needed for an experiment, how many moles of H_2O_2 would be needed?

45. Consider the following equation:

    $$4 NH_3(g) + 5 O_2(g) \rightarrow 4 NO(g) + 6 H_2O(g)$$

    (a) How many liters of oxygen are required to react with 2.5 L NH_3? Both gases are at STP.
    (b) How many grams of water vapor can be produced from 25 L NH_3 if both gases are at STP?
    (c) How many liters of NO can be produced when 25 L O_2 are reacted with 25 L NH_3? All gases are at the same temperature and pressure.

46. Consider the following equation:

    $$C_3H_8(g) + 5 O_2(g) \rightarrow 3 CO_2(g) + 4 H_2O(g)$$

    (a) How many liters of oxygen are required to react with 7.2 L C_3H_8? Both gases are at STP.
    (b) How many grams of CO_2 will be produced from 35 L C_3H_8 if both gases are at STP?
    (c) How many liters of water vapor can be produced when 15 L C_3H_8 are reacted with 15 L O_2? All gases are at the same temperature and pressure.

47. Oxygen gas can be generated by the decomposition of potassium chlorate according to the following equation:

    $$2 KClO_3(s) \rightarrow 2 KCl(s) + 3 O_2(g)$$

    How many liters of oxygen at STP will be produced when 0.525 kg of KCl is also produced?

48. When glucose is burned in a closed container, carbon dioxide gas and water are produced according to the following equation:

    $$C_6H_{12}O_6(s) + 6 O_2(g) \rightarrow 6 CO_2(g) + 6 H_2O(l)$$

    How many liters of CO_2 at STP can be produced when 1.50 kg of glucose are burned?

Additional Exercises

All exercises with blue numbers have answers in the appendix of the text.

49. How is it possible that 1 mole of liquid water occupies a volume of 18 mL, but 1 mole of gaseous water occupies a volume of 22.4 L? (See the photo in Section 12.1.)

50. Explain why it is necessary to add air to a car's tires during the winter.

51. You have a 10-L container filled with 0.5 mol of O_2 gas at a temperature of 30.°C with a pressure of 945 torr.
(a) What will happen to the pressure if the container size is doubled while keeping the temperature and number of moles constant?
(b) What will happen to the pressure when the temperature is doubled while keeping the size of the container and the number of moles constant?
(c) What will happen to the pressure when the amount of O_2 gas is cut in half while keeping the size of the container and the temperature constant?
(d) What will happen to the pressure if 1 mole of N_2 gas is added to the container while keeping the temperature and size of the container the same?

52. Sketch a graph to show each of the following relationships:
(a) P vs. V at constant temperature and number of moles
(b) T vs. V at constant pressure and number of moles
(c) T vs. P at constant volume and number of moles
(d) n vs. V at constant temperature and pressure

53. Why is it dangerous to incinerate an aerosol can?

54. What volume does 1 mol of an ideal gas occupy at standard conditions?

55. Which of these occupies the greatest volume?
(a) 0.2 mol of chlorine gas at 48°C and 80 cm Hg
(b) 4.2 g of ammonia at 0.65 atm and −11°C
(c) 21 g of sulfur trioxide at 55°C and 110 kPa

56. Which of these has the greatest density?
(a) SF_6 at STP
(b) C_2H_6 at room conditions
(c) He at −80°C and 2.15 atm

57. A chemist carried out a chemical reaction that produced a gas. It was found that the gas contained 80.0% carbon and 20.0% hydrogen. It was also noticed that 1500 mL of the gas at STP had a mass of 2.01 g.

(a) What is the empirical formula of the compound?
(b) What is the molecular formula of the compound?
(c) What Lewis structure fits this compound?

58. Three gases were added to the same 2.0-L container. The total pressure of the gases was 790 torr at room temperature (25.0°C). If the mixture contained 0.65 g of oxygen gas, 0.58 g of carbon dioxide, and an unknown amount of nitrogen gas, determine the following:
(a) the total number of moles of gas in the container
(b) the number of grams of nitrogen in the container
(c) the partial pressure of each gas in the mixture

*59. When carbon monoxide and oxygen gas react, carbon dioxide results. If 500. mL of O_2 at 1.8 atm and 15°C are mixed with 500. mL of CO at 800 mm Hg and 60°C, how many milliliters of CO_2 at STP could possibly result?

*60. One of the methods for estimating the temperature at the center of the sun is based on the ideal gas law. If the center is assumed to be a mixture of gases whose average molar mass is 2.0 g/mol, and if the density and pressure are 1.4 g/cm^3 and 1.3×10^9 atm, respectively, calculate the temperature.

*61. A soccer ball of constant volume 2.24 L is pumped up with air to a gauge pressure of 13 lb/in.^{2} at 20.0°C. The molar mass of air is about 29 g/mol.
(a) How many moles of air are in the ball?
(b) What mass of air is in the ball?
(c) During the game, the temperature rises to 30.0°C. What mass of air must be allowed to escape to bring the gauge pressure back to its original value?

*62. A balloon will burst at a volume of 2.00 L. If it is partially filled at 20.0°C and 65 cm Hg to occupy 1.75 L, at what temperature will it burst if the pressure is exactly 1 atm at the time that it breaks?

63. Given a sample of a gas at 27°C, at what temperature would the volume of the gas sample be doubled, the pressure remaining constant?

64. A gas sample at 22°C and 740 torr pressure is heated until its volume is doubled. What pressure would restore the sample to its original volume?

65. A gas occupies 250. mL at 700. torr and 22°C. When the pressure is changed to 500. torr, what temperature (°C) is needed to maintain the same volume?

66. The tires on an automobile were filled with air to 30. psi at 71.0°F. When driving at high speeds, the tires become hot. If the tires have a bursting pressure of 44 psi, at what temperature (°F) will the tires "blow out"?

67. What pressure will 800. mL of a gas at STP exert when its volume is 250. mL at 30°C?

68. How many gas molecules are present in 600. mL of N_2O at 40°C and 400. torr pressure? How many atoms are present? What would be the volume of the sample at STP?

69. An automobile tire has a bursting pressure of 60 lb/in.^{2}. The normal pressure inside the tire is 32 lb/in.^{2}. When traveling at high speeds, a tire can get quite hot. Assuming that the temperature of the tire is 25°C before running it, determine whether the tire will burst when the inside temperature gets to 212°F. Show your calculations.

70. If you prepared a barometer using water instead of mercury, how high would the column of water be at one atmosphere pressure? (Neglect the vapor pressure of water.)

71. How many moles of oxygen are in a 55-L cylinder at 27°C and 2.20×10^3 lb/in.^{2}?

*72. How many moles of Cl_2 are in one cubic meter (1.00 m^3) of Cl_2 gas at STP?

73. At STP, 560. mL of a gas have a mass of 1.08 g. What is the molar mass of the gas?

74. A gas has a density at STP of 1.78 g/L. What is its molar mass?

75. Using the ideal gas law, $PV = nRT$, calculate the following:
    (a) the volume of 0.510 mol of H_2 at 47°C and 1.6 atm pressure
    (b) the number of grams in 16.0 L of CH_4 at 27°C and 600. torr pressure
    *(c) the density of CO_2 at 4.00 atm pressure and −20.0°C
    (d) the molar mass of a gas having a density of 2.58 g/L at 27°C and 1.00 atm pressure.

*76. Acetylene (C_2H_2) and hydrogen fluoride (HF) react to give difluoroethane:

$$C_2H_2(g) + 2\,HF(g) \longrightarrow C_2H_4F_2(g)$$

When 1.0 mol of C_2H_2 and 5.0 mol of HF are reacted in a 10.0-L flask, what will be the pressure in the flask at 0°C when the reaction is complete?

77. What volume of hydrogen at STP can be produced by reacting 8.30 mol of Al with sulfuric acid? The equation is

$$2\,Al(s) + 3\,H_2SO_4(aq) \longrightarrow Al_2(SO_4)_3(aq) + 3\,H_2(g)$$

78. What are the relative rates of effusion of N_2 and He?

*79. (a) What are the relative rates of effusion of CH_4 and He?
    (b) If these two gases are simultaneously introduced into opposite ends of a 100.-cm tube and allowed to diffuse toward each other, at what distance from the helium end will molecules of the two gases meet?

*80. A gas has a percent composition by mass of 85.7% carbon and 14.3% hydrogen. At STP the density of the gas is 2.50 g/L. What is the molecular formula of the gas?

*81. Assume that the reaction

$$2\,CO(g) + O_2(g) \longrightarrow 2\,CO_2(g)$$

goes to completion. When 10. mol of CO and 8.0 mol of O_2 react in a closed 10.-L vessel,
    (a) how many moles of CO, O_2, and CO_2 are present at the end of the reaction?
    (b) what will be the total pressure in the flask at 0°C?

*82. If 250 mL of O_2, measured at STP, are obtained by the decomposition of the $KClO_3$ in a 1.20-g mixture of KCl and $KClO_3$,

$$2\,KClO_3(s) \longrightarrow 2\,KCl(s) + 3\,O_2(g)$$

what is the percent by mass of $KClO_3$ in the mixture?

Challenge Exercises

All exercises with blue numbers have answers in the appendix of the text.

*83. Air has a density of 1.29 g/L at STP. Calculate the density of air on Pikes Peak, where the pressure is 450 torr and the temperature is 17°C.

*84. Consider the arrangement of gases shown below. If the valve between the gases is opened and the temperature is held constant, determine the following:
(a) the pressure of each gas
(b) the total pressure in the system

CO$_2$
$V = 3.0$ L
$P = 150.$ torr

H$_2$
$V = 1.0$ L
$P = 50.$ torr

*85. A steel cylinder contained 50.0 L of oxygen gas under a pressure of 40.0 atm and at a temperature of 25°C. What was the pressure in the cylinder during a storeroom fire that caused the temperature to rise 152°C? (Be careful!)

*86. A balloon has a mass of 0.5 g when completely deflated. When it is filled with an unknown gas, the mass increases to 1.7 g. You notice on the canister of the unknown gas that it occupies a volume of 0.4478 L at a temperature of 50°C. You note the temperature in the room is 25°C. Identify the gas.

Answers to Practice Exercises

12.1 (a) 851 torr, (b) 851 mm Hg
12.2 0.596 atm
12.3 4.92 L
12.4 762 torr
12.5 84.7 L
12.6 861 K (588°C)
12.7 518 mL

12.8 1.4×10^2 g/mol
12.9 0.89 g/L
12.10 0.953 mol
12.11 128 g/mol
12.12 1.56 L O$_2$
12.13 75.0 L O$_2$, 45.0 L CO$_2$, 60.0 L H$_2$O

Properties of Liquids

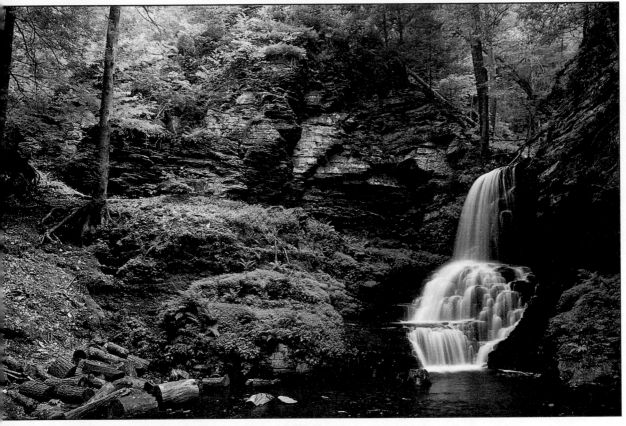

Bushkill Falls and Bridesmaid Falls (lower level) at Delaware Water Gap, Pennsylvania.

Chapter Outline

13.1 What Is a Liquid?

13.2 Evaporation

13.3 Vapor Pressure

13.4 Surface Tension

13.5 Boiling Point

13.6 Freezing Point or Melting Point

13.7 Changes of State

13.8 The Hydrogen Bond

13.9 Hydrates

13.10 Water, a Unique Liquid

13.11 Water Purification

Planet Earth, that magnificient blue sphere we enjoy viewing from space, is spectacular. Over 75% of Earth is covered with water. We are born from it, drink it, bathe in it, cook with it, enjoy its beauty in waterfalls and rainbows, and stand in awe of the majesty of icebergs. Water supports and enhances life.

In chemistry, water provides the medium for numerous reactions. The shape of the water molecule is the basis for hydrogen bonds. These bonds determine the unique properties and reactions of water. The tiny water molecule holds the answers to many of the mysteries of chemical reactions.

13.1   What Is a Liquid?

In the last chapter, we found that gases contain particles that are far apart, in rapid random motion, and essentially independent of each other. The kinetic-molecular theory, along with the ideal gas law, summarizes the behavior of most gases at relatively high temperatures and low pressures.

Solids are obviously very different from gases. Solids contain particles that are very close together; solids have a high density, compress negligibly, and maintain their shape regardless of container. These characteristics indicate large attractive forces between particles. The model for solids is very different from the one for gases.

Liquids, on the other hand, lie somewhere between the extremes of gases and solids. Liquids contain particles that are close together; liquids are essentially incompressible, and have definite volume. These properties are very similar to those of solids. But liquids also take the shape of their containers; this is closer to the model of a gas.

Although liquids and solids show similar properties, they differ tremendously from gases (see Figure 13.1). No simple mathematical relationship, like the ideal gas law, works well for liquids or solids. Instead, these models are directly related to the forces of attraction between molecules. With these general statements in mind, let's consider some specific properties of liquids.

Figure 13.1
The three states of matter. (a) Solid—water molecules are held together rigidly and are very close to each other. (b) Liquid—water molecules are close together but are free to move around and slide over each other. (c) Gas—water molecules are far apart and move freely and randomly.

Solid (Ice)   Liquid (Water)   Gas (Steam)
(a)   (b)   (c)

13.2 Evaporation

When beakers of water, ethyl ether, and ethyl alcohol (all liquids at room temperature) are allowed to stand uncovered, their volumes gradually decrease. The process by which this change takes place is called *evaporation*.

Attractive forces exist between molecules in the liquid state. Not all of these molecules, however, have the same kinetic energy. Molecules that have greater-than-average kinetic energy can overcome the attractive forces and break away from the surface of the liquid to become a gas (see Figure 13.2). **Evaporation,** or **vaporization,** is the escape of molecules from the liquid state to the gas or vapor state.

evaporation
vaporization

Figure 13.2
The high-energy molecules escape from the surface of a liquid in a process known as evaporation.

In evaporation, molecules of greater-than-average kinetic energy escape from a liquid, leaving it cooler than it was before they escaped. For this reason, evaporation of perspiration is one way the human body cools itself and keeps its temperature constant. When volatile liquids such as ethyl chloride (C_2H_5Cl) are sprayed on the skin, they evaporate rapidly, cooling the area by removing heat. The numbing effect of the low temperature produced by evaporation of ethyl chloride allows it to be used as a local anesthetic for minor surgery.

Solids such as iodine, camphor, naphthalene (moth balls), and, to a small extent, even ice will go directly from the solid to the gaseous state, bypassing the liquid state. This change is a form of evaporation and is called **sublimation:**

sublimation

$$\text{liquid} \xrightarrow{\text{evaporation}} \text{vapor}$$

$$\text{solid} \xrightarrow{\text{sublimation}} \text{vapor}$$

13.3 Vapor Pressure

When a liquid vaporizes in a closed system like that shown in Figure 13.3b, some of the molecules in the vapor or gaseous state strike the surface and return to the liquid state by the process of **condensation.** The rate of condensation increases until it's equal to the rate of vaporization. At this point, the space above the liquid is said to be saturated with vapor, and an equilibrium, or steady state, exists between the liquid and the vapor. The equilibrium equation is

condensation

$$\text{liquid} \underset{\text{condensation}}{\overset{\text{evaporation}}{\rightleftharpoons}} \text{vapor}$$

Figure 13.3
(a) Molecules in an open beaker evaporate from the liquid and disperse into the atmosphere. Evaporation will continue until all the liquid is gone. (b) Molecules leaving the liquid are confined to a limited space. With time, the concentration in the vapor phase will increase until an equilibrium between liquid and vapor is established.

(a)　　　　　(b)

This equilibrium is dynamic; both processes—vaporization and condensation—are taking place, even though we cannot see or measure a change. The number of molecules leaving the liquid in a given time interval is equal to the number of molecules returning to the liquid.

At equilibrium, the molecules in the vapor exert a pressure like any other gas. The pressure exerted by a vapor in equilibrium with its liquid is known as the **vapor pressure** of the liquid. The vapor pressure may be thought of as a measure of the "escaping" tendency of molecules to go from the liquid to the vapor state. The vapor pressure of a liquid is independent of the amount of liquid and vapor present, but it increases as the temperature rises. Figure 13.4 illustrates a liquid–vapor equilibrium and the measurement of vapor pressure.

vapor pressure

(a) Evacuated flask　　(b) Water added at 20°C　　(c) Water–vapor equilibrium at 20° C　　(d) Water–vapor equilibrium at 30° C

Figure 13.4
Measurement of the vapor pressure of water at 20°C and 30°C. (a) The system is evacuated. The mercury manometer attached to the flask shows equal pressure in both legs. (b) Water has been added to the flask and begins to evaporate, exerting pressure as indicated by the manometer. (c) When equilibrium is established, the pressure inside the flask remains constant at 17.5 torr. (d) The temperature is changed to 30°C, and equilibrium is reestablished with the vapor pressure at 31.8 torr.

Table 13.1 Vapor Pressure of Water, Ethyl Alcohol, and Ethyl Ether at Various Temperatures

Temperature (°C)	Vapor pressure (torr)		
	Water	Ethyl alcohol	Ethyl ether*
0	4.6	12.2	185.3
10	9.2	23.6	291.7
20	17.5	43.9	442.2
30	31.8	78.8	647.3
40	55.3	135.3	921.3
50	92.5	222.2	1276.8
60	152.9	352.7	1729.0
70	233.7	542.5	2296.0
80	355.1	812.6	2993.6
90	525.8	1187.1	3841.0
100	760.0	1693.3	4859.4
110	1074.6	2361.3	6070.1

*Note that the vapor pressure of ethyl ether at temperatures of 40°C and higher exceeds standard pressure, 760 torr, which indicates that the substance has a low boiling point and should therefore be stored in a cool place in a tightly sealed container.

When equal volumes of water, ethyl ether, and ethyl alcohol are placed in beakers and allowed to evaporate at the same temperature, we observe that the ether evaporates faster than the alcohol, which evaporates faster than the water. This order of evaporation is consistent with the fact that ether has a higher vapor pressure at any particular temperature than ethyl alcohol or water. One reason for this higher vapor pressure is that the attraction is less between ether molecules than between alcohol or water molecules. The vapor pressures of these three compounds at various temperatures are compared in Table 13.1.

Substances that evaporate readily are said to be **volatile.** A volatile liquid has a relatively high vapor pressure at room temperature. Ethyl ether is a very volatile liquid; water is not too volatile; and mercury, which has a vapor pressure of 0.0012 torr at 20°C, is essentially a nonvolatile liquid. Most substances that are normally in a solid state are nonvolatile (solids that sublime are exceptions).

volatile

13.4 Surface Tension

Have you ever observed water and mercury in the form of small drops? These liquids form drops because liquids have *surface tension*. A droplet of liquid that is not falling or under the influence of gravity (as on the space shuttle) will form a sphere. Spheres minimize the ratio of surface area to volume. The molecules within the liquid are attracted to the surrounding liquid molecules, but at the liquid's surface, the attraction is nearly all inward. This pulls the surface into a spherical shape. The resistance of a liquid to an increase in its surface area is called the **surface tension** of the liquid. Substances with large attractive forces between molecules have high surface tensions. The effect of surface tension in water is illustrated by floating a needle on the surface of still water.

Mercury droplets.

surface tension

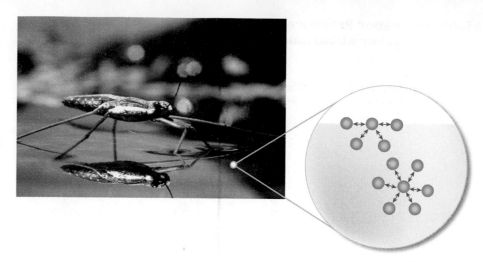

A water strider skims the surface of the water as a result of surface tension. At the molecular level the surface tension results from the net attraction of the water molecules toward the liquid below. In the interior of the water, the forces are balanced in all directions.

Other examples include a water strider walking across a calm pond and water beading on a freshly waxed car. Surface tension is temperature dependent, decreasing with increasing temperature.

capillary action

Liquids also exhibit a phenomenon called **capillary action,** the spontaneous rising of a liquid in a narrow tube. This action results from the *cohesive forces* within the liquid and the *adhesive forces* between the liquid and the walls of the container. If the forces between the liquid and the container are greater than those within the liquid itself, the liquid will climb the walls of the container. For example, consider the California sequoia, a tree that reaches over 200 feet in height. Although water rises only 33 feet in a glass tube (under atmospheric pressure), capillary action causes water to rise from the sequoia's roots to all its parts.

meniscus

The meniscus in liquids is further evidence of cohesive and adhesive forces. When a liquid is placed in a glass cylinder, the surface of the liquid shows a curve called the **meniscus** (see Figure 13.5). The concave shape of water's meniscus shows that the adhesive forces between the glass and water are stronger than the cohesive forces within the water. In a nonpolar substance such as mercury, the meniscus is convex, indicating that the cohesive forces within mercury are greater than the adhesive forces between the glass wall and the mercury.

Figure 13.5
The meniscus of mercury (left) and water (right). The meniscus is the characteristic curve of the surface of a liquid in a narrow tube.

13.5 Boiling Point

The boiling temperature of a liquid is related to its vapor pressure. We've seen that vapor pressure increases as temperature increases. When the internal or vapor pressure of a liquid becomes equal to the external pressure, the liquid boils. (By external pressure we mean the pressure of the atmosphere above the liquid.) The boiling temperature of a pure liquid remains constant as long as the external pressure does not vary.

The boiling point (bp) of water is 100°C at 1 atm pressure. Table 13.1 shows that the vapor pressure of water at 100°C is 760 torr. The significant fact here is that the boiling point is the temperature at which the vapor pressure of the water or other liquid is equal to standard, or atmospheric, pressure at sea level.

boiling point   **Boiling point** is the temperature at which the vapor pressure of a liquid is equal to the external pressure above the liquid.

We can readily see that a liquid has an infinite number of boiling points. When we give the boiling point of a liquid, we should also state the pressure. When we express the boiling point without stating the pressure, we mean it to be the **normal boiling point** at standard pressure (760 torr). Using Table 13.1 again, we see that the normal boiling point of ethyl ether is between 30°C and 40°C, and for ethyl alcohol it is between 70°C and 80°C because for each compound 760 torr lies within these stated temperature ranges. At the normal boiling point, 1 g of a liquid changing to a vapor (gas) absorbs an amount of energy equal to its heat of vaporization (see Table 13.2).

normal boiling point

The boiling point at various pressures can be evaluated by plotting the data of Table 13.1 on the graph in Figure 13.6, where temperature is plotted horizontally along the x-axis and vapor pressure is plotted vertically along the y-axis.

Table 13.2 Physical Properties of Ethyl Chloride, Ethyl Ether, Ethyl Alcohol, and Water

Substance	Boiling point (°C)	Melting point (°C)	Heat of vaporization J/g (cal/g)	Heat of fusion J/g (cal/g)
Ethyl chloride	12.3	−139	387 (92.5)	—
Ethyl ether	34.6	−116	351 (83.9)	—
Ethyl alcohol	78.4	−112	855 (204.3)	104 (24.9)
Water	100.0	0	2259 (540)	335 (80)

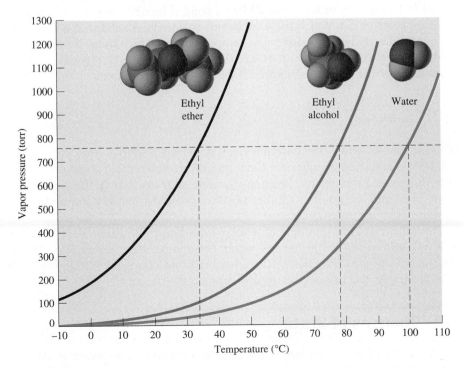

Figure 13.6
Vapor pressure–temperature curves for ethyl ether, ethyl alcohol, and water.

vapor pressure curves   The resulting curves are known as **vapor pressure curves.** Any point on these curves represents a vapor–liquid equilibrium at a particular temperature and pressure. We can find the boiling point at any pressure by tracing a horizontal line from the designated pressure to a point on the vapor pressure curve. From this point, we draw a vertical line to obtain the boiling point on the temperature axis. Three such points are shown in Figure 13.6; they represent the normal boiling points of the three compounds at 760 torr. By reversing this process, you can ascertain at what pressure a substance will boil at a specific temperature. The boiling point is one of the most commonly used physical properties for characterizing and identifying substances.

> **Practice 13.1** _____
>
> Use the graph in Figure 13.6 to determine the boiling points of ethyl ether, ethyl alcohol, and water at 600 torr.
>
> **Practice 13.2** _____
>
> The average atmospheric pressure in Denver is 0.83 atm. What is the boiling point of water in Denver?

13.6 Freezing Point or Melting Point

As heat is removed from a liquid, the liquid becomes colder and colder, until a temperature is reached at which it begins to solidify. A liquid changing into a solid is said to be _freezing,_ or _solidifying._ When a solid is heated continuously, a temperature is reached at which the solid begins to liquefy. A solid that is changing into a liquid is said to be _melting._ The temperature at which the solid phase of a **freezing or melting point**   substance is in equilibrium with its liquid phase is known as the **freezing point** or **melting point** of that substance. The equilibrium equation is

$$\text{solid} \underset{\text{freezing}}{\overset{\text{melting}}{\rightleftharpoons}} \text{liquid}$$

When a solid is slowly and carefully heated so that a solid–liquid equilibrium is achieved and then maintained, the temperature will remain constant as long as both phases are present. The energy is used solely to change the solid to the liquid. The melting point is another physical property that is commonly used for characterizing substances.

The most common example of a solid–liquid equilibrium is ice and water. In a well-stirred system of ice and water, the temperature remains at 0°C as long as both phases are present. The melting point changes only slightly with pressure unless the pressure change is very large.

13.7 Changes of State

The majority of solids undergo two changes of state upon heating. A solid changes to a liquid at its melting point, and a liquid changes to a gas at its boiling point. This warming process can be represented by a graph called a _heating_

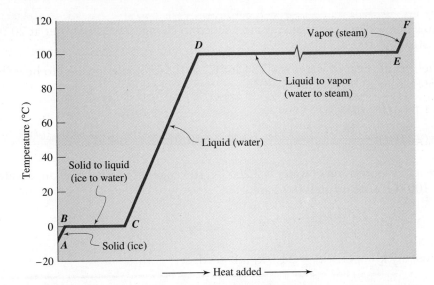

curve (Figure 13.7). This figure shows ice being heated at a constant rate. As energy flows into the ice, the vibrations within the crystal increase and the temperature rises ($A \longrightarrow B$). Eventually, the molecules begin to break free from the crystal and melting occurs ($B \longrightarrow C$). During the melting process all energy goes into breaking down the crystal structure; the temperature remains constant.

The energy required to change exactly one gram of a solid at its melting point into a liquid is called the **heat of fusion.** When the solid has completely melted, the temperature once again rises ($C \longrightarrow D$); the energy input is increasing the molecular motion within the water. At 100°C, the water reaches its boiling point; the temperature remains constant while the added energy is used to vaporize the water to steam ($D \longrightarrow E$). The **heat of vaporization** is the energy required to change exactly one gram of liquid to vapor at its normal boiling point. The attractive forces between the liquid molecules are overcome during vaporization. Beyond this temperature all the water exists as steam and is being heated further ($E \longrightarrow F$).

heat of fusion

heat of vaporization

How many joules of energy are needed to change 10.0 g of ice at 0.00°C to water at 20.0°C?

Example 13.1

Ice will absorb 335 J/g (heat of fusion) in going from a solid at 0°C to a liquid at 0°C. An additional 4.184 J/g°C (specific heat of water) is needed to raise the temperature of the water by 1°C.
   Joules needed to melt the ice:

SOLUTION

$$(10.0\ \cancel{g})\left(\frac{335\ \text{J}}{1\ \cancel{g}}\right) = 3.35 \times 10^3\ \text{J}$$

Joules needed to heat the water from 0.00°C to 20.0°C [(mass) (sp ht)(Δt)]:

$$(10.0\ \cancel{g})\left(\frac{4.184\ \text{J}}{1\ \cancel{g}\cancel{°C}}\right)(20.0\ \cancel{°C}) = 837\ \text{J}$$

Thus, 3350 J + 837 J = 4.19×10^3 J are needed.

Example 13.2 How many kilojoules of energy are needed to change 20.0 g of water at 20.°C to steam at 100.°C?

SOLUTION The specific heat of water is 4.184 J/g°C, so the kilojoules needed to heat the water from 20.°C to 100.°C are:

$$\text{(mass) (sp ht) } (\Delta t) = \text{energy}$$

$$(20.0 \,\cancel{g})\left(\frac{4.184 \,\cancel{J}}{1 \,\cancel{g°C}}\right)\left(\frac{1 \,kJ}{1000 \,\cancel{J}}\right)(100. - 20.0) \,\cancel{°C} = 6.7 \,kJ$$

Heat of vaporization of water is 2.26 kJ/g, so kilojoules needed to change water at 100.°C to steam at 100.°C are:

$$(20.0 \,\cancel{g})\left(\frac{2.26 \,kJ}{1 \,\cancel{g}}\right) = 45.2 \,kJ$$

Thus, 6.7 kJ + 45.2 kJ = 51.9 kJ are needed.

Practice 13.3 _____

How many kilojoules of energy are required to change 50.0 g of ethyl alcohol from 60.0°C to vapor at 78.4°C? The specific heat of ethyl alcohol is 2.138 J/g°C.

13.8 The Hydrogen Bond

Table 13.3 compares the physical properties of H_2O, H_2S, H_2Se, and H_2Te. From this comparison, it is apparent that four physical properties of water—melting point, boiling point, heat of fusion, and heat of vaporization—are extremely high and do not fit the trend relative to the molar masses of the four compounds. For example, if the properties of water followed the progression shown by the other three compounds, we would expect the melting point of water to be below −85°C and the boiling point to be below −60°C.

Why does water exhibit these anomalies? Because liquid water molecules are held together more strongly than other molecules in the same family.

hydrogen bond   The intermolecular force acting between water molecules is called a **hydrogen bond,** which acts like a very weak bond between two polar molecules.

Table 13.3 Physical Properties of Water and Other Hydrogen Compounds of Group 6A Elements

Formula	Color	Molar mass (g/mol)	Melting point (°C)	Boiling point, 1 atm (°C)	Heat of fusion J/g (cal/g)	Heat of vapor ization J/g (cal/g)
H_2O	Colorless	18.02	0.00	100.0	335 (80.0)	2.26×10^3 (540)
H_2S	Colorless	34.09	−85.5	−60.3	69.9 (16.7)	548 (131)
H_2Se	Colorless	80.98	−65.7	−41.3	31 (7.4)	238 (57.0)
H_2Te	Colorless	129.6	−49	−2	—	179 (42.8)

Did you think artificial sweeteners were a product of the post–World War II chemical industry? Not so—many of them have been around a long time, and several of the important ones were discovered quite by accident. In 1878, Ira Remsen was working late in his laboratory and realized he was about to miss a dinner with friends. In his haste to leave the lab, he forgot to wash his hands. Later at dinner he broke a piece of bread and tasted it only to discover that it was very sweet. The sweet taste had to be the chemical he had been working with in the lab. Back at the lab, he isolated saccharin—the first of the artificial sweeteners.

In 1937, Michael Sveda was smoking a cigarette in his laboratory (a very dangerous practice to say the least!). He touched the cigarette to his lips and was surprised by the exceedingly sweet taste. The chemical on his hands turned out to be cyclamate, which soon became a staple of the artificial sweetener industry.

In 1965, James Schlatter was researching antiulcer drugs for the pharmaceutical firm G. D. Searle. In the course of his work, he accidentally ingested a small amount of a preparation and found to his surprise that it had an extremely sweet taste. He had discovered aspartame, a molecule consisting of two amino acids joined together. Since only very small quantities of aspartame are necessary to produce sweetness, it proved to be an excellent low-calorie artificial sweetener. More than 50 different molecules have a sweet taste, and it is difficult a find a single binding site that could interact with all of them.

Our taste receptors are composed of proteins that can form hydrogen bonds with other molecules. The proteins contain —NH and —OH groups (with hydrogen available to bond) as well as C=O groups (providing oxygen for hydrogen bonding). "Sweet molecules" also contain H-bonding groups including —OH, —NH$_2$, and O or N. These molecules not only must have the proper atoms to form hydrogen bonds, they must also contain a hydrophobic region (repels H$_2$O). A new model for binding to a sweetness receptor has been developed at Senomyx in La Jolla, California. The model shows four binding sites that can act independently. Small molecules bind to a pocket on a subunit as shown in the model. Large molecules (such as proteins) bind to a different site above one of the pockets.

Sweet bondage. Model shows how the sweetener aspartame binds to a site on the sweetness receptor's T1R3 subunit. Red and blue are hydrogen-bond donor and acceptor residues, respectively; aspartame is in gold, except for its carboyxlate (red) and ammonium (blue) groups. Model prepared with MOLMOL (*J. Mol. Graphics* 1996, 14, 51). Adapted from *J. Med. Chem.*

A hydrogen bond is formed between polar molecules that contain hydrogen covalently bonded to a small, highly electronegative atom such as fluorine, oxygen, or nitrogen (F—H, O—H, N—H). A hydrogen bond is actually the dipole–dipole attraction between polar molecules containing these three types of polar bonds.

> Compounds that have significant hydrogen-bonding ability are those that contain H covalently bonded to F, O, or N.

Because a hydrogen atom has only one electron, it forms only one covalent bond. When it is bonded to a strong electronegative atom such as oxygen, a hydrogen atom will also be attracted to an oxygen atom of another molecule, forming a dipole–dipole attraction (H bond) between the two molecules. Water has two types of bonds: covalent bonds that exist between hydrogen and oxygen atoms within a molecule and hydrogen bonds that exist between hydrogen and oxygen atoms in *different* water molecules.

Oxygen

Hydrogen

· · · Hydrogen bonds

Figure 13.8
Hydrogen bonding. Water in the liquid and solid states exists as aggregates in which the water molecules are linked together by hydrogen bonds.

Hydrogen bonds are *intermolecular* bonds; that is, they are formed between atoms in different molecules. They are somewhat ionic in character because they are formed by electrostatic attraction. Hydrogen bonds are much weaker than the ionic or covalent bonds that unite atoms to form compounds. Despite their weakness, they are of great chemical importance.

The oxygen atom in water can form two hydrogen bonds—one through each of the unbonded pairs of electrons. Figure 13.8 shows two water molecules linked by a hydrogen bond and eight water molecules linked by hydrogen bonds. A dash (—) is used for the covalent bond and a dotted line (· · · ·) for the hydrogen bond. In water each molecule is linked to others through hydrogen bonds to form a three-dimensional aggregate of water molecules. This intermolecular hydrogen bonding effectively gives water the properties of a much larger, heavier molecule, explaining in part its relatively high melting point, boiling point, heat of fusion, and heat of vaporization. As water is heated and energy absorbed, hydrogen bonds are continually being broken until at 100°C, with the absorption of an additional 2.26 kJ/g, water separates into individual molecules, going into the gaseous state. Sulfur, selenium, and tellurium are not sufficiently electronegative for their hydrogen compounds to behave like water. The lack of hydrogen bonding is one reason H_2S is a gas, not a liquid, at room temperature.

Fluorine, the most electronegative element, forms the strongest hydrogen bonds. This bonding is strong enough to link hydrogen fluoride molecules together as *dimers,* H_2F_2, or as larger $(HF)_n$ molecular units. The dimer structure may be represented in this way:

$$H \diagdown \overset{\displaystyle \overset{..}{\underset{..}{F}:}}{\underset{\underset{\text{— H bond}}{\overset{..}{F}:}}{\overset{..}{\cdot \cdot \cdot}} \overset{.. H}{}}$$

Hydrogen bonding can occur between two different atoms that are capable of forming H bonds. Thus, we may have an O· · · ·H—N or O—H· · · ·N linkage in which the hydrogen atom forming the H bond is between an oxygen and a nitrogen atom. This form of H bond exists in certain types of protein molecules and many biologically active substances.

Would you expect hydrogen bonding to occur between molecules of these substances?   **Example 13.3**

(a)

H—C—C—Ö—H
  H H   H H
ethyl alcohol

(b)

H—C—Ö—C—H
  H H   H H
dimethyl ether

SOLUTION

(a) Hydrogen bonding should occur in ethyl alcohol because one hydrogen atom is bonded to an oxygen atom:

H—C—C—Ö—H⋯:O—C—C—H
  H H   H bond   H H

(b) There is no hydrogen bonding in dimethyl ether because all the hydrogen atoms are bonded only to carbon atoms.

Both ethyl alcohol and dimethyl ether have the same molar mass (46.07 g/mol). Although both compounds have the same molecular formula, C_2H_6O, ethyl alcohol has a much higher boiling point (78.4°C) than dimethyl ether (−23.7°C) because of hydrogen bonding between the alcohol molecules.

Practice 13.4 _____

Would you expect hydrogen bonding to occur between molecules of these substances?

(a)

H—C—C—N—H
  H H H   H H

(b)

H—C—N—C—H
  H H H   H H

(c)

H—C—C——N:
  H H H—C—H
  H H H—C—H
           H

13.9 Hydrates

When certain solutions containing ionic compounds are allowed to evaporate, some water molecules remain as part of the crystalline compound that is left after evaporation is complete. Solids that contain water molecules as part of their crystalline structure are known as **hydrates.** Water in a hydrate is known as **water of hydration** or **water of crystallization.**

hydrate
water of hydration
water of crystallization

Ice could well become the most abundant source of energy in the 21st century. This particular type of ice has unusual properties. Even its existence has been the subject of debate since 1810, when Humphrey Davy synthesized the first hydrate of chlorine. Chemists have long debated whether water could crystallize around a gas to form an unusual state of matter known as a gas hydrate. The material looks just like a chunk of ice left from a winter storm, gray and ugly. But unlike ice, this gas hydrate ice pops and sizzles and a lit match causes it to burst into flames. Left alone, it melts quickly into a puddle of water.

Just what is gas hydrate ice and how does it form? It turns out that gas hydrates form when water and some gases (like propane, ethane, or methane) are present at high pressure and low temperature. These conditions occur in nature at the bottom of the ocean and under permafrost in the Arctic north. Oil companies struggle

Nuggets of gas hydrate burn as they revert to water (or ice) and gas.

with hydrate ice since it plugs pipelines and damages oil rigs in oil fields in the Arctic. A gas hydrate is a gas molecule trapped inside the crystal lattice of water molecules. The most common gas is methane produced from the decomposition of organic matter. Sonar mapping has revealed large gas hydrate deposits in polar regions such as Alaska and Siberia and off the southeastern coast of the United States.

Scientists think that gas hydrates trap huge amounts of natural gas. One cubic meter of methane hydrate can contain the amount of gas that would fill 164 m^3 at room temperature and pressure. If this gas could be brought to the surface economically, it could supply 1000 years of fuel for the United States. As prices increase for energy sources, the costs for liberating gas hydrates become more feasible. At this point, most American research is focused on figuring out how to recover the gas cost-effectively from the hydrates and on transporting the methane to customers. But one day in the not too distant future we could be heating our homes and running our cars by mining gas hydrates from ice!

Formulas for hydrates are expressed by first writing the usual anhydrous (without water) formula for the compound and then adding a dot followed by the number of water molecules present. An example is $BaCl_2 \cdot 2 H_2O$. This formula tells us that each formula unit of this compound contains one barium ion, two chloride ions, and two water molecules. A crystal of the compound contains many of these units in its crystalline lattice.

In naming hydrates, we first name the compound exclusive of the water and then add the term *hydrate,* with the proper prefix representing the number of water molecules in the formula. For example, $BaCl_2 \cdot 2 H_2O$ is called *barium chloride dihydrate.* Hydrates are true compounds and follow the law of definite composition. The molar mass of $BaCl_2 \cdot 2 H_2O$ is 244.2 g/mol; it contains 56.22% barium, 29.03% chlorine, and 14.76% water.

Water molecules in hydrates are bonded by electrostatic forces between polar water molecules and the positive or negative ions of the compound. These forces are not as strong as covalent or ionic chemical bonds. As a result, water of crystallization can be removed by moderate heating of the compound. A partially dehydrated or completely anhydrous compound may result. When $BaCl_2 \cdot 2 H_2O$ is heated, it loses its water at about 100°C:

$$BaCl_2 \cdot 2 H_2O(s) \xrightarrow{100°C} BaCl_2(s) + 2 H_2O(g)$$

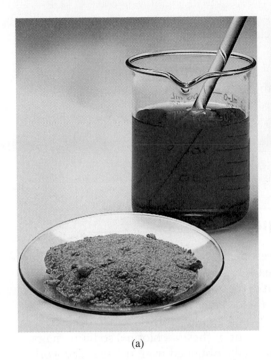

Figure 13.9
(a) When these blue crystals of $CuSO_4 \cdot 5 H_2O$ are dissolved in water, a blue solution forms. (b) The anhydrous crystals of $CuSO_4$ are pale green. When water is added, they immediately change color to blue $CuSO_4 \cdot 5 H_2O$ crystals.

(a)

(b)

When a solution of copper(II) sulfate ($CuSO_4$) is allowed to evaporate, beautiful blue crystals containing 5 moles water per 1 mole $CuSO_4$ are formed (Figure 13.9a). The formula for this hydrate is $CuSO_4 \cdot 5 H_2O$; it is called copper(II) sulfate pentahydrate. When $CuSO_4 \cdot 5 H_2O$ is heated, water is lost, and a pale green-white powder, anhydrous $CuSO_4$, is formed:

$$CuSO_4 \cdot 5 H_2O(s) \xrightarrow{250°C} CuSO_4(s) + 5 H_2O(g)$$

When water is added to anhydrous copper(II) sulfate, the foregoing reaction is reversed, and the compound turns blue again (Figure 13.9b). Because of this outstanding color change, anhydrous copper(II) sulfate has been used as an indicator to detect small amounts of water. The formation of the hydrate is noticeably exothermic.

The formula for plaster of paris is $(CaSO_4)_2 \cdot H_2O$. When mixed with the proper quantity of water, plaster of paris forms a dihydrate and sets to a hard mass. It is therefore useful for making patterns for the production of art objects, molds, and surgical casts. The chemical reaction is

$$(CaSO_4)_2 \cdot H_2O(s) + 3 H_2O(l) \longrightarrow 2 CaSO_4 \cdot 2 H_2O(s)$$

Table 13.4 lists a number of common hydrates.

Table 13.4 Selected Hydrates

Hydrate	Name	Hydrate	Name
$CaCl_2 \cdot 2 H_2O$	calcium chloride dihydrate	$Na_2CO_3 \cdot 10 H_2O$	sodium carbonate decahydrate
$Ba(OH)_2 \cdot 8 H_2O$	barium hydroxide octahydrate	$(NH_4)_2C_2O_4 \cdot H_2O$	ammonium oxalate monohydrate
$MgSO_4 \cdot 7 H_2O$	magnesium sulfate heptahydrate	$NaC_2H_3O_2 \cdot 3 H_2O$	sodium acetate trihydrate
$SnCl_2 \cdot 2 H_2O$	tin(II) chloride dihydrate	$Na_2B_4O_7 \cdot 10 H_2O$	sodium tetraborate decahydrate
$CoCl_2 \cdot 6 H_2O$	cobalt(II) chloride hexahydrate	$Na_2S_2O_3 \cdot 5 H_2O$	sodium thiosulfate pentahydrate

Practice 13.5

Write formulas for

(a) beryllium carbonate tetrahydrate
(b) cadmium permanganate hexahydrate
(c) chromium(III) nitrate nonahydrate
(d) platinum(IV) oxide trihydrate

Practice 13.6

Calculate the percent water in Epsom salts, $MgSO_4 \cdot 7\,H_2O$.

13.10   Water, a Unique Liquid

Water is our most common natural resource. It covers about 75% of Earth's surface. Not only is it found in the oceans and seas, in lakes, rivers, streams, and glacial ice deposits, it is always present in the atmosphere and in cloud formations.

About 97% of Earth's water is in the oceans. This *saline* water contains vast amounts of dissolved minerals. More than 70 elements have been detected in the mineral content of seawater. Only four of these—chlorine, sodium, magnesium, and bromine—are now commercially obtained from the sea. The world's *fresh* water comprises the other 3%, of which about two-thirds is locked up in polar ice caps and glaciers. The remaining fresh water is found in groundwater, lakes, rivers and the atmosphere.

Water is an essential constituent of all living matter. It is the most abundant compound in the human body, making up about 70% of total body mass. About 92% of blood plasma is water; about 80% of muscle tissue is water; and about 60% of a red blood cell is water. Water is more important than food in the sense that we can survive much longer without food than without water.

Physical Properties of Water

Water is a colorless, odorless, tasteless liquid with a melting point of 0°C and a boiling point of 100°C at 1 atm. The heat of fusion of water is 335 J/g (80 cal/g). The heat of vaporization of water is 2.26 kJ/g (540 cal/g). The values for water for both the heat of fusion and the heat of vaporization are high compared with those for other substances; this indicates strong attractive forces between the molecules.

Ice and water exist together in equilibrium at 0°C, as shown in Figure 13.10. When ice at 0°C melts, it absorbs 335 J/g in changing into a liquid; the temperature remains at 0°C. To refreeze the water, 335 J/g must be removed from the liquid at 0°C.

In Figure 13.10, both boiling water and steam are shown to have a temperature of 100°C. It takes 418 J to heat 1 g of water from 0°C to 100°C, but water at its boiling point absorbs 2.26 kJ/g in changing to steam. Although boiling water and steam are both at the same temperature, steam contains considerably more heat per gram and can cause more severe burns than hot water.

The maximum density of water is 1.000 g/mL at 4°C. Water has the unusual property of contracting in volume as it is cooled to 4°C and then expanding when cooled from 4°C to 0°C. Therefore, 1 g of water occupies a volume greater than 1 mL at all temperatures except 4°C. Although most liquids

Figure 13.10
Water equilibrium systems.
In the beaker on the left, ice and
water are in equilibrium at 0°C;
in the beaker on the right, boiling
water and steam are in
equilibrium at 100°C.

contract in volume all the way down to the point at which they solidify, a large
increase (about 9%) in volume occurs when water changes from a liquid at 0°C
to a solid (ice) at 0°C. The density of ice at 0°C is 0.917 g/mL, which means
that ice, being less dense than water, will float in water.

Structure of the Water Molecule

A single water molecule consists of two hydrogen atoms and one oxygen atom.
Each hydrogen atom is bonded to the oxygen atom by a single covalent bond.
This bond is formed by the overlap of the $1s$ orbital of hydrogen with an unpaired $2p$ orbital of oxygen. The average distance between the two nuclei is
known as the *bond length*. The O—H bond length in water is 0.096 nm. The
water molecule is nonlinear and has a bent structure with an angle of about
105° between the two bonds (see Figure 13.11).

Oxygen is the second most electronegative element. As a result, the two covalent OH bonds in water are polar. If the three atoms in a water molecule
were aligned in a linear structure, such as H $\longrightarrow$ O $\longleftarrow$ H, the two polar

Figure 13.11
Diagrams of a water molecule: (a) electron distribution, (b) bond angle and O—H bond length,
(c) molecular orbital structure, and (d) dipole representation.

bonds would be acting in equal and opposite directions and the molecule would be nonpolar. However, water is a highly polar molecule. It therefore does not have a linear structure. When atoms are bonded in a nonlinear fashion, the angle formed by the bonds is called the *bond angle*. In water the HOH bond angle is 105°. The two polar covalent bonds and the bent structure result in a partial negative charge on the oxygen atom and a partial positive charge on each hydrogen atom. The polar nature of water is responsible for many of its properties, including its behavior as a solvent.

Formation of Water

Water is very stable to heat; it decomposes to the extent of only about 1% at temperatures up to 2000°C. Pure water is a nonconductor of electricity. But when a small amount of sulfuric acid or sodium hydroxide is added, the solution is readily decomposed into hydrogen and oxygen by an electric current. Two volumes of hydrogen are produced for each volume of oxygen:

$$2\,H_2O(l) \xrightarrow[\text{H}_2\text{SO}_4 \text{ or NaOH}]{\text{electrical energy}} 2\,H_2(g) + O_2(g)$$

Water is formed when hydrogen burns in air. Pure hydrogen burns very smoothly in air, but mixtures of hydrogen and air or oxygen explode when ignited. The reaction is strongly exothermic:

$$2\,H_2(g) + O_2(g) \longrightarrow 2\,H_2O(g) + 484\,\text{kJ}$$

Water is produced by a variety of other reactions, especially by (1) acid–base neutralizations, (2) combustion of hydrogen-containing materials, and (3) metabolic oxidation in living cells:

1. $HCl(aq) + NaOH(aq) \longrightarrow NaCl(aq) + H_2O(l)$

2. $2\,C_2H_2(g) + 5\,O_2(g) \longrightarrow 4\,CO_2(g) + 2\,H_2O(g) + 1212\,\text{kJ}$
    acetylene

    $CH_4(g) + 2\,O_2(g) \longrightarrow CO_2(g) + 2\,H_2O(g) + 803\,\text{kJ}$
    methane

3. $C_6H_{12}O_6(aq) + 6\,O_2(g) \xrightarrow{\text{enzymes}} 6\,CO_2(g) + 6\,H_2O(l) + 2519\,\text{kJ}$
    glucose

The combustion of acetylene shown in (2) is strongly exothermic and is capable of producing very high temperatures. It is used in oxygen–acetylene torches to cut and weld steel and other metals. Methane is known as natural gas and is commonly used as fuel for heating and cooking. The reaction of glucose with oxygen shown in (3) is the reverse of photosynthesis. It is the overall reaction by which living cells obtain needed energy by metabolizing glucose to carbon dioxide and water.

Reactions of Water with Metals and Nonmetals

The reactions of metals with water at different temperatures show that these elements vary greatly in their reactivity. Metals such as sodium, potassium, and calcium react with cold water to produce hydrogen and a metal hydroxide. A small piece of sodium added to water melts from the heat produced by the reaction, forming a silvery metal ball, which rapidly flits back

Sodium reacts vigorously with water to produce hydrogen gas and sodium hydroxide.

and forth on the surface of the water. Caution must be used when experimenting with this reaction, because the hydrogen produced is frequently ignited by the sparking of the sodium, and it will explode, spattering sodium. Potassium reacts even more vigorously than sodium. Calcium sinks in water and liberates a gentle stream of hydrogen. The equations for these reactions are

$$2\,Na(s) + 2\,H_2O(l) \longrightarrow H_2(g) + 2\,NaOH(aq)$$
$$2\,K(s) + 2\,H_2O(l) \longrightarrow H_2(g) + 2\,KOH(aq)$$
$$Ca(s) + 2\,H_2O(l) \longrightarrow H_2(g) + Ca(OH)_2(aq)$$

Zinc, aluminum, and iron do not react with cold water but will react with steam at high temperatures, forming hydrogen and a metallic oxide. The equations are

$$Zn(s) + H_2O(g) \longrightarrow H_2(g) + ZnO(s)$$
$$2\,Al(s) + 3\,H_2O(g) \longrightarrow 3\,H_2(g) + Al_2O_3(s)$$
$$3\,Fe(s) + 4\,H_2O(g) \longrightarrow 4\,H_2(g) + Fe_3O_4(s)$$

Copper, silver, and mercury are examples of metals that do not react with cold water or steam to produce hydrogen. We conclude that sodium, potassium, and calcium are chemically more reactive than zinc, aluminum, and iron, which are more reactive than copper, silver, and mercury.

Certain nonmetals react with water under various conditions. For example, fluorine reacts violently with cold water, producing hydrogen fluoride and free oxygen. The reactions of chlorine and bromine are much milder, producing what is commonly known as "chlorine water" and "bromine water," respectively. Chlorine water contains HCl, HOCl, and dissolved Cl_2; the free chlorine gives it a yellow-green color. Bromine water contains HBr, HOBr, and dissolved Br_2; the free bromine gives it a reddish-brown color. The equations for these reactions are

$$2\,F_2(g) + 2\,H_2O(l) \longrightarrow 4\,HF(aq) + O_2(g)$$
$$Cl_2(g) + H_2O(l) \longrightarrow HCl(aq) + HOCl(aq)$$
$$Br_2(l) + H_2O(l) \longrightarrow HBr(aq) + HOBr(aq)$$

Reactions of Water with Metal and Nonmetal Oxides

Metal oxides that react with water to form hydroxides are known as **basic anhydrides.** Examples are

basic anhydride

$$CaO(s) + H_2O(l) \longrightarrow Ca(OH)_2(aq)$$
$$\text{calcium hydroxide}$$

$$Na_2O(s) + H_2O(l) \longrightarrow 2\,NaOH(aq)$$
$$\text{sodium hydroxide}$$

Certain metal oxides, such as CuO and Al_2O_3, do not form solutions containing OH^- ions because the oxides are insoluble in water.

Practice 13.7 ————————————————————

Copper does not react with water to produce hydrogen gas. However, hydrogen reacts with copper(II) oxide to produce water. Write a balanced equation for this reaction.

acid anhydride Nonmetal oxides that react with water to form acids are known as **acid anhydrides.** Examples are

$$CO_2(g) + H_2O(l) \rightleftharpoons H_2CO_3(aq)$$
carbonic acid

$$SO_2(g) + H_2O(l) \rightleftharpoons H_2SO_3(aq)$$
sulfurous acid

$$N_2O_5(s) + H_2O(l) \longrightarrow 2\,HNO_3(aq)$$
nitric acid

The word *anhydrous* means "without water." An anhydride is a metal oxide or a nonmetal oxide derived from a base or an oxy-acid by the removal of water. To determine the formula of an anhydride, the elements of water are removed from an acid or base formula until all the hydrogen is removed. Sometimes more than one formula unit is needed to remove all the hydrogen as water. The formula of the anhydride then consists of the remaining metal or nonmetal and the remaining oxygen atoms. In calcium hydroxide, removal of water as indicated leaves CaO as the anhydride:

$$Ca{\overset{O|H}{\underset{\boxed{OH}}{}}} \xrightarrow{\Delta} CaO + H_2O$$

In sodium hydroxide, H_2O cannot be removed from one formula unit, so two formula units of NaOH must be used, leaving Na_2O as the formula of the anhydride:

$$\begin{matrix}NaO\boxed{H}\\ Na\boxed{OH}\end{matrix} \xrightarrow{\Delta} Na_2O + H_2O$$

The removal of H_2O from H_2SO_4 gives the acid anhydride SO_3:

$$H_2SO_4 \xrightarrow{\Delta} SO_3 + H_2O$$

The foregoing are examples of typical reactions of water but are by no means a complete list of the known reactions of water.

13.11 Water Purification

Natural fresh waters are not pure, but contain dissolved minerals, suspended matter, and sometimes harmful bacteria. The water supplies of large cities are usually drawn from rivers or lakes. Such water is generally unsafe to drink without treatment. To make such water safe to drink, it is treated by some or all of the following processes (see Figure 13.12):

1. *Screening.* Removal of relatively large objects, such as trash, fish, and so on.
2. *Flocculation and sedimentation.* Chemicals, usually lime, CaO, and alum (aluminum sulfate), $Al_2(SO_4)_3$, are added to form a flocculent jellylike precipitate of aluminum hydroxide. This precipitate traps most of the fine suspended matter in the water and carries it to the bottom of the sedimentation basin.
3. *Sand filtration.* Water is drawn from the top of the sedimentation basin and passed downward through fine sand filters. Nearly all the remaining suspended matter and bacteria are removed by the sand filters.

Figure 13.12
Typical municipal
water-treatment plant.

4. *Aeration.* Water is drawn from the bottom of the sand filters and is aerated by spraying. The purpose of this process is to remove objectionable odors and tastes.

5. *Disinfection.* In the final stage chlorine gas is injected into the water to kill harmful bacteria before the water is distributed to the public. Ozone is also used to disinfect water. In emergencies, water may be disinfected by simply boiling it for a few minutes.

Water that contains dissolved calcium and magnesium salts is called *hard water.* Ordinary soap does not lather well in hard water; the soap reacts with the calcium and magnesium ions to form an insoluble greasy scum. However, synthetic soaps, known as detergents, have excellent cleaning qualities and do not form precipitates with hard water. Hard water is also undesirable because it causes "scale" to form on the walls of water heaters, tea kettles, coffee pots, and steam irons, which greatly reduces their efficiency.

Many people today find that they want to further purify their drinking and cooking water. Often this is because of objectionable tastes in the municipal drinking water. People add filtration systems to appliances such as refrigerators and coffee pots or use home water filtration systems. People also buy bottled water for drinking and cooking.

Four techniques are used to "soften" hard water:

1. **Distillation** The water is boiled, and the steam formed is condensed into a liquid again, leaving the minerals behind in the distilling vessel. Figure 13.13 illustrates a simple laboratory distillation apparatus. Commercial stills are capable of producing hundreds of liters of distilled water per hour.

2. **Calcium and magnesium precipitation** Calcium and magnesium ions are precipitated from hard water by adding sodium carbonate and lime. Insoluble calcium carbonate and magnesium hydroxide are precipitated and are removed by filtration or sedimentation.

3. **Ion exchange** Hard water is effectively softened as it is passed through a bed or tank of zeolite—a complex sodium aluminum silicate. In this process, sodium ions replace objectionable calcium and magnesium ions, and the water is thereby softened:

$$Na_2(zeolite)(s) + Ca^{2+}(aq) \longrightarrow Ca(zeolite)(s) + 2\,Na^+(aq)$$

The zeolite is regenerated by back-flushing with concentrated sodium chloride solution, reversing the foregoing reaction.

If the drinking water of children contains an optimum amount of fluoride ion, their teeth will be more resistant to decay.

Bottled water provides a convenient source of hydration for athletes.

Figure 13.13
Simple laboratory setup for
distilling liquids.

4. Demineralization Both cations and anions are removed by a two-stage
ion-exchange system. Special synthetic organic resins are used in the ion-
exchange beds. In the first stage metal cations are replaced by hydrogen
ions. In the second stage anions are replaced by hydroxide ions. The hydrogen and hydroxide ions react, and essentially pure, mineral-free water
leaves the second stage.

Our oceans are an enormous source of water, but seawater contains about
3.5 lb of salts per 100 lb of water. This 35,000 ppm of dissolved salts makes seawater unfit for agricultural and domestic uses. Water that contains less than
1000 ppm of salts is considered reasonably good for drinking, and safe drinking water is already being obtained from the sea in many parts of the world.
Continuous research is being done in an effort to make usable water from our
oceans more abundant and economical. See Figure 13.14.

Figure 13.14
Catalina Island, California, gets
most of its water from a
desalinization plant.

Table 13.5 Classification of Water Pollutants

Type of pollutant	Examples
Oxygen-demanding wastes	Decomposable organic wastes from domestic sewage and industrial wastes of plant and animal origin
Infectious agents	Bacteria, viruses, and other organisms from domestic sewage, animal wastes, and animal process wastes
Plant nutrients	Principally compounds of nitrogen and phosphorus
Organic chemicals	Large numbers of chemicals synthesized by industry, pesticides, chlorinated organic compounds
Other minerals and chemicals	Inorganic chemicals from industrial operations, mining, oil field operations, and agriculture
Radioactive substances	Waste products from mining and processing radioactive materials, airborne radioactive fallout, increased use of radioactive materials in hospitals and research
Heat from industry	Large quantities of heated water returned to water bodies from power plants and manufacturing facilities after use for cooling
Sediment from land erosion	Solid matter washed into streams and oceans by erosion, rain, and water runoff

Our highly technical, industrial society has produced many new chemicals that add to our need for water purification. Many of the newer chemicals are not removed or destroyed by the usual water-treatment processes. For example, among the 66 organic compounds found in the drinking water of a major city on the Mississippi River, 3 are labeled slightly toxic, 17 moderately toxic, 15 very toxic, 1 extremely toxic, and 1 supertoxic. Two are known carcinogens (cancer-producing agents), 11 are suspect, and 3 are metabolized to carcinogens. The U.S. Public Health Service classifies water pollutants under eight broad categories. These categories are shown in Table 13.5.

The disposal of hazardous waste products adds to the water purification problem. These substances are unavoidable in the manufacture of many products we consider indispensable today. One common way to dispose of these wastes is to place them in toxic waste dumps. What has been found after many years of disposing of wastes in this manner is that toxic substances have seeped into the groundwater deposits. As a result, many people have become ill, and water wells have been closed until satisfactory methods of detoxifying this water are found. This problem is serious, because one-half the U.S. population gets its drinking water from groundwater. Cleaning up the thousands of industrial dumps and finding and implementing new and safe methods of disposing of wastes is ongoing and costly.

Chapter 13 Review

13.1   What Is a Liquid?

- Properties:
  - Particles are close together yet free to move.
  - Liquids are incompressible.
  - Liquids have definite volume but take the shape of the container.

13.2   Evaporation

KEY TERMS

Evaporation, or vaporization
Sublimation

- During evaporation, molecules of greater-than-average kinetic energy escape from the liquid.
- Sublimation is the evaporation of a solid directly to a gas.

13.3   Vapor Pressure

KEY TERMS

Condensation
Vapor pressure
Volatile

- Gaseous molecules return to the liquid state through condensation.
- At equilibrium the rate of evaporation equals the rate of condensation.
- The pressure exerted by the vapor in equilibrium with its liquid in a closed container is the vapor pressure of the liquid.
- A volatile substance evaporates readily.

13.4   Surface Tension

KEY TERMS

Surface tension
Capillary action
Meniscus

- The resistance of a liquid to an increase in its surface area is the surface tension of the liquid.
- Capillary action is caused by the cohesive forces within the liquid and adhesive forces between the liquid and the walls of the container:
  - If the forces between the liquid and the container are greater than those within the liquid, the liquid will climb the walls of the container.
- A meniscus is evidence of cohesive and adhesive forces:
  - It is concave if the adhesive forces are stronger than the cohesive forces.
  - It is convex if the cohesive forces are stronger than the adhesive forces.

13.5   Boiling Point

KEY TERMS

Boiling point
Normal boiling point
Vapor pressure curves

- The boiling point of a liquid is the temperature at which its vapor pressure equals the atmospheric pressure:
  - At 1 atm the boiling point is called the normal boiling point for a liquid.

13.6   Freezing Point or Melting Point

KEY TERM

Freezing point or melting point

- The temperature at which a solid is in equilibrium with its liquid phase is the freezing point or melting point.

13.7   Changes of State

KEY TERMS

Heat of fusion
Heat of vaporization

- A graph of the warming of a liquid is called a heating curve:
  - Horizontal lines on the heating curve represent changes of state:
    - The energy required to change 1 g of solid to liquid at its melting point is the heat of fusion.
    - The energy required to change 1 g of liquid to gas at its normal boiling point is the heat of vaporization.
    - The energy required to change the phase of a sample (at its melting or boiling point) is

    energy = (mass)(heat of fusion (or vaporization))
  - The energy required to heat molecules without a phase change is determined by

  energy = (mass)(sp ht)(Δt)

13.8   The Hydrogen Bond

KEY TERM

Hydrogen bond

- A hydrogen bond is the dipole–dipole attraction between polar molecules containing any of these types of bonds: F—H, O—H, N—H.

13.9   Hydrates

KEY TERMS

Hydrate
Water of hydration or water crystallization

- Solids that contain water molecules as part of their crystalline structure are called hydrates.
- Formulas of hydrates are given by writing the formula for the anhydrous compound followed by a dot and then the number of water molecules present.
- Water molecules in hydrates are bonded by electrostatic forces, which are weaker than covalent or ionic bonds:
  - Water of hydration can be removed by heating the hydrate to form a partially dehydrated compound or the anhydrous compound.

13.10 Water, a Unique Liquid

KEY TERMS

Basic anhydride
Acid anhydride

- Water is our most common resource, covering 75% of the Earth's surface.
- Water is a colorless, odorless, tasteless liquid with a melting point of 0°C and boiling point of 100°C at 1 atm
- The water molecule consists of 2 H atoms and 1 O atom bonded together at a 105° bond angle, making the molecule polar.
- Water can be formed in a variety of ways including:
  - Burning hydrogen in air
  - Acid–base neutralizations
  - Combustion of hydrogen-containing materials
  - Metabolic oxidation in living cells

- Water undergoes reactions with many metals and nonmetals, depending on their reactivity.
- Water reacts with metal oxides (basic anhydrides) to form bases (hydroxides).
- Water reacts with nonmetal oxides (acid anhydrides) to form acids.

13.11 Water Purification

- Water can be purified at the commercial level by using some or all of the following processes: screening, flocculation and sedimentation, sand filtration, aeration, and disinfection.
- Water can be softened by one of the following techniques: distillation, calcium and magnesium precipitation, ion exchange, or demineralization.

Review Questions

All questions with blue numbers have answers in the appendix of the text.

1. Compare the potential energy of the three states of water shown in Figure 13.10.

2. In what state (solid, liquid, or gas) would H_2S, H_2Se, and H_2Te be at 0°C? (Table 13.3)

3. What property or properties of liquids are similar to solids?

4. What property or properties of liquids are similar to gases?

5. The temperature of the water in the beaker on the hotplate (Figure 13.10) reads 100°C. What is the pressure of the atmosphere?

6. Diagram a water molecule and point out the negative and positive ends of the dipole.

7. If the water molecule were linear, with all three atoms in a straight line rather than in the shape of a V, as shown in Figure 13.11, what effect would this have on the physical properties of water?

8. How do we specify 1, 2, 3, 4, 5, 6, 7, and 8 molecules of water in the formulas of hydrates? (Table 13.4)

9. Would the distillation setup in Figure 13.13 be satisfactory for separating salt and water? for separating ethyl alcohol and water? Explain.

10. If the liquid in the flask in Figure 13.13 is ethyl alcohol and the atmospheric pressure is 543 torr, what temperature will show on the thermometer? (Use Figure 13.6.)

11. If water were placed in both containers in Figure 13.3, would both have the same vapor pressure at the same temperature? Explain.

12. Why doesn't the vapor pressure of a liquid depend on the amount of liquid and vapor present?

13. In Figure 13.3, in which case, (a) or (b), will the atmosphere above the liquid reach a point of saturation?

14. Suppose a solution of ethyl ether and ethyl alcohol is placed in the closed bottle in Figure 13.3. (Use Figure 13.6 for information on the substances.)
    (a) Are both substances present in the vapor?
    (b) If the answer to part (a) is yes, which has more molecules in the vapor?

15. In Figure 13.4, if 50% more water is added in part (b), what equilibrium vapor pressure will be observed in (c)?

16. At approximately what temperature would each of the substances shown in Figure 13.6 boil when the pressure is 30 torr?

17. Use the graph in Figure 13.6 to find the following:
    (a) boiling point of water at 500 torr
    (b) normal boiling point of ethyl alcohol
    (c) boiling point of ethyl ether at 0.50 atm

18. Consider Figure 13.7.
    (a) Why is line BC horizontal? What is happening in this interval?
    (b) What phases are present in the interval BC?
    (c) When heating is continued after point C, another horizontal line, DE, is reached at a higher temperature. What does this line represent?

19. List six physical properties of water.

20. What condition is necessary for water to have its maximum density? What is its maximum density?

21. Account for the fact that an ice–water mixture remains at 0°C until all the ice is melted, even though heat is applied to it.

22. Which contains less energy: ice at 0°C or water at 0°C? Explain.

23. Why does ice float in water? Would ice float in ethyl alcohol ($d = 0.789$ g/mL)? Explain.

24. If water molecules were linear instead of bent, would the heat of vaporization be higher or lower? Explain.

25. The heat of vaporization for ethyl ether is 351 J/g and that for ethyl alcohol is 855 J/g. From this data, which of these compounds has hydrogen bonding? Explain.

26. Would there be more or less H bonding if water molecules were linear instead of bent? Explain.

27. In which condition are there fewer hydrogen bonds between molecules: water at 40°C or water at 80°C? Explain.

28. Which compound

    $H_2NCH_2CH_2NH_2$ or $CH_3CH_2CH_2NH_2$

    would you expect to have the higher boiling point? Explain. (Both compounds have similar molar masses.)

29. Explain why rubbing alcohol warmed to body temperature still feels cold when applied to your skin.

30. The vapor pressure at 20°C for the following substances is

methyl alcohol	96 torr
acetic acid	11.7 torr
benzene	74.7 torr
bromine	173 torr
water	17.5 torr
carbon tetrachloride	91 torr
mercury	0.0012 torr
toluene	23 torr

(a) Arrange these substances in order of increasing rate of evaporation.

(b) Which substance listed has the highest boiling point? the lowest?

31. Suggest a method whereby water could be made to boil at 50°C.

32. Explain why a higher temperature is obtained in a pressure cooker than in an ordinary cooking pot.

33. What is the relationship between vapor pressure and boiling point?

34. On the basis of the kinetic-molecular theory, explain why vapor pressure increases with temperature.

35. Why does water have such a relatively high boiling point?

36. The boiling point of ammonia, NH_3, is −33.4°C and that of sulfur dioxide, SO_2, is −10.0°C. Which has the higher vapor pressure at −40°C?

37. Explain what is occurring physically when a substance is boiling.

38. Explain why HF (bp = 19.4°C) has a higher boiling point than HCl (bp = −85°C), whereas F_2 (bp = −188°C) has a lower boiling point than Cl_2 (bp = −34°C).

39. Why does a boiling liquid maintain a constant temperature when heat is continuously being added?

40. At what specific temperature will ethyl ether have a vapor pressure of 760 torr?

41. What water temperature would you theoretically expect to find at the bottom of a very deep lake? Explain.

42. Is the formation of hydrogen and oxygen from water an exothermic or an endothermic reaction? How do you know?

Paired Exercises

All exercises with blue numbers have answers in the appendix of the text.

1. Write the formulas for the anhydrides of these acids: $HClO_4$, H_2CO_3, H_3PO_4.

2. Write the formulas for the anhydrides of these acids: H_2SO_3, H_2SO_4, HNO_3.

3. Write the formulas for the anhydrides of these bases: LiOH, NaOH, $Mg(OH)_2$.

4. Write the formulas for the anhydrides of these bases: KOH, $Ba(OH)_2$, $Ca(OH)_2$.

5. Complete and balance these equations:

    (a) $Ba(OH)_2 \xrightarrow{\Delta}$

    (b) $CH_3OH + O_2 \xrightarrow{\Delta}$
        methyl alcohol

    (c) $Rb + H_2O \longrightarrow$

    (d) $SnCl_2 \cdot 2H_2O \xrightarrow{\Delta}$

    (e) $HNO_3 + NaOH \longrightarrow$

    (f) $CO_2 + H_2O \longrightarrow$

6. Complete and balance these equations:

    (a) $Li_2O + H_2O \longrightarrow$

    (b) $KOH \xrightarrow{\Delta}$

    (c) $Ba + H_2O \longrightarrow$

    (d) $Cl_2 + H_2O \longrightarrow$

    (e) $SO_3 + H_2O \longrightarrow$

    (f) $H_2SO_3 + KOH \longrightarrow$

7. Name these hydrates:
(a) $BaBr_2 \cdot 2\,H_2O$
(b) $AlCl_3 \cdot 6\,H_2O$
(c) $FePO_4 \cdot 4\,H_2O$

8. Name these hydrates:
(a) $MgNH_4PO_4 \cdot 6\,H_2O$
(b) $FeSO_4 \cdot 7\,H_2O$
(c) $SnCl_4 \cdot 5\,H_2O$

9. Distinguish between deionized water and
(a) hard water
(b) soft water

10. Distinguish between deionized water and
(a) distilled water
(b) natural water

11. In which of the following substances would you expect to find hydrogen bonding? Explain your reasons.
(a) C_2H_6
(b) NH_3
(c) H_2O
(d) HI
(e) C_2H_5OH

12. In which of the following substances would you expect to find hydrogen bonding? Explain your reasons.
(a) HF
(b) CH_4
(c) H_2O_2
(d) CH_3OH
(e) H_2

13. For each of the compounds in Question 11 that forms hydrogen bonds, draw a diagram of the two molecules using a dotted line to indicate where the hydrogen bonding will occur.

14. For each of the compounds in Question 12 that forms hydrogen bonds, draw a diagram of the two molecules using a dotted line to indicate where the hydrogen bonding will occur.

15. Water spreads out on a glass plate. In this example, are adhesive forces or cohesive forces strongest? Explain.

16. Rainex is a product that causes water to "bead" instead of spread when sprayed on car windshields. When the water forms these droplets, are adhesive forces or cohesive forces strongest? Explain.

17. How many moles of compound are in 100. g of $CoCl_2 \cdot 6\,H_2O$?

18. How many moles of compound are in 100. g of $FeI_2 \cdot 4\,H_2O$?

19. How many moles of water can be obtained from 100. g of $CoCl_2 \cdot 6\,H_2O$?

20. How many moles of water can be obtained from 100. g of $FeI_2 \cdot 4\,H_2O$?

21. When a person purchases Epsom salts, $MgSO_4 \cdot 7\,H_2O$, what percent of the compound is water?

22. Calculate the mass percent of water in the hydrate $Al_2(SO_4)_3 \cdot 18\,H_2O$.

23. Sugar of lead, a hydrate of lead(II) acetate, $Pb(C_2H_3O_2)_2$, contains 14.2% H_2O. What is the empirical formula for the hydrate?

24. A 25.0-g sample of a hydrate of $FePO_4$ was heated until no more water was driven off. The mass of anhydrous sample is 16.9 g. What is the empirical formula of the hydrate?

25. How many joules are needed to change 120. g of water at 20.°C to steam at 100.°C?

26. How many joules of energy must be removed from 126 g of water at 24°C to form ice at 0°C?

*27. Suppose 100. g of ice at 0°C are added to 300. g of water at 25°C. Is this sufficient ice to lower the temperature of the system to 0°C and still have ice remaining? Show evidence for your answer.

*28. Suppose 35.0 g of steam at 100.°C are added to 300. g of water at 25°C. Is this sufficient steam to heat all the water to 100.°C and still have steam remaining? Show evidence for your answer.

*29. If 75 g of ice at 0.0°C were added to 1.5 L of water at 75°C, what would be the final temperature of the mixture?

*30. If 9560 J of energy were absorbed by 500. g of ice at 0.0°C, what would be the final temperature?

31. How many grams of water will react with the following?
(a) 1.00 g Na
(b) 1.00 g MgO
(c) 1.00 g N_2O_5

32. How many grams of water will react with the following?
(a) 1.00 mol K
(b) 1.00 mol Ca
(c) 1.00 mol SO_3

Additional Exercises

All exercises with blue numbers have answers in the appendix of the text.

33. You walk out to your car just after a rain shower and notice that small droplets of water are scattered on the hood. How was it possible for the water to form those droplets?

34. Which causes a more severe burn: liquid water at 100°C or steam at 100°C? Why?

35. You have a shallow dish of alcohol set into a tray of water. If you blow across the tray, the alcohol evaporates, while the water cools significantly and eventually freezes. Explain why.

36. Regardless of how warm the outside temperature may be, we always feel cool when stepping out of a swimming pool, the ocean, or a shower. Why is this so?

37. Sketch a heating curve for a substance X whose melting point is 40°C and whose boiling point is 65°C.
    (a) Describe what you will observe as a 60.-g sample of X is warmed from 0°C to 100°C.
    (b) If the heat of fusion of X is 80. J/g, the heat of vaporization is 190. J/g, and if 3.5 J are required to warm 1 g of X each degree, how much energy will be needed to accomplish the change in (a)?

38. For the heating curve of water (see Figure 13.7), why doesn't the temperature increase when ice is melting into liquid water and liquid water is changing into steam?

39. Why does the vapor pressure of a liquid increase as the temperature is increased?

40. At the top of Mount Everest, which is just about 29,000 feet above sea level, the atmospheric pressure is about 270 torr. Use Figure 13.6 to determine the approximate boiling temperature of water on Mount Everest.

41. Explain how anhydrous copper(II) sulfate ($CuSO_4$) can act as an indicator for moisture.

42. Write formulas for magnesium sulfate heptahydrate and disodium hydrogen phosphate dodecahydrate.

43. How can soap make soft water from hard water? What objections are there to using soap for this purpose?

44. What substance is commonly used to destroy bacteria in water?

45. What chemical, other than chlorine or chlorine compounds, can be used to disinfect water for domestic use?

46. Some organic pollutants in water can be oxidized by dissolved molecular oxygen. What harmful effect can result from this depletion of oxygen in the water?

47. Why should you not drink liquids that are stored in ceramic containers, especially unglazed ones?

48. Write the chemical equation showing how magnesium ions are removed by a zeolite water softener.

49. Write an equation to show how hard water containing calcium chloride ($CaCl_2$) is softened by using sodium carbonate (Na_2CO_3).

50. Draw a heating curve and label the different phases and phase changes on the curve for solid H_2S at −95 °C and a pressure of 1 atm going to gaseous H_2S. (*Note:* For H_2S the melting point is −82.9°C and the boiling point is −59.6°C.)

51. How many calories are required to change 225 g of ice at 0°C to steam at 100.°C?

52. The molar heat of vaporization is the number of joules required to change 1 mol of a substance from liquid to vapor at its boiling point. What is the molar heat of vaporization of water?

*53. The specific heat of zinc is 0.096 cal/g°C. Determine the energy required to raise the temperature of 250. g of zinc from room temperature (20.0°C) to 150.°C.

54. How many joules of energy would be liberated by condensing 50.0 mol of steam at 100.0°C and allowing the liquid to cool to 30.0°C?

55. How many kilojoules of energy are needed to convert 100. g of ice at −10.0°C to water at 20.0°C? (The specific heat of ice at −10.0°C is 2.01 J/g°C.)

56. What mass of water must be decomposed to produce 25.0 L of oxygen at STP?

*57. Suppose 1.00 mol of water evaporates in 1.00 day. How many water molecules, on the average, leave the liquid each second?

58. Compare the volume occupied by 1.00 mol of liquid water at 0°C and 1.00 mol of water vapor at STP.

59. A mixture of 80.0 mL of hydrogen and 60.0 mL of oxygen is ignited by a spark to form water.
    (a) Does any gas remain unreacted? Which one, H_2 or O_2?
    (b) What volume of which gas (if any) remains unreacted? (Assume the same conditions before and after the reaction.)

60. A student (with slow reflexes) puts his hand in a stream of steam at 100.°C until 1.5 g of water has condensed. If the water then cools to room temperature (20.0°C), how many joules have been absorbed by the student's hand?

61. Determine which of the following molecules would hydrogen bond with other molecules like it. For those that do not hydrogen bond explain why. For those that hydrogen bond draw a diagram of two molecules using a dotted line to indicate where the hydrogen bonding will occur.
    (a) Br_2 (d) H_2O
    (b) $CH_3—O—H$ (e) H_2S
    (c) $CH_3—O—CH_3$

62. The heat of fusion of a substance is given in units of J/g. The specific heat of a substance is given in units of J/g °C. Why is a temperature factor not needed in the units for heat of fusion?

63. How many joules of energy are required to change 50.0 g Cu from 25.0°C to a liquid at its melting point, 1083 °C?

    Specific heat of Cu = 0.385 J/g °C
    Heat of fusion for Cu = 134 J/g

Challenge Exercises

All exercises with blue numbers have answers in the appendix of the text.

64. Why does a lake freeze from the top down? What significance does this have for life on Earth?

*66. A quantity of sulfuric acid is added to 100. mL of water. The final volume of the solution is 122 mL and has a density of 1.26 g/mL. What mass of acid was added? Assume the density of the water is 1.00 g/mL.

*65. Suppose 150. g of ice at 0.0°C are added to 0.120 L of water at 45°C. If the mixture is stirred and allowed to cool to 0.0°C, how many grams of ice remain?

Answers to Practice Exercises

13.1 28°C, 73°C, 93°C

13.2 approximately 95°C

13.3 44.8 kJ

13.4 (a) yes, (b) yes, (c) no

13.5 (a) $BeCO_3 \cdot 4\,H_2O$
(b) $Cd(MnO_4)_2 \cdot 6\,H_2O$
(c) $Cr(NO_3)_3 \cdot 9\,H_2O$
(d) $PtO_2 \cdot 3\,H_2O$

13.6 51.17% H_2O

13.7 $CuO + H_2 \longrightarrow Cu + H_2O$

CHAPTER 14

Solutions

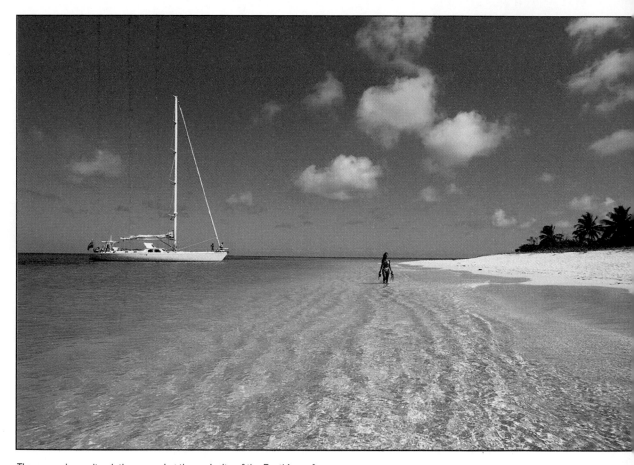

The ocean is a salt solution covering the majority of the Earth's surface.

Chapter Outline

14.1 General Properties of Solutions

14.2 Solubility

14.3 Factors Related to Solubility

14.4 Rate of Dissolving Solids

14.5 Solutions: A Reaction Medium

14.6 Concentration of Solutions

14.7 Colligative Properties of Solutions

14.8 Osmosis and Osmotic Pressure

Most substances we encounter in our daily lives are mixtures. Often they are homogeneous mixtures, which are called *solutions*. Some solutions we commonly encounter are shampoo, soft drinks, or wine. Blood plasma is a complex mixture composed of compounds and ions dissolved in water and proteins suspended in the solution. These solutions all have water as a main component, but many common items, such as air, gasoline, and steel, are also solutions that do not contain water. What are the necessary components of a solution? Why do some substances dissolve, while others do not? What effect does a dissolved substance have on the properties of the solution? Answering these questions is the first step in understanding the solutions we encounter in our daily lives.

14.1   General Properties of Solutions

The term **solution** is used in chemistry to describe a system in which one or more substances are homogeneously mixed or dissolved in another substance. A simple solution has two components: a solute and a solvent. The **solute** is the component that is dissolved or is the least abundant component in the solution. The **solvent** is the dissolving agent or the most abundant component in the solution. For example, when salt is dissolved in water to form a solution, salt is the solute and water is the solvent. Complex solutions containing more than one solute and/or more than one solvent are common.

    The three states of matter—solid, liquid, and gas—give us nine different types of solutions: solid dissolved in solid, solid dissolved in liquid, solid dissolved in gas, liquid dissolved in liquid, and so on. Of these, the most common solutions are solid dissolved in liquid, liquid dissolved in liquid, gas dissolved in liquid, and gas dissolved in gas. Some common types of solutions are listed in Table 14.1.

    A true solution is one in which the particles of dissolved solute are molecular or ionic in size, generally in the range of 0.1 to 1 nm (10^{-8} to 10^{-7} cm). The properties of a true solution are as follows:

1. A mixture of two or more components—solute and solvent—is homogeneous and has a variable composition; that is, the ratio of solute to solvent can be varied.
2. The dissolved solute is molecular or ionic in size.
3. It is either colored or colorless and is usually transparent.
4. The solute remains uniformly distributed throughout the solution and will not settle out with time.
5. The solute can generally be separated from the solvent by purely physical means (for example, by evaporation).

solution

solute

solvent

Table 14.1 Common Types of Solutions

Phase of solution	Solute	Solvent	Example
Gas	gas	gas	air
Liquid	gas	liquid	soft drinks
Liquid	liquid	liquid	antifreeze
Liquid	solid	liquid	salt water
Solid	gas	solid	H_2 in Pt
Solid	solid	solid	brass

We usually consider silicon, a main ingredient in sand and computer chips, to be reasonably inert. So imagine researcher Frederic Mikulec's surprise when he cut a silicon wafer with a diamond scribe and it blew up in his face. Michael Sailor, professor of chemistry and biochemistry at the University of California, San Diego, recalls, "It was just a small explosion, like a cap going off in a cap gun." The event inspired the research team to investigate the process further.

The silicon wafers being used were specially processed to have many tiny holes. Gadolinium nitrate was dissolved in ethanol and added to the porous silicon wafers. After the ethanol evaporated, the gadolinium nitrate remained inside the pores in the silicon chip. Sailor's team discovered that they could explode these chips using a 9-volt battery or by scratching them.

Next the team began to look for ways to use these exploding silicon chips. The scientists added metal ions to the wafers and exploded them. Their results indicated that the elements could be identified by their emission spectra in the flame from the explosion. Currently, groundwater samples must be taken to the lab and analyzed in an emission spectrometer by being mixed with other chemicals and burned. Sailor thinks that these wafers could be used as a sensor for analyzing water contaminants in the field. Imagine being able to test water for heavy metals using an instrument you could hold in the palm of your hand.

Since the silicon wafers explode like gunpowder, these tiny wafers could also be used to operate tiny micromotors or microvehicles. Sailor also thinks that these silicon wafers could replace blasting caps in explosives.

Let's illustrate these properties using water solutions of sugar and of potassium permanganate. We prepare two sugar solutions, the first containing 10 g of sugar added to 100 mL of water and the second containing 20 g of sugar added to 100 mL of water. Each solution is stirred until all the solute dissolves, demonstrating that we can vary the composition of a solution. Every portion of the solution has the same sweet taste because the sugar molecules are uniformly distributed throughout. If confined so that no solvent is lost, the solution will taste and appear the same a week or a month later. A solution cannot be separated into its components by filtering it. But by carefully evaporating the water, we can recover the sugar from the solution.

To observe the dissolving of potassium permanganate ($KMnO_4$), we drop a few crystals of it in a beaker of water. Almost at once, the beautiful purple color of dissolved permanganate ions (MnO_4^-) appears and streams to the bottom of the beaker as the crystals dissolve. After a while, the purple color disperses until it's evenly distributed throughout the solution. This dispersal demonstrates that molecules and ions move about freely and spontaneously (diffuse) in a liquid or solution.

Solution permanency is explained in terms of the kinetic-molecular theory (see Section 12.2). According to the KMT, both the solute and solvent particles (molecules or ions) are in constant random motion. This motion is energetic enough to prevent the solute particles from settling out under the influence of gravity.

Note the beautiful purple trails of $KMnO_4$ as the crystals dissolve.

14.2 Solubility

solubility

The term **solubility** describes the amount of one substance (solute) that will dissolve in a specified amount of another substance (solvent) under stated conditions. For example, 36.0 g of sodium chloride will dissolve in 100 g of water at 20°C. We say then that the solubility of NaCl in water is 36.0 g/100 g H_2O at 20°C. Solubility is often used in a relative way. For instance, we say that a substance is very soluble, moderately soluble, slightly soluble, or insoluble. Although these terms do not accurately indicate how much solute will dissolve, they are frequently used to describe the solubility of a substance qualitatively.

Two other terms often used to describe solubility are *miscible* and *immiscible*. Liquids that are capable of mixing and forming a homogeneous solution are **miscible**; those that do not form solutions or are generally insoluble in each other are **immiscible**. Methyl alcohol and water are miscible in each other in all proportions. Oil and water are immiscible, forming two separate layers when they are mixed, as shown in Figure 14.1.

miscible
immiscible

The general guidelines for the solubility of common ionic compounds (salts) are given in Figure 14.2. These guidelines have some exceptions, but they provide a solid foundation for the compounds considered in this course. The solubilities of over 200 compounds are given in the Solubility Table in Appendix V. Solubility data for thousands of compounds can be found by consulting standard reference sources.*

The term *salt* is used interchangeably with *ionic compound* by many chemists.

The quantitative expression of the amount of dissolved solute in a particular quantity of solvent is known as the **concentration of a solution**. Several methods of expressing concentration are described in Section 14.6.

concentration of a solution

Figure 14.1
An immiscible mixture of oil and water.

Figure 14.2
The solubility of various common ions. Substances containing the ions on the left are generally soluble in cold water, while those substances containing the ions on the right are insoluble in cold water. The arrows point to the exceptions.

---

*Two commonly used handbooks are *Lange's Handbook of Chemistry*, 15th ed. (New York: McGrawHill, 1998), and the *Handbook of Chemistry and Physics*, 85th ed. (Cleveland: Chemical Rubber Co., 2004).

14.3 Factors Related to Solubility

Predicting solubilities is complex and difficult. Many variables, such as size of ions, charge on ions, interaction between ions, interaction between solute and solvent, and temperature, complicate the problem. Because of the factors involved, the general rules of solubility given in Figure 14.2 have many exceptions. However, these rules are useful because they do apply to many of the more common compounds that we encounter in the study of chemistry. Keep in mind that these are rules, not laws, and are therefore subject to exceptions. Fortunately, the solubility of a solute is relatively easy to determine experimentally. Now let's examine the factors related to solubility.

The Nature of the Solute and Solvent

The old adage "like dissolves like" has merit, in a general way. Polar or ionic substances tend to be more compatible with other polar substances. Nonpolar substances tend to be compatible with other nonpolar substances and less miscible with polar substances. Thus ionic compounds, which are polar, tend to be much more soluble in water, which is polar, than in solvents such as ether, hexane, or benzene, which are essentially nonpolar. Sodium chloride, an ionic substance, is soluble in water, slightly soluble in ethyl alcohol (less polar than water), and insoluble in ether and benzene. Pentane, C_5H_{12}, a nonpolar substance, is only slightly soluble in water but is very soluble in benzene and ether.

At the molecular level the formation of a solution from two nonpolar substances, such as hexane and benzene, can be visualized as a process of simple mixing. These nonpolar molecules, having little tendency to either attract or repel one another, easily intermingle to form a homogeneous solution.

Solution formation between polar substances is much more complex. See, for example, the process by which sodium chloride dissolves in water (Figure 14.3). Water molecules are very polar and are attracted to other polar molecules or ions. When salt crystals are put into water, polar water molecules become attracted to the sodium and chloride ions on the crystal surfaces and weaken the attraction between Na^+ and Cl^- ions. The positive end of the water dipole is attracted to the Cl^- ions, and the negative end of the water dipole to the Na^+ ions. The weakened attraction permits the ions to move apart, making room for more water dipoles. Thus the surface ions are surrounded by water molecules, becoming hydrated ions, $Na^+(aq)$ and $Cl^-(aq)$, and slowly diffuse away from the crystals and dissolve in solution:

$$NaCl(crystal) \xrightarrow{\text{H}_2\text{O}} Na^+(aq) + Cl^-(aq)$$

Examination of the data in Table 14.2 reveals some of the complex questions relating to solubility.

The Effect of Temperature on Solubility

Temperature affects the solubility of most substances, as shown by the data in Table 14.2. Most solutes have a limited solubility in a specific solvent at a fixed temperature. For most solids dissolved in a liquid, an increase in temperature

= Water

= Na^+

= Cl^-

Figure 14.3
Dissolution of sodium chloride in water. Polar water molecules are attracted to Na^+ and Cl^- ions in the salt crystal, weakening the attraction between the ions. As the attraction between the ions weakens, the ions move apart and become surrounded by water dipoles. The hydrated ions slowly diffuse away from the crystal to become dissolved in solution.

Table 14.2 Solubility of Alkali Metal Halides in Water

| Salt | Solubility (g salt/100 g H_2O) | |
	0°C	100°C
LiF	0.12	0.14 (at 35°C)
LiCl	67	127.5
LiBr	143	266
LiI	151	481
NaF	4	5
NaCl	35.7	39.8
NaBr	79.5	121
NaI	158.7	302
KF	92.3 (at 18°C)	Very soluble
KCl	27.6	57.6
KBr	53.5	104
KI	127.5	208

results in increased solubility (see Figure 14.4). However, no single rule governs the solubility of solids in liquids with change in temperature. Some solids increase in solubility only slightly with increasing temperature (see NaCl in Figure 14.4); other solids decrease in solubility with increasing temperature (see Li_2SO_4 in Figure 14.4).

On the other hand, the solubility of a gas in water usually decreases with increasing temperature (see HCl and SO_2 in Figure 14.4). The tiny bubbles that form when water is heated are due to the decreased solubility of air at higher temperatures. The decreased solubility of gases at higher temperatures is explained in terms of the KMT by assuming that, in order to dissolve, the gas

Figure 14.4
Solubility of various compounds in water. Solids are shown in red and gases are shown in blue.

molecules must form bonds of some sort with the molecules of the liquid. An increase in temperature decreases the solubility of the gas because it increases the kinetic energy (speed) of the gas molecules and thereby decreases their ability to form "bonds" with the liquid molecules.

The Effect of Pressure on Solubility

Small changes in pressure have little effect on the solubility of solids in liquids or liquids in liquids but have a marked effect on the solubility of gases in liquids. The solubility of a gas in a liquid is directly proportional to the pressure of that gas above the solution. Thus the amount of a gas dissolved in solution will double if the pressure of that gas over the solution is doubled. For example, carbonated beverages contain dissolved carbon dioxide under pressures greater than atmospheric pressure. When a can of carbonated soda is opened, the pressure is immediately reduced to the atmospheric pressure, and the excess dissolved carbon dioxide bubbles out of the solution.

Saturated, Unsaturated, and Supersaturated Solutions

At a specific temperature there is a limit to the amount of solute that will dissolve in a given amount of solvent. When this limit is reached, the resulting solution is said to be *saturated*. For example, when we put 40.0 g of KCl into 100 g of H_2O at 20°C, we find that 34.0 g of KCl dissolve and 6.0 g of KCl remain undissolved. The solution formed is a saturated solution of KCl.

Two processes are occurring simultaneously in a saturated solution. The solid is dissolving into solution and, at the same time, the dissolved solute is crystallizing out of solution. This may be expressed as

$$\text{solute (undissolved)} \rightleftharpoons \text{solute (dissolved)}$$

When these two opposing processes are occurring at the same rate, the amount of solute in solution is constant, and a condition of equilibrium is established between dissolved and undissolved solute. Therefore, a **saturated solution** contains dissolved solute in equilibrium with undissolved solute. Thus, any point on any solubility curve (Figure 14.4) represents a saturated solution of that solute. For example, a solution containing 60 g NH_4Cl per 100 g H_2O is saturated at 70°C.

It's important to state the temperature of a saturated solution, because a solution that is saturated at one temperature may not be saturated at another. If the temperature of a saturated solution is changed, the equilibrium is disturbed, and the amount of dissolved solute will change to reestablish equilibrium.

A saturated solution may be either dilute or concentrated, depending on the solubility of the solute. A saturated solution can be conveniently prepared by dissolving a little more than the saturated amount of solute at a temperature somewhat higher than room temperature. Then the amount of solute in solution will be in excess of its solubility at room temperature, and, when the solution cools, the excess solute will crystallize, leaving the solution saturated. (In this case, the solute must be more soluble at higher temperatures and must not form a supersaturated solution.) Examples expressing the solubility of saturated solutions at two different temperatures are given in Table 14.3.

An **unsaturated solution** contains less solute per unit of volume than does its corresponding saturated solution. In other words, additional solute can be dissolved in an unsaturated solution without altering any other conditions.

Pouring root beer into a glass illustrates the effect of pressure on solubility. The escaping CO_2 produces the foam.

saturated solution

unsaturated solution

Table 14.3 Saturated Solutions at 20°C and 50°C

Solute	Solubility (g solute/100 g H_2O)	
	20°C	50°C
NaCl	36.0	37.0
KCl	34.0	42.6
$NaNO_3$	88.0	114.0
$KClO_3$	7.4	19.3
$AgNO_3$	222.0	455.0
$C_{12}H_{22}O_{11}$	203.9	260.4

The heat released in this hot pack results from the crystallization of a supersaturated solution of sodium acetate.

Consider a solution made by adding 40 g of KCl to 100 g of H_2O at 20°C (see Table 14.3). The solution formed will be saturated and will contain about 6 g of undissolved salt, because the maximum amount of KCl that can dissolve in 100 g of H_2O at 20°C is 34 g. If the solution is now heated and maintained at 50°C, all the salt will dissolve and even more can be dissolved. The solution at 50°C is unsaturated.

In some circumstances, solutions can be prepared that contain more solute than needed for a saturated solution at a particular temperature. These solutions are said to be **supersaturated**. However, we must qualify this definition by noting that a supersaturated solution is unstable. Disturbances, such as jarring, stirring, scratching the walls of the container, or dropping in a "seed" crystal, cause the supersaturation to return to saturation, releasing heat. When a supersaturated solution is disturbed, the excess solute crystallizes out rapidly, returning the solution to a saturated state.

supersaturated solution

Supersaturated solutions are not easy to prepare but may be made from certain substances by dissolving, in warm solvent, an amount of solute greater than that needed for a saturated solution at room temperature. The warm solution is then allowed to cool very slowly. With the proper solute and careful work, a supersaturated solution will result.

Will a solution made by adding 2.5 g of $CuSO_4$ to 10 g of H_2O be saturated or unsaturated at 20°C?

Example 14.1

We first need to know the solubility of $CuSO_4$ at 20°C. From Figure 14.4, we see that the solubility of $CuSO_4$ at 20°C is about 21 g per 100 g of H_2O. This amount is equivalent to 2.1 g of $CuSO_4$ per 10 g of H_2O.

SOLUTION

Since 2.5 g per 10 g of H_2O is greater than 2.1 g per 10 g of H_2O, the solution will be saturated and 0.4 g of $CuSO_4$ will be undissolved.

Practice 14.1 _____

Will a solution made by adding 9.0 g NH_4Cl to 20 g of H_2O be saturated or unsaturated at 50°C?

14.4 Rate of Dissolving Solids

The rate at which a solid dissolves is governed by (1) the size of the solute particles, (2) the temperature, (3) the concentration of the solution, and (4) agitation or stirring. Let's look at each of these conditions:

1. *Particle size.* A solid can dissolve only at the surface that is in contact with the solvent. Because the surface-to-volume ratio increases as size decreases, smaller crystals dissolve faster than large ones. For example, if a salt crystal 1 cm on a side (a surface area of 6 cm^2) is divided into 1000 cubes, each 0.1 cm on a side, the total surface of the smaller cubes is 60 cm^2—a 10-fold increase in surface area (see Figure 14.5).

2. *Temperature.* In most cases, the rate of dissolving of a solid increases with temperature. This increase is due to kinetic effects. The solvent molecules move more rapidly at higher temperatures and strike the solid surfaces more often, causing the rate of dissolving to increase.

3. *Concentration of the solution.* When the solute and solvent are first mixed, the rate of dissolving is at its maximum. As the concentration of the solution increases and the solution becomes more nearly saturated with the solute, the rate of dissolving decreases greatly. The rate of dissolving is graphed in Figure 14.6. Note that about 17 g dissolve in the first five-minute interval, but only about 1 g dissolves in the fourth five-minute interval. Although different solutes show different rates, the rate of dissolving always becomes very slow as the concentration approaches the saturation point.

4. *Agitation or stirring.* The effect of agitation or stirring is kinetic. When a solid is first put into water, it comes in contact only with solvent in its immediate vicinity. As the solid dissolves, the amount of dissolved solute around the solid becomes more and more concentrated, and the rate of dissolving slows down. If the mixture is not stirred, the dissolved solute

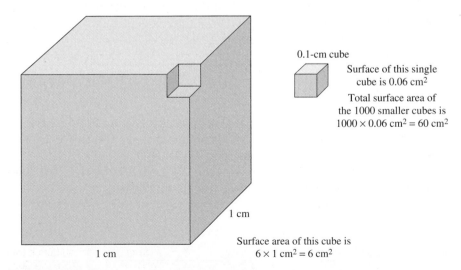

Figure 14.5
Surface area of crystals. A crystal 1 cm on a side has a surface area of 6 cm^2. Subdivided into 1000 smaller crystals, each 0.1 cm on a side, the total surface area is increased to 60 cm^2.

Figure 14.6
Rate of dissolution of a solid solute in a solvent. The rate is maximum at the beginning and decreases as the concentration approaches saturation.

diffuses very slowly through the solution; weeks may pass before the solid is entirely dissolved. Stirring distributes the dissolved solute rapidly through the solution, and more solvent is brought into contact with the solid, causing it to dissolve more rapidly.

14.5 Solutions: A Reaction Medium

Many solids must be put into solution to undergo appreciable chemical reaction. We can write the equation for the double-displacement reaction between sodium chloride and silver nitrate:

$$NaCl + AgNO_3 \longrightarrow AgCl + NaNO_3$$

But suppose we mix solid $NaCl$ and solid $AgNO_3$ and look for a chemical change. If any reaction occurs, it is slow and virtually undetectable. In fact, the crystalline structures of $NaCl$ and $AgNO_3$ are so different that we could separate them by tediously picking out each kind of crystal from the mixture. But if we dissolve the $NaCl$ and $AgNO_3$ separately in water and mix the two solutions, we observe the immediate formation of a white, curdlike precipitate of silver chloride.

Molecules or ions must collide with one another in order to react. In the foregoing example, the two solids did not react because the ions were securely locked within their crystal structures. But when the $NaCl$ and $AgNO_3$ are dissolved, their crystal lattices are broken down and the ions become mobile. When the two solutions are mixed, the mobile Ag^+ and Cl^- ions come into contact and react to form insoluble $AgCl$, which precipitates out of solution. The soluble Na^+ and NO_3^- ions remain mobile in solution but form the crystalline salt $NaNO_3$ when the water is evaporated:

$$NaCl(aq) + AgNO_3(aq) \longrightarrow AgCl(s) + NaNO_3(aq)$$

$$Na^+(aq) + Cl^-(aq) + Ag^+(aq) + NO_3^-(aq) \longrightarrow AgCl(s) + Na^+(aq) + NO_3^-(aq)$$

| sodium chloride solution | silver nitrate solution | silver chloride | sodium nitrate in solution |

The mixture of the two solutions provides a medium or space in which the Ag^+ and Cl^- ions can react. (See Chapter 15 for further discussion of ionic reactions.)

Solutions also function as diluting agents in reactions in which the undiluted reactants would combine with each other too violently. Moreover, a solution of known concentration provides a convenient method for delivering a specific amount of reactant.

14.6 Concentration of Solutions

The concentration of a solution expresses the amount of solute dissolved in a given quantity of solvent or solution. Because reactions are often conducted in solution, it's important to understand the methods of expressing concentration and to know how to prepare solutions of particular concentrations. The concentration of a solution may be expressed qualitatively or quantitatively. Let's begin with a look at the qualitative methods of expressing concentration.

Dilute and Concentrated Solutions

When we say that a solution is *dilute* or *concentrated*, we are expressing, in a relative way, the amount of solute present. One gram of a compound and 2 g of a compound in solution are both dilute solutions when compared with the same volume of a solution containing 20 g of a compound. Ordinary concentrated hydrochloric acid contains 12 mol of HCl per liter of solution. In some laboratories, the dilute acid is made by mixing equal volumes of water and the concentrated acid. In other laboratories the concentrated acid is diluted with two or three volumes of water, depending on its use. The term **dilute solution**, then, describes a solution that contains a relatively small amount of dissolved solute. Conversely, a **concentrated solution** contains a relatively large amount of dissolved solute.

dilute solution

concentrated solution

Mass Percent Solution

The mass percent method expresses the concentration of the solution as the percent of solute in a given mass of solution. It says that for a given mass of solution, a certain percent of that mass is solute. Suppose we take a bottle from the reagent shelf that reads "sodium hydroxide, NaOH, 10%." This statement means that for every 100 g of this solution, 10 g will be NaOH and 90 g will be water. (Note that this amount of solution is 100 g, not 100 mL.) We could also make this same concentration of solution by dissolving 2.0 g of NaOH in 18 g of water. Mass percent concentrations are most generally used for solids dissolved in liquids:

Note that mass percent is independent of the formula for the solute.

$$\text{mass percent} = \frac{\text{g solute}}{\text{g solute + g solvent}} \times 100 = \frac{\text{g solute}}{\text{g solution}} \times 100$$

As instrumentation advances are made in chemistry, our ability to measure the concentration of dilute solutions is increasing as well. In addition to mass percent, chemists now commonly use **parts per million (ppm)**:

parts per million (ppm)

$$\text{parts per million} = \frac{\text{g solute}}{\text{g solute + g solvent}} \times 1,000,000$$

Currently, air and water contaminants, drugs in the human body, and pesticide residues are measured in parts per million.

What is the mass percent of sodium hydroxide in a solution that is made by dissolving 8.00 g NaOH in 50.0 g H_2O?

Example 14.2

SOLUTION

grams of solute(NaOH) = 8.00 g

grams of solvent (H_2O) = 50.0 g

$$\left(\frac{8.00 \text{ g NaOH}}{8.00 \text{ g NaOH} + 50.0 \text{ g } H_2O}\right)100 = 13.8\% \text{ NaOH solution}$$

What masses of potassium chloride and water are needed to make 250. g of 5.00% solution?

Example 14.3

SOLUTION

The percent expresses the mass of the solute:

250. g = total mass of solution

5.00% of 250. g = (0.0500)(250. g) = 12.5 g KCl (solute)

250. g − 12.5 g = 238 g H_2O

Dissolving 12.5 g KCl in 238 g H_2O gives a 5.00% KCl solution.

A 34.0% sulfuric acid solution has a density of 1.25 g/mL. How many grams of H_2SO_4 are contained in 1.00 L of this solution?

Example 14.4

SOLUTION

Since H_2SO_4 is the solute, we first solve the mass percent equation for grams of solute:

$$\text{mass percent} = \left(\frac{\text{g solute}}{\text{g solution}}\right)100$$

$$\text{g solute} = \frac{(\text{mass percent})(\text{g solution})}{100}$$

The mass percent is given, so we need to determine the grams of solution. The mass of the solution can be calculated from the density data. Convert density (g/mL) to grams:

1.00 L = 1.00 × 10³ mL

$$\left(\frac{1.25 \text{ g}}{mL}\right)(1.00 \times 10^3 \text{ mL}) = 1250 \text{ g (mass of solution)}$$

Now we have all the figures to calculate the grams of solute:

$$\text{g solute} = \frac{(34.0 \text{ g } H_2SO_4)(1250 \text{ g})}{100 \text{ g}} = 425 \text{ g } H_2SO_4$$

Thus, 1.00 L of 34.0% H_2SO_4 solution contains 425 g H_2SO_4.

Practice 14.2

What is the mass percent of Na_2SO_4 in a solution made by dissolving 25.0 g Na_2SO_4 in 225.0 g H_2O?

Mass/Volume Percent (m/v)

This method expresses concentration as grams of solute per 100 mL of solution. With this system, a 10.0%-m/v-glucose solution is made by dissolving 10.0 g of glucose in water, diluting to 100 mL, and mixing. The 10.0%-m/v solution could also be made by diluting 20.0 g to 200 mL, 50.0 g to 500 mL, and so on. Of course, any other appropriate dilution ratio may be used:

$$\text{mass/volume percent} = \frac{\text{g solute}}{\text{mL solution}} \times 100$$

Example 14.5 A 3.0% H_2O_2 solution is commonly used as a topical antiseptic to prevent infection. What volume of this solution will contain 10. g of H_2O_2?

SOLUTION First solve the mass/volume percent equation for mL of solution:

$$\text{mL solution} = \frac{(\text{g solute})}{(\text{m/v percent})}(100)$$

$$\text{mL solution} = \frac{(10. \text{ g solute})}{(3.0 \text{ m/v percent})}(100) = 330 \text{ mL}$$

Volume Percent

Solutions that are formulated from two liquids are often expressed as *volume percent* with respect to the solute. The volume percent is the volume of a liquid in 100 mL of solution. The label on a bottle of ordinary rubbing alcohol reads "isopropyl alcohol, 70% by volume." Such a solution could be made by mixing 70 mL of alcohol with water to make a total volume of 100 mL, but we cannot use 30 mL of water, because the two volumes are not necessarily additive:

$$\text{volume percent} = \frac{\text{volume of liquid in question}}{\text{total volume of solution}} \times 100$$

Volume percent is used to express the concentration of alcohol in beverages. Wines generally contain 12% alcohol by volume. This translates into 12 mL of alcohol in each 100 mL of wine. The beverage industry also uses the concentration unit of *proof* (twice the volume percent). Pure ethyl alcohol is 100% and therefore 200 proof. Scotch whiskey is 86 proof, or 43% alcohol.

Molarity

Mass percent solutions do not equate or express the molar masses of the solute in solution. For example, 1000. g of 10.0% NaOH solution contains 100. g NaOH; 1000. g of 10.0% KOH solution contains 100. g KOH. In terms of moles of NaOH and KOH, these solutions contain

$$\text{mol NaOH} = (100. \text{ g NaOH})\left(\frac{1 \text{ mol NaOH}}{40.00 \text{ g NaOH}}\right) = 2.50 \text{ mol NaOH}$$

$$\text{mol KOH} = (100. \text{ g KOH})\left(\frac{1 \text{ mol KOH}}{56.11 \text{ g KOH}}\right) = 1.78 \text{ mol KOH}$$

From these figures, we see that the two 10.0% solutions do not contain the same number of moles of NaOH and KOH. As a result, we find that a 10.0% NaOH solution contains more reactive base than a 10.0% KOH solution.

We need a method of expressing concentration that will easily indicate how many moles of solute are present per unit volume of solution. For this purpose, the concentration known as molarity is used in calculations involving chemical reactions.

A 1-molar solution contains 1 mol of solute per liter of solution. For example, to make a 1-molar solution of sodium hydroxide (NaOH) we dissolve 40.00 g NaOH (1 mol) in water and dilute the solution with more water to a volume of 1 L. The solution contains 1 mol of the solute in 1 L of solution and is said to be 1 molar in concentration. Figure 14.7 illustrates the preparation of a 1-molar solution. Note that the volume of the solute and the solvent together is 1 L.

The concentration of a solution can, of course, be varied by using more or less solute or solvent; but in any case the **molarity** of a solution is the number of moles of solute per liter of solution. The abbreviation for molarity is M. The units of molarity are moles per liter. The expression "2.0 M NaOH" means a 2.0-molar solution of NaOH (2.0 mol, or 80. g, of NaOH dissolved in water to make 1.0 L of solution):

molarity (M)

$$\text{molarity} = M = \frac{\text{number of moles of solute}}{\text{liter of solution}} = \frac{\text{moles}}{\text{liter}}$$

Flasks that are calibrated to contain specific volumes at a particular temperature are used to prepare solutions of a desired concentration. These *volumetric flasks* have a calibration mark on the neck that accurately indicates the measured volume. Molarity is based on a specific volume of solution and therefore will vary slightly with temperature because volume varies with temperature (1000 mL H_2O at 20°C = 1001 mL at 25°C).

(a) Add 1 mole of solute to a 1-liter volumetric flask

(b) Dissolve in solvent

(c) Add solvent to the 1-liter mark and mix thoroughly

Figure 14.7
Preparation of a 1 M solution.

Suppose we want to make 500 mL of 1 M solution. This solution can be prepared by determining the mass of 0.5 mol of the solute and diluting with water in a 500-mL (0.5-L) volumetric flask. The molarity will be

$$M = \frac{0.5 \text{ mol solute}}{0.5 \text{ L solution}} = 1 \text{ molar}$$

You can see that it isn't necessary to have a liter of solution to express molarity. All we need to know is the number of moles of dissolved solute and the volume of solution. Thus, 0.001 mol NaOH in 10 mL of solution is 0.1 M:

$$\left(\frac{0.001 \text{ mol}}{10 \text{ mL}}\right)\left(\frac{1000 \text{ mL}}{1 \text{ L}}\right) = 0.1 \, M$$

When we stop to think that a balance is not calibrated in moles, but in grams, we can incorporate grams into the molarity formula. We do so by using the relationship

$$\text{moles} = \frac{\text{grams of solute}}{\text{molar mass}}$$

Substituting this relationship into our expression for molarity, we get

$$M = \frac{\text{mol}}{\text{L}} = \frac{\text{g solute}}{\text{molar mass solute} \times \text{L solution}}$$

We can now determine the mass of any amount of a solute that has a known formula, dilute it to any volume, and calculate the molarity of the solution using this formula.

Molarities of concentrated acids commonly used in the laboratory:

HCl   12 M
$HC_2H_3O_2$ 17 M
HNO_3 16 M
H_2SO_4 18 M

Example 14.6   What is the molarity of a solution containing 1.4 mol of acetic acid ($HC_2H_3O_2$) in 250. mL of solution?

SOLUTION   By the unit-conversion method, we note that the concentration given in the problem statement is 1.4 mol per 250. mL (mol/mL). Since molarity = mol/L, the needed conversion is

$$\frac{\text{mol}}{\text{mL}} \longrightarrow \frac{\text{mol}}{\text{L}} = M$$

$$\left(\frac{1.4 \text{ mol}}{250. \text{ mL}}\right)\left(\frac{1000 \text{ mL}}{\text{L}}\right) = \frac{5.6 \text{ mol}}{\text{L}} = 5.6 \, M$$

Example 14.7   What is the molarity of a solution made by dissolving 2.00 g of potassium chlorate in enough water to make 150. mL of solution?

SOLUTION   We use the unit-conversion method. The steps in the conversions must lead to units of moles/liter:

$$\frac{\text{g KClO}_3}{\text{mL}} \longrightarrow \frac{\text{g KClO}_3}{\text{L}} \longrightarrow \frac{\text{mol KClO}_3}{\text{L}} = M$$

The data are

mass KClO$_3$ = 2.00 g   molar mass KClO$_3$ = 122.6 g/mol   volume = 150. mL

$$\left(\frac{2.00 \text{ g KClO}_3}{150. \text{ mL}}\right)\left(\frac{1000 \text{ mL}}{L}\right)\left(\frac{1 \text{ mol KClO}_3}{122.6 \text{ g KClO}_3}\right) = \frac{0.109 \text{ mol}}{L} = 0.109 \text{ } M \text{ KClO}_3$$

How many grams of potassium hydroxide are required to prepare 600. mL of 0.450 M KOH solution?   **Example 14.8**

The conversion is   **SOLUTION**

$$\text{milliliters} \longrightarrow \text{liters} \longrightarrow \text{moles} \longrightarrow \text{grams}$$

The data are

$$\text{volume} = 600. \text{ mL} \qquad M = \frac{0.450 \text{ mol}}{L} \qquad \text{molar mass KOH} = \frac{56.11 \text{ g}}{\text{mol}}$$

The calculation is

$$(600. \text{ mL})\left(\frac{1 \text{ L}}{1000 \text{ mL}}\right)\left(\frac{0.450 \text{ mol}}{L}\right)\left(\frac{56.11 \text{ g KOH}}{\text{mol}}\right) = 15.1 \text{ g KOH}$$

Practice 14.3

What is the molarity of a solution made by dissolving 7.50 g of magnesium nitrate [Mg(NO$_3$)$_2$] in enough water to make 25.0 mL of solution?

Practice 14.4

How many grams of sodium chloride are needed to prepare 125 mL of a 0.037 M NaCl solution?

Calculate the number of moles of nitric acid in 325 mL of 16 M HNO$_3$ solution.   **Example 14.9**

Use the equation   **SOLUTION**

$$\text{moles} = \text{liters} \times M$$

Substitute the data given in the problem and solve:

$$\text{moles} = (0.325 \text{ L})\left(\frac{16 \text{ mol HNO}_3}{1 \text{ L}}\right) = 5.2 \text{ mol HNO}_3$$

What volume of 0.250 M solution can be prepared from 16.0 g of potassium carbonate?   **Example 14.10**

We start with 16.0 g K$_2$CO$_3$; we need to find the volume of 0.250 M solution   **SOLUTION**
that can be prepared from this amount of K$_2$CO$_3$. The conversion therefore is

$$\text{g K}_2\text{CO}_3 \longrightarrow \text{mol K}_2\text{CO}_3 \longrightarrow \text{L solution}$$

The data are

$$\text{mass } K_2CO_3 = 16.0 \text{ g} \quad M = \frac{0.250 \text{ mol}}{1 \text{ L}} \quad \text{molar mass } K_2CO_3 = \frac{138.2 \text{ g}}{1 \text{ mol}}$$

$$(16.0 \text{ g } K_2CO_3)\left(\frac{1 \text{ mol } K_2CO_3}{138.2 \text{ g } K_2CO_3}\right)\left(\frac{1 \text{ L}}{0.250 \text{ mol } K_2CO_3}\right) = 0.463 \text{ L } (463 \text{ mL})$$

Thus, 463 mL of 0.250 M solution can be made from 16.0 g K_2CO_3.

Example 14.11   How many milliliters of 2.00 M HCl will react with 28.0 g NaOH?

SOLUTION   **Step 1**   Write and balance the equation for the reaction:

$$HCl(aq) + NaOH(aq) \longrightarrow NaCl(aq) + H_2O(aq)$$

The equation states that 1 mol of HCl reacts with 1 mol of NaOH.

Step 2   Find the number of moles NaOH in 28.0 g NaOH:

g NaOH $\longrightarrow$ mol NaOH

$$(28.0 \text{ g NaOH})\left(\frac{1 \text{ mol}}{40.00 \text{ g}}\right) = 0.700 \text{ mol NaOH}$$

28.0 g NaOH = 0.700 mol NaOH

Step 3   Solve for moles and volume of HCl needed. From Steps 1 and 2, we see that 0.700 mol HCl will react with 0.700 mol NaOH, because the ratio of moles reacting is 1 : 1. We know that 2.00 M HCl contain 2.00 mol HCl per liter, and so the volume that contains 0.700 mol HCl will be less than 1 L:

mol NaOH $\longrightarrow$ mol HCl $\longrightarrow$ L HCl $\longrightarrow$ mL HCl

$$(0.700 \text{ mol NaOH})\left(\frac{1 \text{ mol HCl}}{1 \text{ mol NaOH}}\right)\left(\frac{1 \text{ L HCl}}{2.00 \text{ mol HCl}}\right) = 0.350 \text{ L HCl}$$

$$(0.350 \text{ L HCl})\left(\frac{1000 \text{ mL}}{1 \text{ L}}\right) = 350. \text{ mL HCl}$$

Therefore, 350. mL of 2.00 M HCl contain 0.700 mol HCl and will react with 0.700 mol, or 28.0 g, of NaOH.

Practice 14.5
What volume of 0.035 M AgNO$_3$ can be made from 5.0 g of AgNO$_3$?

Practice 14.6
How many milliliters of 0.50 M NaOH are required to react completely with 25.00 mL of 1.5 M HCl?

We've now examined several ways to measure concentration of solutions quantitatively. A summary of these concentration units is found in Table 14.4.

Table 14.4 Concentration Units for Solutions

Units	Symbol	Definition
Mass percent	% m/m	$\dfrac{\text{mass solute}}{\text{mass solution}} \times 100$
Parts per million	ppm	$\dfrac{\text{mass solute}}{\text{mass solution}} \times 1{,}000{,}000$
Mass/volume percent	% m/v	$\dfrac{\text{mass solute}}{\text{mL solution}} \times 100$
Volume percent	% v/v	$\dfrac{\text{mL solute}}{\text{mL solution}} \times 100$
Molarity	M	$\dfrac{\text{moles solute}}{\text{L solution}}$
Molality	m	$\dfrac{\text{moles solute}}{\text{kg solvent}}$

Molality is covered in Section 14.7

Dilution Problems

Chemists often find it necessary to dilute solutions from one concentration to another by adding more solvent to the solution. If a solution is diluted by adding pure solvent, the volume of the solution increases, but the number of moles of solute in the solution remains the same. Thus the moles/liter (molarity) of the solution decreases. Always read a problem carefully to distinguish between (1) how much solvent must be added to dilute a solution to a particular concentration and (2) to what volume a solution must be diluted to prepare a solution of a particular concentration.

Calculate the molarity of a sodium hydroxide solution that is prepared by mixing 100. mL of 0.20 M NaOH with 150. mL of water. Assume that the volumes are additive.

Example 14.12

This problem is a dilution problem. If we double the volume of a solution by adding water, we cut the concentration in half. Therefore, the concentration of the above solution should be less than 0.10 M. In the dilution, the moles of NaOH remain constant; the molarity and volume change. The final volume is (100. mL + 150. mL), or 250. mL.

SOLUTION

To solve this problem, (1) calculate the moles of NaOH in the original solution and (2) divide the moles of NaOH by the final volume of the solution to obtain the new molarity:

Step 1   Calculate the moles of NaOH in the original solution:

$$M = \dfrac{\text{mol}}{\text{L}} \qquad \text{mol} = \text{L} \times M$$

$$(0.100 \, \cancel{L})\left(\dfrac{0.20 \, \text{mol NaOH}}{1 \, \cancel{L}}\right) = 0.020 \, \text{mol NaOH}$$

Step 2 Solve for the new molarity, taking into account that the total volume of the solution after dilution is 250. mL (0.250 L):

$$M = \frac{0.020 \text{ mol NaOH}}{0.250 \text{ L}} = 0.080 \, M \text{ NaOH}$$

ALTERNATIVE SOLUTION When the moles of solute in a solution before and after dilution are the same, the moles before and after dilution may be set equal to each other. That is,

$$\text{mol}_1 = \text{mol}_2$$

where mol_1 = moles before dilution and mol_2 = moles after dilution. Then

$$\text{mol}_1 = L_1 \times M_1 \qquad \text{mol}_2 = L_2 \times M_2$$
$$L_1 \times M_1 = L_2 \times M_2$$

When both volumes are in the same units, a more general statement can be made:

$$V_1 \times M_1 = V_2 \times M_2$$

For this problem,

$$V_1 = 100. \text{ mL} \qquad\qquad M_1 = 0.20 \, M$$
$$V_2 = 150. \text{ mL} + 100. \text{ mL} \qquad M_2 = \text{(unknown)}$$

Then

$$(100. \text{ mL})(0.20 \, M) = (250. \text{ mL})M_2$$

Solving for M_2, we get

$$M_2 = \frac{(100. \text{ mL})(0.20 \, M)}{250. \text{ mL}} = 0.080 \, M \text{ NaOH}$$

Practice 14.7

Calculate the molarity of a solution prepared by diluting 125 mL of 0.400 M $K_2Cr_2O_7$ with 875 mL of water.

Example 14.13 How many grams of silver chloride will be precipitated by adding sufficient silver nitrate to react with 1500. mL of 0.400 M barium chloride solution?

$$2 \, AgNO_3(aq) + BaCl_2(aq) \longrightarrow 2 \, AgCl(s) + Ba(NO_3)_2(aq)$$

SOLUTION This problem is a stoichiometry problem. The fact that $BaCl_2$ is in solution means that we need to consider the volume and concentration of the solution in order to determine the number of moles of $BaCl_2$ reacting.

Step 1 Determine the number of moles of $BaCl_2$ in 1500. mL of 0.400 M solution:

$$M = \frac{\text{mol}}{\text{L}} \qquad \text{mol} = L \times M \qquad 1500. \text{ mL} = 1.500 \text{ L}$$

$$(1.500 \, L)\left(\frac{0.400 \text{ mol } BaCl_2}{L}\right) = 0.600 \text{ mol } BaCl_2$$

Step 2 Calculate the moles and grams of AgCl:

$$\text{mol BaCl}_2 \longrightarrow \text{mol AgCl} \longrightarrow \text{g AgCl}$$

$$(0.600 \ \cancel{\text{mol BaCl}_2})\left(\frac{2 \ \cancel{\text{mol AgCl}}}{1 \ \cancel{\text{mol BaCl}_2}}\right)\left(\frac{143.4 \ \text{g AgCl}}{\cancel{\text{mol AgCl}}}\right) = 172 \ \text{g AgCl}$$

Practice 14.8

How many grams of lead(II) iodide will be precipitated by adding sufficient $Pb(NO_3)_2$ to react with 750 mL of 0.250 M KI solution?

$$2 \ KI(aq) + Pb(NO_3)_2(aq) \longrightarrow PbI_2(s) + 2 \ KNO_3(aq)$$

14.7 Colligative Properties of Solutions

Two solutions—one containing 1 mol (60.06 g) of urea (NH_2CONH_2) and the other containing 1 mol (342.3 g) of sucrose ($C_{12}H_{22}O_{11}$) each in 1 kg of water—both have a freezing point of $-1.86°C$, not $0°C$ as for pure water. Urea and sucrose are distinct substances, yet they lower the freezing point of the water by the same amount. The only thing apparently common to these two solutions is that each contains 1 mol (6.022×10^{23} molecules) of solute and 1 kg of solvent. In fact, when we dissolve 1 mol of any nonionizable solute in 1 kg of water, the freezing point of the resulting solution is $-1.86°C$.

These results lead us to conclude that the freezing point depression for a solution containing 6.022×10^{23} solute molecules (particles) and 1 kg of water is a constant, namely, $1.86°C$. Freezing point depression is a general property of solutions. Furthermore, the amount by which the freezing point is depressed is the same for all solutions made with a given solvent; that is, each solvent shows a characteristic *freezing point depression constant*. Freezing point depression constants for several solvents are given in Table 14.5.

The solution formed by the addition of a nonvolatile solute to a solvent has a lower freezing point, a higher boiling point, and a lower vapor pressure than that of the pure solvent. These effects are related and are known as colligative

Table 14.5 Freezing Point Depression and Boiling Point Elevation Constants of Selected Solvents

Solvent	Freezing point of pure solvent (°C)	Freezing point depression constant, K_f $\left(\dfrac{°C \ kg \ solvent}{mol \ solute}\right)$	Boiling point of pure solvent (°C)	Boiling point elevation constant, K_b $\left(\dfrac{°C \ kg \ solvent}{mol \ solute}\right)$
Water	0.00	1.86	100.0	0.512
Acetic acid	16.6	3.90	118.5	3.07
Benzene	5.5	5.1	80.1	2.53
Camphor	178	40	208.2	5.95

(a)   (b)

Figure 14.8
Vapor pressure curves of pure water and water solutions, showing (a) boiling point elevation and (b) freezing point depression effects (concentration: 1 mol solute/1 kg water).

colligative properties

properties. The **colligative properties** are properties that depend only on the number of solute particles in a solution, not on the nature of those particles. Freezing point depression, boiling point elevation, and vapor pressure lowering are colligative properties of solutions.

The colligative properties of a solution can be considered in terms of vapor pressure. The vapor pressure of a pure liquid depends on the tendency of molecules to escape from its surface. If 10% of the molecules in a solution are nonvolatile solute molecules, the vapor pressure of the solution is 10% lower than that of the pure solvent. The vapor pressure is lower because the surface of the solution contains 10% nonvolatile molecules and 90% of the volatile solvent molecules. A liquid boils when its vapor pressure equals the pressure of the atmosphere. We can thus see that the solution just described as having a lower vapor pressure will have a higher boiling point than the pure solvent. The solution with a lowered vapor pressure doesn't boil until it has been heated above the boiling point of the solvent (see Figure 14.8a). Each solvent has its own characteristic boiling point elevation constant (Table 14.5). The boiling point elevation constant is based on a solution that contains 1 mol of solute particles per kilogram of solvent. For example, the boiling point elevation constant for a solution containing 1 mol of solute particles per kilogram of water is 0.512°C, which means that this water solution will boil at 100.512°C.

The freezing behavior of a solution can also be considered in terms of lowered vapor pressure. Figure 14.8b shows the vapor pressure relationships of ice, water, and a solution containing 1 mol of solute per kilogram of water. The freezing point of water is at the intersection of the liquid and solid vapor pressure curves (i.e., at the point where water and ice have the same vapor pressure). Because the vapor pressure of the liquid is lowered by the solute, the vapor pressure curve of the solution does not intersect the vapor pressure curve of the solid until the solution has been cooled below the freezing point of pure water. So the solution must be cooled below 0°C in order for it to freeze.

The foregoing discussion dealing with freezing point depressions is restricted to *un-ionized* substances. The discussion of boiling point elevations is restricted to *nonvolatile* and un-ionized substances. The colligative properties of ionized substances are not under consideration at this point; we will discuss them in Chapter 15.

Engine coolant is one application of colligative properties. The addition of coolant to the water in a radiator raises its boiling point and lowers its freezing point.

Sodium chloride or calcium chloride is used to melt ice on snowy streets and highways.

Some practical applications involving colligative properties are (1) use of salt–ice mixtures to provide low freezing temperatures for homemade ice cream, (2) use of sodium chloride or calcium chloride to melt ice from streets, and (3) use of ethylene glycol–water mixtures as antifreeze in automobile radiators (ethylene glycol also raises the boiling point of radiator fluid, thus allowing the engine to operate at a higher temperature).

Both the freezing point depression and the boiling point elevation are directly proportional to the number of moles of solute per kilogram of solvent. When we deal with the colligative properties of solutions, another concentration expression, *molality*, is used. The **molality (*m*)** of a solute is the number of moles of solute per kilogram of solvent:

molality (*m*)

$$m = \frac{\text{mol solute}}{\text{kg solvent}}$$

Note that a lowercase *m* is used for molality concentrations and a capital *M* for molarity. The difference between molality and molarity is that molality refers to moles of solute *per kilogram of solvent*, whereas molarity refers to moles of solute *per liter of solution*. For un-ionized substances, the colligative properties of a solution are directly proportional to its molality.

Molality is independent of volume. It is a mass-to-mass relationship of solute to solvent and allows for experiments, such as freezing point depression and boiling point elevation, to be conducted at variable temperatures.

The following equations are used in calculations involving colligative properties and molality:

$$\Delta t_f = m K_f \qquad \Delta t_b = m K_b \qquad m = \frac{\text{mol solute}}{\text{kg solvent}}$$

m = molality; mol solute/kg solvent

Δt_f = freezing point depression; °C

Δt_b = boiling point elevation; °C

K_f = freezing point depression constant; °C kg solvent/mol solute

K_b = boiling point elevation constant; °C kg solvent/mol solute

Example 14.14   What is the molality (m) of a solution prepared by dissolving 2.70 g CH_3OH in 25.0 g H_2O?

SOLUTION   Since $m = \dfrac{\text{mol solute}}{\text{kg solvent}}$, the conversion is

$$\dfrac{2.70 \text{ g } CH_3OH}{25.0 \text{ g } H_2O} \longrightarrow \dfrac{\text{mol } CH_3OH}{25.0 \text{ g } H_2O} \longrightarrow \dfrac{\text{mol } CH_3OH}{1 \text{ kg } H_2O}$$

The molar mass of CH_3OH is (12.01 + 4.032 + 16.00), or 32.04 g/mol:

$$\left(\dfrac{2.70 \text{ g } CH_3OH}{25.0 \text{ g } H_2O}\right)\left(\dfrac{1 \text{ mol } CH_3OH}{32.04 \text{ g } CH_3OH}\right)\left(\dfrac{1000 \text{ g } H_2O}{1 \text{ kg } H_2O}\right) = \dfrac{3.37 \text{ mol } CH_3OH}{1 \text{ kg } H_2O}$$

The molality is 3.37 m.

Practice 14.9 _____

What is the molality of a solution prepared by dissolving 150.0 g $C_6H_{12}O_6$ in 600.0 g H_2O?

Example 14.15   A solution is made by dissolving 100. g of ethylene glycol ($C_2H_6O_2$) in 200. g of water. What is the freezing point of this solution?

SOLUTION   To calculate the freezing point of the solution, we first need to calculate Δt_f, the change in freezing point. Use the equation

$$\Delta t_f = mK_f = \dfrac{\text{mol solute}}{\text{kg solvent}} \times K_f$$

K_f (for water): $\dfrac{1.86 °C \text{ kg solvent}}{\text{mol solute}}$ (from Table 14.5)

mol solute: $(100. \text{ g } C_2H_6O_2)\left(\dfrac{1 \text{ mol } C_2H_6O_2}{62.07 \text{ g } C_2H_6O_2}\right) = 1.61 \text{ mol } C_2H_6O_2$

kg solvent: $(200. \text{ g } H_2O)\left(\dfrac{1 \text{ kg}}{1000 \text{ g}}\right) = 0.200 \text{ kg } H_2O$

$$\Delta t_f = \left(\dfrac{1.61 \text{ mol } C_2H_6O_2}{0.200 \text{ kg } H_2O}\right)\left(\dfrac{1.86 °C \text{ kg } H_2O}{1 \text{ mol } C_2H_6O_2}\right) = 15.0 °C$$

The freezing point depression, 15.0°C, must be subtracted from 0°C, the freezing point of the pure solvent (water):

freezing point of solution = freezing point of solvent − Δt_f

$$= 0.0 °C - 15.0 °C = -15.0 °C$$

Therefore, the freezing point of the solution is −15.0°C. This solution will protect an automobile radiator down to −15.0°C (5°F).

Ice cream is mainly composed of water (from milk and cream), milk solids, milk fats, and, frequently, various sweeteners (corn syrup or sugar), flavorings, emulsifiers, and stabilizers. But that smooth, creamy, rich flavor and texture are the result of the chemistry of the mixing and freezing process. The rich, smooth texture of great ice cream results from the milk fat. By law, if the carton is labeled "ice cream," it must contain a minimum of 10% milk fat. That carton of ice cream also contains 20%–50% air whipped into the ingredients during the initial mixing process.

H. Douglas Goff, a professor of food science and ice cream expert from Ontario, Canada, says, "There are no real chemical reactions that take place when you make ice cream, but that doesn't mean that there isn't plenty of chemistry." The structure of ice cream contributes greatly to its taste. Tiny air bubbles are formed in the initial whipping process. These bubbles are distributed through a network of fat globules and liquid water. The milk

fat has surface proteins on the globules in the milk to keep the fat dissolved in solution. Ice cream manufacturers destabilize these globules using emulsifiers (such as egg yolks, mono- or diglycerides) and let them come together in larger networks sort of like grape clusters.

Once the mixture is fully whipped, it is cooled to begin the freezing process. But ice cream does not freeze at 0°C even though it is 55–64% water. The freezing point is depressed as a colligative property of the ice cream solution. Once the freezing of the water begins, the concentration of the solution increases, which continues to depress the freezing point. Goff tells us that even at a typical serving temperature (−16°C for most ice cream), only about 72% of the water in the ice cream is frozen. The unfrozen solution keeps the ice cream "scoopable."

One last factor that affects the quality of ice cream is the size of the ice crystals. For very smooth ice cream, tiny crystals are needed. To produce these, the ice cream must freeze very slowly. Large crystals give a coarse, grainy texture. Now, as you savor that premium ice cream cone, you'll know just how colligative properties and the chemistry of freezing helped make it so delicious!

A solution made by dissolving 4.71 g of a compound of unknown molar mass in 100.0 g of water has a freezing point of −1.46°C. What is the molar mass of the compound?

Example 14.16

First substitute the data in $\Delta t_f = mK_f$ and solve for m:

SOLUTION

$\Delta t_f = +1.46$ (since the solvent, water, freezes at 0°C)

$$K_f = \frac{1.86°C\,\text{kg}\,H_2O}{\text{mol solute}}$$

$$1.46°C = mK_f = m \times \frac{1.86°C\,\text{kg}\,H_2O}{\text{mol solute}}$$

$$m = \frac{1.46°\cancel{C} \times \text{mol solute}}{1.86°\cancel{C} \times \text{kg}\,H_2O} = \frac{0.785\,\text{mol solute}}{\text{kg}\,H_2O}$$

Now convert the data, 4.71 g solute/100.0 g H₂O, to g/mol:

$$\left(\frac{4.71\,\text{g}\,\cancel{\text{solute}}}{100.0\,\cancel{\text{g}\,H_2O}}\right)\left(\frac{1000\,\cancel{\text{g}\,H_2O}}{1\,\cancel{\text{kg}\,H_2O}}\right)\left(\frac{1\,\cancel{\text{kg}\,H_2O}}{0.785\,\text{mol}\,\cancel{\text{solute}}}\right) = 60.0\,\text{g/mol}$$

The molar mass of the compound is 60.0 g/mol.

Practice 14.10 _____

What is the freezing point of the solution in Practice Exercise 14.9? What is the boiling point?

14.8 Osmosis and Osmotic Pressure

When red blood cells are put into distilled water, they gradually swell and in time may burst. If red blood cells are put in a 5%-urea (or a 5%-salt) solution, they gradually shrink and take on a wrinkled appearance. The cells behave in this fashion because they are enclosed in semipermeable membranes. A **semipermeable membrane** allows the passage of water (solvent) molecules through it in either direction but prevents the passage of larger solute molecules or ions. When two solutions of different concentrations (or water and a water solution) are separated by a semipermeable membrane, water diffuses through the membrane from the solution of lower concentration into the solution of higher concentration. The diffusion of water, either from a dilute solution or from pure water, through a semipermeable membrane into a solution of higher concentration is called **osmosis**.

A 0.90% (0.15 M) sodium chloride solution is known as a *physiological saline solution* because it is *isotonic* with blood plasma; that is, it has the same concentration of NaCl as blood plasma. Because each mole of NaCl yields about 2 mol of ions when in solution, the solute particle concentration in physiological saline solution is nearly 0.30 M. Five-percent-glucose solution (0.28 M) is also approximately isotonic with blood plasma. Blood cells neither swell nor shrink in an isotonic solution. The cells described in the preceding paragraph swell in water because water is *hypotonic* to cell plasma. The cells shrink in 5%-urea solution because the urea solution is *hypertonic* to the cell plasma. To prevent possible injury to blood cells by osmosis, fluids for intravenous use are usually made up at approximately isotonic concentration.

semipermeable membrane

osmosis

Human red blood cells. Left: In an isotonic solution the concentration is the same inside and outside the cell (0.9% saline). Center: In a hypertonic solution (1.6% saline) water leaves the cells, causing them to crenate (shrink). Right: In a hypotonic solution (0.2% saline) the cells swell as water moves into the cell center. Magnification is 260,000×.

Cross section on
molecular level

Semipermeable
membrane

Rising
solution
level

Thistle
tube

Sugar
solution

Water

- Sugar molecule
- Water molecule

Semipermeable
membrane (cellophane)

Figure 14.9
Laboratory demonstration of
osmosis: As a result of osmosis,
water passes through the
membrane, causing the solution
to rise in the thistle tube.

All solutions exhibit *osmotic pressure*, which is another colligative property. Osmotic pressure is a pressure difference between the system and atmospheric pressure. The osmotic pressure of a system can be measured by applying enough pressure to stop the flow of water due to osmosis in the system. The difference between the applied pressure and atmospheric pressure is the **osmotic pressure**. When pressure greater than the osmotic pressure is applied to a system, the flow of water can be reversed from that of osmosis. This process can be used to obtain useful drinking water from seawater and is known as *reverse osmosis*. Osmotic pressure is dependent only on the concentration of the solute particles and is independent of their nature. The osmotic pressure of a solution can be measured by determining the amount of counterpressure needed to prevent osmosis; this pressure can be very large. The osmotic pressure of a solution containing 1 mol of solute particles in 1 kg of water is about 22.4 atm, which is about the same as the pressure exerted by 1 mol of a gas confined in a volume of 1 L at 0°C.

Osmosis has a role in many biological processes, and semipermeable membranes occur commonly in living organisms. An example is the roots of plants, which are covered with tiny structures called root hairs; soil water enters the plant by osmosis, passing through the semipermeable membranes covering the root hairs. Artificial or synthetic membranes can also be made.

Osmosis can be demonstrated with the simple laboratory setup shown in Figure 14.9. As a result of osmotic pressure, water passes through the cellophane membrane into the thistle tube, causing the solution level to rise. In osmosis, the net transfer of water is always from a less concentrated to a more concentrated solution; that is, the effect is toward equalization of the concentration on both sides of the membrane. Note that the effective movement of water in osmosis is always from the region of *higher water concentration* to the region of *lower water concentration*.

Osmosis can be explained by assuming that a semipermeable membrane has passages that permit water molecules and other small molecules to pass in either direction. Both sides of the membrane are constantly being struck by water molecules in random motion. The number of water molecules crossing the

osmotic pressure

membrane is proportional to the number of water molecule-to-membrane impacts per unit of time. Because the solute molecules or ions reduce the concentration of water, there are more water molecules and thus more water molecule impacts on the side with the lower solute concentration (more dilute solution). The greater number of water molecule-to-membrane impacts on the dilute side thus causes a net transfer of water to the more concentrated solution. Again, note that the overall process involves the net transfer, by diffusion through the membrane, of water molecules from a region of higher water concentration (dilute solution) to one of lower water concentration (more concentrated solution).

This is a simplified picture of osmosis. No one has ever seen the hypothetical passages that allow water molecules and other small molecules or ions to pass through them. Alternative explanations have been proposed, but our discussion has been confined to water solutions. Osmotic pressure is a general colligative property, however, and is known to occur in nonaqueous systems.

Chapter 14 Review

14.1 General Properties of Solutions

KEY TERMS

Solution
Solute
Solvent

- A solution is a homogeneous mixture of two or more substances:
  - Consists of solvent—the dissolving agent—and solute—the component(s) dissolved in the solvent
  - Is a homogeneous mixture
  - Contains molecular or ionic particles
  - Can be colored or colorless
  - Can be separated into solute and solvent by a physical separation process

14.2 Solubility

KEY TERMS

Solubility
Miscible
Immiscible
Concentration of a solution

- Solubility describes the amount of solute that will dissolve in a specified amount of solvent.
- Solubility can also be qualitative.
- General guideline for ionic solubility are:

Soluble		Insoluble
Na^+, K^+, NH_4^+		
Nitrates, NO_3^- Acetates, $C_2H_3O_2^-$		
Chlorides, Cl^- Bromides Br^- Iodides, I^-	except →	Ag^+, Hg_2^{2+}, Pb^{2+}
Sulfates, SO_4^{2-} Ag^+, Ca^{2+} are slightly soluble	except →	Ba^{2+}, Sr^{2+}, Pb^{2+}
NH_4^+ alkali metal cations	← except	Carbonates, CO_3^{2-} Phosphates, PO_4^{3-} Hydroxides, OH^- Sulfides, S^{2-}

- Liquids can also be classified as miscible (soluble in each other) or immiscible (not soluble in each other).
- The concentration of a solution is the quantitative measurement of the amount of solute that is dissolved in a solution.

14.3 Factors Related to Solubility

KEY TERMS

Saturated solution
Unsaturated solution
Supersaturated solution

- Like tends to dissolve like is a general rule for solvents and solutes.
- As temperature increases:
  - Solubility of a solid in a liquid tends to increase.
  - Solubility of a gas in a liquid tends to decrease.
- As pressure increases:
  - Solubility of a solid in a liquid remains constant.
  - Solubility of a gas in a liquid tends to increase.
- At a specific temperature, the amount of solute that can dissolve in a solvent has a limit:
  - Unsaturated solutions contain less solute than the limit.
  - Saturated solutions contain dissolved solute at the limit.
  - Supersaturated solutions contain more solute than the limit and are therefore unstable:
    - If disturbed, the excess solute will precipitate out of solution.

14.4 Rate of Dissolving Solids

- The rate at which a solute dissolves is determined by these factors:
  - Particle size
  - Temperature
  - Concentration of solution
  - Agitation

14.5 Solutions: A Reaction Medium

- Molecules or ions must collide in order to react.
- Solutions provide a medium for the molecules or ions to collide.

14.6 Concentration of Solutions

KEY TERMS

Dilute solution
Concentrated solution
Parts per million (ppm)
Molarity (M)

- Concentrations can be measured in many ways:

Mass percent	% m/m	$\dfrac{\text{mass solute}}{\text{mass solution}} \times 100$
Parts per million	ppm	$\dfrac{\text{mass solute}}{\text{mass solution}} \times 1{,}000{,}000$
Mass/volume percent	% m/v	$\dfrac{\text{mass solute}}{\text{mL solution}} \times 100$
Volume percent	% v/v	$\dfrac{\text{mL solute}}{\text{mL solution}} \times 100$
Molarity	M	$\dfrac{\text{moles solute}}{\text{L solution}}$
Molality	m	$\dfrac{\text{moles solute}}{\text{kg solvent}}$

- Dilution of solutions requires the addition of more solvent to an existing solution:
  - The number of moles in the diluted solution is the same as that in the original solution.
  - $M_1V_1 = M_2V_2$.

14.7 Colligative Properties of Solutions

KEY TERMS

Colligative properties
Molality (m)

- Properties of a solution that depend only on the number of solute particles in solution are called colligative properties:
  - Freezing point depression
    $$\Delta t_f = mK_f$$
  - Boiling point elevation
    $$\Delta t_b = mK_b$$
  - Osmotic pressure
- Molality is used in working with colligative properties.

14.8 Osmosis and Osmotic Pressure

KEY TERMS

Semipermeable membrane
Osmosis
Osmotic pressure

- Osmosis is the diffusion of water through a semipermeable membrane:
  - Occurs from dilute solution to a solution of higher concentration
- Osmosis results in osmotic pressure, which is a colligative property of a solution.

Review Questions

All questions with blue numbers have answers in the appendix of the text.

1. Sketch the orientation of water molecules (a) about a single sodium ion and (b) about a single chloride ion in solution.
2. Estimate the number of grams of sodium fluoride that would dissolve in 100 g of water at 50°C. (Table 14.2)
3. What is the solubility at 25°C of these substances (Figure 14.4)?
   (a) potassium chloride
   (b) potassium chlorate
   (c) potassium nitrate

4. What is different in the solubility trend of the potassium halides compared with that of the lithium halides and the sodium halides? (Table 14.2)

5. What is the solubility, in grams of solute per 100 g of H_2O, of (a) $KClO_3$ at 60°C; (b) HCl at 20°C; (c) Li_2SO_4 at 80°C; and (d) KNO_3 at 0°C? (Figure 14.4)

6. Which substance, KNO_3 or NH_4Cl, shows the greater increase in solubility with increased temperature? (Figure 14.4)

7. Does a 2-molal solution in benzene or a 1-molal solution in camphor show the greater freezing point depression? (Table 14.5)

8. What would be the total surface area if the 1-cm cube in Figure 14.5 were cut into cubes 0.01 cm on a side?

9. At which temperatures—10°C, 20°C, 30°C, 40°C, or 50°C—would you expect a solution made from 63 g of ammonium chloride and 150 g of water to be unsaturated? (Figure 14.4)

10. Explain why the rate of dissolving decreases. (Figure 14.6)

11. Explain how a supersaturated solution of $NaC_2H_3O_2$ can be prepared and proven to be supersaturated.

12. Assume that the thistle tube in Figure 14.9 contains 1.0 M sugar solution and that the water in the beaker has just been replaced by a 2.0 M solution of urea. Would the solution level in the thistle tube continue to rise, remain constant, or fall? Explain.

13. What is a true solution?

14. Name and distinguish between the two components of a solution.

15. Is it always apparent in a solution which component is the solute, for example, in a solution of a liquid in a liquid?

16. Explain why the solute does not settle out of a solution.

17. Is it possible to have one solid dissolved in another? Explain.

18. An aqueous solution of KCl is colorless, $KMnO_4$ is purple, and $K_2Cr_2O_7$ is orange. What color would you expect of an aqueous solution of $Na_2Cr_2O_7$? Explain.

19. Explain why hexane will dissolve benzene but will not dissolve sodium chloride.

20. Some drinks like tea are consumed either hot or cold, whereas others like Coca-Cola are drunk only cold. Why?

21. Why is air considered to be a solution?

22. In which will a teaspoonful of sugar dissolve more rapidly, 200 mL of iced tea or 200 mL of hot coffee? Explain in terms of the KMT.

23. What is the effect of pressure on the solubility of gases in liquids? of solids in liquids?

24. Why do smaller particles dissolve faster than large ones?

25. In a saturated solution containing undissolved solute, solute is continuously dissolving, but the concentration of the solution remains unchanged. Explain.

26. Explain why there is no apparent reaction when crystals of $AgNO_3$ and NaCl are mixed, but a reaction is apparent immediately when solutions of $AgNO_3$ and NaCl are mixed.

27. Why do salt trucks distribute salt over icy roads in the winter?

28. What do we mean when we say that concentrated nitric acid (HNO_3) is 16 molar?

29. Will 1 L of 1 M NaCl contain more chloride ions than 0.5 L of 1 M $MgCl_2$? Explain.

30. Champagne is usually cooled in a refrigerator prior to opening. It's also opened very carefully. What would happen if a warm bottle of champagne were shaken and opened quickly and forcefully?

31. Describe how you would prepare 750 mL of 5.0 M NaCl solution.

32. Explain in terms of the KMT how a semipermeable membrane functions when placed between pure water and a 10% sugar solution.

33. Which has the higher osmotic pressure, a solution containing 100 g of urea (NH_2CONH_2) in 1 kg H_2O or a solution containing 150 g of glucose, $C_6H_{12}O_6$, in 1 kg H_2O?

34. Explain why a lettuce leaf in contact with salad dressing containing salt and vinegar soon becomes wilted and limp whereas another lettuce leaf in contact with plain water remains crisp.

35. A group of shipwreck survivors floated for several days on a life raft before being rescued. Those who had drunk some seawater were found to be suffering the most from dehydration. Explain.

36. Arrange the following bases (in descending order) according to the volume of each that will react with 1 L of 1 M HCl:
   (a) 1 M NaOH
   (b) 1.5 M $Ca(OH)_2$
   (c) 2 M KOH
   (d) 0.6 M $Ba(OH)_2$

37. Explain in terms of vapor pressure why the boiling point of a solution containing a nonvolatile solute is higher than that of the pure solvent.

38. Explain why the freezing point of a solution is lower than the freezing point of the pure solvent.

39. When water and ice are mixed, the temperature of the mixture is 0°C. But if methyl alcohol and ice are mixed, a temperature of -10°C is readily attained. Explain why the two mixtures show such different temperature behavior.

40. Which would be more effective in lowering the freezing point of 500. g of water?
   (a) 100. g of sucrose, $C_{12}H_{22}O_{11}$, or 100. g of ethyl alcohol, C_2H_5OH
   (b) 100. g of sucrose or 20.0 g of ethyl alcohol
   (c) 20.0 g of ethyl alcohol or 20.0 g of methyl alcohol, CH_3OH.

41. What is the difference between molarity and molality?

42. Is the molarity of a 5 m aqueous solution of NaCl greater or less than 5 M? Explain.

Paired Exercises

All exercises with blue numbers have answers in the appendix of the text.

1. Which of the substances listed below are reasonably soluble and which are insoluble in water?
   (See Figure 14.2 or Appendix V.)
   (a) KOH   (d) $AgC_2H_3O_2$
   (b) $NiCl_2$ (e) Na_2CrO_4
   (c) ZnS

2. Which of the substances listed below are reasonably soluble and which are insoluble in water?
   (See Figure 14.2 or Appendix V.)
   (a) PbI_2 (d) $Fe(NO_3)_3$
   (b) $MgCO_3$ (e) $BaSO_4$
   (c) $CaCl_2$

Percent Solution

3. Calculate the mass percent of the following solutions:
   (a) 15.0 g KCl + 100.0 g H_2O
   (b) 2.50 g Na_3PO_4 + 10.0 g H_2O
   (c) 0.20 mol $NH_4C_2H_3O_2$ + 125 g H_2O
   (d) 1.50 mol NaOH in 33.0 mol H_2O

4. Calculate the mass percent of the following solutions:
   (a) 25.0 g $NaNO_3$ in 125.0 g H_2O
   (b) 1.25 g $CaCl_2$ in 35.0 g H_2O
   (c) 0.75 mol K_2CrO_4 in 225 g H_2O
   (d) 1.20 mol H_2SO_4 in 72.5 mol H_2O

5. A chemistry lab experiment requires 25.2 g of silver nitrate. How many grams of a 15.5% by mass solution of silver nitrate should be used?

6. A reaction requires 25.0 g of sodium chloride. How many grams of a 10.0% by mass solution would provide this amount of solute?

7. In 25 g of a 7.5% by mass solution of $CaSO_4$
   (a) how many grams of solute are present?
   (b) how many grams of solvent are present?

8. In 75 g of a 12.0% by mass solution of $BaCl_2$
   (a) how many grams of solute are present?
   (b) how many grams of solvent are present?

9. Calculate the mass/volume percent of a solution made by dissolving 22.0 g of CH_3OH (methanol) in C_2H_5OH (ethanol) to make 100. mL of solution.

10. Calculate the mass/volume percent of a solution made by dissolving 4.20 g of NaCl in H_2O to make 12.5 mL of solution.

11. What is the volume percent of 10.0 mL of CH_3OH (methanol) dissolved in water to a volume of 40.0 mL?

12. What is the volume percent of 2.0 mL of hexane, C_6H_{14}, dissolved in benzene, C_6H_6, to a volume of 9.0 mL?

Molarity

13. Calculate the molarity of the following solutions:
    (a) 0.25 mol of solute in 75.0 mL of solution
    (b) 1.75 mol of KBr in 0.75 L of solution
    (c) 35.0 g of $NaC_2H_3O_2$ in 1.25 L of solution
    (d) 75 g of $CuSO_4 \cdot 5\,H_2O$ in 1.0 L of solution

14. Calculate the molarity of the following solutions:
    (a) 0.50 mol of solute in 125 mL of solution
    (b) 2.25 mol of $CaCl_2$ in 1.50 L of solution
    (c) 275 g $C_6H_{12}O_6$ in 775 mL of solution
    (d) 125 g $MgSO_4 \cdot 7\,H_2O$ in 2.50 L of solution

15. Calculate the number of moles of solute in each of the following solutions:
    (a) 1.5 L of 1.20 M H_2SO_4
    (b) 25.0 mL of 0.0015 M $BaCl_2$
    (c) 125 mL of 0.35 M K_3PO_4

16. Calculate the number of moles of solute in each of the following solutions:
    (a) 0.75 L of 1.50 M HNO_3
    (b) 10.0 mL of 0.75 M $NaClO_3$
    (c) 175 mL of 0.50 M LiBr

17. Calculate the grams of solute in each of the following solutions:
    (a) 2.5 L of 0.75 M K_2CrO_4
    (b) 75.2 mL of 0.050 M $HC_2H_3O_2$
    (c) 250 mL of 16 M HNO_3

18. Calculate the grams of solute in each of the following solutions:
    (a) 1.20 L of 18 M H_2SO_4
    (b) 27.5 mL of 1.50 M $KMnO_4$
    (c) 120 mL of 0.025 M $Fe_2(SO_4)_3$

19. How many milliliters of 0.750 M H_3PO_4 will contain the following?
    (a) 0.15 mol H_3PO_4
    (b) 35.5 g H_3PO_4

20. How many milliliters of 0.250 M NH_4Cl will contain the following?
    (a) 0.85 mol NH_4Cl
    (b) 25.2 g NH_4Cl

Dilution

21. What will be the molarity of the resulting solutions made by mixing the following? Assume that volumes are additive.
(a) 125 mL of 5.0 M H_3PO_4 with 775 mL of H_2O
(b) 250 mL of 0.25 M Na_2SO_4 with 750 mL of H_2O
(c) 75 mL of 0.50 M HNO_3 with 75 mL of 1.5 M HNO_3

23. Calculate the volume of concentrated reagent required to prepare the diluted solutions indicated:
(a) 15 M H_3PO_4 to prepare 750 mL of 3.0 M H_3PO_4
(b) 16 M HNO_3 to prepare 250 mL of 0.50 M HNO_3

25. Calculate the molarity of the solutions made by mixing 125 mL of 6.0 M $HC_2H_3O_2$ with the following:
(a) 525 mL of H_2O
(b) 175 mL of 1.5 M $HC_2H_3O_2$

22. What will be the molarity of the resulting solutions made by mixing the following? Assume that volumes are additive.
(a) 175 mL of 3.0 M H_2SO_4 with 275 mL of H_2O
(b) 350 mL of 0.10 M $CuSO_4$ with 150 mL of H_2O
(c) 50.0 mL of 0.250 M HCl with 25.0 mL of 0.500 M HCl

24. Calculate the volume of concentrated reagent required to prepare the diluted solutions indicated:
(a) 18 M H_2SO_4 to prepare 225 mL of 2.0 M H_2SO_4
(b) 15 M NH_3 to prepare 75 mL of 1.0 M NH_3

26. Calculate the molarity of the solutions made by mixing 175 mL of 3.0 M HCl with the following:
(a) 250 mL of H_2O
(b) 115 mL of 6.0 M HCl

Stoichiometry

27. Use the equation to calculate the following:

$$3\,Ca(NO_3)_2(aq) + 2\,Na_3PO_4(aq) \rightarrow Ca_3(PO_4)_2(s) + 6\,NaNO_3(aq)$$

(a) the moles $Ca_3(PO_4)_2$ produced from 2.7 mol Na_3PO_4
(b) the moles $NaNO_3$ produced from 0.75 mol $Ca(NO_3)_2$
(c) the moles Na_3PO_4 required to react with 1.45 L of 0.225 M $Ca(NO_3)_2$
(d) the grams of $Ca_3(PO_4)_2$ that can be obtained from 125 mL of 0.500 M $Ca(NO_3)_2$
(e) the volume of 0.25 M Na_3PO_4 needed to react with 15.0 mL of 0.50 M $Ca(NO_3)_2$
(f) the molarity (M) of the $Ca(NO_3)_2$ solution when 50.0 mL react with 50.0 mL of 2.0 M Na_3PO_4

28. Use the equation to calculate the following:

$$2\,NaOH(aq) + H_2SO_4(aq) \rightarrow Na_2SO_4(aq) + 2\,H_2O(l),$$

(a) the moles Na_2SO_4 produced from 3.6 mol H_2SO_4
(b) the moles H_2O produced from 0.025 mol NaOH
(c) the moles NaOH required to react with 2.50 L of 0.125 M H_2SO_4
(d) the grams of Na_2SO_4 that can be obtained from 25 mL of 0.050 M NaOH
(e) the volume of 0.250 M H_2SO_4 needed to react with 25.5 mL of 0.750 M NaOH
(f) the molarity (M) of the NaOH solution when 48.20 mL react with 35.72 mL of 0.125 M H_2SO_4

29. Use the equation to calculate the following:

$$2\,KMnO_4(aq) + 16\,HCl(aq) \rightarrow 2\,MnCl_2(aq) + 5\,Cl_2(g) + 8\,H_2O(l) + 2\,KCl(aq)$$

(a) the moles of H_2O that can be obtained from 15.0 mL of 0.250 M HCl
(b) the volume of 0.150 M $KMnO_4$ needed to produce 1.85 mol $MnCl_2$
(c) the volume of 2.50 M HCl needed to produce 125 mL of 0.525 M KCl
(d) the molarity (M) of the HCl solution when 22.20 mL react with 15.60 mL of 0.250 M $KMnO_4$
(e) the liters of Cl_2 gas at STP produced by the reaction of 125 mL of 2.5 M HCl
(f) the liters of Cl_2 gas at STP produced by the reaction of 15.0 mL of 0.750 M HCl and 12.0 mL of 0.550 M $KMnO_4$

30. Use the equation to calculate the following:

$$K_2CO_3(aq) + 2\,HC_2H_3O_2(aq) \rightarrow 2\,KC_2H_3O_2(aq) + H_2O(l) + CO_2(g)$$

(a) the moles of H_2O that can be obtained from 25.0 mL of 0.150 M $HC_2H_3O_2$
(b) the volume of 0.210 M K_2CO_3 needed to produce 17.5 mol $KC_2H_3O_2$
(c) the volume of 1.25 M $HC_2H_3O_2$ needed to react with 75.2 mL 0.750 M K_2CO_3
(d) the molarity (M) of the $HC_2H_3O_2$ solution when 10.15 mL react with 18.50 mL of 0.250 M K_2CO_3
(e) the liters of CO_2 gas at STP produced by the reaction of 105 mL of 1.5 M $HC_2H_3O_2$
(f) the liters of CO_2 gas at STP produced by the reaction of 25.0 mL of 0.350 M K_2CO_3 and 25.0 mL of 0.250 M $HC_2H_3O_2$

Molality and Colligative Properties

31. Calculate the molality of these solutions:
(a) 14.0 g CH_3OH in 100.0 g of H_2O
(b) 2.50 mol of benzene (C_6H_6) in 250 g of hexane (C_6H_{14})

32. Calculate the molality of these solutions:
(a) 1.0 g $C_6H_{12}O_6$ in 1.0 g H_2O
(b) 0.250 mol iodine in 1.0 kg H_2O

33. What is the (a) molality, (b) freezing point, and (c) boiling point of a solution containing 2.68 g of naphthalene ($C_{10}H_8$) in 38.4 g of benzene (C_6H_6)?

34. What is the (a) molality, (b) freezing point, and (c) boiling point of a solution containing 100.0 g of ethylene glycol ($C_2H_6O_2$) in 150.0 g of water?

*35. The freezing point of a solution of 8.00 g of an unknown compound dissolved in 60.0 g of acetic acid is 13.2°C. Calculate the molar mass of the compound.

*36. What is the molar mass of a compound if 4.80 g of the compound dissolved in 22.0 g of H_2O give a solution that freezes at −2.50°C?

Additional Exercises

All exercises with blue *numbers have answers in the appendix of the text.*

37. What happens to salt (NaCl) crystals when they are dissolved in water?

38. What happens to sugar molecules ($C_{12}H_{22}O_{11}$) when they are dissolved in water?

39. Why do sugar and salt behave differently when dissolved in water?

40. Why don't blood cells shrink or swell in an isotonic sodium chloride solution (0.9% saline)?

41. In the picture of dissolving $KMnO_4$ found in Section 14.1, the compound is forming purple streaks as it dissolves. Why?

42. In Figure 14.4, observe the line for KNO_3. Explain why it slopes up from left to right. How does the slope compare to the slopes of the other substances? What does this mean?

43. How many grams of solution, 10.0% NaOH by mass, are required to neutralize 150 mL of a 1.0 M HCl solution?

*44. How many grams of solution, 10.0% NaOH by mass, are required to neutralize 250.0 g of a 1.0 m solution of HCl?

*45. A sugar syrup solution contains 15.0% sugar, $C_{12}H_{22}O_{11}$, by mass and has a density of 1.06 g/mL.
(a) How many grams of sugar are in 1.0 L of this syrup?
(b) What is the molarity of this solution?
(c) What is the molality of this solution?

*46. A solution of 3.84 g C_4H_2N (empirical formula) in 250.0 g of benzene depresses the freezing point of benzene 0.614°C. What is the molecular formula for the compound?

47. Hydrochloric acid (HCl) is sold as a concentrated aqueous solution (12.0 mol/L). If the density of the solution is 1.18 g/mL, determine the molality of the solution.

*48. How many grams of KNO_3 are needed to make 450 mL of a solution that is to contain 5.5 mg/mL of potassium ion? Calculate the molarity of the solution.

49. What mass of 5.50% solution can be prepared from 25.0 g KCl?

50. Given a solution containing 16.10 g $C_2H_6O_2$ in 82.0 g H_2O that has a boiling point of 101.62°C, verify that the boiling point elevation constant K_f for water is 0.512°C kg H_2O/mole solute.

51. Physiological saline (NaCl) solutions used in intravenous injections have a concentration of 0.90% NaCl (mass/volume).
(a) How many grams of NaCl are needed to prepare 500.0 mL of this solution?
(b) How much water must evaporate from this solution to give a solution that is 9.0% NaCl (mass/volume)?

*52. A solution is made from 50.0 g KNO_3 and 175 g H_2O. How many grams of water must evaporate to give a saturated solution of KNO_3 in water at 20°C? (See Figure 14.4.)

53. What volume of 70.0% rubbing alcohol can you prepare if you have only 150 mL of pure isopropyl alcohol on hand?

54. At 20°C, an aqueous solution of HNO_3 that is 35.0% HNO_3 by mass has a density of 1.21 g/mL.
(a) How many grams of HNO_3 are present in 1.00 L of this solution?
(b) What volume of this solution will contain 500. g HNO_3?

*55. What is the molarity of a nitric acid solution if the solution is 35.0% HNO_3 by mass and has a density of 1.21 g/mL?

56. To what volume must a solution of 80.0 g H_2SO_4 in 500.0 mL of solution be diluted to give a 0.10 M solution?

57. A 10.0-mL sample of 16 M HNO_3 is diluted to 500.0 mL. What is the molarity of the final solution?

58. (a) How many moles of hydrogen will be liberated from 200.0 mL of 3.00 M HCl reacting with an excess of magnesium? The equation is

$$Mg(s) + 2\,HCl(aq) \longrightarrow MgCl_2(aq) + H_2(g)$$

(b) How many liters of hydrogen gas (H_2) measured at 27°C and 720 torr will be obtained? (*Hint*: Use the ideal gas law.)

59. Which will be more effective in neutralizing stomach acid, HCl: a tablet containing 1.20 g $Mg(OH)_2$ or a tablet containing 1.00g $Al(OH)_3$? Show evidence for your answer.

60. Which would be more effective as an antifreeze in an automobile radiator? A solution containing
(a) 10 kg of methyl alcohol (CH_3OH) or 10 kg of ethyl alcohol (C_2H_5OH)?
(b) 10 m solution of methyl alcohol or 10 m solution of ethyl alcohol?

* **61.** Automobile battery acid is 38% H_2SO_4 and has a density of 1.29 g/mL. Calculate the molality and the molarity of this solution.

* **62.** What is the (a) molality and (b) boiling point of an aqueous sugar, $C_{12}H_{22}O_{11}$, solution that freezes at $-5.4°C$?

63. A solution of 6.20 g $C_2H_6O_2$ in water has a freezing point of $-0.372°C$. How many grams of H_2O are in the solution?

64. What (a) mass and (b) volume of ethylene glycol ($C_2H_6O_2$, density = 1.11 g/mL) should be added to 12.0 L of water in an automobile radiator to protect it from freezing at $-20°C$? (c) To what temperature Fahrenheit will the radiator be protected?

* **65.** If 150 mL of 0.055 M HNO_3 are needed to completely neutralize 1.48 g of an *impure* sample of sodium hydrogen carbonate (baking soda), what percent of the sample is baking soda?

66. (a) How much water must be added to concentrated sulfuric acid (H_2SO_4) (17.8 M) to prepare 8.4 L of 1.5 M sulfuric acid solution?
(b) How many moles of H_2SO_4 are in each milliliter of the original concentrate?
(c) How many moles are in each milliliter of the diluted solution?

* **67.** How would you prepare a 6.00 M HNO_3 solution if only 3.00 M and 12.0 M solutions of the acid are available for mixing?

* **68.** A 20.0-mL portion of an HBr solution of unknown strength is diluted to exactly 240 mL. If 100.0 mL of this diluted solution requires 88.4 mL of 0.37 M NaOH to achieve complete neutralization, what was the strength of the original HBr solution?

69. When 80.5 mL of 0.642 M $Ba(NO_3)_2$ are mixed with 44.5 mL of 0.743 M KOH, a precipitate of $Ba(OH)_2$ forms. How many grams of $Ba(OH)_2$ do you expect?

70. A 0.25 M solution of lithium carbonate (Li_2CO_3), a drug used to treat manic depression, is prepared.
(a) How many moles of Li_2CO_3 are present in 45.8 mL of the solution?
(b) How many grams of Li_2CO_3 are in 750 mL of the same solution?
(c) How many milliliters of the solution would be needed to supply 6.0 g of the solute?
(d) If the solution has a density of 1.22 g/mL, what is its mass percent?

Challenge Exercises

All exercises with blue numbers have answers in the appendix of the text.

71. When solutions of hydrochloric acid and sodium sulfite react, a salt, water, and sulfur dioxide gas are produced. How many liters of sulfur dioxide gas at 775 torr and 22°C can be produced when 125 mL of 2.50 M hydrochloric acid react with 75.0 mL of 1.75 M sodium sulfite?

***72.** Consider a saturated solution at 20°C made from 5.549 moles of water and an unknown solute. You determine the mass of the container containing the solution to be 563 g. The mass of the empty container is 375 g. Identify the solute.

Answers to Practice Exercises

14.1 unsaturated
14.2 10.0% Na_2SO_4 solution
14.3 2.02 M
14.4 0.27 g NaCl
14.5 0.84 L (840 mL)
14.6 75 mL NaOH solution

14.7 $5.00 \times 10^{-2} M$
14.8 43 g
14.9 1.387 m
14.10 freezing point = $-2.58°C$, boiling point = $100.71°C$

PUTTING IT TOGETHER: Review for Chapters 12–14

Multiple Choice:

Choose the correct answer to each of the following.

1. Which of these statements is *not* one of the principal assumptions of the kinetic-molecular theory for an ideal gas?
   (a) All collisions of gaseous molecules are perfectly elastic.
   (b) A mole of any gas occupies 22.4 L at STP.
   (c) Gas molecules have no attraction for one another.
   (d) The average kinetic energy for molecules is the same for all gases at the same temperature.

2. Which of the following is not equal to 1.00 atm?
   (a) 760. cm Hg   (c) 760. mm Hg
   (b) 29.9 in. Hg   (d) 760. torr

3. If the pressure on 45 mL of gas is changed from 600. torr to 800. torr, the new volume will be
   (a) 60 mL   (c) 0.045 L
   (b) 34 mL   (d) 22.4 L

4. The volume of a gas is 300. mL at 740. torr and 25°C. If the pressure remains constant and the temperature is raised to 100.°C, the new volume will be
   (a) 240. mL   (c) 376 mL
   (b) 1.20 L   (d) 75.0 mL

5. The volume of a dry gas is 4.00 L at 15.0°C and 745 torr. What volume will the gas occupy at 40.0°C and 700. torr?
   (a) 4.63 L   (c) 3.92 L
   (b) 3.46 L   (d) 4.08 L

6. A sample of Cl_2 occupies 8.50 L at 80.0°C and 740. mm Hg. What volume will the Cl_2 occupy at STP?
   (a) 10.7 L   (c) 11.3 L
   (b) 6.75 L   (d) 6.40 L

7. What volume will 8.00 g O_2 occupy at 45°C and 2.00 atm?
   (a) 0.462 L   (c) 9.62 L
   (b) 104 L   (d) 3.26 L

8. The density of NH_3 gas at STP is
   (a) 0.760 g/mL   (c) 1.32 g/mL
   (b) 0.760 g/L   (d) 1.32 g/L

9. The ratio of the relative rate of effusion of methane (CH_4) to sulfur dioxide (SO_2) is
   (a) $^{64}/_{16}$ (c) $^1/_4$
   (b) $^{16}/_{64}$ (d) $^2/_1$

10. Measured at 65°C and 500. torr, the mass of 3.21 L of a gas is 3.5 g. The molar mass of this gas is
    (a) 21 g/mole   (c) 24 g/mole
    (b) 46 g/mole   (d) 130 g/mole

11. Box A contains O_2 (molar mass = 32.0) at a pressure of 200 torr. Box B, which is identical to box A in volume, contains twice as many molecules of CH_4 (molar mass = 16.0) as the molecules of O_2 in box A. The temperatures of the gases are identical. The pressure in box B is
    (a) 100 torr   (c) 400 torr
    (b) 200 torr   (d) 800 torr

12. A 300.-mL sample of oxygen (O_2) is collected over water at 23°C and 725 torr. If the vapor pressure of water at 23°C is 21.0 torr, the volume of dry O_2 at STP is
    (a) 256 mL   (c) 341 mL
    (b) 351 mL   (d) 264 mL

13. A tank containing 0.01 mol of neon and 0.04 mol of helium shows a pressure of 1 atm. What is the partial pressure of neon in the tank?
    (a) 0.8 atm   (c) 0.2 atm
    (b) 0.01 atm   (d) 0.5 atm

14. How many liters of NO_2 (at STP) can be produced from 25.0 g Cu reacting with concentrated nitric acid?

    $Cu(s) + 4 HNO_3(aq) \longrightarrow$
    $$Cu(NO_3)_2(aq) + 2 H_2O(l) + 2 NO_2(g)$$

    (a) 4.41 L   (c) 17.6 L
    (b) 8.82 L   (d) 44.8 L

15. How many liters of butane vapor are required to produce 2.0 L CO_2 at STP?

    $2 C_4H_{10}(g) + 13 O_2(g) \longrightarrow 8 CO_2(g) + 10 H_2O(g)$
    butane

    (a) 2.0 L   (c) 0.80 L
    (b) 4.0 L   (d) 0.50 L

16. What volume of CO_2 (at STP) can be produced when 15.0 g C_2H_6 and 50.0 g O_2 are reacted?

    $2 C_2H_6(g) + 7 O_2(g) \longrightarrow 4 CO_2(g) + 6 H_2O(g)$

    (a) 20.0 L   (c) 35.0 L
    (b) 22.4 L   (d) 5.6 L

17. Which of these gases has the highest density at STP?
    (a) N_2O (c) Cl_2
    (b) NO_2 (d) SO_2

18. What is the density of CO_2 at 25°C and 0.954 atm?
    (a) 1.72 g/L   (c) 0.985 g/L
    (b) 2.04 g/L   (d) 1.52 g/L

19. How many molecules are present in 0.025 mol of H_2 gas?
    (a) 1.5×10^{22} molecules
    (b) 3.37×10^{23} molecules
    (c) 2.40×10^{25} molecules
    (d) 1.50×10^{22} molecules

20. 5.60 L of a gas at STP have a mass of 13.0 g. What is the molar mass of the gas?
    (a) 33.2 g/mol   (c) 66.4 g/mol
    (b) 26.0 g/mol   (d) 52.0 g/mol

21. The heat of fusion of water is
    (a) 4.184 J/g   (c) 2.26 kJ/g
    (b) 335 J/g   (d) 2.26 kJ/mol

22. The heat of vaporization of water is
    (a) 4.184 J/g   (c) 2.26 kJ/g
    (b) 335 J/g   (d) 2.26 kJ/mol

23. The specific heat of water is
    (a) 4.184 J/g°C   (c) 2.26 kJ/g°C
    (b) 335 J/g°C   (d) 18 J/g°C

24. The density of water at 4°C is
(a) 1.0 g/mL
(c) 18.0 g/mL
(b) 80 g/mL
(d) 14.7 lb/in.^{3}

25. SO_2 can be properly classified as a(n)
(a) basic anhydride
(c) anhydrous salt
(b) hydrate
(d) acid anhydride

26. When compared to H_2S, H_2Se, and H_2Te, water is found to have the highest boiling point because it
(a) has the lowest molar mass
(b) is the smallest molecule
(c) has the highest bonding
(d) forms hydrogen bonds better than the others

27. In which of the following molecules will hydrogen bonding be important?
(a) H—F
(c) H—Br

28. Which of the following is an incorrect equation?
(a) $H_2SO_4 + 2\,NaOH \longrightarrow Na_2SO_4 + 2\,H_2O$
(b) $C_2H_6 + O_2 \longrightarrow 2\,CO_2 + 3\,H_2$
(c) $2\,H_2O \xrightarrow[\text{H}_2\text{SO}_4]{\text{electrolysis}} 2\,H_2 + O_2$
(d) $Ca + 2\,H_2O \longrightarrow H_2 + Ca(OH)_2$

29. Which of the following is an incorrect equation?
(a) $C + H_2O(g) \xrightarrow{1000°C} CO(g) + H_2(g)$
(b) $CaO + H_2O \longrightarrow Ca(OH)_2$
(c) $2\,NO_2 + H_2O \longrightarrow 2\,HNO_3$
(d) $Cl_2 + H_2O \longrightarrow HCl + HOCl$

30. How many kilojoules are required to change 85 g of water at 25°C to steam at 100.°C?
(a) 219 kJ
(c) 590 kJ
(b) 27 kJ
(d) 192 kJ

31. A chunk of 0°C ice, mass 145 g, is dropped into 75 g of water at 62°C. The heat of fusion of water is 335 J/g. The result, after thermal equilibrium is attained, will be
(a) 87 g ice and 133 g liquid water, all at 0°C
(b) 58 g ice and 162 g liquid water, all at 0°C
(c) 220 g water at 7°C
(d) 220 g water at 17°C

32. The formula for iron(II) sulfate heptahydrate is
(a) $Fe_2SO_4 \cdot 7\,H_2O$
(c) $FeSO_4 \cdot 7\,H_2O$
(b) $Fe(SO_4)_2 \cdot 6\,H_2O$
(d) $Fe_2(SO_4)_3 \cdot 7\,H_2O$

33. The process by which a solid changes directly to a vapor is called
(a) vaporization
(c) sublimation
(b) evaporation
(d) condensation

34. Hydrogen bonding
(a) occurs only between water molecules
(b) is stronger than covalent bonding
(c) can occur between NH_3 and H_2O
(d) results from strong attractive forces in ionic compounds

35. A liquid boils when
(a) the vapor pressure of the liquid equals the external pressure above the liquid
(b) the heat of vaporization exceeds the vapor pressure
(c) the vapor pressure equals 1 atm
(d) the normal boiling temperature is reached

36. Consider two beakers, one containing 50 mL of liquid A and the other 50 mL of liquid B. The boiling point of A is 90°C and that of B is 72°C. Which of these statements is correct?
(a) A will evaporate faster than B.
(b) B will evaporate faster than A.
(c) Both A and B evaporate at the same rate.
(d) Insufficient data to answer the question.

37. 95.0 g of 0.0°C ice is added to exactly 100. g of water at 60.0°C. When the temperature of the mixture first reaches 0.0°C, the mass of ice still present is
(a) 0.0 g
(c) 10.0 g
(b) 20.0 g
(d) 75.0 g

38. Which of the following is not a general property of solutions?
(a) a homogeneous mixture of two or more substances
(b) variable composition
(c) dissolved solute breaks down to individual molecules
(d) the same chemical composition, the same chemical properties, and the same physical properties in every part

39. If NaCl is soluble in water to the extent of 36.0 g NaCl/100 g H_2O at 20°C, then a solution at 20°C containing 45 g NaCl/150 g H_2O would be
(a) dilute
(c) supersaturated
(b) saturated
(d) unsaturated

40. If 5.00 g NaCl are dissolved in 25.0 g of water, the percent of NaCl by mass is
(a) 16.7%
(c) 0.20%
(b) 20.0%
(d) no correct answer given

41. How many grams of 9.0% $AgNO_3$ solution will contain 5.3 g $AgNO_3$?
(a) 47.7 g
(c) 59 g
(b) 0.58 g
(d) no correct answer given

42. The molarity of a solution containing 2.5 mol of acetic acid ($HC_2H_3O_2$) in 400. mL of solution is
(a) 0.063 M
(c) 0.103 M
(b) 1.0 M
(d) 6.3 M

43. What volume of 0.300 M KCl will contain 15.3 g KCl?
(a) 1.46 L
(c) 61.5 mL
(b) 683 mL
(d) 4.60 L

44. What mass of $BaCl_2$ will be required to prepare 200. mL of 0.150 M solution?
(a) 0.750 g
(c) 6.25 g
(b) 156 g
(d) 31.2 g

Problems 45–47 relate to the reaction

$$CaCO_3 + 2\,HCl \longrightarrow CaCl_2 + H_2O + CO_2$$

45. What volume of 6.0 M HCl will be needed to react with 0.350 mol of $CaCO_3$?
(a) 42.0 mL   (c) 117 mL
(b) 1.17 L   (d) 583 mL

46. If 400. mL of 2.0 M HCl react with excess $CaCO_3$, the volume of CO_2 produced, measured at STP, is
(a) 18 L   (c) 9.0 L
(b) 5.6 L   (d) 56 L

47. If 5.3 g $CaCl_2$ are produced in the reaction, what is the molarity of the HCl used if 25 mL of it reacted with excess $CaCO_3$?
(a) 3.8 M (c) 0.38 M
(b) 0.19 M (d) 0.42 M

48. If 20.0 g of the nonelectrolyte urea $(CO(NH_2)_2)$ is dissolved in 25.0 g of water, the freezing point of the solution will be
(a) $-2.47°C$ (c) $-24.7°C$
(b) $-1.40°C$ (d) $-3.72°C$

49. When 256 g of a nonvolatile, nonelectrolyte unknown were dissolved in 500. g H_2O, the freezing point was found to be $-2.79°C$. The molar mass of the unknown solute is
(a) 357   (c) 768
(b) 62.0   (d) 341

50. How many milliliters of 6.0 M H_2SO_4 must you use to prepare 500. mL of 0.20 M sulfuric acid solution?
(a) 30   (c) 12
(b) 17   (d) 100

51. How many milliliters of water must be added to 200. mL of 1.40 M HCl to make a solution that is 0.500 M HCl?
(a) 360. mL   (c) 140. mL
(b) 560. mL   (d) 280. mL

52. Which procedure is most likely to increase the solubility of most solids in liquids?
(a) stirring
(b) pulverizing the solid
(c) heating the solution
(d) increasing the pressure

53. The addition of a crystal of $NaClO_3$ to a solution of $NaClO_3$ causes additional crystals to precipitate. The original solution was
(a) unsaturated   (c) saturated
(b) dilute   (d) supersaturated

54. Which of these anions will not form a precipitate with silver ions, Ag^+?
(a) Cl^- (c) Br^-
(b) NO_3^- (d) CO_3^{2-}

55. Which of these salts are considered to be soluble in water?
(a) $BaSO_4$ (c) AgI
(b) NH_4Cl (d) PbS

56. A solution of ethyl alcohol and benzene is 40% alcohol by volume. Which statement is correct?
(a) The solution contains 40 mL of alcohol in 100 mL of solution.
(b) The solution contains 60 mL of benzene in 100 mL of solution.
(c) The solution contains 40 mL of alcohol in 100 g of solution.
(d) The solution is made by dissolving 40 mL of alcohol in 60 mL of benzene.

57. Which of the following is not a colligative property?
(a) boiling point elevation
(b) freezing point depression
(c) osmotic pressure
(d) surface tension

58. When a solute is dissolved in a solvent
(a) the freezing point of the solution increases.
(b) the vapor pressure of the solution increases.
(c) the boiling point of the solution increases.
(d) the concentration of the solvent increases.

59. Which of the following solutions will have the lowest freezing point where X is any element or nonelectrolytic compound?
(a) 1.0 mol X in 1 kg H_2O
(b) 2.0 mol X in 1 kg H_2O
(c) 1.2 mol X in 1 kg H_2O
(d) 0.80 mol X in 1 kg H_2O

60. In the process of osmosis, water passes through a semipermeable membrane
(a) from a more concentrated solution to a dilute solution
(b) from a dilute solution to a more concentrated solution
(c) in order to remove a solute from a solution
(d) so that a sugar solution can become sweeter

Free Response Questions

Answer each of the following. Be sure to include your work and explanations in a clear, logical form.

1. Which solution should have a higher boiling point: 215 mL of a 10.0% (m/v) aqueous KCl solution or 224 mL of a 1.10 M aqueous NaCl solution?

2. A glass containing 345 mL of a soft drink (a carbonated beverage) was left sitting out on a kitchen counter. If the CO_2 released at room temperature (25°C) and pressure (1 atm) occupies 1.40 L, at a minimum, what is the concentration (in ppm) of the CO_2 in the original soft drink (assume the density of the original soft drink is 0.965 g/mL).

3. Dina and Murphy were trying to react 100. mL of a 0.10 M HCl solution with KOH. The procedure called for a 10% KOH solution. Dina made a 10% mass/volume solution while Murphy made a 10% by mass solution. (Assume there is no volume change upon dissolving KOH.) Which solution required less volume to fully react with 100. mL of the HCl solution?

4. A flask containing 825 mL of a solution containing 0.355 mole of $CuSO_4$ was left open overnight. The next morning the flask only contained 755 mL of solution.

H H H
O···H–O–H
H H H
O O
H H

(1) →
(2) →
(3) →

(a) What is the concentration (molarity) of the $CuSO_4$ solution remaining in the flask?

(b) Which of the pathways shown below best represents the evaporation of water and why are the others wrong?

H—H(g) + O═O(g)

$H^+(g)$ + $OH^-(g)$

O   O
H   H(g) + H   H(g)

5. Three students at Jamston High—Zack, Gaye, and Lamont—each had the opportunity to travel over spring break. As part of a project, each of them measured the boiling point of pure water at their vacation spot. Zack found a boiling point of 93.9°C, Gaye measured 101.1°C, and Lamont read 100.°C. Which student most likely went snow skiing near Ely, Nevada, and which student most likely went water skiing in Honolulu? From the boiling point information, what can you surmise about the Dead Sea region, the location of the third student's vacation? Explain.

6. Why does a change in pressure of a gas significantly affect its volume whereas a change in pressure on a solid or liquid has negligible effect on their respective volumes? If the accompanying picture represents a liquid at the molecular level, draw what you might expect a solid and a gas to look like.

7. (a) If you filled up three balloons with equal volumes of hydrogen, argon, and carbon dioxide gas, all at the same temperature and pressure, which balloon would weigh the most? the least? Explain.

(b) If you filled up three balloons with equal masses of nitrogen, oxygen, and neon, all to the same volume at the same temperature, which would have the lowest pressure?

8. Ray ran a double-displacement reaction using 0.050 mol $CuCl_2$ and 0.10 mol $AgNO_3$. The resulting white precipitate was removed by filtration. The filtrate was accidentally left open on the lab bench for over a week, and when Ray returned the flask contained solid, blue crystals. Ray weighed the crystals and found they had a mass of 14.775 g. Was Ray expecting this number? If not, what did he expect?

9. Why is it often advantageous or even necessary to run reactions in solution rather than mixing two solids? Would you expect reactions run in the gas phase to be more similar to solutions or to solids? Why?

10. A solution of 5.36 g of a molecular compound dissolved in 76.8 g benzene (C_6H_6) and has a boiling point of 81.48°C. What is the molar mass of the compound?

Boiling point for benzene = 80.1°C
K_b for benzene = 2.53°C kg solvent/mol solute

CHAPTER 15

Acids, Bases, and Salts

Citrus fruits are all acidic substances.

Chapter Outline

15.1 Acids and Bases

15.2 Reactions of Acids

15.3 Reactions of Bases

15.4 Salts

15.5 Electrolytes and Nonelectrolytes

15.6 Dissociation and Ionization of Electrolytes

15.7 Strong and Weak Electrolytes

15.8 Ionization of Water

15.9 Introduction to pH

15.10 Neutralization

15.11 Writing Net Ionic Equations

15.12 Acid Rain

15.13 Colloids

15.14 Properties of Colloids

15.15 Applications of Colloidal Properties

Acids are important chemicals. They are used in cooking to produce the surprise of tartness (from lemons) and to release CO_2 bubbles from leavening agents in baking. Vitamin C is an acid that is an essential nutrient in our diet. Our stomachs release acid to aid in digestion. Excess stomach acid can produce heartburn and indigestion. Bacteria in our mouths produce acids that can dissolve tooth enamel to form cavities. In our recreational activities we are concerned about acidity levels in swimming pools and spas. Acids are essential in the manufacture of detergents, plastics, and storage batteries. The acid–base properties of substances are found in all areas of our lives. In this chapter we consider the properties of acids, bases, and salts.

15.1 Acids and Bases

The word *acid* is derived from the Latin *acidus,* meaning "sour" or "tart," and is also related to the Latin word *acetum,* meaning "vinegar." Vinegar has been around since antiquity as a product of the fermentation of wine and apple cider. The sour constituent of vinegar is acetic acid ($HC_2H_3O_2$). Characteristic properties commonly associated with acids include the following:

1. sour taste
2. the ability to change the color of litmus, a vegetable dye, from blue to red
3. the ability to react with
   • metals such as zinc and magnesium to produce hydrogen gas
   • hydroxide bases to produce water and an ionic compound (salt)
   • carbonates to produce carbon dioxide

These properties are due to the hydrogen ions, H^+, released by acids in a water solution.

Classically, a *base* is a substance capable of liberating hydroxide ions, OH^-, in water solution. Hydroxides of the alkali metals (Group 1A) and alkaline earth metals (Group 2A), such as LiOH, NaOH, KOH, $Ca(OH)_2$, and $Ba(OH)_2$, are the most common inorganic bases. Water solutions of bases are called *alkaline solutions* or *basic solutions.* Some of the characteristic properties commonly associated with bases include the following:

1. bitter or caustic taste
2. a slippery, soapy feeling
3. the ability to change litmus from red to blue
4. the ability to interact with acids

Several theories have been proposed to answer the question "What is an acid and what is a base?" One of the earliest, most significant of these theories was advanced in 1884 by Svante Arrhenius (1859–1927), a Swedish scientist, who stated that "an acid is a hydrogen-containing substance that dissociates to produce hydrogen ions, and a base is a hydroxide-containing substance that dissociates to produce hydroxide ions in aqueous solutions." Arrhenius postulated that the hydrogen ions are produced by the dissociation of acids in water and that the hydroxide ions are produced by the dissociation of bases in water:

$$HA \xrightarrow{\text{H}_2\text{O}} H^+(aq) + A^-(aq)$$
acid

$$MOH \xrightarrow{\text{H}_2\text{O}} M^+(aq) + OH^-(aq)$$
base

An Arrhenius acid solution contains an excess of H^+ ions.
An Arrhenius base solution contains an excess of OH^- ions.

In 1923, the Brønsted–Lowry proton transfer theory was introduced by J. N. Brønsted (1897–1947), a Danish chemist, and T. M. Lowry (1847–1936), an English chemist. This theory states that an acid is a proton donor and a base is a proton acceptor.

A Brønsted–Lowry acid is a proton (H^+) donor.
A Brønsted–Lowry base is a proton (H^+) acceptor.

Consider the reaction of hydrogen chloride gas with water to form hydrochloric acid:

$$HCl(g) + H_2O(l) \longrightarrow H_3O^+(aq) + Cl^-(aq) \qquad (1)$$

In the course of the reaction, HCl donates, or gives up, a proton to form a Cl^- ion, and H_2O accepts a proton to form the H_3O^+ ion. Thus, HCl is an acid and H_2O is a base, according to the Brønsted–Lowry theory.

A hydrogen ion (H^+) is nothing more than a bare proton and does not exist by itself in an aqueous solution. In water H^+ combines with a polar water molecule to form a hydrated hydrogen ion (H_3O^+) commonly called a **hydronium ion**. The H^+ is attracted to a polar water molecule, forming a bond with one of the two pairs of unshared electrons:

$$H^+ + H\!:\!\ddot{\underset{\displaystyle H}{O}}\!: \longrightarrow \left[H\!:\!\ddot{\underset{\displaystyle H}{O}}\!:\!H \right]^+$$

hydronium ion

hydronium ion

Note the electron structure of the hydronium ion. For simplicity we often use H^+ instead of H_3O^+ in equations, with the explicit understanding that H^+ is always hydrated in solution.

When a Brønsted–Lowry acid donates a proton, as illustrated in the equation below, it forms the conjugate base of that acid. When a base accepts a proton, it forms the conjugate acid of that base. A conjugate acid and base are produced as products. The formulas of a conjugate acid–base pair differ by one proton (H^+). Consider what happens when $HCl(g)$ is bubbled through water, as shown by this equation:

conjugate acid–base pair

$$HCl(g) + H_2O(l) \longrightarrow Cl^-(aq) + H_3O^+(aq)$$

conjugate acid–base pair

acid   base   base   acid

The conjugate acid–base pairs are $HCl\!-\!Cl^-$ and $H_3O^+\!-\!H_2O$. The conjugate base of HCl is Cl^-, and the conjugate acid of Cl^- is HCl. The conjugate base of H_3O^+ is H_2O, and the conjugate acid of H_2O is H_3O^+.

Another example of conjugate acid–base pairs can be seen in this equation:

$$NH_4^+ + H_2O \longrightarrow H_3O^+ + NH_3$$

acid   base   acid   base

Here the conjugate acid–base pairs are $NH_4^+\!-\!NH_3$ and $H_3O^+\!-\!H_2O$.

Example 15.1   Write the formula for (a) the conjugate base of H_2O and of HNO_3, and (b) the conjugate acid of SO_4^{2-} and of $C_2H_3O_2^{-}$.

SOLUTION   (a) To write the conjugate base of an acid, remove one proton from the acid formula:

$$H_2O \xrightarrow{-H^+} OH^- \qquad \text{(conjugate base)}$$

$$HNO_3 \xrightarrow{-H^+} NO_3^- \qquad \text{(conjugate base)}$$

Note that, by removing an H^+, the conjugate base becomes more negative than the acid by one minus charge.

(b) To write the conjugate acid of a base, add one proton to the formula of the base:

$$SO_4^{2-} \xrightarrow{+H^+} HSO_4^- \qquad \text{(conjugate acid)}$$

$$C_2H_3O_2^- \xrightarrow{+H^+} HC_2H_3O_2 \qquad \text{(conjugate acid)}$$

In each case the conjugate acid becomes more positive than the base by a +1 charge due to the addition of H^+.

Remember: The difference between an acid or a base and its conjugate is one proton, H^+.

Practice 15.1 _____

Indicate the conjugate base for these acids:
(a) H_2CO_3 (b) HNO_2 (c) $HC_2H_3O_2$

Practice 15.2 _____

Indicate the conjugate acid for these bases:
(a) HSO_4^- (b) NH_3 (c) OH^-

A more general concept of acids and bases was introduced by Gilbert N. Lewis. The Lewis theory deals with the way in which a substance with an unshared pair of electrons reacts in an acid–base type of reaction. According to this theory a base is any substance that has an unshared pair of electrons (electron pair donor), and an acid is any substance that will attach itself to or accept a pair of electrons.

> A Lewis acid is an electron pair acceptor.
> A Lewis base is an electron pair donor.

In the reaction

$$H^+ + \overset{\displaystyle H}{\underset{\displaystyle H}{:\!\ddot{N}\!:\!H}} \longrightarrow \left[\overset{\displaystyle H}{\underset{\displaystyle H}{H\!:\!\ddot{N}\!:\!H}} \right]^+$$

acid   base

Table 15.1 Summary of Acid–Base Definitions

Theory	Acid	Base
Arrhenius	A hydrogen-containing substance that produces hydrogen ions in aqueous solution	A hydroxide-containing substance that produces hydroxide ions in aqueous solution
Brønsted–Lowry	A proton (H^+) donor	A proton (H^+) acceptor
Lewis	Any species that will bond to an unshared pair of electrons (electron pair acceptor)	Any species that has an unshared pair of electrons (electron pair donor)

The H^+ is a Lewis acid and $:NH_3$ is a Lewis base. According to the Lewis theory, substances other than proton donors (e.g., BF_3) behave as acids:

$$\underset{\text{acid}}{\text{F:}\overset{\text{F}}{\underset{\text{F}}{\text{B}}}} + \underset{\text{base}}{:\overset{\text{H}}{\underset{\text{H}}{\text{N}}}\text{:H}} \longrightarrow \text{F:}\overset{\text{F H}}{\underset{\text{F H}}{\text{B:N}}}\text{:H}$$

These three theories, which explain how acid–base reactions occur, are summarized in Table 15.1. We will generally use the theory that best explains the reaction under consideration. Most of our examples will refer to aqueous solutions. Note that in an aqueous acidic solution the H^+ ion concentration is always greater than OH^- ion concentration. And vice versa—in an aqueous basic solution the OH^- ion concentration is always greater than the H^+ ion concentration. When the H^+ and OH^- ion concentrations in a solution are equal, the solution is *neutral*; that is, it is neither acidic nor basic.

15.2 Reactions of Acids

In aqueous solutions, the H^+ or H_3O^+ ions are responsible for the characteristic reactions of acids. The following reactions are in an aqueous medium:

Reaction with Metals Acids react with metals that lie above hydrogen in the activity series of elements to produce hydrogen and an ionic compound (salt) (see Section 17.5):

acid + metal $\longrightarrow$ hydrogen + ionic compound

$2\,HCl(aq) + Ca(s) \longrightarrow H_2(g) + CaCl_2(aq)$

$H_2SO_4(aq) + Mg(s) \longrightarrow H_2(g) + MgSO_4(aq)$

$6\,HC_2H_3O_2(aq) + 2\,Al(s) \longrightarrow 3\,H_2(g) + 2\,Al(C_2H_3O_2)_3(aq)$

Acids such as nitric acid (HNO_3) are oxidizing substances (see Chapter 17) and react with metals to produce water instead of hydrogen. For example,

$3\,Zn(s) + 8\,HNO_3(\text{dilute}) \longrightarrow 3\,Zn(NO_3)_2(aq) + 2\,NO(g) + 4\,H_2O(l)$

Reaction with Bases The interaction of an acid and a base is called a *neutralization reaction*. In aqueous solutions, the products of this reaction are a salt and water:

$$acid + base \longrightarrow salt + water$$
$$HBr(aq) + KOH(aq) \longrightarrow KBr(aq) + H_2O(l)$$
$$2\,HNO_3(aq) + Ca(OH)_2(aq) \longrightarrow Ca(NO_3)_2(aq) + 2\,H_2O(l)$$
$$2\,H_3PO_4(aq) + 3\,Ba(OH)_2(aq) \longrightarrow Ba_3(PO_4)_2(s) + 6\,H_2O(l)$$

Reaction with Metal Oxides This reaction is closely related to that of an acid with a base. With an aqueous acid solution, the products are a salt and water:

$$acid + metal\ oxide \longrightarrow salt + water$$
$$2\,HCl(aq) + Na_2O(s) \longrightarrow 2\,NaCl(aq) + H_2O(l)$$
$$H_2SO_4(aq) + MgO(s) \longrightarrow MgSO_4(aq) + H_2O(l)$$
$$6\,HCl(aq) + Fe_2O_3(s) \longrightarrow 2\,FeCl_3(aq) + 3\,H_2O(l)$$

Reaction with Carbonates Many acids react with carbonates to produce carbon dioxide, water, and an ionic compound:

Carbonic acid (H_2CO_3) is not the product because it is unstable and spontaneously decomposes into water and carbon dioxide.

$$H_2CO_3(aq) \longrightarrow CO_2(g) + H_2O(l)$$
$$acid + carbonate \longrightarrow salt + water + carbon\ dioxide$$
$$2\,HCl(aq) + Na_2CO_3(aq) \longrightarrow 2\,NaCl(aq) + H_2O(l) + CO_2(g)$$
$$H_2SO_4(aq) + MgCO_3(s) \longrightarrow MgSO_4(aq) + H_2O(l) + CO_2(g)$$
$$HCl(aq) + NaHCO_3(aq) \longrightarrow NaCl(aq) + H_2O(l) + CO_2(g)$$

15.3 Reactions of Bases

The OH^- ions are responsible for the characteristic reactions of bases. The following reactions are in an aqueous medium:

Reaction with Acids Bases react with acids to produce a salt and water. See reaction of acids with bases in Section 15.2.

amphoteric

Amphoteric Hydroxides Hydroxides of certain metals, such as zinc, aluminum, and chromium, are **amphoteric;** that is, they are capable of reacting as either an acid or a base. When treated with a strong acid, they behave like bases; when reacted with a strong base, they behave like acids:

Strong acids and bases are discussed in Section 15.7.

$$Zn(OH)_2(s) + 2\,HCl(aq) \longrightarrow ZnCl_2(aq) + 2\,H_2O(l)$$
$$Zn(OH)_2(s) + 2\,NaOH(aq) \longrightarrow Na_2Zn(OH)_4(aq)$$

Reaction of NaOH and KOH with Certain Metals Some amphoteric metals react directly with the strong bases sodium hydroxide and potassium hydroxide to produce hydrogen:

$$base + metal + water \longrightarrow salt + hydrogen$$
$$2\,NaOH(aq) + Zn(s) + 2\,H_2O(l) \longrightarrow Na_2Zn(OH)_4(aq) + H_2(g)$$
$$2\,KOH(aq) + 2\,Al(s) + 6\,H_2O(l) \longrightarrow 2\,KAl(OH)_4(aq) + 3\,H_2(g)$$

Practice 15.3

Write the formulas and names of the acid and base from which these salts are formed.
(a) K_3PO_4 (b) $MgBr_2$ (c) $LiCl$ (d) $FeCO_3$

Try this experiment with your friends: Chill some soft drinks well. Next, pour the soft drink into a glass. Stick your tongue into the liquid and time how long you can keep it there.

What causes the tingling we feel in our mouth (or on our tongue)? Many people believe it is the bubbles of carbon dioxide, but scientists have found that is not the answer. The tingling is caused by "chemisthesis." Bruce Bryant at Monell Chemical Senses Center in Philadelphia says that chemisthesis is a chemically induced sensation that does not involve taste or odor receptors. The tongue-tingling response to the soft drink is caused by production of protons (H^+ ions) released when an enzyme (carbonic anhydrase) acts on CO_2. The H^+ ions acidify nerve endings, producing the sensation of tingling.

Carbon dioxide also stimulates other neurons when the drink is cold. This means that at a constant pressure of CO_2, a cold drink will produce more tingling than a room-temperature drink. If the drink is at room temperature, CO_2

Tongue papillae contain nerve endings that send temperature and tactile information to the brain.

will increase the cool feeling. Chilling a soda increases the effect of the protons on the nerve endings. At the same time, the high concentration of CO_2 in a freshly opened soda makes the cold soda feel even colder. Together these chemical effects make it really painful to keep your tongue in the chilled soda.

Chemisthesis is important for survival, as well as for having fun with friends.

Bryant says, "It tells you that something is chemically impinging on your body, that tissue is in imminent danger. Burrowing animals can sense toxic levels of carbon dioxide and even feel a sting when exposed to those levels."

15.4 Salts

Salts are very abundant in nature. Most of the rocks and minerals of Earth's mantle are salts of one kind or another. Huge quantities of dissolved salts also exist in the oceans. Salts can be considered compounds derived from acids and bases. They consist of positive metal or ammonium ions combined with negative nonmetal ions (OH^- and O^{2-} excluded). The positive ion is the base counterpart and the nonmetal ion is the acid counterpart:

Chemists use the terms *ionic compound* and *salt* interchangeably.

Salts are usually crystalline and have high melting and boiling points.

From a single acid such as hydrochloric acid (HCl), we can produce many chloride compounds by replacing the hydrogen with metal ions (e.g., NaCl, KCl, RbCl, $CaCl_2$, $NiCl_2$). Hence the number of known salts greatly exceeds the number of known acids and bases. If the hydrogen atoms of a binary acid are replaced by a nonmetal, the resulting compound has covalent bonding and is therefore not considered to be ionic (e.g., PCl_3, S_2Cl_2, Cl_2O, NCl_3, ICl).

You may want to review Chapter 6 for nomenclature of acids, bases, and salts.

15.5 Electrolytes and Nonelectrolytes

We can show that solutions of certain substances are conductors of electricity with a simple conductivity apparatus, which consists of a pair of electrodes connected to a voltage source through a light bulb and switch (see Figure 15.1). If the medium between the electrodes is a conductor of electricity, the light bulb will glow when the switch is closed. When chemically pure water is placed in the beaker and the switch is closed, the light does not glow, indicating that water is a virtual nonconductor. When we dissolve a small amount of sugar in the water and test the solution, the light still does not glow, showing that a sugar solution is also a nonconductor. But when a small amount of salt, NaCl, is dissolved in water and this solution is tested, the light glows brightly. Thus the salt solution conducts electricity. A fundamental difference exists between the chemical bonding in sugar and that in salt. Sugar is a covalently bonded (molecular) substance; common salt is a substance with ionic bonds.

electrolyte
nonelectrolyte

Substances whose aqueous solutions are conductors of electricity are called **electrolytes.** Substances whose solutions are nonconductors are known as **nonelectrolytes.** The classes of compounds that are electrolytes are acids, bases, and other ionic compounds (salts). Solutions of certain oxides also are conductors because the oxides form an acid or a base when dissolved in water. One major difference between electrolytes and nonelectrolytes is that electrolytes are capable of producing ions in solution, whereas nonelectrolytes do not have this property. Solutions that contain a sufficient number of ions will conduct an electric current. Although pure water is essentially a nonconductor, many city water supplies contain enough dissolved ionic matter to cause the light to glow dimly when the water is tested in a conductivity apparatus. Table 15.2 lists some common electrolytes and nonelectrolytes.

Acids, bases, and salts are electrolytes.

(a)   (b)   (c)

Figure 15.1
A conductivity apparatus shows the difference in conductivity of solutions. (a) Distilled water does not conduct electricity. (b) Sugar water is a nonelectrolyte. (c) Salt water is a strong electrolyte and conducts electricity.

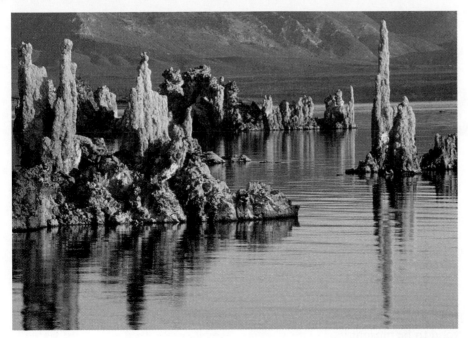

These strange mineral formations called "tufa" exist at Mono Lake, California. Tufa is formed by water bubbling through sand saturated with NaCl, Na_2CO_3, and Na_2SO_4.

Table 15.2 Representative Electrolytes and Nonelectrolytes

Electrolytes		Nonelectrolytes	
H_2SO_4	$HC_2H_3O_2$	$C_{12}H_{22}O_{11}$ (sugar)	CH_3OH (methyl alcohol)
HCl	NH_3	C_2H_5OH (ethyl alcohol)	$CO(NH_2)_2$ (urea)
HNO_3	K_2SO_4	$C_2H_4(OH)_2$ (ethylene glycol)	O_2
NaOH	$NaNO_3$	$C_3H_5(OH)_3$ (glycerol)	H_2O

15.6 Dissociation and Ionization of Electrolytes

Arrhenius received the 1903 Nobel Prize in chemistry for his work on electrolytes. He found that a solution conducts electricity because the solute dissociates immediately upon dissolving into electrically charged particles (ions). The movement of these ions toward oppositely charged electrodes causes the solution to be a conductor. According to his theory, solutions that are relatively poor conductors contain electrolytes that are only partly dissociated. Arrhenius also believed that ions exist in solution whether or not an electric current is present. In other words, the electric current does not cause the formation of ions. Remember that positive ions are cations; negative ions are anions.

We have seen that sodium chloride crystals consist of sodium and chloride ions held together by ionic bonds. **Dissociation** is the process by which the ions of a salt separate as the salt dissolves. When placed in water, the sodium and chloride ions are attracted by the polar water molecules, which surround each ion as it dissolves. In water, the salt dissociates, forming hydrated sodium

dissociation

Figure 15.2
Hydrated sodium and chloride ions. When sodium chloride dissolves in water, each Na^+ and Cl^- ion becomes surrounded by water molecules. The negative end of the water dipole is attracted to the Na^+ ion, and the positive end is attracted to the Cl^- ion.

and chloride ions (see Figure 15.2). The sodium and chloride ions in solution are surrounded by a specific number of water molecules and have less attraction for each other than they had in the crystalline state. The equation representing this dissociation is

$$NaCl(s) + (x + y)\,H_2O \longrightarrow Na^+(H_2O)_x + Cl^-(H_2O)_y$$

A simplified dissociation equation in which the water is omitted but understood to be present is

$$NaCl(s) \longrightarrow Na^+(aq) + Cl^-(aq)$$

Remember that sodium chloride exists in an aqueous solution as hydrated ions, not as NaCl units, even though the formula NaCl (or $Na^+ + Cl^-$) is often used in equations.

The chemical reactions of salts in solution are the reactions of their ions. For example, when sodium chloride and silver nitrate react and form a precipitate of silver chloride, only the Ag^+ and Cl^- ions participate in the reaction. The Na^+ and NO_3^- remain as ions in solution:

$$Ag^+(aq) + Cl^-(aq) \longrightarrow AgCl(s)$$

ionization   **Ionization** is the formation of ions; it occurs as a result of a chemical reaction of certain substances with water. Glacial acetic acid (100% $HC_2H_3O_2$) is a liquid that behaves as a nonelectrolyte when tested by the method described in Section 15.5. But a water solution of acetic acid conducts an electric current (as indicated by the dull-glowing light of the conductivity apparatus). The equation for the reaction with water, which forms hydronium and acetate ions, is

$$\underset{\text{acid}}{HC_2H_3O_2} + \underset{\text{base}}{H_2O} \rightleftharpoons \underset{\text{acid}}{H_3O^+} + \underset{\text{base}}{C_2H_3O_2^-}$$

or, in the simplified equation,

$$HC_2H_3O_2 \rightleftharpoons H^+ + C_2H_3O_2^-$$

In this ionization reaction, water serves not only as a solvent but also as a base according to the Brønsted–Lowry theory.

Hydrogen chloride is predominantly covalently bonded, but when dissolved in water, it reacts to form hydronium and chloride ions:

$$HCl(g) + H_2O(l) \longrightarrow H_3O^+(aq) + Cl^-(aq)$$

When a hydrogen chloride solution is tested for conductivity, the light glows brilliantly, indicating many ions in the solution.

Ionization occurs in each of the preceding two reactions with water, producing ions in solution. The necessity for water in the ionization process can be demonstrated by dissolving hydrogen chloride in a nonpolar solvent such as hexane and testing the solution for conductivity. The solution fails to conduct electricity, indicating that no ions are produced.

The terms *dissociation* and *ionization* are often used interchangeably to describe processes taking place in water. But, strictly speaking, the two are different. In the dissociation of a salt, the salt already exists as ions; when it dissolves in water, the ions separate, or dissociate, and increase in mobility. In the ionization process, ions are produced by the reaction of a compound with water.

15.7 Strong and Weak Electrolytes

Electrolytes are classified as strong or weak depending on the degree, or extent, of dissociation or ionization. **Strong electrolytes** are essentially 100% ionized in solution; **weak electrolytes** are much less ionized (based on comparing 0.1 M solutions). Most electrolytes are either strong or weak, with a few classified as moderately strong or weak. Most salts are strong electrolytes. Acids and bases that are strong electrolytes (highly ionized) are called *strong acids* and *strong bases*. Acids and bases that are weak electrolytes (slightly ionized) are called *weak acids* and *weak bases*.

strong electrolyte
weak electrolyte

For equivalent concentrations, solutions of strong electrolytes contain many more ions than do solutions of weak electrolytes. As a result, solutions of strong electrolytes are better conductors of electricity. Consider two solutions, 1 M HCl and 1 M HC$_2$H$_3$O$_2$. Hydrochloric acid is almost 100% ionized; acetic acid is about 1% ionized. (See Figure 15.3.) Thus, HCl is a strong acid and HC$_2$H$_3$O$_2$ is a weak acid. Hydrochloric acid has about 100 times as many hydronium ions in solution as acetic acid, making the HCl solution much more acidic.

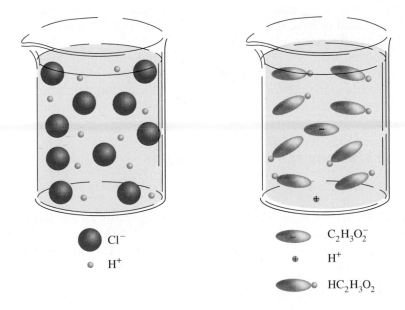

Cl$^-$	
H$^+$	

C$_2$H$_3$O$_2^-$	
H$^+$	
HC$_2$H$_3$O$_2$	

Figure 15.3
HCl solution (left) is 100% ionized, while in HC$_2$H$_3$O$_2$ solution (right) almost all of the solute is in molecular form. HCl is a strong acid, while HC$_2$H$_3$O$_2$ is a weak acid. Note: The water molecules in the solution are not shown in this figure.

$HC_2H_3O_2$

$C_2H_3O_2^-$

We can distinguish between strong and weak electrolytes experimentally using the apparatus described in Section 15.5. A 1 M HCl solution causes the light to glow brilliantly, but a 1 M $HC_2H_3O_2$ solution causes only a dim glow. The strong base sodium hydroxide (NaOH) can be distinguished in a similar fashion from the weak base ammonia (NH_3). The ionization of a weak electrolyte in water is represented by an equilibrium equation showing that both the un-ionized and ionized forms are present in solution. In the equilibrium equation of $HC_2H_3O_2$ and its ions, we say that the equilibrium lies "far to the left" because relatively few hydrogen and acetate ions are present in solution:

$$HC_2H_3O_2(aq) \rightleftharpoons H^+(aq) + C_2H_3O_2^-(aq)$$

We have previously used a double arrow in an equation to represent reversible processes in the equilibrium between dissolved and undissolved solute in a saturated solution. A double arrow ($\rightleftharpoons$) is also used in the ionization equation of soluble weak electrolytes to indicate that the solution contains a considerable amount of the un-ionized compound in equilibrium with its ions in solution. (See Section 16.1 for a discussion of reversible reactions.) A single arrow is used to indicate that the electrolyte is essentially all in the ionic form in the solution. For example, nitric acid is a strong acid; nitrous acid is a weak acid. Their ionization equations in water may be indicated as

$$HNO_3(aq) \xrightarrow{H_2O} H^+(aq) + NO_3^-(aq)$$

$$HNO_2(aq) \xrightleftharpoons{H_2O} H^+(aq) + NO_2^-(aq)$$

Practically all soluble salts, acids (such as sulfuric, nitric, and hydrochloric acids), and bases (such as sodium, potassium, calcium, and barium hydroxides) are strong electrolytes. Weak electrolytes include numerous other acids and bases such as acetic acid, nitrous acid, carbonic acid, and ammonia. The terms *strong acid, strong base, weak acid,* and *weak base* refer to whether an acid or base is a strong or weak electrolyte. A brief list of strong and weak electrolytes is given in Table 15.3.

Electrolytes yield two or more ions per formula unit upon dissociation—the actual number being dependent on the compound. Dissociation is complete or nearly complete for nearly all soluble ionic compounds and for certain other strong electrolytes, such as those given in Table 15.3. The following

Table 15.3 Strong and Weak Electrolytes

Strong electrolytes		Weak electrolytes	
Most soluble salts	$HClO_4$	$HC_2H_3O_2$	$H_2C_2O_4$
H_2SO_4	NaOH	H_2CO_3	H_3BO_3
HNO_3	KOH	HNO_2	HClO
HCl	$Ca(OH)_2$	H_2SO_3	NH_3
HBr	$Ba(OH)_2$	H_2S	HF

are dissociation equations for several strong electrolytes. In all cases, the ions are actually hydrated:

$$NaOH \xrightarrow{H_2O} Na^+(aq) + OH^-(aq) \qquad \text{2 ions in solution per formula unit}$$

$$Na_2SO_4 \xrightarrow{H_2O} 2\,Na^+(aq) + SO_4^{2-}(aq) \qquad \text{3 ions in solution per formula unit}$$

$$Fe_2(SO_4)_3 \xrightarrow{H_2O} 2\,Fe^{3+}(aq) + 3\,SO_4^{2-}(aq)$$
$$\text{5 ions in solution per formula unit}$$

One mole of NaCl will give 1 mol of Na^+ ions and 1 mol of Cl^- ions in solution, assuming complete dissociation of the salt. One mole of $CaCl_2$ will give 1 mol of Ca^{2+} ions and 2 mol of Cl^- ions in solution:

$$NaCl \xrightarrow{H_2O} Na^+(aq) + Cl^-(aq)$$
1 mol   1 mol   1 mol

$$CaCl_2 \xrightarrow{H_2O} Ca^{2+}(aq) + 2\,Cl^-(aq)$$
1 mol   1 mol   2 mol

What is the molarity of each ion in a solution of (a) 2.0 M NaCl and (b) 0.40 M K_2SO_4? Assume complete dissociation.

Example 15.2

(a) According to the dissociation equation,

SOLUTION

$$NaCl \xrightarrow{H_2O} Na^+(aq) + Cl^-(aq)$$
1 mol   1 mol   1 mol

the concentration of Na^+ is equal to that of NaCl: 1 mol NaCl $\longrightarrow$ 1 mol Na^+ and the concentration of Cl^- is also equal to that of NaCl. Therefore the concentrations of the ions in 2.0 M NaCl are 2.0 M Na^+ and 2.0 M Cl^-.

(b) According to the dissociation equation

$$K_2SO_4 \xrightarrow{H_2O} 2\,K^+(aq) + SO_4^{2-}(aq)$$
1 mol   2 mol   1 mol

the concentration of K^+ is twice that of K_2SO_4 and the concentration of SO_4^{2-} is equal to that of K_2SO_4. Therefore the concentrations of the ions in 0.40 M K_2SO_4 are 0.80 M K^+ and 0.40 M SO_4^{2-}.

Practice 15.4

What is the molarity of each ion in a solution of (a) 0.050 M $MgCl_2$ and (b) 0.070 M $AlCl_3$?

Colligative Properties of Electrolyte Solutions

We have learned that when 1 mol of sucrose, a nonelectrolyte, is dissolved in 1000 g of water, the solution freezes at $-1.86°C$. When 1 mol NaCl is dissolved in 1000 g of water, the freezing point of the solution is not $-1.86°C$, as might be expected, but is closer to $-3.72°C$ (-1.86×2). The reason for the lower freezing point is that 1 mol NaCl in solution produces 2 mol of particles ($2 \times 6.022 \times 10^{23}$ ions) in solution. Thus the freezing point depression produced by 1 mol NaCl is essentially equivalent to that produced by 2 mol of a nonelectrolyte. An electrolyte such as $CaCl_2$, which yields three ions in water, gives a freezing point depression of about three times that of a nonelectrolyte. These freezing point data provide additional evidence that electrolytes dissociate when dissolved in water. The other colligative properties are similarly affected by substances that yield ions in aqueous solutions.

Practice 15.5

What is the boiling point of a 1.5 m solution of KCl(aq)?

15.8 Ionization of Water

Pure water is a *very* weak electrolyte, but it does ionize slightly. Two equations commonly used to show how water ionizes are

$$H_2O + H_2O \rightleftharpoons H_3O^+ + OH^-$$
$$\text{acid} \quad\ \text{base} \quad\quad\ \text{acid} \quad\ \text{base}$$

and

$$H_2O \rightleftharpoons H^+ + OH^-$$

The first equation represents the Brønsted–Lowry concept, with water reacting as both an acid and a base, forming a hydronium ion and a hydroxide ion. The second equation is a simplified version, indicating that water ionizes to give a hydrogen and a hydroxide ion. Actually, the proton (H^+) is hydrated and exists as a hydronium ion. In either case equal molar amounts of acid and base are produced so that water is neutral, having neither H^+ nor OH^- ions in excess. The ionization of water at 25°C produces an H^+ ion concentration of 1.0×10^{-7} mol/L and an OH^- ion concentration of 1.0×10^{-7} mol/L. Square brackets, [ ], are used to indicate that the concentration is in moles per liter. Thus [H^+] means the concentration of H^+ is in moles per liter. These concentrations are usually expressed as

$$[H^+] \text{ or } [H_3O^+] = 1.0 \times 10^{-7} \text{ mol/L}$$
$$[OH^-] = 1.0 \times 10^{-7} \text{ mol/L}$$

These figures mean that about two out of every billion water molecules are ionized. This amount of ionization, small as it is, is a significant factor in the behavior of water in many chemical reactions.

15.9 Introduction to pH

The acidity of an aqueous solution depends on the concentration of hydrogen or hydronium ions. The pH scale of acidity provides a simple, convenient, numerical way to state the acidity of a solution. Values on the pH scale are obtained by mathematical conversion of H^+ ion concentrations to pH by the expression

$$pH = -\log[H^+]$$

where $[H^+]$ = H^+ or H_3O^+ ion concentration in moles per liter. The **pH** is defined as the *negative* logarithm of the H^+ or H_3O^+ concentration in moles per liter:

pH

$$pH = -\log[H^+] = -\log(1 \times 10^{-7}) = -(-7) = 7$$

For example, the pH of pure water at 25°C is 7 and is said to be neutral; that is, it is neither acidic nor basic, because the concentrations of H^+ and OH^- are equal. Solutions that contain more H^+ ions than OH^- ions have pH values less than 7, and solutions that contain less H^+ ions than OH^- ions have pH values greater than 7.

> pH < 7.00 is an acidic solution
>
> pH = 7.00 is a neutral solution
>
> pH > 7.00 is a basic solution

When $[H^+] = 1 \times 10^{-5}$ mol/L, pH = 5 (acidic)

When $[H^+] = 1 \times 10^{-9}$ mol/L, pH = 9 (basic)

Instead of saying that the hydrogen ion concentration in the solution is 1×10^{-5} mol/L, it's customary to say that the pH of the solution is 5. The smaller the pH value, the more acidic the solution (see Figure 15.4).

The pH scale, along with its interpretation, is given in Table 15.4, and Table 15.5 lists the pH of some common solutions. Note that a change of only one pH unit means a 10-fold increase or decrease in H^+ ion concentration. For example, a solution with a pH of 3.0 is 10 times more acidic than a solution with a pH of 4.0. A simplified method of determining pH from $[H^+]$ follows:

$$[H^+] = 1 \times 10^{-5} \quad \longleftarrow \quad \text{pH = this number (5)}$$
$$\text{pH = 5}$$

when this number
is exactly 1

$$[H^+] = 2 \times 10^{-5} \quad \longleftarrow \quad \text{pH is between this number and}$$
next lower number (4 and 5)
$$\text{pH = 4.7}$$

when this number
is between 1 and 10

Calculating the pH value for H^+ ion concentrations requires the use of logarithms, which are exponents. The **logarithm** (log) of a number is simply the power to which 10 must be raised to give that number. Thus the log of 100 is 2 ($100 = 10^2$), and the log of 1000 is 3 ($1000 = 10^3$). The log of 500 is 2.70, but you can't determine this value easily without a scientific calculator.

logarithm

Help on using calculators is found in Appendix II.

Figure 15.4
The pH scale of acidity and basicity.

Table 15.4 pH Scale for Expressing Acidity

[H$^+$] (mol/L)	pH	
1 × 10^{-14}	14	↑
1 × 10^{-13}	13	
1 × 10^{-12}	12	Increasing
1 × 10^{-11}	11	basicity
1 × 10^{-10}	10	
1 × 10^{-9}	9	
1 × 10^{-8}	8	
1 × 10^{-7}	7	Neutral
1 × 10^{-6}	6	
1 × 10^{-5}	5	
1 × 10^{-4}	4	
1 × 10^{-3}	3	Increasing
1 × 10^{-2}	2	acidity
1 × 10^{-1}	1	
1 × 10^{0}	0	↓

Table 15.5 The pH of Common Solutions

Solution	pH
Gastric juice	1.0
0.1 M HCl	1.0
Lemon juice	2.3
Vinegar	2.8
0.1 M HC$_2$H$_3$O$_2$	2.9
Orange juice	3.7
Tomato juice	4.1
Coffee, black	5.0
Urine	6.0
Milk	6.6
Pure water (25°C)	7.0
Blood	7.4
Household ammonia	11.0
1 M NaOH	14.0

Remember: Change the sign on your calculator since pH = –log[H$^+$].

Let's determine the pH of a solution with [H$^+$] = 2 × 10^{-5} using a calculator. Enter 2 × 10^{-5} into your calculator and press the log key. The number −4.69... will be displayed. The pH is then

$$pH = -\log[H^+] = -(-4.69\ldots) = 4.7$$

Next we must determine the correct number of significant figures in the logarithm. The rules for logs are different from those we use in other math operations. The number of decimal places for a log must equal the number of significant figures in the original number. Since 2 × 10^{-5} has one significant figure, we should round the log to one decimal place (4.69...) = 4.7.

Example 15.3   What is the pH of a solution with an [H$^+$] of (a) 1.0 × 10^{-11}, (b) 6.0 × 10^{-4}, and (c) 5.47 × 10^{-8}?

SOLUTION

(a) [H$^+$] = 1.0 × 10^{-11}
(2 significant figures)

pH = −log(1.0 × 10^{-11})
pH = 11.00
(2 decimal places)

(b) [H$^+$] = 6.0 × 10^{-4}
(2 significant figures)

log (6.0 × 10^{-4}) = −3.22
pH = −log[H$^+$]
pH = −(−3.22) = 3.22
(2 decimal places)

(c) [H$^+$] = 5.47 × 10^{-8}
(3 significant figures)

log (5.47 × 10^{-8}) = −7.262
pH = −log[H$^+$]
pH = −(−7.262) = 7.262
(3 decimal places)

Practice 15.6_____

What is the pH of a solution with [H$^+$] of (a) 3.9 × 10^{-12} M, (b) 1.3 × 10^{-3} M, and (c) 3.72 × 10^{-6} M?

As our fleet of commercial airplanes ages, scientists are looking for new ways to detect corrosion and stop it *before* it becomes a safety issue. The problem is that planes are subjected to heat, rain, and wind—all factors that can contribute to metal corrosion. How can a maintenance crew detect corrosion on a huge aircraft since many imaging techniques work on only small areas at a time? Imagine a plane that could tell the maintenance crew it needed repair.

Gerald Frankel and Jim Zhang from the Ohio State University have developed a paint that detects changes in pH. Corrosion of metals (in contact with air and water) results in a chemi-

Phenolphthalein paint after eight days.

cal reaction that produces hydroxide ions, which increase pH. Frankel and Zhang made a clear acrylic coating that they mixed with phenolphthalein.

This acid–base indicator turns bright fuschia when the pH increases above 8.0. The scientists can show the "paint" turns pink at corrosion sites as small as 15 μm deep.

Technicians at Wright-Patterson Air Force Base in Ohio think this could lead to a whole new way to detect corrosion. William Mullins says, "You could walk down the vehicle and see there's a pink spot." This method would be especially good at detecting corrosion concealed around rivets and where metals overlap. The only limitation appears to be that the coating must be clear.

The next time your airline maintenance crew sees pink spots, you may be in for a delay or a change of aircraft!

The measurement and control of pH is extremely important in many fields. Proper soil pH is necessary to grow certain types of plants successfully. The pH of certain foods is too acidic for some diets. Many biological processes are delicately controlled pH systems. The pH of human blood is regulated to very close tolerances through the uptake or release of H^+ by mineral ions, such as HCO_3^-, HPO_4^{2-}, and $H_2PO_4^-$, Changes in the pH of the blood by as little as 0.4 pH unit result in death.

Compounds with colors that change at particular pH values are used as indicators in acid–base reactions. For example, phenolphthalein, an organic compound, is colorless in acid solution and changes to pink at a pH of 8.3. When a solution of sodium hydroxide is added to a hydrochloric acid solution containing phenolphthalein, the change in color (from colorless to pink) indicates that all the acid is neutralized. Commercially available pH test paper contains chemical indicators. The indicator in the paper takes on different colors when wetted with solutions of different pH. Thus the pH of a solution can be estimated by placing a drop on the test paper and comparing the color of the test paper with a color chart calibrated at different pH values. Common applications of pH test indicators are the kits used to measure and adjust the pH of swimming pools, hot tubs, and saltwater aquariums. Electronic pH meters are used for making rapid and precise pH determinations (see Figure 15.5).

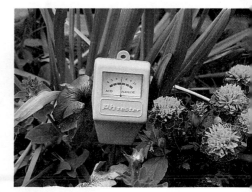
Figure 15.5
The pH of a substance (in this case garden soil) can be measured by using a pH meter.

15.10 Neutralization

The reaction of an acid and a base to form a salt and water is known as **neutralization.** We've seen this reaction before, but now with our knowledge about ions and ionization, let's reexamine the process of neutralization.

Consider the reaction that occurs when solutions of sodium hydroxide and hydrochloric acid are mixed. The ions present initially are Na^+ and OH^- from

neutralization

the base and H^+ and Cl^- from the acid. The products, sodium chloride and water, exist as Na^+ and Cl^- ions and H_2O molecules. A chemical equation representing this reaction is

$$HCl(aq) + NaOH(aq) \longrightarrow NaCl(aq) + H_2O(l)$$

This equation, however, does not show that HCl, NaOH, and NaCl exist as ions in solution. The following total ionic equation gives a better representation of the reaction:

$$(H^+ + Cl^-) + (Na^+ + OH^-) \longrightarrow Na^+ + Cl^- + H_2O(l)$$

spectator ion   This equation shows that the Na^+ and Cl^- ions did not react. These ions are called **spectator ions** because they were present but did not take part in the reaction. The only reaction that occurred was that between the H^+ and OH^- ions. Therefore the equation for the neutralization can be written as this net ionic equation:

$$H^+(aq) + OH^-(aq) \longrightarrow H_2O(l)$$
$$\text{acid} \qquad\quad \text{base} \qquad\qquad \text{water}$$

titration   This simple net ionic equation represents not only the reaction of sodium hydroxide and hydrochloric acid, but also the reaction of any strong acid with any water-soluble hydroxide base in an aqueous solution. The driving force of a neutralization reaction is the ability of an H^+ ion and an OH^- ion to react and form a molecule of water.

The amount of acid, base, or other species in a sample can be determined by **titration,** which measures the volume of one reagent required to react with a measured mass or volume of another reagent. Consider the titration of an acid with a base. A measured volume of acid of unknown concentration is placed in a flask, and a few drops of an indicator solution are added. Base solution of known concentration is slowly added from a buret to the acid until the indicator changes color. The indicator selected is one that changes color when the stoichiometric quantity (according to the equation) of base has been added to the acid. At this

(a)   (b)   (c)

Tritation. (a) Before the titration begins, the liqiuid in the flask is colorless and clear. (b) At the endpoint a pole pink color persists in the flash. (c) After the endpoint, the color of the solution turns bright pink, indicating an excess of base.

point, known as the *end point of the titration,* the titration is complete, and the volume of base used to neutralize the acid is read from the buret. The concentration or amount of acid in solution can be calculated from the titration data and the chemical equation for the reaction. Let's look at some examples.

Example 15.4

Suppose that 42.00 mL of 0.150 M NaOH solution are required to neutralize 50.00 mL of hydrochloric acid solution. What is the molarity of the acid solution?

SOLUTION

The equation for the reaction is

$$NaOH(aq) + HCl(aq) \longrightarrow NaCl(aq) + H_2O(l)$$

In this neutralization NaOH and HCl react in a 1:1 mole ratio. Therefore the moles of HCl in solution are equal to the moles of NaOH required to react with it. First we calculate the moles of NaOH used, and from this value we determine the moles of HCl:

Data: 42.00 mL of 0.150 M NaOH   50.00 mL HCl

molarity of acid = M (unknown)

Determine the moles of NaOH:

M = mol/L   42.00 mL = 0.04200 L

$$(0.04200 \, \cancel{L})\left(\frac{0.150 \, \text{mol NaOH}}{1 \, \cancel{L}}\right) = 0.00630 \, \text{mol NaOH}$$

Since NaOH and HCl react in a 1:1 ratio, 0.00630 mol HCl was present in the 50.00 mL of HCl solution. Therefore, the molarity of the HCl is

$$M = \frac{\text{mol}}{\text{L}} = \frac{0.00630 \, \text{mol HCl}}{0.05000 \, \text{L}} = 0.126 \, M \, \text{HCl}$$

Example 15.5

Suppose that 42.00 mL of 0.150 M NaOH solution are required to neutralize 50.00 mL of H_2SO_4 solution. What is the molarity of the acid solution?

SOLUTION

The equation for the reaction is

$$2 \, NaOH(aq) + H_2SO_4(aq) \longrightarrow Na_2SO_4(aq) + 2 \, H_2O(l)$$

The same amount of base (0.00630 mol NaOH) is used in this titration as in Example 15.4, but the mole ratio of acid to base in the reaction is 1:2. The moles of H_2SO_4 reacted can be calculated:

Data: 42.00 mL of 0.150 M NaOH = 0.00630 mol NaOH

$$(0.00630 \, \cancel{\text{mol NaOH}})\left(\frac{1 \, \text{mol H}_2\text{SO}_4}{2 \, \cancel{\text{mol NaOH}}}\right) = 0.00315 \, \text{mol H}_2\text{SO}_4$$

Therefore 0.00315 mol H_2SO_4 was present in 50.00 mL of H_2SO_4 solution. The molarity of the H_2SO_4 is

$$M = \frac{\text{mol}}{\text{L}} = \frac{0.00315 \, \text{mol H}_2\text{SO}_4}{0.05000 \, \text{L}} = 0.0630 \, M \, \text{H}_2\text{SO}_4$$

Example 15.6　A 25.00-mL sample of H_2SO_4 solution required 14.26 mL of 0.2240 M NaOH for complete neutralization. What is the molarity of the sulfuric acid?

SOLUTION　The equation for the reaction is

$$2\,NaOH(aq) + H_2SO_4(aq) \longrightarrow Na_2SO_4(aq) + 2\,H_2O(l)$$

The moles of NaOH needed are

$$\text{moles NaOH} = (V_{\text{NaOH}})(M_{\text{NaOH}})$$

$$= (0.01426\ \text{L})\left(0.2240\ \frac{\text{mol}}{\text{L}}\right)$$

$$= 0.003194\ \text{mol NaOH}$$

Since the mole ratio of acid to base is $\dfrac{1\,H_2SO_4}{2\,NaOH}$, the moles of acid in the sample are

$$(0.003194\ \text{mol NaOH})\left(\frac{1\,H_2SO_4}{2\,NaOH}\right) = 0.001597\ \text{mol}\ H_2SO_4$$

Now to find the molarity of the sample, we divide the moles of acid by its original volume:

$$\frac{0.001597\ \text{mol}\ H_2SO_4}{0.02500\ \text{L}} = 0.06388\ M\ H_2SO_4$$

Practice 15.7_____

A 50.0-mL sample of HCl required 24.81 mL of 0.1250 M NaOH for neutralization. What is the molarity of the acid?

15.11　Writing Net Ionic Equations

In Section 15.10, we wrote the reaction of hydrochloric acid and sodium hydroxide in three different equations:

1. $HCl(aq) + NaOH(aq) \longrightarrow NaCl(aq) + H_2O(l)$

2. $(H^+ + Cl^-) + (Na^+ + OH^-) \longrightarrow Na^+ + Cl^- + H_2O(l)$

3. $H^+ + OH^- \longrightarrow H_2O$

formula equation
total ionic equation

In the **formula equation** (1), compounds are written in their molecular, or formula, expressions. In the **total ionic equation** (2), compounds are written to show the form in which they are predominantly present: strong electrolytes as ions in solution and nonelectrolytes, weak electrolytes, precipitates, and gases in their molecular forms. In the **net ionic equation** (3), only those molecules or ions that have reacted are included in the equation; ions or molecules that do not react (the spectators) are omitted.

net ionic equation

When balancing equations thus far, we've been concerned only with the atoms of the individual elements. Because ions are electrically charged, ionic equations often end up with a net electrical charge. A balanced equation must have the same net charge on each side, whether that charge is positive, negative, or zero. Therefore, when balancing ionic equations, we must make sure that both the same number of each kind of atom and the same net electrical charge are present on each side.

Here is a list of rules for writing ionic equations:

1. Strong electrolytes in solution are written in their ionic form.

2. Weak electrolytes are written in their molecular form.

3. Nonelectrolytes are written in their molecular form.

4. Insoluble substances, precipitates, and gases are written in their molecular forms.

5. The net ionic equation should include only substances that have undergone a chemical change. Spectator ions are omitted from the net ionic equation.

6. Equations must be balanced, both in atoms and in electrical charge.

Study the following examples. In each one, the formula equation is given. Write the total ionic equation and the net ionic equation for each.

$$HNO_3(aq) + KOH(aq) \longrightarrow KNO_3(aq) + H_2O(l)$$
formula equation

$$(H^+ + NO_3^-) + (K^+ + OH^-) \longrightarrow K^+ + NO_3^- + H_2O$$
total ionic equation

$$H^+ + OH^- \longrightarrow H_2O$$
net ionic equation

Example 15.7

SOLUTION

The HNO_3, KOH, and KNO_3 are soluble, strong electrolytes. The K^+ and NO_3^- ions are spectator ions, have not changed, and are not included in the net ionic equation. Water is a nonelectrolyte and is written in the molecular form.

$$2\,AgNO_3(aq) + BaCl_2(aq) \longrightarrow 2\,AgCl(s) + Ba(NO_3)_2(aq)$$
formula equation

$$(2\,Ag^+ + 2\,NO_3^-) + (Ba^{2+} + 2\,Cl^-) \longrightarrow 2\,AgCl(s) + Ba^{2+} + 2\,NO_3^-$$
total ionic equation

$$Ag^+ + Cl^- \longrightarrow AgCl(s)$$
net ionic equation

Example 15.8

SOLUTION

Although AgCl is an ionic compound, it is written in the un-ionized form on the right side of the ionic equations because most of the Ag^+ and Cl^- ions are no longer in solution but have formed a precipitate of AgCl. The Ba^{2+} and NO_3^- ions are spectator ions.

Example 15.9

$$Na_2CO_3(aq) + H_2SO_4(aq) \longrightarrow Na_2SO_4(aq) + H_2O(l) + CO_2(g)$$
formula equation

SOLUTION

$$(2\,Na^+ + CO_3^{2-}) + (2\,H^+ + SO_4^{2-}) \longrightarrow 2\,Na^+ + SO_4^{2-} + H_2O(l) + CO_2(g)$$
total ionic equation

$$CO_3^{2-} + 2\,H^+ \longrightarrow H_2O(l) + CO_2(g)$$
net ionic equation

Carbon dioxide (CO_2) is a gas and evolves from the solution; Na^+ and SO_4^{2-} are spectator ions.

Example 15.10

$$HC_2H_3O_2(aq) + NaOH(aq) \longrightarrow NaC_2H_3O_2(aq) + H_2O(l)$$
formula equation

SOLUTION

$$HC_2H_3O_2 + (Na^+ + OH^-) \longrightarrow Na^+ + C_2H_3O_2^- + H_2O$$
total ionic equation

$$HC_2H_3O_2 + OH^- \longrightarrow C_2H_3O_2^- + H_2O$$
net ionic equation

Acetic acid, $HC_2H_3O_2$, a weak acid, is written in the molecular form, but sodium acetate, $NaC_2H_3O_2$, a soluble salt, is written in the ionic form. The Na^+ ion is the only spectator ion in this reaction. Both sides of the net ionic equation have a -1 electrical charge.

Example 15.11

$$Mg(s) + 2\,HCl(aq) \longrightarrow MgCl_2(aq) + H_2(g)$$
formula equation

SOLUTION

$$Mg + (2\,H^+ + 2\,Cl^-) \longrightarrow Mg^{2+} + 2\,Cl^- + H_2(g)$$
total ionic equation

$$Mg + 2\,H^+ \longrightarrow Mg^{2+} + H_2(g)$$
net ionic equation

The net electrical charge on both sides of the equation is $+2$.

Example 15.12

$$H_2SO_4(aq) + Ba(OH)_2(aq) \longrightarrow BaSO_4(s) + 2\,H_2O(l)$$
formula equation

SOLUTION

$$(2\,H^+ + SO_4^{2-}) + (Ba^{2+} + 2\,OH^-) \longrightarrow BaSO_4(s) + 2\,H_2O(l)$$
total ionic equation

$$2\,H^+ + SO_4^{2-} + Ba^{2+} + 2\,OH^- \longrightarrow BaSO_4(s) + 2\,H_2O(l)$$
net ionic equation

Barium sulfate ($BaSO_4$) is a highly insoluble salt. If we conduct this reaction using the conductivity apparatus described in Section 15.5, the light glows brightly at first but goes out when the reaction is complete because almost no ions are left in solution. The $BaSO_4$ precipitates out of solution, and water is a nonconductor of electricity.

Practice 15.8_____

Write the net ionic equation for
$$3\,H_2S(aq) + 2\,Bi(NO_3)_3(aq) \longrightarrow Bi_2S_3(s) + 6\,HNO_3(aq)$$

15.12 Acid Rain

Acid rain is defined as any atmospheric precipitation that is more acidic than usual. The increase in acidity might be from natural or industrial sources. Rain acidity varies throughout the world and across the United States. The pH of rain is generally lower in the eastern United States and higher in the West. Unpolluted rain has a pH of 5.6, and so is slightly acidic. This acidity results from the dissolution of carbon dioxide in the water producing carbonic acid:

$$CO_2(g) + H_2O(l) \rightleftharpoons H_2CO_3(aq) \rightleftharpoons H^+(aq) + HCO_3^-(aq)$$

The general process involves the following steps:

1. emission of nitrogen and sulfur oxides into the air
2. transportation of these oxides throughout the atmosphere
3. chemical reactions between the oxides and water, forming sulfuric acid (H_2SO_4) and nitric acid (HNO_3)
4. rain or snow, which carries the acids to the ground

The oxides may also be deposited directly on a dry surface and become acidic when normal rain falls on them.

Acid rain is not a new phenomenon. Rain was probably acidic in the early days of our planet as volcanic eruptions, fires, and decomposition of organic matter released large volumes of nitrogen and sulfur oxides into the atmosphere. Use of fossil fuels, especially since the Industrial Revolution, has made significant changes in the amounts of pollutants being released into the atmosphere. As increasing amounts of fossil fuels have been burned, more and more sulfur and nitrogen oxides have poured into the atmosphere, thus increasing the acidity of rain.

This relief work on St. Bartholomew's Church in New York shows the destructive power of acid rain. The marble is being dissolved slowly over time.

Acid rain affects a variety of factors in our environment. For example, freshwater plants and animals decline significantly when rain is acidic; large numbers of fish and plants die when acidic water from spring thaws enters the lakes. Acidic rainwater leaches aluminum from the soil into lakes, where the aluminum compounds adversely affect the gills of fish. In addition to leaching aluminum from the soil, acid rain also causes other valuable minerals, such as magnesium and calcium, to dissolve and run into lakes and streams. It can also dissolve the waxy protective coat on plant leaves, making them vulnerable to attack by bacteria and fungi.

In our cities, acid rain is responsible for extensive and continuing damage to buildings, monuments, and statues. It reduces the durability of paint and promotes the deterioration of paper, leather, and cloth. In short, we are just beginning to explore the effects of acid rain on human beings and on our food chain.

15.13 Colloids

When we add sugar to a flask of water and shake it, the sugar dissolves and forms a clear homogeneous *solution*. When we do the same experiment with very fine sand and water, the sand particles form a *suspension,* which settles when the shaking stops. When we repeat the experiment again using ordinary cornstarch, we find that the starch does not dissolve in cold water. But if the mixture is heated and stirred, the starch forms a cloudy, opalescent *dispersion.* This dispersion does not appear to be clear and homogeneous like the sugar solution, yet it is not obviously heterogeneous and does not settle like the sand suspension. In short, its properties are intermediate between those of the sugar solution and those of the sand suspension. The starch dispersion is actually a *colloid,* a name derived from the Greek *kolla,* meaning "glue," and was coined by the English scientist Thomas Graham in 1861.

colloid   As it is now used, the word **colloid** means a dispersion in which the dispersed particles are larger than the solute ions or molecules of a true solution and smaller than the particles of a mechanical suspension. The term does not imply a gluelike quality, although most glues are colloidal materials. The size of colloidal particles ranges from a lower limit of about 1 nm (10^{-7} cm) to an upper limit of about 1000 nm (10^{-4} cm). There are eight types of colloids, which are summarized in Table 15.6.

Table 15.6 Types of Colloidal Dispersions

Type	Name	Examples
Gas in liquid	foam	whipped cream, soapsuds
Gas in solid	solid foam	Styrofoam, foam rubber, pumice
Liquid in gas	liquid aerosol	fog, clouds
Liquid in liquid	emulsion	milk, vinegar in oil salad dressing, mayonnaise
Liquid in solid	solid emulsion	cheese, opals, jellies
Solid in gas	solid aerosol	smoke, dust in air
Solid in liquid	sol	india ink, gold sol
Solid in solid	solid sol	tire rubber, certain gems (e.g., rubies)

Metal foams are a new class of materials that may revolutionize the car industry. Automotive makers know that one key to making cars that are fuel efficient is to decrease weight. Up till now, that weight reduction meant higher cost (in materials like titanium and aluminum) and problems in crash testing (since light vehicles don't absorb energy well).

German automotive supplier Willhelm Karmann (whose company manufactured the Volkswagen Karmann-Ghia) developed an aluminum foam composite material with some amazing properties. Parts made of this foam composite weigh 30–50% less than an equivalent steel part and are 10 times stiffer. The material is so light it floats in water and although it costs 20–25% more than steel, it could be used for as much as 20% of a compact car. Since the surface of the new material is not smooth, it would likely be used in structural areas of the car (firewalls, roof panels, luggage compartment walls, etc.)

How is aluminum foam made? Two layers of aluminum sheet and a middle powder layer (made of titanium metal hydride and aluminum powder) are rolled together under very high pressure to make a single flat sheet. This sheet metal is then processed in traditional ways to make a variety of 3-D shapes. Then the sheet is placed in a 1148°F oven for two minutes. This quick bake allows the Al metal to melt and mix with H_2 gas (released from the titanium hydride), making foam. The sheet rises just like a cake (increasing five to seven times in thickness). When the foam cools it is a rigidly formed 3-D structure between two aluminum skins.

Not only is the new aluminum foam part much lighter than its steel counterpart; it also performs well in crash tests. One of these days you may climb into a car that is really a foam colloid—with great fuel economy and crash resistance.

The fundamental difference between a colloidal dispersion and a true solution is the size, not the nature, of the particles. The solute particles in a solution are usually single ions or molecules that may be hydrated to varying degrees. Colloidal particles are usually aggregations of ions or molecules. However, the molecules of some polymers, such as proteins, are large enough to be classified as colloidal particles when in solution. To fully appreciate the differences in relative sizes, the volumes (not just the linear dimensions) of colloidal particles and solute particles must be compared. The difference in volumes can be approximated by assuming that the particles are spheres. A large colloidal particle has a diameter of about 500 nm, whereas a fair-sized ion or molecule has a diameter of about 0.5 nm. Thus the diameter of the colloidal particle is about 1000 times that of the solute particle. Because the volumes of spheres are proportional to the cubes of their diameters, we can calculate that the volume of a colloidal particle can be up to a billion ($10^3 \times 10^3 \times 10^3 = 10^9$) times greater than that of a solution particle.

15.14 Properties of Colloids

In 1827, while observing a strongly illuminated aqueous suspension of pollen under a high-powered microscope, Robert Brown (1773–1858) noted that the pollen grains appeared to have a trembling, erratic motion. He later

Brownian movement

determined that this erratic motion is not confined to pollen but is characteristic of colloidal particles in general. This random motion of colloidal particles is called **Brownian movement.** We can readily observe such movement by confining cigarette smoke in a small transparent chamber and illuminating it with a strong beam of light at right angles to the optical axis of the microscope. The smoke particles appear as tiny randomly moving lights because the light is reflected from their surfaces. This motion is due to the continual bombardment of the smoke particles by air molecules. Since Brownian movement can be seen when colloidal particles are dispersed in either a gaseous or a liquid medium, it affords nearly direct visual proof that matter at the molecular level is moving randomly, as postulated by the kinetic-molecular theory.

When an intense beam of light is passed through an ordinary solution and viewed at an angle, the beam passing through the solution is hardly visible. A beam of light, however, is clearly visible and sharply outlined when it is passed

Tyndall effect

through a colloidal dispersion. This phenomenon is known as the **Tyndall effect.** The Tyndall effect, like Brownian movement, can be observed in nearly all colloidal dispersions. It occurs because the colloidal particles are large enough to scatter the rays of visible light. The ions or molecules of true solutions are too small to scatter light and therefore do not exhibit a noticeable Tyndall effect.

Another important characteristic of colloids is that the particles have relatively huge surface areas. We saw in Section 14.4 that the surface area is increased 10-fold when a 1-cm cube is divided into 1000 cubes with sides of 0.1 cm. When a 1-cm cube is divided into colloidal-size cubes measuring 10^{-6} cm, the combined surface area of all the particles becomes a million times greater than that of the original cube.

A beam of lights is visible in a colloid solution (left) but not in a true solution. The size of the particles in the colloid are large enough to scatter the light beam making it visible.

Colloidal particles become electrically charged when they adsorb ions on their surfaces. *Adsorption* should not be confused with *absorption*. Adsorption refers to the adhesion of molecules or ions to a surface, whereas absorption refers to the taking in of one material by another material. Adsorption occurs because the atoms or ions at the surface of a particle are not completely surrounded by other atoms or ions as are those in the interior. Consequently, these surface atoms or ions attract and adsorb ions or polar molecules from the dispersion medium onto the surfaces of the colloidal particles. This property is directly related to the large surface area presented by the many tiny particles.

15.15 Applications of Colloidal Properties

Activated charcoal has an enormous surface area, approximately 1 million square centimeters per gram in some samples. Hence, charcoal is very effective in selectively adsorbing the polar molecules of some poisonous gases and is therefore used in gas masks. Charcoal can be used to adsorb impurities from liquids as well as from gases, and large amounts are used to remove substances that have objectionable tastes and odors from water supplies. In sugar refineries, activated charcoal is used to adsorb colored impurities from the raw sugar solutions.

Colloidal particles become electrically charged when they adsorb ions on their surface.

A process widely used for dust and smoke control in many urban and industrial areas was devised by an American, Frederick Cottrell (1877–1948). The Cottrell process takes advantage of the fact that the particulate matter in dust and smoke is electrically charged. Air to be cleaned of dust or smoke is passed between electrode plates charged with a high voltage. Positively charged particles are attracted to, neutralized, and thereby precipitated at the negative electrodes. Negatively charged particles are removed in the same fashion at the positive electrodes. Large Cottrell units are fitted with devices for automatic removal of precipitated material. Small units, designed for removing dust and pollen from air in the home, are now on the market. Unfortunately, Cottrell units remove only particulate matter; they cannot remove gaseous pollutants such as carbon monoxide, sulfur dioxide, and nitrogen oxides.

Thomas Graham found that a parchment membrane would allow the passage of true solutions but would prevent the passage of colloidal dispersions. Dissolved solutes can be removed from colloidal dispersions through the use of such a membrane by a process called **dialysis.** The membrane itself is called a *dialyzing membrane.* Artificial membranes are made from such materials as parchment paper, collodion, or certain kinds of cellophane. Dialysis can be demonstrated by putting a colloidal starch dispersion and some copper(II) sulfate solution in a parchment paper bag and suspending it in running water. In a few hours, the blue color of the copper(II) sulfate has disappeared, and only the starch dispersion remains in the bag.

dialysis

A life-saving application of dialysis has been the development of artificial kidneys. The blood of a patient suffering from partial kidney failure is passed through the artificial kidney machine for several hours, during which time the soluble waste products are removed by dialysis.

Chapter 15 Review

15.1 Acids and Bases

KEY TERM

Hydronium ion

- Characteristic properties of acids include:
  - They taste sour.
  - They change litmus from blue to red.
  - They react with:
    - Metals to form hydrogen gas
    - Hydroxide bases to form water and a salt
    - Carbonates to produce CO_2
- Characteristic properties of bases include:
  - They taste bitter or caustic.
  - They have a slippery feeling.
  - They change litmus from red to blue.
  - They interact with acids to form water and a salt.
- Arrhenius definition of acids and bases:
  - Acids contain excess H^+ ions in aqueous solutions.
  - Bases contain excess OH^- ions in aqueous solutions.
- Brønsted–Lowry definition of acids and bases:
  - Acids are proton donors.
  - Bases are proton acceptors.
- Lewis definition of acids and bases:
  - Acids are electron pair acceptors.
  - Bases are electron pair donors.

15.2 Reactions of Acids

- Acids react with metals above hydrogen in the activity series to form hydrogen gas and a salt.
- Acids react with bases to form water and a salt (neutralization reaction).
- Acids react with metal oxides to form water and salt.
- Acids react with carbonates to form water, a salt, and carbon dioxide.

15.3 Reactions of Bases

KEY TERM

Amphoteric

- Bases react with acids to form water and a salt (neutralization reaction).
- Some amphoteric metals react with NaOH or KOH to form hydrogen and a salt.

15.4 Salts

- Salts can be considered to be derived from the reaction of an acid and a base.
- Salts are crystalline and have high melting and boiling points.

15.5 Electrolytes and Nonelectrolytes

KEY TERMS

Electrolyte
Nonelectrolyte

- A substance whose aqueous solution conducts electricity is called an electrolyte:
  - Aqueous solution contains ions.
- A substance whose aqueous solution does not conduct electricity is called a nonelectrolyte:
  - Aqueous solution contains molecules, not ions.

15.6 Dissociation and Ionization of Electrolytes

KEY TERMS

Dissociation

Ionization

- Dissociation is the process by which the ions of a salt separate as the salt dissolves.
- Ionization is the formation of ions that occurs as a result of a chemical reaction with water.

15.7 Strong and Weak Electrolytes

KEY TERMS

Strong electrolyte

Weak electrolyte

- Strong electrolytes are 100% ionized in solution:
  - Strong acids and strong bases
- Weak electrolytes are much less ionized in solution:
  - Weak acids and bases
- Colligative properties of electrolyte solutions depend on the number of particles produced during the ionization of the electrolyte.

15.8 Ionization of Water

- Water can self-ionize to form H^+ and OH^- ions.
- Concentrations of ions in water at 25°C:
  - $[H^+] = 1.0 \times 10^{-7}$
  - $[OH^-] = 1.0 \times 10^{-7}$

15.9 Introduction to pH

KEY TERMS

pH

Logarithm

- $pH = -\log[H^+]$:
  - $pH < 7$ for acidic solution
  - $pH = 7$ in neutral solution
  - $pH > 7$ for basic solution
- The number of decimal places in a logarithm equals the number of significant figures in the original number.

15.10 Neutralization

KEY TERMS

Neutralization

Spectator ion

Titration

- The reaction of an acid and a base to form water and a salt is called neutralization:
  - The general equation for a neutralization reaction is
    $$H^+(aq) + OH^-(aq) \longrightarrow H_2O(l)$$
- The quantitative study of a neutralization reaction is called a titration.

15.11 Writing Net Ionic Equations

KEY TERMS

Formula equation

Total ionic equation

Net ionic equation

- In a formula equation the compounds are written in their molecular, or formula, expressions.
- In the total ionic equation compounds are written as ions if they are strong electrolytes in solution and as molecules if they are precipitates, nonelectrolytes, or weak electrolytes in solution.
- The net ionic equation shows only the molecules or ions that have changed:
  - The spectators (nonreactors) are omitted.

15.12 Acid Rain

- Acid rain is any atmospheric precipitation that is more acidic that usual.
- Acid rain causes significant damage and destruction to our environment.

15.13 Colloids

KEY TERM

Colloid

- A colloid is a dispersion containing particles between 1 nm and 1000 nm.
- Colloid particles are usually aggregates of ions or molecules.
- The difference between a colloid and a true solution is the size of the particles.

15.14 Properties of Colloids

KEY TERMS

Brownian movement

Tyndall effect

- The random motion of particles in a colloid is called Brownion movement.
- Colloid particles cause light from an intense beam to be scattered clearly, showing the path of the light through the colloid.

15.15 Application of Colloidal Properties

KEY TERM

Dialysis

- Activated charcoal can be used to adsorb impurities from liquids and gases.
- Dialysis is based on the idea that a true solution will pass through a parchment membrane but a colloid will not.

Review Questions

All questions with blue *numbers have answers in the appendix of the text.*

1. Since a hydrogen ion and a proton are identical, what differences exist between the Arrhenius and Brønsted–Lowry definitions of an acid? (Table 15.1)

2. According to Figure 15.1, what type of substance must be in solution for the bulb to light?

3. Which of the following classes of compounds are electrolytes: acids, alcohols, bases, salts? (Table 15.2)

4. What two differences are apparent in the arrangement of water molecules about the hydrated ions as depicted in Figure 15.2?

5. The pH of a solution with a hydrogen ion concentration of 0.003 M is between what two whole numbers? (Table 15.4)

6. Which is more acidic, tomato juice or blood? (Table 15.5)

7. Use the three acid–base theories (Arrhenius, Brønsted–Lowry, and Lewis) to define an acid and a base.

8. For each acid–base theory referred to in Question 7, write an equation illustrating the neutralization of an acid with a base.

9. Write the Lewis structure for the (a) bromide ion, (b) hydroxide ion, and (c) cyanide ion. Why are these ions considered to be bases according to the Brønsted–Lowry and Lewis acid–base theories?

10. Into what three classes of compounds do electrolytes generally fall?

11. Name each compound listed in Table 15.3.

12. A solution of HCl in water conducts an electric current, but a solution of HCl in hexane does not. Explain this behavior in terms of ionization and chemical bonding.

13. How do ionic compounds exist in their crystalline structure? What occurs when they are dissolved in water?

14. An aqueous methyl alcohol, CH_3OH, solution does not conduct an electric current, but a solution of sodium hydroxide, NaOH, does. What does this information tell us about the OH group in the alcohol?

15. Why does molten NaCl conduct electricity?

16. Explain the difference between dissociation of ionic compounds and ionization of molecular compounds.

17. Distinguish between strong and weak electrolytes.

18. Explain why ions are hydrated in aqueous solutions.

19. What is the main distinction between water solutions of strong and weak electrolytes?

20. What are the relative concentrations of $H^+(aq)$ and $OH^-(aq)$ in (a) a neutral solution, (b) an acid solution, and (c) a basic solution?

21. Write the net ionic equation for the reaction of a strong acid with a water-soluble hydroxide base in an aqueous solution.

22. The solubility of HCl gas in water, a polar solvent, is much greater than its solubility in hexane, a nonpolar solvent. How can you account for this difference?

23. Pure water, containing equal concentrations of both acid and base ions, is neutral. Why?

24. Indicate the fundamental difference between a colloidal dispersion and a true solution.

25. Explain the process of dialysis, giving a practical application in society.

26. A solution with a pH of 7 is neutral. A solution with a pH less than 7 is acidic. A solution with a pH greater than 7 is basic. What do these statements mean?

27. How does acid rain form?

28. Explain the purpose of a titration.

Paired Exercises

All exercises with blue *numbers have answers in the appendix of the text.*

1. Identify the conjugate acid–base pairs in the following equations:
   (a) $HC_2H_3O_2 + H_2SO_4 \rightleftharpoons H_2C_2H_3O_2^+ + HSO_4^-$
   (b) The two-step ionization of sulfuric acid,
   $$H_2SO_4 + H_2O \longrightarrow H_3O^+ + HSO_4^-$$
   $$HSO_4^- + H_2O \rightleftharpoons H_3O^+ + SO_4^{2-}$$
   (c) $HClO_4 + H_2O \longrightarrow H_3O^+ + ClO_4^-$
   (d) $CH_3O^- + H_3O^+ \longrightarrow CH_3OH + H_2O$

2. Identify the conjugate acid–base pairs in the following equations:
   (a) $HCl + NH_3 \longrightarrow NH_4^+ + Cl^-$
   (b) $HCO_3^- + OH^- \rightleftharpoons CO_3^{2-} + H_2O$
   (c) $HCO_3^- + H_3O^+ \rightleftharpoons H_2CO_3 + H_2O$
   (d) $HC_2H_3O_2 + H_2O \rightleftharpoons H_3O^+ + C_2H_3O_2^-$

3. Complete and balance these equations:
    (a) $Zn(s) + HCl(aq) \longrightarrow$
    (b) $Al(OH)_3(s) + H_2SO_4(aq) \longrightarrow$
    (c) $Na_2CO_3(aq) + HC_2H_3O_2(aq) \longrightarrow$
    (d) $MgO(s) + HI(aq) \longrightarrow$
    (e) $Ca(HCO_3)_2(s) + HBr(aq) \longrightarrow$
    (f) $KOH(aq) + H_3PO_4(aq) \longrightarrow$

4. Complete and balance these equations:
    (a) $Fe_2O_3(s) + HBr(aq) \longrightarrow$
    (b) $Al(s) + H_2SO_4(aq) \longrightarrow$
    (c) $NaOH(aq) + H_2CO_3(aq) \longrightarrow$
    (d) $Ba(OH)_2(s) + HClO_4(aq) \longrightarrow$
    (e) $Mg(s) + HClO_4(aq) \longrightarrow$
    (f) $K_2O(s) + HI(aq) \longrightarrow$

5. For each of the formula equations in Question 3, write total and net ionic equations.

6. For each of the formula equations in Question 4, write total and net ionic equations.

7. Which of these compounds are electrolytes? Consider each substance to be mixed with water.
    (a) HCl
    (b) CO_2
    (c) $CaCl_2$
    (d) $C_{12}H_{22}O_{11}$ (sugar)
    (e) C_3H_7OH (rubbing alcohol)
    (f) CCl_4 (insoluble)

8. Which of these compounds are electrolytes? Consider each substance to be mixed with water.
    (a) $NaHCO_3$ (baking soda)
    (b) N_2 (insoluble gas)
    (c) $AgNO_3$
    (d) $HCOOH$ (formic acid)
    (e) $RbOH$
    (f) K_2CrO_4

9. Calculate the molarity of the ions present in these salt solutions. Assume each salt to be 100% dissociated:
    (a) $0.015\ M\ NaCl$
    (b) $4.25\ M\ NaKSO_4$
    (c) $0.20\ M\ CaCl_2$
    (d) 22.0 g KI in $500.$ mL of solution

10. Calculate the molarity of the ions present in these salt solutions. Assume each salt to be 100% dissociated:
    (a) $0.75\ M\ ZnBr_2$
    (b) $1.65\ M\ Al_2(SO_4)_3$
    (c) $900.$ g $(NH_4)_2SO_4$ in 20.0 L of solution
    (d) 0.0120 g $Mg(ClO_3)_2$ in 1.00 mL of solution

11. In Exercise 9, how many grams of each ion would be present in $100.$ mL of each solution?

12. In Exercise 10, how many grams of each ion would be present in $100.$ mL of each solution?

13. Calculate the $[H^+]$ for:
    (a) a solution with a pH = 8.5
    (b) pure water
    (c) a solution with a pH = 2.5

14. Calculate the $[H^+]$ for:
    (a) a solution with a pH = 2.4
    (b) a solution with a pH = 10.0
    (c) tap water with a pH = 6.4

15. What is the molar concentration of all ions present in a solution prepared by mixing the following?
    (Neglect the concentration of H^+ and OH^- from water and assume that volumes of solutions are additive.)
    (a) 30.0 mL of $1.0\ M\ NaCl$ and 40.0 mL of $1.0\ M\ NaCl$
    (b) 30.0 mL of $1.0\ M\ HCl$ and 30.0 mL of $1.0\ M\ NaOH$
    (c) 100.0 mL of $0.40\ M\ KOH$ and 100.0 mL of $0.80\ M\ HCl$

16. What is the molar concentration of all ions present in a solution prepared by mixing the following?
    (Neglect the concentration of H^+ and OH^- from water and assume volumes of solutions are additive.)
    (a) 100.0 mL of $2.0\ M\ KCl$ and 100.0 mL of $1.0\ M\ CaCl_2$
    (b) 35.0 mL of $0.20\ M\ Ba(OH)_2$ and 35.0 mL of $0.20\ M\ H_2SO_4$
    (c) 1.00 L of $1.0\ M\ AgNO_3$ and $500.$ mL of $2.0\ M\ NaCl$

17. Given the data for the following separate titrations, calculate the molarity of the HCl:

	mL HCl	Molarity HCl	mL NaOH	Molarity NaOH
(a)	40.13	M	37.70	0.728
(b)	19.00	M	33.66	0.306
(c)	27.25	M	18.00	0.555

18. Given the data for the following separate titrations, calculate the molarity of the NaOH:

	mL HCl	Molarity HCl	mL NaOH	Molarity NaOH
(a)	37.19	0.126	31.91	M
(b)	48.04	0.482	24.02	M
(c)	13.13	1.425	39.39	M

19. Rewrite the following unbalanced equations, changing them into balanced net ionic equations (assume that all reactions are in water solution):

(a) $K_2SO_4(aq) + Ba(NO_3)_2(aq) \longrightarrow$
$$KNO_3(aq) + BaSO_4(s)$$

(b) $CaCO_3(s) + HCl(aq) \longrightarrow$
$$CaCl_2(aq) + CO_2(g) + H_2O(l)$$

(c) $Mg(s) + HC_2H_3O_2(aq) \longrightarrow$
$$Mg(C_2H_3O_2)_2(aq) + H_2(g)$$

20. Rewrite the following unbalanced equations, changing them into balanced net ionic equations (assume that all reactions are in water solution):

(a) $H_2S(g) + CdCl_2(aq) \longrightarrow CdS(s) + HCl(aq)$

(b) $Zn(s) + H_2SO_4(aq) \longrightarrow ZnSO_4(aq) + H_2(g)$

(c) $AlCl_3(aq) + Na_3PO_4(aq) \longrightarrow$
$$AlPO_4(s) + NaCl(aq)$$

21. For each of the given pairs, determine which solution is more acidic. All are water solutions. Explain your answer.

(a) $1\ M$ HCl or $1\ M$ H_2SO_4?

(b) $1\ M$ HCl or $1\ M$ $HC_2H_3O_2$?

22. For each of the given pairs, determine which solution is more acidic. All are water solutions. Explain your answer.

(a) $1\ M$ HCl or $2\ M$ HCl?

(b) $1\ M$ HNO_3 or $1\ M$ H_2SO_4?

23. What volume (in milliliters) of $0.245\ M$ HCl will neutralize 10.0 g $Al(OH)_3$? The equation is

$$3\,HCl(aq) + Al(OH)_3(s) \longrightarrow AlCl_3(aq) + 3\,H_2O(l)$$

24. What volume (in milliliters) of $0.245\ M$ HCl will neutralize 50.0 mL of $0.100\ M$ $Ca(OH)_2$? The equation is

$$2\,HCl(aq) + Ca(OH)_2(aq) \longrightarrow CaCl_2(aq) + 2\,H_2O(l)$$

*25. A 0.200-g sample of impure NaOH requires 18.25 mL of $0.2406\ M$ HCl for neutralization. What is the percent of NaOH in the sample?

*26. A batch of sodium hydroxide was found to contain sodium chloride as an impurity. To determine the amount of impurity, a 1.00-g sample was analyzed and found to require 49.90 mL of $0.466\ M$ HCl for neutralization. What is the percent of NaCl in the sample?

*27. What volume of H_2 gas, measured at 27°C and 700. torr, can be obtained by reacting 5.00 g of zinc metal with 100. mL of $0.350\ M$ HCl? The equation is

$$Zn(s) + 2\,HCl(aq) \longrightarrow ZnCl_2(aq) + H_2(g)$$

*28. What volume of H_2 gas, measured at 27°C and 700. torr, can be obtained by reacting 5.00 g of zinc metal with 200. mL of $0.350\ M$ HCl? The equation is

$$Zn(s) + 2\,HCl(aq) \longrightarrow ZnCl_2(aq) + H_2(g)$$

29. Calculate the pH of solutions having these H^+ ion concentrations:

(a) $0.01\ M$

(b) $1.0\ M$

(c) $6.5 \times 10^{-9}\ M$

30. Calculate the pH of solutions having these H^+ ion concentrations:

(a) $1 \times 10^{-7}\ M$

(b) $0.50\ M$

(c) $0.00010\ M$

31. Calculate the pH of

(a) orange juice, $3.7 \times 10^{-4}\ M\,H^+$

(b) vinegar, $2.8 \times 10^{-3}\ M\,H^+$

32. Calculate the pH of

(a) black coffee, $5.0 \times 10^{-5}\ M\,H^+$

(b) limewater, $3.4 \times 10^{-11}\ M\,H^+$

33. Determine whether each of the following is a strong acid, weak acid, strong base, or weak base. Then write an equation describing the process that occurs when the substance is dissolved in water.

(a) NH_3

(b) HCl

(c) KOH

(d) $HC_2H_3O_2$

34. Determine whether each of the following is a strong acid, weak acid, strong base, or weak base. Then write an equation describing the process that occurs when the substance is dissolved in water.

(a) $H_2C_2O_4$

(b) $Ba(OH)_2$

(c) $HClO_4$

(d) HBr

Additional Exercises

All exercises with blue *numbers have answers in the appendix of the text.*

35. Determine whether each of the following describes a substance that is acidic, basic, or neutral.

(a) $[H^+] = 1 \times 10^{-4}$

(b) Phenolphthalein turns pink.

(c) pH = 8

(d) $[H^+] = 1 \times 10^{-9}$

(e) $[OH^-] = 1 \times 10^{-7}$

(f) Blue litmus paper turns red.

36. Draw pictures similar to those found in Figure 15.2 to show what happens to the following compounds when each is dissolved in water.
    (a) $CaCl_2$ (b) KF   (c) $AlBr_3$

37. What is the concentration of Ca^{2+} ions in a solution of CaI_2 having an I^- ion concentration of 0.520 M?

38. If 29.26 mL of 0.430 M HCl neutralize 20.40 mL of $Ba(OH)_2$ solution, what is the molarity of the $Ba(OH)_2$ solution? The reaction is

    $$Ba(OH)_2(aq) + 2\,HCl(aq) \longrightarrow BaCl_2(aq) + 2\,H_2O(l)$$

39. A 1 m solution of acetic acid, $HC_2H_3O_2$, in water freezes at a lower temperature than a 1 m solution of ethyl alcohol, C_2H_5OH, in water. Explain.

40. At the same cost per pound, which alcohol, CH_3OH or C_2H_5OH, would be more economical to purchase as an antifreeze for your car? Why?

41. How does a hydronium ion differ from a hydrogen ion?

42. Arrange, in decreasing order of freezing points, 1 m aqueous solutions of HCl, $HC_2H_3O_2$, $C_{12}H_{22}O_{11}$ (sucrose), and $CaCl_2$. (List the one with the highest freezing point first.)

43. At 100°C the H^+ concentration in water is about 1×10^{-6} mol/L, about 10 times that of water at 25°C. At which of these temperatures is
    (a) the pH of water the greater?
    (b) the hydrogen ion (hydronium ion) concentration the higher?
    (c) the water neutral?

44. What is the relative difference in H^+ concentration in solutions that differ by one pH unit?

45. What is the mole percent of a 1.00 m aqueous solution?

46. A sample of pure sodium carbonate with a mass of 0.452 g was dissolved in water and neutralized with 42.4 mL of hydrochloric acid. Calculate the molarity of the acid:

    $$Na_2CO_3(aq) + 2\,HCl(aq) \longrightarrow$$
    $$2\,NaCl(aq) + CO_2(g) + H_2O(l)$$

47. What volume (mL) of 0.1234 M HCl is needed to neutralize 2.00 g $Ca(OH)_2$?

48. How many grams of KOH are required to neutralize 50.00 mL of 0.240 M HNO_3?

49. Two drops (0.1 mL) of 1.0 M HCl are added to water to make 1.0 L of solution. What is the pH of this solution if the HCl is 100% ionized?

50. What volume of concentrated (18.0 M) sulfuric acid must be used to prepare 50.0 L of 5.00 M solution?

*51. If 3.0 g NaOH are added to 500. mL of 0.10 M HCl, will the resulting solution be acidic or basic? Show evidence for your answer.

*52. If 380 mL of 0.35 M $Ba(OH)_2$ are added to 500. mL of 0.65 M HCl, will the mixture be acidic or basic? Find the pH of the resulting solution.

*53. If 50.00 mL of 0.2000 M HCl are titrated with 0.2000 M NaOH, find the pH of the solution after the following amounts of base have been added:
    (a) 0.000 mL   (d) 49.00 mL   (f) 49.99 mL
    (b) 10.00 mL   (e) 49.90 mL   (g) 50.00 mL
    (c) 25.00 mL
    Plot your answers on a graph with pH on the y-axis and mL NaOH on the x-axis.

54. Sulfuric acid reacts with NaOH:
    (a) Write a balanced equation for the reaction producing Na_2SO_4.
    (b) How many milliliters of 0.10 M NaOH are needed to react with 0.0050 mol H_2SO_4?
    (c) How many grams of Na_2SO_4 will also form?

*55. A 10.0-mL sample of HNO_3 was diluted to a volume of 100.00 mL. Then, 25 mL of that diluted solution were needed to neutralize 50.0 mL of 0.60 M KOH. What was the concentration of the original nitric acid?

56. The pH of a solution of a strong acid was determined to be 3. If water is then added to dilute this solution, would the pH change? Why or why not? Could enough water ever be added to raise the pH of an acid solution above 7?

Challenge Exercises

All exercises with blue numbers have answers in the appendix of the text.

*57. An HCl solution has a pH of -0.300. What volume of water must be added to 200 mL of this solution to change the pH to -0.150?

*58. Lactic acid (found in sour milk) has an empirical formula of $HC_3H_5O_3$. A 1.0-g sample of lactic acid required 17.0 mL of 0.65 M NaOH to reach the end point of a titration. What is the molecular formula for lactic acid?

Answers to Practice Exercises

15.1 (a) HCO_3^-, (b) NO_2^-, (c) $C_2H_3O_2^-$
15.2 (a) H_2SO_4, (b) NH_4^+, (c) H_2O
15.3 (a) KOH, potassium hydroxide; H_3PO_4, phosphoric acid
    (b) $Mg(OH)_2$, magnesium hydroxide; HBr, hydrobromic acid
    (c) LiOH, lithium hydroxide; HCl, hydrochloric acid
    (d) $Fe(OH)_2$, iron(II) hydroxide; H_2CO_3, carbonic acid
15.4 (a) 0.050 M Mg^{2+}, 0.10 M Cl^-,
    (b) 0.070 M Al^{3+}, 0.21 M Cl^-
15.5 approximately 101.54°C
15.6 (a) 11.41,   (b) 2.89,   (c) 5.429
15.7 0.0620 M HCl
15.8 $3\,H_2S(aq) + 2\,Bi^{3+}(aq) \longrightarrow Bi_2S_3(s) + 6\,H^+(aq)$

Chemical Equilibrium

A coral reef is a system in ecological equilibrium with the ocean surrounding it.

Chapter Outline

16.1 Reversible Reactions

16.2 Rates of Reaction

16.3 Chemical Equilibrium

16.4 Le Châtelier's Principle

16.5 Effect of Concentration on Equilibrium

16.6 Effect of Volume on Equilibrium

16.7 Effect of Temperature on Equilibrium

16.8 Effect of Catalysts on Equilibrium

16.9 Equilibrium Constants

16.10 Ion Product Constant for Water

16.11 Ionization Constants

16.12 Solubility Product Constant

16.13 Acid–Base Properties of Salts

16.14 Buffer Solutions: The Control of pH

Thus far, we've considered chemical change as proceeding from reactants to products. Does that mean that the change then stops? No, but often it appears to be the case at the macroscopic level. A solute dissolves until the solution becomes saturated. Once a solid remains undissolved in a container, the system appears to be at rest. The human body is a marvelous chemical factory, yet from day to day it appears to be quite the same. For example, the blood remains at a constant pH, even though all sorts of chemical reactions are taking place. Another example is a terrarium, which can be watered and sealed for long periods of time with no ill effects. Or an antacid, which absorbs excess stomach acid and does *not* change the pH of the stomach. In all of these cases, reactions are proceeding, even though visible signs of chemical change are absent. Similarly, when a system is at equilibrium, chemical reactions are dynamic at the molecular level. In this chapter, we will consider chemical systems as they approach equilibrium conditions.

16.1 Reversible Reactions

In the preceding chapters, we treated chemical reactions mainly as reactants changing to products. However, many reactions do not go to completion. Some reactions do not go to completion because they are reversible; that is, when the products are formed, they react to produce the starting reactants.

We've encountered reversible systems before. One is the vaporization of a liquid by heating and its subsequent condensation by cooling:

$$\text{liquid} + \text{heat} \longrightarrow \text{vapor}$$
$$\text{vapor} + \text{cooling} \longrightarrow \text{liquid}$$

The conversion between nitrogen dioxide, NO_2, and dinitrogen tetroxide, N_2O_4, shows us visible evidence of the reversibility of a reaction. The NO_2 is a reddish-brown gas that changes with cooling to N_2O_4, a colorless gas. The reaction is reversible by heating N_2O_4:

NO_2

N_2O_4

$$2\,NO_2(g) \xrightarrow{\text{cooling}} N_2O_4(g)$$
$$N_2O_4(g) \xrightarrow{\text{heating}} 2\,NO_2(g)$$

These two reactions may be represented by a single equation with a double arrow, $\rightleftharpoons$, to indicate that the reactions are taking place in both directions at the same time:

$$2\,NO_2(g) \rightleftharpoons N_2O_4(g)$$

This reversible reaction can be demonstrated by sealing samples of NO_2 in two tubes and placing one tube in warm water and the other in ice water (see Figure 16.1).

reversible chemical reaction

A **reversible chemical reaction** is one in which the products formed react to produce the original reactants. Both the forward and reverse reactions occur simultaneously. The forward reaction is called *the reaction to the right*, and the reverse reaction is called *the reaction to the left*. A double arrow is used in the equation to indicate that the reaction is reversible.

Figure 16.1
Reversible reaction of NO_2 and N_2O_4. More of the dark brown molecules are visible in the heated container on the right than in the room-temperature tube on the left.

16.2 Rates of Reaction

Every reaction has a rate, or speed, at which it proceeds. Some are fast, and some are extremely slow. The study of reaction rates and reaction mechanisms is known as **chemical kinetics.**

The rate of a reaction is variable and depends on the concentration of the reacting species, the temperature, the presence of catalysts, and the nature of the reactants. Consider the hypothetical reaction

$$A + B \longrightarrow C + D \quad \text{(forward reaction)}$$
$$C + D \longrightarrow A + B \quad \text{(reverse reaction)}$$

in which a collision between A and B is necessary for a reaction to occur. The rate at which A and B react depends on the concentration, or the number of A and B molecules present; it will be fastest, for a fixed set of conditions, when they are first mixed (as shown by the height of the red line in Figure 16.2).

chemical kinetics

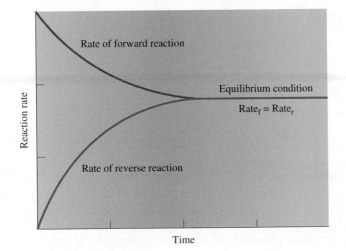

Figure 16.2
The rates of the forward and reverse reactions become equal at some point in time. The forward reaction rate (red) decreases as a result of decreasing amounts of reactants. The reverse reaction rate (blue) starts at zero and increases as the amount of product increases. When the two rates become equal (purple), a state of chemical equilibrium has been reached.

As the reaction proceeds, the number of A and B molecules available for reaction decreases, and the rate of reaction slows down (seen as the red line flattens in Figure 16.2). If the reaction is reversible, the speed of the reverse reaction is zero at first (blue line in Figure 16.2) and gradually increases as the concentrations of C and D increase. As the number of A and B molecules decreases, the forward rate slows down because A and B cannot find one another as often in order to accomplish a reaction. To counteract this diminishing rate of reaction, an excess of one reagent is often used to keep the reaction from becoming impractically slow. Collisions between molecules may be compared to video games. When many objects are on the screen, collisions occur frequently; but if only a few objects are present, collisions can usually be avoided.

16.3 Chemical Equilibrium

equilibrium

chemical equilibrium

Any system at **equilibrium** represents a dynamic state in which two or more opposing processes are taking place at the same time and at the same rate. A chemical equilibrium is a dynamic system in which two or more opposing chemical reactions are going on at the same time and at the same rate. When the rate of the forward reaction is exactly equal to the rate of the reverse reaction, a condition of **chemical equilibrium** exists (see purple line in Figure 16.2). The concentrations of the products and the reactants are not changing, and the system appears to be at a standstill because the products are reacting at the same rate at which they are being formed.

> Chemical equilibrium:
> rate of forward reaction = rate of reverse reaction

A saturated salt solution is in a condition of equilibrium:

$$NaCl(s) \rightleftharpoons Na^+(aq) + Cl^-(aq)$$

At equilibrium, salt crystals are continuously dissolving, and Na^+ and Cl^- ions are continuously crystallizing. Both processes are occurring at the same rate.

The ionization of weak electrolytes is another chemical equilibrium system:

$$HC_2H_3O_2(aq) + H_2O(l) \rightleftharpoons H_3O^+(aq) + C_2H_3O_2^-(aq)$$

In this reaction, the equilibrium is established in a 1 M solution when the forward reaction has gone about 1%—that is, when only 1% of the acetic acid molecules in solution have ionized. Therefore, only a relatively few ions are present, and the acid behaves as a weak electrolyte.

The reaction represented by

$$H_2(g) + I_2(g) \xrightleftharpoons{700\ K} 2\ HI(g)$$

provides another example of chemical equilibrium. Theoretically, 1.00 mol of hydrogen should react with 1.00 mol of iodine to yield 2.00 mol of hydrogen iodide. Actually, when 1.00 mol H_2 and 1.00 mol I_2 are reacted at 700 K, only 1.58 mol HI is present when equilibrium is attained. Since 1.58 is 79% of the theoretical yield of 2.00 mol HI, the forward reaction is only 79% complete at

Dentists have understood for more than 20 years what causes cavities, but, until now, there have been only a limited number of over-the-counter products to help us avoid our dates with the drill. Bacteria in the mouth break down sugars remaining in the mouth after eating. Acids produced during this process slip through tooth enamel, dissolving minerals below the surface in a process called demineralization. Saliva works to rebuild teeth by adding calcium and phosphate back in a process called remineralization. Under ideal conditions (assuming that you brush after eating), these two processes form an equilibrium.

Unfortunately, bacteria in plaque (resulting from not brushing) shift the equilibrium toward demineralization (shown in the figure) and a cavity can begin to form. Scientists realized that fluoride encourages remineralization in teeth by replacing hydroxyl ions in nature's calcium phosphate (hydroxyapatite). The substitution changes the hydroxyapatite to fluorapatite, which is more acid resistant.

The good news is that scientists have now figured out how to shift the equilibrium between demineralization and remineralization toward remineralization. This new understanding has resulted in products that actually stop cavities and prevent the need for a filling. Scientists now know that adding calcium and phosphate to the mouth regularly speeds up remineralization of the teeth. The largest problem in doing it this way is to figure out how to dispense the calcium and phosphate through the surface enamel gradually.

If a calcium ion finds a phosphate ion before this happens, a precipitate forms that can't penetrate the enamel. Researchers have developed new products such as chewing gum and sticky substances (see photo) that can be applied between the teeth. These products work better than brushing or rinsing because they last longer—a brushing or rinsing lasts only a minute.

Several types of remineralizing chewing gums are currently available,

Normal Conditions

Remineralization

Demineralization

Cavity-Forming Conditions

Remineralization

Plaque

Demineralization
Calcium and Phosphate

Remineralizing Therapies

Remineralization
Fluoride, Calcium, and Phosphate

Demineralization

including Trident Advantage and Trident for Kids. The calcium and phosphate ions are stabilized by using a system that mimics nature. Researchers looked at how newborns get their large supply of calcium and phosphate. Eric Reynolds of the University of Melbourne in Australia says, "Nature's evolved this system in milk to carry very high levels of calcium and phosphate in a highly bioavailable form." They used casein (a milk protein) to produce microscopic particles of stabilized calcium phosphate that diffuse through the surface of teeth. When these particles were placed into the chewing gum and chewed, the results were better than hoped. In Japan, researchers at Tokyo's FAP Dental Institute have designed a toothpaste that repairs cavities smaller than 50 μm. The toothpaste grows nanocrystals of hydroxyapatite, which is treated with fluoride right on the cavity. The synthetic enamel not only repairs the tiny cavity as you brush it but also strengthens the natural enamel. Brushing can now repair cavities and prevent new ones at the same time.

Wedge

These sticky wedges dissolve slowly to release fluoride, calcium, and phosphate to teeth.

equilibrium. The equilibrium mixture will also contain 0.21 mol each of unreacted H_2 and I_2 (1.00 mol − 0.79 mol = 0.21 mol):

$$H_2(g) + I_2(g) \xrightarrow{700 \text{ K}} 2\,HI(g)$$

This equation represents the condition if the reaction were 100% complete; 2.00 mol HI would be formed and no H_2 and I_2 would be left unreacted.

$$H_2(g) + I_2(g) \underset{}{\overset{700 \text{ K}}{\rightleftharpoons}} 2\,HI(g)$$

0.21 mol   0.21 mol   1.58 mol

This equation represents the actual equilibrium attained starting with 1.00 mol each of H_2 and I_2. It shows that the forward reaction is only 79% complete.

16.4 Le Châtelier's Principle

In 1888, the French chemist Henri Le Châtelier (1850–1936) set forth a simple, far-reaching generalization on the behavior of equilibrium systems. This generalization, known as **Le Châtelier's principle,** states:

Le Châtelier's principle

> If a stress is applied to a system in equilibrium, the system will respond in such a way as to relieve that stress and restore equilibrium under a new set of conditions.

The application of Le Châtelier's principle helps us predict the effect of changing conditions in chemical reactions. We will examine the effect of changes in concentration, temperature, and volume.

16.5 Effect of Concentration on Equilibrium

The manner in which the rate of a chemical reaction depends on the concentration of the reactants must be determined experimentally. Many simple, one-step reactions result from a collision between two molecules or ions. The rate of such one-step reactions can be altered by changing the concentration of the reactants or products. An increase in concentration of the reactants provides more individual reacting species for collisions and results in an increase in the rate of reaction.

An equilibrium is disturbed when the concentration of one or more of its components is changed. As a result, the concentration of all species will change, and a new equilibrium mixture will be established. Consider the hypothetical equilibrium represented by the equation

$$A + B \rightleftharpoons C + D$$

where A and B react in one step to form C and D. When the concentration of B is increased, the following results occur:

1. The rate of the reaction to the right (forward) increases. This rate is proportional to the concentration of A times the concentration of B.
2. The rate to the right becomes greater than the rate to the left.
3. Reactants A and B are used faster than they are produced; C and D are produced faster than they are used.
4. After a period of time, rates to the right and left become equal, and the system is again in equilibrium.
5. In the new equilibrium the concentration of A is less, and the concentrations of B, C, and D are greater than in the original equilibrium. *Conclusion:* The equilibrium has shifted to the right.

Concentration	Change
[H_2]	?
[I_2]	increase
[HI]	?

Applying this change in concentration to the equilibrium mixture of 1.00 mol of hydrogen and 1.00 mol of iodine from Section 16.3, we find that, when an additional 0.20 mol I_2 is added, the yield of HI (based on H_2) is 85% (1.70 mol) instead of 79%. Here is how the two systems compare after the new equilibrium mixture is reached:

Original equilibrium	New equilibrium
1.00 mol H_2 + 1.00 mol I_2	1.00 mol H_2 + 1.20 mol I_2
Yield: 79% HI	Yield: 85% HI (based on H_2)
Equilibrium mixture contains:	Equilibrium mixture contains:
1.58 mol HI	1.70 mol HI
0.21 mol H_2	0.15 mol H_2
0.21 mol I_2	0.35 mol I_2

Analyzing this new system, we see that, when 0.20 mol I_2 is added, the equilibrium shifts to the right to counteract the increase in I_2 concentration. Some of the H_2 reacts with added I_2 and produces more HI, until an equilibrium mixture is established again. When I_2 is added, the concentration of I_2 increases, the concentration of H_2 decreases, and the concentration of HI increases.

Concentration	Change
$[H_2]$	decrease
$[I_2]$	increase
[HI]	increase

Practice 16.1

Use a chart like those in the margin to show what would happen to the concentrations of each substance in the system

$$H_2(g) + I_2(g) \rightleftharpoons 2\,HI(g)$$

upon adding (a) more H_2 and (b) more HI.

The equation

$$Fe^{3+}(aq) + SCN^-(aq) \rightleftharpoons Fe(SCN)^{2+}(aq)$$

pale yellow   colorless   red

represents an equilibrium that is used in certain analytical procedures as an indicator because of the readily visible, intense red color of the complex $Fe(SCN)^{2+}$ ion. A very dilute solution of iron(III) (Fe^{3+}) and thiocyanate (SCN^-) is light red. When the concentration of either Fe^{3+} or SCN^- is increased, the equilibrium shift to the right is observed by an increase in the intensity of the color, resulting from the formation of additional $Fe(SCN)^{2+}$.

If either Fe^{3+} or SCN^- is removed from solution, the equilibrium will shift to the left, and the solution will become lighter in color. When Ag^+ is added to the solution, a white precipitate of silver thiocyanate (AgSCN) is formed, thus removing SCN^- ion from the equilibrium:

$$Ag^+(aq) + SCN^-(aq) \rightleftharpoons AgSCN(s)$$

The system accordingly responds to counteract the change in SCN^- concentration by shifting the equilibrium to the left. This shift is evident by a decrease in the intensity of the red color due to a decreased concentration of $Fe(SCN)^{2+}$.

Now consider the effect of changing the concentrations in the equilibrium mixture of chlorine water. The equilibrium equation is

$$Cl_2(aq) + 2\,H_2O(l) \rightleftharpoons HOCl(aq) + H_3O^+(aq) + Cl^-(aq)$$

The variation in concentrations and the equilibrium shifts are tabulated in the following table. An X in the second or third column indicates that the reagent is increased or decreased. The fourth column indicates the direction of the equilibrium shift.

| Reagent | Concentration | | Equilibrium shift |
	Increase	Decrease	
Cl_2	—	X	Left
H_2O	X	—	Right
HOCl	X	—	Left
H_3O^+	—	X	Right
Cl^-	X	—	Left

Consider the equilibrium in a 0.100 M acetic acid solution:

$$HC_2H_3O_2(aq) + H_2O(l) \rightleftharpoons H_3O^+(aq) + C_2H_3O_2^-(aq)$$

In this solution, the concentration of the hydronium ion (H_3O^+), which is a measure of the acidity, is 1.34×10^{-3} mol/L, corresponding to a pH of 2.87. What will happen to the acidity when 0.100 mol of sodium acetate ($NaC_2H_3O_2$) is added to 1 L of 0.100 M acetic acid ($HC_2H_3O_2$)? When $NaC_2H_3O_2$ dissolves, it dissociates into sodium ions (Na^+) and acetate ions ($C_2H_3O_2^-$). The acetate ion from the salt is a common ion to the acetic acid equilibrium system and increases the total acetate ion concentration in the solution. As a result the equilibrium shifts to the left, decreasing the hydronium ion concentration and lowering the acidity of the solution. Evidence of this decrease in acidity is shown by the fact that the pH of a solution that is 0.100 M in $HC_2H_3O_2$ and 0.100 M in $NaC_2H_3O_2$ is 4.74. The pH of several different solutions of $HC_2H_3O_2$ and $NaC_2H_3O_2$ is shown in the table that follows. Each time the acetate ion is increased, the pH increases, indicating a further shift in the equilibrium toward un-ionized acetic acid.

Concentration	Change
$[HC_2H_3O_2]$	increase
$[H_3O^+]$	decrease
$[C_2H_3O_2^-]$	increase

Solution	pH
1 L 0.100 M $HC_2H_3O_2$	2.87
1 L 0.100 M $HC_2H_3O_2$ + 0.100 mol $NaC_2H_3O_2$	4.74
1 L 0.100 M $HC_2H_3O_2$ + 0.200 mol $NaC_2H_3O_2$	5.05
1 L 0.100 M $HC_2H_3O_2$ + 0.300 mol $NaC_2H_3O_2$	5.23

In summary, we can say that when the concentration of a reagent on the left side of an equation is increased, the equilibrium shifts to the right. When the concentration of a reagent on the right side of an equation is increased, the equilibrium shifts to the left. In accordance with Le Châtelier's principle, the equilibrium always shifts in the direction that tends to reduce the concentration of the added reactant.

Practice 16.2

Aqueous chromate ion, CrO_4^{2-}, exists in equilibrium with aqueous dichromate ion, $Cr_2O_7^{2-}$, in an acidic solution. What effect will (a) increasing the dichromate ion and (b) adding HCl have on the equilibrium?

$$2\,CrO_4^{2-}(aq) + 2\,H^+(aq) \rightleftharpoons Cr_2O_7^{2-}(aq) + H_2O(l)$$

16.6 Effect of Volume on Equilibrium

Changes in volume significantly affect the reaction rate only when one or more of the reactants or products is a gas and the reaction is run in a closed container. In these cases the effect of decreasing the volume of the reacting gases is equivalent to increasing their concentrations. In the reaction

$$CaCO_3(s) \overset{\Delta}{\rightleftharpoons} CaO(s) + CO_2(g)$$

calcium carbonate decomposes into calcium oxide and carbon dioxide when heated about 825°C. Decreasing the volume of the container speeds up the reverse reaction and causes the equilibrium to shift to the left. Decreasing the volume increases the concentration of CO_2, the only gaseous substance in the reaction.

If the volume of the container is decreased, the pressure of the gas will increase. In a system composed entirely of gases, this decrease in the volume of the container will cause the reaction and the equilibrium to shift to the side that contains the smaller number of molecules. To clarify what's happening: When the container volume is decreased, the pressure in the container is increased. The system tries to lower this pressure by reducing the number of molecules. Let's consider an example that shows these effects.

Prior to World War I, Fritz Haber (1868–1934) invented the first major process for the fixation of nitrogen. In this process nitrogen and hydrogen are reacted together in the presence of a catalyst at moderately high temperature and pressure to produce ammonia:

$$N_2(g) + 3H_2(g) \rightleftharpoons 2NH_3(g) + 92.5\,kJ$$

| 1 mol | 3 mol | 2 mol |
| 1 volume | 3 volumes | 2 volumes |

Haber received the Nobel Prize in chemistry for this process in 1918.

The left side of the equation in the Haber process represents 4 mol of gas combining to give 2 mol of gas on the right side of the equation. A decrease in the volume of the container shifts the equilibrium to the right. This decrease in volume results in a higher concentration of both reactants and products. The equilibrium shifts to the right toward fewer molecules.

Gaseous ammonia is often used to add nitrogen to the fields before planting and during early growth.

When the total number of gaseous molecules on both sides of an equation is the same, a change in volume does not cause an equilibrium shift. The following reaction is an example:

$N_2(g)$	+	$O_2(g)$	$\rightleftharpoons$	$2\,NO(g)$
1 mol		1 mol		2 mol
1 volume		1 volume		2 volumes
6.022×10^{23} molecules		6.022×10^{23} molecules		$2(6.022 \times 10^{23})$ molecules

When the volume of the container is decreased, the rate of both the forward and the reverse reactions will increase because of the higher concentrations of N_2, O_2, and NO. But the equilibrium will not shift because the number of molecules is the same on both sides of the equation and the effects on concentration are the same on both forward and reverse rates.

Example 16.1 What effect would a decrease in volume of the container have on the position of equilibrium in these reactions?
(a) $2\,SO_2(g) + O_2(g) \rightleftharpoons 2\,SO_3(g)$
(b) $H_2(g) + Cl_2(g) \rightleftharpoons 2\,HCl(g)$
(c) $N_2O_4(g) \rightleftharpoons 2\,NO_2(g)$

SOLUTION
(a) The equilibrium will shift to the right because the substance on the right has a smaller number of moles than those on the left.
(b) The equilibrium position will be unaffected because the moles of gases on both sides of the equation are the same.
(c) The equilibrium will shift to the left because $N_2O_4(g)$ represents the smaller number of moles.

Practice 16.3

What effect would a decrease in the container's volume have on the position of the equilibrium in these reactions?

(a) $2\,NO(g) + Cl_2(g) \rightleftharpoons 2\,NOCl(g)$
(b) $COBr_2(g) \rightleftharpoons CO(g) + Br_2(g)$

16.7 Effect of Temperature on Equilibrium

When the temperature of a system is raised, the rate of reaction increases because of increased kinetic energy and more frequent collisions of the reacting species. In a reversible reaction, the rate of both the forward and the reverse reactions is increased by an increase in temperature; however, the reaction that absorbs heat increases to a greater extent, and the equilibrium shifts to favor that reaction.

An increase in temperature generally increases the rate of reaction. Molecules at elevated temperatures have more kinetic energy; their collisions are thus more likely to result in a reaction.

When heat is applied to a system in equilibrium, the reaction that absorbs heat is favored. When the process, as written, is endothermic, the forward reaction is increased. When the reaction is exothermic, the reverse reaction is

High temperatures can cause the destruction or decomposition of the reactants or products.

favored. In this sense heat may be treated as a reactant in endothermic reactions or as a product in exothermic reactions. Therefore temperature is analogous to concentration when applying Le Châtelier's principle to heat effects on a chemical reaction.

Hot coke (C) is a very reactive element. In the reaction

$$C(s) + CO_2(g) + heat \rightleftharpoons 2\,CO(g)$$

very little if any CO is formed at room temperature. At 1000°C, the equilibrium mixture contains about an equal number of moles of CO and CO_2. Since the reaction is endothermic, the equilibrium is shifted to the right at higher temperatures.

When phosphorus trichloride reacts with dry chlorine gas to form phosphorus pentachloride, the reaction is exothermic:

$$PCl_3(l) + Cl_2(g) \rightleftharpoons PCl_5(s) + 88\,kJ$$

Heat must continuously be removed during the reaction to obtain a good yield of the product. According to Le Châtelier's principle, heat will cause the product, PCl_5, to decompose, re-forming PCl_3 and Cl_2. The equilibrium mixture at 200°C contains 52% PCl_5, and at 300°C it contains 3% PCl_5, verifying that heat causes the equilibrium to shift to the left.

Light sticks. The chemical reaction that produces light in these light sticks is endothermic. Placing the light stick in hot water (right) favors this reaction, producing a brighter light than when the light stick is in ice water (left).

What effect would an increase in temperature have on the position of the equilibrium in these reactions?

Example 16.2

$$4\,HCl(g) + O_2(g) \rightleftharpoons 2\,H_2O(g) + 2\,Cl_2(g) + 95.4\,kJ \quad (1)$$
$$H_2(g) + Cl_2(g) \rightleftharpoons 2\,HCl(g) + 185\,kJ \quad (2)$$
$$CH_4(g) + 2\,O_2(g) \rightleftharpoons CO_2(g) + 2\,H_2O(g) + 890\,kJ \quad (3)$$
$$N_2O_4(g) + 58.6\,kJ \rightleftharpoons 2\,NO_2(g) \quad (4)$$
$$2\,CO_2(g) + 566\,kJ \rightleftharpoons 2\,CO(g) + O_2(g) \quad (5)$$
$$H_2(g) + I_2(g) + 51.9\,kJ \rightleftharpoons 2\,HI(g) \quad (6)$$

Reactions 1, 2, and 3 are exothermic; an increase in temperature will cause the equilibrium to shift to the left. Reactions 4, 5, and 6 are endothermic; an increase in temperature will cause the equilibrium to shift to the right.

SOLUTION

Practice 16.4 _____

What effect would an increase in temperature have on the position of the equilibrium in these reactions?

(a) $2\,SO_2(g) + O_2(g) \rightleftharpoons 2\,SO_3(g) + 198\,kJ$
(b) $H_2(g) + CO_2(g) + 41\,kJ \rightleftharpoons H_2O(g) + CO(g)$

16.8 Effect of Catalysts on Equilibrium

A **catalyst** is a substance that influences the rate of a chemical reaction and can be recovered essentially unchanged at the end of the reaction. A catalyst does not shift the equilibrium of a reaction; it affects only the speed at which the equilibrium is reached. It does this by lowering the activation energy for the

catalyst

Figure 16.3
Energy diagram for an exothermic reaction. Energy is put into the reaction (activation energy) to initiate the process. In the reaction shown, all of the activation energy and the net energy are released as the reaction proceeds to products. Note that the presence of a catalyst lowers the activation energy but does not change the energies of the reactants or the products.

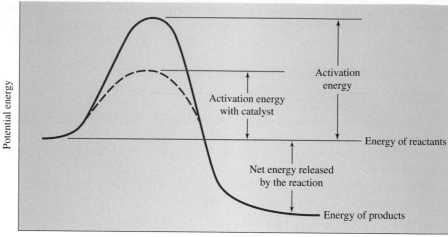

activation energy

reaction (see Figure 16.3). **Activation energy** is the minimum energy required for the reaction to occur. A catalyst speeds up a reaction by lowering the activation energy while not changing the energies of reactants or products. If a catalyst does not affect the equilibrium, then it follows that it must affect the rate of both the forward and the reverse reactions equally.

The reaction between phosphorus trichloride and sulfur is highly exothermic, but it's so slow that very little product, thiophosphoryl chloride, is obtained, even after prolonged heating. When a catalyst such as aluminum chloride is added, the reaction is complete in a few seconds:

$$PCl_3(l) + S(s) \xrightarrow{AlCl_3} PSCl_3(l)$$

The lab preparation of oxygen uses manganese dioxide as a catalyst to increase the rates of decomposition of both potassium chlorate and hydrogen peroxide:

$$2\,KClO_3(s) \xrightarrow[\Delta]{MnO_2} 2\,KCl(s) + 3\,O_2(g)$$

$$2\,H_2O_2(aq) \xrightarrow{MnO_2} 2\,H_2O(l) + O_2(g)$$

Catalysts are extremely important to industrial chemistry. Hundreds of chemical reactions that are otherwise too slow to be of practical value have been put to commercial use once a suitable catalyst was found. And in the area of biochemistry, catalysts are of supreme importance because nearly all chemical reactions in all forms of life are completely dependent on biochemical catalysts known as *enzymes*.

16.9 Equilibrium Constants

In a reversible chemical reaction at equilibrium, the concentrations of the reactants and products are constant. At equilibrium, the rates of the forward and reverse reactions are equal, and an equilibrium constant expression can be written relating the products to the reactants. For the general reaction

$$a\mathrm{A} + b\mathrm{B} \rightleftharpoons c\mathrm{C} + d\mathrm{D}$$

at a given temperature, the equilibrium constant expression can be written as

$$K_{eq} = \frac{[C]^c[D]^d}{[A]^a[B]^b}$$

where K_{eq}, the **equilibrium constant**, is constant at a particular temperature. The quantities in brackets are the concentrations of each substance in moles per liter. The superscript letters a, b, c, and d are the coefficients of the substances in the balanced equation. According to convention, we place the concentrations of the products (the substances on the right side of the equation as written) in the numerator and the concentrations of the reactants in the denominator.

equilibrium constant, K_{eq}

Note: The exponents are the same as the coefficients in the balanced equation.

Example 16.3

Write equilibrium constant expressions for

(a) $3 H_2(g) + N_2(g) \rightleftharpoons 2 NH_3(g)$
(b) $CO(g) + 2 H_2(g) \rightleftharpoons CH_3OH(g)$

SOLUTION

(a) The only product, NH_3, has a coefficient of 2. Therefore, the numerator will be $[NH_3]^2$. Two reactants are present: H_2, with a coefficient of 3, and N_2, with a coefficient of 1. The denominator will thus be $[H_2]^3[N_2]$. The equilibrium constant expression is

$$K_{eq} = \frac{[NH_3]^2}{[H_2]^3[N_2]}$$

(b) For this equation, the numerator is $[CH_3OH]$ and the denominator is $[CO][H_2]^2$. The equilibrium constant expression is

$$K_{eq} = \frac{[CH_3OH]}{[CO][H_2]^2}$$

Practice 16.5

Write equilibrium constant expressions for

(a) $2 N_2O_5(g) \rightleftharpoons 4 NO_2(g) + O_2(g)$
(b) $4 NH_3(g) + 3 O_2(g) \rightleftharpoons 2 N_2(g) + 6 H_2O(g)$

The magnitude of an equilibrium constant indicates the extent to which the forward and reverse reactions take place. When K_{eq} is greater than 1, the amount of products at equilibrium is greater than the amount of reactants. When K_{eq} is less than 1, the amount of reactants at equilibrium is greater than the amount of products. A very large value for K_{eq} indicates that the forward reaction goes essentially to completion. A very small K_{eq} means that the reverse reaction goes nearly to completion and that the equilibrium is far to the left (toward the reactants). Consider the following two examples:

$$H_2(g) + I_2(g) \rightleftharpoons 2 HI(g) \qquad K_{eq} = 54.8 \text{ at } 425°C$$

$$COCl_2(g) \rightleftharpoons CO(g) + Cl_2(g) \qquad K_{eq} = 7.6 \times 10^{-4} \text{ at } 400°C$$

In the first example, K_{eq} indicates that more product than reactant exists at equilibrium.

In the second equation, K_{eq} indicates that $COCl_2$ is stable and that very little decomposition to CO and Cl_2 occurs at 400°C. The equilibrium is far to the left.

When the molar concentrations of all species in an equilibrium reaction are known, the K_{eq} can be calculated by substituting the concentrations into the equilibrium constant expression.

Example 16.4

Calculate K_{eq} for the following reaction based on concentrations of $PCl_5 = 0.030$ mol/L, $PCl_3 = 0.97$ mol/L, and $Cl_2 = 0.97$ mol/L at 300°C.

$$PCl_5(g) \rightleftharpoons PCl_3(g) + Cl_2(g)$$

SOLUTION

First write the K_{eq} expression; then substitute the respective concentrations into this equation and solve:

$$K_{eq} = \frac{[PCl_3][Cl_2]}{[PCl_5]} = \frac{(0.97)(0.97)}{(0.030)} = 31$$

Remember: Units are not
included for K_{eq}.

This K_{eq} is considered to be a fairly large value, indicating that at 300°C the decomposition of PCl_5 proceeds far to the right.

Practice 16.6 _____

Calculate the K_{eq} for this reaction. Is the forward or the reverse reaction favored?

$$2\,NO(g) + O_2(g) \rightleftharpoons 2\,NO_2(g)$$

when $[NO] = 0.050\,M$, $[O_2] = 0.75\,M$, and $[NO_2] = 0.25\,M$.

16.10 Ion Product Constant for Water

We've seen that water ionizes to a slight degree. This ionization is represented by these equilibrium equations:

$$H_2O + H_2O \rightleftharpoons H_3O^+ + OH^- \tag{1}$$
$$H_2O \rightleftharpoons H^+ + OH^- \tag{2}$$

Equation 1 is the more accurate representation of the equilibrium because free protons (H^+) do not exist in water. Equation 2 is a simplified and often-used representation of the water equilibrium. The actual concentration of H^+ produced in pure water is minute and amounts to only 1.00×10^{-7} mol/L at 25°C. In pure water,

$$[H^+] = [OH^-] = 1.00 \times 10^{-7}\,mol/L$$

since both ions are produced in equal molar amounts, as shown in equation 2.

The $H_2O \rightleftharpoons H^+ + OH^-$ equilibrium exists in water and in all water solutions. A special equilibrium constant called the **ion product constant for water, K_w**, applies to this equilibrium. The constant K_w is defined as the product of the H^+ ion concentration and the OH^- ion concentration, each in moles per liter:

ion product constant
for water, K_w

$$K_w = [H^+][OH^-]$$

The numerical value of K_w is 1.00×10^{-14}, since for pure water at 25°C,

$$K_w = [H^+][OH^-] = (1.00 \times 10^{-7})(1.00 \times 10^{-7}) = 1.00 \times 10^{-14}$$

Table 16.1 Relationship of H^+ and OH^- Concentrations in Water Solutions

$[H^+]$	$[OH^-]$	K_w	pH	pOH
1.00×10^{-2}	1.00×10^{-12}	1.00×10^{-14}	2.00	12.00
1.00×10^{-4}	1.00×10^{-10}	1.00×10^{-14}	4.00	10.00
2.00×10^{-6}	5.00×10^{-9}	1.00×10^{-14}	5.70	8.30
1.00×10^{-7}	1.00×10^{-7}	1.00×10^{-14}	7.00	7.00
1.00×10^{-9}	1.00×10^{-5}	1.00×10^{-14}	9.00	5.00

The value of K_w for all water solutions at 25°C is the constant 1.00×10^{-14}. It is important to realize that as the concentration of one of these ions, H^+ or OH^-, increases, the other decreases. However, the product of $[H^+]$ and $[OH^-]$ always equals 1.00×10^{-14}. This relationship can be seen in the examples shown in Table 16.1. If the concentration of one ion is known, the concentration of the other can be calculated from the K_w expression.

$$K_w = [H^+][OH^-] \qquad [H^+] = \frac{K_w}{[OH^-]} \qquad [OH^-] = \frac{K_w}{[H^+]}$$

What is the concentration of (a) H^+ and (b) OH^- in a 0.001 M HCl solution? Remember that HCl is 100% ionized.

Example 16.5

(a) Since all the HCl is ionized, $H^+ = 0.001$ mol/L, or 1×10^{-3} mol/L:

SOLUTION

$$\text{HCl} \longrightarrow \text{H}^+ + \text{Cl}^-$$
$$\qquad\quad 0.001\,M \quad 0.001\,M$$

$$[H^+] = 1 \times 10^{-3}\,\text{mol/L}$$

(b) To calculate the $[OH^-]$ in this solution, use the following equation and substitute the values for K_w and $[H^+]$:

$$[OH^-] = \frac{K_w}{[H^+]}$$

$$[OH^-] = \frac{1.00 \times 10^{-14}}{1 \times 10^{-3}} = 1 \times 10^{-11}\,\text{mol/L}$$

Practice 16.7

Determine the $[H^+]$ and $[OH^-]$ in

(a) $5.0 \times 10^{-5}\,M\,\text{HNO}_3$ (b) $2.0 \times 10^{-6}\,M\,\text{KOH}$

What is the pH of a 0.010 M NaOH solution? Assume that NaOH is 100% ionized.

Example 16.6

Since all the NaOH is ionized, $[OH^-] = 0.010$ mol/L or 1.0×10^{-2} mol/L.

SOLUTION

$$\text{NaOH} \longrightarrow \text{Na}^+ + \text{OH}^-$$
$$\qquad\quad\;\; 0.010\,M \quad 0.010\,M$$

To find the pH of the solution, we first calculate the H^+ ion concentration. Use the following equation and substitute the values for K_w and $[OH^-]$:

$$[H^+] = \frac{K_w}{[OH^-]} = \frac{1.00 \times 10^{-14}}{1.0 \times 10^{-2}} = 1.0 \times 10^{-12}\,\text{mol/L}$$

$$pH = -\log[H^+] = -\log(1.0 \times 10^{-12}) = 12.00$$

Practice 16.8

Determine the pH for the following solutions:

(a) $5.0 \times 10^{-5}\,M\,HNO_3$ (b) $2.0 \times 10^{-6}\,M\,KOH$

Just as pH is used to express the acidity of a solution, pOH is used to express the basicity of an aqueous solution. The pOH is related to the OH^- ion concentration in the same way that the pH is related to the H^+ ion concentration:

$$pOH = -\log[OH^-]$$

Thus a solution in which $[OH^-] = 1.0 \times 10^{-2}$, as in Example 16.6, will have $pOH = 2.00$.

In pure water, where $[H^+] = 1.00 \times 10^{-7}$ and $[OH^-] = 1.00 \times 10^{-7}$, the pH is 7.0 and the pOH is 7.0. The sum of the pH and pOH is always 14.0:

$$pH + pOH = 14.00$$

In Example 16.6, the pH can also be found by first calculating the pOH from the OH^- ion concentration and then subtracting from 14.00.

$$pH = 14.00 - pOH = 14.00 - 2.00 = 12.00$$

Table 16.1 summarizes the relationship between $[H^+]$ and $[OH^-]$ in water solutions.

16.11 Ionization Constants

In addition to K_w, several other equilibrium constants are commonly used. Strong acids are essentially 100% ionized. Weak acids are only slightly ionized. Let's consider the equilibrium constant for acetic acid in solution. Because it is a weak acid, an equilibrium is established between molecular $HC_2H_3O_2$ and its ions in solution:

$$HC_2H_3O_2(aq) \rightleftharpoons H^+(aq) + C_2H_3O_2^-(aq)$$

The ionization constant expression is the concentration of the products divided by the concentration of the reactants:

$$K_a = \frac{[H^+][C_2H_3O_2^-]}{[HC_2H_3O_2]}$$

acid ionization constant, K_a The constant is called the **acid ionization constant, K_a**, a special type of equilibrium constant. The concentration of water in the solution is large compared to other concentrations and does not change appreciably. It is therefore part of the constant K_{eq}.

At 25°C, a 0.100 M $HC_2H_3O_2$ solution is 1.34% ionized and has an $[H^+]$ of 1.34×10^{-3} mol/L. From this information we can calculate the ionization constant for acetic acid.

A 0.100 M solution initially contains 0.100 mol of acetic acid per liter. Of this 0.100 mol, only 1.34%, or 1.34×10^{-3} mol, is ionized, which gives an $[H^+] = 1.34 \times 10^{-3}$ mol/L. Because each molecule of acid that ionizes yields one H^+ and one $C_2H_3O_2^-$, the concentration of $C_2H_3O_2^-$ ions is also 1.34×10^{-3} mol/L. This ionization leaves $0.100 - 0.00134 = 0.099$ mol/L of un-ionized acetic acid.

Acid	Initial concentration (mol/L)	Equilibrium concentration (mol/L)
$[HC_2H_3O_2]$	0.100	0.099
$[H^+]$	0	0.00134
$[C_2H_3O_2^-]$	0	0.00134

Substituting these concentrations in the equilibrium expression, we obtain the value for K_a:

$$K_a = \frac{[H^+][C_2H_3O_2^-]}{[HC_2H_3O_2]} = \frac{(1.34 \times 10^{-3})(1.34 \times 10^{-3})}{(0.099)} = 1.8 \times 10^{-5}$$

The K_a for acetic acid, 1.8×10^{-5}, is small and indicates that the position of the equilibrium is far toward the un-ionized acetic acid. In fact, a 0.100 M acetic acid solution is 99% un-ionized.

Once the K_a for acetic acid is established, it can be used to describe other systems containing H^+, $C_2H_3O_2^-$, and $HC_2H_3O_2$ in equilibrium at 25°C. The ionization constants for several other weak acids are listed in Table 16.2.

Table 16.2 Ionization Constants (K_a) of Weak Acids at 25°C

Acid	Formula	K_a	Acid	Formula	K_a
Acetic	$HC_2H_3O_2$	1.8×10^{-5}	Hydrocyanic	HCN	4.0×10^{-10}
Benzoic	$HC_7H_5O_2$	6.3×10^{-5}	Hypochlorous	HClO	3.5×10^{-8}
Carbolic (phenol)	HC_6H_5O	1.3×10^{-10}	Nitrous	HNO_2	4.5×10^{-4}
Cyanic	HCNO	2.0×10^{-4}	Hydrofluoric	HF	6.5×10^{-4}
Formic	$HCHO_2$	1.8×10^{-4}			

What is the $[H^+]$ in a 0.50 M $HC_2H_3O_2$ solution? The ionization constant, K_a, for $HC_2H_3O_2$ is 1.8×10^{-5}.

Example 16.7

To solve this problem, first write the equilibrium equation and the K_a expression:

SOLUTION

$$HC_2H_3O_2 \rightleftharpoons H^+ + C_2H_3O_2^- \qquad K_a = \frac{[H^+][C_2H_3O_2^-]}{[HC_2H_3O_2]} = 1.8 \times 10^{-5}$$

We know that the initial concentration of $HC_2H_3O_2$ is 0.50 M. We also know from the ionization equation that one $C_2H_3O_2^-$ is produced for every H^+ produced; that is, the $[H^+]$ and the $[C_2H_3O_2^-]$ are equal. To solve, let $Y = [H^+]$, which also equals the $[C_2H_3O_2^-]$. The un-ionized $[HC_2H_3O_2]$ remaining will then be $0.50 - Y$, the starting concentration minus the amount that ionized:

$$[H^+] = [C_2H_3O_2^-] = Y \qquad [HC_2H_3O_2] = 0.50 - Y$$

	Initial	Equilibrium
$[H^+]$	0	Y
$[C_2H_3O_2^-]$	0	Y
$[HC_2H_3O_2]$	0.5	$0.5 - Y$

Substituting these values into the K_a expression, we obtain

$$K_a = \frac{(Y)(Y)}{0.50 - Y} = \frac{Y^2}{0.50 - Y} = 1.8 \times 10^{-5}$$

The quadratic equation is
$$y = \frac{-b \pm \sqrt{b^2 - 4ac}}{2a} \text{ for the}$$
equation $ay^2 + by + c = 0$

An exact solution of this equation for Y requires the use of a mathematical equation known as the *quadratic equation*. However, an approximate solution is obtained if we assume that Y is small and can be neglected compared with 0.50. Then $0.50 - Y$ will be equal to approximately 0.50. The equation now becomes

$$\frac{Y^2}{0.50} = 1.8 \times 10^{-5}$$

$$Y^2 = 0.50 \times 1.8 \times 10^{-5} = 0.90 \times 10^{-5} = 9.0 \times 10^{-6}$$

Taking the square root of both sides of the equation, we obtain

$$Y = \sqrt{9.0 \times 10^{-6}} = 3.0 \times 10^{-3} \,\text{mol/L}$$

Thus, the $[H^+]$ is approximately 3.0×10^{-3} mol/L in a 0.50 M $HC_2H_3O_2$ solution. The exact solution to this problem, using the quadratic equation, gives a value of 2.99×10^{-3} mol/L for $[H^+]$, showing that we were justified in neglecting Y compared with 0.50.

Practice 16.9 _____

Calculate the hydrogen ion concentration in (a) 0.100 M hydrocyanic acid (HCN) solution and (b) 0.0250 M carbolic acid (HC_6H_5O) solution.

Example 16.8   Calculate the percent ionization in a 0.50 M $HC_2H_3O_2$ solution.

SOLUTION   The percent ionization of a weak acid, $HA(aq) \rightleftharpoons H^+(aq) + A^-(aq)$, is found by dividing the concentration of the H^+ or A^- ions at equilibrium by the initial concentration of HA and multiplying by 100. For acetic acid,

$$\frac{\text{concentration of } [H^+] \text{ or } [C_2H_3O_2^-]}{\text{initial concentration of } [HC_2H_3O_2]} \times 100 = \text{percent ionized}$$

To solve this problem, we first need to calculate the $[H^+]$. This calculation has already been done for a 0.50 M $HC_2H_3O_2$ solution:

$$[H^+] = 3.0 \times 10^{-3} \,\text{mol/L in a } 0.50 \, M \text{ solution} \quad \text{(from Example 16.7)}$$

This $[H^+]$ represents a fractional amount of the initial 0.50 M $HC_2H_3O_2$. Therefore

$$\frac{3.0 \times 10^{-3} \,\text{mol/L}}{0.50 \,\text{mol/L}} \times 100 = 0.60\% \text{ ionized}$$

A 0.50 M $HC_2H_3O_2$ solution is 0.60% ionized.

Practice 16.10 _____

Calculate the percent ionization for
(a) 0.100 M hydrocyanic acid (HCN)
(b) 0.0250 M carbolic acid (HC_6H_5O)

16.12 Solubility Product Constant

The **solubility product constant, K_{sp}**, is the equilibrium constant of a slightly soluble salt. To evaluate K_{sp}, consider this example. The solubility of AgCl in water is 1.3×10^{-5} mol/L at 25°C. The equation for the equilibrium between AgCl and its ions in solution is

solubility product constant, K_{sp}

$$AgCl(s) \rightleftharpoons Ag^+(aq) + Cl^-(aq)$$

The equilibrium constant expression is

$$K_{eq} = \frac{[Ag^+][Cl^-]}{[AgCl(s)]}$$

The amount of solid AgCl does not affect the equilibrium system provided that some is present. In other words, the concentration of solid AgCl is constant whether 1 mg or 10 g of the salt are present. Therefore the product obtained by multiplying the two constants K_{eq} and [AgCl(s)] is also a constant. This is the solubility product constant, K_{sp}:

$$K_{eq} \times [AgCl(s)] = [Ag^+][Cl^-] = K_{sp}$$
$$K_{sp} = [Ag^+][Cl^-]$$

The K_{sp} is equal to the product of the $[Ag^+]$ and the $[Cl^-]$, each in moles per liter. When 1.3×10^{-5} mol/L of AgCl dissolves, it produces 1.3×10^{-5} mol/L each of Ag^+ and Cl^-. From these concentrations, the K_{sp} can be calculated:

$$[Ag^+] = 1.3 \times 10^{-5} \text{ mol/L} \qquad [Cl^-] = 1.3 \times 10^{-5} \text{ mol/L}$$
$$K_{sp} = [Ag^+][Cl^-] = (1.3 \times 10^{-5})(1.3 \times 10^{-5}) = 1.7 \times 10^{-10}$$

Once the K_{sp} value for AgCl is established, it can be used to describe other systems containing Ag^+ and Cl^-.

The K_{sp} expression does not have a denominator. It consists only of the concentrations (mol/L) of the ions in solution. As in other equilibrium expressions, each of these concentrations is raised to a power that is the same number as its coefficient in the balanced equation. Here are equilibrium equations and the K_{sp} expressions for several other substances:

$$AgBr(s) \rightleftharpoons Ag^+(aq) + Br^-(aq) \qquad K_{sp} = [Ag^+][Br^-]$$
$$BaSO_4(s) \rightleftharpoons Ba^{2+}(aq) + SO_4^{2-}(aq) \qquad K_{sp} = [Ba^{2+}][SO_4^{2-}]$$
$$Ag_2CrO_4(s) \rightleftharpoons 2\,Ag^+(aq) + CrO_4^{2-}(aq) \qquad K_{sp} = [Ag^+]^2[CrO_4^{2-}]$$
$$CuS(s) \rightleftharpoons Cu^{2+}(aq) + S^{2-}(aq) \qquad K_{sp} = [Cu^{2+}][S^{2-}]$$
$$Mn(OH)_2(s) \rightleftharpoons Mn^{2+}(aq) + 2\,OH^-(aq) \qquad K_{sp} = [Mn^{2+}][OH^-]^2$$
$$Fe(OH)_3(s) \rightleftharpoons Fe^{3+}(aq) + 3\,OH^-(aq) \qquad K_{sp} = [Fe^{3+}][OH^-]^3$$

Table 16.3 lists K_{sp} values for these and several other substances.

When the product of the molar concentration of the ions in solution (each raised to its proper power) is greater than the K_{sp} for that substance, precipitation will occur. If the ion product is less than the K_{sp} value, no precipitation will occur.

Table 16.3 Solubility Product Constants (K_{sp}) at 25°C

Compound	K_{sp}	Compound	K_{sp}
AgCl	1.7×10^{-10}	CaF$_2$	3.9×10^{-11}
AgBr	5.2×10^{-13}	CuS	8.5×10^{-45}
AgI	8.5×10^{-17}	Fe(OH)$_3$	6.1×10^{-38}
AgC$_2$H$_3$O$_2$	2.1×10^{-3}	PbS	3.4×10^{-28}
Ag$_2$CrO$_4$	1.9×10^{-12}	PbSO$_4$	1.3×10^{-8}
BaCrO$_4$	8.5×10^{-11}	Mn(OH)$_2$	2.0×10^{-13}
BaSO$_4$	1.5×10^{-9}		

Example 16.9

Write K_{sp} expressions for AgI and PbI_2, both of which are slightly soluble salts.

SOLUTION   First write the equilibrium equations:

$$AgI(s) \rightleftharpoons Ag^+(aq) + I^-(aq)$$

$$PbI_2(s) \rightleftharpoons Pb^{2+}(aq) + 2\,I^-(aq)$$

Since the concentration of the solid crystals is constant, the K_{sp} equals the product of the molar concentrations of the ions in solution. In the case of PbI_2, the $[I^-]$ must be squared:

$$K_{sp} = [Ag^+][I^-] \qquad K_{sp} = [Pb^{2+}][I^-]^2$$

Example 16.10

The K_{sp} value for lead sulfate is 1.3×10^{-8}. Calculate the solubility of $PbSO_4$ in grams per liter.

SOLUTION   First write the equilibrium equation and the K_{sp} expression:

$$PbSO_4 \rightleftharpoons Pb^{2+}(aq) + SO_4^{2-}(aq)$$

$$K_{sp} = [Pb^{2+}][SO_4^{2-}] = 1.3 \times 10^{-8}$$

Since the lead sulfate that is in solution is completely dissociated, the $[Pb^{2+}]$ or $[SO_4^{2-}]$ is equal to the solubility of $PbSO_4$ in moles per liter. Let

$$Y = [Pb^{2+}] = [SO_4^{2-}]$$

Substitute Y into the K_{sp} equation and solve:

$$[Pb^{2+}][SO_4^{2-}] = (Y)(Y) = 1.3 \times 10^{-8}$$
$$Y^2 = 1.3 \times 10^{-8}$$
$$Y = 1.1 \times 10^{-4}\,mol/L$$

The solubility of $PbSO_4$ therefore is $1.1 \times 10^{-4}\,mol/L$. Now convert mol/L to g/L:

1 mol of $PbSO_4$ has a mass of $(207.2\,g + 32.07\,g + 64.00\,g)\ 303.3\,g$

$$\left(\frac{1.1 \times 10^{-4}\ \cancel{mol}}{L}\right)\left(\frac{303.3\,g}{\cancel{mol}}\right) = 3.3 \times 10^{-2}\,g/L$$

The solubility of $PbSO_4$ is $3.3 \times 10^{-2}\,g/L$.

Practice 16.11 _____

Write the K_{sp} expression for
(a) $Cr(OH)_3$ (b) $Cu_3(PO_4)_2$

Practice 16.12 _____

The K_{sp} value for CuS is 8.5×10^{-45}. Calculate the solubility of CuS in grams per liter.

An ion added to a solution already containing that ion is called a *common ion*. When a common ion is added to an equilibrium solution of a weak electrolyte or a slightly soluble salt, the equilibrium shifts according to Le Châtelier's principle. For example, when silver nitrate ($AgNO_3$) is added to a saturated solution of AgCl,

$$AgCl(s) \rightleftharpoons Ag^+ + Cl^-$$

the equilibrium shifts to the left due to the increase in the $[Ag^+]$. As a result, the $[Cl^-]$ and the solubility of AgCl decreases. The AgCl and $AgNO_3$ have the common ion Ag^+. A shift in the equilibrium position upon addition of an ion already contained in the solution is known as the **common ion effect.**

common ion effect

Silver nitrate is added to a saturated AgCl solution until the $[Ag^+]$ is $0.10\ M$. What will be the $[Cl^-]$ remaining in solution?

Example 16.11

This is an example of the common ion effect. The addition of $AgNO_3$ puts more Ag^+ in solution; the Ag^+ combines with Cl^- and causes the equilibrium to shift to the left, reducing the $[Cl^-]$ in solution. After the addition of Ag^+ to the mixture, the $[Ag^+]$ and $[Cl^-]$ in solution are no longer equal.

SOLUTION

We use the K_{sp} to calculate the $[Cl^-]$ remaining in solution. The K_{sp} is constant at a particular temperature and remains the same no matter how we change the concentration of the species involved:

$$K_{sp} = [Ag^+][Cl^-] = 1.7 \times 10^{-10} \qquad [Ag^+] = 0.10\ mol/L$$

We then substitute the $[Ag^+]$ into the K_{sp} expression and calculate the $[Cl^-]$:

$$[0.10][Cl^-] = 1.7 \times 10^{-10}$$

$$[Cl^-] = \frac{1.7 \times 10^{-10}}{0.10} = 1.7 \times 10^{-9}\ mol/L$$

This calculation shows a 10,000-fold reduction of Cl^- ions in solution. It illustrates that Cl^- ions may be quantitatively removed from solution with an excess of Ag^+ ions.

Practice 16.13 _____

Sodium sulfate (Na_2SO_4) is added to a saturated solution of $BaSO_4$ until the concentration of the sulfate ion is $2.0 \times 10^{-2}\ M$. What will be the concentration of the Ba^{2+} ions remaining in solution?

16.13 Acid–Base Properties of Salts

hydrolysis   **Hydrolysis** is the term used for the general reaction in which a water molecule is split. For example, the net ionic hydrolysis reaction for a sodium acetate solution is

$$C_2H_3O_2^-(aq) + HOH(l) \rightleftharpoons HC_2H_3O_2(aq) + OH^-(aq)$$

In this reaction the water molecule is split, with the H^+ combining with $C_2H_3O_2^-$ to give the weak acid $HC_2H_3O_2$ and the OH^- going into solution, making the solution more basic.

Salts that contain an ion of a weak acid undergo hydrolysis. For example, a $0.10\ M$ NaCN solution has a pH of 11.1. The hydrolysis reaction that causes this solution to be basic is

$$CN^-(aq) + HOH(l) \rightleftharpoons HCN(aq) + OH^-(aq)$$

If a salt contains the ion of a weak base, the ion produces an acidic solution by transferring a proton to water. An example is ammonium chloride, which produces the NH_4^+ and Cl^- in solution. The NH_4^+ dissociates to give an acidic solution:

$$NH_4^+(aq) + H_2O(l) \rightleftharpoons NH_3(aq) + H_3O^+(aq)$$

The ions of a salt derived from a strong acid and a strong base, such as NaCl, do not undergo hydrolysis and thus form neutral solutions. Table 16.4 lists the ionic composition of various salts and the nature of the aqueous solutions that they form.

Table 16.4   Ionic Composition of Salts and the Nature of the Aqueous Solutions They Form

Type of salt	Nature of aqueous solution	Examples
Weak base–strong acid	Acidic	NH_4Cl, NH_4NO_3
Strong base–weak acid	Basic	$NaC_2H_3O_2$, K_2CO_3
Weak base–weak acid	Depends on the salt	$NH_4C_2H_3O_2$, NH_4NO_2
Strong base–strong acid	Neutral	$NaCl$, KBr

Practice 16.14 _____

Indicate whether these salts would produce an acidic, basic, or neutral aqueous solution:
(a) KCN   (b) $NaNO_3$ (c) NH_4Br

16.14 Buffer Solutions: The Control of pH

The control of pH within narrow limits is critically important in many chemical applications and vitally important in many biological systems. For example, human blood must be maintained between pH 7.35 and 7.45 for the efficient transport of oxygen from the lungs to the cells. This narrow pH range is maintained by buffer systems in the blood.

buffer solution   A **buffer solution** resists changes in pH when diluted or when small amounts of acid or base are added. Two common types of buffer solutions are (1) a weak acid mixed with a salt of its conjugate base and (2) a weak base mixed with a salt of its conjugate acid.

The transport of oxygen and carbon dioxide between the lungs and tissues is a complex process that involves several reversible reactions, each of which behaves in accordance with Le Châtelier's principle.

The binding of oxygen to hemoglobin is a reversible reaction. The oxygen molecule must attach to the hemoglobin (Hb) and then later detach. The equilibrium equation for this reaction can be written:

$$Hb + O_2 \rightleftharpoons HbO_2$$

In the lungs the concentration of oxygen is high and favors the forward reaction. Oxygen quickly binds to the hemoglobin until it is saturated with oxygen.

In the tissues the concentration of oxygen is lower and in accordance with Le Châtelier's principle: The equilibrium position shifts to the left and the hemoglobin releases oxygen to the tissues. Approximately 45% of the oxygen diffuses out of the capillaries into the tissues, where it may be picked up by *myoglobin*, another carrier molecule.

Myoglobin functions as an oxygen-storage molecule, holding the oxygen until it is required in the energy-producing portions of the cell. The reaction between myoglobin (Mb) and oxygen can be written as an equilibrium reaction:

$$Mb + O_2 \rightleftharpoons MbO_2$$

The hemoglobin and myoglobin equations are very similar, so what accounts for the transfer of the oxygen from the hemoglobin to the myoglobin? Although both equilibria involve similar interactions, the affinity between oxy-

Oxygen and carbon dioxide are exchanged in the red blood cells when they are in capillaries.

gen and hemoglobin is different from the affinity between myoglobin and oxygen. In the tissues the position of the hemoglobin equilibrium is such that it is 55% saturated with oxygen, whereas the myoglobin is at 90% oxygen saturation. Under these conditions hemoglobin will release oxygen, while myoglobin will bind oxygen. Thus oxygen is loaded onto hemoglobin in the lungs and unloaded in the tissue's cells.

Carbon dioxide produced in the cells must be removed from the tissues. Oxygen-depleted hemoglobin molecules accomplish this by becoming car-

riers of carbon dioxide. The carbon dioxide does not bind at the heme site as the oxygen does, but rather at one end of the protein chain. When carbon dioxide dissolves in water, some of the CO_2 reacts to release hydrogen ions:

$$CO_2 + H_2O \rightleftharpoons HCO_3^- + H^+$$

To facilitate the removal of CO_2 from the tissues, this equilibrium needs to be moved toward the right. This shift is accomplished by the removal of H^+ from the tissues by the hemoglobin molecule. The deoxygenated hemoglobin molecule can bind H^+ ions as well as CO_2. In the lungs this whole process is reversed so the CO_2 is removed from the hemoglobin and exhaled.

Molecules that are similar in structure to the oxygen molecule can become involved in competing equilibria. Hemoglobin is capable of binding with carbon monoxide (CO), nitrogen monoxide (NO), and cyanide (CN^-). The extent of the competition depends on the affinity. Since these molecules have a greater affinity for hemoglobin than oxygen, they will effectively displace oxygen from hemoglobin. For example,

$$HbO_2 + CO \rightleftharpoons HbCO + O_2$$

Since the affinity of hemoglobin for CO is 150 times stronger than its affinity for oxygen, the equilibrium position lies far to the right. This explains why CO is a poisonous substance and why oxygen is administered to victims of CO poisoning. The hemoglobin molecules can only transport oxygen if the CO is released and the oxygen shifts the equilibrium toward the left.

The action of a buffer system can be understood by considering a solution of acetic acid and sodium acetate. The weak acid ($HC_2H_3O_2$) is mostly un-ionized and is in equilibrium with its ions in solution. The sodium acetate is completely ionized:

$$HC_2H_3O_2(aq) \rightleftharpoons H^+(aq) + C_2H_3O_2^-(aq)$$
$$NaC_2H_3O_2(aq) \longrightarrow Na^+(aq) + C_2H_3O_2^-(aq)$$

Because the sodium acetate is completely ionized, the solution contains a much higher concentration of acetate ions than would be present if only acetic acid were in solution. The acetate ion represses the ionization of acetic acid and

The aquariums at the MGM Grand Hotel are a buffer system.

also reacts with water, causing the solution to have a higher pH (be more basic) than an acetic acid solution (see Section 16.5). Thus, a 0.1 M acetic acid solution has a pH of 2.87, but a solution that is 0.1 M in acetic acid and 0.1 M in sodium acetate has a pH of 4.74. This difference in pH is the result of the common ion effect.

A buffer solution has a built-in mechanism that counteracts the effect of adding acid or base. Consider the effect of adding HCl or NaOH to an acetic acid–sodium acetate buffer. When a small amount of HCl is added, the acetate ions of the buffer combine with the H^+ ions from HCl to form un-ionized acetic acid, thus neutralizing the added acid and maintaining the approximate pH of the solution. When NaOH is added, the OH^- ions react with acetic acid to neutralize the added base and thus maintain the approximate pH. The equations for these reactions are

$$H^+(aq) + C_2H_3O_2^-(aq) \rightleftharpoons HC_2H_3O_2(aq)$$

$$OH^-(aq) + HC_2H_3O_2(aq) \rightleftharpoons H_2O(l) + C_2H_3O_2^-(aq)$$

Data comparing the changes in pH caused by adding HCl and NaOH to pure water and to an acetic acid–sodium acetate buffer solution are shown in Table 16.5. Notice that adding 0.01 mol HCl or 0.01 mol NaOH to 100 mL water changes the pH of the solution by 5 units. But adding the same amount of HCl or NaOH to a buffer solution causes a change of only 0.08 or 0.09 pH units. So buffers really protect the pH of a solution.

Buffers cannot withstand the addition of large amounts of acids or bases. They overpower the capacity of the buffer to absorb acid or base. The maximum buffering effect occurs when the weak acid or base and its conjugate are of equal molar concentrations.

The human body has a number of buffer systems. One of these, the hydrogen carbonate–carbonic acid buffer, $HCO_3^- — H_2CO_3$, maintains the blood plasma at a pH of 7.4. The phosphate system, $HPO_4^{2-} — H_2PO_4^-$, is an important buffer in the red blood cells, as well as in other places in the body.

Table 16.5 Changes in pH Caused by the Addition of HCl and NaOH

Solution	pH	Change in pH
H_2O (1000 mL)	7	—
H_2O + 0.010 mol HCl	2	5
H_2O + 0.010 mol NaOH	12	5
Buffer solution (1000 mL)		
0.10 M $HC_2H_3O_2$ + 0.10 M $NaC_2H_3O_2$	4.74	—
Buffer + 0.010 mol HCl	4.66	0.08
Buffer + 0.010 mol NaOH	4.83	0.09

Chapter 16 Review

16.1 Reversible Reactions

KEY TERM

Reversible chemical reaction

- In a reversible chemical reaction, the products formed react to produce the original reaction mixture.
- The forward reaction is called the reaction to the right.
- The reverse reaction is called the reaction to the left.

16.2 Rates of Reaction

KEY TERM

Chemical kinetics

- The rate of a reaction is variable and depends on:
  - Concentration of reacting species
  - Temperature
  - Presence of catalysts
  - Nature of the reactants

16.3 Chemical Equilibrium

KEY TERMS

Equilibrium
Chemical equilibrium

- A system at equilibrium is dynamic.
- At chemical equilibrium, the rate of the forward reaction equals the rate of the reverse reaction.

16.4 Le Châtelier's Principle

KEY TERM

Le Châtelier's principle

- If stress is applied to a system at equilibrium, the system will respond in such a way as to relive the stress and restore equilibrium under a new set of conditions.

16.5 Effect of Concentration on Equilibrium

- When the concentration of a reactant is increased, the equilibrium shifts toward the right.
- When the concentration of a product on the right side of the equation is increased, the equilibrium shifts to the left.
- If the concentration of a reactant is decreased, the equilibrium shifts toward the side where the reactant that is decreased is.

16.6 Effect of Volume on Equilibrium

- In gaseous reactions:
  - A decrease in volume causes the equilibrium position to shift toward the side of the reaction with the fewest molecules.
  - If the number of molecules is the same on both sides of a reaction, a decrease in volume has no effect on the equilibrium position.

16.7 Effect of Temperature on Equilibrium

- When heat is added to a reaction, the side of the equation that absorbs heat is favored:
  - If the reaction is endothermic, the forward reaction increases.
  - If the reaction is exothermic, the reverse reaction increases.

16.8 Effect of Catalysts on Equilibrium

KEY TERMS

Catalyst
Activation energy

- A catalyst does not shift the equilibrium of a reaction; it only affects the speed at which equilibrium is reached.
- The activation energy for a reaction is the minimum energy required for the reaction to occur:
  - A catalyst lowers the activation energy for a reaction by providing a different pathway for the reaction.

16.9 Equilibrium Constants

KEY TERM

Equilibrium constant, K_{eq}

- For the general reaction $a\text{A} + b\text{B} \rightleftharpoons c\text{C} + d\text{D}$

$$K_{eq} = \frac{[\text{C}]^c[\text{D}]^d}{[\text{A}]^a[\text{B}]^b}$$

- The magnitude of K_{eq} indicates the extent of the reaction:
  - If $K_{eq} > 1$:Amount of products is greater than reactants at equilibrium.
  - If $K_{eq} < 1$:Amount of reactants is less than products at equilibrium.

16.10 Ion Product Constant for Water

KEY TERM

Ion product constant for water, K_w

- $K_w = [\text{H}^+][\text{OH}^-]$
- $[\text{H}^+] = \dfrac{K_w}{[\text{OH}^-]}$
- $[\text{OH}^-] = \dfrac{K_w}{[\text{H}^+]}$

16.11 Ionization Constants

KEY TERM

Acid ionization constant, K_a

- The ionization constant can be used in the equilibrium expression to calculate the concentration of a reactant, the percent ionization of a substance, or the pH of a weak acid.

16.12 Solubility Product Constant

KEY TERMS

Solubility product constant, K_{sp}
Common ion effect

- The solubility product constant is used to calculate the solubility of a slightly soluble salt in water.

- The solubility product constant can also be used to determine whether or not precipitation will occur in a solution:
  - If the product of the molar concentrations of the ions in solution is greater than the K_{sp}, precipitation will occur.
- A shift in the equilibrium position upon addition of an ion already contained in the solution is known as the common ion effect.

16.13 Acid–Base Properties of Salts

KEY TERM

Hydrolysis

- Salts of weak acids and weak bases can react with water to cause hydrolysis and form solutions that are not neutral.

Type of salt	Nature of aqueous solution	Examples
Weak base–strong acid	Acidic	NH_4Cl, NH_4NO_3
Strong base–weak acid	Basic	$NaC_2H_3O_2$, K_2CO_3
Weak base–weak acid	Depends on the salt	$NH_4C_2H_3O_2$, NH_4NO_2
Strong base–strong acid	Neutral	$NaCl$, KBr

16.14 Buffer Solutions: The Control of pH

KEY TERM

Buffer solution

- A buffer solution could contain:
  - A weak acid and a salt of its conjugate base
  - A weak base and a salt of its conjugate acid
- A buffer solution resists a change in pH by neutralizing small amounts of acid or base added to it.

Review Questions

All questions with blue numbers have answers in the appendix of the text.

1. How would you expect the two tubes in Figure 16.1 to appear if both are at 25°C?

2. Is the reaction $N_2O_4 \rightleftharpoons 2 NO_2$ exothermic or endothermic? (Figure 16.1)

3. Why is heat treated as a reactant in an endothermic process and as a product in an exothermic process?

4. At equilibrium how do the forward and reverse reaction rates compare? (Figure 16.2)

5. Why don't free protons (H^+) exist in water?

6. For each solution in Table 16.1, what is the sum of the pH plus the pOH? What would be the pOH of a solution whose pH is −1?

7. Of the acids listed in Table 16.2, which ones are stronger than acetic acid and which are weaker?

8. Tabulate the relative order of molar solubilities of $AgCl$, $AgBr$, AgI, $AgC_2H_3O_2$, $PbSO_4$, $BaSO_4$, $BaCrO_4$, and PbS. List the most soluble first. (Table 16.3)

9. Which compound in the following pairs has the greater molar solubility? (Table 16.3)
   (a) $Mn(OH)_2$ or Ag_2CrO_4
   (b) $BaCrO_4$ or Ag_2CrO_4

10. Explain how the acetic acid–sodium acetate buffer system maintains its pH when 0.010 mol of HCl is added to 1 L of the buffer solution. (Table 16.5)

11. Explain why a precipitate of NaCl forms when HCl gas is passed into a saturated aqueous solution of NaCl.

12. Why does the rate of a reaction usually increase when the concentration of one of the reactants is increased?

13. If pure hydrogen iodide, HI, is placed in a vessel at 700 K, will it decompose? Explain.

14. Why does an increase in temperature cause the rate of reaction to increase?

15. What does a catalyst do?

16. Describe how equilibrium is reached when the substances A and B are first mixed and react as

    $$A + B \rightleftharpoons C + D$$

17. With dilution, aqueous solutions of acetic acid $(HC_2H_3O_2)$ show increased ionization. For example, a 1.0 M solution of acetic acid is 0.42% ionized, whereas a 0.10 M solution is 1.34% ionized. Explain the behavior using the ionization equation and equilibrium principles.

18. A 1.0 M solution of acetic acid ionizes less and has a higher concentration of H^+ ions than a 0.10 M acetic acid solution. Explain this behavior. (See Question 17 for data.)

19. What would cause two separate samples of pure water to have slightly different pH values?

20. Why are the pH and pOH equal in pure water?

21. Explain why silver acetate is more soluble in nitric acid than in water. (*Hint:* Write the equilibrium equation first and then consider the effect of the acid on the acetate ion.) What would happen if hydrochloric acid were used in place of nitric acid?

22. Dissolution of sodium acetate $(NaC_2H_3O_2)$ in pure water gives a basic solution. Why? (*Hint:* A small amount of $HC_2H_3O_2$ is formed.)

23. Describe why the pH of a buffer solution remains almost constant when a small amount of acid or base is added to it.

24. Describe the similarities and differences between K_a, K_b, K_w and K_{sp}.

Paired Exercises

All exercises with blue numbers have answers in the appendix of the text.

1. Express these reversible systems in equation form:
   (a) a mixture of ice and liquid water at 0°C
   (b) crystals of Na_2SO_4 in a saturated aqueous solution of Na_2SO_4

2. Express these reversible systems in equation form:
   (a) liquid water and vapor at 100°C in a pressure cooker
   (b) a closed system containing boiling sulfur dioxide, SO_2

3. Consider the following system at equilibrium:

   $$4\,NH_3(g) + 3\,O_2(g) \rightleftharpoons 2\,N_2(g) + 6\,H_2O(g) + 1531\ kJ$$

   (a) Is the reaction exothermic or endothermic?
   (b) If the system's state of equilibrium is disturbed by the addition of O_2, in which direction, left or right, must the reaction occur to reestablish equilibrium? After the new equilibrium has been established, how will the final molar concentrations of NH_3, O_2, N_2, and H_2O compare (increase or decrease) with their concentrations before the addition of the O_2?

4. Consider the following system at equilibrium:

   $$4\,NH_3(g) + 3\,O_2(g) \rightleftharpoons 2\,N_2(g) + 6\,H_2O(g) + 1531\ kJ$$

   (a) If the system's state of equilibrium is disturbed by the addition of N_2, in which direction, left or right, must the reaction occur to reestablish equilibrium? After the new equilibrium has been established, how will the final molar concentrations of NH_3, O_2, N_2, and H_2O compare (increase or decrease) with their concentrations before the addition of the N_2?
   (b) If the system's state of equilibrium is disturbed by the addition of heat, in which direction will the reaction occur, left or right, to reestablish equilibrium?

5. Consider the following system at equilibrium:

$N_2(g) + 3 H_2(g) \rightleftharpoons 2 NH_3(g) + 92.5 \, kJ$

Complete the table that follows. Indicate changes in moles by entering I, D, N, or ? in the table. (I = increase, D = decrease, N = no change, ? = insufficient information to determine.)

Change of stress imposed on the system at equilibrium	Direction of reaction, left or right, to reestablish equilibrium	Change in number of moles		
		N_2	H_2	NH_3
(a) Add N_2				
(b) Remove H_2				
(c) Decrease volume of reaction vessel				
(d) Increase temperature				

6. Consider the following system at equilibrium:

$N_2(g) + 3 H_2(g) \rightleftharpoons 2 NH_3(g) + 92.5 \, kJ$

Complete the table that follows. Indicate changes in moles by entering I, D, N, or ? in the table. (I = increase, D = decrease, N = no change, ? = insufficient information to determine.)

Change of stress imposed on the system at equilibrium	Direction of reaction, left or right, to reestablish equilibrium	Change in number of moles		
		N_2	H_2	NH_3
(a) Add NH_3				
(b) Increase volume of reaction vessel				
(c) Add catalyst				
(d) Add both H_2 and NH_3				

7. For the following equations, tell in which direction, left or right, the equilibrium will shift when these changes are made: The temperature is increased, the pressure is increased by decreasing the volume of the reaction vessel, and a catalyst is added.

(a) $3 O_2(g) + 271 \, kJ \rightleftharpoons 2 O_3(g)$
(b) $CH_4(g) + Cl_2(g) \rightleftharpoons CH_3Cl(g) + HCl(g) + 110 \, kJ$
(c) $2 NO(g) + 2 H_2(g) \rightleftharpoons N_2(g) + 2 H_2O(g) + 665 \, kJ$

8. For the following equations, tell in which direction, left or right, the equilibrium will shift when these changes are made: The temperature is increased, the pressure is increased by decreasing the volume of the reaction vessel, and a catalyst is added.

(a) $2 SO_3(g) + 197 \, kJ \rightleftharpoons 2 SO_2(g) + O_2(g)$
(b) $4 NH_3(g) + 3 O_2(g) \rightleftharpoons 2 N_2(g) + 6 H_2O(g) + 1531 \, kJ$
(c) $OF_2(g) + H_2O(g) \rightleftharpoons O_2(g) + 2 HF(g) + 318 \, kJ$

9. Utilizing Le Châtelier's principle, indicate the shift (if any) that would occur to

$C_2H_6(g) + heat \rightleftharpoons C_2H_4(g) + H_2(g)$

(a) if the concentration of hydrogen gas is decreased
(b) if the temperature is lowered
(c) if a catalyst is added

10. Utilizing Le Châtelier's principle, indicate the shift (if any) that would occur to

$C_2H_6(g) + heat \rightleftharpoons C_2H_4(g) + H_2(g)$

(a) if C_2H_6 is removed from the system
(b) if the volume of the container is increased
(c) if the temperature is raised

11. Write the equilibrium constant expression for these reactions:

(a) $2 NO_2(g) + 7 H_2(g) \rightleftharpoons 2 NH_3(g) + 4 H_2O(g)$
(b) $H_2CO_3(aq) \rightleftharpoons H^+(aq) + HCO_3^-(aq)$
(c) $2 COF_2(g) \rightleftharpoons CO_2(g) + CF_4(g)$

12. Write the equilibrium constant expression for these reactions:

(a) $H_3PO_4(aq) \rightleftharpoons H^+(aq) + H_2PO_4^-(aq)$
(b) $CS_2(g) + 4 H_2(g) \rightleftharpoons CH_4(g) + 2 H_2S(g)$
(c) $4 NO_2(g) + O_2(g) \rightleftharpoons 2 N_2O_5(g)$

13. Write the solubility product expression, K_{sp}, for these substances:

(a) AgCl   (c) $Zn(OH)_2$
(b) $PbCrO_4$ (d) $Ca_3(PO_4)_2$

14. Write the solubility product expression, K_{sp}, for these substances:

(a) $MgCO_3$ (c) $Tl(OH)_3$
(b) CaC_2O_4 (d) $Pb_3(AsO_4)_2$

15. What effect will decreasing the $[H^+]$ of a solution have on (a) pH, (b) pOH, (c) $[OH^-]$, and (d) K_w?

16. What effect will increasing the $[H^+]$ of a solution have on (a) pH, (b) pOH, (c) $[OH^-]$, and (d) K_w?

17. Decide whether these salts form an acidic, basic, or neutral solution when dissolved in water:

(a) $CaBr_2$ (c) NaCN
(b) $Al(NO_3)_3$ (d) K_3PO_4

18. Decide whether these salts form an acidic, basic, or neutral solution when dissolved in water:

(a) NH_4Cl (c) $CuSO_4$
(b) $NaC_2H_3O_2$ (d) KI

19. Write hydrolysis equations for aqueous solutions of these salts:
    (a) KNO_2 (b) $Mg(C_2H_3O_2)_2$

20. Write hydrolysis equations for aqueous solutions of these salts:
    (a) NH_4NO_3 (b) Na_2SO_3

21. Write hydrolysis equations for these ions:
    (a) HCO_3^- (b) NH_4^+

22. Write hydrolysis equations for these ions:
    (a) OCl^- (b) ClO_2^-

23. One of the important pH-regulating systems in the blood consists of a carbonic acid–sodium hydrogen carbonate buffer:

    $$H_2CO_3(aq) \rightleftharpoons H^+(aq) + HCO_3^-(aq)$$
    $$NaHCO_3(aq) \longrightarrow Na^+(aq) + HCO_3^-(aq)$$

    Explain how this buffer resists changes in pH when excess acid, H^+, gets into the bloodstream.

24. One of the important pH-regulating systems in the blood consists of a carbonic acid–sodium hydrogen carbonate buffer:

    $$H_2CO_3(aq) \rightleftharpoons H^+(aq) + HCO_3^-(aq)$$
    $$NaHCO_3(aq) \longrightarrow Na^+(aq) + HCO_3^-(aq)$$

    Explain how this buffer resists changes in pH when excess base, OH^-, gets into the bloodstream.

25. Calculate (a) the $[H^+]$, (b) the pH, and (c) the percent ionization of a 0.25 M solution of $HC_2H_3O_2$. ($K_a = 1.8 \times 10^{-5}$)

26. Calculate (a) the $[H^+]$, (b) the pH, and (c) the percent ionization of a 0.25 M solution of phenol, HC_6H_5O. ($K_a = 1.3 \times 10^{-10}$)

27. A 0.025 M solution of a weak acid, HA, is 0.45% ionized. Calculate the ionization constant, K_a, for the acid.

28. A 0.500 M solution of a weak acid, HA, is 0.68% ionized. Calculate the ionization constant, K_a, for the acid.

29. Calculate the percent ionization and pH of solutions of $HC_2H_3O_2$ ($K_a = 1.8 \times 10^{-5}$) having the following molarities: (a) 1.0 M, (b) 0.10 M, and (c) 0.010 M.

30. Calculate the percent ionization and pH of solutions of $HClO$ ($K_a = 3.5 \times 10^{-8}$) having the following molarities: (a) 1.0 M, (b) 0.10 M, and (c) 0.010 M.

*31. A 0.37 M solution of a weak acid (HA) has a pH of 3.7. What is the K_a for this acid?

*32. A 0.23 M solution of a weak acid (HA) has a pH of 2.89. What is the K_a for this acid?

33. A student needs a sample of 1.0 M NaOH for a laboratory experiment. Calculate the $[H^+]$, $[OH^-]$, pH, and pOH of this solution.

34. A laboratory cabinet contains a stock solution of 3.0 M HNO_3. Calculate the $[H^+]$, $[OH^-]$, pH, and pOH of this solution.

35. Calculate the pH and the pOH of these solutions:
    (a) 0.250 M HBr
    (b) 0.333 M KOH
    (c) 0.895 M $HC_2H_3O_2$ ($K_a = 1.8 \times 10^{-5}$)

36. Calculate the pH and the pOH of these solutions:
    (a) 0.0010 M NaOH
    (b) 0.125 M HCl
    (c) 0.0250 M HC_6H_5O ($K_a = 1.3 \times 10^{-10}$)

37. Calculate the $[OH^-]$ in these solutions:
    (a) $[H^+] = 1.0 \times 10^{-4}$ (b) $[H^+] = 2.8 \times 10^{-6}$

38. Calculate the $[OH^-]$ in these solutions:
    (a) $[H^+] = 4.0 \times 10^{-9}$ (b) $[H^+] = 8.9 \times 10^{-2}$

39. Calculate the $[H^+]$ in these solutions:
    (a) $[OH^-] = 6.0 \times 10^{-7}$ (b) $[OH^-] = 1 \times 10^{-8}$

40. Calculate the $[H^+]$ in these solutions:
    (a) $[OH^-] = 4.5 \times 10^{-6}$ (b) $[OH^-] = 7.3 \times 10^{-4}$

41. Given the following solubility data, calculate the solubility product constant for each substance:
    (a) $BaSO_4$, 3.9×10^{-5} mol/L
    (b) Ag_2CrO_4, 7.8×10^{-5} mol/L
    (c) $CaSO_4$, 0.67 g/L
    (d) $AgCl$, 0.0019 g/L

42. Given the following solubility data, calculate the solubility product constant for each substance:
    (a) ZnS, 3.5×10^{-12} mol/L
    (b) $Pb(IO_3)_2$, 4.0×10^{-5} mol/L 2.6×10^{-13}
    (c) Ag_3PO_4, 6.73×10^{-3} g/L
    (d) $Zn(OH)_2$, 2.33×10^{-4} g/L 5.13×10^{-17}

43. Calculate the molar solubility for these substances:
    (a) CaF_2, $K_{sp} = 3.9 \times 10^{-11}$
    (b) $Fe(OH)_3$, $K_{sp} = 6.1 \times 10^{-38}$

44. Calculate the molar solubility for these substances:
    (a) $PbSO_4$, $K_{sp} = 1.3 \times 10^{-8}$
    (b) $BaCrO_4$, $K_{sp} = 8.5 \times 10^{-11}$

45. For each substance in Exercise 43, calculate the solubility in grams per 100. mL of solution.

46. For each substance in Exercise 44, calculate the solubility in grams per 100. mL of solution.

47. Solutions containing 100. mL of 0.010 M Na_2SO_4 and 100. mL of 0.001 M $Pb(NO_3)_2$ are mixed. Show by calculation whether or not a precipitate will form. Assume that the volumes are additive. (K_{sp} for $PbSO_4 = 1.3 \times 10^{-8}$)

48. Solutions containing 50.0 mL of 1.0×10^{-4} M $AgNO_3$ and 100. mL of 1.0×10^{-4} M NaCl are mixed. Show by calculation whether or not a precipitate will form. Assume the volumes are additive. (K_{sp} for AgCl $= 1.7 \times 10^{-10}$)

49. How many moles of AgBr will dissolve in 1.0 L of 0.10 M NaBr? ($K_{sp} = 5.2 \times 10^{-13}$ for AgBr)

50. How many moles of AgBr will dissolve in 1.0 L of 0.10 M $MgBr_2$? ($K_{sp} = 5.2 \times 10^{-13}$ for AgBr)

51. Calculate the [H^+] and the pH of a buffer solution that is 0.20 M in $HC_2H_3O_2$ and contains sufficient sodium acetate to make the [$C_2H_3O_2^-$] equal to 0.10 M. (K_a for $HC_2H_3O_2 = 1.8 \times 10^{-5}$)

52. Calculate the [H^+] and the pH of a buffer solution that is 0.20 M in $HC_2H_3O_2$ and contains sufficient sodium acetate to make the [$C_2H_3O_2^-$] equal to 0.20 M. (K_a for $HC_2H_3O_2 = 1.8 \times 10^{-5}$)

53. When 1.0 mL of 1.0 M HCl is added to 50. mL of 1.0 M NaCl, the [H^+] changes from $1 \times 10^{-7} M$ to $2.0 \times 10^{-2} M$. Calculate the initial pH and the pH change in the solution.

54. When 1.0 mL of 1.0 M HCl is added to 50. mL of a buffer solution that is 1.0 M in $HC_2H_3O_2$ and 1.0 M in $NaC_2H_3O_2$, the [H^+] changes from $1.8 \times 10^{-5} M$ to $1.9 \times 10^{-5} M$. Calculate the initial pH and the pH change in the solution.

Additional Exercises

All exercises with blue *numbers have answers in the appendix of the text.*

55. In a K_{sp} expression the concentration of solid salt is not included. Why?

56. The energy diagram in Figure 16.3 is for an exothermic reaction. How can you tell?

57. Sketch the energy diagram for an endothermic reaction using a solid line. Sketch what would happen to the energy profile if a catalyst is added to the reaction using a dotted line.

58. What is the maximum number of moles of HI that can be obtained from a reaction mixture containing 2.30 mol I_2 and 2.10 mol H_2?

59. (a) How many moles of HI are produced when 2.00 mol H_2 and 2.00 mol I_2 are reacted at 700 K? (Reaction is 79% complete.)
   (b) Addition of 0.27 mol I_2 to the system increases the yield of HI to 85%. How many moles of H_2, I_2, and HI are now present?
   (c) From the data in part (a), calculate K_{eq} for the reaction at 700 K.

*60. After equilibrium is reached in the reaction of 6.00 g H_2 with 200. g I_2 at 500. K, analysis shows that the flask contains 64.0 g of HI. How many moles of H_2, I_2, and HI are present in this equilibrium mixture?

61. What is the equilibrium constant of the reaction
$$PCl_3(g) + Cl_2(g) \rightleftharpoons PCl_5(g)$$
if a 20.-L flask contains 0.10 mol PCl_3, 1.50 mol Cl_2, and 0.22 mol PCl_5?

62. If the rate of a reaction doubles for every 10°C rise in temperature, how much faster will the reaction go at 100°C than at 30°C?

63. Would you believe that an NH_4Cl solution is acidic? Calculate the pH of a 0.30 M NH_4Cl solution.
($NH_4^+ \rightleftharpoons NH_3 + H^+$, $K_{eq} = 5.6 \times 10^{-10}$)

64. Calculate the pH of an acetic acid buffer composed of $HC_2H_3O_2$ $(0.30\,M)$ and $C_2H_3O_2^-$ $(0.20\,M)$.

65. Solutions of 50.0 mL of 0.10 M $BaCl_2$ and 50.0 mL of 0.15 M Na_2CrO_4 are mixed, forming a precipitate of $BaCrO_4$. Calculate the concentration of Ba^{2+} that remains in solution.

66. Calculate the ionization constant for the given acids. Each acid ionizes as follows: HA $\rightleftharpoons$ H^+ + A^-.

Acid	Acid concentration	[H^+]
Hypochlorous, HOCl	0.10 M	5.9×10^{-5} mol/L
Propanoic, $HC_3H_5O_2$	0.15 M	1.4×10^{-3} mol/L
Hydrocyanic, HCN	0.20 M	8.9×10^{-6} mol/L

67. The K_{sp} of CaF_2 is 3.9×10^{-11}. Calculate (a) the molar concentrations of Ca^{2+} and F^- in a saturated solution and (b) the grams of CaF_2 that will dissolve in 500. mL of water.

68. Calculate whether or not a precipitate will form in the following mixed solutions:
   (a) 100 mL of 0.010 M Na_2SO_4 and 100 mL of 0.001 M $Pb(NO_3)_2$
   (b) 50.0 mL of $1.0 \times 10^{-4} M$ $AgNO_3$ and 100. mL of $1.0 \times 10^{-4} M$ NaCl
   (c) 1.0 g $Ca(NO_3)_2$ in 150 mL H_2O and 250 mL of 0.01 M NaOH
   $$K_{sp}PbSO_4 = 1.3 \times 10^{-8}$$
   $$K_{sp}AgCl = 1.7 \times 10^{-10}$$
   $$K_{sp}Ca(OH)_2 = 1.3 \times 10^{-6}$$

69. If $BaCl_2$ is added to a saturated $BaSO_4$ solution until the $[Ba^{2+}]$ is 0.050 M,
   (a) what concentration of SO_4^{2-} remains in solution?
   (b) how much $BaSO_4$ remains dissolved in 100. mL of the solution? ($K_{sp} = 1.5 \times 10^{-9}$ for $BaSO_4$)

70. The K_{sp} for $PbCl_2$ is 2.0×10^{-5}. Will a precipitate form when 0.050 mol $Pb(NO_3)_2$ and 0.010 mol NaCl are in 1.0 L solution? Show evidence for your answer.

71. Suppose the concentration of a solution is 0.10 M Ba^{2+} and 0.10 M Sr^{2+}. Which sulfate, $BaSO_4$ or $SrSO_4$, will precipitate first when a dilute solution of H_2SO_4 is added dropwise to the solution? Show evidence for your answer. ($K_{sp} = 1.5 \times 10^{-9}$ for $BaSO_4$ and $K_{sp} = 3.5 \times 10^{-7}$ for $SrSO_4$)

72. Calculate the K_{eq} for the reaction

$$SO_2(g) + O_2(g) \rightleftharpoons SO_3(g)$$

when the equilibrium concentrations of the gases at 530°C are $[SO_3] = 11.0\,M$, $[SO_2] = 4.20\,M$, and $[O_2] = 0.60 \times 10^{-3}\,M$.

73. If it takes 0.048 g BaF_2 to saturate 15.0 mL of water, what is the K_{sp} of BaF_2?

74. The K_{eq} for the formation of ammonia gas from its elements is 4.0. If the equilibrium concentrations of nitrogen gas and hydrogen gas are both 2.0 M, what is the equilibrium concentration of the ammonia gas?

75. The K_{sp} of $SrSO_4$ is 7.6×10^{-7}. Should precipitation occur when 25.0 mL of $1.0 \times 10^{-3}\,M$ $SrCl_2$ solution are mixed with 15.0 mL of $2.0 \times 10^{-3}\,M$ Na_2SO_4? Show proof.

*76. The solubility of Hg_2I_2 in H_2O is 3.04×10^{-7} g/L. The reaction $Hg_2I_2 \rightleftharpoons Hg_2^{2+} + 2\,I^-$ represents the equilibrium. Calculate the K_{sp}.

77. Under certain circumstances, when oxygen gas is heated, it can be converted into ozone according to the following reaction equation:

$$3\,O_2(g) + heat \rightleftharpoons 2\,O_3(g)$$

Name three different ways that you could increase the production of the ozone.

78. One day in a laboratory, some water spilled on a table. In just a few minutes the water had evaporated. Some days later, a similar amount of water spilled again. This time, the water remained on the table after 7 or 8 hours. Name three conditions that could have changed in the lab to cause this difference.

79. For the reaction $CO(g) + H_2O(g) \rightleftharpoons CO_2(g) + H_2(g)$ at a certain temperature, K_{eq} is 1. At equilibrium would you expect to find
   (a) only CO and H_2
   (b) mostly CO_2 and H_2
   (c) about equal concentrations of CO and H_2O, compared to CO_2 and H_2
   (d) mostly CO and H_2O
   (e) only CO and H_2O
   Explain your answer briefly.

80. Write the equilibrium constant expressions for these reactions:
   (a) $3\,O_2(g) \rightleftharpoons 2\,O_3(g)$
   (b) $H_2O(g) \rightleftharpoons H_2O(l)$
   (c) $MgCO_3(s) \rightleftharpoons MgO(s) + CO_2(g)$
   (d) $2\,Bi^{3+}(aq) + 3\,H_2S(aq) \rightleftharpoons Bi_2S_3(s) + 6\,H^+(aq)$

81. Reactants A and B are mixed, each initially at a concentration of 1.0 M. They react to produce C according to this equation:

$$2\,A + B \rightleftharpoons C$$

When equilibrium is established, the concentration of C is found to be 0.30 M. Calculate the value of K_{eq}.

82. At a certain temperature, K_{eq} is 2.2×10^{-3} for the reaction

$$2\,ICl(g) \rightleftharpoons I_2(g) + Cl_2(g)$$

Now calculate the K_{eq} value for the reaction

$$I_2(g) + Cl_2(g) \rightleftharpoons 2\,ICl(g)$$

83. One drop of 1 M OH^- ion is added to a 1 M solution of HNO_2. What will be the effect of this addition on the equilibrium concentration of the following?

$$HNO_2(aq) \rightleftharpoons H^+(aq) + NO_2^-(aq)$$

   (a) $[OH^-]$ (c) $[NO_2^-]$
   (b) $[H^+]$ (d) $[HNO_2]$

84. How many grams of $CaSO_4$ will dissolve in 600. mL of water? ($K_{sp} = 2.0 \times 10^{-4}$ for $CaSO_4$)

85. A student found that 0.098 g of PbF_2 was dissolved in 400. mL of saturated PbF_2. What is the K_{sp} for the lead(II) fluoride?

Challenge Exercises

All exercises with blue numbers have answers in the appendix of the text.

*86. All a snowbound skier had to eat were walnuts! He was carrying a bag holding 12 dozen nuts. With his mittened hands, he cracked open the shells. Each nut that was opened resulted in one kernel and two shell halves. When he tired of the cracking and got ready to do some eating, he discovered he had 194 total pieces (whole nuts, shell halves, and kernels). What is the K_{eq} for this reaction?

*87. At 500°C, the reaction

$$SO_2(g) + NO_2(g) \rightleftharpoons NO(g) + SO_3(g)$$

has $K_{eq} = 81$. What will be the equilibrium concentrations of the four gases if the two reactants begin with equal concentrations of 0.50 M?

Answers to Practice Exercises

16.1 (a)

Concentration	Change
$[H_2]$	increase
$[I_2]$	decrease
$[HI]$	increase

(b)

Concentration	Change
$[H_2]$	increase
$[I_2]$	increase
$[HI]$	increase

16.2 (a) equilibrium shifts to the left;
(b) equilibrium shifts to the right

16.3 (a) equilibrium shifts to the right;
(b) equilibrium shifts to the left

16.4 (a) equilibrium shifts to the left;
(b) equilibrium shifts to the right

16.5 (a) $K_{eq} = \dfrac{[NO_2]^4[O_2]}{[N_2O_5]^2}$;   (b) $K_{eq} = \dfrac{[N_2]^2[H_2O]^6}{[NH_3]^4[O_2]^3}$

16.6 $K_{eq} = 33$; the forward reaction is favored

16.7 (a) $[H^+] = 5.0 \times 10^{-5}$ $[OH^-] = 2.0 \times 10^{-10}$
(b) $[H^+] = 5.0 \times 10^{-9}$ $[OH^-] = 2.0 \times 10^{-6}$

16.8 (a) 4.30   (b) 8.30

16.9 (a) 6.3×10^{-6} (b) 1.8×10^{-6}

16.10 (a) 6.3×10^{-3}% ionized
(b) 7.2×10^{-3}% ionized

16.11 (a) $K_{sp} = [Cr^{3+}][OH^-]^3$
(b) $K_{sp} = [Cu^{2+}]^3[PO_4^{3-}]^2$

16.12 8.8×10^{-21} g/L

16.13 7.5×10^{-8} mol/L

16.14 (a) basic
(b) neutral
(c) acidic

CHAPTER 17

Oxidation–Reduction

The Golden Gate Bridge in San Francisco requires frequent painting to prevent oxidation of the metal in the sea air.

Chapter Outline

17.1 Oxidation Number

17.2 Oxidation–Reduction

17.3 Balancing Oxidation–Reduction Equations

17.4 Balancing Ionic Redox Equations

17.5 Activity Series of Metals

17.6 Electrolytic and Voltaic Cells

The variety of oxidation–reduction reactions that affect us every day is amazing. Our society runs on batteries—in our calculators, laptop computers, cars, toys, radios, televisions, and more. We paint iron railings and galvanize nails to combat corrosion. We electroplate jewelry and computer chips with very thin coatings of gold or silver. We bleach our clothes using chemical reactions that involve electron transfer. We test for glucose in urine or alcohol in the breath with reactions that show vivid color changes. Plants turn energy into chemical compounds through a series of reactions called photosynthesis. These reactions all involve the transfer of electrons between substances in a chemical process called *oxidation–reduction*.

17.1   Oxidation Number

The oxidation number of an atom (sometimes called its *oxidation state*) represents the number of electrons lost, gained, or unequally shared by an atom. Oxidation numbers can be zero, positive, or negative. An oxidation number of zero means the atom has the same number of electrons assigned to it as there are in the free neutral atom. A positive oxidation number means the atom has fewer electrons assigned to it than in the neutral atom, and a negative oxidation number means the atom has more electrons assigned to it than in the neutral atom.

The oxidation number of an atom that has lost or gained electrons to form an ion is the same as the positive or negative charge of the ion. (See Table 17.1.) In the ionic compound NaCl, the oxidation numbers are clearly $+1$ for the Na^+ ion and -1 for the Cl^- ion. The Na^+ ion has one less electron than the neutral Na atom, and the Cl^- ion has one more electron than the neutral Cl atom. In $MgCl_2$, two electrons have transferred from the Mg atom to the two Cl atoms; the oxidation number of Mg is $+2$.

In covalently bonded substances, where electrons are shared between two atoms, oxidation numbers are assigned by an arbitrary system based on relative electronegativities. For symmetrical covalent molecules such as H_2 and Cl_2, each atom is assigned an oxidation number of zero because the bonding pair of electrons is shared equally between two like atoms, neither of which is more electronegative than the other:

$$H\!:\!H \qquad :\!\ddot{\underset{..}{Cl}}\!:\!\ddot{\underset{..}{Cl}}\!:$$

When the covalent bond is between two unlike atoms, the bonding electrons are shared unequally because the more electronegative element has a greater attraction for them. In this case the oxidation numbers are determined by assigning both electrons to the more electronegative element.

Thus in compounds with covalent bonds such as NH_3 and H_2O,

the pairs of electrons are unequally shared between the atoms and are attracted toward the more electronegative elements, N and O. This causes the N and O atoms to be relatively negative with respect to the H atoms. At the same

Table 17.1   Oxidation Numbers for Common Ions

Ion	Oxidation number
H^+	$+1$
Na^+	$+1$
K^+	$+1$
Li^+	$+1$
Ag^+	$+1$
Cu^{2+}	$+2$
Ca^{2+}	$+2$
Ba^{2+}	$+2$
Fe^{2+}	$+2$
Mg^{2+}	$+2$
Zn^{2+}	$+2$
Al^{3+}	$+3$
Fe^{3+}	$+3$
Cl^-	-1
Br^-	-1
F^-	-1
I^-	-1
S^{2-}	-2
O^{2-}	-2

time, it causes the H atoms to be relatively positive with respect to the N and O atoms. In H_2O, both pairs of shared electrons are assigned to the O atom, giving it two electrons more than the neutral O atom, and each H atom is assigned one electron less than the neutral H atom. Therefore, the oxidation number of the O atom is -2, and the oxidation number of each H atom is $+1$. In NH_3, the three pairs of shared electrons are assigned to the N atom, giving it three electrons more than the neutral N atom, and each H atom has one electron less than the neutral atom. Therefore, the oxidation number of the N atom is -3, and the oxidation number of each H atom is $+1$.

Assigning correct oxidation numbers to elements is essential for balancing oxidation–reduction equations. The **oxidation number** or **oxidation state** of an element is an integer value assigned to each element in a compound or ion that allows us to keep track of electrons associated with each atom. Oxidation numbers have a variety of uses in chemistry—from writing formulas to predicting properties of compounds and assisting in the balancing of oxidation–reduction reactions in which electrons are transferred.

As a starting point, the oxidation number of an uncombined element, regardless of whether it is monatomic or diatomic, is zero. Rules for assigning oxidation numbers are summarized in Table 17.2. Many elements have multiple oxidation numbers; for example, nitrogen

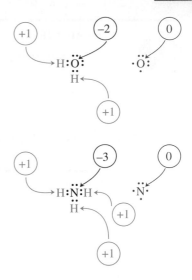

Oxidation number
Oxidation state

	N_2	N_2O	NO	N_2O_3	NO_2	N_2O_5	NO_3^-
Oxidation number	0	+1	+2	+3	+4	+5	+5

Use the following steps to find the oxidation number for an element within a compound:

Step 1	Write the oxidation number of each known atom below the atom in the formula.
Step 2	Multiply each oxidation number by the number of atoms of that element in the compound.
Step 3	Write an expression indicating the sum of all the oxidation numbers in the compound. *Remember:* The sum of the oxidation numbers in a compound must equal zero.

The oxidation number for Cu metal is 0 while the oxidation number for Cu^{2+} ions in the crystal is $+2$.

Table 17.2 Rules for Assigning Oxidation Number

1. All elements in their free state (uncombined with other elements) have an oxidation number of zero (e.g., Na, Cu, Mg, H_2, O_2, Cl_2, N_2).
2. H is $+1$, except in metal hydrides, where it is -1 (e.g., NaH, CaH_2).
3. O is -2, except in peroxides, where it is -1, and in OF_2, where it is $+2$.
4. The metallic element in an ionic compound has a positive oxidation number.
5. In covalent compounds the negative oxidation number is assigned to the most electronegative atom.
6. The algebraic sum of the oxidation numbers of the elements in a compound is zero.
7. The algebraic sum of the oxidation numbers of the elements in a polyatomic ion is equal to the charge of the ion.

Example 17.1   Determine the oxidation number for carbon in carbon dioxide:

SOLUTION   **CO₂**

Step 1   −2
Step 2   (−2)2
Step 3   C + (−4) = 0
Step 4   C = +4 (oxidation number for carbon)

Example 17.2   Determine the oxidation number for sulfur in sulfuric acid:

SOLUTION   **H₂SO₄**

Step 1   +1   −2
Step 2   2(+1) = +2   4(−2) = −8
Step 3   +2 + S+(−8) = 0
Step 4   S = +6 (oxidation number for sulfur)

> **Practice 17.1**
> Determine the oxidation number of (a) S in Na₂SO₄, (b) As in K₃AsO₄, and (c) C in CaCO₃.

   Oxidation numbers in a polyatomic ion (ions containing more than one atom) are determined in a similar fashion, except that in a polyatomic ion the sum of the oxidation numbers must equal the charge on the ion instead of zero.

Example 17.3   Determine the oxidation number for manganese in the permanganate ion MnO₄⁻:

SOLUTION   **MnO₄⁻**

Step 1   −2
Step 2   (−2)4
Step 3   Mn + (−8) = −1 (the charge on the ion)
             Mn = +7 (oxidation number for manganese)

Example 17.4   Determine the oxidation number for carbon in the oxalate ion C₂O₄²⁻:

SOLUTION   **C₂O₄²⁻**

                         −2
Step 1   (−2)4
Step 2   2C + (−8) = −2 (the charge on the ion)
Step 3   2C = +6
             C = +3 (oxidation number for C)

Determine the oxidation numbers of (a) N in NH_4^+, (b) Cr in $Cr_2O_7^{2-}$, and (c) P in PO_4^{3-}.

Determine the oxidation number of each element in (a) KNO_3 and (b) SO_4^{2-}.

Example 17.5

SOLUTION

(a) Potassium is a Group 1A metal; therefore, it has an oxidation number of $+1$. The oxidation number of each O atom is -2 (Table 17.2, Rule 3). Using these values and the fact that the sum of the oxidation numbers of all the atoms in a compound is zero, we can determine the oxidation number of N:

$$KNO_3$$
$$+1 + N + 3(-2) = 0$$
$$N = +6 - 1 = +5$$

The oxidation numbers are K, $+1$; N, $+5$; O, -2.

(b) Because SO_4^{2-} is an ion, the sum of oxidation numbers of the S and the O atoms must be -2, the charge of the ion. The oxidation number of each O atom is -2 (Table 17.2, Rule 3). Then

$$SO_4^{2-}$$
$$S + 4(-2) = -2, \quad S - 8 = -2$$
$$S = -2 + 8 = +6$$

The oxidation numbers are S, $+6$; O, -2.

Determine the oxidation number of each element in these species:
(a) $BeCl_2$ (b) $HClO$ (c) H_2O_2 (d) NH_4^+ (e) BrO_3^-

17.2 Oxidation–Reduction

Oxidation–reduction, also known as **redox**, is a chemical process in which the oxidation number of an element is changed. The process may involve the complete transfer of electrons to form ionic bonds or only a partial transfer or shift of electrons to form covalent bonds.

oxidation-reduction
redox

   Oxidation occurs whenever the oxidation number of an element increases as a result of losing electrons. Conversely, **reduction** occurs whenever the oxidation number of an element decreases as a result of gaining electrons. For example, a change in oxidation number from $+2$ to $+3$ or from -1 to 0 is oxidation; a change from $+5$ to $+2$ or from -2 to -4 is reduction (see Figure 17.1). Oxidation and reduction occur simultaneously in a chemical reaction; one cannot take place without the other.

oxidation
reduction

Figure 17.1
Oxidation and reduction. Oxidation results in an increase in the oxidation number, and reduction results in a decrease in the oxidation number.

Many combination, decomposition, and single-displacement reactions involve oxidation–reduction. Let's examine the combustion of hydrogen and oxygen from this point of view:

$$2\,H_2 + O_2 \longrightarrow 2\,H_2O$$

Both reactants, hydrogen and oxygen, are elements in the free state and have an oxidation number of zero. In the product (water), hydrogen has been oxidized to $+1$ and oxygen reduced to -2. The substance that causes an increase in the oxidation state of another substance is called an **oxidizing agent.** The substance that causes a decrease in the oxidation state of another substance is called a **reducing agent.** In this reaction the oxidizing agent is free oxygen, and the reducing agent is free hydrogen. In the reaction

oxidizing agent

reducing agent

$$Zn(s) + H_2SO_4(aq) \longrightarrow ZnSO_4(aq) + H_2(g)$$

metallic zinc is oxidized, and hydrogen ions are reduced. Zinc is the reducing agent, and hydrogen ions, the oxidizing agent. Electrons are transferred from the zinc metal to the hydrogen ions. The electron transfer is more clearly expressed as

$$Zn^0 + 2\,H^+ + SO_4^{2-} \longrightarrow Zn^{2+} + SO_4^{2-} + H_2^0$$

In the reaction between Zn and H_2SO_4 the Zn is oxidized while hydrogen is reduced.

Oxidation:	**Increase in oxidation number**
	Loss of electrons
Reduction:	**Decrease in oxidation number**
	Gain of electrons

The oxidizing agent is reduced and gains electrons. The reducing agent is oxidized and loses electrons. The transfer of electrons is characteristic of all redox reactions.

17.3 Balancing Oxidation–Reduction Equations

Many simple redox equations can be balanced readily by inspection, or by trial and error:

$$Na + Cl_2 \longrightarrow NaCl \qquad \text{(unbalanced)}$$
$$2\,Na + Cl_2 \longrightarrow 2\,NaCl \quad \text{(balanced)}$$

Balancing this equation is certainly not complicated. But as we study more complex reactions and equations such as

$$P + HNO_3 + H_2O \longrightarrow NO + H_3PO_4 \qquad \text{(unbalanced)}$$
$$3\,P + 5\,HNO_3 + 2\,H_2O \longrightarrow 5\,NO + 3\,H_3PO_4 \quad \text{(balanced)}$$

the trial-and-error method of balancing equations takes an unnecessarily long time.

One systematic method for balancing oxidation–reduction equations is based on the transfer of electrons between the oxidizing and reducing agents. Consider the first equation again:

The superscript 0 shows that the oxidation number is 0 for elements in their uncombined state.

$$Na^0 + Cl_2^0 \longrightarrow Na^+Cl^- \quad \text{(unbalanced)}$$

In this reaction, sodium metal loses one electron per atom when it changes to a sodium ion. At the same time chlorine gains one electron per atom. Because chlorine is diatomic, two electrons per molecule are needed to form a chloride ion from each atom. These electrons are furnished by two sodium atoms. Stepwise, the reaction may be written as two half-reactions, the oxidation half-reaction and the reduction half-reaction:

$$
\begin{aligned}
2\,Na^0 &\longrightarrow 2\,Na^+ + 2\,e^- \qquad \text{oxidation half-reaction} \\
Cl_2^0 + 2\,e^- &\longrightarrow \qquad\qquad 2\,Cl^- \quad\;\; \text{reduction half-reaction} \\
\hline
Cl_2^0 + 2\,Na^0 &\longrightarrow 2\,Na^+Cl^-
\end{aligned}
$$

When the two half-reactions, each containing the same number of electrons, are added together algebraically, the electrons cancel out. In this reaction there are no excess electrons; the two electrons lost by the two sodium atoms are utilized by chlorine. In all redox reactions the loss of electrons by the reducing agent must equal the gain of electrons by the oxidizing agent. Here sodium is oxidized and chlorine is reduced. Chlorine is the oxidizing agent; sodium is the reducing agent.

In the following examples, we use the change-in-oxidation-number method, a system for balancing more complicated redox equations.

Balance the equation

$$Sn + HNO_3 \longrightarrow SnO_2 + NO_2 + H_2O \quad \text{(unbalanced)}$$

Example 17.6

Step 1 Assign oxidation numbers to each element to identify the elements being oxidized and those being reduced. Write the oxidation numbers below each element to avoid confusing them with ionic charge:

SOLUTION

$$
\begin{array}{ccccccc}
Sn & + & HNO_3 & \longrightarrow & SnO_2 & + & NO_2 & + & H_2O \\
\downarrow & & \swarrow\;\downarrow\;\searrow & & \downarrow\;\searrow & & \downarrow\;\searrow & & \downarrow\;\downarrow \\
0 & & +1\;+5\;-2 & & +4\;-2 & & +4\;-2 & & +1\;-2
\end{array}
$$

Note that the oxidation numbers of Sn and N have changed.

Step 2 Now write two new equations (half-reactions), using only the elements that change in oxidation number. Then add electrons to bring the equations into electrical balance. One equation represents the oxidation step; the other represents the reduction step. *Remember*: Oxidation produces electrons; reduction uses electrons.

$$Sn^0 \longrightarrow Sn^{4+} + 4\,e^- \quad \text{(oxidation)}$$
$\qquad Sn^0$ loses 4 electrons

$$N^{5+} + 1\,e^- \longrightarrow N^{4+} \quad \text{(reduction)}$$
$\qquad N^{5+}$ gains 1 electron

Step 3 Multiply the two equations by the smallest whole numbers that will make the electrons lost by oxidation equal to the number of electrons gained by reduction. In this reaction the oxidation step is multiplied by 1 and the reduction step by 4. The equations become

$$Sn^0 \longrightarrow Sn^{4+} + 4\,e^- \quad \text{(oxidation)}$$
$\qquad Sn^0$ loses 4 electrons

$$4\,N^{5+} + 4\,e^- \longrightarrow 4\,N^{4+} \quad \text{(reduction)}$$

$4\,N^{5+}$ gains 4 electrons

We have now established the ratio of the oxidizing to the reducing agent as being four atoms of N to one atom of Sn.

Step 4 Transfer the coefficient in front of each substance in the balanced oxidation–reduction equations to the corresponding substance in the original equation. We need to use 1 Sn, 1 SnO$_2$, 4 HNO$_3$, and 4 NO$_2$:

$$Sn + 4\,HNO_3 \longrightarrow SnO_2 + 4\,NO_2 + H_2O \quad \text{(unbalanced)}$$

Step 5 In the usual manner, balance the remaining elements that are not oxidized or reduced to give the final balanced equation:

$$Sn + 4\,HNO_3 \longrightarrow SnO_2 + 4\,NO_2 + 2\,H_2O \quad \text{(balanced)}$$

In balancing the final elements, we must not change the ratio of the elements that were oxidized and reduced.

Finally, check to ensure that both sides of the equation have the same number of atoms of each element. The final balanced equation contains 1 atom of Sn, 4 atoms of N, 4 atoms of H, and 12 atoms of O on each side.

Because each new equation presents a slightly different problem and because proficiency in balancing equations requires practice, let's work through two more examples.

Example 17.7   Balance the equation

$$I_2 + Cl_2 + H_2O \longrightarrow HIO_3 + HCl \quad \text{(unbalanced)}$$

SOLUTION   **Step 1** Assign oxidation numbers:

$$I_2 + Cl_2 + H_2O \longrightarrow HIO_3 + HCl$$

$$\quad 0 \quad\quad 0 \quad\quad +1 \;-2 \quad\quad +1\;+5\;-2 \quad +1\;-1$$

The oxidation numbers of I$_2$ and Cl$_2$ have changed, I$_2$ from 0 to +5, and Cl$_2$ from 0 to −1.

Step 2 Write the oxidation and reduction steps. Balance the number of atoms and then balance the electrical charge using electrons:

$$I_2 \longrightarrow 2\,I^{5+} + 10\,e^- \quad \text{(oxidation)} \quad (10\,e^-\text{ are needed to balance}$$

I$_2$ loses 10 electrons   the +10 charge)

$$Cl_2 + 2\,e^- \longrightarrow 2\,Cl^- \quad \text{(reduction)} \quad (2\,e^-\text{ are needed to balance}$$

Cl$_2$ gains 2 electrons   the −2 charge)

Step 3 Adjust loss and gain of electrons so that they are equal. Multiply the oxidation step by 1 and the reduction step by 5:

$$I_2 \longrightarrow 2\,I^{5+} + 10\,e^- \quad \text{(oxidation)}$$

I$_2$ loses 10 electrons

$$5\,Cl_2 + 10\,e^- \longrightarrow 10\,Cl^- \quad \text{(reduction)}$$

5 Cl$_2$ gain 10 electrons

Step 4 Transfer the coefficients from the balanced redox equations into the original equation. We need to use 1 I$_2$, 2 HIO$_3$, 5 Cl$_2$, and 10 HCl:

$$I_2 + 5\,Cl_2 + H_2O \longrightarrow 2\,HIO_3 + 10\,HCl \quad \text{(unbalanced)}$$

Step 5 Balance the remaining elements, H and O:

$$I_2 + 5\,Cl_2 + 6\,H_2O \longrightarrow 2\,HIO_3 + 10\,HCl \quad \text{(balanced)}$$

Check: The final balanced equation contains 2 atoms of I, 10 atoms of Cl, 12 atoms of H, and 6 atoms of O on each side.

Balance the equation

$$K_2Cr_2O_7 + FeCl_2 + HCl \longrightarrow CrCl_3 + KCl + FeCl_3 + H_2O \quad \text{(unbalanced)}$$

Step 1 Assign oxidation numbers (Cr and Fe have changed):

$$K_2Cr_2O_7 + FeCl_2 + HCl \longrightarrow CrCl_3 + KCl + FeCl_3 + H_2O$$

$$\begin{array}{ccccccc} +1\ +6\ -2 & +2\ -1 & +1\ -1 & +3\ -1 & +1\ -1 & +3\ -1 & +1\ -2 \end{array}$$

Example 17.8

SOLUTION

Step 2 Write the oxidation and reduction steps. Balance the number of atoms and then balance the electrical charge using electrons:

$$Fe^{2+} \longrightarrow Fe^{3+} + 1\,e^- \quad \text{(oxidation)}$$

Fe^{2+} loses 1 electron

$$2\,Cr^{6+} + 6\,e^- \longrightarrow 2\,Cr^{3+} \quad \text{(reduction)}$$

$2\,Cr^{6+}$ gain 6 electrons

Step 3 Balance the loss and gain of electrons. Multiply the oxidation step by 6 and the reduction step by 1 to equalize the transfer of electrons.

$$6\,Fe^{2+} \longrightarrow 6\,Fe^{3+} + 6\,e^- \quad \text{(oxidation)}$$

$6\,Fe^{2+}$ lose 6 electrons

$$2\,Cr^{6+} + 6\,e^- \longrightarrow 2\,Cr^{3+} \quad \text{(reduction)}$$

$2\,Cr^{6+}$ gain 6 electrons

Step 4 Transfer the coefficients from the balanced redox equations into the original equation. (Note that one formula unit of $K_2Cr_2O_7$ contains two Cr atoms.) We need to use $1\,K_2Cr_2O_7$, $2\,CrCl_3$, $6\,FeCl_2$, and $6\,FeCl_3$:

$$K_2Cr_2O_7 + 6\,FeCl_2 + HCl \longrightarrow$$
$$2\,CrCl_3 + KCl + 6\,FeCl_3 + H_2O \quad \text{(unbalanced)}$$

Step 5 Balance the remaining elements in the order K, Cl, H, O:

$$K_2Cr_2O_7 + 6\,FeCl_2 + 14\,HCl \longrightarrow$$
$$2\,CrCl_3 + 2\,KCl + 6\,FeCl_3 + 7\,H_2O \quad \text{(balanced)}$$

Check: The final balanced equation contains 2 K atoms, 2 Cr atoms, 7 O atoms, 6 Fe atoms, 26 Cl atoms, and 14 H atoms on each side.

Practice 17.4 _____

Balance these equations using the change-in-oxidation-number method:

(a) $HNO_3 + S \longrightarrow NO_2 + H_2SO_4 + H_2O$

(b) $CrCl_3 + MnO_2 + H_2O \longrightarrow MnCl_2 + H_2CrO_4$

(c) $KMnO_4 + HCl + H_2S \longrightarrow KCl + MnCl_2 + S + H_2O$

17.4 Balancing Ionic Redox Equations

The main difference between balancing ionic redox equations and molecular redox equations is in how we handle the ions. In the ionic redox equations, besides having the same number of atoms of each element on both sides of the final equation, we must also have equal net charges. In assigning oxidation numbers, we must therefore remember to consider the ionic charge.

Several methods are used to balance ionic redox equations, including, with slight modification, the oxidation-number method just shown for molecular equations. But the most popular method is probably the ion–electron method.

The ion–electron method uses ionic charges and electrons to balance ionic redox equations. Oxidation numbers are not formally used, but it is necessary to determine what is being oxidized and what is being reduced. The method is as follows:

Step 1 Write the two half-reactions that contain the elements being oxidized and reduced using the entire formula of the ion or molecule.

Step 2 Balance the elements other than oxygen and hydrogen.

Step 3 Balance oxygen and hydrogen.
Acidic solution: For reactions in acidic solution, use H^+ and H_2O to balance oxygen and hydrogen. For each oxygen needed, use one H_2O. Then add H^+ as needed to balance the hydrogen atoms.
Basic solution: For reactions in alkaline solutions, first balance as though the reaction were in an acidic solution, using Steps 1–3. Then add as many OH^- ions to each side of the equation as there are H^+ ions in the equation. Now combine the H^+ and OH^- ions into water (for example, $4\,H^+$ and $4\,OH^-$ give $4\,H_2O$). Rewrite the equation, canceling equal numbers of water molecules that appear on opposite sides of the equation.

Step 4 Add electrons (e^-) to each half-reaction to bring them into electrical balance.

Step 5 Since the loss and gain of electrons must be equal, multiply each half-reaction by the appropriate number to make the number of electrons the same in each half-reaction.

Step 6 Add the two half-reactions together, canceling electrons and any other identical substances that appear on opposite sides of the equation.

Example 17.9   Balance this equation using the ion–electron method:

$$MnO_4^- + S^{2-} \longrightarrow Mn^{2+} + S^0 \quad \text{(acidic solution)}$$

SOLUTION   **Step 1**   Write two half-reactions, one containing the element being oxidized and the other the element being reduced (use the entire molecule or ion):

$$S^{2-} \longrightarrow S^0 \quad \text{(oxidation)}$$
$$MnO_4^- \longrightarrow Mn^{2+} \quad \text{(reduction)}$$

Step 2   Balance elements other than oxygen and hydrogen (accomplished in Step 1: 1 S and 1 Mn on each side).

Oxidation–reduction reactions are the basis for many interesting applications. Consider photochromic glass, which is used for lenses in light-sensitive glasses. These lenses, manufactured by the Corning Glass Company, can change from transmitting 85% of light to transmitting only 22% of light when exposed to bright sunlight.

Photochromic glass is composed of linked tetrahedrons of silicon and oxygen atoms jumbled in a disorderly array, with crystals of silver chloride caught between the silica tetrahedrons. When the glass is clear, the visible light passes right through the molecules. The glass absorbs ultraviolet light, however, and this energy triggers an oxidation–reduction reaction between Ag^+ and Cl^-:

$$Ag^+ + Cl^- \xrightarrow{\text{UV light}} Ag^0 + Cl^0$$

To prevent the reaction from reversing itself immediately, a few ions of Cu^+ are incorporated into the silver chloride crystal. These Cu^+ ions react with the newly formed chlorine atoms:

$$Cu^+ + Cl^0 \longrightarrow Cu^{2+} + Cl^-$$

The silver atoms move to the surface of the crystal and form small colloidal clusters of silver metal. This metallic silver absorbs visible light, making the lens appear dark (colored).

As the glass is removed from the light, the Cu^{2+} ions slowly move to the surface of the crystal, where they interact with the silver metal:

$$Cu^{2+} + Ag^0 \longrightarrow Cu^+ + Ag^+$$

The glass clears as the silver ions rejoin chloride ions in the crystals.

An oxidation–reduction reaction causes these photochromic glasses to change from light to dark in bright sunlight.

Step 3 Balance O and H. Remember the solution is acidic. The oxidation requires neither O nor H, but the reduction equation needs $4\,H_2O$ on the right and $8\,H^+$ on the left:

$$S^{2-} \longrightarrow S^0$$
$$8\,H^+ + MnO_4^- \longrightarrow Mn^{2+} + 4\,H_2O$$

Step 4 Balance each half-reaction electrically with electrons:

$$S^{2-} \longrightarrow S^0 + 2\,e^-$$
net charge = −2 on each side

$$5\,e^- + 8\,H^+ + MnO_4^- \longrightarrow Mn^{2+} + 4\,H_2O$$
net charge = +2 on each side

Step 5 Equalize loss and gain of electrons. In this case, multiply the oxidation equation by 5 and the reduction equation by 2:

$$5\,S^{2-} \longrightarrow 5\,S^0 + 10\,e^-$$
$$10\,e^- + 16\,H^+ + 2\,MnO_4^- \longrightarrow 2\,Mn^{2+} + 8\,H_2O$$

Step 6 Add the two half-reactions together, canceling the $10\,e^-$ from each side, to obtain the balanced equation:

$$5\,S^{2-} \longrightarrow 5\,S^0 + \cancel{10\,e^-}$$
$$\underline{\cancel{10\,e^-} + 16\,H^+ + 2\,MnO_4^- \longrightarrow 2\,Mn^{2+} + 8\,H_2O}$$
$$16\,H^+ + 2\,MnO_4^- + 5\,S^{2-} \longrightarrow 2\,Mn^{2+} + 5\,S^0 + 8\,H_2O \quad \text{(balanced)}$$

Check: Both sides of the equation have a charge of +4 and contain the same number of atoms of each element.

Example 17.10   Balance this equation:

$$CrO_4^{2-} + Fe(OH)_2 \longrightarrow Cr(OH)_3 + Fe(OH)_3 \quad \text{(basic solution)}$$

SOLUTION   **Step 1**   Write the two half-reactions:

$$Fe(OH)_2 \longrightarrow Fe(OH)_3 \quad \text{(oxidation)}$$
$$CrO_4^{2-} \longrightarrow Cr(OH)_3 \quad \text{(reduction)}$$

Step 2   Balance elements other than H and O (accomplished in Step 1).

Step 3   Remember the solution is basic. Balance O and H as though the solution were acidic. Use H_2O and H^+. To balance O and H in the oxidation equation, add 1 H_2O on the left and 1 H^+ on the right side:

$$Fe(OH)_2 + H_2O \longrightarrow Fe(OH)_3 + H^+$$

Add 1 OH^- to each side:

$$Fe(OH)_2 + H_2O + OH^- \longrightarrow Fe(OH)_3 + H^+ + OH^-$$

Combine H^+ and OH^- as H_2O and rewrite, canceling H_2O on each side:

$$Fe(OH)_2 + \cancel{H_2O} + OH^- \longrightarrow Fe(OH)_3 + \cancel{H_2O}$$

$$\boxed{Fe(OH)_2 + OH^- \longrightarrow Fe(OH)_3} \quad \text{(oxidation)}$$

To balance O and H in the reduction equation, add 1 H_2O on the right and 5 H^+ on the left:

$$CrO_4^{2-} + 5\,H^+ \longrightarrow Cr(OH)_3 + H_2O$$

Add 5 OH^- to each side:

$$CrO_4^{2-} + 5\,H^+ + 5\,OH^- \longrightarrow Cr(OH)_3 + H_2O + 5\,OH^-$$

Combine $5\,H^+ + 5\,OH^- \longrightarrow 5\,H_2O$:

$$CrO_4^{2-} + 5\,H_2O \longrightarrow Cr(OH)_3 + H_2O + 5\,OH^-$$

Rewrite, canceling 1 H_2O from each side:

$$\boxed{CrO_4^{2-} + 4\,H_2O \longrightarrow Cr(OH)_3 + 5\,OH^-} \quad \text{(reduction)}$$

Step 4   Balance each half-reaction electrically with electrons:

$$Fe(OH)_2 + OH^- \longrightarrow Fe(OH)_3 + e^-$$
$$\text{(balanced oxidation equation)}$$
$$CrO_4^{2-} + 4\,H_2O + 3\,e^- \longrightarrow Cr(OH)_3 + 5\,OH^-$$
$$\text{(balanced reduction equation)}$$

Step 5   Equalize the loss and gain of electrons. Multiply the oxidation reaction by 3:

$$3\,Fe(OH)_2 + 3\,OH^- \longrightarrow 3\,Fe(OH)_3 + 3\,e^-$$
$$CrO_4^{2-} + 4\,H_2O + 3\,e^- \longrightarrow Cr(OH)_3 + 5\,OH^-$$

Step 6   Add the two half-reactions together, canceling the $3\,e^-$ and $3\,OH^-$ from each side of the equation:

$$3\,Fe(OH)_2 + 3\,OH^- \longrightarrow 3\,Fe(OH)_3 + \cancel{3\,e^-}$$

$$\underline{CrO_4^{2-} + 4\,H_2O + \cancel{3\,e^-} \longrightarrow Cr(OH)_3 + 5\,OH^-}$$

$$CrO_4^{2-} + 3\,Fe(OH)_2 + 4\,H_2O \longrightarrow Cr(OH)_3 + 3\,Fe(OH)_3 + 2\,OH^- \quad \text{(balanced)}$$

Check: Each side of the equation has a charge of -2 and contains the same number of atoms of each element.

Practice 17.5 _____

Balance these equations using the ion–electron method:

(a) $I^- + NO_2^- \longrightarrow I_2 + NO$ (acidic solution)

(b) $Cl_2 + IO_3^- \longrightarrow IO_4^- + Cl^-$ (basic solution)

(c) $AuCl_4^- + Sn^{2+} \longrightarrow Sn^{4+} + AuCl + Cl^-$

   Ionic equations can also be balanced using the change-in-oxidation-number method shown in Example 17.6. To illustrate this method, let's use the equation from Example 17.10.

Balance this equation using the change-in-oxidation-number method:

$$CrO_4^{2-} + Fe(OH)_2 \longrightarrow Cr(OH)_3 + Fe(OH)_3 \quad \text{(basic solution)}$$

Example 17.11

Steps 1 and 2   Assign oxidation numbers and balance the charges with electrons: **SOLUTION**

$$Cr^{6+} + 3\,e^- \longrightarrow Cr^{3+} \quad \text{(reduction)}$$
$$Cr^{6+} \text{ gains } 3\,e^-$$

$$Fe^{2+} \longrightarrow Fe^{3+} + e^- \quad \text{(oxidation)}$$
$$Fe^{2+} \text{ loses } 1\,e^-$$

Step 3   Equalize the loss and gain of electrons, and then multiply the oxidation step by 3:

$$Cr^{6+} + 3\,e^- \longrightarrow Cr^{3+} \quad \text{(reduction)}$$
$$Cr^{6+} \text{ gains } 3\,e^-$$

$$3\,Fe^{2+} \longrightarrow 3\,Fe^{3+} + 3\,e^- \quad \text{(oxidation)}$$
$$3\,Fe^{2+} \text{ loses } 3\,e^-$$

Step 4   Transfer coefficients back to the original equation:

$$CrO_4^{2-} + 3\,Fe(OH)_2 \longrightarrow Cr(OH)_3 + 3\,Fe(OH)_3$$

Step 5   Balance electrically. Because the solution is basic, use OH^- to balance charges. The charge on the left side is -2 and on the right side is 0. Add $2\,OH^-$ ions to the right side of the equation:

$$CrO_4^{2-} + 3\,Fe(OH)_2 \longrightarrow Cr(OH)_3 + 3\,Fe(OH)_3 + 2\,OH^-$$

Adding $4\,H_2O$ to the left side balances the equation:

$$CrO_4^{2-} + 3\,Fe(OH)_2 + 4\,H_2O \longrightarrow$$
$$Cr(OH)_3 + 3\,Fe(OH)_3 + 2\,OH^- \quad \text{(balanced)}$$

Check: Each side of the equation has a charge of -2 and contains the same number of atoms of each element.

Figure 17.2
A coil of copper placed in a silver nitrate solution forms silver crystals on the wire. The pale blue of the solution indicates the presence of copper ions.

Table 17.3 Activity Series of Metals

Ease of oxidation ↑

$$K \longrightarrow K^+ + e^-$$
$$Ba \longrightarrow Ba^{2+} + 2e^-$$
$$Ca \longrightarrow Ca^{2+} + 2e^-$$
$$Na \longrightarrow Na^+ + e^-$$
$$Mg \longrightarrow Mg^{2+} + 2e^-$$
$$Al \longrightarrow Al^{3+} + 3e^-$$
$$Zn \longrightarrow Zn^{2+} + 2e^-$$
$$Cr \longrightarrow Cr^{3+} + 3e^-$$
$$Fe \longrightarrow Fe^{2+} + 2e^-$$
$$Ni \longrightarrow Ni^{2+} + 2e^-$$
$$Sn \longrightarrow Sn^{2+} + 2e^-$$
$$Pb \longrightarrow Pb^{2+} + 2e^-$$
$$\mathbf{H_2 \longrightarrow 2H^+ + 2e^-}$$
$$Cu \longrightarrow Cu^{2+} + 2e^-$$
$$As \longrightarrow As^{3+} + 3e^-$$
$$Ag \longrightarrow Ag^+ + e^-$$
$$Hg \longrightarrow Hg^{2+} + 2e^-$$
$$Au \longrightarrow Au^{3+} + 3e^-$$

activity series of metals

Practice 17.6 _____

Balance these equations using the change-in-oxidation-number method:

(a) $Zn \longrightarrow Zn(OH)_4^{2-} + H_2$ (basic solution)
(b) $H_2O_2 + Sn^{2+} \longrightarrow Sn^{4+}$ (acidic solution)
(c) $Cu + Cu^{2+} \longrightarrow Cu_2O$ (basic solution)

17.5 Activity Series of Metals

Knowledge of the relative chemical reactivities of the elements helps us predict the course of many chemical reactions. For example, calcium reacts with cold water to produce hydrogen, and magnesium reacts with steam to produce hydrogen. Therefore, calcium is considered a more reactive metal than magnesium:

$$Ca(s) + 2H_2O(l) \longrightarrow Ca(OH)_2(aq) + H_2(g)$$
$$Mg(s) + \underset{\text{steam}}{H_2O(g)} \longrightarrow MgO(s) + H_2(g)$$

The difference in their activity is attributed to the fact that calcium loses its two valence electrons more easily than magnesium and is therefore more reactive and/or more readily oxidized than magnesium.

When a coil of copper is placed in a solution of silver nitrate ($AgNO_3$), free silver begins to plate out on the copper. (See Figure 17.2.) After the reaction has continued for some time, we can observe a blue color in the solution, indicating the presence of copper(II) ions. The equations are

$$Cu^0(s) + 2AgNO_3(aq) \longrightarrow 2Ag^0(s) + Cu(NO_3)_2(aq)$$
$$Cu^0(s) + 2Ag^+(aq) \longrightarrow 2Ag^0(s) + Cu^{2+}(aq) \qquad \text{(net ionic equation)}$$
$$Cu^0(s) \longrightarrow Cu^{2+}(aq) + 2e^- \qquad \text{(oxidation of } Cu^0\text{)}$$
$$Ag^+(aq) + e^- \longrightarrow Ag^0(s) \qquad \text{(reduction of } Ag^+\text{)}$$

If a coil of silver is placed in a solution of copper(II) nitrate, $Cu(NO_3)_2$, no reaction is visible.

$$Ag^0(s) + Cu(NO_3)_2(aq) \longrightarrow \text{no reaction}$$

In the reaction between Cu and $AgNO_3$, electrons are transferred from Cu^0 atoms to Ag^+ ions in solution. Copper has a greater tendency than silver to lose electrons, so an electrochemical force is exerted upon silver ions to accept electrons from copper atoms. When an Ag^+ ion accepts an electron, it is reduced to an Ag^0 atom and is no longer soluble in solution. At the same time, Cu^0 is oxidized and goes into solution as Cu^{2+} ions. From this reaction, we can conclude that copper is more reactive than silver.

Metals such as sodium, magnesium, zinc, and iron that react with solutions of acids to liberate hydrogen are more reactive than hydrogen. Metals such as copper, silver, and mercury that do not react with solutions of acids to liberate hydrogen are less reactive than hydrogen. By studying a series of reactions such as these, we can list metals according to their chemical activity, placing the most active at the top and the least active at the bottom. This list is called the **activity series of metals.** Table 17.3 lists some of the common metals in the series. The arrangement corresponds to the ease with which the elements are oxidized or lose electrons, with the most easily oxidizable element listed first. More extensive tables are available in chemistry reference books.

The general principles governing the arrangement and use of the activity series are as follows:

1. The reactivity of the metals listed decreases from top to bottom.
2. A free metal can displace the ion of a second metal from solution, provided that the free metal is above the second metal in the activity series.
3. Free metals above hydrogen react with nonoxidizing acids in solution to liberate hydrogen gas.
4. Free metals below hydrogen do not liberate hydrogen from acids.
5. Conditions such as temperature and concentration may affect the relative position of some of these elements.

Here are two examples using the activity series of metals.

Will zinc metal react with dilute sulfuric acid?

Example 17.12

From Table 17.3, we see that zinc is above hydrogen; therefore, zinc atoms will lose electrons more readily than hydrogen atoms. Hence zinc atoms will reduce hydrogen ions from the acid to form hydrogen gas and zinc ions. In fact, these reagents are commonly used for the laboratory preparation of hydrogen. The equation is

SOLUTION

$$Zn(s) + H_2SO_4(aq) \longrightarrow ZnSO_4(aq) + H_2(g)$$

$$Zn(s) + 2\,H^+(aq) \longrightarrow Zn^{2+}(aq) + H_2(g) \qquad \text{(net ionic equation)}$$

Will a reaction occur when copper metal is placed in an iron(II) sulfate solution?

Example 17.13

No, copper lies below iron in the series, loses electrons less easily than iron, and therefore will not displace iron(II) ions from solution. In fact, the reverse is true. When an iron nail is dipped into a copper(II) sulfate solution, it becomes coated with free copper. The equations are

SOLUTION

$$Cu(s) + FeSO_4(aq) \longrightarrow \text{no reaction}$$

$$Fe(s) + CuSO_4(aq) \longrightarrow FeSO_4(aq) + Cu(s)$$

From Table 17.3, we may abstract the following pair in their relative position to each other:

$$Fe \longrightarrow Fe^{2+} + 2\,e^-$$

$$Cu \longrightarrow Cu^{2+} + 2\,e^-$$

According to Principle 2 on the use of the activity series, we can predict that free iron will react with copper(II) ions in solution to form free copper metal and iron(II) ions in solution:

$$Fe(s) + Cu^{2+}(aq) \longrightarrow Fe^{2+}(aq) + Cu(s) \qquad \text{(net ionic equation)}$$

Practice 17.7 _____

Indicate whether these reactions will occur:

(a) Sodium metal is placed in dilute hydrochloric acid.
(b) A piece of lead is placed in magnesium nitrate solution.
(c) Mercury is placed in a solution of silver nitrate.

17.6 Electrolytic and Voltaic Cells

electrolysis
electrolytic cell

The process in which electrical energy is used to bring about chemical change is known as **electrolysis.** An **electrolytic cell** uses electrical energy to produce a chemical reaction. The use of electrical energy has many applications in industry—for example, in the production of sodium, sodium hydroxide, chlorine, fluorine, magnesium, aluminum, and pure hydrogen and oxygen, and in the purification and electroplating of metals.

What happens when an electric current is passed through a solution? Let's consider a hydrochloric acid solution in a simple electrolytic cell, as shown in Figure 17.3. The cell consists of a source of direct current (a battery) connected to two electrodes that are immersed in a solution of hydrochloric acid.

cathode
anode

The negative electrode is called the **cathode** because cations are attracted to it. The positive electrode is called the **anode** because anions are attracted to it. The cathode is attached to the negative pole and the anode to the positive pole of the battery. The battery supplies electrons to the cathode.

When the electric circuit is completed, positive hydronium ions (H_3O^+) migrate to the cathode, where they pick up electrons and evolve as hydrogen gas. At the same time the negative chloride ions (Cl^-) migrate to the anode, where they lose electrons and evolve as chlorine gas.

Reaction at the cathode:

$$H_3O^+ + 1\,e^- \longrightarrow H^0 + H_2O \quad \text{(reduction)}$$
$$H^0 + H^0 \longrightarrow H_2$$

Reaction at the anode:

$$Cl^- \longrightarrow Cl^0 + 1\,e^- \qquad \text{(oxidation)}$$
$$Cl^0 + Cl^0 \longrightarrow Cl_2$$
$$2\,HCl(aq) \xrightarrow{\text{electrolysis}} H_2(g) + Cl_2(g) \quad \text{(net reaction)}$$

Note that oxidation–reduction has taken place. Chloride ions lost electrons (were oxidized) at the anode, and hydronium ions gained electrons (were reduced) at the cathode.

Figure 17.3
During the electrolysis of a hydrochloric acid solution, positive hydronium ions are attracted to the cathode, where they gain electrons and form hydrogen gas. Chloride ions migrate to the anode, where they lose electrons and form chlorine gas. The equation for this process is
$2\,HCl(aq) \longrightarrow H_2(g) + Cl_2(g)$.

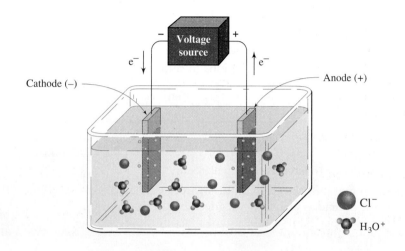

Oxidation always occurs at the anode and reduction at the cathode.

When concentrated sodium chloride solutions (brines) are electrolyzed, the products are sodium hydroxide, hydrogen, and chlorine. The overall reaction is

$$2\,Na^+(aq) + 2\,Cl^-(aq) + 2\,H_2O(l) \xrightarrow{\text{electrolysis}}$$
$$2\,Na^+(aq) + 2\,OH^-(aq) + H_2(g) + Cl_2(g)$$

The net ionic equation is

$$2\,Cl^-(aq) + 2\,H_2O(l) \longrightarrow 2\,OH^-(aq) + H_2(g) + Cl_2(g)$$

During electrolysis, Na^+ ions move toward the cathode and Cl^- ions move toward the anode. The anode reaction is similar to that of hydrochloric acid; the chlorine is liberated:

$$2\,Cl^-(aq) \longrightarrow Cl_2(g) + 2\,e^-$$

Even though Na^+ ions are attracted by the cathode, the facts show that hydrogen is liberated there. No evidence of metallic sodium is found, but the area around the cathode tests alkaline from the accumulated OH^- ions. The reaction at the cathode is

$$2\,H_2O(l) + 2\,e^- \longrightarrow H_2(g) + 2\,OH^-(aq)$$

If electrolysis is allowed to continue until all the chloride is reacted, the solution remaining will contain only sodium hydroxide, which on evaporation yields solid NaOH. Large amounts of sodium hydroxide and chlorine are made by this process.

When molten sodium chloride (without water) is subjected to electrolysis, metallic sodium and chlorine gas are formed:

$$2\,Na^+(l) + 2\,Cl^-(l) \xrightarrow{\text{electrolysis}} 2\,Na(s) + Cl_2(g)$$

An important electrochemical application is the electroplating of metals. Electroplating is the art of covering a surface or an object with a thin adherent electrodeposited metal coating. Electroplating is done for protection of the surface of the base metal or for a purely decorative effect. The layer deposited is surprisingly thin, varying from as little as 5×10^{-5} cm to 2×10^{-3} cm, depending on the metal and the intended use. The object to be plated is set up as the cathode and is immersed in a solution containing ions of the plating metal. When an electric current passes through the solution, metal ions that migrate to the cathode are reduced, depositing on the object as the free metal. In most cases, the metal deposited on the object is replaced in the solution by using an anode of the same metal. The following equations show the chemical changes in the electroplating of nickel:

Reaction at the cathode: $Ni^{2+}(aq) + 2\,e^- \longrightarrow Ni(s)$ (Ni plated out on an object)

Reaction at the anode: $Ni(s) \longrightarrow Ni^{2+}(aq) + 2\,e^-$ (Ni replenished in solution)

Metals commonly used in commercial electroplating are copper, nickel, zinc, lead, cadmium, chromium, tin, gold, and silver.

Jewelry and eyeglasses are electroplated with rhodium to prevent tarnishing.

In the electrolytic cell shown in Figure 17.3, electrical energy from the voltage source is used to bring about nonspontaneous redox reactions. The hydrogen and chlorine produced have more potential energy than was present in the hydrochloric acid before electrolysis.

Conversely, some spontaneous redox reactions can be made to supply useful amounts of electrical energy. When a piece of zinc is put in a copper(II) sulfate solution, the zinc quickly becomes coated with metallic copper. We expect this coating to happen because zinc is above copper in the activity series; copper(II) ions are therefore reduced by zinc atoms:

$$Zn^0(s) \ + \ Cu^{2+}(aq) \ \longrightarrow \ Zn^{2+}(aq) \ + \ Cu^0(s)$$

This reaction is clearly a spontaneous redox reaction, but simply dipping a zinc rod into a copper(II) sulfate solution will not produce useful electric current. However, when we carry out this reaction in the cell shown in Figure 17.4, an electric current is produced. The cell consists of a piece of zinc immersed in a zinc sulfate solution and connected by a wire through a voltmeter to a piece of copper immersed in copper(II) sulfate solution. The two solutions are connected by a salt bridge. Such a cell produces an electric current and a potential of about 1.1 volts when both solutions are 1.0 M in concentration. A cell that produces electric current from a spontaneous chemical reaction is called a **voltaic cell.** A voltaic cell is also known as a *galvanic cell*.

voltalc cell

The driving force responsible for the electric current in the zinc–copper cell originates in the great tendency of zinc atoms to lose electrons relative to the tendency of copper(II) ions to gain electrons. In the cell shown in Figure 17.4, zinc atoms lose electrons and are converted to zinc ions at the zinc electrode surface; the electrons flow through the wire (external circuit) to the copper electrode. Here copper(II) ions pick up electrons and are reduced to copper atoms, which plate out on the copper electrode. Sulfate ions flow from the $CuSO_4$ solution via the salt bridge into the $ZnSO_4$ solution (internal circuit) to complete the circuit.

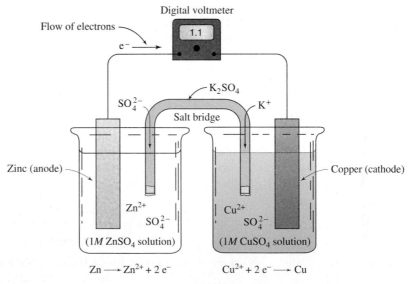

Figure 17.4
Zinc–copper voltaic cell. The cell has a potential of 1.1 volts when $ZnSO_4$ and $CuSO_4$ solutions are 1.0 M. The salt bridge provides electrical contact between the two half-cells.

The equations for the reactions of this cell are

anode	$Zn^0(s) \longrightarrow Zn^{2+}(aq) + 2\,e^-$	(oxidation)
cathode	$Cu^{2+}(aq) + 2\,e^- \longrightarrow Cu^0(s)$	(reduction)
net ionic	$Zn^0(s) + Cu^{2+}(aq) \longrightarrow Zn^{2+}(aq) + Cu^0(s)$	
overall	$Zn(s) + CuSO_4(aq) \longrightarrow ZnSO_4(aq) + Cu(s)$	

The redox reaction, the movement of electrons in the metallic or external part of the circuit, and the movement of ions in the solution or internal part of the circuit of the zinc–copper cell are very similar to the actions that occur in the electrolytic cell of Figure 17.3. The only important difference is that the re-actions of the zinc–copper cell are spontaneous. This spontaneity is the crucial difference between all voltaic and electrolytic cells.

> Voltaic cells use chemical reactions to produce electrical energy.
> Electrolytic cells use electrical energy to produce chemical reactions.

Calculators, watches, radios, cell phones, and portable CD players are powered by small efficient voltaic cells called *dry cell batteries*. These batteries are called dry cells because they do not contain a liquid electrolyte (like the voltaic cells discussed earlier). Dry cell batteries are found in several different versions.

The *acid*-type dry cell battery contains a zinc inner case that functions as the anode. A carbon (graphite) rod runs through the center and is in contact with the zinc case at one end and a moist paste of solid MnO_2, NH_4Cl, and car-bon that functions as the cathode. (See Figure 17.5a.) The cell produces about 1.5 volts. The *alkaline*-type dry cell battery is the same as the acid type except the NH_4Cl is replaced by either KOH or NaOH. These dry cells typically last longer because the zinc anode corrodes more slowly in basic conditions. A third type of dry cell is the *zinc–mercury* cell shown in Figure 17.5b. The reactions occurring in this cell are

anode	$Zn^0 + 2\,OH^- \longrightarrow ZnO + H_2O + 2\,e^-$	(oxidation)
cathode	$HgO + H_2O + 2\,e^- \longrightarrow Hg^0 + 2\,OH^-$	(reduction)
net ionic	$Zn^0 + Hg^{2+} \longrightarrow Zn^{2+} + Hg^0$	
overall	$Zn^0 + HgO \longrightarrow ZnO + Hg^0$	

(a)

(b)

Figure 17.5
(a) A common acid-type dry cell.
(b) Diagram of an alkaline zinc–mercury cell.

Figure 17.6
Cross-sectional diagram of a lead storage battery cell.

To offset the relatively high initial cost, this cell (1) provides current at a very steady potential of about 1.5 volts; (2) has an exceptionally long service life—that is, high energy output to weight ratio; (3) is completely self-contained; and (4) can be stored for relatively long periods of time when not in use.

An automobile storage battery is an energy reservoir. The charged battery acts as a voltaic cell and through chemical reactions furnishes electrical energy to operate the starter, lights, radio, and so on. When the engine is running, a generator or alternator produces and forces an electric current through the battery and, by electrolytic chemical action, restores it to the charged condition.

The cell unit consists of a lead plate filled with spongy lead and a lead(IV) oxide plate, both immersed in dilute sulfuric acid solution, which serves as the electrolyte (see Figure 17.6). When the cell is discharging, or acting as a voltaic cell, these reactions occur:

Pb plate (anode): $Pb^0 \longrightarrow Pb^{2+} + 2\,e^-$ (oxidation)

PbO$_2$ plate (cathode): $PbO_2 + 4\,H^+ + 2\,e^- \longrightarrow Pb^{2+} + 2\,H_2O$ (reduction)

Net ionic redox
reaction: $Pb^0 + PbO_2 + 4\,H^+ \longrightarrow 2\,Pb^{2+} + 2\,H_2O$

Precipitation reaction
on plates: $Pb^{2+}(aq) + SO_4^{2-}(aq) \longrightarrow PbSO_4(s)$

Because lead(II) sulfate is insoluble, the Pb^{2+} ions combine with SO_4^{2-} ions to form a coating of $PbSO_4$ on each plate. The overall chemical reaction of the cell is

$$Pb(s) + PbO_2(s) + 2\,H_2SO_4(aq) \xrightarrow[\text{cycle}]{\text{discharge}} 2\,PbSO_4(s) + 2\,H_2O(l)$$

The cell can be recharged by reversing the chemical reaction. This reversal is accomplished by forcing an electric current through the cell in the opposite direction. Lead sulfate and water are reconverted to lead, lead(IV) oxide, and sulfuric acid:

$$2\,PbSO_4(s) + 2\,H_2O(l) \xrightarrow[\text{cycle}]{\text{charge}} Pb(s) + PbO_2(s) + 2\,H_2SO_4(aq)$$

The electrolyte in a lead storage battery is a 38% by mass sulfuric acid solution having a density of 1.29 g/mL. As the battery is discharged, sulfuric acid is removed, thereby decreasing the density of the electrolyte solution. The state of

Scientists are constantly trying to make longer-lasting environmentally friendly batteries to fuel our many electronic devices. Now a new type of alkaline batteries called "super-iron" batteries have been developed by chemists at Technion-Israel Institute of Technology in Israel. A traditional cell of this type is shown on p. 455. In the new battery, the heavy manganese dioxide cathode is replaced with "super iron." (See accompanying diagram.) This special type of iron compound contains iron(VI) in compounds such as K_2FeO_4 or $BaFeO_4$. Iron typically has an oxida-tion state of +2 or +3. In super iron, each iron atom is missing six electrons instead of the usual 2 or 3. This allows the battery to store 1505 J more energy than other alkaline batteries. When the battery is used (or the cell discharges), the following reaction occurs:

$$2\,MFeO_4 + 3\,Zn \longrightarrow$$
$$Fe_2O_3 + ZnO + 2\,MZnO_2$$
$$(M = K_2 \text{ or } Ba)$$

The iron compounds used in this battery are much less expensive than the current MnO_2 compounds and the products are more environmentally friendly (Fe_2O_3 is a form of rust).

Manganese dioxide molecules in conventional batteries can only each accept 1 electron while iron(VI) compounds can absorb 3 electrons each. The super-iron compounds are highly conductive, which means the super-iron battery will work well in our high-drain-rate electronic items. The accompanying graph shows a comparison between a conventional battery and a super-iron battery (both AAA). The conventional AAA battery lasts less than half as long as a super-iron AAA battery. Just think how many more CDs you could play in that time without changing the batteries!

Super-iron battery

Electrolyte (aqueous KOH) — Super-iron cathode — Separator — Case, cathode current collector — Cover — Zinc paste anode — Anode current collector

charge or discharge of the battery can be estimated by measuring the density (or specific gravity) of the electrolyte solution with a hydrometer. When the density has dropped to about 1.05 g/mL, the battery needs recharging.

In a commerical battery, each cell consists of a series of cell units of alternating lead–lead(IV) oxide plates separated and supported by wood, glass wool, or fiberglass. The energy storage capacity of a single cell is limited, and its electrical potential is only about 2 volts. Therefore a bank of six cells is connected in series to provide the 12-volt output of the usual automobile battery.

Practice 17.8

Consider the reaction

$$Sn^{2+}(aq) + Cu(s) \longrightarrow Sn(s) + Cu^{2+}(aq)$$

(a) Which metal is oxidized and which is reduced?
(b) Write the reaction occurring at the anode and at the cathode.

Chapter 17 Review

17.1 Oxidation Number
KEY TERM
Oxidation number or oxidation state

• To assign an oxidation number:

1. All elements in their free state (uncombined with other elements) have an oxidation number of zero (e.g., Na, Cu, Mg, H_2, O_2, Cl_2, N_2).
2. H is +1, except in metal hydrides, where it is −1 (e.g., NaH, CaH_2).
3. O is −2, except in peroxides, where it is −1, and in OF_2, where it is +2.
4. The metallic element in an ionic compound has a positive oxidation number.
5. In covalent compounds the negative oxidation number is assigned to the most electronegative atom.
6. The algebraic sum of the oxidation numbers of the elements in a compound is zero.
7. The algebraic sum of the oxidation numbers of the elements in a polyatomic ion is equal to the charge of the ion.

17.2 Oxidation–Reduction
KEY TERMS
Oxidation–reduction or redox
Oxidation
Reduction
Oxidizing agent
Reducing agent

• Oxidation–reduction is a chemical process in which electrons are transferred from one atom to another to change the oxidation number of the atom.
• When the oxidation number increases, oxidation occurs, resulting in the loss of electrons.
• When the oxidation number decreases, reduction occurs, resulting in the gain of electrons.

17.3 Balancing Oxidation–Reduction Equations
• Trial and error or inspection methods.
• Can be balanced by writing and balancing the half-reactions for the overall reaction.
• Can be balanced by the change-in-oxidation-number method.

17.4 Balancing Ionic Redox Equations
• To balance equations that are ionic, charge must also be balanced (in addition to atoms and ions).
• The ion–electron method for balancing equations is:

Step 1 Write the two half-reactions that contain the elements being oxidized and reduced using the entire formula of the ion or molecule.
Step 2 Balance the elements other than oxygen and hydrogen.
Step 3 Balance oxygen and hydrogen.
Acidic solution: For reactions in acidic solution, use H^+ and H_2O to balance oxygen and hydrogen. For each oxygen needed, use one H_2O. Then add H^+ as needed to balance the hydrogen atoms.
Basic solution: For reactions in alkaline solutions, first balance as though the reaction were in an acidic solution, using Steps 1–3. Then add as many OH^- ions to each side of the equation as there are H^+ ions in the equation. Now combine the H^+ and OH^- ions into water (for example, $4\,H^+$ and $4\,OH^-$ give $4\,H_2O$). Rewrite the equation, canceling equal numbers of water molecules that appear on opposite sides of the equation.
Step 4 Add electrons (e^-) to each half-reaction to bring them into electrical balance.
Step 5 Since the loss and gain of electrons must be equal, multiply each half-reaction by the appropriate number to make the number of electrons the same in each half-reaction.
Step 6 Add the two half-reactions together, canceling electrons and any other identical substances that appear on opposite sides of the equation.

17.5 Activity Series of Metals
KEY TERM
Activity series of metals

• The activity series lists metals from most to least reactive.
• A free metal can displace anything lower on the activity series.

17.6 Electrolytic and Voltaic Cells
KEY TERMS
Electrolysis
Electrolytic cell
Cathode
Anode
Voltaic cell

• Electrolysis is the process of using electricity to bring about chemical change.
• A typical electrolytic cell is shown next:

• Oxidation always occurs at the anode and reduction at the cathode.
• A cell that produces electric current from a spontaneous chemical reaction is a voltaic or galvanic cell.

Review Questions

All questions with blue numbers have answers in the appendix of the text.

1. Why do we say that the more active a metal is, the more easily it will be oxidized?

2. In the equation
$$I_2 + 5\,Cl_2 + 6\,H_2O \longrightarrow 2\,HIO_3 + 10\,HCl$$
   (a) has iodine been oxidized or has it been reduced?
   (b) has chlorine been oxidized or has it been reduced? (Figure 17.1)

3. Which element of each pair is more active? (Table 17.3)
   (a) Ag or Al   (b) Na or Ba   (c) Ni or Cu

4. Will the following combinations react in aqueous solution? (Table 17.3)
   (a) $Zn(s) + Cu^{2+}(aq)$ (e) $Ba(s) + FeCl_2(aq)$
   (b) $Ag(s) + H^+(aq)$ (f) $Pb(s) + NaCl(aq)$
   (c) $Sn(s) + Ag^+(aq)$ (g) $Ni(s) + Hg(NO_3)_2(aq)$
   (d) $As(s) + Mg^{2+}(aq)$ (h) $Al(s) + CuSO_4(aq)$

5. Why doesn't a silver wire placed in a solution of copper(II) nitrate ($Cu(NO_3)_2$) produce a reaction?

6. The reaction between powdered aluminum and iron(III) oxide (in the thermite process) producing molten iron is very exothermic.
   (a) Write the equation for the chemical reaction that occurs.
   (b) Explain in terms of Table 17.3 why a reaction occurs.
   (c) Would you expect a reaction between powdered iron and aluminum oxide?
   (d) Would you expect a reaction between powdered aluminum and chromium(III) oxide?

7. Write equations for the chemical reaction of aluminum, chromium, gold, iron, copper, magnesium, mercury, and zinc with dilute solutions of (a) hydrochloric acid and (b) sulfuric acid. If a reaction will not occur, write "no reaction" as the product. (Table 17.3)

8. What is the difference between an oxidation number for an atom in an ionic compound and an oxidation number for an atom in a covalently bonded compound?

9. State the charge and purpose of the anode and the cathode in an electrolytic or volatic cell.

10. A $NiCl_2$ solution is placed in the apparatus shown in Figure 17.3, instead of the HCl solution shown. Write equations for the following:
   (a) the anode reaction
   (b) the cathode reaction
   (c) the net electrochemical reaction

11. What is the major distinction between the reactions occurring in Figures 17.3 and 17.4?

12. In the cell shown in Figure 17.4,
   (a) what would be the effect of removing the voltmeter and connecting the wires shown coming to the voltmeter?
   (b) what would be the effect of removing the salt bridge?

13. Why are oxidation and reduction said to be complementary processes?

14. When molten $CaBr_2$ is electrolyzed, calcium metal and bromine are produced. Write equations for the two half-reactions that occur at the electrodes. Label the anode half-reaction and the cathode half-reaction.

15. Why is direct current used instead of alternating current in the electroplating of metals?

16. What property of lead(IV) oxide and lead(II) sulfate makes it unnecessary to have salt bridges in the cells of a lead storage battery?

17. Explain why the density of the electrolyte in a lead storage battery decreases during the discharge cycle.

18. In one type of alkaline cell used to power devices such as portable radios, Hg^{2+} ions are reduced to metallic mercury when the cell is being discharged. Does this reduction occur at the anode or the cathode? Explain.

19. Differentiate between an electrolytic cell and a voltaic cell.

20. Why is a porous barrier or a salt bridge necessary in some voltaic cells?

Paired Exercises

All exercises with blue numbers have answers in the appendix of the text.

1. What is the oxidation number of the underlined element in each compound?
(a) $\underline{Na}Cl$
(b) $Fe\underline{Cl}_3$
(c) $\underline{Pb}O_2$
(d) $Na\underline{N}O_3$
(e) $H_2\underline{S}O_3$
(f) $\underline{N}H_4Cl$

2. What is the oxidation number of the underlined element in each compound?
(a) $K\underline{Mn}O_4$
(b) $\underline{I}_2$
(c) $\underline{N}H_3$
(d) $K\underline{Cl}O_3$
(e) $K_2\underline{Cr}O_4$
(f) $K_2\underline{Cr}_2O_7$

3. What is the oxidation number of the underlined elements?
(a) $\underline{S}^{2-}$
(b) $\underline{N}O_2^-$
(c) $Na_2\underline{O}_2$
(d) $\underline{Bi}^{3+}$

4. What is the oxidation number of the underlined elements?
(a) $\underline{O}_2$
(b) $\underline{As}O_4^{3-}$
(c) $Fe(\underline{O}H)_3$
(d) $\underline{I}O_3^-$

5. In the following half-reactions, which element is changing oxidation state? Is the half-reaction an oxidation or a reduction? Supply the proper number of electrons to each side to balance each equation.
(a) $Zn^{2+} \longrightarrow Zn$
(b) $2\,Br^- \longrightarrow Br_2$
(c) $MnO_4^- + 8\,H^+ \longrightarrow Mn^{2+} + 4\,H_2O$
(d) $Ni \longrightarrow Ni^{2+}$

6. In the following half-reactions, which element is changing oxidation state? Is the half-reaction an oxidation or a reduction? Supply the proper number of electrons to each side to balance each equation.
(a) $SO_3^{2-} + H_2O \longrightarrow SO_4^{2-} + 2\,H^+$
(b) $NO_3^- + 4\,H^+ \longrightarrow NO + 2\,H_2O$
(c) $S_2O_4^{2-} + 2\,H_2O \longrightarrow 2\,SO_3^{2-} + 4\,H^+$
(d) $Fe^{2+} \longrightarrow Fe^{3+}$

7. In the following unbalanced equations, identify
(a) the oxidized element and the reduced element
(b) the oxidizing agent and the reducing agent
1. $Cr + HCl \longrightarrow CrCl_3 + H_2$
2. $SO_4^{2-} + I^- + H^+ \longrightarrow H_2S + I_2 + H_2O$

8. In the following unbalanced equations, identify
(a) the oxidized element and the reduced element
(b) the oxidizing agent and the reducing agent
1. $AsH_3 + Ag^+ + H_2O \longrightarrow H_3AsO_4 + Ag + H^+$
2. $Cl_2 + NaBr \longrightarrow NaCl + Br_2$

9. Determine whether the following oxidation–reduction reactions are balanced correctly. If they are not, provide the correct balanced reaction.

(a) unbalanced:
$CH_4(g) + O_2(g) \longrightarrow CO_2(g) + H_2O(g)$
balanced:
$CH_4(g) + 2\,O_2(g) \longrightarrow CO_2(g) + 2\,H_2O(g)$

(b) unbalanced:
$NO^{2-}(aq) + Al(s) \longrightarrow NH_3(g) + AlO_2^-(aq)$
(basic solution)
balanced: $2\,H_2O(l) + Al(s) + NO^{2-}(aq) \longrightarrow$
$AlO_2^-(aq) + NH_3(g) + OH^-(aq)$

(c) unbalanced:
$Mg(s) + HCl(aq) \longrightarrow Mg^{2+}(aq) + Cl^-(aq) + H_2(g)$
balanced: $Mg(s) + 2\,HCl(aq) \longrightarrow$
$Mg^{2+}(aq) + Cl^-(aq) + H_2(g)$

(d) unbalanced: $CH_3OH(aq) + Cr_2O_7^{2-}(aq) \longrightarrow$
$CH_2O(aq) + Cr^{3+}(aq)$ (acidic solution)
balanced:
$3\,CH_3OH(aq) + 14\,H^+(aq) + Cr_2O_7^{2-}(aq) \longrightarrow$
$2\,Cr^{3+}(aq) + 3\,CH_2O(aq) + 7\,H_2O(l) + 6\,H^+(aq)$

10. Determine whether the following oxidation–reduction reactions are balanced correctly. If they are not, provide the correct balanced reaction.

(a) unbalanced:
$MnO_2(s) + Al(s) \longrightarrow Mn(s) + Al_2O_3(s)$
balanced:
$MnO_2(s) + 2\,Al(s) \longrightarrow Mn(s) + Al_2O_3(s)$

(b) unbalanced:
$Cu(s) + Ag^+(aq) \longrightarrow Cu^{2+}(aq) + Ag(s)$
balanced:
$Cu(s) + 2\,Ag^+(aq) \longrightarrow Cu^{2+}(aq) + 2\,Ag(s)$

(c) unbalanced:
$Br^-(aq) + MnO_4^-(aq) \longrightarrow Br_2(l) + Mn^{2+}(aq)$
(acidic solution)
balanced:
$16\,H^+(aq) + 10\,Br^-(aq) + 2\,MnO_4^-(aq) \longrightarrow$
$5\,Br_2(l) + 2\,Mn^{2+}(aq) + 8\,H_2O(l)$

(d) unbalanced:
$MnO_4^-(aq) + S^{2-}(aq) \longrightarrow MnS(s) + S(s)$
(basic solution)
balanced: $8\,H^+(aq) + MnO_4^-(aq) + S^{2-}(aq) \longrightarrow$
$S(s) + MnS(s) + 4\,H_2O(l)$

11. Balance these equations using the change-in-oxidation-number method:

(a) $Zn + S \longrightarrow ZnS$

(b) $AgNO_3 + Pb \longrightarrow Pb(NO_3)_2 + Ag$

(c) $Fe_2O_3 + CO \longrightarrow Fe + CO_2$

(d) $H_2S + HNO_3 \longrightarrow S + NO + H_2O$

(e) $MnO_2 + HBr \longrightarrow MnBr_2 + Br_2 + H_2O$

12. Balance these equations using the change-in-oxidation-number method.

(a) $Cl_2 + KOH \longrightarrow KCl + KClO_3 + H_2O$

(b) $Ag + HNO_3 \longrightarrow AgNO_3 + NO + H_2O$

(c) $CuO + NH_3 \longrightarrow N_2 + Cu + H_2O$

(d) $PbO_2 + Sb + NaOH \longrightarrow PbO + NaSbO_2 + H_2O$

(e) $H_2O_2 + KMnO_4 + H_2SO_4 \longrightarrow$
$\quad\quad\quad O_2 + MnSO_4 + K_2SO_4 + H_2O$

13. Balance these ionic redox equations using the ion–electron method. These reactions occur in acidic solution.

(a) $Zn + NO_3^- \longrightarrow Zn^{2+} + NH_4^+$

(b) $NO_3^- + S \longrightarrow NO_2 + SO_4^{2-}$

(c) $PH_3 + I_2 \longrightarrow H_3PO_2 + I^-$

(d) $Cu + NO_3^- \longrightarrow Cu^{2+} + NO$

*(e) $ClO_3^- + Cl^- \longrightarrow Cl_2$

14. Balance these ionic redox equations using the ion–electron method. These reactions occur in acidic solution.

(a) $ClO_3^- + I^- \longrightarrow I_2 + Cl^-$

(b) $Cr_2O_7^{2-} + Fe^{2+} \longrightarrow Cr^{3+} + Fe^{3+}$

(c) $MnO_4^- + SO_2 \longrightarrow Mn^{2+} + SO_4^{2-}$

(d) $H_3AsO_3 + MnO_4^- \longrightarrow H_3AsO_4 + Mn^{2+}$

*(e) $Cr_2O_7^{2+} + H_3AsO_3 \longrightarrow Cr^{3+} + H_3AsO_4$

15. Balance these ionic redox equations using the ion–electron method. These reactions occur in basic solution.

(a) $Cl_2 + IO_3^- \longrightarrow Cl^- + IO_4^-$

(b) $MnO_4^- + ClO_2^- \longrightarrow MnO_2 + ClO_4^-$

(c) $Se \longrightarrow Se^{2-} + SeO_3^{2-}$

*(d) $Fe_3O_4 + MnO_4^- \longrightarrow Fe_2O_3 + MnO_2$

*(e) $BrO^- + Cr(OH)_4^- \longrightarrow Br^- + CrO_4^{2-}$

16. Balance these ionic redox equations using the ion–electron method. These reactions occur in basic solution.

(a) $MnO_4^- + SO_3^{2-} \longrightarrow MnO_2 + SO_4^{2-}$

(b) $ClO_2 + SbO_2^- \longrightarrow ClO_2^- + Sb(OH)_6^-$

(c) $Al + NO_3^- \longrightarrow NH_3 + Al(OH)_4^-$

*(d) $P_4 \longrightarrow HPO_3^{2-} + PH_3$

*(e) $Al + OH^- \longrightarrow Al(OH)_4^- + H_2$

17. Balance these reactions:

(a) $IO_3^-(aq) + I^-(aq) \longrightarrow I_2(aq)$ (acid solution)

(b) $Mn^{2+}(aq) + S_2O_8^{2-}(aq) \longrightarrow MnO_4^-(aq) + SO_4^{2-}(aq)$
(acid solution)

(c) $Co(NO_2)_6^{3-}(aq) + MnO_4^-(aq) \longrightarrow$
$\quad Co^{2+}(aq) + Mn^{2+}(aq) + NO_3^-(aq)$ (acid solution)

18. Balance these reactions:

(a) $Mo_2O_3(s) + MnO_4^-(aq) \longrightarrow MoO_3(s) + Mn^{2+}(aq)$
(acid solution)

(b) $BrO^-(aq) + Cr(OH)_4^-(aq) \longrightarrow$
$\quad\quad\quad Br^-(aq) + CrO_4^{2-}(aq)$ (basic solution)

(c) $S_2O_3^{2-}(aq) + MnO_4^-(aq) \longrightarrow$
$\quad\quad\quad SO_4^{2-}(aq) + MnO_2(s)$ (basic solution)

Additional Exercises

All exercises with blue numbers have answers in the appendix of the text.

19. Draw a picture of an electrolytic cell made from an aqueous HBr solution.

20. The chemical reactions taking place during discharge in a lead storage battery are

$Pb + SO_4^{2-} \longrightarrow PbSO_4$

$PbO_2 + SO_4^{2-} + 4 H^+ \longrightarrow PbSO_4 + 2 H_2O$

(a) Complete each half-reaction by supplying electrons.

(b) Which reaction is oxidation and which is reduction?

(c) Which reaction occurs at the anode of the battery?

21. Use this unbalanced redox equation

$KMnO_4 + HCl \longrightarrow KCl + MnCl_2 + H_2 + Cl_2$

to indicate

(a) the oxidizing agent

(b) the reducing agent

(c) the number of electrons that are transferred per mole of oxidizing agent

22. Brass is an alloy of zinc and copper. When brass is in contact with salt water, it corrodes as the zinc dissolves from the alloy, leaving almost pure copper. Explain why the zinc is preferentially dissolved.

23. How many moles of NO gas will be formed by the reaction of 25.0 g of silver with nitric acid?

$Ag + HNO_3 \longrightarrow AgNO_3 + NO + H_2O$ (acid solution)

24. What volume of chlorine gas, measured at STP, must react with excess KOH to form 0.300 mol $KClO_3$?

$Cl_2 + KOH \longrightarrow KCl + KClO_3 + H_2O$ (acid solution)

25. What mass of $KMnO_4$ is needed to react with 100. mL H_2O_2 solution? ($d = 1.031$ g/mL, 9.0% H_2O_2 by mass)

$H_2O_2 + KMnO_4 + H_2SO_4 \longrightarrow$
$\quad\quad\quad O_2 + MnSO_4 + K_2SO_4 + H_2O$ (acid solution)

*26. What volume of 0.200 M $K_2Cr_2O_7$ will be required to oxidize 5.00 g H_3AsO_3?

$$Cr_2O_7^{2-} + H_3AsO_3 \longrightarrow Cr^{3+} + H_3AsO_4 \quad \text{(acid solution)}$$

*27. What volume of 0.200 M $K_2Cr_2O_7$ will be required to oxidize the Fe^{2+} ion in 60.0 mL of 0.200 M $FeSO_4$ solution?

$$Cr_2O_7^{2-} + Fe^{2+} \longrightarrow Cr^{3+} + Fe^{3+} \quad \text{(acid solution)}$$

28. How many moles of H_2 can be produced from 100.0 g Al according to this reaction?

$$Al + OH^- \longrightarrow Al(OH)_4^- + H_2 \quad \text{(basic solution)}$$

29. There is something incorrect about these half-reactions:
(a) $Cu^+ + e^- \longrightarrow Cu^{2+}$　　(b) $Pb^{2+} + e^{2-} \longrightarrow Pb$

Identify what is wrong, and correct it.

30. Why can oxidation *never* occur without reduction?

31. The following observations were made concerning metals A, B, C, and D.
(a) When a strip of metal A is placed in a solution of B^{2+} ions, no reaction is observed.
(b) Similarly, A in a solution containing C^+ ions produces no reaction.
(c) When a strip of metal D is placed in a solution of C^+ ions, black metallic C deposits on the surface of D, and the solution tests positively for D^{2+} ions.
(d) When a piece of metallic B is placed in a solution of D^{2+} ions, metallic D appears on the surface of B and B^{2+} ions are found in the solution.

Arrange the ions—A^+, B^{2+}, C^+, and D^{2+}—in order of their ability to attract electrons. List them in order of increasing ability.

32. Tin normally has oxidation numbers of 0, +2, and +4. Which of these species can be an oxidizing agent, which can be a reducing agent, and which can be both? In each case, what product would you expect as the tin reacts?

33. Manganese is an element that can exist in numerous oxidation states. In each of these compounds, identify the oxidation number of the manganese. Which compound would you expect to be the best oxidizing agent and why?
(a) $Mn(OH)_2$　(c) MnO_2　　(e) $KMnO_4$
(b) MnF_3　　(d) K_2MnO_4

34. Which equations represent oxidations?
(a) $Mg \longrightarrow Mg^{2+}$　　(c) $KMnO_4 \longrightarrow MnO_2$
(b) $SO_2 \longrightarrow SO_3$　　(d) $Cl_2O_3 \longrightarrow Cl^-$

*35. In the following equation, note the reaction between manganese(IV) oxide and bromide ions:

$$MnO_2 + Br^- \longrightarrow Br_2 + Mn^{2+}$$

(a) Balance this redox reaction in acidic solution.
(b) How many grams of MnO_2 would be needed to produce 100.0 mL of 0.05 M Mn^{2+}?
(c) How many liters of bromine vapor at 50°C and 1.4 atm would also result?

36. Use the table shown to complete the following reactions. If no reaction occurs, write NR:
(a) $F_2 + Cl^- \longrightarrow$
(b) $Br_2 + Cl^- \longrightarrow$
(c) $I_2 + Cl^- \longrightarrow$
(d) $Br_2 + I^- \longrightarrow$

Activity
ease of reduction ↑ F_2
Cl_2
Br_2
I_2

37. Manganese metal reacts with HCl to give hydrogen gas and the Mn^{2+} ion in solution. Write a balanced equation for the reaction.

38. If zinc is allowed to react with dilute nitric acid, zinc is oxidized to the 2+ ion, while the nitrate ion can be reduced to ammonium, NH_4^+. Write a balanced equation for the reaction in acidic solution.

39. In the following equations, identify the
(a) atom or ion oxidized
(b) atom or ion reduced
(c) oxidizing agent
(d) reducing agent
(e) change in oxidation number associated with each oxidizing process
(f) change in oxidation number associated with each reducing process

1. $C_3H_8 + O_2 \longrightarrow CO_2 + H_2O$
2. $HNO_3 + H_2S \longrightarrow NO + S + H_2O$
3. $CuO + NH_3 \longrightarrow N_2 + H_2O + Cu$
4. $H_2O_2 + Na_2SO_3 \longrightarrow Na_2SO_4 + H_2O$
5. $H_2O_2 \longrightarrow H_2O + O_2$

40. In the galvanic cell shown in the diagram, a strip of silver is placed in a solution of silver nitrate, and a strip of lead is placed in a solution of lead(II) nitrate. The two beakers are connected with a salt bridge. Determine
(a) the anode
(b) the cathode
(c) where oxidation occurs
(d) where reduction occurs
(e) in which direction electrons flow through the wire
(f) in which direction ions flow through the solution

Challenge Exercises

All exercises with blue numbers have answers in the appendix of the text.

*41. A sample of crude potassium iodide was analyzed using this reaction (not balanced):

$$I^- + SO_4^{2-} \longrightarrow I_2 + H_2S \qquad \text{(acid solution)}$$

If a 4.00-g sample of crude KI produced 2.79 g of iodine, what is the percent purity of the KI?

*42. What volume of NO gas, measured at 28°C and 744 torr, will be formed by the reaction of 0.500 mol Ag reacting with excess nitric acid?

$$Ag + HNO_3 \longrightarrow AgNO_3 + NO + H_2O \quad \text{(acid solution)}$$

Answers to Practice Exercises

17.1 (a) S = +6, (b) As = +5, (c) C = +4

17.2 (*Note:* H = +1 even though it comes second in the formula: N is a nonmetal.)
(a) N = -3, (b) Cr = +6, (c) P = +5

17.3 (a) Be = +2; Cl = -1, (b) H = +1; Cl = +1; O = -2, (c) H = +1; O = -1, (d) N = -3; H = +1, (e) Br = +5; O = -2

17.4 (a) $6\,HNO_3 + S \longrightarrow 6\,NO_2 + H_2SO_4 + 2\,H_2O$
(b) $2\,CrCl_3 + 3\,MnO_2 + 2\,H_2O \longrightarrow 3\,MnCl_2 + 2\,H_2CrO_4$
(c) $2\,KMnO_4 + 6\,HCl + 5\,H_2S \longrightarrow$
$$2\,KCl + 2\,MnCl_2 + 5\,S + 8\,H_2O$$

17.5 (a) $4\,H^+ + 2\,I^- + 2\,NO_2^- \longrightarrow I_2 + 2\,NO + 2\,H_2O$
(b) $2\,OH^- + Cl_2 + IO_3^- \longrightarrow IO_4^- + H_2O + 2\,Cl^-$
(c) $AuCl_4^- + Sn^{2+} \longrightarrow Sn^{4+} + AuCl + 3\,Cl^-$

17.6 (a) $Zn + 2\,H_2O + 2\,OH^- \longrightarrow Zn(OH)_4^{2-} + H_2$
(b) $H_2O_2 + Sn^{2+} + 2\,H^+ \longrightarrow Sn^{4+} + 2\,H_2O$
(c) $Cu + Cu^{2+} + 2\,OH^- \longrightarrow Cu_2O + H_2O$

17.7 (a) yes, (b) no, (c) no

17.8 (a) Cu is oxidized; Sn is reduced
(b) anode: $Cu(s) \longrightarrow Cu^{2+}(aq) + 2e^-$
cathode: $Sn^{2+}(aq) + 2e^- \longrightarrow Sn(s)$

PUTTING IT TOGETHER: Review for Chapters 15–17

Multiple Choice:
Choose the correct answer to each of the following.

1. When the reaction
   $$Al + HCl \longrightarrow$$
   is completed and balanced, this term appears in the balanced equation:
   (a) $3\,HCl$ (b) $AlCl_2$ (c) $3\,H_2$ (d) $4\,Al$

2. When the reaction
   $$CaO + HNO_3 \longrightarrow$$
   is completed and balanced, this term appears in the balanced equation:
   (a) H_2 (b) $2\,H_2$ (c) $2\,CaNO_3$ (d) H_2O

3. When the reaction
   $$H_3PO_4 + KOH \longrightarrow$$
   is completed and balanced, this term appears in the balanced equation:
   (a) H_3PO_4 (c) KPO_4
   (b) $6\,H_2O$ (d) $3\,KOH$

4. When the reaction
   $$HCl + Cr_2(CO_3)_3 \longrightarrow$$
   is completed and balanced, this term appears in the balanced equation:
   (a) Cr_2Cl (b) $3\,HCl$ (c) $3\,CO_2$ (d) H_2O

5. Which of these is not a salt?
   (a) $K_2Cr_2O_7$ (c) $Ca(OH)_2$
   (b) $NaHCO_3$ (d) $Na_2C_2O_4$

6. Which of these is not an acid?
   (a) H_3PO_4 (b) H_2S (c) H_2SO_4 (d) NH_3

7. Which of these is a weak electrolyte?
   (a) NH_4OH (c) K_3PO_4
   (b) $Ni(NO_3)_2$ (d) $NaBr$

8. Which of these is a nonelectrolyte?
   (a) $HC_2H_3O_2$ (c) $KMnO_4$
   (b) $MgSO_4$ (d) CCl_4

9. Which of these is a strong electrolyte?
   (a) H_2CO_3 (c) NH_4OH
   (b) HNO_3 (d) H_3BO_3

10. Which of these is a weak electrolyte?
    (a) $NaOH$ (c) $HC_2H_3O_2$
    (b) $NaCl$ (d) H_2SO_4

11. A solution has an H^+ concentration of $3.4 \times 10^{-5}\,M$. The pH is
    (a) 4.47   (c) 3.53
    (b) 5.53   (d) 5.47

12. A solution with a pH of 5.85 has an H^+ concentration of
    (a) $7.1 \times 10^{-5}\,M$ (c) $3.8 \times 10^{-4}\,M$
    (b) $7.1 \times 10^{-6}\,M$ (d) $1.4 \times 10^{-6}\,M$

13. If 16.55 mL of 0.844 M NaOH are required to titrate 10.00 mL of a hydrochloric acid solution, the molarity of the acid solution is
    (a) 0.700 M (c) 1.40 M
    (b) 0.510 M (d) 0.255 M

14. What volume of 0.462 M NaOH is required to neutralize 20.00 mL of 0.391 M HNO_3?
    (a) 23.6 mL   (c) 9.03 mL
    (b) 16.9 mL   (d) 11.8 mL

15. 25.00 mL of H_2SO_4 solution requires 18.92 mL of 0.1024 M NaOH for complete neutralization. The molarity of the acid is
    (a) 0.1550 M (c) 0.07750 M
    (b) 0.03875 M (d) 0.06765 M

16. Dilute hydrochloric acid is a typical acid, as shown by its
    (a) color   (c) solubility
    (b) odor   (d) taste

17. What is the pH of a 0.00015 M HCl solution?
    (a) 4.0   (c) between 3 and 4
    (b) 2.82   (d) no correct answer given

18. The chloride ion concentration in 300. mL of 0.10 M $AlCl_3$ is
    (a) 0.30 M (c) 0.030 M
    (b) 0.10 M (d) 0.90 M

19. The amount of $BaSO_4$ that will precipitate when 100. mL of 0.10 M $BaCl_2$ and 100. mL of 0.10 M Na_2SO_4 are mixed is
    (a) 0.010 mol   (c) 23 g
    (b) 0.10 mol   (d) no correct answer given

20. The freezing point of a 0.50 m NaCl aqueous solution will be about what?
    (a) $-1.86°C$ (c) $-2.79°C$
    (b) $-0.93°C$ (d) no correct answer given

21. The equation
    $$HC_2H_3O_2 + H_2O \rightleftharpoons H_3O^+ + C_2H_3O_2^-$$
    implies that
    (a) If you start with 1.0 mol $HC_2H_3O_2$, 1.0 mol H_3O^+ and 1.0 mol $C_2H_3O_2^-$ will be produced.
    (b) An equilibrium exists between the forward reaction and the reverse reaction.
    (c) At equilibrium, equal molar amounts of all four will exist.
    (d) The reaction proceeds all the way to the products and then reverses, going all the way back to the reactants.

22. If the reaction $A + B \rightleftharpoons C + D$ is initially at equilibrium and then more A is added, which of the following is not true?
    (a) More collisions of A and B will occur; the rate of the forward reaction will thus be increased.
    (b) The equilibrium will shift toward the right.
    (c) The moles of B will be increased.
    (d) The moles of D will be increased.

23. What will be the H^+ concentration in a 1.0 M HCN solution? ($K_a = 4.0 \times 10^{-10}$)
    (a) $2.0 \times 10^{-5} M$ (c) $4.0 \times 10^{-10} M$
    (b) $1.0 M$ (d) $2.0 \times 10^{-10} M$

24. What is the percent ionization of HCN in Exercise 23?
    (a) 100%   (c) 2.0×10^{-3}%
    (b) 2.0×10^{-8}%   (d) 4.0×10^{-8}%

25. If $[H^+] = 1 \times 10^{-5} M$, which of the following is not true?
    (a) pH = 5   (c) $[OH^-] = 1 \times 10^{-5} M$
    (b) pOH = 9   (d) The solution is acidic.

26. If $[H^+] = 2.0 \times 10^{-4} M$, then $[OH^-]$ will be
    (a) $5.0 \times 10^{-9} M$ (c) $2.0 \times 10^{-4} M$
    (b) $3.70 M$ (d) $5.0 \times 10^{-11} M$

27. The solubility product of $PbCrO_4$ is 2.8×10^{-13}. The solubility of $PbCrO_4$ is
    (a) $5.3 \times 10^{-7} M$ (c) $7.8 \times 10^{-14} M$
    (b) $2.8 \times 10^{-13} M$ (d) $1.0 M$

28. The solubility of AgBr is $7.2 \times 10^{-7} M$. The value of the solubility product is
    (a) 7.2×10^{-7} (c) 5.2×10^{-48}
    (b) 5.2×10^{-13} (d) 5.2×10^{-15}

29. Which of these solutions would be the best buffer solution?
    (a) $0.10 M$ $HC_2H_3O_2$ + $0.10 M$ $NaC_2H_3O_2$
    (b) $0.10 M$ HCl
    (c) $0.10 M$ HCl + $0.10 M$ NaCl
    (d) pure water

30. For the reaction $H_2(g) + I_2(g) \rightleftharpoons 2 HI(g)$, at 700 K, $K_{eq} = 56.6$. If an equilibrium mixture at 700 K was found to contain 0.55 M HI and 0.21 M H_2, the I_2 concentration must be
    (a) $0.046 M$ (c) $22 M$
    (b) $0.025 M$ (d) $0.21 M$

31. The equilibrium constant for the reaction
    $2 A + B \rightleftharpoons 3 C + D$ is
    (a) $\dfrac{[C]^3[D]}{[A]^2[B]}$ (c) $\dfrac{[3C][D]}{[2A][B]}$

    (b) $\dfrac{[2A][B]}{[3C][D]}$ (d) $\dfrac{[A]^2[B]}{[C]^3[D]}$

32. In the equilibrium represented by
    $N_2(g) + O_2(g) \rightleftharpoons 2 NO_2(g)$
    as the pressure is increased, the amount of NO_2 formed
    (a) increases   (c) remains the same
    (b) decreases   (d) increases and decreases irregularly

33. Which factor will not increase the concentration of ammonia as represented by this equation?
    $3 H_2(g) + N_2(g) \rightleftharpoons 2 NH_3(g) + 92.5\,kJ$
    (a) increasing the temperature
    (b) increasing the concentration of N_2
    (c) increasing the concentration of H_2
    (d) increasing the pressure

34. If HCl(g) is added to a saturated solution of AgCl, the concentration of Ag^+ in solution
    (a) increases
    (b) decreases
    (c) remains the same
    (d) increases and decreases irregularly

35. The solubility of $CaCO_3$ at 20°C is 0.013 g/L. What is the K_{sp} for $CaCO_3$?
    (a) 1.3×10^{-8} (c) 1.7×10^{-8}
    (b) 1.3×10^{-4} (d) 1.7×10^{-4}

36. The K_{sp} for $BaCrO_4$ is 8.5×10^{-11}. What is the solubility of $BaCrO_4$ in grams per liter?
    (a) 9.2×10^{-6} (c) 2.3×10^{-3}
    (b) 0.073 (d) 8.5×10^{-11}

37. What will be the $[Ba^{2+}]$ when 0.010 mol Na_2CrO_4 is added to 1.0 L of saturated $BaCrO_4$ solution? See Exercise 36 for K_{sp}.
    (a) $8.5 \times 10^{-11} M$ (c) $9.2 \times 10^{-6} M$
    (b) $8.5 \times 10^{-9} M$ (d) $9.2 \times 10^{-4} M$

38. Which would occur if a small amount of sodium acetate crystals, $NaC_2H_3O_2$, were added to 100 mL of 0.1 M $HC_2H_3O_2$ at constant temperature?
    (a) The number of acetate ions in the solution would decrease.
    (b) The number of acetic acid molecules would decrease.
    (c) The number of sodium ions in solution would decrease.
    (d) The H^+ concentration in the solution would decrease.

39. If the temperature is decreased for the endothermic reaction
    $A + B \rightleftharpoons C + D$
    which of the following is true?
    (a) The concentration of A will increase.
    (b) No change will occur.
    (c) The concentration of B will decrease.
    (d) The concentration of D will increase.

40. In K_2SO_4, the oxidation number of sulfur is
    (a) +2   (c) +6
    (b) +4   (d) −2

41. In $Ba(NO_3)_2$, the oxidation number of N is
    (a) +5   (c) +4
    (b) −3   (d) −1

42. In the reaction
    $H_2S + 4 Br_2 + 4 H_2O \longrightarrow H_2SO_4 + 8 HBr$
    the oxidizing agent is
    (a) H_2S (c) H_2O
    (b) Br_2 (d) H_2SO_4

43. In the reaction
    $VO_3^- + Fe^{2+} + 4 H^+ \longrightarrow VO^{2+} + Fe^{3+} + 2 H_2O$
    the element reduced is
    (a) V   (c) O
    (b) Fe   (d) H

Questions 44–46 pertain to the activity series below.

K   Ca   Mg   Al   Zn   Fe   H   Cu   Ag

44. Which of these pairs will not react in water solution?
(a) Zn, $CuSO_4$
(b) Cu, $Al_2(SO_4)_3$
(c) Fe, $AgNO_3$
(d) Mg, $Al_2(SO_4)_3$

45. Which element is the most easily oxidized?
(a) K
(b) Mg
(c) Zn
(d) Cu

46. Which element will reduce Cu^{2+} to Cu but will not reduce Zn^{2+} to Zn?
(a) Fe
(b) Ca
(c) Ag
(d) Mg

47. In the electrolysis of fused (molten) $CaCl_2$, the product at the negative electrode is
(a) Ca^{2+}
(b) Cl^-
(c) Cl_2
(d) Ca

48. In its reactions, a free element from Group 2A in the periodic table is most likely to
(a) be oxidized
(b) be reduced
(c) be unreactive
(d) gain electrons

49. In the partially balanced redox equation
$$3\,Cu + HNO_3 \longrightarrow 3\,Cu(NO_3)_2 + 2\,NO + H_2O$$
the coefficient needed to balance H_2O is
(a) 8
(b) 6
(c) 4
(d) 3

50. Which reaction does not involve oxidation–reduction?
(a) burning sodium in chlorine
(b) chemical union of Fe and S
(c) decomposition of $KClO_3$
(d) neutralization of NaOH with H_2SO_4

51. How many moles of Fe^{2+} can be oxidized to Fe^{3+} by 2.50 mol Cl_2 according to this equation?
$$Fe^{2+} + Cl_2 \longrightarrow Fe^{3+} + Cl^-$$
(a) 2.50 mol
(b) 5.00 mol
(c) 1.00 mol
(d) 22.4 mol

52. How many grams of sulfur can be produced in this reaction from 100. mL of 6.00 M HNO_3?
$$HNO_3 + H_2S \longrightarrow S + NO + H_2O$$
(a) 28.9 g
(b) 19.3 g
(c) 32.1 g
(d) 289 g

53. Which of these ions can be reduced by H_2?
(a) Hg^{2+}
(b) Sn^{2+}
(c) Zn^{2+}
(d) K^+

54. Which of the following is *not* true of a zinc–mercury cell?
(a) It provides current at a steady potential.
(b) It has a short service life.
(c) It is self-contained.
(d) It can be stored for long periods of time.

Balancing Oxidation–Reduction Equations

Balance each equation.

55. $P + HNO_3 \longrightarrow HPO_3 + NO + H_2O$

56. $MnSO_4 + PbO_2 + H_2SO_4 \longrightarrow$
$$HMnO_4 + PbSO_4 + H_2O$$

57. $Cr_2O_7^{2-} + Cl^- \longrightarrow Cr^{3+} + Cl_2$ (acidic solution)

58. $MnO_4^- + AsO_3^{3-} \longrightarrow Mn^{2+} + AsO_4^{3-}$ (acidic solution)

59. $S^{2-} + Cl_2 \longrightarrow SO_4^{2-} + Cl^-$ (basic solution)

60. $Zn + NO_3^- \longrightarrow Zn(OH)_4^{2-} + NH_3$ (basic solution)

61. $KOH + Cl_2 \longrightarrow KCl + KClO + H_2O$ (basic solution)

62. $As + ClO_3^- \longrightarrow H_3AsO_3 + HClO$ (acidic solution)

63. $MnO_4^- + Cl^- \longrightarrow Mn^{2+} + Cl_2$ (acidic solution)

64. $H_2O_2 + Cl_2O_7 \longrightarrow ClO_2^- + O_2$ (basic solution)

Free Response Questions

Answer each of the following. Be sure to include your work and explanations in a clear, logical form.

1. You are investigating the properties of two new metallic elements found on Pluto, Bz and Yz. Bz reacts with aqueous HCl. However, the formula for the compound it forms with chlorine is $YzCl_2$. Write the balanced reaction that should occur if a galvanic cell is set up with Bz and Yz electrodes in solutions containing the metallic ions.

2. Suppose that 25 mL of an iron(II) nitrate solution are added to a beaker containing aluminum metal. Assume the reaction went to completion with no excess reagents. The solid iron produced was removed by filtration.
(a) Write a balanced redox equation for the reaction.
(b) For which solution, the initial iron(II) nitrate or the solution after the solid iron was filtered out, would the freezing point be lower? Explain your answer.

3. 50. mL of 0.10 M HCl solution are poured equally into two flasks.
(a) What is the pH of the HCl solution?
(b) Next, 0.050 mol Zn is added to Flask A and 0.050 mol Cu is added to Flask B. Determine the pH of each solution after approximately 20 minutes.

4. (a) Write a balanced acid–base reaction that produces Na_2S.
(b) If Na_2S is added to an aqueous solution of H_2S ($K_a = 9.1 \times 10^{-8}$), will the pH of the solution rise or fall? Explain.

5. (a) Would you expect a reaction to take place between HCN (aq) and $AgNO_3$ (aq) (K_{sp} for AgCN = 5.97×10^{-17})? Explain, and if a reaction occurs, write the net ionic equation.
(b) If NaCN is added to distilled water, would you expect the solution to be acidic, basic, or neutral? Explain using any chemical equations that may be appropriate.

6. For each set of beakers below, draw a picture you might expect to see when the contents of the beakers are mixed together and allowed to react.

(a)

Na$_2$SO$_4$ (aq)   Pb(NO$_3$)$_2$ (aq)

(b)

K$_2$SO$_4$ (aq)   CuCl$_2$ (aq)

7. The picture below represents the equilibrium condition of 2 A$_3$X $\rightleftharpoons$ 2 A$_2$X + A$_2$.

(a) What is the equilibrium constant?
(b) Does the equilibrium lie to the left or to the right?
(c) Do you think the reaction is a redox reaction? Explain your answer.

8. The picture below represents the equilibrium condition of the reaction X$_2$ + 2G $\rightleftharpoons$ X$_2$G$_2$.

(a) What is the equilibrium constant?
(b) If the ratio of reactants to products increased when the temperature was raised, was the reaction exothermic or endothermic?
(c) Provide a logical explanation for why the equilibrium shifts to the right when the pressure is increased at constant temperature.

9. The hydroxide ion concentration of a solution is 3.4×10^{-10} M. Balance the following equation:

$$Fe^{2+}(aq) + MnO_4^-(aq) \longrightarrow Fe^{3+}(aq) + Mn^{2+}(aq)$$

CHAPTER 18

Nuclear Chemistry

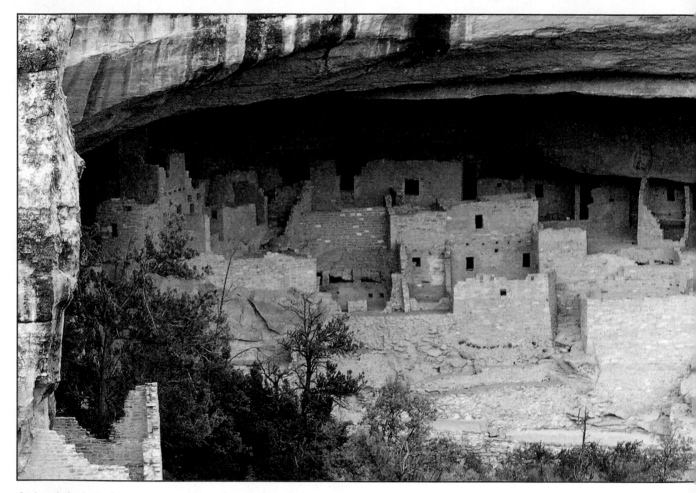

Carbon dating is used to determine the age of civilizations that lived in Cliff Palace, at Mesa Verde, Colorado.

Chapter Outline

18.1 Discovery of Radioactivity

18.2 Natural Radioactivity

18.3 Alpha Particles, Beta Particles, and Gamma Rays

18.4 Radioactive Disintegration Series

18.5 Transmutation of Elements

18.6 Artificial Radioactivity

18.7 Measurement of Radioactivity

18.8 Nuclear Fission

18.9 Nuclear Power

18.10 The Atomic Bomb

18.11 Nuclear Fusion

18.12 Mass–Energy Relationship in Nuclear Reactions

18.13 Transuranium Elements

18.14 Biological Effects of Radiation

The nucleus of the atom is a source of tremendous energy. Harnessing this energy has enabled us to fuel power stations, treat cancer, and preserve food. We use isotopes in medicine to diagnose illness and to detect minute quantities of drugs or hormones. Researchers use radioactive tracers to sequence the human genome. We also use nuclear processes to detect explosives in luggage and to establish the age of objects such as human artifacts and rocks. In this chapter we'll consider properties of atomic nuclei and their applications in our lives.

18.1 Discovery of Radioactivity

In 1895, Wilhelm Conrad Roentgen (1845-1923) made an important breakthrough that eventually led to the discovery of radioactivity. Roentgen discovered X-rays when he observed that a vacuum discharge tube enclosed in a thin, black cardboard box caused a nearby piece of paper coated with the salt barium platinocyanide to glow with a brilliant phosphorescence. From this and other experiments, he concluded that certain rays, which he called X-rays, were emitted from the discharge tube, penetrated the box, and caused the salt to glow. Roentgen's observations that X-rays could penetrate other bodies and affect photographic plates led to the development of X-ray photography.

X-ray technology opened the door for a new world of diagnosis and treatment.

Shortly after this discovery, Antoine Henri Becquerel (1852–1908) attempted to show a relationship between X-rays and the phosphorescence of uranium salts. In one experiment he wrapped a photographic plate in black paper, placed a sample of uranium salt on it, and exposed it to sunlight. The developed photographic plate showed that rays emitted from the salt had penetrated the paper. When Becquerel attempted to repeat the experiment, the sunlight was intermittent, so he placed the entire setup in a drawer. Several days later he developed the photographic plate, expecting to find it only slightly affected. To his amazement, he found an intense image on the plate. He repeated the experiment in total darkness and obtained the same results, proving that the uranium salt emitted rays that affected the photographic plate without its being exposed to sunlight. Thus did the discovery of radioactivity come about, a combination of numerous experiments by the finest minds of the day—and serendipity. Becquerel later showed that the rays coming from uranium are able to ionize air and are also capable of penetrating thin sheets of metal.

The name *radioactivity* was coined two years later (in 1898) by Marie Curie. **Radioactivity** is the spontaneous emission of particles and/or rays from the nucleus of an atom. Elements having this property are said to be radioactive.

radioactivity

In 1898, Marie Sklodowska Curie (1867–1934) and her husband Pierre Curie (1859–1906) turned their research interests to radioactivity. In a short time, the Curies discovered two new elements, polonium and radium, both of which are radioactive. To confirm their work on radium, they processed 1 ton of pitchblende residue ore to obtain 0.1 g of pure radium chloride, which they used to make further studies on the properties of radium and to determine its atomic mass.

In 1899, Ernest Rutherford began to investigate the nature of the rays emitted from uranium. He found two particles, which he called *alpha* and *beta particles*. Soon he realized that uranium, while emitting these particles, was changing into another element. By 1912, over 30 radioactive isotopes were known, and many more are known today. The *gamma ray,* a third type of emission from radioactive materials similar to an X-ray, was discovered by Paul Villard (1860–1934) in 1900. Rutherford's description of the nuclear atom led scientists to attribute the phenomenon of radioactivity to reactions taking place in the nuclei of atoms.

The symbolism and notation we described for isotopes in Chapter 5 are also very useful in nuclear chemistry:

For example, $^{238}_{92}U$ represents a uranium isotope with an atomic number of 92 and a mass number of 238. This isotope is also designated as U-238, or uranium-238, and contains 92 protons and 146 neutrons. The protons and neutrons collectively are known as **nucleons.** The mass number is the total number of nucleons in the nucleus. Table 18.1 shows the isotopic notations for several particles associated with nuclear chemistry.

nucleon

When we speak of isotopes, we mean atoms of the same element with different masses, such as $^{16}_{8}O$, $^{17}_{8}O$, $^{18}_{8}O$. In nuclear chemistry we use the term **nuclide** to mean any isotope of any atom. Thus, $^{16}_{8}O$ and $^{235}_{92}U$ are referred to as nuclides. Nuclides that spontaneously emit radiation are referred to as *radionuclides.*

nuclide

Table 18.1 Isotopic Notation for Several Particles (and Small Isotopes) Associated with Nuclear Chemistry

Particle	Symbol	Z Atomic number	A Mass number
Neutron	$^{1}_{0}n$	0	1
Proton	$^{1}_{1}H$	1	1
Beta particle (electron)	$^{0}_{-1}e$	−1	0
Positron (positive electron)	$^{0}_{+1}e$	1	0
Alpha particle (helium nucleus)	$^{4}_{2}He$	2	4
Deuteron (heavy hydrogen nucleus)	$^{2}_{1}H$	1	2

18.2 Natural Radioactivity

radioactive decay

Radioactive elements continuously undergo **radioactive decay,** or disintegration, to form different elements. The chemical properties of an element are associated with its electronic structure, but radioactivity is a property of the nucleus. Therefore, neither ordinary changes of temperature and pressure nor the chemical or physical state of an element has any effect on its radioactivity.

The principal emissions from the nuclei of radionuclides are known as alpha particles, beta particles, and gamma rays. Upon losing an alpha or beta particle, the radioactive element changes into a different element. We will explain this process in detail later.

Each radioactive nuclide disintegrates at a specific and constant rate, which is expressed in units of half-life. The **half-life** ($t_{1/2}$) is the time required for one-half of a specific amount of a radioactive nuclide to disintegrate. The half-lives of the elements range from a fraction of a second to billions of years. To illustrate, suppose we start with 1.0 g of $^{226}_{88}Ra$ ($t_{1/2} = 1620$ years):

half-life

$$1.0\,g\,^{226}_{88}Ra \xrightarrow[1620\ \text{years}]{t_{1/2}} 0.50\,g\,^{226}_{88}Ra \xrightarrow[1620\ \text{years}]{t_{1/2}} 0.25\,g\,^{226}_{88}Ra$$

The half-lives of the various radioisotopes of the same element differ dramatically. Half-lives for certain isotopes of radium, carbon, and uranium are listed in Table 18.2.

Element	$t_{1/2}$
$^{238}_{92}U$	4.5×10^9 years
$^{226}_{88}Ra$	1620 years
$^{15}_{6}C$	2.4 seconds

Table 18.2 Half-Lives for Radium, Carbon, and Uranium Isotopes

Isotope	Half-life	Isotope	Half-life
Ra-223	11.7 days	C-14	5730 years
Ra-224	3.64 days	C-15	2.4 seconds
Ra-225	14.8 days	U-235	7.1×10^8 years
Ra-226	1620 years	U-238	4.5×10^9 years
Ra-228	6.7 years		

Example 18.1

The half-life of $^{131}_{53}I$ is 8 days. How much $^{131}_{53}I$ from a 32-g sample remains after five half-lives?

SOLUTION

Using the following graph, we can find the number of grams of ^{131}I remaining after one half-life:

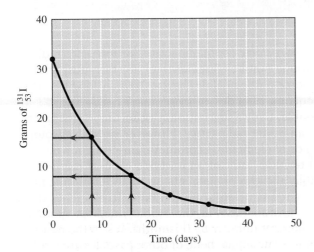

Trace a perpendicular line from 8 days on the *x*-axis to the line on the graph.

Now trace a horizontal line from this point on the plotted line to the y-axis and read the corresponding grams of ^{131}I. Continue this process for each half-life, adding 8 days to the previous value on the x-axis:

Half-lives	0	1	2	3	4	5
Number of days		8	16	24	32	40
Amount remaining	32 g	16 g	8 g	4 g	2 g	1 g

Starting with 32 g, 1 g ^{131}I remains after five half-lives (40 days).

Example 18.2   In how many half-lives will 10.0 g of a radioactive nuclide decay to less than 10% of its original value?

SOLUTION   We know that 10% of the original amount is 1.0 g. After the first half-life, half the original material remains and half has decayed (5.00 g). After the second half-life, one-fourth of the original material remains (i.e., one-half of 5.00 g). This progression continues, reducing the quantity remaining by half for each half-life that passes.

Half-lives	0	1	2	3	4
Percent remaining	100%	50%	25%	12.5%	6.25%
Amount remaining	10.0 g	5.00 g	2.50 g	1.25 g	0.625 g

Therefore the amount remaining will be less than 10% sometime between the third and the fourth half-lives.

Practice 18.1 _____

The half-life of $^{14}_{6}$C is 5730 years. How much $^{14}_{6}$C will remain after six half-lives in a sample that initially contains 25.0 g?

Nuclides are said to be either *stable* (nonradioactive) or *unstable* (radioactive). Elements that have atomic numbers greater than 83 (bismuth) are naturally radioactive, although some of the nuclides have extremely long half-lives. Some of the naturally occurring nuclides of elements 81, 82, and 83 are radioactive, and some are stable. Only a few naturally occurring elements that have atomic numbers less than 81 are radioactive. However, no stable isotopes of element 43 (technetium) or of element 61 (promethium) are known.

Radioactivity is believed to be a result of an unstable ratio of neutrons to protons in the nucleus. Stable nuclides of elements up to about atomic number 20 generally have about a 1 : 1 neutron-to-proton ratio. In elements above number 20, the neutron-to-proton ratio in the stable nuclides gradually increases to about 1.5 : 1 in element number 83 (bismuth). When the neutron-to-proton ratio is too high or too low, alpha, beta, or other particles are emitted to achieve a more stable nucleus.

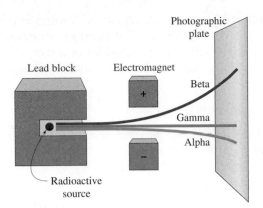

Figure 18.1
The effect of an electromagnetic field on alpha particles, beta particles, and gamma rays. Lighter beta particles are deflected considerably more than alpha particles. Alpha and beta particles are deflected in opposite directions. Gamma radiation is not affected by the electromagnetic field.

18.3 Alpha Particles, Beta Particles, and Gamma Rays

The classical experiment proving that alpha and beta particles are oppositely charged was performed by Marie Curie (see Figure 18.1). She placed a radioactive source in a hole in a lead block and positioned two poles of a strong electromagnet so that the radiations that were given off passed between them. The paths of three different kinds of radiation were detected by means of a photographic plate placed some distance beyond the electromagnet. The lighter beta particles were strongly deflected toward the positive pole of the electromagnet; the heavier alpha particles were less strongly deflected and in the opposite direction. The uncharged gamma rays were not affected by the electromagnet and struck the photographic plates after traveling along a path straight out of the lead block.

Alpha Particles

An **alpha particle** (α) consists of two protons and two neutrons, has a mass of about 4 amu, and has a charge of $+2$. It is a helium nucleus that is usually given one of the following symbols: α or $^{4}_{2}\text{He}$. When an alpha particle is emitted from the nucleus, a different element is formed. The atomic number of the new element is 2 less, and the mass is 4 amu less, than that of the starting element.

alpha particle

Loss of an alpha particle from the nucleus results in
    loss of 4 in the mass number (A)
    loss of 2 in the atomic number (Z)

For example, when $^{238}_{92}\text{U}$ loses an alpha particle, $^{234}_{90}\text{Th}$ is formed, because two neutrons and two protons are lost from the uranium nucleus. This disintegration may be written as a nuclear equation:

$$^{238}_{92}\text{U} \longrightarrow {}^{234}_{90}\text{Th} + \alpha \quad \text{or} \quad {}^{238}_{92}\text{U} \longrightarrow {}^{234}_{90}\text{Th} + {}^{4}_{2}\text{He}$$

For the loss of an alpha particle from $^{226}_{88}\text{Ra}$, the equation is

$$^{226}_{88}\text{Ra} \longrightarrow {}^{222}_{86}\text{Rn} + {}^{4}_{2}\text{He} \quad \text{or} \quad {}^{226}_{88}\text{Ra} \longrightarrow {}^{222}_{86}\text{Rn} + \alpha$$

A nuclear equation, like a chemical equation, consists of reactants and products and must be balanced. To have a balanced nuclear equation, the sum of the mass numbers (superscripts) on both sides of the equation must be equal and the sum of the atomic numbers (subscripts) on both sides of the equation must be equal:

sum of mass numbers equals 226

$$^{226}_{88}\text{Ra} \longrightarrow {}^{222}_{86}\text{Rn} + {}^{4}_{2}\text{He}$$

sum of atomic numbers equals 88

What new nuclide will be formed when $^{230}_{90}\text{Th}$ loses an alpha particle? The new nuclide will have a mass of 226 amu and will contain 88 protons, so its atomic number is 88. Locate the corresponding element on the periodic table—in this case, $^{226}_{88}\text{Ra}$ or radium-226.

sum of mass numbers is 230

$$^{230}_{90}\text{Th} \longrightarrow {}^{226}_{88}? + {}^{4}_{2}\text{He}$$

sum of atomic numbers is 90

Beta Particles

beta particle

The **beta particle** (β) is identical in mass and charge to an electron; its charge is −1. Both a beta particle and a proton are produced by the decomposition of a neutron:

$$^{1}_{0}\text{n} \longrightarrow {}^{1}_{0}\text{p} + {}^{0}_{-1}\text{e}$$

The beta particle leaves, and the proton remains in the nucleus. When an atom loses a beta particle from its nucleus, a different element is formed that has essentially the same mass but an atomic number that is 1 greater than that of the starting element. The beta particle is written as β or $^{0}_{-1}\text{e}$.

Loss of a beta particle from the nucleus results in

no change in the mass number (A)

increase of 1 in the atomic number (Z)

Examples of equations in which a beta particle is lost are

$$^{234}_{90}\text{Th} \longrightarrow {}^{234}_{91}\text{Pa} + \beta$$

$$^{234}_{91}\text{Pa} \longrightarrow {}^{234}_{92}\text{U} + {}^{0}_{-1}\text{e}$$

$$^{210}_{82}\text{Pb} \longrightarrow {}^{210}_{83}\text{Bi} + \beta$$

Gamma Rays

gamma ray

Gamma rays (γ) are photons of energy. A gamma ray is similar to an X-ray but is more energetic. It has no electrical charge and no measurable mass. Gamma rays are released from the nucleus in many radioactive changes along with either alpha or beta particles. Gamma radiation does not result in a change of atomic number or the mass of an element.

Loss of a gamma ray from the nucleus results in

no change in mass number (A) or atomic number (Z)

Example 18.3

(a) Write an equation for the loss of an alpha particle from the nuclide $^{194}_{78}Pt$.
(b) What nuclide is formed when $^{228}_{88}Ra$ loses a beta particle from its nucleus?

SOLUTION

(a) Loss of an alpha particle, $^{4}_{2}He$, results in a decrease of 4 in the mass number and a decrease of 2 in the atomic number:

Mass of new nuclide: $A - 4$ or $194 - 4 = 190$
Atomic number of new nuclide: $Z - 2$ or $78 - 2 = 76$

Looking up element number 76 on the periodic table, we find it to be osmium, Os. The equation then is

$$^{194}_{78}Pt \longrightarrow {}^{190}_{76}Os + {}^{4}_{2}He$$

(b) The loss of a beta particle from a $^{228}_{88}Ra$ nucleus means a gain of 1 in the atomic number with no essential change in mass. The new nuclide will have an atomic number of $(Z + 1)$, or 89, which is actinium, Ac:

$$^{228}_{88}Ra \longrightarrow {}^{228}_{89}Ac + {}^{0}_{-1}e$$

Example 18.4

What nuclide will be formed when $^{214}_{82}Pb$ successively emits two beta particles, then one alpha particle from its nucleus? Write successive equations showing these changes.

SOLUTION

The changes brought about in the three steps outlined are as follows:

β loss: Increase of 1 in the atomic number; no change in mass
β loss: Increase of 1 in the atomic number; no change in mass
α loss: Decrease of 2 in the atomic number; decrease of 4 in the mass

The equations are

$$^{214}_{82}Pb \longrightarrow {}^{214}_{83}X + \beta \longrightarrow {}^{214}_{84}X + \beta \longrightarrow {}^{210}_{82}X + \alpha$$

where X stands for the new nuclide formed. Looking up each of these elements by their atomic numbers, we rewrite the equations

$$^{214}_{82}Pb \xrightarrow{\beta} {}^{214}_{83}Bi \xrightarrow{\beta} {}^{214}_{84}Po \xrightarrow{\alpha} {}^{210}_{82}Pb$$

Practice 18.2
What nuclide will be formed when $^{222}_{86}Rn$ emits an alpha particle?

The ability of radioactive rays to pass through various objects is in proportion to the speed at which they leave the nucleus. Gamma rays travel at the velocity of light (186,000 miles per second) and are capable of penetrating several inches of lead. The velocities of beta particles are variable, the fastest being about nine-tenths the velocity of light. Alpha particles have velocities less than

Figure 18.2
Relative penetrating ability of alpha, beta, and gamma radiation. (a) Thin sheet of paper; (b) thin sheet of aluminum; (c) 5-cm lead block.

one-tenth the velocity of light. Figure 18.2 illustrates the relative penetrating power of these rays. A few sheets of paper will stop alpha particles; a thin sheet of aluminum will stop both alpha and beta particles; and a 5-cm block of lead will reduce, but not completely stop, gamma radiation. In fact, it is difficult to stop all gamma radiation. Table 18.3 summarizes the properties of alpha, beta, and gamma radiation.

Table 18.3 Characteristics of Nuclear Radiation

Radiation	Symbol	Mass (amu)	Electrical charge	Velocity	Composition	Ionizing power
Alpha	α, ^4_2He	4	$+2$	Variable, less than 10% the speed of light	He nucleus	High
Beta	β, $^0_{-1}\text{e}$	$\dfrac{1}{1837}$	-1	Variable, up to 90% the speed of light	Identical to an electron	Moderate
Gamma	γ	0	0	Speed of light	Photons or electromagnetic waves of energy	Almost none

18.4 Radioactive Disintegration Series

These three series begin with the elements uranium, thorium, and actinium.

The naturally occurring radioactive elements with a higher atomic number than lead (Pb) fall into three orderly disintegration series. Each series proceeds from one element to the next by the loss of either an alpha or a beta particle, finally ending in a nonradioactive nuclide. The uranium series starts with $^{238}_{92}\text{U}$ and ends with $^{206}_{82}\text{Pb}$. The thorium series starts with $^{232}_{90}\text{Th}$ and ends with $^{208}_{82}\text{Pb}$. The actinium series starts with $^{235}_{92}\text{U}$ and ends with $^{207}_{82}\text{Pb}$. A fourth series, the neptunium series, starts with the synthetic element $^{241}_{94}\text{Pu}$ and ends with the stable bismuth nuclide $^{209}_{83}\text{Bi}$. The uranium series is shown in Figure 18.3. Gamma radiation, which accompanies alpha and beta radiation, is not shown.

By using these series and the half-lives of their members, scientists have been able to approximate the age of certain geologic deposits. This approximation is done by comparing the amount of $^{238}_{92}\text{U}$ with the amount of $^{206}_{82}\text{Pb}$ and other nuclides in the series that are present in a particular geologic formation. Rocks found in Canada and Finland have been calculated to be about 3.0×10^9 (3 billion) years old. Some meteorites have been determined to be 4.5×10^9 years old.

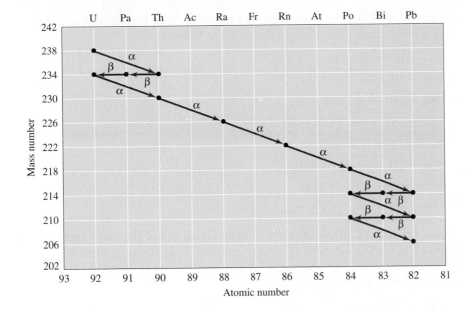

Figure 18.3
The uranium disintegration series. $^{238}_{92}U$ decays by a series of alpha (α) and beta (β) emissions to the stable nuclide $^{206}_{82}Pb$.

Practice 18.3

What nuclides are formed when $^{238}_{92}U$ undergoes the following decays?

(a) an alpha particle and a beta particle
(b) three alpha particles and two beta particles

18.5 Transmutation of Elements

Transmutation is the conversion of one element into another by either natural or artificial means. Transmutation occurs spontaneously in natural radioactive disintegrations. Alchemists tried for centuries to convert lead and mercury into gold by artificial means, but transmutation by artificial means was not achieved until 1919, when Ernest Rutherford succeeded in bombarding the nuclei of nitrogen atoms with alpha particles and produced oxygen nuclides and protons. The nuclear equation for this transmutation can be written as

$$^{14}_{7}N + \alpha \longrightarrow \,^{17}_{8}O + \,^{1}_{1}H \quad \text{or} \quad ^{14}_{7}N + \,^{4}_{2}He \longrightarrow \,^{17}_{8}O + \,^{1}_{1}H$$

It is believed that the alpha particle enters the nitrogen nucleus, forming $^{18}_{9}F$ as an intermediate, which then decomposes into the products.

Rutherford's experiments opened the door to nuclear transmutations of all kinds. Atoms were bombarded by alpha particles, neutrons, protons, deuterons ($^{2}_{1}H$), electrons, and so forth. Massive instruments were developed for accelerating these particles to very high speeds and energies to aid their penetration of the nucleus. The famous cyclotron was developed by E. O. Lawrence (1901–1958) at the University of California; later instruments include the Van de Graaf electrostatic generator, the betatron, and the electron and proton

transmutation

Electrons and positrons are accelerated along the tunnels at the Stanford Linear Accelerator Center.

synchrotrons. With these instruments many nuclear transmutations became possible. Equations for a few of these are as follows:

$$^{7}_{3}\text{Li} + {}^{1}_{1}\text{H} \longrightarrow 2\,{}^{4}_{2}\text{He}$$

$$^{40}_{18}\text{Ar} + {}^{1}_{1}\text{H} \longrightarrow {}^{40}_{19}\text{K} + {}^{1}_{0}\text{n}$$

$$^{23}_{11}\text{Na} + {}^{1}_{1}\text{H} \longrightarrow {}^{23}_{12}\text{Mg} + {}^{1}_{0}\text{n}$$

$$^{114}_{48}\text{Cd} + {}^{2}_{1}\text{H} \longrightarrow {}^{115}_{48}\text{Cd} + {}^{1}_{1}\text{H}$$

$$^{2}_{1}\text{H} + {}^{2}_{1}\text{H} \longrightarrow {}^{3}_{1}\text{H} + {}^{1}_{1}\text{H}$$

$$^{209}_{83}\text{Bi} + {}^{2}_{1}\text{H} \longrightarrow {}^{210}_{84}\text{Po} + {}^{1}_{0}\text{n}$$

$$^{16}_{8}\text{O} + {}^{1}_{0}\text{n} \longrightarrow {}^{13}_{6}\text{C} + {}^{4}_{2}\text{He}$$

$$^{238}_{92}\text{U} + {}^{12}_{6}\text{C} \longrightarrow {}^{244}_{98}\text{Cf} + 6\,{}^{1}_{0}\text{n}$$

On November 1, 2004, the IUPAC officially approved the name roentgenium, symbol Rg, for the element of atomic number 111. Roentgenium was named for Wilhelm Conrad Roentgen, who discovered X-rays in 1895. The nuclear equation for its formation is

$$^{209}_{83}\text{Bi} + {}^{64}_{28}\text{Ni} \longrightarrow {}^{272}_{111}\text{Rg} + {}^{1}_{0}\text{n}$$

18.6 Artificial Radioactivity

Irene Joliot-Curie (daughter of Pierre and Marie Curie) and her husband Frederic Joliot-Curie observed that when aluminum-27 is bombarded with alpha particles, neutrons and positrons (positive electrons) are emitted as part of the products. When the source of alpha particles is removed, neutrons cease to be produced, but positrons continue to be emitted. This observation suggested that the neutrons and positrons come from two separate reactions. It also indicated that a product of the first reaction is radioactive. After further investigation, they discovered that, when aluminum-27 is bombarded with alpha particles, phosphorus-30 and neutrons are produced. Phosphorus-30 is radioactive, has a half-life of 2.5 minutes, and decays to silicon-30 with the emission of a positron. The equations for these reactions are

$$^{27}_{13}\text{Al} + {}^{4}_{2}\text{He} \longrightarrow {}^{30}_{15}\text{P} + {}^{1}_{0}\text{n}$$

$$^{30}_{15}\text{P} \longrightarrow {}^{30}_{14}\text{Si} + {}^{0}_{+1}\text{e}$$

artificial radioactivity
induced radioactivity

The radioactivity of nuclides produced in this manner is known as **artificial radioactivity** or **induced radioactivity.** Artificial radionuclides behave like natural radioactive elements in two ways: They disintegrate in a definite fashion and they have a specific half-life. The Joliot-Curies received the Nobel Prize in chemistry in 1935 for the discovery of artificial, or induced, radioactivity.

18.7 Measurement of Radioactivity

Radiation from radioactive sources is so energetic that it is called *ionizing radiation.* When it strikes an atom or a molecule, one or more electrons are knocked off, and an ion is created. The Geiger counter, an instrument commonly used to detect and measure radioactivity, depends on this property. The instrument consists of a Geiger-Müller detecting tube and a counting device.

The detector tube is a pair of oppositely charged electrodes in an argon gas-filled chamber fitted with a thin window. When radiation, such as a beta particle, passes through the window into the tube, some argon is ionized, and a momentary pulse of current (discharge) flows between the electrodes. These current pulses are electronically amplified in the counter and appear as signals in the form of audible clicks, flashing lights, meter deflections, or numerical readouts (Figure 18.4).

The amount of radiation that an individual encounters can be measured by a film badge. This badge contains a piece of photographic film in a light-proof holder and is worn in areas where radiation might be encountered. The silver grains in the film will darken when exposed to radiation. The badges are processed after a predetermined time interval to determine the amount of radiation the wearer has been exposed to.

A scintillation counter is used to measure radioactivity for biomedical applications. A scintillator contains molecules that emit light when exposed to ionizing radiation. A light-sensitive detector counts the flashes and converts them to a numerical readout.

The *curie* is the unit used to express the amount of radioactivity produced by an element. One **curie (Ci)** is defined as the quantity of radioactive material giving 3.7×10^{10} disintegrations per second. The basis for this figure is pure radium, which has an activity of 1 Ci/g. Because the curie is such a large quantity, the millicurie and microcurie, representing one-thousandth and one-millionth of a curie, respectively, are more practical and more commonly used.

The curie only measures radioactivity emitted by a radionuclide. Different units are required to measure exposure to radiation. The **roentgen (R)** quantifies exposure to gamma or X-rays; 1 roentgen is defined as the amount of radiation required to produce 2.1×10^9 ions/cm³ of dry air. The **rad (radiation absorbed dose)** is defined as the amount of radiation that provides 0.01 J of energy per kilogram of matter. The amount of radiation absorbed will change depending on the type of matter. The roentgen and the rad are numerically similar; 1 roentgen of gamma radiation provides 0.92 rad in bone tissue.

Neither the rad nor the roentgen indicates the biological damage caused by radiation. One rad of alpha particles has the ability to cause 10 times more damage than 1 rad of gamma rays or beta particles. Another unit, **rem (roentgen equivalent to man)**, takes into account the degree of biological effect caused by the type of radiation exposure; 1 rem is equal to the dose in rads multiplied by a factor specific to the form of radiation. The factor is 10 for alpha particles and 1 for both beta particles and gamma rays. Units of radiation are summarized in Table 18.4.

Figure 18.4
Geiger–Müller survey meter.

curie (Ci)

roentgen (R)

**rad
(radiation absorbed dose)**

**rem
(roentgen equivalent
to man)**

Table 18.4 Radiation Units

Unit	Measure	Equivalent
curie (Ci)	rate of decay of a radioactive substance	$1\,Ci = 3.7 \times 10^{10}$ disintegrations/sec
roentgen (R)	exposure based on the quantity of ionization produced in air	$1\,R = 2.1 \times 10^9$ ions/cm³
rad	absorbed dose of radiation	$1\,rad = 0.01$ J/kg matter
rem	radiation dose equivalent	$1\,rem = 1\,rad \times$ factor
gray (Gy) (SI unit)	energy absorbed by tissue	$1\,Gy = 1$ J/kg tissue ($1\,Gy = 100\,rad$)

18.8 Nuclear Fission

nuclear fission

In **nuclear fission** a heavy nuclide splits into two or more intermediate-sized fragments when struck in a particular way by a neutron. The fragments are called *fission products*. As the atom splits, it releases energy and two or three neutrons, each of which can cause another nuclear fission. The first instance of nuclear fission was reported in January 1939 by the German scientists Otto Hahn (1879–1968) and Fritz Strassmann (1902–1980). Detecting isotopes of barium, krypton, cerium, and lanthanum after bombarding uranium with neutrons led scientists to believe that the uranium nucleus had been split.

Characteristics of nuclear fission are as follows:

1. Upon absorption of a neutron, a heavy nuclide splits into two or more smaller nuclides (fission products).
2. The mass of the nuclides formed ranges from about 70 to 160 amu.
3. Two or more neutrons are produced from the fission of each atom.
4. Large quantities of energy are produced as a result of the conversion of a small amount of mass into energy.
5. Most nuclides produced are radioactive and continue to decay until they reach a stable nucleus.

One process by which this fission takes place is illustrated in Figure 18.5. When a heavy nucleus captures a neutron, the energy increase may be sufficient to cause deformation of the nucleus until the mass finally splits into two fragments, releasing energy and usually two or more neutrons.

In a typical fission reaction, a $^{235}_{92}U$ nucleus captures a neutron and forms unstable $^{236}_{92}U$. This $^{236}_{92}U$ nucleus undergoes fission, quickly disintegrating into two fragments, such as $^{139}_{56}Ba$ and $^{94}_{36}Kr$, and three neutrons. The three neutrons in turn may be captured by three other $^{235}_{92}U$ atoms, each of which undergoes

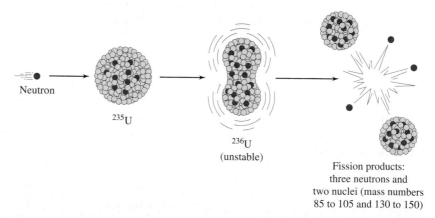

Neutron

^{235}U

^{236}U
(unstable)

Fission products:
three neutrons and
two nuclei (mass numbers
85 to 105 and 130 to 150)

Figure 18.5
The fission process. When a neutron is captured by a heavy nucleus, the nucleus becomes more unstable. The more energetic nucleus begins to deform, resulting in fission. Two nuclear fragments and three neutrons are produced by this fission process.

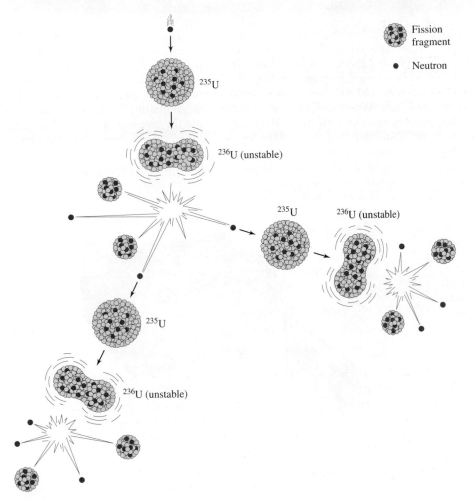

Figure 18.6
Fission and chain reaction of $^{235}_{92}$U. Each fission produces two major fission fragments and three neutrons, which may be captured by other $^{235}_{92}$U nuclei, continuing the chain reaction.

fission, producing nine neutrons, and so on. A reaction of this kind, in which the products cause the reaction to continue or magnify, is known as a **chain reaction.** For a chain reaction to continue, enough fissionable material must be present so that each atomic fission causes, on average, at least one additional fission. The minimum quantity of an element needed to support a self-sustaining chain reaction is called the **critical mass.** Since energy is released in each atomic fission, chain reactions provide a steady supply of energy. (A chain reaction is illustrated in Figure 18.6.) Two of the many possible ways in which $^{235}_{92}$U may fission are shown by these nuclear equations:

chain reaction

critical mass

$$^{235}_{92}\text{U} + ^{1}_{0}\text{n} \longrightarrow ^{139}_{56}\text{Ba} + ^{94}_{36}\text{Kr} + 3\,^{1}_{0}\text{n}$$

$$^{235}_{92}\text{U} + ^{1}_{0}\text{n} \longrightarrow ^{144}_{54}\text{Xe} + ^{90}_{38}\text{Sr} + 2\,^{1}_{0}\text{n}$$

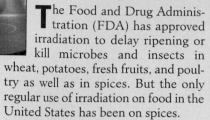

The Food and Drug Administration (FDA) has approved irradiation to delay ripening or kill microbes and insects in wheat, potatoes, fresh fruits, and poultry as well as in spices. But the only regular use of irradiation on food in the United States has been on spices.

Now the FDA has approved irradiation of red meat, and new legislation allows the labels about irradiated food to be in much smaller type. The general public has not yet accepted the use of radiation to reduce microbes and help preserve food. Why? Many people believe that irradiating food makes it radioactive. This is not true; in fact, the most that irradiation of food does is to produce compounds similar to those created by cooking and also to reduce the vitamin content of some food.

How does irradiation of food work? The radioactive source currently used is cobalt-60. It is contained in a concrete cell (see diagram) with 6-foot-thick walls. Inside the cell is a pool of water with racks of thin Co-60 rods suspended above. When the rods are not being used, they are submerged in the water, which absorbs the gamma radiation. When food to be irradiated moves into the cells, the Co-60 rods are lifted from the water, and the boxes of food move among them on a conveyor being irradiated from all sides.

Scientists who have investigated the process say that food irradiation is safe. The Center for Disease Control and Prevention in Atlanta estimates that food-borne illness causes as many as 9000 deaths per year. Irradiation provides one way to reduce microbial contamination in food. It does not solve the problem of careless handling of food by processors, and long-term studies on humans have not yet been concluded regarding irradiated food supplies. Once again, we face the issue of balancing benefit and risk.

Concrete shielding

Conveyor

Radioactive chamber

18.9 Nuclear Power

Nearly all electricity for commercial use is produced by machines consisting of a turbine linked by a drive shaft to an electrical generator. The energy required to run the turbine can be supplied by falling water, as in hydroelectric power plants, or by steam, as in thermal power plants.

The world's demands for energy, largely from fossil fuels, is heavy. At the present rates of consumption, the estimated world supply of fossil fuels is sufficient for only a few centuries. Although the United States has large coal and oil shale deposits, it currently imports over 40% of its oil supply. We clearly need to develop alternative energy sources. At present, uranium is the most productive alternative energy source, and about 17% of the electrical energy used in the United States is generated from power plants using uranium fuel.

A nuclear power plant is a thermal power plant in which heat is produced by a nuclear reactor instead of by combustion of fossil fuel. The major components of a nuclear reactor are

1. an arrangement of nuclear fuel, called the reactor core
2. a control system, which regulates the rate of fission and thereby the rate of heat generation
3. a cooling system, which removes the heat from the reactor and also keeps the core at the proper temperature

Figure 18.7
Diagram of a nuclear power plant. Heat produced by the reactor core is carried by a cooling fluid to a steam generator, which turns a turbine that produces electricity.

One type of reactor uses metal slugs containing uranium enriched from the normal 0.7% U-235 to about 3% U-235. The self-sustaining fission reaction is moderated, or controlled, by adjustable control rods containing substances that slow down and capture some of the neutrons produced. Ordinary water, heavy water, and molten sodium are typical coolants used. Energy obtained from nuclear reactions in the form of heat is used in the production of steam to drive turbines for generating electricity. (See Figure 18.7.)

The potential dangers of nuclear power were tragically demonstrated by the accidents at Three Mile Island, Pennsylvania (1979), and Chernobyl in the former U.S.S.R. (1986). Both accidents resulted from the loss of coolant to the reactor core. The reactors at Three Mile Island were covered by concrete containment buildings and therefore released a relatively small amount of radioactive material into the atmosphere. But because the Soviet Union did not require containment structures on nuclear power plants, the Chernobyl accident resulted in 31 deaths and the resettlement of 135,000 people. The release of large quantities of I-131, Cs-134, and Cs-137 appears to be causing long-term health problems in that exposed population.

Another problem with nuclear power plants is their radioactive waste production. The United States has debated where and how to permanently store nuclear wastes. In July 2002, an underground site was authorized at Yucca Mountain, Nevada, about 90 miles north of Las Vegas. The spent nuclear fuel will remain radioactive for thousands of years.

In the United States, reactors designed for commercial power production use uranium oxide, U_3O_8, that is enriched with the relatively scarce fissionable U-235 isotope. Because the supply of U-235 is limited, a new type of reactor known as the *breeder reactor* has been developed. Breeder reactors produce additional fissionable material at the same time that the fission reaction is occurring. In a breeder reactor, excess neutrons convert nonfissionable iso-topes, such as U-238 or Th-232, to fissionable isotopes, Pu-239 or U-233, as shown below:

$$\ce{^{238}_{92}U + ^{1}_{0}n -> ^{239}_{92}U ->[\beta] ^{239}_{93}Np ->[\beta] ^{239}_{94}Pu}$$

$$\ce{^{232}_{90}Th + ^{1}_{0}n -> ^{233}_{90}Th ->[\beta] ^{233}_{91}Pa ->[\beta] ^{233}_{92}U}$$

These transmutations make it possible to greatly extend the supply of fuel for nuclear reactors. No breeder reactors are presently in commercial operation in the United States, but a number of them are being operated in Europe and Great Britain.

18.10 The Atomic Bomb

The atomic bomb is a fission bomb; it operates on the principle of a very fast chain reaction that releases a tremendous amount of energy. An atomic bomb and a nuclear reactor both depend on self-sustaining nuclear fission chain reactions. The essential difference is that in a bomb the fission is "wild," or uncontrolled, whereas in a nuclear reactor the fission is moderated and care-fully controlled. A minimum critical mass of fissionable material is needed for a bomb; otherwise a major explosion will not occur. When a quantity smaller than the critical mass is used, too many neutrons formed in the fission step escape without combining with another nucleus, and a chain reaction does not occur. Therefore the fissionable material of an atomic bomb must be stored as two or more subcritical masses and brought together to form the critical mass at the desired time of explosion. The temperature developed in an atomic bomb is believed to be about 10 million degrees Celsius.

The nuclides used in atomic bombs are U-235 and Pu-239. Uranium deposits contain about 0.7% of the U-235 isotope, the remainder being U-238. Uranium-238 does not undergo fission except with very high-energy neutrons. It was discovered, however, that U-238 captures a low-energy neutron without undergoing fission and that the product, U-239, changes to Pu-239 (plutonium) by a beta-decay process. Plutonium-239 readily undergoes fission upon capture of a neutron and is therefore useful for nuclear weapons. The equations for the nuclear transformations are

$$\ce{^{238}_{92}U + ^{1}_{0}n -> ^{239}_{92}U ->[\beta] ^{239}_{93}Np ->[\beta] ^{239}_{94}Pu}$$

The hazards of an atomic bomb explosion include not only shock waves from the explosive pressure and tremendous heat, but also intense radiation in the form of alpha particles, beta particles, gamma rays, and ultraviolet rays. Gamma rays and X-rays can penetrate deeply into the body, causing burns, sterilization, and gene mutation, which can adversely affect future generations. Both radioactive fission products and unfissioned material are present after the explosion. If the bomb explodes near the ground, many tons of dust are lift-

The mushroom cloud is a signature of uncontrolled fission in an atomic bomb.

ed into the air. Radioactive material adhering to this dust, known as *fallout,* is spread by air currents over wide areas of the land and constitutes a lingering source of radiation hazard.

Today nuclear war is probably the most awesome threat facing civilization. Only two rather primitive fission-type atom bombs were used to destroy the Japanese cities of Hiroshima and Nagasaki and bring World War II to an early end. The threat of nuclear war is increased by the fact that the number of nations possessing nuclear weapons is steadily increasing.

18.11 Nuclear Fusion

The process of uniting the nuclei of two light elements to form one heavier nucleus is known as **nuclear fusion.** Such reactions can be used for producing energy, because the masses of the two nuclei that fuse into a single nucleus are greater than the mass of the nucleus formed by their fusion. The mass differential is liberated in the form of energy. Fusion reactions are responsible for the tremendous energy output of the sun. Thus, aside from relatively small amounts from nuclear fission and radioactivity, fusion reactions are the ultimate source of our energy, even the energy from fossil fuels. They are also responsible for the devastating power of the thermonuclear, or hydrogen, bomb.

nuclear fusion

Fusion reactions require temperatures on the order of tens of millions of degrees for initiation. Such temperatures are present in the sun but have been produced only momentarily on Earth. For example, the hydrogen, or fusion, bomb is triggered by the temperature of an exploding fission bomb. Two typical fusion reactions are

$$\underset{\text{tritium}}{^{3}_{1}\text{H}} \quad + \quad \underset{\text{deuterium}}{^{2}_{1}\text{H}} \quad \longrightarrow \quad {^{4}_{2}}\text{He} + {^{1}_{0}}\text{n} + \text{energy}$$

$$\underset{\substack{3.0150 \\ \text{amu}}}{^{3}_{1}\text{H}} \quad + \quad \underset{\substack{1.0079 \\ \text{amu}}}{^{1}_{1}\text{H}} \quad \longrightarrow \quad \underset{\substack{4.0026 \\ \text{amu}}}{^{4}_{2}\text{He}} + \text{energy}$$

The total mass of the reactants in the second equation is 4.0229 amu, which is 0.0203 amu greater than the mass of the product. This difference in mass is manifested in the great amount of energy liberated.

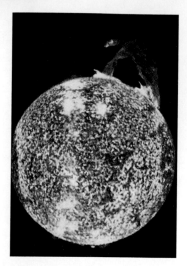

Solar flares such as these are indications of fusion reactions occurring at temperatures of millions of degrees.

A great deal of research in the United States and in other countries, especially the former Soviet Union, has focused on controlled nuclear fusion reactions. The goal of controlled nuclear fusion has not yet been attained, although the required ignition temperature has been reached in several devices. Evidence to date leads us to believe that we can develop a practical fusion power reactor. Fusion power, if we can develop it, will be far superior to fission power for the following reasons:

1. Virtually infinite amounts of energy are possible from fusion. Uranium supplies for fission power are limited, but heavy hydrogen, or deuterium (the most likely fusion fuel), is abundant. It is estimated that the deuterium in a cubic mile of seawater used as fusion fuel could provide more energy than the petroleum reserves of the entire world.

2. From an environmental viewpoint, fusion power is much "cleaner" than fission power because fusion reactions (in contrast to uranium and plutonium fission reactions) do not produce large amounts of long-lived and dangerously radioactive isotopes.

18.12 Mass-Energy Relationship in Nuclear Reactions

Large amounts of energy are released in nuclear reactions; thus significant amounts of mass are converted to energy. We stated earlier that the amount of mass converted to energy in chemical changes is insignificant compared to the amount of energy released in a nuclear reaction. In fission reactions, about 0.1% of the mass is converted into energy. In fusion reactions as much as 0.5% of the mass may be changed into energy. The Einstein equation, $E = mc^2$, can be used to calculate the energy liberated, or available, when the mass loss is known. For example, in the fusion reaction

$$\underset{7.016 \text{ g}}{^{7}_{3}\text{Li}} \quad + \quad \underset{1.008 \text{ g}}{^{1}_{1}\text{H}} \quad \longrightarrow \quad \underset{4.003 \text{ g}}{^{4}_{2}\text{He}} \quad + \quad \underset{4.003 \text{ g}}{^{4}_{2}\text{He}} \quad + \quad \text{energy}$$

the mass difference between the reactants and products ($8.024 \text{ g} - 8.006 \text{ g}$) is 0.018 g. The energy equivalent to this amount of mass is 1.62×10^{12} J. By comparison, this is more than 4 million times greater than the 3.9×10^{5} J of energy obtained from the complete combustion of 12.01 g (1 mol) of carbon.

The mass of a nucleus is actually less than the sum of the masses of the protons and neutrons that make up that nucleus. The difference between the mass of the protons and the neutrons in a nucleus and the mass of the nucleus is known as the **mass defect.** The energy equivalent to this difference in mass is known as the **nuclear binding energy.** This energy is the amount that would be required to break a nucleus into its individual protons and neutrons. The higher the binding energy, the more stable the nucleus. Elements of intermediate atomic masses have high binding energies. For example, iron (element number 26) has a very high binding energy and therefore a very stable nucleus. Just as electrons attain less energetic and more stable arrangements through ordinary chemical reactions, neutrons and protons attain less energetic and more stable arrangements through nuclear fission or fusion reactions. Thus, when uranium undergoes fission, the products have less mass (and greater binding energy) than the original uranium. In like manner, when hydrogen and

mass defect
nuclear binding energy

lithium fuse to form helium, the helium has less mass (and greater binding energy) than the hydrogen and lithium. It is this conversion of mass to energy that accounts for the very large amounts of energy associated with both fission and fusion reactions.

Calculate the mass defect and the nuclear binding energy for an α particle (helium nucleus).

Example 18.5

Data: proton mass = 1.0073 g/mol

neutron mass = 1.0087 g/mol

α mass = 4.0015 g/mol

$1.0 \text{ g} = 9.0 \times 10^{13} \text{ J}$

First calculate the sum of the individual parts of an α particle (2 protons +2 neutrons).

SOLUTION

2 protons: $2 \times 1.0073 \text{ g/mol} = 2.0146 \text{ g/mol}$

2 neutrons: $2 \times 1.0087 \text{ g/mol} = \underline{2.0174 \text{ g/mol}}$

4.0320 g/mol

The mass defect is the difference between the mass of the α particle and its component parts:

4.0320 g/mol

$\underline{4.0015 \text{ g/mol}}$

0.0305 g/mol (mass defect)

To find the nuclear binding energy, convert the mass defect to its energy equivalent:

$(0.0305 \text{ g/mol})(9.0 \times 10^{13} \text{ J/g}) = 2.7 \times 10^{12} \text{ J/mol} (6.5 \times 10^{11} \text{ cal})$

18.13 Transuranium Elements

The elements that follow uranium on the periodic table and that have atomic numbers greater than 92 are known as the **transuranium elements.** They are synthetic radioactive elements; none of them occur naturally.

transuranium element

The first transuranium element, number 93, was discovered in 1939 by Edwin M. McMillan (1907–1991) at the University of California while he was investigating the fission of uranium. He named it neptunium for the planet Neptune. In 1941, element 94, plutonium, was identified as a beta-decay product of neptunium:

$$^{238}_{93}\text{Np} \longrightarrow {}^{238}_{94}\text{Pu} + {}^{0}_{-1}\text{e}$$

$$^{239}_{93}\text{Np} \longrightarrow {}^{239}_{94}\text{Pu} + {}^{0}_{-1}\text{e}$$

Plutonium is one of the most important fissionable elements known today.

Since 1964, the discoveries of eight new transuranium elements, numbers 104–111, have been announced. These elements have been produced in minute quantities by high-energy particle accelerators.

18.14 Biological Effects of Radiation

ionizing radiation

Radiation with energy to dislocate bonding electrons and create ions when passing through matter is classified as **ionizing radiation.** Alpha particles, beta particles, gamma rays, and X-rays fall into this classification. Ionizing radiation can damage or kill living cells and can be particularly devastating when it strikes the cell nuclei and affects molecules involved in cell reproduction. The effects of radiation on living organisms fall into these general categories: (1) acute or short-term effects, (2) long-term effects, and (3) genetic effects.

Acute Radiation Damage

High levels of radiation, especially from gamma rays or X-rays, produce nausea, vomiting, and diarrhea. The effect has been likened to a sunburn throughout the body. If the dosage is high enough, death will occur in a few days. The damaging effects of radiation appear to be centered in the nuclei of the cells, and cells that are undergoing rapid cell division are most susceptible to damage. It is for this reason that cancers are often treated with gamma radiation from a Co-60 source. Cancerous cells multiply rapidly and are destroyed by a level of radiation that does not seriously damage normal cells.

Long–Term Radiation Damage

Protracted exposure to low levels of any form of ionizing radiation can weaken an organism and lead to the onset of malignant tumors, even after fairly long time delays. The largest exposure to synthetic sources of radiation is from X-rays. Evidence suggests that the lives of early workers in radioactivity and X-ray technology may have been shortened by long-term radiation damage.

Strontium-90 isotopes are present in the fallout from atmospheric testing of nuclear weapons. Strontium is in the same periodic-table group as calcium, and its chemical behavior is similar to that of calcium. Hence when foods contaminated with Sr-90 are eaten, Sr-90 ions are laid down in the bone tissue along with ordinary calcium ions. Strontium-90 is a beta emitter with a half-life of 28 years. Blood cells manufactured in bone marrow are affected by the radiation from Sr-90. Hence there is concern that Sr-90 accumulation in the environment may cause an increase in the incidence of leukemia and bone cancers.

Genetic Effects

The information needed to create an individual of a particular species, be it a bacterial cell or a human being, is contained within the nucleus of a cell. This genetic information is encoded in the structure of DNA (deoxyribonucleic acid) molecules, which make up genes. The DNA molecules form precise duplicates of themselves when cells divide, thus passing genetic information from one generation to the next. Radiation can damage DNA molecules. If the damage is not severe enough to prevent the individual from reproducing, a mutation may result. Most mutation-induced traits are undesirable. Unfortunately, if the bearer of the altered genes survives to reproduce, these traits are passed along to succeeding generations. In other words, the genetic effects of increased radiation exposure are found in future generations, not only in the present generation.

Because radioactive rays are hazardous to health and living tissue, special precautions must be taken in designing laboratories and nuclear reactors, in disposing of waste materials, and in monitoring the radiation exposure of people working in this field.

Imagine viewing a living process as it is occurring. This was the dream of many scientists in the past as they tried to extract this knowledge from dead tissue. Today, because of innovations in nuclear chemistry, this dream is a common, everyday occurrence.

Compounds containing a radionuclide are described as being *labeled*, or *tagged*. These compounds undergo their normal chemical reactions, but their location can be detected because of their radioactivity. When such compounds are given to a plant or an animal, the movement of the nuclide can be traced through the organism by the use of a Geiger counter or other detecting device.

In an early use of the tracer technique, the pathway by which CO_2 becomes fixed into carbohydrate ($C_6H_{12}O_6$) during photosynthesis was determined. The net equation for photosynthesis is

$$6\,CO_2 + 6\,H_2O \longrightarrow C_6H_{12}O_6 + 6\,O_2$$

Radioactive $^{14}CO_2$ was injected into a colony of green algae, and the algae were then placed in the dark and killed at selected time intervals. When the radioactive compounds were separated by paper chromatography and analyzed, the results elucidated a series of light-independent photosynthetic reactions.

Biological research using tracer techniques have determined

1. the rate of phosphate uptake by plants, using radiophosphorus
2. the flow of nutrients in the digestive tract using radioactive barium compounds.
3. the accumulation of iodine in the thyroid gland, using radioactive iodine
4. the absorption of iron by the hemoglobin of the blood, using radioactive iron

In chemistry, uses for tracers are unlimited. The study of reaction mechanisms, the measurement of the rates of chemical reactions, and the determi-

A radioactive tracer is injected into this patient and absorbed by the brain. The PET scanner detects photons emitted by the tracer and produces an image used in medical diagnosis and research.

nation of physical constants are just a few of the areas of application.

Radioactive tracers are commonly used in medical diagnosis. The radionuclide must be effective at a low concentration and have a short half-life to reduce the possibility of damage to the patient.

Radioactive iodine (I-131) is used to determine thyroid function, where the body concentrates iodine. In this process a small amount of radioactive potassium or sodium iodide is ingested. A detector is focused on the thyroid gland and measures the amount of iodine in the gland. This picture is then compared to that of a normal thyroid to detect any differences.

Doctors examine the heart's pumping performance and check for evidence of obstruction in coronary arteries by *nuclear scanning*. The radionuclide Tl-201, when injected into the bloodstream, lodges in healthy heart muscle. Thallium-201 emits gamma radiation, which is detected by a special imaging device called a *scintillation camera*. The data obtained are simultaneously trans-

lated into pictures by a computer. With this technique doctors can observe whether heart tissue has died after a heart attack and whether blood is flowing freely through the coronary passages.

One of the most recent applications of nuclear chemistry is the use of positron emission tomography (PET) in the measurement of dynamic processes in the body, such as oxygen use or blood flow. In this application, a compound is made that contains a positron-emitting nuclide such as C-11, O-15, or N-13. The compound is injected into the body, and the patient is placed in an instrument that detects the positron emission. A computer produces a three-dimensional image of the area.

PET scans have been used to locate the areas of the brain involved in epileptic seizures. Glucose tagged with C-11 is injected, and an image of the brain is produced. Since the brain uses glucose almost exclusively for energy, diseased areas that use glucose at a rate different than normal tissue can then be identified.

Chapter 18 Review

18.1   Discovery of Radioactivity
KEY TERMS

Radioactivity
Nucleon
Nuclide

- Radioactivity is the spontaneous emission of particles and energy from the nucleus of an atom.
- Protons and neutrons are known as nucleons.
- In nuclear chemistry isotopes are also known as nuclides.

18.2   Natural Radioactivity
KEY TERMS

Radioactivite decay
Half-life

- Radioactive elements undergo radioactive decay to form different elements.
- Radioactivity is not affected by changes in temperature, pressure, or the state of the element.
- Principal emissions:
  - Alpha particles
  - Beta particles
  - Gamma rays
- The half-life of a nuclide is the time required for $1/2$ of a specific amount of the nuclide to disintegrate.

18.3   Alpha Particles, Beta Particles, and Gamma Rays
KEY TERMS

Alpha particle
Beta particle
Gamma ray

- Alpha particles:
  - Consist of 2 protons and 2 neutrons with mass of 4 amu and charge of +2
  - Loss of an alpha particle from the nucleus results in:
    - Loss of 4 in mass number
    - Loss of 2 in atomic number
  - High ionizing power, low penetration
- Beta particles:
  - Same as an electron
  - Loss of a beta particle from the nucleus results in:
    - No change in mass number
    - Increase of 1 in atomic number
  - Moderate ionizing power and penetration
- Gamma rays:
  - Photons of energy
  - Loss of a gamma ray from the nucleus results in:
    - No change in mass number (A)
    - No change in atomic number (Z)
  - Almost no ionizing power, high penetration

18.4   Radioactive Disintegration Series
- As element undergo disintegration, they eventually become stable after the loss of a series of particles and energy.
- The disintegration series can be used to determine the age of geologic deposits.

18.5   Transmutation of Elements
KEY TERM

Transmutation

- Transmutation of an element can occur spontaneously or artificially.

18.6   Artificial Radioactivity
KEY TERM

Artificial radioactivity or induced radioactivity

- Artificial radionuclides behave like natural radioactive elements in two ways:
  - They disintegrate in a definite fashion.
  - They have a specific half-life.

18.7   Measurement of Radioactivity
KEY TERMS

Curie (Ci)
Roentgen (R)
Rad (radiation absorbed dose)
Rem (roentgen equivalent to man)

- Radioactivity can be measured in several ways:
  - Geiger–Müller counter
  - Film badge
  - Scintillation counter
- Units for measuring radiation:
  - Gray
  - Curie
  - Roentgen
  - Rad
  - Rem

18.8   Nuclear Fission
KEY TERMS

Nuclear fission
Chain reaction
Critical mass

- Characteristics of fission:
  - A heavy nuclide splits into two or more smaller nuclides.
  - The mass of nuclides formed is 70 to 160 amu.
  - Two or more neutrons are produced per fission.
  - Large quantities of energy are released.
  - Most nuclides produced are radioactive and continue to decay until they reach stability.

18.9 Nuclear Power

- A nuclear power plant is a thermal plant with a nuclear fission source of energy.
- Major components of a nuclear reactor:
  - Reactor core
  - Control system
  - Cooling system

18.10 The Atomic Bomb

- Out-of-control fission reaction

18.11 Nuclear Fusion

KEY TERM

Nuclear fusion

- Nuclear fusion is the process of uniting 2 light nuclei to form 1 heavier nucleus.
- Nuclear fusion requires extreme temperatures for ignition.

18.12 Mass–Energy Relationship in Nuclear Reactions

KEY TERM

Mass defect
Nuclear binding energy

- The mass defect is the difference between the actual mass of an atom and the calculated mass of the protons and neutrons in the nucleus of that atom.
- The nuclear binding energy is the energy equivalent to the mass defect. $E = mc^2$ is used to determine the energy liberated in a nuclear reaction.

18.13 Transuranium Elements

KEY TERM

Transuranium elements

- Transuranium elements are the elements that follow uranium on the periodic table.

18.14 Biological Effects of Radiation

KEY TERM

Ionizing radiation

- Effects of radiation fall into 3 general categories:
  - Acute or short-term
  - Long-term
  - Genetic

Review Questions

All questions with blue *numbers have answers in the appendix of the text.*

1. To afford protection from radiation injury, which kind of radiation requires (a) the most shielding? (b) the least shielding?
2. Why is an alpha particle deflected less than a beta particle in passing through an electromagnetic field?
3. Name three pairs of nuclides that might be obtained by fissioning U-235 atoms.
4. Identify these people and their associations with the early history of radioactivity:
   (a) Antoine Henri Becquerel
   (b) Marie and Pierre Curie
   (c) Wilhelm Roentgen
   (d) Ernest Rutherford
   (e) Otto Hahn and Fritz Strassmann
5. Why is the radioactivity of an element unaffected by the usual factors that affect the rate of chemical reactions, such as ordinary changes of temperature and concentration?
6. Distinguish between the terms *isotope* and *nuclide*.
7. The half-life of Pu-244 is 76 million years. If Earth's age is about 5 billion years, discuss the feasibility of finding this nuclide as a naturally occurring nuclide.
8. Tell how alpha, beta, and gamma radiation are distinguished from the standpoint of
   (a) charge
   (b) relative mass
   (c) nature of particle or ray
   (d) relative penetrating power
9. Distinguish between natural and artificial radioactivity.
10. What is a radioactive disintegration series?
11. Briefly discuss the transmutation of elements.
12. Stable Pb-208 is formed from Th-232 in the thorium disintegration series by successive α, β, β, α, α, α, α, β, β, α particle emissions. Write the symbol (including mass and atomic number) for each nuclide formed in this series.
13. The nuclide Np-237 loses a total of seven alpha particles and four beta particles. What nuclide remains after these losses?
14. Bismuth-211 decays by alpha emission to give a nuclide that in turn decays by beta emission to yield a stable nuclide. Show these two steps with nuclear equations.
15. What were Otto Hahn and Fritz Strassmann's contribution to nuclear physics?
16. What is a breeder reactor? Explain how it accomplishes the "breeding."
17. What is the essential difference between the nuclear reactions in a nuclear reactor and those in an atomic bomb?
18. Why must a certain minimum amount of fissionable material be present before a self-supporting chain reaction can occur?
19. What are mass defect and nuclear binding energy?
20. Explain why radioactive rays are classified as ionizing radiation.

21. Give a brief description of the biological hazards associated with radioactivity.

22. Strontium-90 has been found to occur in radioactive fallout. Why is there so much concern about this radionuclide being found in cow's milk? (Half-life of Sr-90 is 28 years.)

23. What is a radioactive tracer? How is it used?

24. Describe the radiocarbon method for dating archaeological artifacts.

25. How might radioactivity be used to locate a leak in an underground pipe?

26. Anthropologists have found bones whose age suggests that the human line may have emerged in Africa as much as 4 million years ago. If wood or charcoal were found with such bones, would C-14 dating be useful in dating the bones? Explain.

27. What are the disadvantages of nuclear power?

28. What are the hazards associated with an atomic bomb explosion?

29. What types of elements undergo fission? How about fusion?

30. Using Figure 18.7, describe how nuclear power is generated.

Paired Exercises

All exercises with blue numbers have answers in the appendix of the text.

1. Indicate the number of protons, neutrons, and nucleons in these nuclei:
   (a) $^{35}_{17}Cl$ (b) $^{226}_{88}Ra$

2. Indicate the number of protons, neutrons, and nucleons in these nuclei:
   (a) $^{235}_{92}U$ (b) $^{82}_{35}Br$

3. How are the mass and the atomic number of a nucleus affected by the loss of an alpha particle?

4. How are the mass and the atomic number of a nucleus affected by the loss of a beta particle?

5. Write nuclear equations for the alpha decay of
   (a) $^{218}_{85}At$ (b) $^{221}_{87}Fr$

6. Write nuclear equations for the alpha decay of
   (a) $^{192}_{78}Pt$ (b) $^{210}_{84}Po$

7. Write nuclear equations for the beta decay of
   (a) $^{14}_{6}C$ (b) $^{137}_{55}Cs$

8. Write nuclear equations for the beta decay of
   (a) $^{239}_{93}Np$ (b) $^{90}_{38}Sr$

9. Determine the type of emission or emissions (alpha, beta, or gamma) that occurred in the following transitions:
   (a) $^{226}_{88}Ra$ to $^{222}_{86}Rn$
   (b) $^{222}_{88}Ra$ to $^{222}_{87}Fr$ to $^{222}_{87}Fr$
   (c) $^{238}_{92}U$ to $^{238}_{93}Np$

10. Determine the type of emission or emissions (alpha, beta, or gamma) that occurred in the following transitions:
    (a) $^{210}_{82}Pb$ to $^{210}_{82}Pb$
    (b) $^{234}_{91}Pa$ to $^{230}_{89}Ac$ to $^{230}_{90}Th$
    (c) $^{234}_{90}Th$ to $^{230}_{88}Ra$ to $^{230}_{88}Ra$

11. Write a nuclear equation for the conversion of $^{13}_{6}C$ to $^{14}_{6}C$.

12. Write a nuclear equation for the conversion of $^{30}_{15}P$ to $^{30}_{14}Si$.

13. Complete and balance these nuclear equations by supplying the missing particles:
    (a) $^{66}_{29}Cu \longrightarrow {}^{66}_{30}Zn + \underline{\hspace{1cm}}$
    (b) $^{0}_{-1}e + \underline{\hspace{1cm}} \longrightarrow {}^{7}_{3}Li$
    (c) $^{27}_{13}Al + {}^{4}_{2}He \longrightarrow {}^{30}_{14}Si + \underline{\hspace{1cm}}$
    (d) $^{85}_{37}Rb + \underline{\hspace{1cm}} \longrightarrow {}^{82}_{35}Br + {}^{4}_{2}He$

14. Complete and balance these nuclear equations by supplying the missing particles:
    (a) $^{27}_{13}Al + {}^{4}_{2}He \longrightarrow {}^{30}_{15}P + \underline{\hspace{1cm}}$
    (b) $^{27}_{14}Si \longrightarrow {}^{0}_{+1}e + \underline{\hspace{1cm}}$
    (c) $\underline{\hspace{1cm}} + {}^{2}_{1}H \longrightarrow {}^{13}_{7}N + {}^{1}_{0}n$
    (d) $\underline{\hspace{1cm}} \longrightarrow {}^{82}_{36}Kr + {}^{0}_{-1}e$

15. Strontium-90 has a half-life of 28 years. If a 1.00-mg sample was stored for 112 years, what mass of Sr-90 would remain?

16. Strontium-90 has a half-life of 28 years. If a sample was tested in 1980 and found to be emitting 240 counts/min, in what year would the same sample be found to be emitting 30 counts/min? How much of the original Sr-90 would be left?

17. By what series of emissions does $^{230}_{90}Th$ disintegrate to $^{218}_{84}Po$?

18. By what series of emissions does $^{222}_{88}Ra$ disintegrate to $^{210}_{83}Bi$?

*19. Consider the fission reaction

$$^{235}_{92}U + ^{1}_{0}n \longrightarrow ^{94}_{38}Sr + ^{139}_{54}Xe + 3\,^{1}_{0}n + energy$$

Calculate the following using this mass data (1.0 g is equivalent to 9.0×10^{13} J):

U-235 = 235.0439 amu   Sr-94 = 93.9154 amu

Xe-139 = 138.9179 amu   n = 1.0087 amu

(a) the energy released in joules for a single event (one uranium atom splitting)
(b) the energy released in joules per mole of uranium splitting
(c) the percentage of mass lost in the reaction

20. Consider the fusion reaction

$$^{1}_{1}H + ^{2}_{1}H \longrightarrow ^{3}_{2}He + energy$$

Calculate the following using this mass data (1.0 g is equivalent to 9.0×10^{13} J):

$^{1}_{1}H$ = 1.00794 amu

$^{2}_{1}H$ = 2.01410 amu

$^{3}_{2}H$ = 3.01603 amu

(a) the energy released in joules per mole of He-3 formed
(b) the percentage of mass lost in the reaction

Additional Exercises

All exercises with blue numbers have answers in the appendix of the text.

21. The Th-232 disintegration series starts with $^{232}_{90}Th$ and emits the following rays successively: α, β, β, α, α, α, β, α, β, α. The series ends with the stable $^{208}_{82}Pb$. Write the formula for each nuclide in the series.

22. The half-life of Ra-224 is 3.64 days. How long will it take for 7/8 of an 8.0-g sample to disappear?

23. When the nuclide $^{249}_{98}Cf$ was bombarded with $^{15}_{7}N$, four neutrons and a new transuranium element were formed. Write the nuclear equation for this transmutation.

24. What percent of the mass of the nuclide $^{226}_{88}Ra$ is neutrons? electrons?

25. If radium costs $50,000 a gram, how much will 0.0100 g of $^{226}RaCl_2$ cost if the price is based only on the radium content?

26. An archaeological specimen was analyzed and found to be emitting only 25% as much C-14 radiation per gram of carbon as newly cut wood. How old is this specimen?

27. Barium-141 is a beta emitter. What is the half-life if a 16.0-g sample of the nuclide decays to 0.500 g in 90 minutes?

*28. Calculate (a) the mass defect and (b) the binding energy of $^{7}_{3}Li$ using the mass data:

$^{7}_{3}Li$ = 7.0160 g n = 1.0087 g

p = 1.0073 g e^- = 0.00055 g

1.0 g ≡ 9.0×10^{13} J (from $E = mc^2$)

*29. In the disintegration series $^{235}_{92}U \longrightarrow ^{207}_{82}Pb$, how many alpha and beta particles are emitted?

30. List three devices used for radiation detection and explain their operation.

31. The half-life of I-123 is 13 hours. If 10 mg of I-123 are administered to a patient, how much I-123 remains after 3 days and 6 hours?

32. Clearly distinguish between fission and fusion. Give an example of each.

33. Starting with 1 g of a radioactive isotope whose half-life is 10 days, sketch a graph showing the pattern of decay for that material. On the x-axis, plot time (you may want to simply show multiples of the half-life), and on the y-axis, plot mass of material remaining. Then after completing the graph, explain why a sample never really gets to the point where *all* of its radioactivity is considered to be gone.

34. Identify each missing product (name the element and give its atomic number and mass number) by balancing the following nuclear equations:

(a) $^{235}U + ^{1}_{0}n \longrightarrow ^{143}Xe + 3\,^{1}_{0}n +$ _____

(b) $^{235}U + ^{1}_{0}n \longrightarrow ^{102}Y + 3\,^{1}_{0}n +$ _____

(c) $^{14}N + ^{1}_{0}n \longrightarrow ^{1}H +$ _____

35. Consider these reactions:

(a) $H_2O(l) \longrightarrow H_2O(g)$

(b) $2\,H_2(g) + O_2(g) \longrightarrow 2\,H_2O(g)$

(c) $^{2}_{1}H + ^{2}_{1}H \longrightarrow ^{3}_{1}H + ^{1}_{1}H$

The following energy values belong to one of these equations:

energy$_1$ 115.6 kcal released

energy$_2$ 10.5 kcal absorbed

energy$_3$ 7.5×10^7 kcal released

Match the equation to the energy value and briefly explain your choices.

36. When $^{235}_{92}U$ is struck by a neutron, the unstable isotope $^{236}_{92}U$ results. When that daughter isotope undergoes fission, there are numerous possible products. If strontium-90 and three neutrons are the results of one such fission, what is the other product?

37. Write balanced nuclear equations for

(a) beta emission by $^{29}_{12}Mg$

(b) alpha emission by $^{150}_{60}Nd$

(c) positron emission by $^{72}_{33}As$

38. How much of a sample of cesium-137 ($t_{1/2}$ = 30 years) must have been present originally if, after 270 years, 15.0 g remain?

39. Suppose that the existence of element 114 were confirmed and reported. What element would this new substance fall beneath on the periodic table? Would it be a metal? What typical ion might you expect it to form in solution?

40. Cobalt-60 has a half-life of 5.26 years. If 1.00 g of ^{60}Co were allowed to decay, how many grams would be left after
  (a) one half-life?
  (b) two half-lives?
  (c) four half-lives?
  (d) ten half-lives?

41. Write balanced equations to show these changes:
  (a) alpha emission by boron-11
  (b) beta emission by strontium-88
  (c) neutron absorption by silver-107
  (d) proton emission by potassium-41
  (e) electron absorption by antimony-116

42. The $^{14}_{6}$C content of an ancient piece of wood was found to be one-sixteenth of that in living trees. How many years old is this piece of wood if the half-life of carbon-14 is 5730 years?

43. The curie is equal to 3.7×10^{10} disintegration/sec, and the becquerel is equivalent to just 1 disintegration/sec. Suppose a hospital has a 150-g radioactive source with an activity of 1.24 Ci. What is its activity in becquerels?

Challenge Exercises

All exercises with blue numbers have answers in the appendix of the text.

*44. Rubidium-87, a beta emitter, is the product of positron emission. Identify
  (a) the product of rubidium-87 decay
  (b) the precursor of rubidium-87

*45. Potassium-42 is used to locate brain tumors. Its half-life is 12.5 hours. Starting with 15.4 mg, what fraction will remain after 100 hours? If it was necessary to have at least 1 μg for a particular procedure, could you hold the original sample for 200 hours before using it?

Answers to Practice Exercises

18.1 0.391 g

18.2 $^{218}_{84}$Po

18.3 (a) $^{234}_{91}$Pa

  (b) $^{226}_{88}$Ra

Introduction to Organic Chemistry

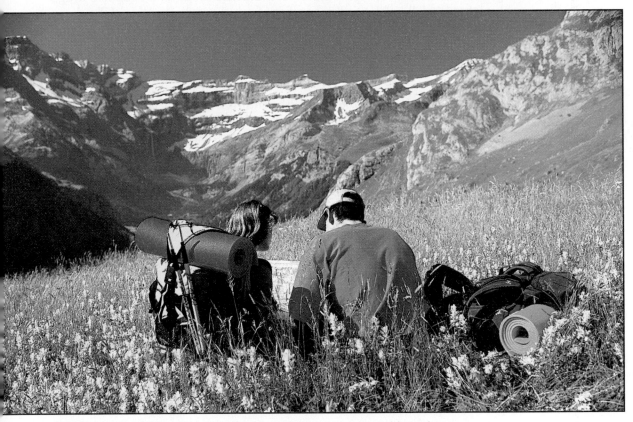

Most of the gear used by outdoor enthusiasts is made from polymers containing carbon.

Chapter Outline

19.1 The Beginnings of Organic Chemistry

19.2 The Carbon Atom

19.3 Hydrocarbons

19.4 Alkanes

19.5 Structural Formulas and Isomerism

19.6 Naming Alkanes

19.7 Alkenes and Alkynes

19.8 Naming Alkenes and Alkynes

19.9 Reactions of Alkenes

19.10 Aromatic Hydrocarbons

19.11 Naming Aromatic Compounds

19.12 Hydrocarbon Derivatives

19.13 Alkyl Halides

19.14 Alcohols

19.15 Naming Alcohols

19.16 Ethers

19.17 Aldehydes and Ketones

19.18 Naming Aldehydes and Ketones

19.19 Carboxylic Acids

19.20 Esters

19.21 Polymers—Macromolecules

Numerous substances throughout nature incorporate silicon or carbon in their structures. Silicon is the staple of the geologist—it combines with oxygen in various ways to produce silica and a family of compounds known as the silicates. These compounds form the chemical foundation of most sand, rocks, and soil. In the living world, carbon combines with hydrogen, oxygen, nitrogen, and sulfur to form millions of compounds.

The petroleum and the polymer products we find indispensable are two of the many industries that depend on carbon chemistry. Synthetic fibers (clothing and carpeting) and plastics (containers, compact discs, computer terminals, and pens) are made from carbon compounds. Cold remedies, cleaning products, nutrients for space travel, convenience foods, and countless drugs, both legal and illegal, have all come about from our understanding of the chemistry of carbon.

19.1 The Beginnings of Organic Chemistry

During the late 18th and the early 19th centuries, the fact that compounds obtained from animal and vegetable sources defied the established rules for inorganic compounds baffled chemists. They knew that compound formation in inorganic compounds is due to a simple attraction between positively and negatively charged elements, and usually only one, or at most a few, compounds could be made from a given group of two or three elements. But one group of four elements—carbon, hydrogen, oxygen, and nitrogen—gave rise to a large number of remarkably stable compounds.

No organic compounds had ever been synthesized from inorganic substances and chemists had no other explanation for the complexities of organic compounds, so they believed these compounds were formed by a "vital force."

vital–force theory The **vital-force theory** held that organic substances could originate only from living material. In 1828, the results of a simple experiment by German chemist Friedrich Wöhler (1800–1882) proved to be the end of this theory. As he attempted to prepare ammonium cyanate (NH_4CNO) by heating cyanic acid, HCNO, and ammonia, NH_3, Wöhler obtained a white crystalline substance he identified as urea, $H_2N—CO—NH_2$. Wöhler knew urea to be an authentic organic substance because it had been isolated from urine. The implications of Wöhler's results were not immediately recognized, but the fact that one organic compound had been isolated from inorganic compounds changed the face of chemistry forever.

With Wöhler's work, it was apparent that no vital force other than skill and knowledge was needed to make organic chemicals in the laboratory. Today the

organic chemistry branch of chemistry that deals with carbon compounds, **organic chemistry,** does not imply that these compounds must originate from living matter. A few special kinds of carbon compounds (e.g., carbon oxides, metal carbides, and metal carbonates) are excluded from the organic classification because their chemistry is more closely related to that of inorganic substances.

The field of organic chemistry is vast; it includes not only all living organisms but also a great many other materials that we use daily. Foodstuffs (fats, proteins, carbohydrates); fuels; fabrics; wood and paper products; paints and varnishes; plastics; dyes; soaps and detergents; cosmetics; medicinals; and rubber products—all are organic materials.

The sources of organic compounds are carbon-containing raw materials—petroleum and natural gas, coal, carbohydrates, fats, and oils. In the United States we produce about 250 billion pounds of organic chemicals from these sources, which amounts to about 1100 pounds per year for every man, woman, and child. About 90% of this 250 billion pounds comes from petroleum and natural gas. Because world reserves of petroleum and natural gas are finite, we will sometime have to rely on other sources to make the vast amount of organic substances that we depend on. Fortunately, we know how to synthesize many organic compounds from sources other than petroleum, although at a much greater expense.

19.2 The Carbon Atom

The carbon atom is central to all organic compounds. The atomic number of carbon is 6, and its electron structure is $1s^2 2s^2 2p^2$. Two stable isotopes of carbon exist: C-12 and C-13. In addition, carbon has several radioactive isotopes, C-14 being the most widely known of these because of its use in radiocarbon dating.

A carbon atom usually forms four covalent bonds. The most common geometric arrangement of these bonds is tetrahedral (see Figure 19.1). In this structure, the four covalent bonds are not planar about the carbon atom but are directed toward the corners of a regular tetrahedron. (A tetrahedron is a solid figure with four sides.) The angle between these tetrahedral bonds is 109.5°.

With four valence electrons, the carbon atom ($\cdot \overset{\cdot}{\underset{\cdot}{C}} \cdot$) forms four single covalent bonds by sharing electrons with other atoms. The structures of methane and carbon tetrachloride illustrate this point:

$$H : \overset{\cdot\cdot}{\underset{\cdot\cdot}{C}} : H \qquad H - \overset{|}{\underset{|}{C}} - H \qquad Cl : \overset{\cdot\cdot}{\underset{\cdot\cdot}{C}} : Cl \qquad Cl - \overset{|}{\underset{|}{C}} - Cl$$

methane   carbon tetrachloride

Actually, these compounds have a tetrahedral shape with bond angles of 109.5° (see Figure 19.2), but the bonds are often drawn at right angles. In methane each bond is formed by the sharing of electrons between a carbon and a hydrogen atom.

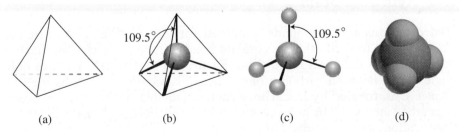

(a)   (b)   (c)   (d)

Figure 19.1
Tetrahedral structure of carbon. (a) A regular tetrahedron; (b) a carbon atom within a regular tetrahedron; (c) a methane molecule, CH_4; (d) spacefilling model of methane, CH_4.

Figure 19.2
Spacefilling models of CH_4 and CCl_4.

Carbon–carbon bonds are formed because carbon atoms can share electrons with other carbon atoms. One, two, or three pairs of electrons can be shared between two carbon atoms, forming a single, double, or triple bond, respectively:

$$\cdot\overset{\cdot}{C}:\overset{\cdot}{C}\cdot \qquad \cdot\overset{\cdot}{C}::\overset{\cdot}{C}\cdot \qquad \cdot C:::C\cdot$$

$$\cdot\overset{\cdot}{C}-\overset{\cdot}{C}\cdot \qquad \cdot\overset{\cdot}{C}=\overset{\cdot}{C}\cdot \qquad \cdot C\equiv C\cdot$$

single bond   double bond   triple bond

Each dash represents a covalent bond. Carbon, more than any other element, has the ability to form chains of covalently bonded atoms. This bonding ability is the main reason for the large number of organic compounds. Three examples are shown here. It's easy to see how, through this bonding ability, long chains of carbon atoms form by linking one carbon atom to another through covalent bonds:

$$\cdot\overset{\cdot}{C}:\overset{\cdot}{C}:\overset{\cdot}{C}\cdot \qquad \cdot\overset{\cdot}{C}-\overset{\cdot}{C}-\overset{\cdot}{C}\cdot \qquad \overset{\overset{\cdot}{C}}{\underset{109.5°}{C}}\overset{\cdot}{C}$$

three carbon atoms bonded by single bonds

seven-carbon chain

ten carbon atoms bonded together

Carbon forms so many different compounds that a system for grouping the molecules is necessary. Organic molecules are classified according to structural features. The members of each class of compounds contain a characteristic atom or group of atoms called a **functional group.** Molecules with the same functional group share similarities in structure that result in similar chemical properties. Thus, we need study only a few members of a particular class of compounds to be able to predict the behavior of other molecules in that class. In this brief introduction to organic chemistry we will consider two categories of compounds: hydrocarbons and hydrocarbon derivatives.

functional group

19.3 Hydrocarbons

hydrocarbon

saturated hydrocarbon

unsaturated hydrocarbon

Hydrocarbons are compounds composed entirely of carbon and hydrogen atoms bonded to each other by covalent bonds. These molecules are further classified as saturated or unsaturated. **Saturated hydrocarbons** have only single bonds between carbon atoms. These hydrocarbons are classified as *alkanes.* **Unsaturated hydrocarbons** contain a double or triple bond between two carbon atoms and include *alkenes, alkynes,* and *aromatic compounds.* These classifications are summarized in Figure 19.3.

Fossil fuels—natural gas, petroleum, and coal—are the principal sources of hydrocarbons. Natural gas is primarily methane with small amounts of ethane, propane, and butane. Petroleum is a mixture of hydrocarbons from which

Rudolf Diesel chose peanut oil for his fuel when he first invented the diesel engine in 1892. While diesel engines now power many of our vehicles (farm equipment, trucks, barges), the diesel fuel currently in use is a mixture of thousands of hydrocarbons obtained from crude oil. Diesel fuel is known for its crude and dirty smoke-containing noxious gases and particles of sulfur, organic compounds, and soot.

Today scientists and environmentalists are beginning to return to a vegetable-based renewable fuel. It turns out that vegetable oils can be converted to esters, which are less viscous and make good fuels. These new fuels are called biodiesel, and they can be made by using virtually any vegetable oil including soybean oil or used restaurant oil.

Most biodiesel used today is a blend with regular diesel fuel. Diesel engines can run on biodiesel with little or no engine modification. Biodiesel has been in common use in Europe for years. All diesel in France is 5% biodiesel. Eu-

ropean governments have maintained policies to reduce their dependence on foreign oil. In the United States, some city and government fleets use biodiesel. A number of states, including Maine, Nevada, California, and Hawaii, even have biodiesel available to

the public at pumps. The popularity of biodiesel is increasing significantly, as shown in the accompanying graph.

The Veggie Van, designed by Joshua Tickell, has run more than 25,000 miles on biodiesel. Tickell even collected his fuel on the road, stopping at fast-food places for gallons of used oil for a fill-up. Tickell says, "It's instantly available renewable energy, one we can begin using today and use for the next 10 to 15 years until higher technology comes to fruition." So hang onto that used French fry oil—you may soon use it to run your car!

Gallons of Biodiesel Distributed

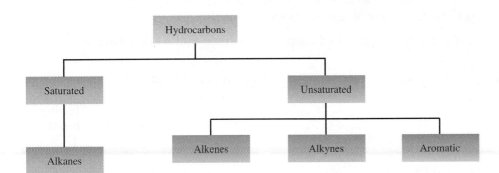

Figure 19.3
General classification of hydrocarbons.

gasoline, kerosene, fuel oil, lubricating oil, paraffin wax, and petrolatum (themselves mixtures of hydrocarbons) are separated. Coal tar, a volatile by-product of the steel industry's process of making coke from coal, is the source of many valuable chemicals, including the aromatic hydrocarbons benzene, toluene, and naphthalene.

19.4 Alkanes

alkane

The **alkanes,** also known as *paraffins* or *saturated hydrocarbons,* are straight- or branched-chain hydrocarbons with only single covalent bonds between the carbon atoms. We will study the alkanes in some detail because many other classes of organic compounds are derivatives of these substances. Be sure to learn the names of the first 10 members of the alkane series, because they are the basis for naming other classes of compounds.

Methane, CH_4, is the first member of the alkane series. Members having two, three, and four carbon atoms are ethane, propane, and butane, respectively. The first four alkanes have common names and must be memorized, but the next six names are derived from Greek numbers. The names and formulas of the first 10 alkanes are given in Table 19.1.

Successive compounds in the alkane series differ from each other in composition by one carbon and two hydrogen atoms (CH_2). When each member of a series differs from the next member by a CH_2 group, the series is called a **homologous series.** The members of a homologous series are similar in structure but differ in formula. All common classes of organic compounds exist in homologous series, which can be represented by a general formula. For open-chain alkanes the general formula is C_nH_{2n+2}, where n corresponds to the number of carbon atoms in the molecule. The formulas of specific alkanes are easily determined from this general formula. Thus for pentane, $n = 5$ and $2n + 2 = 12$, so its formula is C_5H_{12}. For hexadecane, a 16-carbon alkane, the formula is $C_{16}H_{34}$.

One single reaction of alkanes has inspired people to explore equatorial jungles, to endure the heat and sandstorms of the deserts of Africa and the Middle East, to mush across the frozen Arctic, and to drill holes in Earth more than 30,000 feet deep! The substance is oil, and the reaction is combustion with oxygen to produce heat energy. Combustion reactions overshadow all other reactions of alkanes in economic importance. For example, note the heat generated when methane reacts with oxygen:

$$CH_4(g) + 2\,O_2(g) \longrightarrow CO_2(g) + 2\,H_2O(g) + 802.5\,kJ\ (191.8\,kcal)$$

Learn the names of the alkanes in Table 19.1; they form the stem of many other names.

homologous series

Oil contains alkanes as well as other hydrocarbons.

Table 19.1 Names, Formulas, and Physical Properties of Straight–Chain Alkanes

Name	Molecular formula C_nH_{2n+2}	Condensed structural formula	Boiling point (°C)	Melting point (°C)
Methane	CH_4	CH_4	−161	−183
Ethane	C_2H_6	CH_3CH_3	−88	−172
Propane	C_3H_8	$CH_3CH_2CH_3$	−45	−187
Butane	C_4H_{10}	$CH_3CH_2CH_2CH_3$	−0.5	−138
Pentane	C_5H_{12}	$CH_3CH_2CH_2CH_2CH_3$	36	−130
Hexane	C_6H_{14}	$CH_3CH_2CH_2CH_2CH_2CH_3$	69	−95
Heptane	C_7H_{16}	$CH_3CH_2CH_2CH_2CH_2CH_2CH_3$	98	−90
Octane	C_8H_{18}	$CH_3CH_2CH_2CH_2CH_2CH_2CH_2CH_3$	125	−57
Nonane	C_9H_{20}	$CH_3CH_2CH_2CH_2CH_2CH_2CH_2CH_2CH_3$	151	−54
Decane	$C_{10}H_{22}$	$CH_3CH_2CH_2CH_2CH_2CH_2CH_2CH_2CH_2CH_3$	174	−30

Thermal energy is converted to mechanical and electrical energy all over the world. But combustion reactions are not usually of great interest to organic chemists, because carbon dioxide and water are the only chemical products of complete combustion. Aside from their combustibility, alkanes are limited in reactivity.

19.5 Structural Formulas and Isomerism

The properties of an organic substance are dependent on its molecular structure. By structure, we mean the way in which the atoms bond within the molecule. The majority of organic compounds are made from relatively few elements—carbon, hydrogen, oxygen, nitrogen, and the halogens. In these compounds, carbon has four bonds to each atom, nitrogen three bonds, oxygen two bonds, and hydrogen and the halogens one bond to each atom:

$$
-\overset{|}{\underset{|}{C}}- \quad H- \quad -O- \quad -\overset{|}{N}- \quad Cl- \quad Br- \quad I- \quad F-
$$

The combustion of alkanes forms the basis for the gas and petroleum industry.

Alkane molecules contain only carbon–carbon and carbon–hydrogen bonds. Each carbon atom is joined to four other atoms by four single covalent bonds. These bonds are separated by angles of 109.5° (corresponding to those angles formed by lines drawn from the center of a regular tetrahedron to its corners). Alkane molecules are essentially nonpolar. Because of this low polarity, these molecules have very little intermolecular attraction and therefore relatively low boiling points compared with other organic compounds of similar molar mass.

The three-dimensional character of atoms and molecules is difficult to portray without models or computer-generated drawings. Methane and ethane are shown here in Lewis structure and line structure form:

$$
\begin{array}{cc}
\overset{\displaystyle H}{\underset{\displaystyle \ddot{H}}{H:\!\overset{..}{C}\!:H}} & H-\overset{\displaystyle H}{\underset{\displaystyle H}{\overset{|}{C}}{\!}-H \\
\text{methane} &
\end{array}
\qquad
\begin{array}{cc}
\overset{\displaystyle H\ H}{\underset{\displaystyle \ddot{H}\ \ddot{H}}{H:\!\overset{..}{C}\!:\!\overset{..}{C}\!:H}} & H-\overset{\displaystyle H\ \ H}{\underset{\displaystyle H\ \ H}{\overset{|\ \ \ |}{C}-\overset{}{C}}}{\!}-H \\
& \text{ethane}
\end{array}
$$

Lewis diagrams can be drawn for alkanes, but these molecules are usually represented by replacing bonding electron pairs with single lines.

To write the correct structural formula for propane (C_3H_8), the next member of the alkane series, we need to place each atom in the molecule. An alkane contains only single bonds, so each carbon atom must be bonded to four other atoms by either C—C or C—H bonds. Hydrogen must be bonded to only one carbon atom by a C—H bond, since C—H—C bonds do not occur, and an H—H bond would simply represent a hydrogen molecule. Thus the only possible structure for propane is

$$
H-\overset{\displaystyle H\ \ \ H\ \ \ H}{\underset{\displaystyle H\ \ \ H\ \ \ H}{\overset{|\ \ \ \ |\ \ \ \ |}{C}-\overset{}{C}-\overset{}{C}}}{\!}-H
$$

propane

However, it's possible to write two structural formulas corresponding to the molecular formula C_4H_{10} (butane). Two C_4H_{10} compounds with these structural formulas actually exist:

The butane with the unbranched carbon chain is called *normal butane* (abbreviated *n*-butane); it boils at 0.5°C and melts at −138.3°C. The branched-chain butane is called 2-methylpropane; it boils at −11.7°C and melts at −159.5°C. These differences in physical properties are sufficient to establish that the two compounds, though they have the same molecular formula, are different substances. The structural arrangements of the atoms in methane, ethane, propane, butane, and 2-methylpropane are shown in Figure 19.4.

Figure 19.4
Spacefilling models illustrating structural formulas of methane, ethane, propane, butane, and 2-methylpropane. Condensed structural formulas are shown above the names, line sructures to the left of the names.

This phenomenon of two or more compounds having the same molecular formula, but different structural arrangements, is called **isomerism.** The individual compounds are called **isomers.** Butane, C_4H_{10}, shown above, has two isomers. Isomerism is common among organic compounds and is another reason for the large number of known compounds. There are 3 isomers of pentane, 5 isomers of hexane, 9 isomers of heptane, 18 isomers of octane, 35 isomers of nonane, and 75 isomers of decane. The phenomenon of isomerism is a compelling reason for using structural formulas.

isomerism
isomers

> Isomers are compounds with the same molecular formula but different structural formulas.

To save time and space, condensed structural formulas, in which the atoms and groups attached to a carbon atom are written to the right of that carbon atom, are often used. For example, the condensed structural formula for pentane is $CH_3CH_2CH_2CH_2CH_3$ or $CH_3(CH_2)_3CH_3$. Some condensed structural formulas are shown in Figure 19.4.

Let's interpret the condensed structural formula for propane:

$$\overset{1}{C}H_3\overset{2}{C}H_2\overset{3}{C}H_3$$

Carbon-1 has three hydrogen atoms attached to it and is bonded to C-2, which has two hydrogen atoms on it and is bonded to C-3. Carbon-3 has three hydrogen atoms bonded to it.

Pentane (C_5H_{12}) has three isomers. Write their structural formulas and their condensed structural formulas.

Example 19.1

In a problem of this kind, it's best to first write the carbon skeleton with the longest continuous carbon chain—in this case, five carbon atoms. We complete the structure by attaching hydrogen atoms around each carbon atom so that each carbon atom has four bonds. The carbon atoms at the ends of the chain need three hydrogen atoms. The three inner carbon atoms each need two hydrogen atoms to give them four bonds:

SOLUTION

$$C-C-C-C-C$$

$$\begin{matrix} H & H & H & H & H \\ | & | & | & | & | \\ H-C-C-C-C-C-H \\ | & | & | & | & | \\ H & H & H & H & H \end{matrix}$$

$$CH_3CH_2CH_2CH_2CH_3$$

For the next isomer, we write a four-carbon chain and attach the fifth carbon atom to either of the middle carbon atoms (don't use the end ones):

$$\begin{matrix} & C & & & C \\ & | & & & | \\ C-C-C-C & & & C-C-C-C \end{matrix}$$

These structures represent the same compound.

Now add the 12 hydrogen atoms to complete the structure:

$$CH_3CH_2CHCH_3 \quad \text{or} \quad CH_3CH_2CH(CH_3)_2$$

For the third isomer, write a three-carbon chain, attach the remaining two carbon atoms to the central carbon atom, and complete the structure by adding the 12 hydrogen atoms:

$$CH_3CCH_3 \quad \text{or} \quad C(CH_3)_4$$

Practice 19.1

Write condensed structural formulas for the isomer of hexane, C_6H_{14}.

19.6 Naming Alkanes

In the early years of organic chemistry, each new compound was given a name, usually by the person who had isolated or synthesized it. Names were not systematic but did carry some information—often about the origin of the substance. Wood alcohol (methanol), for example, was so named because it was obtained by destructive distillation or pyrolysis of wood.

It soon became apparent that a naming system was needed, and, in 1892, such a system was proposed and adopted. In its present form, the International Union of Pure and Applied Chemistry (IUPAC) system is generally unambiguous and internationally accepted. However, a great many well-established common names and abbreviations (such as TNT and DDT) have continued to be used because of their brevity and/or convenience. So you need a knowledge of both the IUPAC system and the common names.

alkyl group To name organic compounds using the IUPAC system, you must recognize certain common alkyl groups. **Alkyl groups** have the general formula C_nH_{2n+1} (one less hydrogen atom than the corresponding alkane). The name of the group is formed from the name of the corresponding alkane by simply dropping -*ane* and substituting a -*yl* ending. The names and formulas of selected alkyl groups are given in Table 19.2. The letter "R" is often used in formulas to represent any of the possible alkyl groups:

$$R = C_nH_{2n+1} (\text{any alkyl group})$$

Table 19.2 Names and Formulas of Selected Alkyl Groups

Formula	Name	Formula	Name
CH_3-	methyl	CH_3CH- with CH_3 above	isopropyl
CH_3CH_2-	ethyl		
$CH_3CH_2CH_2-$	propyl	CH_3CHCH_2- with CH_3 above	isobutyl
$CH_3CH_2CH_2CH_2-$	butyl		
$CH_3(CH_2)_3CH_2-$	pentyl	CH_3CH_2CH- with CH_3 above	sec-butyl (secondary butyl)
$CH_3(CH_2)_4CH_2-$	hexyl		
$CH_3(CH_2)_5CH_2-$	heptyl		
$CH_3(CH_2)_6CH_2-$	octyl	CH_3C- with CH_3 above and CH_3 below	tert-butyl (t-butyl) (tertiary butyl)
$CH_3(CH_2)_7CH_2-$	nonyl		
$CH_3(CH_2)_8CH_2-$	decyl		

The following IUPAC rules are all that are needed to name a great many alkanes. In later sections these rules will be extended to cover other classes of compounds, but advanced texts or references must be consulted for the complete system.

Rule 1 Select the longest continuous chain of carbon atoms as the parent compound, and consider all alkyl groups attached to it as branch chains or substituents that have replaced hydrogen atoms of the parent hydrocarbon. If two chains of equal length are found, use the chain that has the larger number of substituents attached to it as the parent compound. The alkane's name consists of the parent compound's name prefixed by the names of the alkyl groups attached to it.

Rule 2 Number the carbon atoms in the parent carbon chain starting from the end closest to the first carbon atom that has an alkyl or other group substituted for a hydrogen atom. If the first substituent from each end is on the same-numbered carbon, go to the next substituent to determine which end of the chain to start numbering.

Rule 3 Name each alkyl group and designate its position on the parent carbon chain by a number (e.g., 2-methyl means a methyl group attached to C-2).

Rule 4 When the same alkyl-group branch chain occurs more than once, indicate this repetition by a prefix (*di-*, *tri-*, *tetra-*, and so forth) written in front of the alkyl-group name (e.g., *dimethyl* indicates two methyl groups). The numbers indicating the alkyl-group positions are separated by a comma and followed by a hyphen and are placed in front of the name (e.g., 2,3-dimethyl).

Rule 5 When several different alkyl groups are attached to the parent compound, list them in alphabetical order (e.g., ethyl before methyl in 3-ethyl-4-methyloctane). Prefixes are not included in alphabetical ordering (ethyl comes before dimethyl).

Let's use the IUPAC system to name this compound:

$$\overset{4}{CH_3}-\overset{3}{CH_2}-\overset{2}{CH}-\overset{1}{CH_3} \quad or \quad \overset{1}{CH_3}-\overset{2}{CH}-\overset{3}{CH_2}-\overset{4}{CH_3}$$
$$\underset{CH_3}{|} \qquad\qquad\qquad \underset{CH_3}{|}$$

2-methylbutane

The longest continuous chain contains four carbon atoms. Therefore, we use the parent compound name, butane. The methyl group, CH_3-, attached to C-2 is named as a prefix to butane, the "2-" indicating the point of its attachment on the butane chain.

How would we write the structural formula for 2-methylpentane? Its name tells us how. The parent compound, pentane, contains five carbons. We write and number the five-carbon skeleton of pentane, put a methyl group on C-2 (because of the "2-methyl" in the name), and add hydrogens to give each carbon four bonds:

$$\overset{5}{C}-\overset{4}{C}-\overset{3}{C}-\overset{2}{C}-\overset{1}{C} \quad \overset{5}{C}-\overset{4}{C}-\overset{3}{C}-\overset{2}{C}-\overset{1}{C} \quad CH_3-CH_2-CH_2-CH-CH_3$$
$$\underset{CH_3}{|} \qquad\qquad\qquad\qquad \underset{CH_3}{|}$$

2-methylpentane

Could this compound be called 4-methylpentane? No; in the IUPAC system, the parent carbon chain is numbered starting from the end *nearest* the branch chain.

It is very important to understand that the *sequence* of carbon atoms and groups—not the way the sequence is written—determines the name of a compound. These formulas all represent 2-methylpentane; note that carbon numbering does not have to follow a straight line:

$$\overset{1}{CH_3}-\overset{2}{CH}-\overset{3}{CH_2}-\overset{4}{CH_2}-\overset{5}{CH_3} \qquad \overset{5}{CH_3}-\overset{4}{CH_2}-\overset{3}{CH_2}-\overset{2}{CH}-\overset{1}{CH_3}$$
$$\underset{CH_3}{|} \qquad\qquad\qquad\qquad\qquad\qquad \overset{CH_3}{|}$$

$$\overset{1}{CH_3}-\overset{2}{CH}-\overset{3}{CH_2} \qquad\qquad \overset{CH_3}{\underset{2}{|}}$$
$$\underset{4}{|} \qquad\qquad\qquad\qquad CH \qquad CH_2$$
$$CH_2-\overset{5}{CH_3} \qquad CH_3 \quad CH_2 \quad CH_3$$

The following formulas and names demonstrate other aspects of the IUPAC nomenclature system:

$$\overset{4}{CH_3}-\overset{3}{CH}-\overset{2}{CH}-\overset{1}{CH_3} \qquad\qquad \overset{CH_3}{\underset{2}{|}}$$
$$\underset{CH_3}{|}\ \underset{CH_3}{|} \qquad\qquad \overset{4}{CH_3}-\overset{3}{CH_2}-\overset{2}{C}-\overset{1}{CH_3}$$
$$\qquad\qquad\qquad\qquad\qquad \underset{CH_3}{|}$$

2,3-dimethylbutane   2,2-dimethylbutane

In 2,3-dimethylbutane, the longest carbon chain is four, indicating a butane; "dimethyl" indicates two methyl groups; "2,3-" means that one CH_3 is on C-2 and one is on C-3. In 2,2-dimethylbutane, both methyl groups are on the same carbon atom; both numbers are required.

$$\begin{array}{ccccccc} & & & CH_3 & & CH_3 & \\ {}_7 & {}_6 & {}_5 & {}_4| & {}_3 & {}_2| & {}_1 \\ CH_3 & -CH-CH_2- & CH- & CH-CH-CH_3 \\ & | & & | \\ & CH_3 & & CH_3 \end{array}$$

2,3,4,6-tetramethylheptane (not 2,4,5,6-)

Note that this molecule is numbered from right to left.

$$\begin{array}{cccc} & {}_2 & | & \\ & CH_2-CH_3 & \\ & {}_3| & {}_4 & {}_5 & {}_6 \\ CH_3- & CH-CH_2-CH_2-CH_3 \end{array}$$

3-methylhexane

The longest continuous chain in 3-methylhexane has six carbons.

In the next structure, the longest carbon chain is eight. The groups attached or substituted for hydrogen on the octane chain are named in alphabetical order.

$$\begin{array}{ccccccc} & & & CH_2-CH_3 & & & \\ {}_8 & {}_7 & {}_6 & {}_5 & {}_4| & {}_3 & {}_2 & {}_1 \\ CH_3- & CH_2-CH_2- & CH_2-C & -CH-CH-CH_3 \\ & & & | & | & | \\ & & & CH_3 & Cl & CH_3 \end{array}$$

3-chloro-4-ethyl-2,4-dimethyloctane

Write the formulas for (a) 3-ethylpentane and (b) 2,2,4-trimethylpentane.

Example 19.2

SOLUTION

(a) The name *pentane* indicates a five-carbon chain. Write five connecting carbon atoms and number them. Attach an ethyl group, CH_3CH_2-, to C-3. Now add hydrogen atoms to give each carbon atom four bonds: C-1 and C-5 each need three hydrogen atoms; C-2 and C-4 each need two hydrogen atoms; and C-3 needs one hydrogen atom:

$$\begin{array}{ccccc} {}_1 & {}_2 & {}_3 & {}_4 & {}_5 \\ C- & C- & C- & C- & C \end{array} \qquad \begin{array}{ccccc} {}_1 & {}_2 & {}_3 & {}_4 & {}_5 \\ C- & C- & C- & C- & C \\ & & | & & \\ & & CH_2CH_3 \end{array} \qquad \begin{array}{c} CH_3CH_2CHCH_2CH_3 \\ | \\ CH_2CH_3 \end{array}$$

3-ethylpentane

(b) Pentane indicates a five-carbon chain. Write five connecting carbon atoms and number them. There are three methyl groups, CH_3-, in the compound (trimethyl), two attached to C-2 and one attached to C-4. Attach these three methyl groups to their respective carbon atoms. Now add hydrogen atoms to give each carbon atom four bonds. Thus C-1 and C-5 each need three hydrogen atoms; C-2 does not need any hydrogen atoms; C-3 needs two hydrogen atoms; and C-4 needs one hydrogen atom. The formula is complete:

$$\begin{array}{ccccc} {}_1 & {}_2 & {}_3 & {}_4 & {}_5 \\ C- & C- & C- & C- & C \end{array} \qquad \begin{array}{ccccc} & CH_3 & & CH_3 & \\ {}_1 & {}_2| & {}_3 & {}_4| & {}_5 \\ C- & C- & C- & C- & C \\ & | & & & \\ & CH_3 \end{array} \qquad \begin{array}{c} CH_3 \quad CH_3 \\ | \qquad | \\ CH_3CCH_2CHCH_3 \\ | \\ CH_3 \end{array}$$

2,2,4-trimethylpentane

Example 19.3   Name these compounds:

(a) $CH_3CH_2CH_2CH_2CHCH_3$ (b) $CH_3CH_2CH_2CHCH_2CHCH_3$
                    | | |
                   CH_3 CH_3CH_2 CH_2CH_3

SOLUTION   (a) The longest continuous carbon chain contains six carbon atoms (Rule 1). Thus the parent name is hexane. Number the carbon chain from right to left so that the methyl group attached to C-2 is given the lowest possible number (Rule 2). With a methyl group on C-2, the name of the compound is 2-methylhexane (Rule 3).

(b) The longest continuous carbon chain contains eight carbon atoms:

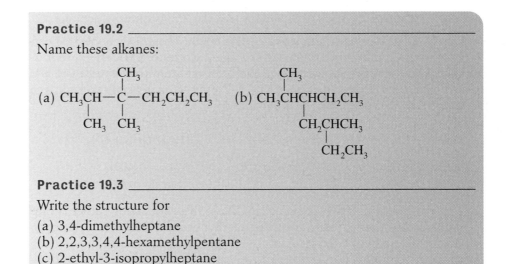

The parent name is octane. As the chain is numbered, a methyl group is on C-3 and an ethyl group is on C-5. The name of the compound is 5-ethyl-3-methyloctane. Note that ethyl is named before methyl (alphabetical order) (Rule 5).

Practice 19.2 _____

Name these alkanes:

                   CH_3 CH_3
                    | |
(a) CH_3CH—C—$CH_2CH_2CH_3$ (b) $CH_3CHCHCH_2CH_3$
          | | |
        CH_3 CH_3 CH_2CHCH_3
                                                        |
                                                     CH_2CH_3

Practice 19.3 _____

Write the structure for

(a) 3,4-dimethylheptane
(b) 2,2,3,3,4,4-hexamethylpentane
(c) 2-ethyl-3-isopropylheptane

19.7 Alkenes and Alkynes

Alkenes and alkynes are classified as unsaturated hydrocarbons. They are said to be unsaturated because, unlike alkanes, their molecules do not contain the maximum possible number of hydrogen atoms. **Alkenes** have two fewer hydrogen atoms, and **alkynes** have four fewer hydrogen atoms than alkanes with a comparable number of carbon atoms. Alkenes contain at least one double bond between adjacent carbon atoms, while alkynes contain at least one triple bond between adjacent carbon atoms.

alkene
alkyne

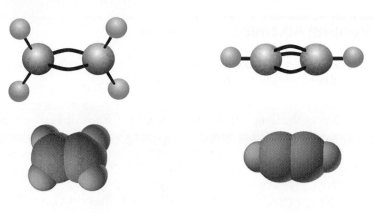

ethylene (common name)
ethene (IUPAC)

acetylene (common name)
ethyne (IUPAC)

Figure 19.5
Ball-and-stick and spacefilling
models for ethylene and acetylene.

Alkenes contain a carbon–carbon double bond.
Alkynes contain a carbon–carbon triple bond.

The simplest alkene is ethylene (or ethene), $CH_2{=}CH_2$, and the simplest alkyne is acetylene (or ethyne), $CH{\equiv}CH$ (Figure 19.5). Ethylene and acetylene are the first members of a homologous series (e.g., $CH_2{=}CH_2$, $CH_3CH{=}CH_2$, and $CH_3CH_2CH{=}CH_2$). Huge quantities of alkenes are made by cracking and dehydrogenating alkanes during the processing of crude oil. These alkenes are used to manufacture motor fuels, polymers, and petrochemicals. Alkene molecules, like those of alkanes, have very little polarity. Hence the physical properties of alkenes are similar to those of the corresponding saturated hydrocarbons.

Remember: In a homologous series the formulas of successive members differ by increments of CH_2.

General formula for alkenes: C_nH_{2n}
General formula for alkynes: C_nH_{2n-2}

Table 19.3 gives the names and formulas for several alkenes and alkynes.

Table 19.3 Names and Formulas for Several Alkenes and Alkynes

Formula	IUPAC name
$CH_2{=}CH_2$	ethene
$CH_3CH{=}CH_2$	propene
$CH_3CH_2CH{=}CH_2$	1-butene
$CH_3CH{=}CHCH_3$	2-butene

$CH_3C{=}CH_2$
    |
   CH_3
	2-methylpropene

$CH{\equiv}CH$	ethyne
$CH_3C{\equiv}CH$	propyne
$CH_3CH_2C{\equiv}CH$	1-butyne
$CH_3C{\equiv}CCH_3$	2-butyne

Acetylene is used for welding steel girders.

19.8 Naming Alkenes and Alkynes

The names of alkenes and alkynes are derived from the corresponding alkanes. To name an alkene (or alkyne) by the IUPAC system,

Step 1	Select the longest continuous carbon–carbon chain that contains a double or triple bond.
Step 2	Name this parent compound as you would an alkane but change the *-ane* ending to *-ene* for an alkene or to *-yne* for an alkyne; thus, propane changes to propene or propyne:

$$CH_3CH_2CH_3 \qquad CH_3CH{=}CH_2 \qquad CH_3C{\equiv}CH$$

propane   propene   propyne

Step 3	Number the carbon chain of the parent compound starting with the end nearer to the double or triple bond. Use the smaller of the two numbers on the double- or triple-bonded carbon atoms to indicate the position of the double or triple bond. Place this number in front of the alkene or alkyne name; 2-butene means that the carbon–carbon double bond is between C-2 and C-3.
Step 4	Branched chains and other groups are treated as in naming alkanes. Name the substituent group, and designate its position on the parent chain with a number.

Study the following examples of named alkenes and alkynes:

$$\overset{4}{C}H_3\overset{3}{C}H_2\overset{2}{C}H{=}\overset{1}{C}H_2$$
1-butene

$$\overset{1}{C}H_3\overset{2}{C}H{=}\overset{3}{C}H\overset{4}{C}H_3$$
2-butene

$$\begin{array}{c} \overset{6}{C}H_3\overset{5}{C}H_2\overset{4}{C}H_2 \\ | \\ \overset{3}{C}H\overset{2}{C}H{=}\overset{1}{C}H_2 \\ | \\ CH_3CH_2CH_2 \end{array}$$
3-propyl-1-hexene

$$\overset{4}{C}H_3\overset{3}{C}H_2\overset{2}{C}{\equiv}\overset{1}{C}H$$
1-butyne

$$\overset{1}{C}H_3{-}\overset{2}{C}{\equiv}\overset{3}{C}{-}\overset{4}{C}H{-}\overset{5}{C}H{-}\overset{6}{C}H_3$$
with CH$_3$ groups on C-4 and C-5
4,5-dimethyl-2-hexyne

To write a structural formula from a systematic name, the naming process is reversed. For example, how would we write the structural formula for 4-methyl-2-pentene? The name indicates (1) five carbons in the longest chain, (2) a double bond between C-2 and C-3, and (3) a methyl group on C-4.

Write five carbon atoms in a row. Place a double bond between C-2 and C-3, and place a methyl group on C-4. Now add hydrogen atoms to give each carbon atom four bonds. Carbons 1 and 5 each need three hydrogen atoms; C-2, C-3, and C-4 each need one hydrogen atom.

$$\overset{1}{C}{-}\overset{2}{C}{=}\overset{3}{C}{-}\overset{4}{C}{-}\overset{5}{C}$$
with CH$_3$ on C-4
carbon skeleton

$$CH_3CH{=}CHCHCH_3$$
with CH$_3$ branch
4-methyl-2-pentene

Write structural formulas for (a) 7-methyl-2-octene and (b) 3-hexyne.

Example 19.4

SOLUTION

(a) Octene, like octane, indicates an eight-carbon chain. The chain contains a double bond between C-2 and C-3 and a methyl group on C-7. Write eight carbon atoms in a row, place a double bond between C-2 and C-3, and place a methyl group on C-7. Now add hydrogen atoms to give each carbon atom four bonds:

$$\overset{1}{C}-\overset{2}{C}=\overset{3}{C}-\overset{4}{C}-\overset{5}{C}-\overset{6}{C}-\overset{7}{C}-\overset{8}{C}$$
$$\qquad\qquad\qquad\quad |$$
$$\qquad\qquad\qquad\quad CH_3$$

carbon skeleton

$$CH_3CH=CHCH_2CH_2CHCH_3$$
$$\qquad\qquad\qquad\qquad |$$
$$\qquad\qquad\qquad\qquad CH_3$$

7-methyl-2-octene

Remember: A double bond counts as 2 bonds.

(b) The stem *hex-* indicates a six-carbon chain; the suffix *-yne* indicates a carbon–carbon triple bond; the number 3 locates the triple bond between C-3 and C-4. Write six carbon atoms in a row and place a triple bond between C-3 and C-4. Now add hydrogen atoms to give each carbon atom four bonds; C-3 and C-4 don't need hydrogen atoms:

$$\overset{1}{C}-\overset{2}{C}-\overset{3}{C}\equiv\overset{4}{C}-\overset{5}{C}-\overset{6}{C}$$

$$CH_3CH_2C\equiv CCH_2CH_3$$

3-hexyne

Practice 19.4

Name these compounds:

(a) $CH_3CHCH=CHCHCH_3$
$\quad\;\; |\qquad\qquad\;\; |$
$\quad\;\; CH_3\qquad\quad CH_2CH_3$

(b) $CH_3C\equiv CCHCH_2CH_2CH_3$
$\qquad\qquad\quad |$
$\qquad\qquad\quad CH_2CH_3$

19.9 Reactions of Alkenes

The alkenes are much more reactive than the corresponding alkanes. This greater reactivity is due to the carbon–carbon double bonds. In our brief overview, we will limit our discussion to the most common reaction of alkenes.

Addition

In organic chemistry, a reaction in which two substances join to produce one compound is called an **addition reaction.** Addition at the carbon–carbon double bond is the most common reaction of alkenes. Hydrogen, halogens (Cl_2 or Br_2), hydrogen halides, and water are some of the reagents that can be added to unsaturated hydrocarbons. Ethylene, for example, reacts in this fashion:

addition reaction

$$CH_2{=}CH_2 \;+\; H_2 \xrightarrow[\text{1 atm}]{\text{Pt} \;\; 25°C} CH_3{-}CH_3$$

ethylene   ethane

$$CH_2{=}CH_2 \;+\; Br{-}Br \longrightarrow BrCH_2{-}CH_2Br$$

1,2-dibromoethane

$\xrightarrow[\text{1 atm}]{\text{Pt } 25°C}$ **indicates the catalyst, Pt, and other necessary conditions for the reaction.**

Visible evidence of the Br_2 addition is the disappearance of the reddish-brown color of bromine as it reacts.

$$CH_2{=}CH_2 \ + \ HCl \ \longrightarrow \ CH_3CH_2Cl$$
chloroethane

The H^+ indicates that the reaction is carried out under acidic conditions.

$$CH_2{=}CH_2 \ + \ HOH \ \xrightarrow{\ H^+\ } \ CH_3CH_2OH$$
ethanol
(ethyl alcohol)

Note that the double bond is broken and the unsaturated alkene molecules become saturated by an addition reaction. Reactions of this kind can occur with almost any molecule that contains a carbon–carbon double bond.

19.10 Aromatic Hydrocarbons

aromatic compound

Benzene and all substances with structures and chemical properties resembling benzene are classified as **aromatic compounds.** The word *aromatic* originally referred to the rather pleasant odor many of these substances possess, but this meaning has been dropped. Benzene, the parent substance of the aromatic hydrocarbons, was first isolated by Michael Faraday in 1825. Its correct molecular formula, C_6H_6, was established a few years later, but finding a reasonable structural formula that would account for the properties of benzene was difficult.

Finally, in 1865, August Kekulé proposed that the carbon atoms in a benzene molecule are arranged in a six-membered ring with one hydrogen atom bonded to each carbon atom and with three carbon–carbon double bonds:

benzene

A turning point in the history of chemistry, Kekulé's structures mark the beginning of our understanding of structure in aromatic compounds. His formulas do have one serious shortcoming, though: They represent benzene and related substances as highly unsaturated compounds. Yet benzene does not readily undergo addition reactions like a typical alkene. For example, benzene does not decolorize bromine solutions rapidly. Instead, the chemical behavior of benzene resembles that of an alkane. Its typical reactions are the substitution type, where a hydrogen atom is replaced by some other group:

$$C_6H_6 + Cl_2 \xrightarrow{\ Fe\ } C_6H_5Cl + HCl$$

Modern theory suggests that the benzene molecule is a hybrid of the two Kekulé structures shown earlier.

For convenience, chemists usually write the structure of benzene as one or the other of these abbreviated forms:

A   B

In both representations, it is understood that there is a carbon atom and a hydrogen atom at each corner of the hexagon. The classical Kekulé structure is represented by A; the modern molecular orbital structure is represented by B. These hexagonal structures are also used to represent the structural formulas of benzene derivatives—that is, substances in which one or more hydrogen atoms in the ring have been replaced by other atoms or groups. Chlorobenzene (C_6H_5Cl), for example, is written in this fashion:

chlorobenzene, C_6H_5Cl

This notation indicates that the chlorine atom has replaced a hydrogen atom and is bonded directly to a carbon atom in the ring. Thus, the correct formula for chlorobenzene is C_6H_5Cl, not C_6H_6Cl.

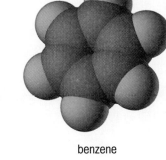

benzene

19.11 Naming Aromatic Compounds

A substituted benzene is derived by replacing one or more of benzene's hydrogen atoms with another atom or group of atoms. Thus a monosubstituted benzene has the formula C_6H_5G, where G is the group replacing a hydrogen atom. Because all the hydrogen atoms in benzene are equivalent, it doesn't matter which hydrogen is replaced by the monosubstituted group.

Monosubstituted Benzenes

Some monosubstituted benzenes are named by adding the name of the substituent group as a prefix to the word *benzene*. The name is written as one word. Here are some examples:

nitrobenzene   ethylbenzene   chlorobenzene   bromobenzene

– NO_2 is a nitro group.

Certain monosubstituted benzenes have special names that should be learned. These are parent names for further substituted compounds:

toluene   phenol   benzoic acid   aniline
(methylbenzene)   (hydroxybenzene)   (aminobenzene)

– NH_2 is an amino group.

The C_6H_5— group is known as the phenyl group (pronounced *fen-ill*). The name *phenyl* is used for compounds that cannot be easily named as benzene derivatives. For example, the following compounds are named as derivatives of alkanes:

3-chloro-2-phenylpentane   diphenylmethane

Disubstituted Benzenes

When two substituent groups replace two hydrogen atoms in a benzene molecule, three isomers are possible. The prefixes *ortho-*, *meta-*, and *para-* (abbreviated *o-*, *m-*, and *p-*) are used to name these disubstituted benzenes. In the *ortho*-compound the substituents are located on adjacent carbon atoms, in the *meta*-compound they are one carbon apart, and in the *para*-compound they are on opposite sides of the ring. When the two groups are different, name them alphabetically followed by the word *benzene*.

The dichlorobenzenes ($C_6H_4Cl_2$) illustrate this method of naming. The three isomers have different physical properties, indicating that they are different substances. Note that the *para*-isomer is a solid and the other two are liquids at room temperature.

Moth crystals or mothballs used to be made of naphthalene. Today many are *para*-dichlorobenzene.

ortho-dichlorobenzene   *meta*-dichlorobenzene   *para*-dichlorobenzene
(1,2-dichlorobenzene)   (1,3-dichlorobenzene)   (1,4-dichlorobenzene)
mp 17.2°C, bp 180.4°C   mp 24.8°C, bp 172°C   mp 53.1°C, bp 174.4°C

When one substituent corresponds to a monosubstituted benzene with a special name, the monosubstituted compound becomes the parent name for the disubstituted compound. In the following examples, the parent compounds are phenol, aniline, toluene, and benzoic acid:

o-nitrophenol   *p*-bromoaniline   *m*-nitrotoluene   *p*-aminobenzoic acid

TNT was used
to demolish the
Kingdome
in Seattle.

Tri- and Polysubstituted Benzenes

When a benzene ring has more than two substituents, the carbon atoms in the ring are numbered. Numbering starts at one of the substituted groups and goes either clockwise or counterclockwise, but it must be done in the direction that gives the lowest possible numbers to the substituent groups. When the compound is named as a derivative of the special parent compound, the substituent of the parent compound is considered to be on C-1 of the ring (the CH_3 group is on C-1 in 2,4,6-trinitrotoluene). The following examples illustrate this system:

1,3,5-trinitrobenzene

1,2,4-tribromobenzene
(not 1,4,6-)

2,4,6-trinitrotoluene
(TNT)

5-bromo-2-chlorophenol

Example 19.5   Write formulas and names for all possible isomers of (a) chloronitrobenzene, $C_6H_4Cl(NO_2)$, and (b) tribromobenzene, $C_6H_3Br_3$.

SOLUTION   (a) The name and formula indicate a chloro-group, Cl, and a nitro-group, NO_2, attached to a benzene ring. There are six positions in which to place these two groups. They can be *ortho-*, *meta-*, or *para-* to each other.

o-chloronitrobenzene   *m*-chloronitrobenzene   *p*-chloronitrobenzene

(b) For tribromobenzene, start by placing the three bromo-groups in the 1, 2, and 3 positions, then the 1, 2, and 4 positions, and so on, until all the possible isomers are formed. Name each isomer to check that no duplicate formulas have been written:

1,2,3-tribromobenzene   1,2,4-tribromobenzene   1,3,5-tribromobenzene

Tribromobenzene has only three isomers. However, if one Br is replaced by a Cl, six isomers are possible.

Practice 19.5

Name these compounds:

(a) CH_2CH_3 (b)   OH   (c) NH_2

          Cl   Br   Br NO_2

19.12 Hydrocarbon Derivatives

Hydrocarbon derivatives are compounds that can be synthesized from a hydrocarbon. These derivatives contain not only carbon and hydrogen, but such additional elements as oxygen, nitrogen, or a halogen. The compounds in each class have similarities in structure and properties. We will consider the classes of hydrocarbon derivatives shown in Table 19.4, which is divided into two sections of different color. The compounds in the first section don't contain a C=O group in their molecules, while those in the second section all contain a C=O group. **carbonyl group**   A carbon atom double bonded to an oxygen atom is called a **carbonyl group.**

Table 19.4 Classes of Hydrocarbon Derivatives

Class	General formula	Structure of functional group	Sample structural formula	Name IUPAC	Common
Alkyl halide	R*X	—X X = F, Cl, Br, I	CH_3Cl CH_3CH_2Cl	chloromethane chloroethane	methyl chloride ethyl chloride
Alcohol	ROH	—OH	CH_3OH CH_3CH_2OH	methanol ethanol	methyl alcohol ethyl alcohol
Ether	R—O—R	R—O—R	CH_3—O—CH_3 CH_3CH_2—O—CH_2CH_3	methoxymethane ethoxyethane	dimethyl ether diethyl ether
Aldehyde	R—C=O (with H)	—C=O (with H)	H—C=O (with H) CH_3—C=O (with H)	methanal ethanal	formaldehyde acetaldehyde
Ketone	R—C—R (with O double bond)	R—C—R (with O double bond)	CH_3—C—CH_3 (with O) CH_3—C—CH_2CH_3 (with O)	propanone 2-butanone	acetone methyl ethyl ketone
Carboxylic acid	R—C(=O)OH	—C(=O)OH	HCOOH CH_3COOH	methanoic acid ethanoic acid	formic acid acetic acid
Ester	R—C(=O)OR	—C(=O)OR	$HCOOCH_3$ CH_3COOCH_3	methyl methanoate methyl ethanoate	methyl formate methyl acetate

*The letter "R" is used to indicate any of the many possible alkyl groups.

19.13 Alkyl Halides

Alkanes react with a halogen in ultraviolet light to produce an alkyl halide, R—X. *Halogenation* is a general term for the substitution of a halogen atom for a hydrogen atom.

$$RH + X_2 \xrightarrow[\text{light}]{\text{UV}} RX + HX \qquad (X = Cl \text{ or } Br)$$

When a specific halogen is used, the name reflects this. For example, when chlorine is the halogen, the process is called *chlorination*.

$$CH_3CH_3 + Cl_2 \xrightarrow[\text{light}]{\text{UV}} CH_3CH_2Cl + HCl$$
chloroethane

The reaction yields alkyl halides, RX, which are useful as intermediates for the manufacture of other substances.

A well-known reaction of methane and chlorine is shown by the equation

$$CH_4 + Cl_2 \xrightarrow[\text{light}]{\text{UV}} \underset{\text{chloromethane}}{CH_3Cl} + HCl$$

According to IUPAC rules, organic halides are named by giving the halogen substituent as a prefix of the parent alkane (e.g., CH_3Cl is chloromethane).

Alkyl halides are primarily used as industrial solvents. Carbon tetrachloride, CCl_4, was once used extensively in the dry-cleaning process, but it has been replaced because of its toxicity and carcinogenic effects. (See the Chemistry in Action on p. 532.)

Organic halides are also used as anesthetics. Chloroform, $CHCl_3$, was once a popular anesthetic. It's no longer used, however, because it is harmful to the respiratory system. Instead, such compounds as halothane ($CF_3CHClBr$) are being used.

Composed of carbon, hydrogen, fluorine, and chlorine, chlorinated fluoro-carbons (CFCs) are alkyl halides used in aerosol propellents and refrigerants. CFCs have been replaced by other compounds because they generate chlorine atoms in the upper atmosphere, which deplete the ozone layer (see Section 12.15). Many aerosols are now propelled by carbon dioxide. Different refrigerant molecules (called hydrofluorocarbons, HFCs) are being used in refrigerators and air conditioners.

19.14 Alcohols

alcohol   **Alcohols** are organic compounds whose molecules contain a hydroxyl ($-OH$) group covalently bonded to a saturated carbon atom. Thus if we substitute an $-OH$ for an H in CH_4, we get CH_3OH, methyl alcohol. The functional group of the alcohols is $-OH$. The general formula for alcohols is ROH, where R is an alkyl or a substituted alkyl group.

Alcohols differ from metal hydroxides in that they do not dissociate or ionize in water. The $-OH$ group is attached to the carbon atom by a covalent bond, not by an ionic bond. The alcohols form a homologous series, with methanol, CH_3OH, as the first member of the series. Models of the structural arrangements of the atoms in methanol and ethanol are shown in Figure 19.6.

primary alcohol   **Alcohols** are classified as **primary** (1°), **secondary** (2°), or **tertiary** (3°),
secondary alcohol   depending on whether the carbon atom to which the $-OH$ group is attached
tertiary alcohol   is bonded to one, two, or three other carbon atoms, respectively. Generalized formulas for 1°, 2°, and 3° alcohols are

Examples of these alcohols are listed in Table 19.5. Methanol, CH_3OH, is grouped with the primary alcohols.

Molecular structures with more than one $-OH$ group bonded to a single carbon atom are generally not stable. But an alcohol molecule can contain two or more $-OH$ groups if each $-OH$ is bonded to a different carbon atom.

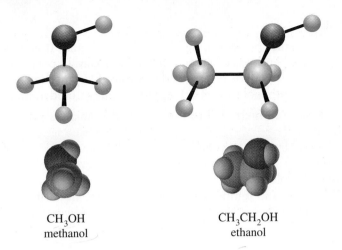

CH$_3$OH
methanol

CH$_3$CH$_2$OH
ethanol

Figure 19.6
Models illustrating structural
formulas of methanol and ethanol.

Accordingly, alcohols are also classified as monohydroxy, dihydroxy, trihydroxy, and so on, on the basis of the number of hydroxyl groups per molecule. **Polyhydroxy alcohol** is a general term for an alcohol that has more than one —OH group per molecule.

polyhydroxy alcohol

Table 19.5 Names and Classifications of Alcohols

Class	Formula	IUPAC name	Common name*	Boiling point (°C)
Primary	CH$_3$OH	methanol	Methyl alcohol	65.0
Primary	CH$_3$CH$_2$OH	ethanol	Ethyl alcohol	78.5
Primary	CH$_3$CH$_2$CH$_2$OH	1-propanol	*n*-propyl alcohol	97.4
Primary	CH$_3$CH$_2$CH$_2$CH$_2$OH	1-butanol	*n*-butyl alcohol	118
Primary	CH$_3$(CH$_2$)$_3$CH$_2$OH	1-pentanol	*n*-pentyl alcohol	138
Primary	CH$_3$(CH$_2$)$_6$CH$_2$OH	1-octanol	*n*-octyl alcohol	195
Primary	CH$_3$CHCH$_2$OH $\quad\mid$ $\quad$CH$_3$	2-methyl-1-propanol	isobutyl alcohol	108
Secondary	CH$_3$CHCH$_3$ $\quad\mid$ $\quad$OH	2-propanol	isopropyl alcohol	82.5
Secondary	CH$_3$CH$_2$CHCH$_3$ $\qquad\mid$ $\qquad$OH	2-butanol	*sec*-butyl alcohol	91.5
Tertiary	$\quad$CH$_3$ $\quad\mid$ CH$_3$—C—OH $\quad\mid$ $\quad$CH$_3$	2-methyl-2-propanol	*tert*-butyl alcohol	82.9
Dihydroxy	HOCH$_2$CH$_2$OH	1,2-ethanediol	ethylene glycol	197
Trihydroxy	HOCH$_2$CHCH$_2$OH $\qquad\mid$ $\qquad$OH	1,2,3-propanetriol	glycerol or glycerine	290

*The abbreviations *n, sec,* and *tert* stand for normal, secondary, and tertiary, respectively.

Methanol

Methanol is a volatile (bp 65°C), highly flammable liquid. It is poisonous and capable of causing blindness or death if taken internally. Exposure to methanol vapors for even short periods of time is dangerous. Despite the hazards, over 8 billion pounds of methanol (3.6×10^9 kg) are manufactured annually and used

- in the conversion of methanol to formaldehyde, which is primarily used to make polymers
- in the manufacture of other chemicals, especially various kinds of esters
- to denature ethyl alcohol (rendering it unfit as a beverage)
- as an industrial solvent
- as an inexpensive and temporary antifreeze for radiators (it isn't a satisfactory permanent antifreeze because its boiling point is lower than that of water)

Ethanol

Ethanol is without doubt the most widely known alcohol. Huge quantities of this substance are prepared by fermentation, using starch and sugar as the raw materials. Conversion of simple sugars to ethanol is accomplished by the yeast enzyme zymase:

$$C_6H_{12}O_6 \xrightarrow{\text{zymase}} 2\,CH_3CH_2OH + 2\,CO_2$$
$$\text{glucose} \qquad\qquad\qquad \text{ethanol}$$

Ethanol is economically significant as

- an intermediate in the manufacture of other chemicals such as acetaldehyde, acetic acid, ethyl acetate, and diethyl ether
- a solvent for many organic substances (e.g., pharmaceuticals, perfumes, flavorings)
- an ingredient in alcoholic beverages
- an alternative fuel for automobiles

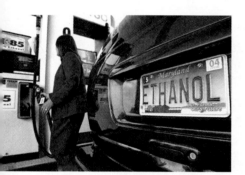

Ethanol is now an alternative fuel used in flexible-fuel vehicles.

Ethanol acts physiologically as a food, a drug, and a poison. It's a food in the limited sense that the body metabolizes small amounts of it to carbon dioxide and water with the production of energy. In moderate quantities, ethanol causes drowsiness and depresses brain functions so that activities requiring skill and judgment (such as automobile driving) are impaired. In larger quantities, it causes nausea, vomiting, impaired perception, and lack of coordination. If a very large amount is consumed, unconsciousness and ultimately death may occur.

For industrial use, ethanol is often denatured (rendered unfit for drinking) by adding small amounts of methanol and other denaturants that are extremely difficult to remove. Denaturing is required by the federal government to protect the beverage alcohol tax source. Special tax-free use permits are issued to scientific and industrial users who require pure ethanol for nonbeverage uses.

Other widely used alcohols include the following:

- Isopropyl alcohol (2-propanol) is the principal ingredient in rubbing alcohol formulations.
- Ethylene glycol (ethanediol) is the main compound in antifreezes; it is also used in the manufacture of synthetic fibers and in the paint industry.
- Glycerol (1,2,3-propanetriol), also known as glycerine, is a syrupy, sweet-tasting liquid used in the manufacture of polymers and explosives, as an emollient in cosmetics, as a humectant in tobacco, and as a sweetener.

19.15 Naming Alcohols

If you know how to name alkanes, it's easy to name alcohols by the IUPAC system. Unfortunately, several alcohols are also known by common names, so it's often necessary to know both names for a given alcohol. The common name is usually formed from the name of the alkyl group that is attached to the —OH group, followed by the word *alcohol*. (See Table 19.5.) To name an alcohol by the IUPAC system,

Step 1	Select the longest continuous chain of carbon atoms containing the hydroxyl group.
Step 2	Number the carbon atoms in this chain so that the one bonded to the —OH group has the lowest possible number.
Step 3	Form the parent alcohol name by replacing the final -*e* of the corresponding alkane name by -*ol*. When isomers are possible, locate the position of the —OH by placing the number (hyphenated) of the carbon atom to which the —OH is bonded immediately before the parent alcohol name.
Step 4	Name each alkyl branch chain (or other group) and designate its position by number.

Study the following examples of this naming system and also those shown in Table 19.5.

$$\overset{3}{C}H_3 - \overset{2}{C}H_2 - \overset{1}{C}H_2OH$$

1-propanol

$$\overset{1}{C}H_3 - \overset{2}{C}H - \overset{3}{C}H_3$$
$$\vert$$
$$OH$$

2-propanol
(isopropyl alcohol)

$$\overset{4}{C}H_3 - \overset{3}{C}H - \overset{2}{C}H_2 - \overset{1}{C}H_2OH$$
$$\vert$$
$$CH_3$$

3-methyl-1-butanol

$$\overset{2}{H}OCH_2 - \overset{1}{C}H_2OH$$

1,2-ethanediol
(ethylene glycol)

Name the following alcohol by the IUPAC system:

$$CH_3CH_2CHCH_2CHCH_3$$
$$\qquad\quad \vert \qquad\quad \vert$$
$$\qquad\quad CH_3 \quad\; OH$$

Example 19.6

Step 1	The longest continuous carbon chain containing the —OH group has six carbon atoms.
Step 2	This carbon chain is numbered from right to left so that the carbon bonded to the —OH group has the smallest possible number. In this case the —OH is on C-2:

$$\overset{6}{C} - \overset{5}{C} - \overset{4}{C} - \overset{3}{C} - \overset{2}{C} - \overset{1}{C}$$
$$\qquad\qquad \vert \qquad\quad \vert$$
$$\qquad\qquad CH_3 \quad\; OH$$

SOLUTION

Step 3   The name of the six-carbon alkane is hexane. Replace the final -e in hexane by -ol, forming the name hexanol. Since the —OH is on C-2, place a 2- before hexanol to give it the parent alcohol name of 2-hexanol.

Step 4   A methyl group, —CH$_3$, is located on C-4. Therefore, the full name of the compound is 4-methyl-2-hexanol.

Example 19.7   Write the structural formula of 3,3-dimethyl-2-hexanol.

SOLUTION   **Step 1**   The 2-hexanol refers to a six-carbon chain with an —OH group on C-2. Write the skeleton structure with an —OH on C-2. Now place the two methyl groups (3,3-dimethyl) on C-3:

$$
\begin{array}{cccccc}
1 & 2 & 3 & 4 & 5 & 6 \\
C-&C-&C-&C-&C-&C \\
 & | & & & & \\
 & OH & & & &
\end{array}
\qquad
\begin{array}{cccccc}
 & & CH_3 & & & \\
1 & 2 & 3| & 4 & 5 & 6 \\
C-&C-&C-&C-&C-&C \\
 & | & | & & & \\
 & HO & CH_3 & & &
\end{array}
$$

Step 2   Finally, add hydrogen atoms to give each carbon atom four bonds:

$$
\begin{array}{c}
CH_3 \\
| \\
CH_3CH-C-CH_2CH_2CH_3 \\
| \quad\quad | \\
HO \quad CH_3
\end{array}
$$

3,3-dimethyl-2-hexanol

Practice 19.6

Name these alcohols and classify each as primary, secondary, or tertiary:

(a) $CH_3CH_2CHCH_2CH_2CHCH_3$
        | |
        OH   CH$_3$

(b)
$$
\begin{array}{c}
OH \\
| \\
CH_3CCH_2CHCH_3 \\
| \quad\quad | \\
CH_3 \quad CH_3
\end{array}
$$

19.16 Ethers

ethanol

dimethyl ether

ether   **Ethers** have the general formula ROR′. The two groups, R and R′, may be derived from saturated, unsaturated, or aromatic hydrocarbons and, for a given ether, may be alike or different. Table 19.6 lists structural formulas and names for different ethers.

Saturated ethers have little chemical reactivity, but because they readily dissolve a great many organic substances, ethers are often used as solvents in both laboratory and manufacturing operations.

Alcohols (ROH) and ethers (ROR′) are isomeric, having the same molecular formula but different structural formulas. For example, the molecular formula for ethanol and dimethyl ether is C_2H_6O, but the structural formulas are

CH_3CH_2OH CH_3-O-CH_3
    ethanol   dimethyl ether

Table 19.6 Names and Structural Formulas of Ethers

Name*	Formula	Boiling point (°C)
Dimethyl ether (methoxymethane)	CH_3-O-CH_3	−24
Methyl ethyl ether (methoxyethane)	$CH_3CH_2-O-CH_3$	8
Diethyl ether (ethoxyethane)	$CH_3CH_2-O-CH_2CH_3$	35
Ethyl isopropyl ether (2-ethoxypropane)	$CH_3CH_2-O-CHCH_3$ $\quad\quad\quad\quad\quad\ |$ $\quad\quad\quad\quad\quad CH_3$	54
Divinyl ether	$CH_2=CH-O-CH=CH_2$	39
Anisole (methoxybenzene)	⬡ — OCH_3	154

*The IUPAC name is in parentheses.

These two molecules are extremely different in both physical and chemical properties. Ethanol boils at 78.3°C, and dimethyl ether boils at −23.7°C. Ethanol is capable of intermolecular hydrogen bonding and therefore has a much higher boiling point. In addition, it also has a greater solubility in water than dimethyl ether.

Naming Ethers

Like alcohols, individual ethers can have several names. Once widely used as an anesthetic, the ether with the formula $CH_3CH_2-O-CH_2CH_3$ is called diethyl ether, ethyl ether, ethoxyethane, or simply ether. Common names of ethers are formed from the names of the groups attached to the oxygen atom followed by the word *ether*:

$$CH_3 - \boxed{O} - CH_3$$
methyl ether methyl
dimethyl ether

$$CH_3 - \boxed{O} - CH_2CH_3$$
methyl ether ethyl
methyl ethyl ether

In the IUPAC system, ethers are named as alkoxy, RO—, derivatives of the alkane corresponding to the longest carbon–carbon chain in the molecule. To name an ether by this system, use the following steps:

Step 1 Select the longest carbon–carbon chain and label it with the name of the corresponding alkane.

Step 2 Change the *-yl* ending of the other hydrocarbon group to *-oxy* to obtain the alkoxy group name. For example, CH_3O- is called *methoxy*.

alkyl + oxy = alkoxy

Step 3   Combine the two names from Steps 1 and 2, giving the alkoxy name first, to form the ether name:

$$CH_3 - O - CH_2CH_3$$

This is the longest carbon–carbon chain, so call it *ethane*.

This is the other group; modify its name to *methoxy* and combine with ethane to obtain the IUPAC name of the ether, *methoxyethane*.

Thus,

$CH_3CH_2 - O - CH_2CH_3$ is ethoxyethane.

$CH_3CH_2CH_2 - O - CH_2CH_2CH_2CH_3$ is propoxybutane.

19.17   Aldehydes and Ketones

carbonyl group

aldehyde
ketone

The aldehydes and ketones are closely related classes of compounds. Their structures contain the **carbonyl group,** $\diagdown$C$=$O, a carbon double bonded to oxygen. **Aldehydes** have at least one hydrogen atom bonded to the carbonyl group, while **ketones** have two alkyl (R) or aromatic (Ar) groups bonded to the carbonyl group:

$$\underset{\text{aldehydes}}{R-\overset{\overset{\displaystyle O}{\|}}{C}-H \qquad Ar-\overset{\overset{\displaystyle O}{\|}}{C}-H} \qquad \underset{\text{ketones}}{R-\overset{\overset{\displaystyle O}{\|}}{C}-R \qquad R-\overset{\overset{\displaystyle O}{\|}}{C}-Ar \qquad Ar-\overset{\overset{\displaystyle O}{\|}}{C}-Ar}$$

In a linear expression, the aldehyde group is often written as CHO or CH$=$O. For example,

$$CH_3CHO \quad \text{is equivalent to} \quad CH_3\overset{\overset{\displaystyle O}{\|}}{C}-H$$

In the linear expression of a ketone, the carbonyl group is written as CO; thus

$$CH_3COCH_3 \quad \text{is equivalent to} \quad CH_3\overset{\overset{\displaystyle O}{\|}}{C}CH_3$$

formaldehyde

acetone

Formaldehyde is the most widely used aldehyde. It is a poisonous, irritating gas that is very soluble in water. Marketed as a 40% aqueous solution called *formalin,* the largest use of this chemical is in the manufacture of polymers. About 2.1 billion pounds (9.6×10^8 kg) of formaldehyde are manufactured annually in the United States. Formaldehyde vapors are intensely irritating to the mucous membranes, and ingestion may cause severe abdominal pains, leading to coma and death.

Acetone and methyl ethyl ketone are widely used organic solvents. Acetone, in particular, is used in very large quantities for this purpose. The U.S. production of acetone is about 1.9 billion pounds (8.7×10^8 kg) annually. It is used as a solvent in the manufacture of drugs, chemicals, and explosives; for removal of paints, varnishes, and fingernail polish; and as a solvent in the plastics industry. Methyl ethyl ketone (MEK) is also widely used as a solvent, especially for lacquers.

19.18 Naming Aldehydes and Ketones

Aldehydes

The IUPAC names of aldehydes are obtained by dropping the final -e and adding -al to the name of the parent hydrocarbon (i.e., the longest carbon–carbon chain carrying the —CHO group). The aldehyde carbon is always at the beginning of the carbon chain, is understood to be carbon number 1, and does not need to be numbered. The first member of the homologous series, $H_2C=O$, is methanal. The name *methanal* is derived from the hydrocarbon methane, which contains one carbon atom. The second member of the series is ethanal, the third member of the series is propanal, and so on:

$$CH_4 \qquad H-\overset{\overset{\displaystyle O}{\|}}{C}-H \qquad CH_3CH_3 \qquad CH_3\overset{\overset{\displaystyle O}{\|}}{C}-H$$

methane   methanal   ethane   ethanal
(from methane + -al)   (from ethane + -al)

The longest carbon chain containing the aldehyde group is the parent compound. Other groups attached to this chain are numbered and named as we have done previously. For example,

$$CH_3CH_2\overset{\overset{\displaystyle CH_3}{|}}{C}HCH_2CH_2\overset{\overset{\displaystyle O}{\|}}{C}-H$$

4-methylhexanal

Common names for some aldehydes are widely used. The common names for the aldehydes are derived from the common names of the carboxylic acids (see Table 19.7). The -ic acid or -oic acid ending of the acid name is dropped and is replaced with the suffix -aldehyde. Thus the name of the one-carbon acid, formic acid, becomes formaldehyde for the one-carbon aldehyde:

$$H-\overset{\overset{\displaystyle O}{\|}}{C}-OH \qquad H-\overset{\overset{\displaystyle O}{\|}}{C}-H \qquad CH_3\overset{\overset{\displaystyle O}{\|}}{C}-OH \qquad CH_3\overset{\overset{\displaystyle O}{\|}}{C}-H$$

formic acid   formaldehyde   acetic acid   acetaldehyde

Ketones

The IUPAC name of a ketone is derived from the name of the alkane corresponding to the longest carbon chain that contains the ketone–carbonyl group. The parent name is formed by changing the -e ending of the alkane to -one. If the chain is longer than four carbons, it's numbered so that the carbonyl carbon has the smallest number possible, and this number is prefixed to the name of the ketone. Other groups bonded to the parent chain are named and numbered as previously indicated for hydrocarbons and alcohols. For example,

$$CH_3-\overset{\overset{\displaystyle O}{\|}}{C}-CH_3 \qquad \overset{5}{C}H_3\overset{4}{C}H_2\overset{3}{C}H_2-\overset{\overset{\displaystyle O}{\|}}{\underset{2}{C}}-\overset{1}{C}H_3 \qquad \overset{1}{C}H_3\overset{2}{C}H_2-\overset{\overset{\displaystyle O}{\|}}{\underset{3}{C}}-\overset{4}{C}H\overset{5}{C}H_2\overset{6}{C}H_3$$

propanone   2-pentanone $\overset{\overset{\displaystyle |}{}}{C}H_3$

4-methyl-3-hexanone

Note that in 4-methyl-3-hexanone the carbon chain is numbered from left to right to give the ketone group the lowest possible number.

An alternative non-IUPAC method commonly used to name simple ketones is to list the names of the alkyl or aromatic groups attached to the carbonyl carbon together with the word *ketone*.

$$
\underset{\substack{\uparrow \\ \text{methyl} \qquad\qquad \text{methyl} \\ \text{(dimethyl ketone, acetone)}}}{CH_3 - \overset{\overset{\displaystyle O}{\|}}{C} - CH_3}
\qquad
\underset{\substack{\uparrow \quad\; \uparrow \qquad\quad \uparrow \\ \text{methyl ketone} \quad \text{ethyl} \\ \text{(methyl ethyl ketone, MEK)}}}{CH_3 - \overset{\overset{\displaystyle O}{\|}}{C} - CH_2CH_3}
\qquad
\underset{\substack{\uparrow \quad\;\; \uparrow \qquad \uparrow \\ \text{methyl ketone} \quad \text{isopropyl} \\ \text{(methyl isopropyl ketone)}}}{CH_3 - \overset{\overset{\displaystyle O}{\|}}{C} - \overset{\overset{\displaystyle CH_3}{|}}{C}HCH_3}
$$

Two of the most widely used ketones are commonly known by non-IUPAC names: Propanone is called acetone, and butanone is known as methyl ethyl ketone, or MEK.

Example 19.8 Write the formulas and the names for the straight-chain five- and six-carbon aldehydes.

SOLUTION The IUPAC names are based on the five- and six-carbon alkanes. Drop the *-e* of the alkane name and add the suffix *-al*. Pentane, C_5, becomes pentanal and hexane, C_6, becomes hexanal.

$$
\underset{\text{pentanal}}{CH_3CH_2CH_2CH_2\overset{\overset{\displaystyle O}{\|}}{C} - H}
\qquad\qquad
\underset{\text{hexanal}}{CH_3CH_2CH_2CH_2CH_2\overset{\overset{\displaystyle O}{\|}}{C} - H}
$$

Example 19.9 Give two names for these ketones:

$$
\text{(a)} \quad CH_3CH_2\overset{\overset{\displaystyle O}{\|}}{C}CH_2\overset{\overset{\displaystyle CH_3}{|}}{C}HCH_3
\qquad\qquad
\text{(b)} \quad CH_3CH_2CH_2\overset{\overset{\displaystyle O}{\|}}{C} - \bigcirc
$$

SOLUTION (a) The parent carbon chain that contains the carbonyl group has six carbons. Number this chain from the end nearest to the carbonyl group. The ketone group is on C-3, and a methyl group is on C-5. The six-carbon alkane is hexane. Drop the *-e* from hexane and add *-one* to give the parent name hexanone. Prefix the name hexanone with 3- to locate the ketone group and with 5-methyl- to locate the methyl group. The name is 5-methyl-3-hexanone. The common name is ethyl isobutyl ketone since the C=O has an ethyl group and an isobutyl group bonded to it.

(b) The longest chain has four carbons. The parent ketone name is butanone, derived by dropping the *-e* of butane and adding *-one*. The butanone has a phenyl group attached to C-1. The IUPAC name is therefore 1-phenyl-1-butanone. The common name is phenyl *n*-propyl ketone, since the C=O group has a phenyl and an *n*-propyl group bonded to it.

Practice 19.7

Write the IUPAC names for these molecules:

(a)
$$\text{H}-\overset{\displaystyle \overset{\text{O}}{\|}}{\text{C}}-\underset{\displaystyle \underset{\text{CH}_3}{|}}{\text{CH}}\text{CH}_2\underset{\displaystyle \underset{\text{CH}_2\text{CH}_3}{|}}{\text{CH}_2}$$

(b)
$$\text{CH}_3\text{CH}_2\text{CH}_2\underset{\displaystyle \underset{\text{CH}_3}{|}}{\text{CH}}\text{CH}_2\underset{\displaystyle \underset{\text{CH}_3}{|}}{\text{CH}}\overset{\displaystyle \overset{\text{O}}{\|}}{\text{C}}\text{CH}_3$$

19.19 Carboxylic Acids

Organic acids, known as **carboxylic acids,** are characterized by the functional group called a **carboxyl group.** The carboxyl group is represented in the following ways:

carboxylic acid
carboxyl group

$$-\overset{\displaystyle \overset{\text{O}}{\|}}{\text{C}}-\text{OH} \quad \text{or} \quad -\text{COOH} \quad \text{or} \quad -\text{CO}_2\text{H}$$

Open-chain carboxylic acids form a homologous series. The carboxyl group is always at the beginning of a carbon chain, and the carbon atom in this group is understood to be C-1 in naming the compound.

To name a carboxylic acid by the IUPAC system, first identify the longest carbon chain including the carboxyl group. Then form the acid name by dropping the *-e* from the corresponding parent hydrocarbon name and adding *-oic acid*. Thus, the names corresponding to the C-1, C-2, and C-3 acids are methanoic acid, ethanoic acid, and propanoic acid. These names are, of course, derived from methane, ethane, and propane:

CH_4	methane	HCOOH	methanoic acid
CH_3CH_3	ethane	CH_3COOH	ethanoic acid
$CH_3CH_2CH_3$	propane	CH_3CH_2COOH	propanoic acid

The IUPAC system is neither the only nor the most generally used method of naming acids. Organic acids are usually known by common names. Methanoic, ethanoic, and propanoic acids are commonly called formic, acetic, and propionic acids, respectively. These names usually refer to a natural source of the acid and are not really systematic. Formic acid was named from the Latin word *formica,* meaning "ant." This acid contributes to the stinging sensation of ant bites. Acetic acid is found in vinegar and is so named from the Latin word for vinegar. The name of butyric acid is derived from the Latin term for butter, since it is a constituent of butterfat. Many of the carboxylic acids, principally those having even numbers of carbon atoms ranging from 4 to about 20, exist in combined form in plant and animal fats. These acids are called *saturated fatty acids.* Table 19.7 lists the IUPAC and common names of the more important saturated acids.

The simplest aromatic acid is benzoic acid. *Ortho*-hydroxybenzoic acid is known as salicylic acid, the basis for many salicylate drugs such as aspirin. There are three methylbenzoic acids, known as *o-, m-,* and *p*-toluic acids.

The bark of a willow tree is a natural source of salicylic acid.

benzoic acid 　　salicylic acid 　　acetylsalicylic acid 　　p-toluic acid
　　　　　　　(o-hydroxybenzoic acid) 　　(aspirin)

Table 19.7 Formulas and Names of Saturated Carboxylic Acids

Formula	IUPAC name	Common name
HCOOH	methanoic acid	formic acid
CH_3COOH	ethanoic acid	acetic acid
CH_3CH_2COOH	propanoic acid	propionic acid
$CH_3(CH_2)_2COOH$	butanoic acid	butyric acid
$CH_3(CH_2)_3COOH$	pentanoic acid	valeric acid
$CH_3(CH_2)_4COOH$	hexanoic acid	caproic acid
$CH_3(CH_2)_6COOH$	octanoic acid	caprylic acid
$CH_3(CH_2)_8COOH$	decanoic acid	capric acid
$CH_3(CH_2)_{10}COOH$	dodecanoic acid	lauric acid
$CH_3(CH_2)_{12}COOH$	tetradecanoic acid	myristic acid
$CH_3(CH_2)_{14}COOH$	hexadecanoic acid	palmitic acid
$CH_3(CH_2)_{16}COOH$	octadecanoic acid	stearic acid
$CH_3(CH_2)_{18}COOH$	eicosanoic acid	arachidic acid

19.20 Esters

ester　Carboxylic acids react with alcohols in an acidic medium to form esters. **Esters** have the general formula RCOOR′ where R′ can be any type of saturated, unsaturated, or aromatic hydrocarbon group. The functional group of the ester is —COOR′:

ester 　　functional group of an ester

The reaction of acetic acid and ethyl alcohol is shown as an example of the group of reactions called *esterification*. In addition to the ester, a molecule of water is formed as a product.

acetic acid 　　ethyl alcohol 　　ethyl acetate
(ethanoic acid) 　　(ethanol) 　　(ethyl ethanoate)

Esters are alcohol derivatives of carboxylic acids. They are named in much the same way as salts. The alcohol part (R′ in OR′) is named first, followed by the name of the acid modified to end in -ate. The -ic ending of the organic acid name is replaced by the ending -ate. Thus in the IUPAC system, ethanoic acid becomes ethanoate. In the common names, acetic acid becomes acetate. To name an ester, be sure to recognize the portion of the ester molecule that comes from the acid and the portion that comes from the alcohol. In the general formula for an ester, the RC=O comes from the acid and the R′O comes from the alcohol:

$$
\underbrace{R-C}_{\text{acid}}\overset{\displaystyle O}{\underset{\|}{}}\underbrace{O-R'}_{\text{alcohol}} \qquad \underbrace{CH_3C}_{\substack{\text{acetic}\\\text{acid}}}\overset{\displaystyle O}{\underset{\|}{}}\underbrace{OCH_3}_{\substack{\text{methyl}\\\text{alcohol}}}
$$

methyl acetate
(methyl ethanoate)

Esters occur naturally in many varieties of plant life. They often have pleasant, fragrant, fruity odors and are used as flavoring and scenting agents. Esters are insoluble in water but soluble in alcohol. Table 19.8 lists selected esters.

Practice 19.8

Give two names for

(a) $CH_3CH_2\overset{\displaystyle O}{\underset{\|}{C}}-OCH_3$

(b) $CH_3CH_2CH_2\overset{\displaystyle O}{\underset{\|}{C}}-OCH_2CH_3$

(c) $CH_3(CH_2)_{10}\overset{\displaystyle O}{\underset{\|}{C}}-OCH_3$

Table 19.8 Odors and Flavors of Selected Esters

Formula	IUPAC name	Common name	Odor or flavor
$CH_3COOCH_2CH_2CHCH_3$ 　　　　　　　CH_3	isopentyl ethanoate	isoamyl acetate	banana, pear
$CH_3CH_2CH_2COOCH_2CH_3$	ethyl butanoate	ethyl butyrate	pineapple
$HCOOCH_2CHCH_3$ 　　　　　CH_3	isobutyl methanoate	isobutyl formate	raspberry
$CH_3COOCH_2(CH_2)_6CH_3$	octyl ethanoate	n-octyl acetate	orange
⬡—$COOCH_3$ ⬡—OH	methyl-2-hydroxy benzoate	methyl salicylate	wintergreen

19.21 Polymers—Macromolecules

Some very large molecules (macromolecules) exist in nature, containing tens of thousands of atoms. Some of these, such as starch, glycogen, cellulose, proteins, and DNA, have molar masses in the millions and are central to many of our life processes. Synthetic macromolecules touch every phase of modern living. Today it's hard to imagine a world without polymers. Textiles for clothing, carpeting, and draperies; shoes; toys; automobile parts; construction materials; synthetic rubber; chemical equipment; medical supplies; cooking utensils; synthetic leather; recreational equipment—the list goes on. The vast majority of these polymeric materials are based on petroleum. Petroleum is nonreplaceable; therefore our dependence on polymers is another good reason for not squandering our limited world supply of petroleum.

polymerization
polymer
monomer
copolymer

The process of forming very large, high-molar-mass molecules from smaller units is called **polymerization.** The large molecule, or unit, is called the **polymer** and the small repeating unit, the **monomer.** Polymers containing more than one kind of monomer are called **copolymers.** The term *polymer* is derived from the Greek word *polumerēs*, meaning "having many parts." Ethylene is a monomer and polyethylene is a polymer. Because of their large size, polymers are often called *macromolecules*. Some synthetic polymers are called *plastics*. The word *plastic* means "capable of being molded, or pliable."

Polyethylene is an example of a synthetic polymer. Ethylene, derived from petroleum, is made to react with itself to form polyethylene (or polythene).

Polyethylene is a long-chain hydrocarbon made from many ethylene units:

$$n\,CH_2{=}CH_2 \longrightarrow -CH_2CH_2[CH_2CH_2]_nCH_2CH_2CH_2CH_2-$$

ethylene   polyethylene

Synthetic polymers are used to form commercial products, such as curtains, carpets, and disposable diapers.

A typical polyethylene molecule is made up of about 2500–25,000 ethylene molecules joined in a continuous structure. Over 30 billion pounds of polyethylene are produced annually in the United States. Its uses are as varied as any single substance known and include chemical equipment, packaging material, electrical insulation, films, industrial protective clothing, and toys.

Ethylene derivatives, in which one or more hydrogen atoms have been replaced by other atoms or groups, can also be polymerized. Many of our commercial synthetic polymers are made from such modified ethylene monomers. For example, $CH_2{=}CH{-}$ is a vinyl group and $CH_2{=}CHCl$ is vinyl chloride. Vinyl chloride can be polymerized to polyvinylchloride (PVC). The names, structures, and some uses for several of these polymers are given in Table 19.9

Table 19.9 Polymers Derived from Modified Ethylene Monomers

Monomer	Polymer	Uses
$CH_2{=}CH_2$ ethylene	${-}(CH_2{-}CH_2{)}_{\overline{n}}$ polyethylene	Packing material, molded articles, containers, toys
$CH_2{=}CH$ $\vert$ CH_3 propylene	$\left({-}CH_2{-}CH{-}\atop \quad\ \vert\ \right)_{n}$ CH_3 polypropylene	Textile fibers, molded articles, lightweight ropes, autoclavable biological equipment
$CH_2{=}CH$ $\vert$ Cl vinyl chloride	$\left({-}CH_2{-}CH{-}\atop \quad\ \vert\ \right)_{n}$ Cl polyvinyl chloride	Garden hoses, pipes, molded articles, floor tile, electrical insulation, vinyl leather
$CH_2{=}CH$ $\vert$ CN acrylonitrile	$\left({-}CH_2{-}CH{-}\atop \quad\ \vert\ \right)_{n}$ CN Orlon, Acrilan	Textile fibers
$CF_2{=}CF_2$ tetrafluoroethylene	${-}(CF_2{-}CF_2{)}_{\overline{n}}$ Teflon	Gaskets, valves, insulation, heat-resistant and chemical-resistant coatings, linings for pots and pans
$CH_2{=}CH$ ⬡ styrene	$\left({-}CH_2{-}CH{-}\right)$ ⬡ $_n$ polystyrene	Molded articles, Styrofoam, insulation, toys, disposable food containers
$CH_2{=}C{-}CH_3$ $\qquad \vert$ $\qquad C{-}O{-}CH_3$ $\qquad \Vert$ $\qquad O$ methylmethacrylate	CH_3 $\left({-}CH_2{-}C{-}\atop\right)_{n}$ $O{=}C$ OCH_3 Lucite, Plexiglas (acrylic resins)	Contact lenses, clear sheets for windows and optical uses, molded articles, automobile finishes

Dry cleaning has been used for many years to clean clothes that can't be placed in water (such as silk, rayon, wool) without severe damage or shrinking. The dry-cleaning process involves no water. Instead, clothes are treated for stains and then washed in perchloroethylene (per-clene), an organic liquid. Unfortunately perclene is a health hazard and is classified as both an air pollutant and environmental contaminant. This means big trouble for small dry cleaners who now must dispose of it as hazardous waste and also worry about health risks to employees. Consequently, researchers are looking for a solvent that can replace perclene for dry cleaning.

One interesting alternative is to use carbon dioxide for dry cleaning. You may think of CO_2 as a greenhouse gas, the gas you breathe in and out, or the solid called dry ice. One thing's for sure—we don't often consider CO_2 as a liquid. Scientists have already started to consider liquid CO_2 as a dry-cleaning alternative. Why? There is no shortage of CO_2 in the world. It can be collected from the waste generated in industry.

No special disposal procedures are needed since it can be recycled and leaks would be nontoxic to humans and the environment in small quantities.

● ◑ ◐ CO_2

▭ Fats, oils, waxes

〰〰 Polymer detergent

Copolymers act as detergents, dissolving in CO_2 and attracting and trapping organic substances in a micelle.

The largest obstacle for the dry-cleaning process is that CO_2 doesn't dissolve most polymers, oils, waxes, and substances that get caught in clothes. Of course, neither does water; it requires the help of detergents. So scientists have now developed polymers that can be used as detergents in liquid CO_2. Joseph DeSimone (University of North Carolina at Chapel Hill) made detergent polymers called copolymers, which are really just two polymers joined together. One end is soluble in CO_2; the other attracts oils and waxes. The new detergent polymers form micelles that trap the oils and waxes within the micelle and dissolve in CO_2.

Now machines that look like bank vaults can clean 50–70 pounds of clothes per load with CO_2 and about 5 ounces of polymer detergent. One extra bonus—since the machine operates at room temperature, stains don't require pretreating! Leather and suede can also be cleaned using CO_2. Someday soon you may walk into your local dry-cleaning shop and get your clothes CO_2 cleaned!

Chapter 19 Review

19.1 The Beginnings of Organic Chemistry
KEY TERMS
Vital-force theory
Organic chemistry

- Originally organic compounds were thought to be formed by a "vital force."
- Organic chemistry today is the chemistry of carbon compounds:
  - This does not imply that the compounds come from living matter.
  - A few carbon compounds are considered to be inorganic since their chemistry matches the chemistry of inorganic compounds better.

19.2 The Carbon Atom
KEY TERM
Functional group

- Carbon is central to all organic compounds:
  - Usually forms four covalent bonds
  - Has tetrahedral structure
  - Can also form double bonds and triple bonds

19.3 Hydrocarbons
KEY TERMS
Hydrocarbon
Saturated hydrocarbon
Unsaturated hydrocarbon

- Hydrocarbons are composed of carbon and hydrogen:
  - Saturated—contain only single bonds
  - Unsaturated—contain a double or triple bond between carbon atoms

19.4 Alkanes
KEY TERMS
Alkane
Homologous series

- Saturated hydrocarbons.
- General formula is C_nH_{2n+2}.
- Most important reaction is combustion to form water, carbon dioxide, and heat.

19.5 Structural Formulas and Isomerism
KEY TERMS
Isomerism
Isomers

- Two or more compounds that have the same molecular formula but different structural arrangements are isomers.
- Formulas can be written to show all the bonds or condensed, using the groups attached to a carbon atom to save space.

19.6 Naming Alkanes
KEY TERM
Alkyl group

- An alkyl group has the general formula C_nH_{2n+1}.
- To name an alkane:
  - Find the longest continuous carbon chain.
  - Count the carbons in the parent chain starting from the end closest to a branch point.
  - Name each alkyl group and designate its position on the parent chain by a number.
  - If the same alkyl group appears multiple times, use a prefix on the alkyl name to indicate how many are in the compound.
  - If multiple different alkyl groups are on the parent chain, list them alphabetically (prefixes do not count).

19.7 Alkenes and Alkynes
KEY TERMS
Alkene
Alkyne

- Alkenes:
  - Contain a carbon–carbon double bond
  - General formula: C_nH_{2n}
- Alkynes:
  - Contain a carbon–carbon triple bond
  - General formula: C_nH_{2n-2}

19.8 Naming Alkenes and Alkynes
- To name an alkene or alkyne:
  - Find the longest continuous carbon chain containing a double or triple carbon–carbon bond.

- Number of the carbon atoms in the parent chain starting with the end closest to the double or triple bond and name the parent chain as you would for an alkane, but change the ending to -ene or -yne.
- Name each alkyl group and designate its position on the parent chain by a number.
- If the same alkyl group appears multiple times, use a prefix on the alkyl name to indicate how many are in the compound.
- If multiple different alkyl groups are on the parent chain, list them alphabetically (prefixes do not count).

19.9 Reactions of Alkenes
KEY TERM
Addition reaction

- The most common reaction for alkenes and alkynes is addition:
  - Occurs at the carbon–carbon double or triple bond
  - Hydrogen, halogens, hydrogen halides, or water can be added across the double bond.
  - Complete addition results in a saturated compound.

19.10 Aromatic Hydrocarbons
KEY TERM
Aromatic compound

- Compounds with structures and chemical properties resembling benzene are called aromatic compounds.

19.11 Naming Aromatic Compounds
- Substituted benzenes have the general formula of C_6H_5G.
- Three major classes:
  - Monosubstituted:
    - Named by adding substituent name as prefixes to *benzene*.
  - Disubstituted:
    - *Ortho-*, *meta-*, *para-* used as prefixes for the substituent groups to show the position of the substituents.
    - Alternatively, numbers can be used to show the positions.
    - They are named by adding substituent names as a prefix to *benzene*.
  - Tri- and polysubstituted:
    - Atoms in the benzene ring are numbered and the substituents start on the C-1 atom.
    - They are named by adding substituent names as prefixes to *benzene*.

19.12 Hydrocarbon Derivatives
KEY TERM
Carbonyl group

- These are compounds that can be synthesized from a hydrocarbon.
- They can contain additional elements such as O, N, or a halogen.
- See Table 19.4 for a list of derivatives.

19.13 Alkyl Halides

- $RH + X_2 \xrightarrow[\text{light}]{\text{UV}} RX + HX$ (X = Cl or Br)

19.14 Alcohols

KEY TERMS

Alcohol
Primary alcohol
Secondary alcohol
Tertiary alcohol
Polyhydroxy alcohol

- Compounds that contain a hydroxyl group bonded to a saturated carbon atom
- Classification:
  - Primary (1°)
  - Secondary (2°)
  - Tertialry (3°)

19.15 Naming Alcohols

- To name an alcohol:
  - Select the longest continuous chain of C atoms containing the alcohol group.
  - Number the C atoms in this chain so the —OH group has the lowest number.
  - Form the parent alcohol name by replacing the -e on *alkane* with -ol.
  - Name any alkyl branches and show their positions by the corresponding numbers.

19.16 Ethers

KEY TERM

Ether

- Ethers are isomers of alcohols.
- Naming ethers:
  - Select the longest continuous chain of C atoms and name it as the corresponding alkane.
  - Change the -yl ending on the other hydrocarbon group to -oxy to get the alkoxy group name.
  - Combine the two names, giving the alkoxy name first, to form the ether name.

19.17 Aldehydes and Ketones

KEY TERMS

Carbonyl group
Aldehyde
Ketone

- Aldehydes and ketones both contain a carbonyl group:
  - Aldehydes have a hydrogen bonded to the carbonyl.
  - Ketones have two alkyl or aromatic groups bonded to the carbonyl carbon atom.

19.18 Naming Aldehydes and Ketones

- Naming aldehydes:
  - Drop the final -e and add -al to the name of the parent hydrocarbon.
- Naming ketones:
  - Name the alkane corresponding to the longest chain containing the ketone group by changing the -e to -one.
  - If the chain is longer than four C atoms, make the carbonyl the lowest number possible.
  - Name groups attached to the parent chain as for hydrocarbons and alcohols.

19.19 Carboxyl Acid

KEY TERMS

Carboxyl group
Carboxyl acid

- The carboxyl group is represented in the following ways:

$$-\overset{\overset{\displaystyle O}{\|}}{C}-OH \qquad -COOH \qquad -CO_2H$$

- To name a carboxylic acid:
  - Find the longest carbon chain including the carboxyl group.
  - Drop the -e from the corresponding parent hydrocarbon name and add -oic acid.
- Other systems are used to name carboxylic acids.

19.20 Esters

KEY TERM

Ester

- Carboxylic acids react with alcohols to form esters.
- General formula RCOOR'
- Esters are named like salts:
  - Alcohol part is named first.
  - Name of the acid is given second, with the acid ending modified to -ate.

methyl acetate
(methyl ethanoate)

19.21 Polymers—Macromolecules

KEY TERMS

Polymerization
Polymer
Monomer
Copolymer

- Polymers are called macromolecules.
- Polymers are composed of small repeating units called monomers or copolymers.

Review Questions

All questions with blue numbers have answers in the appendix of the text.

1. What bonding characteristic of carbon is primarily responsible for the existence of so many organic compounds?

2. What is the most common geometric arrangement of covalent carbon bonds? Illustrate this structure.

3. In addition to single bonds, what other types of bonds can carbon atoms form? Give examples.

4. Draw a Lewis structure for
   (a) a single carbon atom
   (b) molecules of methane, ethene, and ethyne

5. Write the names and draw the structural formulas for the first 10 normal alkanes.

6. Write the names and draw the structural formulas for all possible alkyl groups, C_nH_{2n+1}, containing from one to four carbon atoms.

7. Which one of these compounds belongs to a different homologous series than the others?
   (a) C_2H_4 (c) C_6H_{14}
   (b) CH_4 (d) C_5H_{12}

8. Which word does not belong with the others?
   (a) alkane   (d) ethane
   (b) paraffin   (e) ethylene
   (c) saturated   (f) pentane

9. What is the single most important reaction of alkanes?

10. Draw the structure for vinyl acetylene, which has the formula C_4H_4 and contains one double bond and one triple bond.

11. An open-chain hydrocarbon has the formula C_6H_8. What possible combinations of carbon–carbon double bonds and/or carbon–carbon triple bonds can be in this compound?

12. Why is ethylene glycol (1,2-ethanediol) superior to methyl alcohol (methanol) as an antifreeze for automobile radiators?

13. Alcohols are considered to be toxic to the human body—with ethanol, the alcohol in alcoholic beverages, being the least toxic one. What are the hazards of ingesting (a) methanol and (b) ethanol?

14. What is wrong with the name 1-methylpentane?

Paired Exercises

All exercises with blue numbers have answers in the appendix of the text.

1. Which of the following are organic compounds?
   (a) CH_3CH_3
   (b) NH_3
   (c) CH_3SCH_3
   (d) CH_3CH_2OH
   (e) $Mg(OH)_2$

2. Which of the following are organic compounds?
   (a) H_2SO_4
   (b) CH_3NH_2
   (c) Na_2CO_3
   (d) $CH_3CH_2CH_2CH_2CH_3$
   (e) $CH_3CH_2OCH_3$

3. For each organic compound in Question 1, explain why it is classified as an organic compound.

4. For each organic compound in Question 2, explain why it is classified as an organic compound.

5. Name these normal alkyl groups:
   (a) $C_5H_{11}-$
   (b) $C_7H_{15}-$

6. Name these normal alkyl groups:
   (a) $C_8H_{17}-$
   (b) $C_{10}H_{21}-$

7. Write condensed structural formulas for the five isomers of hexane.

8. Write condensed structural formulas for the nine isomers of heptane.

9. Give IUPAC names for the following:
   (a) CH_3CH_2Cl
   (b) $CH_3CHClCH_3$
   (c) $(CH_3)_2CHCH_2Cl$

10. Give IUPAC names for the following:
    (a) $CH_3CH_2CH_2Cl$
    (b) $(CH_3)_3CCl$
    (c) $CH_3CHClCH_2CH_3$

11. Give IUPAC names for these compounds:
    (a) CH$_3$CHCH$_2$CHCH$_2$CH$_3$
            | |
            CH$_3$ CH$_2$CH$_3$

    (b) CH$_3$CHCH$_2$CHCH$_2$—CHCH$_2$CH$_3$
            | | |
            CH$_3$ CH$_2$CH$_3$ CH$_2$CH$_3$

12. Give IUPAC names for these compounds:
    (a) CH$_3$CHCH$_2$CH$_2$CHCH$_3$
            | |
            CH$_2$ CH$_2$
            | |
            CH$_3$ CH$_3$

    (b) CH$_3$CH$_2$CH$_2$CHCH$_2$CH$_3$
                        |
                        CH$_3$CHCH$_3$

13. Draw structural formulas of these compounds:
(a) 2,4-dimethylpentane
(b) 2,2-dimethylpentane
(c) 3-isopropyloctane

14. Draw structural formulas of these compounds:
(a) 4-ethyl-2-methylhexane
(b) 4-*tert*-butylheptane
(c) 4-ethyl-7-isopropyl-2,4,8-trimethyldecane

15. One name in each pair is incorrect. Draw structures corresponding to each name and indicate which name is incorrect.
(a) 2-methylbutane and 3-methylbutane
(b) 2-ethylbutane and 3-methylpentane

16. One name in each pair is incorrect. Draw structures corresponding to each name and indicate which name is incorrect.
(a) 2-dimethylbutane and 2,2-dimethylbutane
(b) 2,4-dimethylhexane and 2-ethyl-4-methyl-pentane

17. Draw structural formulas for all the isomers of
(a) CH_3Br
(b) CH_2Cl_2
(c) C_2H_5Cl
(d) C_3H_7Br

18. Draw structural formulas for all the isomers of
(a) C_4H_9I
(b) $C_3H_6Cl_2$
(c) C_3H_6BrCl
(d) $C_4H_8Cl_2$

19. Using alkenes and any other necessary inorganic reagents, write reactions showing the formation of
(a) $CH_3CH_2CHClCH_2Cl$
(b) CH_3CH_2Br

20. Using alkenes and any other necessary inorganic reagents, write reactions showing the formation of
(a) $CH_3CHBrCH_2Br$
(b) $CH_3CHBrCH_3$

21. Draw structures containing two carbon atoms for the following classes of compounds:
(a) alkene
(b) alkyne
(c) alkyl halide
(d) alcohol

22. Draw structures containing two carbon atoms for the following classes of compounds:
(a) ether
(b) aldehyde
(c) carboxylic acid
(d) ester

23. Give the IUPAC name for each compound in Exercise 21.

24. Give the IUPAC name for each compound in Exercise 22.

25. Draw structural formulas for the following:
(a) chloromethane
(b) vinyl chloride
(c) chloroform
(d) 1,1-dibromoethene

26. Draw structural formulas for the following:
(a) hexachloroethane
(b) iodoethyne
(c) 6-bromo-3-methyl-3-hexene-1-yne
(d) 1,2-dibromoethene

27. Draw structural formulas for the following:
(a) 2,5-dimethyl-3-hexene
(b) 2-ethyl-3-methyl-1-pentene
(c) 4-methyl-2-pentene

28. Draw structural formulas for the following:
(a) 1,2-diphenylethene
(b) 3-pentene-1-yne
(c) 3-phenyl-1-butyne

29. Name these compounds:

(a) $\underset{\displaystyle \quad\ \ |}{CH_3CH=CCH_2CH_2CH_3}$
$\quad\quad\quad CH_3$

(b) $\underset{\displaystyle |\qquad\ |}{CH_3C=C-CH_3}$
$\quad\ \ CH_3\ \ CH_3$

30. Name these compounds:

(a) $\underset{\displaystyle \quad\quad |}{CH_3CH_2CHCH=CH_2}$
$\quad\quad\quad CH$
$\quad\quad H_3C\quad CH_3$

(b) $\underset{\displaystyle \qquad\quad\ \ |}{CH_3CH_2CH=CCH_2CH_3}$
$\quad\quad\quad\quad CH_3$

31. Complete these reactions and name the products:
(a) $CH_2{=}CHCH_3 + Br_2 \longrightarrow$
(b) $CH_2{=}CH_2 + HBr \longrightarrow$
(c) $CH_3CH{=}CHCH_3 + H_2 \xrightarrow[\text{1 atm}]{\text{Pt 25°C}}$

32. Complete these reactions and name the products:
(a) $CH_2{=}CH_2 + H_2O \xrightarrow{H^+}$
(b) $CH{\equiv}CH + 2Br_2 \longrightarrow$
(c) $CH_2{=}CH_2 + H_2 \xrightarrow[\text{1 atm}]{\text{Pt 25°C}}$

33. Name these compounds:

(a) OH

(b) CH_3

(c) COOH

(d) NH_2

(e) Cl, Cl

(f) Cl, Cl

34. Name these compounds:

(a) Cl, Cl

(b) OH, NO_2

(c) CH_3, Br, Br, Br

(d) CH_2CH_3

(e) CH_3, OH

(f) OH, $C=O$, CH_3, NO_2

35. Draw structural formulas for
(a) benzene
(b) toluene
(c) benzoic acid
(d) aniline

36. Draw structural formulas for
(a) phenol
(b) *o*-bromochlorobenzene
(c) 1,3-dichloro-5-nitrobenzene
(d) *m*-dinitrobenzene

37. Draw structural formulas for
(a) ethylbenzene
(b) 1,3,5-tribromobenzene

38. Draw structural formulas for
(a) *tert*-butylbenzene
(b) 1,1-diphenylethane

39. Draw structural formulas and write the IUPAC names for all isomers of trichlorobenzene ($C_6H_3Cl_3$).

40. Draw structural formulas and write the IUPAC names for all isomers of dichlorobromobenzene ($C_6H_3Cl_2Br$).

41. Write the IUPAC name for these alcohols. Classify each as primary, secondary, or tertiary alcohols.

(a) $CH_3CH_2CHCH_3$
      |
      OH

(b) $CH_3CHCH_2CH_2CHOH$
      | |
      CH_3 CH_3

42. Write the IUPAC name for these alcohols. Classify each as primary, secondary, or tertiary alcohols.

(a) $CH_3CHCHCH_2CH_2CHCH_3$
            |
            OH
      | |
      CH_2CH_3 CH_2CH_3

(b) $CH_2CH_2CHCH_2CH_2CH_3$
      | |
      OH $CH—CH_3$
             |
             CH_3

43. Draw structural formulas for
(a) 2-pentanol
(b) isopropyl alcohol

44. Draw structural formulas for
(a) 2,2-dimethyl-1-heptanol
(b) 1,3 propanediol

45. Name these aldehydes:

(a) $CH_2=O$

(b) $CH_3CH_2CH_2\overset{O}{\overset{\|}{C}}—H$

(c) $CH_3CHCH_2\overset{O}{\overset{\|}{C}}—H$
      |
      CH_3

46. Name these aldehydes:

(a) $\overset{O}{\overset{\|}{C}}—H$

(b) $O=CCH_2CH_2\overset{O}{\overset{\|}{C}}—H$
      |
      H

(c) $CH_3CHCH_2\overset{O}{\overset{\|}{C}}—H$
      |
      OH

47. Name these ketones:

(a) CH_3COCH_3

(b) $CH_3CH_2COCH_3$

(c)

48. Name these ketones:

(a)

(b)

(c)

49. Name these acids:

(a) $CH_3CHBrCOOH$

(b) $CH_2\!=\!CHCH_2COOH$

(c) $CH_3CH_2CH_2COOH$

(d)

50. Name these acids:

(a)

(b)

(c) $CH_3CH_2CH_2CH_2COOH$

(d)

51. Draw structural formulas for these esters:
(a) ethyl formate
(b) methyl ethanoate
(c) isopropyl propanoate

52. Draw structural formulas for these esters:
(a) *n*-nonyl acetate
(b) ethyl benzoate
(c) methyl salicylate

53. Write IUPAC names for these esters:

(a)

(b)

(c)

54. Write IUPAC names for these esters:

(a)

(b)

(c)

55. Complete these equations:

(a) $CH_3COOH + NaOH \longrightarrow$

(b)

56. Complete these equations:

(a) $CH_3COOH + KOH \longrightarrow$

(b)

57. Ethylene and its derivatives are the most common monomers for polymers. Write formulas for these ethylene-based polymers:
(a) polyethylene
(b) polyvinyl chloride

58. Ethylene and its derivatives are the most common monomers for polymers. Write formulas for these ethylene-based polymers:
(a) polyacrylonitrile
(b) Teflon

59. For the following structures, determine whether the given name is correct. If it is incorrect, provide the correct name for the structure.

(a) 3-methylbutane

$$CH_3$$
$$CH_3CHCH_2CH_3$$

(b) *ortho*-dibromobenzene

Br

Br

(c) 2-methyl-1-butyne

$$HC{\equiv}CCHCH_3$$
$$CH_3$$

(d) pentanol

$$OH$$
$$CH_3CH_2CHCH_2CH_3$$

(e) butanoic acid

$$O$$
$$CH_3CH_2CH_2C{-}OH$$

60. For the following structures, determine whether the given name is correct. If it is incorrect, provide the correct name for the structure.

(a) 2-chlorobutane

$$Cl$$
$$CH_3CH_2CHCH_3$$

(b) 2-butaneketone

$$O$$
$$CH_3CCH_2CH_3$$

(c) 2-methyl-3-pentene

$$CH_3CH{=}CHCHCH_3$$
$$CH_3$$

(d) propyl methanoate

$$O$$
$$CH_3CH_2COCH_3$$

(e) 2-ethylbutane

$$CH_3CH_2CHCH_3$$
$$CH_2CH_3$$

61. Classify the structures in Exercise 59 as one of the following: hydrocarbon, alkene, alkyne, aromatic, alkyl halide, alcohol, ester, carboxylic acid.

62. Classify the structures in Exercise 60 as one of the following: hydrocarbon, alkene, alkyne, ether, alkyl halide, ester, aldehyde, ketone.

Additional Exercises

All exercises with blue *numbers have answers in the appendix of the text.*

63. Why do small alkanes have low boiling points?

64. Draw all possible structures for:
(a) C_6H_{14}
(b) C_5H_{10}
(c) C_7H_{16}

65. Draw the structural formulas and write the IUPAC names for all isomers of
(a) pentyne C_5H_8,
(b) hexyne, C_6H_{10}.

66. Eight open-chain isomeric alcohols have the formula $C_5H_{11}OH$.
(a) Draw the structural formula and write the IUPAC name for each of these alcohols.
(b) Indicate which of these isomers are primary, secondary, and tertiary alcohols.

67. What is the molar mass of an open-chain saturated alcohol containing 30 carbon atoms? This alcohol is present in beeswax as an ester.

68. Write the common name and structure of the ether that is isomeric with
(a) 1-propanol
(b) ethanol
(c) isopropyl alcohol.

69. Give the IUPAC and common names and structures of all ethers having the molecular formula $C_5H_{12}O$.

70. Draw structural formulas for propanal and propanone. From these formulas, do you think that aldehydes and ketones are isomeric with each other? Show evidence and substantiate your answer by testing with a four-carbon aldehyde and ketone.

71. Give the common and IUPAC names for the first five straight-chain carboxylic acids.

72. Draw the structural formulas and write the IUPAC names for all the isomers of hexanoic acid, $CH_3(CH_2)_4COOH$.

73. Write equations for the preparation of these esters:
(a) ethyl formate
(b) methyl propanoate
(c) *n*-propyl benzoate

74. (a) Draw a structural formula showing the polymer that can be formed from the following monomers (show four units): (1) propylene, (2) 1-butene, (3) 2-butene
(b) How many ethylene units are in a polyethylene molecule that has a molar mass of 35,000?

75. How many different dibromobenzenes are there? How many tribromobenzenes? Show structures.

76. A compound has the following composition:

C = 24.3%

H = 4.1%

Cl = 71.7%

If 140.3 mL of the vapor at 100°C and 740 mm Hg pressure have a mass of 0.442 g, what is the molecular formula of the compound? Draw possible structures of the compound.

77. A compound was found to contain 24 g of carbon, 4 g of hydrogen, and 32 g of oxygen. The molar mass of the compound is 60. Which of these structures is correct for that compound?
(a) C_2H_5OH
(b) CH_3CHO
(c) CH_3OCH_3
(d) COH
(e) CH_3COOH

78. Write formulas for
(a) ethyl alcohol
(b) iodomethane
(c) 2-chloropentane
(d) decane
(e) *tert*-butyl alcohol

79. Write equations for the reactions in acid solution between
(a) methanol and formic acid
(b) 1-butanol and butanoic acid
(c) 1-hexanol and hexanoic acid

Name the esters produced in each case.

80. Draw the structural formula for
(a) a six-carbon alkane with methyl branches on two adjacent carbons
(b) a six-carbon alcohol with an —OH group not at the end of the chain
(c) a three-carbon organic (carboxylic) acid
(d) a four-carbon aldehyde
(e) the ester formed when a four-carbon alcohol reacts with a three-carbon acid

Answers to Practice Exercises

19.1 (a) $CH_3CH_2CH_2CH_2CH_2CH_3$

(b) $CH_3CH_2CH_2CHCH_3$
              $|$
             CH_3

              CH_3
              $|$
(c) $CH_3CH_2CHCH_2CH_3$

              CH_3 CH_3
              $|$ $|$
(d) $CH_3CH_2CCH_3$ $CH_3CHCHCH_3$
              $|$ $|$
             CH_3 CH_3

19.2 (a) 2,3,3-trimethylhexane
(b) 3-ethyl-2,5-dimethylheptane

19.3
              CH_3 CH_3
              $|$ $|$
(a) $CH_3CH_2CH— CH—CH_2CH_2CH_3$

         CH_3 CH_3 CH_3
         $|$ $|$ $|$
(b) $CH_3C—C—CCH_3$
         $|$ $|$ $|$
         CH_3 CH_3 CH_3

                          CH_3CHCH_3
                          $|$
(c) $CH_3CH_2—CH—CH_2CH_2CH_2CH_3$
                    $|$
                 CH_2CH_3

19.4 (a) 2,5-dimethyl-3-heptene
(b) 4-ethyl-2-octyne

19.5 (a) *m*-chloroethylbenzene
(b) 3,5-dibromophenol
(c) *o*-nitroaniline

19.6 (a) secondary, 6-methyl-3-heptanol
(b) tertiary, 2,4-dimethyl-2-pentanol

19.7 (a) 2-methylhexanal
(b) 3,5-dimethyl-2-octanone

19.8 (a) methyl propanoate
methyl propionate
(b) ethyl butanoate
ethyl butyrate
(c) methyl dodecanoate
methyl laurate

Introduction to Biochemistry

The energy to vacation, breathe, and live our busy lives comes from chemical reactions performed by biomolecules.

Chapter Outline

20.1 Chemistry in Living Organisms

20.2 Carbohydrates

20.3 Lipids

20.4 Amino Acids and Proteins

20.5 Enzymes

20.6 Nucleic Acids

20.7 DNA and Genetics

The study of life has long fascinated us—it's probably the most intriguing of all scientific studies, but the answer to our question "What is life?" still eludes us. The chemical bases for certain fundamental biological processes are well understood; we know how sunlight and carbon dioxide are converted into food. We can pinpoint specific genes and identify their function. We even use DNA to "fingerprint" suspects at crime scenes. But still we search for an answer to "What is life?"

Common molecular principles underscore the diversity of life, which enable us to produce essential substances, such as insulin for diabetics, using bacteria as the manufacturing machine. Other substances, such as the human growth hormone, are being genetically engineered as well.

Biochemistry is more important than ever to the practice of medicine and the study of nutrition. Today we routinely measure the severity of a heart attack by the levels of certain enzymes in the blood. The roles of cholesterol, fats, and trace elements in our diet are current topics of great interest.

20.1 Chemistry in Living Organisms

Chemical substances present in living organisms—from microbes to humans—range in complexity from water and simple salts to DNA (deoxyribonucleic acid) molecules containing tens of thousands of atoms. Four of the chemical elements, hydrogen, carbon, nitrogen, and oxygen, make up approximately 95% of the mass of living matter. Small amounts of sulfur, phosphorus, calcium, sodium, potassium, chlorine, magnesium, and iron, together with trace amounts of many other elements such as copper, manganese, zinc, cobalt, and iodine, are also found in living organisms. The human body consists of about 60% water, with some tissues having a water content as high as 80%.

biochemistry   **Biochemistry** is the branch of chemistry concerned with the chemical reactions occurring in living organisms. Its scope includes such processes as growth, digestion, metabolism, and reproduction.

The four major classes of biomolecules upon which all life depends are carbohydrates, lipids, proteins, and nucleic acids. Each kind of living organism has an amazing ability to select and synthesize a large portion of the many complicated molecules needed for its existence. In fact, the processes carried out in a living organism are similar to those of a highly automated, smoothly running chemical factory. But unlike chemical factories, living organisms are able to expand (grow), repair damage (if not too severe), and, finally, reproduce themselves.

20.2 Carbohydrates

carbohydrate   Chemically, **carbohydrates** are polyhydroxy aldehydes or polyhydroxy ketones or substances that yield these compounds when hydrolyzed. The name *carbohydrate* was given to this class of compounds many years ago by French scientists who called them *hydrates de carbone* because their empirical formulas approximated $(C \cdot H_2O)_n$. However, the hydrogen and oxygen do not actually exist as water or in hydrate form, as we have seen in such compounds as $BaCl_2 \cdot 2 H_2O$. Empirical formulas used to represent carbohydrates are $C_x(H_2O)_y$ and $(CH_2O)_n$.

Carbohydrates, also known as saccharides, occur naturally in plants and are one of the three principal classes of animal food. The other two classes of foods are lipids (fats) and proteins. Plants are able to synthesize carbohydrates by the photosynthetic process. Animals are incapable of this synthesis and are dependent on the plant kingdom for their source of carbohydrates. Animals are, however, capable of converting plant carbohydrate into glycogen (animal carbohydrate) and storing this glycogen throughout the body as a reserve energy source. The amount of energy available from carbohydrates is about 17 kJ/g (4 kcal/g).

Carbohydrates exist as sugars, starches, and cellulose. The simplest of these are the sugars. The common names for sugars end in -ose (e.g., glucose, sucrose, maltose). Carbohydrates are classified as monosaccharides, disaccharides, oligosaccharides, and polysaccharides according to the number of monosaccharide units linked in a molecule.

Monosaccharides

Monosaccharides are carbohydrates that cannot be hydrolyzed to simpler carbohydrate units. They are often called simple sugars, the most common of which is glucose. Monosaccharides containing three-, four-, five-, and six-carbon atoms are called *trioses*, *tetroses*, *pentoses*, and *hexoses*, respectively. Monosaccharides that contain an aldehyde group on one carbon atom and a hydroxyl group on each of the other carbon atoms are called *aldoses*. *Ketoses* are monosaccharides that contain a ketone group on one carbon atom and a hydroxyl group on each of the other carbons.

The carbohydrates in these hay rolls are storage molecules for energy from the sun.

monosaccharide

triose $C_3H_6O_3$
tetrose $C_4H_8O_4$
pentose $C_5H_{10}O_5$
hexose $C_6H_{12}O_6$

aldehyde   hydroxyl   ketone

We can write structural formulas for many monosaccharides, but only a limited number are of biological importance. Sixteen different isomeric aldohexoses of formula $C_6H_{12}O_6$ are known; glucose and galactose are the most important of these. Most sugars exist predominantly in a cyclic structure, forming a five- or six-membered ring. For example, glucose exists as a ring in which C-1 is bonded through an oxygen atom to C-5:

$$H-C=O$$
$$H-C-OH$$
$$H-C-OH$$
$$CH_2OH$$

erythrose
(an aldotetrose)

open-chain form
of glucose

or

cyclic form of glucose

Table 20.1	Relative Sweetness of Sugars
Fructose	100
Sucrose	58
Glucose	43
Maltose	19
Galactose	19
Lactose	9.2
Invert sugar	75

Glucose Glucose ($C_6H_{12}O_6$) is the most important of the monosaccharides. An aldohexose found in the free state in plants and animal tissues, glucose is commonly known as *dextrose* or *grape sugar* and is a component of the disaccharides sucrose, maltose, and lactose; it is also the monomer of the polysaccharides starch, cellulose, and glycogen. Among the common sugars, glucose is of intermediate sweetness (see Table 20.1).

Glucose is the key sugar of the body and is carried by the bloodstream to all body parts. The concentration of glucose in the blood is normally 80–100 mg per 100 mL of blood. Because glucose is the most abundant carbohydrate in the blood, it is also sometimes known as *blood sugar*. Glucose requires no digestion and therefore may be given intravenously to patients who cannot take food by mouth. Glucose is found in the urine of those who have diabetes mellitus (sugar diabetes), a condition called *glycosuria*.

Galactose Galactose ($C_6H_{12}O_6$) is also an aldohexose and occurs along with glucose in the disaccharide lactose and in many oligo- and polysaccharides, such as pectin, gums, and mucilages. It is an isomer of glucose, differing only in the spatial arrangement of the —H and —OH groups around C-4. Galactose is synthesized in the mammary glands to make the lactose of milk and is less than half as sweet as glucose.

Glucose

$$H-C=O$$
$$H-\overset{|}{\underset{|}{C}}-OH \quad 2$$
$$HO-\overset{|}{\underset{|}{C}}-H \quad 3$$
$$HO-\overset{|}{\underset{|}{C}}-H \quad 4$$
$$H-\overset{|}{\underset{|}{C}}-OH \quad 5$$
$$\underset{6}{CH_2OH}$$

galactose
(an aldohexose)

cyclic form of galactose

A severe inherited disease, called galactosemia, is the inability of infants to metabolize galactose. The galactose concentration increases markedly in the blood and also appears in the urine. Galactosemia causes vomiting, diarrhea, enlargement of the liver, and often mental retardation. If not recognized within a few days after birth, it can lead to death. But if diagnosis is made early and lactose is excluded from the diet, the symptoms disappear and normal growth is resumed. Sometimes, as people age, their intestines stop producing lactase enzyme. Instead, lactose is metabolized by bacteria in the large intestine, creating gas and intestinal discomfort. This condition is called lactose intolerance and is treated by taking lactase tablets after meals.

Fructose Fructose ($C_6H_{12}O_6$), also known as levulose, is a ketohexose and occurs in fruit juices, honey, and (along with glucose) as a constituent of the disaccharide sucrose. Fructose is the major constituent of the polysaccharide inulin, a starchlike substance present in many plants, such as dahlia tubers, chicory roots, and Jerusalem artichokes. Fructose is the sweetest of all the sugars, being more than twice as sweet as glucose. This sweetness accounts for the sweet taste of honey; the enzyme invertase, which is present in bees, splits sucrose into glucose and fructose. Fructose is metabolized directly but is also readily converted to glucose in the liver.

fructose
(a ketohexose)

cyclic form of fructose

Ribose Ribose ($C_5H_{10}O_5$) is an aldopentose and is present in adenosine triphosphate (ATP), one of the chemical energy carriers in the body. Ribose and one of its derivatives, 2-deoxyribose, are also important components of the nucleic acids RNA and DNA, respectively—the genetic information carriers in the body.

ribose
(an aldopentose)

cyclic form of ribose

2-deoxyribose

Disaccharides

Disaccharides are carbohydrates whose molecules yield two molecules of the same or of different monosaccharides when hydrolyzed. The three disaccharides that are especially important from a biological viewpoint are sucrose, lactose, and maltose.

disaccharide

Sucrose, $C_{12}H_{22}O_{11}$, which is commonly known as *table sugar,* is found in the free state throughout the plant kingdom. Sugar cane contains 15–20% sucrose, and sugar beets contain 10–17%. Maple syrup and sorghum are also good sources of sucrose.

Lactose ($C_{12}H_{22}O_{11}$), or *milk sugar,* is found free in nature mainly in the milk of mammals. Human milk contains about 6.7% lactose, and cow's milk contains about 4.5%.

Maltose ($C_{12}H_{22}O_{11}$) is found in sprouting grain but occurs much less commonly (in nature) than either sucrose or lactose. Maltose is prepared commercially by the partial hydrolysis of starch, catalyzed either by enzymes or by dilute acids.

Disaccharides are not used directly in the body but are first hydrolyzed to monosaccharides. Disaccharides yield two monosaccharide molecules when hydrolyzed in the laboratory at elevated temperatures in the presence of hydrogen ions (acids) as catalysts. In biological systems, enzymes (biochemical catalysts) carry out the reaction. A different enzyme is required for the hydrolysis of each of the three disaccharides:

$$\text{sucrose} + \text{water} \xrightarrow{\text{H}^+ \text{ or sucrase}} \text{glucose} + \text{fructose}$$

$$\text{lactose} + \text{water} \xrightarrow{\text{H}^+ \text{ or lactase}} \text{galactose} + \text{glucose}$$

$$\text{maltose} + \text{water} \xrightarrow{\text{H}^+ \text{ or maltase}} \text{glucose} + \text{glucose}$$

The structure of a disaccharide is derived from two monosaccharide molecules by the elimination of a water molecule between them. In maltose, for example, the two monosaccharides are glucose. The water molecule is eliminated between the OH group on C-1 of one glucose unit and the OH group on C-4 of the other glucose unit. Thus, the two glucose units are joined through an oxygen atom at C-1 and C-4.

Lactose is a disaccharide found in breast milk.

maltose,
a disaccharide

glucose unit   glucose unit

Sucrose consists of a glucose unit and a fructose unit linked through an oxygen atom from C-1 on glucose to C-2 on fructose. In lactose, the linkage is from C-1 of galactose through an oxygen atom to C-4 of glucose.

Sucrose is the sugar we commonly use in our kitchens for baking.

glucose unit

fructose unit

sucrose, a disaccharide

Polysaccharides

Polysaccharides are also called *complex carbohydrates* and can be hydrolyzed to a large number of monosaccharide units. The molar masses of polysaccharides range up to 1 million or more. Three of the most important polysaccharides are starch, glycogen, and cellulose.

Starch is a polymer of glucose and is found mainly in the seeds, roots, and tubers of plants. Corn, wheat, potatoes, rice, and cassava are the chief sources of starch whose principal use is for food.

Glycogen is the reserve carbohydrate of the animal kingdom and is often called *animal starch*. Glycogen is formed in the body by polymerization of glucose and is stored especially in the liver and in muscle tissue. Glycogen also occurs in some insects and lower plants including fungi and yeasts.

Cellulose, like starch and glycogen, is also a polymer of glucose. It differs from starch and glycogen in the manner in which the cyclic glucose units are linked to form chains. Cellulose is the most abundant organic substance found in nature. It is the chief structural component of plants and wood. Cotton fibers are almost pure cellulose, and wood, after removal of moisture, consists of about 50% cellulose. An important substance in the textile and paper industries, cellulose is also used to make rayon fibers, photographic film, guncotton, and cellophane. Humans cannot utilize cellulose as food because they lack the necessary enzymes to hydrolyze it to glucose. However, cellulose is an important source of bulk in the diet.

The digestion and metabolism of carbohydrates is a complex biochemical process. It starts in the mouth, where the enzyme amylase in the saliva begins the hydrolysis of starch to maltose and temporarily stops in the stomach, where hydrochloric acid deactivates the enzyme. Digestion continues in the intestines, where the hydrochloric acid is neutralized and pancreatic enzymes complete the hydrolysis to maltose. The enzyme maltase then catalyzes the digestion of maltose to glucose:

$$\text{starch} \xrightarrow{\text{amylase}} \text{dextrins} \xrightarrow{\text{amylase}} \text{maltose} \xrightarrow{\text{maltase}} \text{glucose}$$

Other specific enzymes in the intestines convert sucrose and lactose to monosaccharides. Glucose is absorbed through the intestinal walls into the bloodstream, where it is transported to the cells to be used for energy. Excess glucose is rapidly removed by the liver and muscle tissue, where it is polymerized and stored as glycogen. As the body calls for it, glycogen is converted back to glucose, which is ultimately oxidized to carbon dioxide and water with the release of energy. This energy is used by the body for maintenance, growth, and other normal functions.

20.3 Lipids

Lipids are a group of oily, greasy organic substances found in living organisms that are water insoluble but soluble in organic solvents, such as diethyl ether, benzene, and chloroform. Unlike carbohydrates, lipids share no common chemical structure. The most abundant lipids are the fats and oils, which make up one of the three important classes of foods.

polysaccharide

Cotton is a complex carbohydrate made up of almost pure cellulose.

Microvilli (red) in the epithelium of the human duodenum are responsible for the absorption of food. The fingerlike projections maximize absorption capacity.

lipids

fats
oils
triacylglycerols or
triglycerides

Fats and **oils** are esters of glycerol and predominantly long-chain fatty acids (carboxylic acids). Fats and oils are also called **triacylglycerols** or **triglycerides,** since each molecule is derived from one molecule of glycerol and three molecules of fatty acid:

$$
\begin{array}{c}
\text{glycerol} \\
\text{portion} \rightarrow
\end{array}
\left.
\begin{array}{l}
CH_2-O-\overset{\displaystyle O}{\overset{\|}{C}}-R \\[2mm]
CH-O-\overset{\displaystyle O}{\overset{\|}{C}}-R' \\[2mm]
CH_2-O-\overset{\displaystyle O}{\overset{\|}{C}}-R''
\end{array}
\right|
\qquad
\begin{array}{l}
CH_2-O-\overset{\displaystyle O}{\overset{\|}{C}}-C_{17}H_{35} \\[2mm]
CH-O-\overset{\displaystyle O}{\overset{\|}{C}}-C_{15}H_{31} \\[2mm]
CH_2-O-\overset{\displaystyle O}{\overset{\|}{C}}-C_{11}H_{23}
\end{array}
$$

general formula
for a triacylglycerol

typical triacylglycerol
containing three different
fatty acids

The formulas of triacylglycerol molecules vary for the following reasons:

1. The length of the fatty acid chain may vary from 4 to 20 carbons, but the number of carbon atoms in the chain is nearly always even.
2. Each fatty acid may be saturated, or it may be unsaturated and contain one, two, or three carbon–carbon double bonds.
3. An individual triacylglycerol may, and frequently does, contain three different fatty acids.

The most abundant saturated fatty acids in fats and oils are lauric, myristic, palmitic, and stearic acids (see Table 19.7). The most abundant unsaturated acids in fats and oils contain 18 carbon atoms and have one, two, or three carbon–carbon double bonds. Their formulas are

$$CH_3(CH_2)_7CH=CH(CH_2)_7COOH$$
oleic acid

$$CH_3(CH_2)_4CH=CHCH_2CH=CH(CH_2)_7COOH$$
linoleic acid

$$CH_3CH_2CH=CHCH_2CH=CHCH_2CH=CH(CH_2)_7COOH$$
linolenic acid

Other significant unsaturated fatty acids are palmitoleic acid (with 16 carbons) and arachidonic acid (with 20 carbons).

$$CH_3(CH_2)_5CH=CH(CH_2)_7COOH \qquad CH_3(CH_2)_4(CH=CHCH_2)_4CH_2CH_2COOH$$
palmitoleic acid   arachidonic acid

Three unsaturated fatty acids—linoleic, linolenic, and arachidonic—are essential for animal nutrition and must be supplied in the diet. Diets lacking these fatty acids lead to impaired growth and reproduction, and such skin disorders as eczema and dermatitis. We don't require fats in our diet except as a source of these three fatty acids.

Fat cells like these can result in clogged arteries, but they also provide a major reserve of potential energy.

Practice 20.1

Write the structure of a triacylglycerol that contains one unit of palmitic acid, stearic acid, and oleic acids. How many other structures are possible containing these three acids?

Table 20.2 Fatty Acid Composition of Selected Fats and Oils

Fat or oil	Fatty acid (%)				
	Myristic acid	Palmitic acid	Stearic acid	Oleic acid	Linoleic acid
Animal fat					
Butter*	7–10	23–26	10–13	30–40	4–5
Lard	1–2	28–30	12–18	41–48	6–7
Vegetable oil					
Olive	0–1	5–15	1–4	49–84	4–12
Peanut	—	6–9	2–6	50–70	13–26
Corn	0–2	7–11	3–4	43–49	34–42
Cottonseed	0–2	19–24	1–2	23–33	40–48
Soybean	0–2	6–10	2–4	21–29	50–59
Linseed**	—	4–7	2–5	9–38	3–43

*Butyric acid, 3–4%
**Linolenic acid, 25–58%

The major physical difference between fats and oils is that fats are solid and oils are liquid at room temperature. Since the glycerol part of the structure is the same for a fat and an oil, the difference must be due to the fatty acid end of the molecule. Fats contain a higher proportion of saturated fatty acids, whereas oils contain higher amounts of unsaturated fatty acids. The term *polyunsaturated* means that the molecules of a particular product each contain several double bonds.

Fats and oils are obtained from natural sources. In general, fats come from animal sources and oils from vegetable sources. Olive, cottonseed, corn, soybean, linseed, and other oils are obtained from the fruit or seed of their respective vegetable sources. Table 20.2 shows the major constituents of several fats and oils.

Olive oil harvested from these olives is a healthy, monounsatured oil.

Fats are an important energy source for humans and normally account for about 25–50% of caloric intake. When oxidized to carbon dioxide and water, fats supply about 39 kJ/g (9.3 kcal/g), which is more than twice the amount obtained from carbohydrates and proteins.

Fats are digested in the small intestine, where they are first emulsified by the bile salts and then hydrolyzed to di- and monoglycerides, fatty acids, and glycerol. The fatty acids pass through the intestinal wall and are coated with a protein to increase solubility in the blood. They are then transported to various parts of the body where they are broken down in a series of enzyme-catalyzed reactions for the production of energy. Part of the hydrolyzed fat is converted back into fat in the adipose tissue and stored as a reserve source of energy. These fat deposits also function to insulate against loss of heat as well as to protect vital organs against mechanical injury.

Solid fats are preferable to oils for the manufacture of soaps and for use as certain food products. Hydrogenation of oils to make them solid is carried out on a large commercial scale. In this process hydrogen, bubbled through hot oil containing a finely dispersed nickel catalyst, adds to the carbon–carbon double bonds of the oil to saturate the double bonds and form fats. In practice, only some of the double bonds are allowed to become saturated (partially hydrogenated vegetable oil). The product that is marketed as solid "shortening" is used for cooking and baking. Oils and fats are also partially hydrogenated to improve their shelf life. Rancidity in fats and oils results from air oxidation at points of unsaturation, producing low-molar-mass aldehydes and acids of disagreeable odor and flavor.

Soap is made by hydrolyzing fats or oils with aqueous NaOH. This hydrolysis process, called *saponification,* requires 3 moles NaOH per mole of fat:

$$
\begin{array}{l}
\underset{\text{a fat}}{
\begin{array}{l}
\mathrm{CH_2-O-\overset{\displaystyle \|}{\underset{O}{C}}-R} \\[2mm]
\mathrm{CH-O-\overset{\displaystyle \|}{\underset{O}{C}}-R'} \\[2mm]
\mathrm{CH_2-O-\overset{\displaystyle \|}{\underset{O}{C}}-R''}
\end{array}}
+ 3\,\mathrm{NaOH} \longrightarrow
\underset{\text{glycerol}}{
\begin{array}{l}
\mathrm{CH_2-OH} \\[2mm]
\mathrm{CH-OH} \\[2mm]
\mathrm{CH_2-OH}
\end{array}}
+
\underset{\text{soap}}{
\begin{array}{l}
\mathrm{Na^+O^- -\overset{\displaystyle \overset{O}{\|}}{C}-R} \\[2mm]
\mathrm{Na^+O^- -\overset{\displaystyle \overset{O}{\|}}{C}-R'} \\[2mm]
\mathrm{Na^+O^- -\overset{\displaystyle \overset{O}{\|}}{C}-R''}
\end{array}}
$$

The most common soaps are the sodium salts of long-chain fatty acids, such as sodium stearate ($NaOOCC_{17}H_{35}$), sodium palmitate ($NaOOCC_{15}H_{31}$), and sodium oleate ($NaOOCC_{17}H_{33}$).

Other principal classes of lipids, besides fats and oils, are phospholipids, glycolipids, and steroids (see Figure 20.1). The phospholipids are found in all animal and vegetable cells and are abundant in the brain, the spinal cord, egg yolk, and liver. Glycolipids (cerebrosides) contain a long-chain alcohol called *sphingosine.* They contain no glycerol but do contain a monosaccharide (usually galactose). Glycolipids are found in many different tissues but, as the name *cerebroside* indicates, occur in large quantities in brain tissue.

Steroids all have a four-fused carbocyclic-ring system (as in cholesterol) with various side groups attached to the rings. Cholesterol is the most abundant steroid in the body. It occurs in the brain, the spinal column, and nervous tissue, and it is the principal constituent of gallstones. The body synthesizes about 1 g of cholesterol per day, while about 0.3 g per day is ingested in the average diet. The major sources of cholesterol in the diet are meat, liver, and egg yolk. The cholesterol level in the blood generally rises with a person's age and body weight. High blood-level cholesterol is associated with atherosclerosis (hardening of the arteries), which results in reduced flow of blood and high blood pressure. Cholesterol is needed by the body to synthesize other steroids, some of which regulate male and female sexual characteristics. Many synthetic birth control pills are modified steroids that interfere with a woman's normal conception cycle.

Ring structure of steroids

20.4 Amino Acids and Proteins

Proteins are the third important class of foodstuffs. Some common foods with high (over 10%) protein content are gelatin, fish, beans, nuts, cheese, eggs, poultry, and meat. Proteins are present in all body tissue. They can form

a lecithin (a phospholipid)

a cerebroside (a glycolipid)

cholesterol (a steroid)

Norlutin
(a birth control pill)

Phospholipids and glycolipids are common in the brain.

Figure 20.1
Formulas for a phospholipid, a glycolipid, and steroids.

structural elements, such as hair, fingernails, wool, and silk. Proteins also function as enzymes that regulate the countless chemical reactions taking place in every living organism. About 15% of the human body weight is protein. Chemically, proteins are polymers of amino acids with molar masses ranging up to more than 50 million.

Amino acids are carboxylic acids that contain an amino ($-NH_2$) group attached to C-2 (the alpha carbon) and are thus called *α-amino acids*. They also contain another variable group, R. The R group represents any of the various

amino acid

groups that make up the specific amino acids. For example, when R is H—, the amino acid is glycine; when R is CH_3—, the amino acid is alanine; when R is $CH_3SCH_2CH_2$—, the amino acid is methionine.

α-amino acid

Some amino acids have two amino groups and some contain two acid groups.

Practice 20.2 _____

Write formulas for α-aminobutyric acid and β-aminobutyric acid.

There are approximately 200 different known amino acids in nature. Some are found in only one particular species of plant or animal, others in only a few lifeforms. But 20 of these amino acids are found in almost all proteins. Furthermore, these same 20 amino acids are used by all forms of life in the synthesis of proteins; their names, formulas, and abbreviations are given in Table 20.3.

Table 20.3 Common Amino Acids Derived from Proteins

Name	Abbreviation	Formula
Alanine	Ala	$CH_3CHCOOH$ with NH_2 below
Arginine	Arg	$NH_2{-}C{-}NH{-}CH_2CH_2CH_2CHCOOH$ with $\parallel NH$ and NH_2 below
Asparagine	Asn	$NH_2C{-}CH_2CHCOOH$ with $\parallel O$ and NH_2 below
Aspartic acid	Asp	$HOOCCH_2 CHCOOH$ with NH_2 below
Cysteine	Cys	$HSCH_2CHCOOH$ with NH_2 below
Glutamic acid	Glu	$HOOCCH_2 CH_2CHCOOH$ with NH_2 below
Glutamine	Gln	$NH_2 CCH_2CH_2CHCOOH$ with $\parallel O$ and NH_2 below

Name	Abbreviation	Formula
Glycine	Gly	$HCHCOOH$ $\|$ NH_2
Histidine	His	(imidazole ring) $N=CH$, $HC\;\;\;C-CH_2CHCOOH$ with NH_2; ring N–H
Isoleucine*	Ile	$CH_3CH_2CH-CHCOOH$ $CH_3\;\;\;NH_2$
Leucine*	Leu	$(CH_3)_2CHCH_2CHCOOH$ NH_2
Lysine*	Lys	$NH_2CH_2CH_2CH_2CH_2CHCOOH$ NH_2
Methionine*	Met	$CH_3SCH_2CH_2CHCOOH$ NH_2
Phenylalanine*	Phe	(phenyl)$-CH_2CHCOOH$ with NH_2
Proline	Pro	(pyrrolidine ring, N–H)$-COOH$
Serine	Ser	$HOCH_2CHCOOH$ NH_2
Threonine*	Thr	$CH_3CH-CHCOOH$ $OH\;\;\;NH_2$
Tryptophan*	Trp	(indole ring) $C-CH_2CHCOOH$, CH, ring N–H, with NH_2
Tyrosine	Tyr	$HO-$(phenyl)$-CH_2CHCOOH$ with NH_2
Valine*	Val	$(CH_3)_2CHCHCOOH$ NH_2

*Amino acids essential in human nutrition.

Eight are considered essential amino acids, since the human body is *not* capable of synthesizing them. Therefore they must be supplied in our diets if we are to enjoy normal health.

protein   **Proteins** are polymeric substances that yield primarily amino acids on hydrolysis. The bond connecting the amino acids in a protein is commonly called

peptide linkage   a **peptide linkage** or peptide bond. If we combine two glycine molecules by eliminating a water molecule between the amino group of one and the carboxyl group of the second glycine, we form a compound containing the amide structure and the peptide linkage. The compound containing the two amino acid groups is called a *dipeptide:*

glycine   glycine   glycylglycine (Gly-Gly)
                                  (a dipeptide)

The product formed from two glycine molecules is called *glycylglycine* (abbreviated Gly-Gly). Note that the molecule still has a free amino group at one end and a free carboxyl group at the other end. The formation of Gly-Gly is considered the first step in the synthesis of a protein, since each end of the molecule is capable of joining to another amino acid. We can thus visualize the formation of a protein by joining a great many amino acids in this fashion. Another example showing a tripeptide (three amino acids linked together) follows. This compound contains two peptide linkages:

tyrosine   alanine   glycine

tyrosylalanylglycine (a tripeptide)(Tyr-Ala-Gly)

There are five other tripeptide combinations of these three amino acids using only one unit of each amino acid. Peptides containing up to about 40–50

polypeptide   amino acid units in a chain are **polypeptides.** Still longer chains of amino acids are proteins.

The amino acid units in a peptide are called *amino acid residues* or simply residues. (They no longer are amino acids because they have lost an H atom from their amino groups and an OH from their carboxyl groups.) In linear peptides, one end of the chain has a free amino group and the other end a free carboxyl group. The amino-group end is called the *N-terminal residue* and the other end the *C-terminal residue:*

$$\overset{1\quad2\quad3\quad4\quad5\quad6\quad7}{\text{Ala-Pro-Tyr-Met-Gly-Lys-Gly}}$$

N-terminal residue   C-terminal residue

The sequence of amino acids in a chain is numbered starting with the N-terminal residue, which is usually written to the left with the C-terminal residue at the right.

Peptides are named as acyl derivatives of the C-terminal amino acid, with the C-terminal unit keeping its complete name. The -ine ending of all but the C-terminal amino acid is changed to -yl, and these are listed in the order in which they appear, starting with the N-terminal amino acid:

alanyl   tyrosyl   glycine

Ala-Tyr-Gly

Thus Ala-Tyr-Gly is called *alanyltyrosylglycine,* and Arg-Gln-His-Ala is arginylglutamylhistidylalanine.

Practice 20.3 _____

Write full names for
(a) Pro-Tyr   (b) Gln-His-Leu

Practice 20.4 _____

Draw the complete structure for
(a) Ala-Ser   (b) Gly-Thr-Met

Many small, naturally occurring polypeptides have significant biochemical functions. The amino acid sequences of two of these, oxytocin and vasopressin, are shown in Figure 20.2. Oxytocin controls uterine contractions during labor in childbirth and also causes contraction of the smooth muscles of the mam-

$$\overset{1\qquad\quad3\qquad\qquad\qquad\qquad 8}{\text{Cy-Tyr-Ile-Gln-Asn-Cy-Pro-Leu-Gly-NH}_2}$$
$$\text{—S—S—}$$

oxytocin

$$\overset{1\qquad\quad3\qquad\qquad\qquad\qquad 8}{\text{Cy-Tyr-Phe-Gln-Asn-Cy-Pro-Arg-Gly-NH}_2}$$
$$\text{—S—S—}$$

vasopressin

Figure 20.2
Amino acid sequences of oxytocin and vasopressin. The difference in only two amino acids (shown in color) in these two compounds results in very different physiological activity. The *C*-terminal amino acid has an amide structure (Gly-NH$_2$) instead of —COOH.

mary gland, resulting in milk secretion. Vasopressin in high concentration raises the blood pressure and has been used in surgical shock treatment for this purpose. Vasopressin is also an antidiuretic, regulating the excretion of fluid by the kidneys. The absence of vasopressin leads to diabetes insipidus. This condition, which is characterized by excretion of up to 30 L of urine per day, may be controlled by administering vasopressin or its derivatives. Oxytocin and vasopressin are similar nonapeptides, differing only at positions 3 and 8.

Determining the sequence of the amino acids in even one protein molecule was once a formidable task. The amino acid sequence of beef insulin was announced in 1955 by the British biochemist Frederick Sanger (b. 1918). Finding the sequence of this structure required several years of effort by Sanger's team. He was awarded the 1958 Nobel Prize in chemistry for this work. As a result of this work, automated amino acid sequencers have been developed that can determine the sequence of an average-sized protein in a few days. Beef insulin consists of 51 amino acid units in two polypeptide chains. The two chains are connected by disulfide linkages ($-S-S-$) of two cysteine residues at two different sites. The structure is shown in Figure 20.3. Insulins from other animals, including humans, differ slightly by one, two, or three amino acid residues.

Protein digestion takes place in the stomach and the small intestine. Here, digestive enzymes hydrolyze proteins to smaller peptides and amino acids, which pass through the walls of the intestines, are absorbed by the blood, and are transported to the liver and other tissues of the body. The body does not store free amino acids. They are used to synthesize

- proteins to replace and repair body tissues
- other nitrogen-containing substances, such as certain hormones and heme
- nucleic acids
- enzymes that control the synthesis of other necessary products, such as carbohydrates and fats

Figure 20.3
Amino acid sequence of beef insulin.

CHEMISTRY IN ACTION ● The Taste of Umami

The four primary tastes were long ago established as sour, sweet, salty, and bitter. A fifth taste has been added to the group. It was discovered by Kikunae Ikeda, a chemistry professor at Imperial University of Tokyo. The taste is called umami, after the Japanese word *umai*, meaning "delicious." Ikeda isolated and identified glutamate as the substance Japanese chefs used to make food taste better. We know glutamate as MSG (monosodium glutamate), sold in North America as Accent. It is the sodium salt of the amino acid

glutamic acid (see structure in Table 20.3) and is a fine white crystal that looks similar to salt.

Glutamate is found in lots of natural foods, including meat, poultry, vegetables, fish, cheese, and tomatoes. The flavor-enhancing property of glutamate occurs when it is in its free or uncombined form (not bound into protein). Many Asian seasonings give umami taste to foods. These include soy sauce and fish sauce. The Australian product Vegemite also has a strong umami taste.

Long ago MSG was extracted from a seaweed called *Laminaria japonica*

and other plants. Now MSG is produced by fermenting molasses (from sugar cane or beets) or food starch from tapioca. The fermentation product is filtered, purified, and converted to the sodium salt—monosodium glutamate.

Some people are sensitive to MSG, with reactions including migraines and chest pains. Monosodium glutamate has undergone extensive testing and research in the food industry, and the FDA has classified it as "generally recognized as safe." It is widely used to add flavor and umami to many foods.

Proteins are catabolized (degraded) to carbon dioxide, water, and urea. Urea, containing the protein nitrogen, is eliminated from the body in the urine.

Carbohydrates and fats are used primarily to supply heat and energy to the body. Proteins, on the other hand, are used mainly to repair and replace worn-out tissue. Tissue proteins are continuously being broken down and resynthesized. Therefore, protein must be continually supplied to the body in the diet. It is nothing short of amazing how organisms pick out the desired amino acids from the bloodstream and put them together in proper order to synthesize a needed protein.

20.5 Enzymes

Enzymes are the catalysts of biochemical reactions. Most enzymes are proteins, and they catalyze nearly all of the reactions that occur in living cells. Uncatalyzed reactions that may require hours of boiling in the presence of a strong acid or a strong base can occur in a fraction of a second in the presence of the proper enzyme at room temperature and nearly neutral pH. This process is all the more remarkable when we realize that enzymes do not actually cause chemical reactions. They act as catalysts by greatly lowering the activation energy of specific biochemical reactions. The lowered activation energy permits these reactions to proceed at high speed at body temperature.

Louis Pasteur (1822–1895) was one of the first scientists to study enzyme catalyzed reactions. He believed that living yeasts or bacteria were required for these reactions, which he called fermentations—for example, the conversion of glucose to alcohol by yeasts. In 1897, Eduard Büchner (1860–1917) made a cell-free filtrate that contained enzymes prepared by grinding yeast cells with very fine sand. The enzymes in this filtrate converted glucose to alcohol, thus proving that the presence of living cells is not required for enzyme activity. For this work Büchner received the Nobel Prize in chemistry in 1907.

Each organism contains thousands of enzymes. Some enzymes are simple proteins consisting only of amino acid units. Others are conjugated and consist of a protein part, or *apoenzyme,* and a nonprotein part, or *coenzyme.* Both parts

enzyme

557

are essential, and a functioning enzyme consisting of both the protein and non-protein parts is called a *holoenzyme*.

apoenzyme + coenzyme = holoenzyme

Often the coenzyme is a vitamin, and the same coenzyme may be associated with many different enzymes.

For some enzymes, an inorganic component, such as a metal ion (e.g., Ca^{2+}, Mg^{2+}, or Zn^{2+}), is required. This inorganic component is an *activator*. From the standpoint of function, an activator is analogous to a coenzyme, but inorganic components are not called coenzymes.

Another remarkable property of enzymes is their specificity of reaction; that is, a certain enzyme will catalyze the reaction of a specific type of substance. For example, the enzyme maltase catalyzes the reaction of maltose and water to form glucose. Maltase has no effect on the other two common disaccharides sucrose and lactose. Each of these sugars requires a specific enzyme—sucrase to hydrolyze sucrose, lactase to hydrolyze lactose. (See the hydrolysis equations in Section 20.2.)

substrate   The substance acted on by an enzyme is called the **substrate.** Sucrose is the substrate of the enzyme sucrase. Common names for enzymes are formed by adding the suffix *-ase* to the root of the substrate name. Note, for example, the derivations of maltase, sucrase, and lactase from maltose, sucrose, and lactose. Many enzymes, especially digestive enzymes, have common names such as pepsin, rennin, trypsin, and so on. These names have no systematic significance.

Enzymes act according to the following general sequence. Enzyme (E) and substrate (S) combine to form an enzyme–substrate intermediate (E–S). This intermediate decomposes to give the product (P) and regenerate the enzyme:

E + S ⟷ E–S ⟶ E + P

For the hydrolysis of maltose, the sequence is

maltase + maltose ⇌ maltase–maltose
   E   S   E–S

maltase–maltose + H_2O ⟶ maltase + 2 glucose
    E–S   E   P

Enzyme specificity is believed to be due to the particular shape of a small part of the enzyme, its active site, which exactly fits a complementary-shaped part of the substrate (see Figure 20.4). This interaction is analogous to a lock

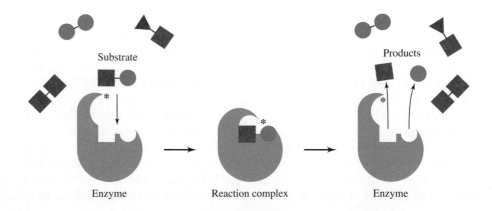

Figure 20.4
Enzyme–substrate interaction illustrating specificity of an enzyme by the lock-and-key model.

and key; the substrate is the lock and the enzyme, the key. Just as a key opens only the lock it fits, the enzyme acts only on a molecule that fits its particular shape. When the substrate and the enzyme come together, they form a substrate–enzyme complex unit. The substrate, activated by the enzyme in the complex, reacts to form the products, regenerating the enzyme.

A more recent model of the enzyme–substrate catalytic site, known as the induced-fit model, visualizes a flexible site of enzyme–substrate attachment, with the substrate inducing a change in the enzyme shape to fit the shape of the substrate. This model allows for the possibility that in some cases the enzyme might wrap itself around the substrate and so form the correct shape of lock and key. Thus, the enzyme does not need to have an exact preformed catalytic site to match the substrate.

20.6 Nucleic Acids

Explaining how hereditary material duplicates itself was a baffling problem for biologists. This explanation and the answer to the question "Why are the offspring of a given species undeniably of that species?" eluded biologists for many years. Many thought the chemical basis for heredity lay in the structure of proteins. But they couldn't find how protein reproduced itself. The answer to the hereditary problem was finally found in the structure of the nucleic acids.

The unit structure of all living things is the cell. Suspended in the nucleus of cells are chromosomes, which consist largely of proteins and nucleic acids. The nucleic acids and proteins are intimately associated in complexes called *nucleoproteins*. Nucleic acids either contain the sugar deoxyribose or they contain the sugar ribose. Accordingly, they are called deoxyribonucleic acid (DNA) and ribonucleic acid (RNA). DNA was discovered in 1869 by Swiss physiologist Friedrich Miescher (1844–1895), who extracted it from the nuclei of cells.

DNA is a polymeric substance made up of thousands of units called nucleotides. The fundamental components of the nucleotides in DNA are phosphoric acid, deoxyribose (a pentose sugar), and the four nitrogen-containing bases adenine, thymine, guanine, and cytosine (abbreviated as A, T, G, and C). Phosphoric acid is obtained from minerals in the diet; deoxyribose is synthesized in the body from glucose; and the four nitrogen bases are made in the body from amino acids. The formulas for these compounds are given in Figure 20.5.

A **nucleotide** in DNA consists of one of the four bases linked to a deoxyribose sugar, which in turn is linked to a phosphate group. Each nucleotide has the following sequence:

| Phosphate | — | Deoxyribose | — | Base |

The structures for a single nucleotide and a segment of a polynucleotide (DNA) are shown in Figure 20.6.

In 1953, the American biologist James D. Watson (b. 1928) and the British physicist Francis H. C. Crick (1916–2004) announced their double-stranded helix structure for DNA. This concept was a milestone in the history of biology, and in 1962 Watson and Crick, along with Maurice H. F. Wilkin (1916–2004), who did the brilliant X-ray diffraction studies on DNA, were awarded the Nobel Prize in medicine and physiology.

Figure 20.5
Fundamental components of nucleotides in DNA.

(a) Adenine deoxyribonucleotide

To another phosphate

(b) Four nucleotide units of a DNA strand   To another phosphate

Figure 20.6
(a) A single nucleotide, adenine deoxyribonucleotide. (b) A segment of one strand of deoxyribonucleic acid (DNA) showing four nucleotides, including those of adenine (A), cytosine (C), guanine (G), and thymine (T).

The structure of DNA, according to Watson and Crick, consists of two polymeric strands of nucleotides in the form of a double helix, with both nucleotide strands coiled around the same axis (see Figure 20.7). Along each strand are alternate phosphate and deoxyribose units with one of the four bases adenine, guanine, cytosine, or thymine attached to deoxyribose as a side group. The double helix is held together by hydrogen bonds extending from the base on one strand of the double helix to a complementary base on the other strand. Watson and Crick furthermore ascertained that adenine is always hydrogen-bonded to thymine and guanine is always hydrogen-bonded to cytosine. Previous analytical work by others substantiated this concept of complementary bases by showing that the molar ratio of adenine

P = Phosphate
D = Deoxyribose
A = Adenine
T = Thymine
C = Cytosine
G = Guanine

The chromosomes in this
mammal cell are in the final stage
of cell division. They contain nucleic
acids and proteins replicated
for the new cell.

**In DNA, adenosine is always
bonded to thymine.**

Figure 20.7
Double-stranded helix structure of DNA. Adenine is hydrogen-bonded to thymine (A∷ T) and
cytosine is hydrogen-bonded to Guanine (C∷∷ G).

to thymine in DNA is approximately 1 : 1 and that of guanine to cytosine is
also approximately 1 : 1.

The structure of DNA has been compared to a ladder twisted into a double
helix, with the rungs of the ladder perpendicular to the twisted railings. The
phosphate and deoxyribose units alternate along the two railings of the ladder,
and two nitrogen bases form each rung of the ladder. The DNA structure is
illustrated in Figure 20.7.

For any individual of any species, the sequence of base pairs and the length
of the nucleotide chains in DNA molecules contain the coded messages that determine all of the characteristics of that individual. In this sense the DNA molecule is a template that stores information for recall as needed. DNA contains
the genetic code of life, which is passed on from one generation to another.

RNA

RNA is a polymer of nucleotides, but it differs from DNA in that (1) it is single-stranded; (2) it contains the pentose sugar ribose instead of deoxyribose; and (3) it contains uracil instead of thymine as one of its four nitrogen bases (see Figure 20.8).

Practice 20.5

Write the formula for the adenine ribonucleotide.

transcription

uracil

$HOCH_2$

ribose

Figure 20.8
Components of RNA that differ from DNA.

Transcription is the process by which DNA directs the synthesis of RNA. The nucleotide sequence of only one strand of DNA is transcribed into a single strand of RNA. This transcription occurs in a complementary fashion. Where there is a guanine base in DNA, a cytosine base will occur in RNA. Cytosine is transcribed to guanine, thymine to adenine, and adenine to uracil. Thymine occurs only in DNA and uracil only in RNA.

The main function of RNA is to direct the synthesis of proteins. RNA is produced in the cell nucleus but performs its function outside of the nucleus. Three kinds of RNA are produced directly from DNA: messenger RNA (*m*RNA), transfer RNA (*t*RNA), and ribosomal RNA (*r*RNA). Messenger RNA contains bases in the exact order transcribed from a strip of the DNA master code. The base sequence on the *m*RNA in turn establishes the sequence of amino acids that comprise a specific protein. The relatively small transfer RNA molecules bring specific amino acids to the site of protein synthesis. There is at least one different *t*RNA for each amino acid. The actual site of protein synthesis is a ribosome, which is composed of *r*RNA and protein. The function of the *r*RNA is not completely understood. However, the ribosome is believed to move along the *m*RNA chain and to aid in the polymerization of amino acids in the order prescribed by the base sequence of the *m*RNA chain. The flow of genetic information is usually in one direction, from DNA to RNA to proteins (Figure 20.9).

Figure 20.9
The process of cellular genetic information.

20.7 DNA and Genetics

Heredity is the process by which the physical and mental characteristics of parents are transferred to their offspring. For this process to occur, the material responsible for genetic transfer must be able to make exact copies of itself. The design for replication is built into Watson and Crick's DNA structure, first by the nature of its double-helical structure and second by the complementary nature of its nitrogen bases, where adenine will bond only to thymine and guanine only to cytosine. The DNA double helix unwinds, or "unzips," into two separate strands at the hydrogen bonds between the bases. Each strand then serves as a template combining only with the proper free nucleotides to produce two identical replicas of itself. This replication of DNA occurs in the cell just before the cell divides, thereby giving each daughter cell the full genetic code of the original cell. This process is illustrated in Figure 20.10.

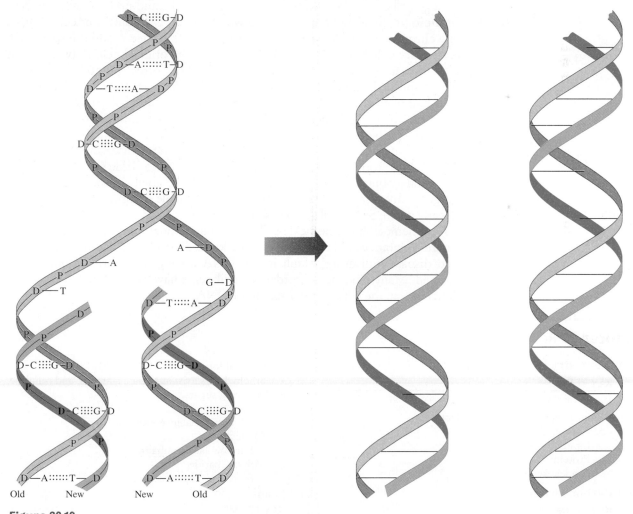

Figure 20.10
Replication of DNA. The two helices unwind, separating at the point of the hydrogen bonds. Each strand then serves as a template, recombining with the proper nucleotides to duplicate itself as a double-stranded helix.

As we have indicated, DNA is an integral part of the chromosomes. Each species carries a specific number of chromosomes in the nucleus of each of its cells. The number of chromosomes varies with different species. Humans have 23 pairs, or 46 chromosomes. Chromosomes are long, threadlike bodies composed of nucleic acids and proteins that contain the basic units of heredity called genes. **genes** **Genes** are segments of the DNA chain that contain the codes for the formation of polypeptides and RNAs. Hundreds of genes can exist along a DNA chain.

mitosis In ordinary cell division, known as **mitosis,** each DNA molecule forms a duplicate by uncoiling to single strands. Each strand then assembles the complementary portion from available free nucleotides to form a duplicate of the original DNA molecule. After cell division is completed, each daughter gene contains the genetic material that corresponds exactly to that present in the original cell before division.

However, in almost all higher forms of life, reproduction takes place by union of the male sperm with the female egg. Cell splitting to form the sperm cell and the egg cell occurs by a different and more complicated process called *meiosis.* **meiosis** In **meiosis,** the genetic material is divided so that each daughter cell receives one chromosome from each pair. After meiosis, the egg cell and the sperm cell each carry only half of the chromosomes from its original cell. Between them they form a new cell that once again contains the correct number of chromosomes and all the hereditary characteristics of the species. Thus the offspring derives half of its genetic characteristics from the father and half from the mother.

Nature is not 100% perfect. Occasionally, DNA replication is not perfect, or a section of the DNA molecule is damaged by X-rays, radioactivity, or drugs, and a mutant organism is produced. For example, in the disease sickle-cell anemia, a large proportion of the red blood cells form into sickle shapes instead of the usual globular shape. This irregularity limits the ability of the blood to carry oxygen and causes the person to be weak and unable to fight infection, leading to early death. Sickle-cell anemia is due to one misplaced amino acid in the structure of hemoglobin. The sickle-cell-producing hemoglobin has a valine residue where a glutamic acid residue should be located. Sickle-cell anemia is an inherited disease indicating a fault in the DNA coding transmitted from parent to child. Many biological disorders and ailments have been traced directly to a deficiency in the genetic information of DNA.

Chapter 20 Review

20.1 Chemistry in Living Organisms
KEY TERM

Biochemistry

- Four major classes of biomolecules:
  - Carbohydrates
  - Lipids
  - Proteins
  - Nucleic acids

20.2 Carbohydrates
KEY TERMS

Carbohydrate
Monosaccharide
Disaccharide
Polysaccharide

- Empirical formulas: $C_x(H_2O)$ or $(CH_2O)_n$.
- Carbohydrates exist as sugars, starches, and cellulose.
- Classifying sugars:
  - Monosaccharide
    - Glucose, galactose, fructose, ribose
  - Disaccharide
    - Sucrose, lactose, maltose
  - Polysaccharide
    - Starch, glycogen, cellulose

20.3 Lipids
KEY TERMS

Lipids
Fats
Oils
Triacylglycerols or triglycerides

- Lipids share no common chemical structure.
- Classes of lipids:
  - Fats and oils
  - Phospholipids
  - Glycolipids
  - Steroids

20.4 Amino Acids and Proteins

KEY TERMS

Amino acids
Protein
Peptide linkage
Polypeptide

- Form various structural elements, and enzymes
- Proteins are polymers that break down mainly into amino acids:
  - Amino acids are connected in a protein by a peptide linkage.
  - Peptides containing up to about 40–50 amino acids are called polypeptides.
  - Longer chains are proteins.

20.5 Enzymes

KEY TERMS

Enzyme
Substrate

- Enzymes catalyze biochemical reactions.
- Enzymes are classified as:
  - Simple—only amino acid units
  - Conjugated—protein part and nonprotein part

- Substance acted on by the enzyme is called the substrate.
- Specificity of enzyme action is due to the shape of a small part of the enzyme molecule.

20.6 Nucleic Acids

KEY TERMS

DNA
Nucleotide
RNA
Transcription

- Components of DNA:
  - Phosphoric acid
  - Deoxyribose
  - Four nitrogen-containing bases
    - Adenine
    - Thymine
    - Guanine
    - Cytosine

- A nucleotide is a base linked to deoxyribose (or ribose) sugar linked to a phosphate group.
- The structure of DNA is a double helix composed of 2 complementary strands joined by hydrogen bonds.
- RNA is a polymer of nucleotides but differs from DNA:
  - Single stranded
  - Contains ribose instead of deoxyribose
  - Contains uracil in place of the base thymine
- Transcription is the process by which DNA directs the synthesis of RNA.

20.7 DNA and Genetics

KEY TERMS

Genes
Mitosis
Meiosis

- Genes are segments of DNA that contain codes for the formation of polypeptides and RNAs.

- In mitosis a DNA molecule uncoils and reassembles to form two new duplicate double helixes. Each new cell contains a full complement of DNA.
- In meiosis the union of the male sperm and the female egg forms a new cell that carries half of the chromosomes from the male and half from the female.

Review Questions

All questions with blue numbers have answers in the appendix of the text.

1. What are the three principal classes of animal food?

2. What are the components of chromosomes?

3. In a linear peptide, what are the N-terminal residue and the C-terminal residue?

4. What is the difference between saturated fats and unsaturated fats?

5. Of the sugars listed in Table 20.1, which is the sweetest disaccharide? Which is the sweetest monosaccharide? (Invert sugar is a mixture of glucose and fructose, so it should not be considered.)

6. Are the fatty acids in vegetable oils more saturated or unsaturated than those in animal fats? Explain your answer. (Table 20.2)

7. Which of the common amino acids have more than one carboxyl group? Which have more than one amino group? (Table 20.3)

8. How many disulfide linkages are there in each molecule of beef insulin? (Figure 20.3)

9. In the four nucleotide units of DNA shown in Figure 20.6, which of these components are part of the backbone chain, and which are off to the side: the nitrogen bases, the deoxyribose, and the phosphoric acid?

10. In the double-stranded helix structure of DNA, which bases are always hydrogen-bonded to the following: cytosine, thymine, adenine, and guanine? (Figure 20.7)

11. Life is dependent on four major classes of biomolecules. What are they?

12. Indicate the three types of carbohydrates. Which is the simplest?

13. What is an aldose, an aldotetrose, a ketose, a ketohexose? Give an example of each.

14. Classify the following as a monosaccharide, disaccharide, or polysaccharide: glucose, sucrose, maltose, fructose, cellulose, lactose, glycogen, galactose, starch, and ribose.

15. Draw structural formulas in the open-chain form for ribose, glucose, fructose, and galactose.

16. Draw structural formulas in the cyclic form for ribose, glucose, fructose, and galactose.

17. State the properties and the sources of ribose, glucose, fructose, and galactose.

18. The molecular formula for lactic acid is $C_3H_6O_3$, and its structural formula is $CH_3CH(OH)COOH$. Is this compound a carbohydrate? Explain.

19. What is the monosaccharide composition of
    (a) sucrose
    (b) maltose
    (c) lactose
    (d) starch
    (e) cellulose
    (f) glycogen

20. Draw structural formulas in the cyclic form for sucrose and maltose.

21. If the most common monosaccharides have the formula $C_6H_{12}O_6$, why do the resulting disaccharides have the formula $C_{12}H_{22}O_{11}$, rather than $C_{12}H_{24}O_{12}$?

22. Write equations, using structural formulas in the cyclic form, for the hydrolysis of
    (a) sucrose
    (b) maltose
    What enzymes catalyze these reactions?

23. Discuss the similarities and differences between starch and cellulose.

24. In what form is carbohydrate stored in the body?

25. Discuss, in simple terms, the metabolism of carbohydrates in the human body.

26. State the natural sources of sucrose, maltose, lactose, and starch.

27. Invert sugar, obtained by the hydrolysis of sucrose to an equal-molar mixture of fructose and glucose, is commonly used as a sweetener in commercial food preparations. Why is invert sugar sweeter than the original sucrose?

28. What is galactosemia and how does it affect humans?

29. What properties of molecules cause them to be classified as lipids?

30. Write structural formulas for glycerol, stearic acid, palmitic acid, oleic acid, and linoleic acid.

31. Distinguish both chemically and physically between a fat and a vegetable oil.

32. What is a triacylglycerol? Give an example.

33. Write the structure for tristearin, a fat in which all the fatty acid units are stearic acid.

34. Write the structure of a triacylglycerol that contains one unit each of linoleic, stearic, and oleic acids. How many other formulas are possible in which the triacylglycerol contains one unit each of these acids?

35. Write equations for the saponification of
    (a) tripalmitin (a fat in which all the fatty acids are palmitic acid)
    (b) the triacylglycerol of Question 34.
    Which product(s) are soaps?

36. How can vegetable oils be solidified? What is the advantage of solidifying these oils?

37. What functions do fats have in the human body?

38. Which fatty acids are essential to human diets?

39. Draw the structural formula of cholesterol.

40. Draw the ring structure that is common to all steroids.

41. List six foods that are major sources of proteins.

42. What functional groups are present in amino acids?

43. Why are the amino acids of proteins called α-amino acids?

44. In Table 20.3, which amino acids (a) are aromatic, (b) contain sulfur, (c) contain an —OH group?

45. Write the full structure for the two possible dipeptides containing glycine and phenylalanine.

46. Write structures for
(a) glycylglycine
(b) glycylglycylalanine
(c) leucylmethionylglycylserine

47. Using amino acid abbreviations, write all the possible tripeptides containing one unit each of glycine, phenylalanine, and leucine.

48. The artificial sweetener aspartame is a dipeptide made from two amino acids, aspartic acid and phenylalanine. Write two possible structures for this dipeptide.

49. What are essential amino acids? Write the names of the amino acids that are essential for humans.

50. When proteins are eaten by a human, what are the metabolic fates of the protein material?

51. Why should protein be continually included in a balanced diet?

52. Write structural formulas for the compounds that make up DNA.

53. (a) What are the three units that make up a nucleotide?
(b) List the components of the four types of nucleotides found in DNA.
(c) Write the structure and name of one of these nucleotides.

54. Briefly describe the structure of DNA as proposed by Watson and Crick.

55. What is the role of hydrogen bonding in the structure of DNA?

56. Explain the concept of complementary bases and how it relates to DNA.

57. A segment of a DNA strand has a base sequence of C-G-A-T-T-G-C-A. What is the base sequence of the other complementary strand of the double helix?

58. Explain the replication process of DNA.

59. Briefly discuss the relationship of DNA to genetics.

60. What are the three differences between DNA and RNA in terms of structure?

61. Distinguish between cell division in mitosis and in meiosis.

62. Specifically, where does protein synthesis occur?

63. What is the relationship between proteins and amino acids?

64. In Figure 20.7, what do the red dotted lines represent?

65. What are enzymes and what is their role in the body?

66. What is meant by specificity of an enzyme?

67. In the polypeptide Tyr-Gly-His-Phe-Val, identify the N-terminal and the C-terminal residues.

68. How are polypeptides numbered? Number the polypeptide in Question 67.

69. What is the name of the bond that bonds one α-amino acid to another in a protein?

70. Differentiate between the lock-and-key and induced-fit models for enzyme function.

71. What is meant by enzyme specificity?

72. What is the function of enzymes in the body?

73. Use molecular structure to explain the following:
(a) Fructose is a ketone.
(b) Glucose is an aldehyde.

74. What is the simplest empirical formula for a carbohydrate?

75. How can you discern the difference between an amino acid and an ordinary carboxylic acid?

***76.** What change in oxidation state is experienced by the carbon in the combustion of sugar, $C_{12}H_{22}O_{11}$, to CO_2?

77. Write the formula for the triacylglycerol formed from
(a) glycerol and butanoic acid
(b) glycerol and one molecule each of stearic ($C_{17}H_{35}COOH$), palmitic ($C_{15}H_{31}COOH$), and myristic ($C_{13}H_{27}COOH$) acids

78. The molar mass of cellulose is approximately 6.0×10^5 g/mol and the molar mass of a soluble starch is on the order of 4.0×10^3 g/mol. The monomer unit in both of these molecules has the empirical formula $C_6H_{10}O_5$. The units are about 5.0×10^{-10} m long. About how many units occur in each molecule, and how long are the molecules of cellulose and starch as a result?

Answers to Practice Exercises

20.1

$$CH_2-O-\overset{\overset{O}{\|}}{C}-(CH_2)_{14}CH_3$$
$$CH-O-\overset{\overset{O}{\|}}{C}-(CH_2)_{16}CH_3$$
$$CH_2-O-\overset{\overset{O}{\|}}{C}-(CH_2)_7CH=CH(CH_2)_7CH_3$$

20.2 α-aminobutyric acid

$$CH_3CH_2\underset{\underset{NH_2}{|}}{CH}COOH$$

β-aminobutyric acid

$$CH_3\underset{\underset{NH_2}{|}}{CH}CH_2COOH$$

20.3 (a) prolyltyrosine
(b) glycylhistidylleucine

20.4 (a)

$$CH_3\underset{\underset{NH_2}{|}}{C}\overset{\overset{O}{\|}}{H}C-NH\underset{\overset{|}{CH_2OH}}{CH}COOH$$

(b)

$$NH_2CH_2\overset{\overset{O}{\|}}{C}-NH\underset{\underset{CH_3CHOH}{|}}{CH}\overset{\overset{O}{\|}}{C}NH\underset{\underset{CH_2CH_2SCH_3}{|}}{CH}COOH$$

20.5 This nucleotide contains ribose.

PUTTING IT TOGETHER: Review for Chapters 18–20

Multiple Choice:
Choose the correct answer to each of the following.

1. If $^{238}_{92}U$ loses an alpha particle, the resulting nuclide is
   (a) $^{237}_{92}U$
   (c) $^{238}_{93}Np$
   (b) $^{234}_{90}Th$
   (d) $^{236}_{90}Th$

2. If $^{210}_{82}Pb$ loses a beta particle, the resulting nuclide is
   (a) $^{209}_{83}Bi$
   (c) $^{206}_{80}Hg$
   (b) $^{210}_{81}Ti$
   (d) $^{210}_{83}Bi$

3. In the equation $^{209}_{83}Bi + ? \longrightarrow ^{210}_{84}Po + ^{1}_{0}n$, the missing bombarding particle is
   (a) $^{2}_{1}H$
   (c) $^{4}_{2}He$
   (b) $^{1}_{0}n$
   (d) $^{0}_{-1}e$

4. Which of the following is not a characteristic of nuclear fission?
   (a) Upon absorption of a proton, a heavy nucleus splits into two or more smaller nuclei.
   (b) Two or more neutrons are produced from the fission of each atom.
   (c) Large quantities of energy are produced.
   (d) Most nuclei formed are radioactive.

5. The half-life of Sn-121 is 10 days. If you started with 40 g of this isotope, how much would you have left 30 days later?
   (a) 10 g
   (c) 15 g
   (b) none
   (d) 5 g

6. $^{241}_{94}Pu$ successively emits $\beta, \alpha, \alpha, \beta, \alpha, \alpha$. At that point, the nuclide has become
   (a) $^{225}_{94}Pu$
   (c) $^{207}_{84}Po$
   (b) $^{225}_{88}Ra$
   (d) $^{219}_{84}Po$

7. Calculate the nuclear binding energy of $^{56}_{26}Fe$.

   Mass data:

   $^{56}_{26}Fe = 55.9349$ g/mol

   n = 1.0087 g/mol $e^- = 0.00055$ g/mol

   p = 1.0073 g/mol $1.0\,g = 9.0 \times 10^{13}\,J$

   (a) 4.8×10^{13} J/mol
   (c) 0.5302 g/mol
   (b) 56.4651 g/mol
   (d) 4.9×10^{15} J/mol

8. The radioactivity ray with the greatest penetrating ability is
   (a) alpha
   (c) gamma
   (b) beta
   (d) proton

9. In a nuclear reaction,
   (a) mass is lost
   (b) mass is gained
   (c) mass is converted into energy
   (d) energy is converted into mass

10. As the temperature of a radionuclide increases, its half-life
    (a) increases
    (c) remains the same
    (b) decreases
    (d) fluctuates

11. The nuclide that has the longest half-life is
    (a) $^{238}_{92}U$
    (c) $^{234}_{90}Th$
    (b) $^{210}_{82}Pb$
    (d) $^{222}_{88}Ra$

12. Which of the following is not a unit of radiation?
    (a) curie
    (c) rod
    (b) roentgen
    (d) rem

13. When $^{235}_{92}U$ is bombarded by a neutron, the atom can fission into
    (a) $^{124}_{53}I + ^{109}_{47}Ag + 2\,^{1}_{0}n$
    (b) $^{124}_{50}Sn + ^{110}_{42}Mo + 2\,^{1}_{0}n$
    (c) $^{134}_{56}Ba + ^{128}_{36}Xe + 2\,^{1}_{0}n$
    (d) $^{90}_{38}Sr + ^{143}_{58}Ce + 2\,^{1}_{0}n$

14. In the nuclear equation
    $^{45}_{21}Sc + ^{1}_{0}n \longrightarrow X + ^{1}_{1}H$
    the nuclide X that is formed is
    (a) $^{45}_{22}Ti$
    (c) $^{46}_{22}Ti$
    (b) $^{45}_{20}Ca$
    (d) $^{45}_{20}K$

15. What type of radiation is a very energetic form of photon?
    (a) alpha
    (c) gamma
    (b) beta
    (d) positron

16. When $^{239}_{92}U$ decays to $^{239}_{93}Np$, what particle is emitted?
    (a) positron
    (c) alpha particle
    (b) neutron
    (d) beta particle

17. The roentgen is the unit of radiation that measures
    (a) an absorbed dose of radiation
    (b) exposure to X-rays
    (c) the dose from a different type of radiation
    (d) the rate of decay of a radioactive substance

18. Which of the following is not a correct name for the alkane shown with it?
    (a) C_2H_6, ethane
    (c) C_7H_{16}, heptane
    (b) C_5H_{12}, propane
    (d) $C_{10}H_{22}$, decane

19. The structural formula of *o*-xylene is

20. The ester

$$\underset{\text{O}\quad\text{CH}_3}{\overset{\parallel\quad\;\;|}{CH_3C-O-CHCH_3}}$$

can be made from which alcohol and carboxylic acid?

(a) $CH_3\overset{\overset{\text{O}}{\parallel}}{C}-OH,\quad CH_3CH_2CH_2OH$

(b) $CH_3\overset{\overset{\text{O}}{\parallel}}{C}-OH,\quad HO-\overset{\overset{\text{CH}_3}{|}}{C}HCH_2OH$

(c) $H-\overset{\overset{\text{O}}{\parallel}}{C}-OH,\quad CH_3\overset{\overset{\text{OH}}{|}}{C}HCH_3$

(d) $CH_3\overset{\overset{\text{O}}{\parallel}}{C}-OH,\quad CH_3\overset{\overset{\text{OH}}{|}}{C}HCH_3$

21. The product of the reaction

$CH_3CH=CH_2 + H_2O + H^+ \longrightarrow$ is a(n)

(a) alcohol   (c) alkyne
(b) aldehyde   (d) carboxylic acid

22. The number of isomers of butyl alcohol, C_4H_9OH, is
(a) 2   (b) 3   (c) 4   (d) 5

23. What is the correct name for this compound?

$$CH_3-\overset{\overset{\text{CH}_3}{|}}{C}H-CH-\overset{\overset{\text{CH}_3}{|}}{C}H-CH=CH_2$$
$$\underset{CH_3-CH-CH_3}{\overset{|}{}}$$

(a) isobutane
(b) 2,4-methyl-3-propyl-5-hexene
(c) 3,5-dimethyl-4-isopropyl-1-hexene
(d) 4,4-diisopropylhexene

24. Which of these acids is named incorrectly?
(a) $CH_3CH_2CH_2COOH$, butyric acid
(d) HCOOH, formic acid
(c) CH_3CH_2COOH, propic acid
(d) CH_3COOH, acetic acid

25. With acid as a catalyst, ethanol and formic acid will react to form

(a) $CH_3\overset{\overset{\text{O}}{\parallel}}{C}-OH$

(b) $H-\overset{\overset{\text{O}}{\parallel}}{C}-O-CH_2CH_3$

(c) $CH_3\overset{\overset{\text{O}}{\parallel}}{C}-O-CH_3$

(d) $CH_3\overset{\overset{\text{O}}{\parallel}}{C}-O-\overset{\overset{\text{O}}{\parallel}}{C}CH_3$

26. The number of isomers of C_6H_{14} is
(a) 3   (b) 5   (c) 6   (d) 8

27. An open-chain hydrocarbon of formula C_6H_8 can have in its formula
(a) one carbon–carbon double bond
(b) two carbon–carbon double bonds
(c) one carbon–carbon triple bond
(d) one carbon–carbon double bond and one carbon–carbon triple bond

28. The reaction

$CH_2=CH_2 + Br_2 \longrightarrow CH_2BrCH_2Br$

represents
(a) dehalogenation   (c) addition
(b) substitution   (d) dehydration

29. The general formula for a ketone is
(a) RCHO   (c) RCOOR
(b) ROR   (d) R_2CO

30. Which of the following cannot be an aromatic compound?
(a) C_6H_5OH (c) C_6H_{14}
(b) C_6H_6 (d) $C_6H_5CH_3$

31. What is the correct name for this compound?

(a) *m*-dinitrophenol   (c) 3,5-dinitrophenol
(b) 2,4-dinitrophenol   (d) 1,3-dinitrophenol

32. Which of these pairs are not isomers?
(a) CH_3OCH_2Cl and CH_2ClCH_2OH
(b) CH_3CH_2CHO and $CH_3OCH_2CH_3$
(c) $CH_3OCH_2OCH_3$ and $CH_3CH(OH)CH_2OH$
(d) $C_6H_4(CH_3)_2$ and $C_6H_5CH_2CH_3$

33. The reaction of $CH_3CH=CHCH_3 + HBr$ produces
(a) $CH_3CH_2CH_2CH_3 + Br_2$
(b) $CH_3CHBrCHBrCH_3 + H_2$
(c) $CH_3CHBrCH_2CH_3$
(d) $CH_3CH_2CH_2CH_2Br$

34. Polyvinyl chloride is a polymer of
(a) $CH_2=CCl_2$
(b) $CF_2=CF_2$
(c) $CH_2=CHCl$
(d) $C_6H_5CH=CHCl$

35. What is the correct name for this compound?

$$\underset{\qquad\quad\text{H}}{\overset{\text{O}\qquad\text{CH}_3}{\overset{\parallel\qquad\;|}{CH_3C-O-\underset{|}{C}-CH_3}}}$$

(a) acetyl-2-propanoate
(b) propyl acetate
(c) ethyl isopropylate
(d) isopropyl ethanoate

36. Teflon is a polymer of
(a) $CH_2=CCl_2$ (c) $CH_2=CHCl$
(b) $CF_2=CF_2$ (d) $C_6H_5CH=CHCl$

37. Sugars are members of a group of compounds with the general name
(a) carbohydrates (c) proteins
(b) lipids (d) steroids

38. The products formed when maltose is hydrolyzed are
(a) glucose and fructose
(b) glucose and galactose
(c) glucose and glucose
(d) galactose and fructose

39. Which is *not* true of starch?
(a) It is a polysaccharide.
(b) It is hydrolyzed to maltose.
(c) It is composed of glucose units.
(d) It is not digestible by humans.

40. Lactose is
(a) a monosaccharide
(b) a disaccharide composed of galactose and glucose
(c) a disaccharide composed of two glucose units
(d) a decomposition product of starch

41. Which is *not* true of glucose?
(a) It is a monosaccharide.
(b) It is a component of sucrose, maltose, lactose, starch, glycogen, and cellulose.
(c) It is a ketohexose.
(d) It is the main source of energy for the body.

42. The sweetest of the common sugars is
(a) fructose (c) glucose
(b) sucrose (d) maltose

43. Which of the following is *not* true?
(a) Lactose is sweeter than sucrose.
(b) Glycogen is known as animal starch.
(c) Cellulose is the most abundant organic substance in nature.
(d) Lactose is a disaccharide known as milk sugar.

44. Which is *not* formed in the saponification of a fat?
(a) glycerol
(b) amino acids
(c) soap
(d) a metal salt of a long-chain fatty acid

45. Which of these lipids does *not* contain a glycerol unit as part of its structure?
(a) a fat (c) a glycolipid
(b) a phospholipid (d) an oil

46. An α-amino acid always contains
(a) an amino group on the carbon atom adjacent to the carboxyl group
(b) a carboxyl group at each end of the molecule
(c) two amino groups
(d) alternating amino and carboxyl groups

47. Which of these amino acids contains sulfur?
(a) alanine (c) cysteine
(b) histidine (d) glycine

48. A compound containing ten amino acid molecules linked together is called a
(a) protein (c) deca-amino acid
(b) polypeptide (d) nucleotide

49. Which of the following is *not* a correct statement about DNA and RNA?
(a) DNA contains deoxyribose, whereas RNA contains ribose.
(b) Both DNA and RNA are polymers made up of nucleotides.
(c) DNA directs the synthesis of proteins and RNA directs the genetic code of life.
(d) DNA exists as a double helix, whereas RNA exists as a single helix.

50. Which of these bases is found in RNA but not in DNA?
(a) thymine (c) guanine
(b) adenine (d) uracil

51. Complementary base pairs in DNA are linked through the formation of
(a) phosphate ester bonds
(b) peptide linkages
(c) hydrogen bonds
(d) ionic bonds

52. In a DNA double helix, hydrogen bonding occurs between
(a) adenine and thymine
(b) thymine and guanine
(c) adenine and uracil
(d) cytosine and thymine

53. Which of these scientists did *not* receive the Nobel Prize for the structure of DNA?
(a) Crick
(b) Watson
(c) Sanger
(d) Wilkins

54. The substance acted on by an enzyme is called a(n)
(a) catalyst (c) coenzyme
(b) apoenzyme (d) substrate

55. The process during which a cell splits to form a sperm or an egg cell is called
(a) mitosis (c) translation
(b) meiosis (d) transcription

Free Response Questions

Answer each of the following. Be sure to include your work and explanations in a clear, logical form.

1. The nuclide $^{223}_{87}$Fr emits three particles losing 8 mass units and 3 atomic number units. Propose a radioactive decay series and write the symbol for the resulting nuclide.

2. Biological molecules such as methionine enkephalin, leucine enkephalin, and β-endorphin begin with a common tetrapeptide tyrosyl-glycyl-glycyl-phenylalanine. Draw the tetrapeptide sequence and write its abbreviated form.

3. What functional groups are in prostaglandin A_2?

prostaglandin A_2

4. Which of the following molecules would you expect to be significantly soluble in water? Why?

$CH_3CH_2CH_2OH$, $CH_3CH_2CH_2CHO$

$CH_3CH_2COCH_3$, $CH_3CH_2CH_2CH_3$

(*Hint:* Remember that "like dissolves like.")

5. Which of these compounds are structural isomers of each other?

methyl propyl ether, butanoic acid, acetone, propene, 2-butanol, phenol, methyl propanoate

6. (a) Indicate which two amino acids could form an ester bond using their R group side chains.

(b) Draw the product of cholesterol reacting with bromine.

cholesterol

(c) Is cholesterol soluble in water? How do you know?

7. (a) Carboxypeptidase A is a digestive enzyme. A zinc ion is necessary for enzyme activity. What is the zinc ion called?

(b) Some RNA strands also show enzymatic activity but DNA does not. What are two structural differences between RNA and DNA?

(c) Do you think we have radioactive isotopes in our bodies? Explain.

8. (a) For each OH group in open-chain glucose, indicate if it is primary, secondary, or tertiary.

(b) Does tyrosine contain an *ortho-*, *meta-*, or *para-*substituted aromatic ring?

(c) How many amino acids are in a pentapeptide? How many molecules of water were lost when it was formed?

(d) Based on what you know about the hydrolysis of disaccharides, if a person has high blood sugar levels, do you think it would be better for them to cut down on lactose or maltose? Why?

9. How many half-lives are required to change 96 g of a sample of a radioactive isotope to 1.5 g over approximately 24 days? What is the approximate half-life of this isotope?

10. (a) What type of process is $^6_3Li + ^1_0n \longrightarrow ^3_1H + ^4_2He$?

(b) How is this process different from radioactive decay?

(c) Could nuclear fission be classified as the same type process?

Mathematical Review

Multiplication Multiplication is a process of adding any given number or quantity to itself a certain number of times. Thus, 4 times 2 means 4 added two times, or 2 added together four times, to give the product 8. Various ways of expressing multiplication are

$$ab \qquad a \times b \qquad a \cdot b \qquad a(b) \qquad (a)(b)$$

Each of these expressions means a times b, or a multiplied by b, or b times a.

When $a = 16$ and $b = 24$, we have $16 \times 24 = 384$.

The expression $°F = (1.8 \times °C) + 32$ means that we are to multiply 1.8 times the Celsius degrees and add 32 to the product. When $°C$ equals 50,

$$°F = (1.8 \times 50) + 32 = 90 + 32 = 122°F$$

The result of multiplying two or more numbers together is known as the *product*.

Division The word *division* has several meanings. As a mathematical expression, it is the process of finding how many times one number or quantity is contained in another. Various ways of expressing division are

$$a \div b \qquad \frac{a}{b} \qquad a/b$$

Each of these expressions means a divided by b.

When $a = 15$ and $b = 3$, $\frac{15}{3} = 5$.

The number above the line is called the *numerator;* the number below the line is the *denominator*. Both the horizontal and the slanted (/) division signs also mean "per." For example, in the expression for density, we determine the mass per unit volume:

$$\text{density} = \text{mass/volume} = \frac{\text{mass}}{\text{volume}} = \text{g/mL}$$

The diagonal line still refers to a division of grams by the number of milliliters occupied by that mass. The result of dividing one number into another is called the *quotient*.

Fractions and Decimals A fraction is an expression of division, showing that the numerator is divided by the denominator. A *proper fraction* is one in which the numerator is smaller than the denominator. In an *improper fraction*, the numerator is the larger number. A decimal or a decimal fraction is a proper fraction in which the

Proper fraction	Decimal fraction	Proper fraction
$\dfrac{1}{8}$ =	0.125 =	$\dfrac{125}{1000}$
$\dfrac{1}{10}$ =	0.1 =	$\dfrac{1}{10}$
$\dfrac{3}{4}$ =	0.75 =	$\dfrac{75}{100}$
$\dfrac{1}{100}$ =	0.01 =	$\dfrac{1}{100}$
$\dfrac{1}{4}$ =	0.25 =	$\dfrac{25}{100}$

denominator is some power of 10. The decimal fraction is determined by carrying out the division of the proper fraction. Examples of proper fractions and their decimal fraction equivalents are shown in the accompanying table.

Addition of Numbers with Decimals To add numbers with decimals, we use the same procedure as that used when adding whole numbers, but we always line up the decimal points in the same column. For example, add 8.21 + 143.1 + 0.325:

$$
\begin{array}{r}
8.21 \\
+143.1 \\
+\ \ \ 0.325 \\
\hline
151.635
\end{array}
$$

When adding numbers that express units of measurement, we must be certain that the numbers added together all have the same units. For example, what is the total length of three pieces of glass tubing: 10.0 cm, 125 mm, and 8.4 cm? If we simply add the numbers, we obtain a value of 143.4, but we are not certain what the unit of measurement is. To add these lengths correctly, first change 125 mm to 12.5 cm. Now all the lengths are expressed in the same units and can be added:

$$
\begin{array}{r}
10.0\ \text{cm} \\
12.5\ \text{cm} \\
8.4\ \text{cm} \\
\hline
30.9\ \text{cm}
\end{array}
$$

Subtraction of Numbers with Decimals To subtract numbers containing decimals, we use the same procedure as for subtracting whole numbers, but we always line up the decimal points in the same column. For example, subtract 20.60 from 182.49:

$$
\begin{array}{r}
182.49 \\
-\ \ 20.60 \\
\hline
161.89
\end{array}
$$

When subtracting numbers that are measurements, be certain that the measurements are in the same units. For example, subtract 22 cm from 0.62 m. First change m to cm, then do the subtraction.

$$
(0.62\ \text{m})\left(\frac{100\ \text{cm}}{\text{m}}\right) = 62\ \text{cm} \qquad
\begin{array}{r}
62\ \ \text{cm} \\
-22\ \ \text{cm} \\
\hline
40.\ \text{cm}
\end{array}
$$

Multiplication of Numbers with Decimals To multiply two or more numbers together that contain decimals, we first multiply as if they were whole numbers. Then, to locate the decimal point in the product, we add together the number of digits to the right of the decimal in all the numbers multiplied together. The product should have this same number of digits to the right of the decimal point.

If a number is a measurement, the answer must be adjusted to the correct number of significant figures.

Multiply 2.05 × 2.05 = 4.2025   (total of four digits to the right of the decimal). Here are more examples:

14.25 × 6.01 × 0.75 = 64.231875   (six digits to the right of the decimal)

39.26 × 60 = 2355.60   (two digits to the right of the decimal)

Division of Numbers with Decimals To divide numbers containing decimals, we first relocate the decimal points of the numerator and denominator by moving them to the right as many places as needed to make the denominator a whole number. (Move the decimal of both the numerator and the denominator the same amount and in the same direction.) For example,

$$\frac{136.94}{4.1} = \frac{1369.4}{41}$$

The decimal point adjustment in this example is equivalent to multiplying both numerator and denominator by 10. Now we carry out the division normally, locating the decimal point immediately above its position in the dividend:

$$41\overline{)1269.4}^{\,33.4} \qquad \frac{0.441}{26.25} = \frac{44.1}{2625} = 2625\overline{)44.1000}^{\,0.0168}$$

These examples are guides to the principles used in performing the various mathematical operations illustrated. Every student of chemistry should learn to use a calculator for solving mathematical problems (see Appendix II). The use of a calculator will save many hours of doing tedious calculations. After solving a problem, the student should check for errors and evaluate the answer to see if it is logical and consistent with the data given.

Algebraic Equations Many mathematical problems that are encountered in chemistry fall into the following algebraic forms. Solutions to these problems are simplified by first isolating the desired term on one side of the equation. This rearrangement is accomplished by treating both sides of the equation in an identical manner until the desired term is isolated.

(a) $a = \dfrac{b}{c}$

To solve for b, multiply both sides of the equation by c:

$$a \times c = \frac{b}{\cancel{c}} \times \cancel{c}$$

$$b = a \times c$$

To solve for c, multiply both sides of the equation by $\dfrac{c}{a}$:

$$\cancel{a} \times \frac{c}{\cancel{a}} = \frac{b}{\cancel{c}} \times \frac{\cancel{c}}{a}$$

$$c = \frac{b}{a}$$

(b) $\dfrac{a}{b} = \dfrac{c}{d}$

To solve for a, multiply both sides of the equation by b:

$$\frac{a}{\cancel{b}} \times \cancel{b} = \frac{c}{d} \times b$$

$$a = \frac{c \times b}{d}$$

To solve for b, multiply both sides of the equation by $\dfrac{b \times d}{c}$:

$$\frac{a}{\cancel{b}} \times \frac{\cancel{b} \times d}{c} = \frac{\cancel{c}}{\cancel{d}} \times \frac{b \times \cancel{d}}{\cancel{c}}$$

$$\frac{a \times d}{c} = b$$

(c) $a \times b = c \times d$

To solve for a, divide both sides of the equation by b:

$$\frac{a \times \cancel{b}}{\cancel{b}} = \frac{c \times d}{b}$$

$$a = \frac{c \times d}{b}$$

(d) $\dfrac{b - c}{a} = d$

To solve for b, first multiply both sides of the equation by a:

$$\frac{\cancel{a}(b - c)}{\cancel{a}} = d \times a$$

$$b - c = d \times a$$

Then add c to both sides of the equation:

$$b - \cancel{c} + \cancel{c} = d \times a + c$$

$$b = (d \times a) + c$$

When $a = 1.8, c = 32$, and $d = 35$,

$$b = (35 \times 1.8) + 32 = 63 + 32 = 95$$

Expression of Large and Small Numbers   In scientific measurement and calculations, we often encounter very large and very small numbers—for example, 0.00000384 and 602,000,000,000,000,000,000,000. These numbers are troublesome to write and awkward to work with, especially in calculations. A convenient method of expressing these large and small numbers in a simplified form is by means of exponents, or powers, of 10. This method of expressing numbers is known as **scientific, or exponential, notation.**

An **exponent** is a number written as a superscript following another number. Exponents are often called *powers* of numbers. The term *power* indicates how many times the number is used as a factor. In the number 10^2, 2 is the exponent, and the number means 10 squared, or 10 to the second power, or $10 \times 10 = 100$. Three other examples are

$$3^2 = 3 \times 3 = 9$$
$$3^4 = 3 \times 3 \times 3 \times 3 = 81$$
$$10^3 = 10 \times 10 \times 10 = 1000$$

For ease of handling, large and small numbers are expressed in powers of 10. Powers of 10 are used because multiplying or dividing by 10 coincides with moving the decimal point in a number by one place. Thus, a number multiplied by 10^1 would move the decimal point one place to the right; 10^2, two places to the right; 10^{-2}, two places to the left. To express a number in powers of 10, we move the decimal point in the original number to a new position, placing it so that the number is a value between 1 and 10. This new decimal number is multiplied by 10 raised to the proper power. For example, to write the number 42,389 in exponential form, the decimal point is placed between the 4 and the 2 (4.2389), and the number is multiplied by 10^4; thus, the number is 4.2389×10^4:

$$42{,}389 = 4.2389 \times 10^4$$
4 3 2 1

scientific, or exponential, notation
exponent

The exponent of 10 (4) tells us the number of places that the decimal point has been moved from its original position. If the decimal point is moved to the left, the exponent is a positive number; if it is moved to the right, the exponent is a negative number. To express the number 0.00248 in exponential notation (as a power of 10), the decimal point is moved three places to the right; the exponent of 10 is -3, and the number is 2.48×10^{-3}.

$$0.00248 = 2.48 \times 10^{-3}$$
$$\underset{1\,2\,3}{\underbrace{\qquad}}$$

Study the following examples of changing a number to scientific notation.

$$1237 = 1.237 \times 10^3$$
$$988 = 9.88 \times 10^2$$
$$147.2 = 1.472 \times 10^2$$
$$2{,}200{,}000 = 2.2 \times 10^6$$
$$0.0123 = 1.23 \times 10^{-2}$$
$$0.00005 = 5 \times 10^{-5}$$
$$0.000368 = 3.68 \times 10^{-4}$$

Exponents in multiplication and division The use of powers of 10 in multiplication and division greatly simplifies locating the decimal point in the answer. In multiplication, first change all numbers to powers of 10, then multiply the numerical portion in the usual manner, and finally add the exponents of 10 algebraically, expressing them as a power of 10 in the product. In multiplication, the exponents (powers of 10) are added algebraically.

$$10^2 \times 10^3 = 10^{(2+3)} = 10^5$$
$$10^2 \times 10^2 \times 10^{-1} = 10^{(2+2-1)} = 10^3$$

Multiply: (40,000)(4200)
Change to powers of 10: $(4 \times 10^4)(4.2 \times 10^3)$
Rearrange: $(4 \times 4.2)(10^4 \times 10^3)$
$16.8 \times 10^{(4+3)}$
16.8×10^7 or 1.68×10^8 (Answer)

Multiply: (380)(0.00020)
$(3.80 \times 10^2)(2.0 \times 10^{-4})$
$(3.80 \times 2.0)(10^2 \times 10^{-4})$
$7.6 \times 10^{(2-4)}$
7.6×10^{-2} or   0.076   (Answer)

Multiply: (125)(284)(0.150)
$(1.25 \times 10^2)(2.84 \times 10^2)(1.50 \times 10^{-1})$
$(1.25)(2.84)(1.50)(10^2 \times 10^2 \times 10^{-1})$
$5.325 \times 10^{(2+2-1)}$
5.33×10^3 (Answer)

In division, after changing the numbers to powers of 10, move the 10 and its exponent from the denominator to the numerator, changing the sign of the exponent. Carry out the division in the usual manner, and evaluate the power of 10. Change the sign(s) of the exponent(s) of 10 in the denominator, and move the 10 and its exponent(s) to the numerator. Then add all the exponents of 10 together. For example,

$$\frac{10^5}{10^3} = 10^5 \times 10^{-3} = 10^{(5-3)} = 10^2$$

$$\frac{10^3 \times 10^4}{10^{-2}} = 10^3 \times 10^4 \times 10^2 = 10^{(3+4+2)} = 10^9$$

Significant Figures in Calculations The result of a calculation based on experimental measurements cannot be more precise than the measurement that has the greatest uncertainty. (See Section 2.4 for additional discussion.)

Addition and subtraction The result of an addition or subtraction should contain no more digits to the right of the decimal point than are contained in the quantity that has the least number of digits to the right of the decimal point.

Perform the operation indicated and then round off the number to the proper number of significant figures:

(a) 142.8   g
     18.843 g
     36.42 g
     198.063 g
     198.1   g   (Answer)

(b)   93.45 mL
     −18.0 mL
      75.45 mL
      75.5 mL   (Answer)

(a) The answer contains only one digit after the decimal point since 142.8 contains only one digit after the decimal point.

(b) The answer contains only one digit after the decimal point since 18.0 contains one digit after the decimal point.

Multiplication and division In calculations involving multiplication or division, the answer should contain the same number of significant figures as the measurement that has the least number of significant figures. In multiplication or division, the position of the decimal point has nothing to do with the number of significant figures in the answer. Study the following examples:

Round off to

$(2.05)(2.05) = 4.2025$ 4.20

$(18.48)(5.2) = 96.096$ 96

$(0.0126)(0.020) = 0.000252$ or

$(1.26 \times 10^{-2})(2.0 \times 10^{-2}) = 2.520 \times 10^{-4}$ 2.5×10^{-4}

$\dfrac{1369.4}{41} = 33.4$ 33

$\dfrac{2268}{4.20} = 540.$ 540.

Dimensional Analysis Many problems of chemistry can be readily solved by dimensional analysis using the factor-label or conversion-factor method. Dimensional analysis involves the use of proper units of dimensions for all factors that are multiplied, divided, added, or subtracted in setting up and solving a problem. Dimensions are physical quantities such as length, mass, and time, which are expressed in such units as centimeters, grams, and seconds, respectively. In solving a problem, we treat these units mathematically just as though they were numbers, which gives us an answer that contains the correct dimensional units.

A measurement or quantity given in one kind of unit can be converted to any other kind of unit having the same dimension. To convert from one kind of unit to another, the original quantity or measurement is multiplied or divided by a conversion factor. The key to success lies in choosing the correct conversion factor. This general method of calculation is illustrated in the following examples.

Suppose we want to change 24 ft to inches. We need to multiply 24 ft by a conversion factor containing feet and inches. Two such conversion factors can be written relating inches to feet:

$\dfrac{12\text{ in.}}{1\text{ ft}}$ or $\dfrac{1\text{ ft}}{12\text{ in.}}$

We choose the factor that will mathematically cancel feet and leave the answer in inches. Note that the units are treated in the same way we treat numbers, multiplying or dividing as required. Two possibilities then arise to change 24 ft to inches:

$$(24 \text{ ft})\left(\frac{12 \text{ in.}}{1 \text{ ft}}\right) \quad \text{or} \quad (24 \text{ ft})\left(\frac{1 \text{ ft}}{12 \text{ in.}}\right)$$

In the first case (the correct method), feet in the numerator and the denominator cancel, giving us an answer of 288 in. In the second case, the units of the answer are ft²/in., the answer being 2.0 ft²/in. In the first case, the answer is reasonable because it is expressed in units having the proper dimensions. That is, the dimension of length expressed in feet has been converted to length in inches according to the mathematical expression

$$\text{ft} \times \frac{\text{in.}}{\text{ft}} = \text{in.}$$

In the second case, the answer is not reasonable because the units (ft²/in.) do not correspond to units of length. The answer is therefore incorrect. The units are the guiding factor for the proper conversion.

The reason we can multiply 24 ft times 12 in./ft and not change the value of the measurement is that the conversion factor is derived from two equivalent quantities. Therefore, the conversion factor 12 in./ft is equal to unity. When you multiply any factor by 1, it does not change the value:

$$12 \text{ in.} = 1 \text{ ft} \quad \text{and} \quad \frac{12 \text{ in.}}{1 \text{ ft}} = 1$$

Convert 16 kg to milligrams. In this problem it is best to proceed in this fashion:

$$\text{kg} \longrightarrow \text{g} \longrightarrow \text{mg}$$

The possible conversion factors are

$$\frac{1000 \text{ g}}{1 \text{ kg}} \quad \text{or} \quad \frac{1 \text{ kg}}{1000 \text{ g}} \qquad \frac{1000 \text{ mg}}{1 \text{ g}} \quad \text{or} \quad \frac{1 \text{ g}}{1000 \text{ mg}}$$

We use the conversion factor that leaves the proper unit at each step for the next conversion. The calculation is

$$(16 \text{ kg})\left(\frac{1000 \text{ g}}{1 \text{ kg}}\right)\left(\frac{1000 \text{ mg}}{1 \text{ g}}\right) = 1.6 \times 10^7 \text{ mg}$$

Regardless of application, the basis of dimensional analysis is the use of conversion factors to organize a series of steps in the quest for a specific quantity with a specific unit.

Graphical Representation of Data   A graph is often the most convenient way to present or display a set of data. Various kinds of graphs have been devised, but the most common type uses a set of horizontal and vertical coordinates to show the relationship of two variables. It is called an x–y graph because the data of one variable are represented on the horizontal or x-axis (abscissa) and the data of the other variable are represented on the vertical or y-axis (ordinate). See Figure I.1.

As a specific example of a simple graph, let us graph the relationship between Celsius and Fahrenheit temperature scales. Assume that initially we have only the information in the table next to Figure I.2.

On a set of horizontal and vertical coordinates (graph paper), scale off at least 100 Celsius degrees on the x-axis and at least 212 Fahrenheit degrees on the y-axis. Locate and mark the three points corresponding to the three temperatures given and draw a line connecting these points (see Figure I.2).

Figure I.1

°C	°F
0	32
50	122
100	212

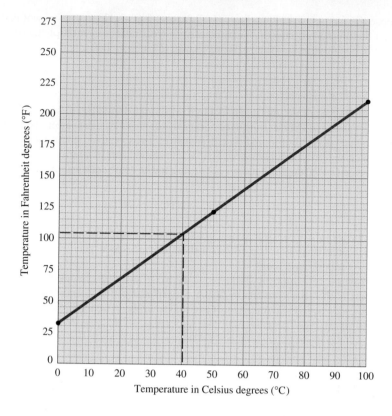

Figure I.2

Here is how a point is located on the graph: Using the (50°C, 122°F) data, trace a vertical line up from 50°C on the x-axis and a horizontal line across from 122°F on the y-axis and mark the point where the two lines intersect. This process is called *plotting*. The other two points are plotted on the graph in the same way. (*Note:* The number of degrees per scale division was chosen to give a graph of convenient size. In this case, there are 5 Fahrenheit degrees per scale division and 2 Celsius degrees per scale division.)

The graph in Figure I.2 shows that the relationship between Celsius and Fahrenheit temperature is that of a straight line. The Fahrenheit temperature corresponding to any given Celsius temperature between 0 and 100° can be determined from the graph. For example, to find the Fahrenheit temperature corresponding to 40°C, trace a perpendicular line from 40°C on the x-axis to the line plotted on the graph. Now trace a horizontal line from this point on the plotted line to the y-axis and read the corresponding Fahrenheit temperature (104°F). See the dashed lines in Figure I.2. In turn, the Celsius temperature corresponding to any Fahrenheit temperature between 32° and 212° can be determined from the graph by tracing a horizontal line from the Fahrenheit temperature to the plotted line and reading the corresponding temperature on the Celsius scale directly below the point of intersection.

The mathematical relationship of Fahrenheit and Celsius temperatures is expressed by the equation $°F = (1.8 \times °C) + 32$. Figure I.2 is a graph of this equation. Because the graph is a straight line, it can be extended indefinitely at either end. Any desired Celsius temperature can be plotted against the corresponding Fahrenheit temperature by extending the scales along both axes as necessary.

Figure I.3 is a graph showing the solubility of potassium chlorate in water at various temperatures. The solubility curve on this graph was plotted from the data in the table next to the graph.

In contrast to the Celsius–Fahrenheit temperature relationship, there is no simple mathematical equation that describes the exact relationship between temperature and the

Temperature (°C)	Solubility (g KClO₃/ 100 g water)
10	5.0
20	7.4
30	10.5
50	19.3
60	24.5
80	38.5

Figure I.3

solubility of potassium chlorate. The graph in Figure I.3 was constructed from experimentally determined solubilities at the six temperatures shown. These experimentally determined solubilities are all located on the smooth curve traced by the unbroken-line portion of the graph. We are therefore confident that the unbroken line represents a very good approximation of the solubility data for potassium chlorate over the temperature range from 10 to 80°C. All points on the plotted curve represent the composition of saturated solutions. Any point below the curve represents an unsaturated solution.

The dashed-line portions of the curve are *extrapolations;* that is, they extend the curve above and below the temperature range actually covered by the plotted solubility data. Curves such as this one are often extrapolated a short distance beyond the range of the known data, although the extrapolated portions may not be highly accurate. Extrapolation is justified only in the absence of more reliable information.

The graph in Figure I.3 can be used with confidence to obtain the solubility of $KClO_3$ at any temperature between 10° and 80°C, but the solubilities between 0° and 10°C and between 80° and 100°C are less reliable. For example, what is the solubility of $KClO_3$ at 55°C, at 40°C, and at 100°C?

First draw a perpendicular line from each temperature to the plotted solubility curve. Now trace a horizontal line to the solubility axis from each point on the curve and read the corresponding solubilities. The values that we read from the graph are

40°C   14.2 g $KClO_3$/100 g water

55°C   22.0 g $KClO_3$/100 g water

100°C   59   g $KClO_3$/100 g water

Of these solubilities, the one at 55°C is probably the most reliable because experimental points are plotted at 50° and at 60°C. The 40°C solubility value is a bit less reliable because the nearest plotted points are at 30° and 50°C. The 100°C solubility is the least reliable of the three values because it was taken from the extrapolated part of the curve, and the nearest plotted point is 80°C. Actual handbook solubility values are 14.0 and 57.0 g of $KClO_3$/100 g water at 40°C and 100°C, respectively.

The graph in Figure I.3 can also be used to determine whether a solution is saturated or unsaturated. For example, a solution contains 15 g of $KClO_3$/100 g of water and is at a temperature of 55°C. Is the solution saturated or unsaturated? *Answer:* The solution is unsaturated because the point corresponding to 15 g and 55°C on the graph is below the solubility curve; all points below the curve represent unsaturated solutions.

Using a Scientific Calculator

A calculator is useful for most calculations in this book. You should obtain a scientific calculator, that is, one that has at least the following function keys on its keyboard.

Addition $\boxed{+}$

Second function $\boxed{F}$, $\boxed{INV}$, $\boxed{Shift}$

Subtraction $\boxed{-}$

Change sign $\boxed{+/-}$

Multiplication $\boxed{\times}$

Exponential number $\boxed{Exp}$

Division $\boxed{\div}$

Logarithm $\boxed{Log}$

Equals $\boxed{=}$

Antilogarithm $\boxed{10^x}$

Not all calculators use the same symbolism for these function keys, nor do all calculators work in the same way. The following discussion may not pertain to your particular calculator. Refer to your instruction manual for variations from the function symbols shown above and for the use of other function keys.

Some keys have two functions, upper and lower. In order to use the upper (second) function, the second function key $\boxed{F}$ must be pressed in order to activate the desired upper function after entering the number.

The display area of the calculator shows the numbers entered and often shows more digits in the answer than should be used. Therefore, the final answer should be rounded to reflect the proper number of significant figures of the calculations.

The second function key may have a different designation on your calculator.

Addition and Subtraction To add numbers using your calculator,

1. Enter the first number to be added followed by the plus key $\boxed{+}$.

2. Enter the second number to be added followed by the plus key $\boxed{+}$.

3. Repeat Step 2 for each additional number to be added, except the last number.

4. After the last number is entered, press the equal key $\boxed{=}$. You should now have the answer in the display area.

5. When a number is to be subtracted, use the minus key $\boxed{-}$ instead of the plus key.

As an example, to add $16.0 + 1.223 + 8.45$, enter 16.0 followed by the $\boxed{+}$ key; then enter 1.223 followed by the $\boxed{+}$ key; then enter 8.45 followed by the $\boxed{=}$ key. The display shows 25.673, which is rounded to the answer 25.7.

Examples of Addition and Subtraction

Calculation	Enter in sequence	Display	Rounded answer
a. 12.0 + 16.2 + 122.3	12.0 $+$ 16.2 $+$ 122.3 $=$	150.5	150.5
b. 132 − 62 + 141	132 $-$ 62 $+$ 141 $=$	211	211
c. 46.23 + 13.2	46.23 $+$ 13.2 $=$	59.43	59.4
d. 129.06 + 49.1 − 18.3	129.06 $+$ 49.1 $-$ 18.3 $=$	159.86	159.9

Multiplication   To multiply numbers using your calculator,

1. Enter the first number to be multiplied followed by the multiplication key $\times$.
2. Enter the second number to be multiplied followed by the multiplication key $\times$.
3. Repeat Step 2 for all other numbers to be multiplied except the last number.
4. Enter the last number to be multiplied followed by the equal key $=$. You now have the answer in the display area.

Round off to the proper number of significant figures.

As an example, to calculate (3.25)(4.184)(22.2), enter 3.25 followed by the $\times$ key; then enter 4.184 followed by the $\times$ key; then enter 22.2 followed by the $=$ key. The display shows 301.8756, which is rounded to the answer 302.

Examples of Multiplication

Calculation	Enter in sequence	Display	Rounded answer
a. 12 × 14 × 18	12 $\times$ 14 $\times$ 18 $=$	3024	3.0×10^3
b. 122 × 3.4 × 60.	122 $\times$ 3.4 $\times$ 60. $=$	24888	2.5×10^4
c. 0.522 × 49.4 × 6.33	0.522 $\times$ 49.4 $\times$ 6.33 $=$	163.23044	163

Division   To divide numbers using your calculator,

1. Enter the numerator followed by the division key $\div$.
2. Enter the denominator followed by the equal key to give the answer.
3. If there is more than one denominator, enter each denominator followed by the division key except for the last number, which is followed by the equal key.

As an example, to calculate $\dfrac{126}{12}$, enter 126 followed by the $\div$ key; then enter 12 followed by the $=$ key. The display shows 10.5, which is rounded to the answer 11.

Examples of Division

Calculation	Enter in sequence	Display	Rounded answer
a. $\dfrac{142}{25}$	142 $\boxed{\div}$ 25 $\boxed{=}$	5.68	5.7
b. $\dfrac{0.422}{5.00}$	0.422 $\boxed{\div}$ 5.00 $\boxed{=}$	0.0844	0.0844
c. $\dfrac{124}{0.022 \times 3.00}$	124 $\boxed{\div}$ 0.022 $\boxed{\div}$ 3.00 $\boxed{=}$	1878.7878	1.9×10^3

Exponents In scientific measurements and calculations, we often encounter very large and very small numbers. To express these large and small numbers conveniently, we use exponents, or powers, of 10. A number in exponential form is treated like any other number; that is, it can be added, subtracted, multiplied, or divided.

To enter an exponential number into your calculator, first enter the nonexponential part of the number and then press the exponent key $\boxed{\text{Exp}}$, followed by the exponent. For example, to enter 4.9×10^3, enter 4.94, then press $\boxed{\text{Exp}}$, and then press 3. When the exponent of 10 is a negative number, press the Change of Sign key $\boxed{+/-}$ after entering the exponent. For example, to enter 4.94×10^{-3}, enter in sequence 4.94 $\boxed{\text{Exp}}$ 3 $\boxed{+/-}$. In most calculators, the exponent will appear in the display a couple of spaces after the nonexponent part of the number—for example, 4.94   03 or 4.94   −03.

Examples Using Exponential Numbers

Calculation	Enter in sequence	Display	Rounded answer
a. $(4.94 \times 10^3)(21.4)$	4.94 $\boxed{\text{Exp}}$ 3 $\boxed{\times}$ 21.4 $\boxed{=}$	105716	1.06×10^5
b. $(1.42 \times 10^4)(2.88 \times 10^{-5})$	1.42 $\boxed{\text{Exp}}$ 4 $\boxed{\times}$ 2.88 $\boxed{\text{Exp}}$ 5 $\boxed{+/-}$ $\boxed{=}$	0.40896	0.409
c. $\dfrac{8.22 \times 10^{-5}}{5.00 \times 10^7}$	8.22 $\boxed{\text{Exp}}$ 5 $\boxed{+/-}$ $\boxed{\div}$ 5.00 $\boxed{\text{Exp}}$ 7 $\boxed{=}$	1.644 −12	1.64×10^{-12}

Logarithms The logarithm of a number is the power (exponent) to which some base number must be raised to give the original number. The most commonly used base number is 10. The base number that we use is 10. For example, the log of 100 is 2.0 (log 100 = $10^{2.0}$). The log of 200 is 2.3 (log 200 = $10^{2.3}$). Logarithms are used in chemistry to calculate the pH of an aqueous acidic solution. The answer (log) should contain the same number of significant figures to the right of the decimal as is in the original number. Thus, log 100 = 2.0, but log 100. is 2.000.

The log key on most calculators is a function key. To determine the log using your calculator, enter the number and then press the log function key. For example, to determine the log of 125, enter 125 and then the $\boxed{\text{Log}}$ key. The answer is 2.097.

Examples Using Logarithms
Determine the log of the following:

	Enter in sequence	Display	Rounded answer
a. 42	42 [Log]	1.6232492	1.62
b. 1.62×10^5	1.62 [Exp] 5 [Log]	5.209515	5.210
c. 6.4×10^{-6}	6.4 [Exp] 6 [+/−] [Log]	−5.19382	−5.19

Antilogarithms (Inverse Logarithms) An antilogarithm is the number from which the logarithm has been calculated. It is calculated using the [10^x] key on your calculator. For example, to determine the antilogarithm of 2.891, enter 2.891 into your calculator and then press the second function key followed by the [10^x] key: 2.891 [F] [10^x].
The display shows 778.03655, which rounds to the answer 778.

Examples Using Antilogarithms
Determine the antilogarithm of the following:

	Enter in sequence	Display	Rounded answer
a. 1.628	1.628 [F] [10^x]	42.461956	42.5
b. 7.086	7.086 [F] [10^x]	12189896	1.22×10^7
c. −6.33	6.33 [+/−] [F] [10^x]	4.6773514 −07	4.7×10^{-7}

Additional Practice Problems*

Problem	Display	Rounded answer
1. $143.5 + 14.02 + 1.202$	158.722	158.7
2. $72.06 - 26.92 - 49.66$	-4.52	-4.52
3. $2.168 + 4.288 - 1.62$	4.836	4.84
4. $(12.3)(22.8)(1.235)$	346.3434	346
5. $(2.42 \times 10^6)(6.08 \times 10^{-4})(0.623)$	916.65728	917
6. $\dfrac{(46.0)(82.3)}{19.2}$	197.17708	197
7. $\dfrac{0.0298}{243}$	1.2263374 -04	1.23×10^{-4}
8. $\dfrac{(5.4)(298)(760)}{(273)(1042)}$	4.2992554	4.3
9. $(6.22 \times 10^6)(1.45 \times 10^3)(9.00)$	8.1171 10	8.12×10^{10}
10. $\dfrac{(1.49 \times 10^6)(1.88 \times 10^6)}{6.02 \times 10^{23}}$	4.6531561 -12	4.65×10^{-12}
11. $\log 245$	2.389166	2.389
12. $\log 6.5 \times 10^{-6}$	-5.1870866	-5.19
13. $\log 24 \times \log 34$	2.1137644	2.11
14. antilog 6.34	2187761.6	2.2×10^6
15. antilog -6.34	4.5708818 -07	4.6×10^{-7}

*Only the problem, the display, and the rounded answer are given.

Units of Measurement

Physical Constants

Constant	Symbol	Value
Atomic mass unit	amu	1.6606×10^{-27} kg
Avogadro's number	N	6.022×10^{23} particles/mol
Gas constant	R (at STP)	0.08205 L atm/K mol
Mass of an electron	m_e	9.109×10^{-28} g
		5.486×10^{-4} amu
Mass of a neutron	m_n	1.675×10^{-27} kg
		1.00866 amu
Mass of a proton	m_p	1.673×10^{-27} kg
		1.00728 amu
Speed of light	c	2.997925×10^8 m/s

SI Units and Conversion Factors

Length

SI unit: meter (m)

1 meter	= 1.0936 yards
	= 100 centimeters
	= 1000 millimeters
1 centimeter	= 0.3937 inch
1 inch	= 2.54 centimeters (exactly)
1 kilometer	= 0.62137 mile
1 mile	= 5280 feet
	= 1.609 kilometers
1 angstrom	= 10^{-10} meter

Mass

SI unit: kilogram (kg)

1 kilogram	= 1000 grams
	= 2.20 pounds
1 gram	= 1000 milligrams
1 pound	= 453.59 grams
	= 0.45359 kilogram
	= 16 ounces
1 ton	= 2000 pounds
	= 907.185 kilograms
1 ounce	= 28.3 grams
1 atomic mass unit	= 1.6606×10^{-27} kilogram

Volume

SI unit: cubic meter (m^3)

1 liter	= 10^{-3} m^3
	= 1 dm^3
	= 1.0567 quarts
	= 1000 milliliters
1 gallon	= 4 quarts
	= 8 pints
	= 3.785 liters
1 quart	= 32 fluid ounces
	= 0.946 liter
1 fluid ounce	= 29.6 milliliters

Temperature

SI unit: kelvin (K)

$$0 \text{ K} = -273.15°C$$
$$= -459.67°F$$
$$\text{K} = °C + 273.15$$
$$°C = \frac{°F - 32}{1.8}$$
$$°F = 1.8(°C) + 32$$
$$°C = \frac{5}{9}(°F - 32)$$

Energy

SI unit: joule (J)

1 joule	= 1 kg m^2/s^2
	= 0.23901 calorie
1 calorie	= 4.184 joules

Pressure

SI unit: pascal (Pa)

1 pascal	= 1 kg/m s^2
1 atmosphere	= 101.325 kilopascals
	= 760 torr (mm Hg)
	= 14.70 pounds per square inch (psi)

Vapor Pressure of Water at Various Temperatures

Temperature (°C)	Vapor pressure (torr)	Temperature (°C)	Vapor pressure (torr)
0	4.6	26	25.2
5	6.5	27	26.7
10	9.2	28	28.3
15	12.8	29	30.0
16	13.6	30	31.8
17	14.5	40	55.3
18	15.5	50	92.5
19	16.5	60	149.4
20	17.5	70	233.7
21	18.6	80	355.1
22	19.8	90	525.8
23	21.2	100	760.0
24	22.4	110	1074.6
25	23.8		

APPENDIX V

Solubility Table

	F^-	Cl^-	Br^-	I^-	O^{2-}	S^{2-}	OH^-	NO_3^-	CO_3^{2-}	SO_4^{2-}	$C_2H_3O_2^-$
H^+	aq	aq	aq	aq	aq	sl.aq	aq	aq	sl.aq	aq	aq
Na^+	aq	aq	aq	aq	aq	aq	aq	aq	aq	aq	aq
K^+	aq	aq	aq	aq	aq	aq	aq	aq	aq	aq	aq
NH_4^+	aq	aq	aq	aq	—	aq	aq	aq	aq	aq	aq
Ag^+	aq	I	I	I	I	I	—	aq	I	I	I
Mg^{2+}	I	aq	aq	aq	I	d	I	aq	I	aq	aq
Ca^{2+}	I	aq	aq	aq	I	d	I	aq	I	I	aq
Ba^{2+}	I	aq	aq	aq	sl.aq	d	sl.aq	aq	I	I	aq
Fe^{2+}	sl.aq	aq	aq	aq	I	I	I	aq	sl.aq	aq	aq
Fe^{3+}	I	aq	aq	—	I	I	I	aq	I	aq	I
Co^{2+}	aq	aq	aq	aq	I	I	I	aq	I	aq	aq
Ni^{2+}	sl.aq	aq	aq	aq	I	I	I	aq	I	aq	aq
Cu^{2+}	sl.aq	aq	aq	—	I	I	I	aq	I	aq	aq
Zn^{2+}	sl.aq	aq	aq	aq	I	I	I	aq	I	aq	aq
Hg^{2+}	d	aq	I	I	I	I	I	aq	I	d	aq
Cd^{2+}	sl.aq	aq	aq	aq	I	I	I	aq	I	aq	aq
Sn^{2+}	aq	aq	aq	sl.aq	I	I	I	aq	I	aq	aq
Pb^{2+}	I	I	I	I	I	I	I	aq	I	I	aq
Mn^{2+}	sl.aq	aq	aq	aq	I	I	I	aq	I	aq	aq
Al^{3+}	I	aq	aq	aq	I	d	I	aq	—	aq	aq

Key: aq = soluble in water
 sl.aq = slightly soluble in water
  I = insoluble in water (less than 1 g/100 g H_2O)
  d = decomposes in water

Answers to Selected Review Questions and Exercises

Chapter 1

Review Questions

3. Six

8. Mercury and water.

9. Air.

15. (a) sugar, a compound and (c) gold, an element

Exercises

2. Two states are present; solid and liquid.

4. The maple leaf represents a heterogeneous mixture.

6. (a) homogeneous
   (b) homogeneous
   (c) heterogeneous
   (d) heterogeneous

Chapter 2

Review Questions

2. 7.6 cm

6. 0.789 g/mL < ice < 0.91 g/mL

8. $d = m/V$ specific gravity $= \dfrac{d_{\text{substance}}}{d_{\text{water}}}$

Exercises

2. (a) 1000 meters = 1 kilometer
   (c) 0.000001 liter = 1 microliter
   (e) 0.001 liter = 1 milliliter

4. (a) mg   (e) Å
   (c) m

6. (a) not significant.   (e) significant
   (c) not significant.

8. (a) 40.0 (3 sig fig)
   (b) 0.081 (2 sig fig)
   (c) 129,042 (6 sig fig)
   (d) 4.090×10^{-3} (4 sig fig)

10. (a) 8.87
    (b) 21.3
    (c) 130.(1.30×10^{2})
    (d) 2.00×10^{6}

12. (a) 4.56×10^{-2}
    (b) 4.0822×10^{3}
    (c) 4.030×10^{1}
    (d) 1.2×10^{7}
    (e) 2.49×10^{-4}

14. (a) 28.1
    (c) 4.0×10^{1}

16. (a) $\frac{1}{4}$
    (b) $\frac{5}{8}$
    (c) $1\frac{2}{3}$ or $\frac{5}{3}$
    (d) $\frac{8}{9}$
    (c) 22

18. (a) 1.0×10^{2}
    (b) 4.6 mL

20. (a) 4.5×10^{8} Å
    (c) 8.0×10^{6} mm
    (e) 6.5×10^{5} mg
    (g) 468 mL

22. (a) 117 ft
    (c) 7.4×10^{4} mm^{3}
    (e) 75.7 L

24. 12 mi/hr

26. 8.33 gr

28. 79 days

30. 2.54×10^{-3} lb

32. 5.94×10^{3} cm/s

34. $4500

36. 3.0×10^{5} straws

38. 160 L

40. 4×10^{5} m^{2}

42. 6 gal

44. 113°F Summer!

46. (a) 90.°F
    (b) −22.6°C
    (c) 546 K
    (d) −300 F

48. −11.4°C = 11.4°F

50. −297°F

52. 3.12 g/mL

54. 1.28 g/mL

56. 3.40×10^2 g

58. A graduated cylinder would be the best choice for adding 100 mL of solvent to a reaction. While the volumetric flask is also labeled 100 mL, volumetric flasks are typically used for doing dilutions. The other three pieces of glassware could also be used, but they hold smaller volumes, so it would take a longer time to measure out 100 mL. Also, because you would have to repeat the measurement many times using the other glassware, there is a greater chance for error.

60. 26 mL

62. 7.0 lb

64. Yes, 116.5 L additional solution

66. $-15°C > 4.5°F$

68. 20.9 lb

70. 5.1×10^3 L

72. 16.4 cm^3

74. 0.965 g/mL

76. ethyl alcohol, because it has the lower density

78. 54.3 mL

82. 76.9 g

Chapter 3

Review Questions

2. (a) Ag   (e) Fe
     (c) H   (g) Mg

4. The symbol of an element represents the element itself.

7. sodium   neon
      fluorine   helium
      nickel   calcium
      zinc   chlorine

10. 86 metals, 7 metalloids, 18 nonmetals

12. 1 metal   0 metalloids   5 nonmetals

14. (a) iodine   (b) bromine

19. H$_2$–hydrogen   Cl$_2$–chlorine
      N$_2$–nitrogen   Br$_2$–bromine
      O$_2$–oxygen   I$_2$–iodine
      F$_2$–fluorine

Exercises

2. (c) H$_2$ (g) ClF
     (f) NO

4. (a) magnesium, bromine
     (c) hydrogen, nitrogen, oxygen
     (e) aluminum, phosphorus, oxygen

6. (a) AlBr$_3$ (c) PbCrO$_4$

8. (a) 1 atom Al, 3 atoms Br
     (c) 12 atoms C, 22 atoms H, 11 atoms O

10. (a) 2 atoms   (e) 17 atoms
      (c) 9 atoms

12. (a) 2 atoms H   (e) 8 atoms H
      (c) 12 atoms H

14. (a) mixture   (e) mixture
      (c) pure substance

16. (c) compound   (d) element

18. (a) compound   (c) mixture
      (b) compound

20. (a) compound   (c) mixture

22. No. The only common liquid elements (at room temperature) are mercury and bromine.

24. 72% solids

25. A physical change is reversible. Therefore, boil the salt–water solution. The water will evaporate and leave the salt behind.

29. (a) 1 carbon atom and 1 oxygen atom; total number of atoms = 2
      (c) 1 hydrogen atom, 1 nitrogen atom, and 3 oxygen atoms; total number of atoms = 5
      (e) 1 calcium atom, 2 nitrogen atoms, and 6 oxygen atoms; total number of atoms = 9

32. 40 atoms H

34. (a) magnesium, manganese, molybdenum, mendelevium, mercury
      (c) sodium, potassium, iron, silver, tin, antimony

36. 420 atoms

39. (a) NaCl   (g) C$_6$H$_{12}$O$_6$
      (c) K$_2$O   (i) Cr(NO$_3$)$_3$
      (e) K$_3$PO$_4$

Chapter 4

Review Questions

2. liquid

4. Water disappears. Gas appears above each electrode and as bubbles in the solution.

6. A new substance is always formed during a chemical change, but never formed during physical changes.

8. Water.

Exercises

2. (a) physical   (e) physical
     (b) physical   (f) physical
     (c) chemical   (g) chemical
     (d) physical   (h) chemical

4. The copper wire, like the platinum wire, changed to a glowing red color when heated. Upon cooling, a new substance, black copper(II) oxide, had appeared.

6. Reactant: water
     Products: hydrogen, oxygen

8. (a) physical   (e) chemical
   (c) chemical
10. (a) potential energy   (e) potential energy
    (c) kinetic energy
12. the transformation of kinetic energy to thermal energy
14. (a) +   (b) −   (c) −   (d) +   (e) −
16. 2.2×10^3 J
18. 5.58×10^3 J
20. 5.03×10^{-2} J/g °C
22. 5°C
26. sp. ht. = 1.02 J/g °C
28. 3.0×10^2 cal
31. 44 g coal
33. 654°C
40. 8 g fat
43. A chemical change has occurred. Hydrogen molecules and oxygen molecules have combined to form water molecules.

Chapter 5

Review Questions

1. (a) copper, 29   (e) zinc, 30
   (c) phosphorus, 15
5. Isotopic notation $^A_Z E$
   Z represents the atomic number
   A represents the mass number

Exercises

2. The formula for hydrogen peroxide is H_2O_2. There are two atoms of oxygen for every two atoms of hydrogen. The molar mass of oxygen is 16.00 g, and the molar mass of hydrogen is 1.01 g. For hydrogen peroxide, the total mass of hydrogen is 2.016 g and the total mass of oxygen is 32.00 g, for a ratio of hydrogen to oxygen of approximately 2 : 32. or 1 : 16. Therefore, there is 1 gram of hydrogen for every 16 grams of oxygen.
4. (a) The nucleus of the atom contains most of the mass.
   (c) The atom is mostly empty space.
6. The nucleus of an atom contains nearly all of its mass.
8. Electrons:
   Dalton—Electrons are not part of his model.
   Thomson—Electrons are scattered throughout the positive mass of matter in the atom.
   Rutherford—Electrons are located out in space away from the central positive mass.
   Positive matter:
   Dalton—No positive matter in his model.
   Thomson—Positive matter is distributed throughout the atom.
   Rutherford—Positive matter is concentrated in a small central nucleus.

10. Yes. The mass of the isotope $^{12}_6$C, 12, is an exact number. The mass of other isotopes are not exact numbers.
12. Three isotopes of hydrogen have the same number of protons and electrons, but differ in the number of neutrons.
14. All five isotopes have nuclei that contain 30 protons and 30 electrons. The numbers of neutrons are:

Isotope mass number	Neutrons
64	34
66	36
67	37
68	38
70	40

16. (a) $^{109}_{47}$Ag   (c) $^{57}_{26}$Fe
18. (a) $^{25}_{12}$Mg   (c) $^{122}_{50}$Sn
20. (a) 35   (c) 45
22. 24.31 amu
24. 6.716 amu
26. 1.0×10^5 : 1.0
28. (a) These two atoms are isotopes.
29. 1.9×10^8 enlargement
31. ^{210}Bi has the largest number of neutrons (127).
33. 131 amu
40. (a) 0.02554%   (c) 0.02741%
42. The electron region is the area around the nucleus where electrons are most likely to be located.

Chapter 6

Review Questions

1. (a) $NaClO_3$ (e) $Zn(HCO_3)_2$
   (c) $Sn(C_2H_3O_2)_2$
3. (a) HBrO   hypobromous acid
       $HBrO_2$ bromous acid
       $HBrO_3$ bromic acid
       $HBrO_4$ perbromic acid
5. Chromium(III) compounds
   (a) $Cr(OH)_3$ (g) $CrPO_4$
   (c) $Cr(NO_2)_3$ (i) Cr_2O_3
   (e) $Cr_2(CO_3)_3$
7. (a) Metals are located in groups IA (except for hydrogen), IIA, IIIB–IIB and atomic numbers 13, 31, 49, 50, 81, 82, 83, lanthanides and actinides.
   (c) The transition metals are located in groups IIIB–IIB in the center of the periodic table. The lanthanides and actinides are located below the main body of the periodic table.

Exercises

2. (a) BaO   (d) $BeBr_2$
   (c) $AlCl_3$ (f) Mg_3P_2

4. Cl^- HSO_4^-
   Br^- HSO_3^-
   F^- CrO_4^{2-}
   I^- CO_3^{2-}
   CN^- HCO_3^-

O^{2-} $C_2H_3O_2^-$
OH^- ClO_3^-
S^{2-} MnO_4^-
SO_4^{2-} $C_2O_4^{2-}$

6. (a) calcium hydroxide
   (c) sulfur
   (d) sodium hydrogen carbonate
   (f) potassium carbonate

8.

Ion	SO_4^{2-}	OH^-	AsO_4^{3-}	$C_2H_3O_2^-$	CrO_4^{2-}
NH_4^+	$(NH_4)_2SO_4$	NH_4OH	$(NH_4)_3AsO_4$	$NH_4C_2H_3O_2$	$(NH_4)_2CrO_4$
Ca^{2+}	$CaSO_4$	$Ca(OH)_2$	$Ca_3(AsO_4)_2$	$Ca(C_2H_3O_2)_2$	$CaCrO_4$
Fe^{3+}	$Fe_2(SO_4)_3$	$Fe(OH)_3$	$FeAsO_4$	$Fe(C_2H_3O_2)_3$	$Fe_2(CrO_4)_3$
Ag^+	Ag_2SO_4	$AgOH$	Ag_3AsO_4	$AgC_2H_3O_2$	Ag_2CrO_4
Cu^{2+}	$CuSO_4$	$Cu(OH)_2$	$Cu_3(AsO_4)_2$	$Cu(C_2H_3O_2)_2$	$CuCrO_4$

10. NH_4^+ compounds: ammonium sulfate, ammonium hydroxide, ammonium arsenate, ammonium acetate, ammonium chromate.

   Ca^{2+} compounds: calcium sulfate, calcium hydroxide, calcium arsenate, calcium acetate, calcium chromate.

   Fe^{3+} compounds: iron(III) sulfate, iron(III) hydroxide, iron(III) arsenate, iron(III) acetate, iron(III) chromate.

   Ag^+ compounds: silver sulfate, silver hydroxide, silver arsenate, silver acetate, silver chromate.

   Cu^{2+} compounds: copper(II) sulfate, copper(II) hydroxide, copper(II) arsenate, copper(II) acetate, copper(II) chromate.

12. (a) sodium nitrate, $NaNO_3$
    (c) barium hydroxide, $Ba(OH)_2$
    (e) silver carbonate, Ag_2CO_3
    (g) potassium nitrite, KNO_2

14. (a) potassium oxide   (e) sodium phosphate
    (c) calcium iodide   (g) zinc nitrate

16. (a) $SnBr_4$ (d) $Hg(NO_2)_2$
    (c) $Fe_2(CO_3)_3$ (f) $Fe(C_2H_3O_2)_2$

18. (a) $HC_2H_3O_2$ (d) H_3BO_3
    (c) H_2S (f) $HClO$

20. (a) phosphoric acid   (f) nitric acid
    (c) iodic acid   (g) hydroiodic acid
    (d) hydrochloric acid

22. (a) Na_2CrO_4 (i) $NaClO$
    (c) $Ni(C_2H_3O_2)_2$ (k) $Cr_2(SO_3)_3$
    (e) $Pb(NO_3)_2$ (m) $Na_2C_2O_4$
    (g) $Mn(OH)_2$

24. (a) calcium hydrogen sulfate
    (c) tin(II) nitrite
    (e) potassium hydrogen carbonate
    (f) bismuth(III) arsenate
    (h) ammonium monohydrogen phosphate
    (j) potassium permanganate

28. (a) sulfate   (e) hydroxide
    (c) nitrate

30. (a) K_2SO_4 (e) KOH
    (c) KNO_3

32. Formula: KCl   Name: potassium chloride

34. (a) $AgNO_3 + NaCl \longrightarrow AgCl + NaNO_3$
    (c) $KOH + H_2SO_4 \longrightarrow K_2SO_4 + H_2O$

37. $Li_3Fe(CN)_6$
    $AlFe(CN)_6$
    $Zn_3[Fe(CN)_6]_2$

Chapter 7

Review Questions

2. A mole of gold has a higher mass than a mole of potassium.
4. A mole of gold atoms contains more electrons than a mole of potassium atoms.
7. 6.022×10^{23}
9. (a) 6.022×10^{23} (e) 32.00
   (c) 1.204×10^{24}
10. 6.022×10^{23} molecules in one molar mass of H_2SO_4.
    4.215×10^{24} atoms in one molar mass of H_2SO_4.

Exercises

2. Molar masses
   (a) NaOH   40.00
   (c) Cr_2O_3 152.0
   (e) $Mg(HCO_3)_2$ 146.3
   (g) $C_6H_{12}O_6$ 180.2
   (i) $BaCl_2 \cdot 2\,H_2O$ 244.2

6. (a) 0.0417 g H_2SO_4 (c) 0.122 g Ti
   (b) 11 g CCl_4 (d) 8.0×10^{-7} g S

8. (a) 1.05×10^{24} molecules Cl_2
   (b) 1.6×10^{23} molecules C_2H_6O
   (c) 1.64×10^{23} molecules CO_2
   (d) 3.75×10^{24} molecules CH_4

10. (a) 1.3×10^2 atoms N_2O_5
 (b) 6.02×10^{24} atoms Au
 (c) 2.67×10^{24} atoms BF_3
 (d) 3.85×10^{22} atoms U
12. (a) 3.271×10^{-22} g Au
 (b) 3.952×10^{-22} g U
 (c) 2.828×10^{-23} g NH_3
 (d) 1.795×10^{-22} g $C_6H_4(NH_2)_2$
14. (a) 0.886 mol S
 (b) 42.8 mol NaCl
 (c) 1.05×10^{24} atoms Mg
 (d) 9.47 mol Br_2
16. One mole of ammonia contains
 (a) 6.022×10^{23} molecules NH_3
 (b) 6.022×10^{23} N atoms
 (c) 1.807×10^{24} H atoms
 (d) 2.409×10^{24} atoms
18. (a) 6.0×10^{24} atoms O
 (b) 5.46×10^{24} atoms O
 (c) 5.0×10^{18} atoms O
20. (a) 1.27 g Cl (c) 2.74 g H
 (b) 9.25×10^{-2} g H
22. (a) 47.98% Zn (e) 23.09% Fe
  52.02% Cl   17.37% N
 (c) 12.26% Mg   59.53% O
  31.24% P
  56.48% O
24. (a) KCl 47.55% Cl (c) $SiCl_4$ 83.46% Cl
 (b) $BaCl_2$ 34.05% Cl (d) LiCl 83.63% Cl
  highest % Cl is LiCl
  lowest % Cl is $BaCl_2$
26. 24.2% C
 4.04% H
 71.72% Cl
28. (a) $KClO_3$ lower % Cl
 (b) $KHSO_4$ higher % S
 (c) Na_2CrO_4 lower % Cr
30. (a) KBr (c) $AgNO_3$
32. Empirical formulas
 (a) CuCl (e) $BaCr_2O_7$
 (c) Cr_2S_3
34. V_2O_5
36. $HgCl_2$
38. The molecular formula is $C_6H_{12}O_6$.
40. The molecular formula is $C_4H_8O_2$.
42. 40.0% C, 6.73% H, 53.3% O. Empirical formula is CH_2O, molecular formula is $C_6H_{12}O_6$.
44. $Al_2(SiO_3)_3$
46. 5.88 g Na
48. 4×10^{24} molecules

50. 9.9×10^{13} dollars/person
52. (a) 10.3 cm^3 (b) 2.18 cm
54. 41.58 g Fe_2S_3
56. 1.0 ton of iron ore contains 2×10^4 g Fe
58. (a) 76.98% Hg (c) 17.27% N
62. 42.10% C
 6.480% H
 51.42% O
64. 4.77 g O in 8.50 g $Al_2(SO_4)_3$
66. The empirical formula is $C_4H_4CaO_6$
68. 1.910×10^{16} years
70. 1.529×10^{-7} g $C_3H_8O_3$
72. $Co_3Mo_2C_{11}$
74. The empirical formula is C_5H_5N.
76. (a) CH_2O (e) $C_6H_2Cl_2O$
 (c) CH_2O

Chapter 8

Review Questions

1. The purpose of balancing chemical equations is to conform to the Law of Conservation of Mass.
5. The charts are one way of keeping track of the number of atoms of each element on the reactant side of a chemical equation and on the product side of an equation. The top row in a chart gives the number and types of atoms on the reactant side and the bottom row gives the number and types of atoms on the product side of a chemical equation.
6. The symbols indicate whether a substance is a solid, a liquid, a gas, or is in an aqueous solution. A solid is indicated by (s), a liquid by (l), a gas by (g), and a aqueous solution by (aq).
8. A combustion reaction is an exothermic process (usually burning) done in the presence of oxygen.

Exercises

2. (a) endothermic (e) exothermic
 (c) endothermic
4. (a) $H_2 + Br_2 \longrightarrow 2\,HBr$ combination
 (c) $Ba(ClO_3)_2 \longrightarrow BaCl_2 + 3\,O_2$ decomposition
 (e) $2\,H_2O_2 \longrightarrow 2\,H_2O + O_2$ decomposition
6. A metal and a nonmetal can react to form a salt; also an acid plus a base.
8. (a) $2\,SO_2 + O_2 \longrightarrow 2\,SO_3$
 (c) $2\,Na + 2\,H_2O \longrightarrow 2\,NaOH + H_2$
 (e) $Bi_2S_3 + 6\,HCl \longrightarrow 2\,BiCl_3 + 3\,H_2S$
 (g) $2\,LiAlH_4 \xrightarrow{\Delta} 2\,LiH + 2\,Al + 3\,H_2$
 (i) $2\,K_3PO_4 + 3\,BaCl_2 \longrightarrow 6\,KCl + Ba_3(PO_4)_2$

10. (a) $2 Cu + S \xrightarrow{\Delta} Cu_2S$

(c) $2 Ag_2O \xrightarrow{\Delta} 4 Ag + O_2$

(e) $Ni_3(PO_4)_2 + 3 H_2SO_4 \longrightarrow 3 NiSO_4 + 2 H_3PO_4$

(g) $3 AgNO_3 + AlCl_3 \longrightarrow 3 AgCl + Al(NO_3)_3$

12. (a) $CuSO_4(aq) + 2 KOH(aq) \longrightarrow$
$$Cu(OH)_2(s) + K_2SO_4(aq)$$

(c) $3 NaHCO_3(s) + H_3PO_4(aq) \longrightarrow$
$$Na_3PO_4(aq) + 3 H_2O(l) + 3 CO_2(g)$$

14. (a) $Cu(s) + FeCl_3(aq) \longrightarrow$ no reaction

(c) $2 Al(s) + 6 HBr(aq) \longrightarrow 3 H_2(g) + 2 AlBr_3(aq)$

16. (a) $SO_2 + H_2O \longrightarrow H_2SO_3$

(c) $Ca + 2 H_2O \longrightarrow Ca(OH)_2 + H_2$

18. (a) $C + O_2 \longrightarrow CO_2$

(c) $CuBr_2 + Cl_2 \longrightarrow CuCl_2 + Br_2$

(e) $2 NaNO_3 \xrightarrow{\Delta} 2 NaNO_2 + O_2$

20. (a) 2 mol of Na react with 1 mol of Cl_2 to produce 2 mol of NaCl and release 822 kJ of energy. Exothermic

(b) 1 mol of PCl_5 absorbs 92.9 kJ of energy to produce 1 mol of PCl_3 and 1 mol of Cl_2. Endothermic

22. (a) $2 Al + 3 I_2 \longrightarrow 2 AlI_3$ + heat

(b) $4 CuO + CH_4$ + heat $\longrightarrow$
$$4 Cu + CO_2 + 2 H_2O$$

24. (a) single displacement,
$Ni(s) + Pb(NO_3)_2(aq) \longrightarrow Pb(s) + Ni(NO_3)_2(aq)$

(c) decomposition, $2 HgO(s) \longrightarrow 2 Hg(l) + O_2(g)$

25. (a) change in color and texture of the bread

(b) change in texture of the white and the yoke

(c) the flame (combustion), change in matchhead, odor

26. $P_4O_{10} + 12 HClO_4 \longrightarrow 6 Cl_2O_7 + 4 H_3PO_4$
58 atoms O on each side

32. (a) $4 K + O_2 \longrightarrow 2 K_2O$

(c) $CO_2 + H_2O \longrightarrow H_2CO_3$

34. (a) $Zn + H_2SO_4 \longrightarrow H_2 + ZnSO_4$

(c) $Mg + 2 AgNO_3 \longrightarrow Mg(NO_3)_2 + 2 Ag$

35. (a) $ZnCl_2 + 2 KOH \longrightarrow Zn(OH)_2 + 2 KCl$

(c) $3 Ca(OH)_2 + 2 H_3PO_4 \longrightarrow$
$$6 H_2O + Ca_3(PO_4)_2$$

(e) $Ba(OH)_2 + 2 HNO_3 \longrightarrow 2 H_2O + Ba(NO_3)_2$

38. (a) $2 C_2H_6 + 7 O_2 \longrightarrow 4 CO_2 + 6 H_2O$

(c) $C_7H_{16} + 11 O_2 \longrightarrow 7 CO_2 + 8 H_2O$

Chapter 9

Review Questions

2. In order to convert grams to moles, the molar mass of the compound under consideration needs to be determined.

4. (a) correct   (e) correct
   (c) correct

6. You can calculate the percent yield of a chemical reaction by dividing the actual yield by the theoretical yield and multiplying by one hundred.

Exercises

2. (a) 25.0 mol $NaHCO_3$ (c) 16 mol CO_2

4. (a) 1.31 g $NiSO_4$ (e) 18 g K_2CrO_4
   (c) 373 g Bi_2S_3

6. HCl

8. (a) $\dfrac{3 \text{ mol } CaCl_2}{1 \text{ mol } Ca_3(PO_4)_2}$ (d) $\dfrac{1 \text{ mol } Ca_3(PO_4)_2}{2 \text{ mol } H_3PO_4}$

   (c) $\dfrac{3 \text{ mol } CaCl_2}{2 \text{ mol } H_3PO_4}$ (f) $\dfrac{2 \text{ mol } H_3PO_4}{6 \text{ mol } HCl}$

10. 2.80 mol Cl_2

12. (a) 8.33 mol H_2O (b) 0.800 mol $Al(OH)_3$

14. 19.7 g $Zn_3(PO_4)_2$

16. 117 g H_2O, 271 g Fe

18. (a) 0.500 mol Fe_2O_3 (e) 0.871 mol O_2
    (c) 6.20 mol SO_2

20. (a)

   Li

   I

neither is limiting

(b)

   Ag

   Cl

Ag is limiting

22. (a) Oxygen is the limiting reactant.

   ○ Nitrogen   ● Oxygen

   Before   After

(b) Water is the limiting reactant.

   Before   After

24. (a) H_2S is the limiting reactant and $Bi(NO_3)_3$ is in excess.
(b) H_2O is the limiting reactant and Fe is in excess.

26. (a) 0.313 mol H_2O.
(b) 0.521 mol H_2O.
(c) 3.3 mol CO_2; C_3H_6 will be left over.

28. Zinc

30. 95.0% yield of Cu

32. 77.8% CaC_2

36. (a) 0.19 mol KO_2 (b) 28 g O_2/hr

38. 65g O_2

40. (a) 0.16 mol $MnCl_2$.   (e) 11 g $KMnO_4$
(c) 82% yield

42. beaker weighs 24.835 g

44. 22.5 KCl; 77.5% KNO_3

48. (a) 3.2×10^2 g C_2H_5OH
(b) 1.10×10^3 g $C_6H_{12}O_6$

50. 3.7×10^2 kg Li_2O

52. 223.0 g H_2SO_4; 72.90% yield

54. 67.2% $KClO_3$

Chapter 10

Review Questions

1. An electron orbital is a region in space around the nucleus of an atom where an electron is most probably found.

3. The valence shell is the outermost energy level of an atom.

8. $1s$, $2s$, $2p$, $3s$, $3p$, $4s$, $3d$, $4p$

9. s–2 electrons per shell
p–6 electrons per shell after the first energy level
d–10 electrons per shell after the second energy level

13. 3 is the third energy level
d indicates an energy sublevel
7 indicates the number of electrons in the d sublevel

15. Elements in the s-block all have one or two electrons in their outermost energy level. These valence electrons are located in an s-orbital.

18. The greatest number of elements in any period is 32. The 6th period has this number of elements.

20. Pairs of elements which are out of sequence with respect to atomic masses are: Ar and K; Co and Ni; Te and I; Th and Pa; U and Np; Pu and Am; Sg and Bh; Hs and Mt.

Exercises

2. (a) F   9 protons
(b) Ag   47 protons
(c) Br   35 protons
(d) Sb   51 protons

4. (a) Cl $1s^2 2s^2 2p^6 3s^2 3p^5$
(c) Li $1s^2 2s^1$
(e) I $1s^2 2s^2 2p^6 3s^2 3p^6 4s^2 3d^{10} 4p^6 5s^2 4d^{10} 5p^5$

6. Bohr said that a number of orbits were available for electrons, each corresponding to an energy level. When an electron falls from a higher energy orbit to a lower orbit, energy is given off as a specific wavelength of light. Only those energies in the visible range are seen in the hydrogen spectrum. Each line corresponds to a change from one orbit to another.

8. 32 electrons in the fourth energy level

10. (a) (14p 14n) $2e^- 8e^- 4e^-$ $^{28}_{14}Si$

(c) (18p 22n) $2e^- 8e^- 8e^-$ $^{40}_{18}Ar$

(e) (15p 16n) $2e^- 8e^- 5e^-$ $^{31}_{15}P$

12. (a) Li $1s^2 2s^1$
(c) Zn $1s^2 2s^2 2p^6 3s^2 3p^6 4s^2 3d^{10}$
(e) K $1s^2 2s^2 2p^6 3s^2 3p^6 4s^1$

14. (a) Sc   (c) Sn

16.

Atomic number	Electron structure
(a) 9	$[He]2s^2 2p^5$
(c) 31	$[Ar]4s^2 3d^{10} 4p^1$
(d) 39	$[Kr]5s^2 4d^1$
(f) 10	$[He]2s^2 2p^6$

18. (a) Phosphorus (P)

(c) Calcium (Ca)

(e) Potassium (K)

20. (a) Cl

(c) Ni

(e) Ba

22. (a) (13p 14n) $2e^- 8e^- 3e^-$ $^{27}_{13}Al$

    (b) (22p 26n) $2e^- 8e^- 10e^- 2e^-$ $^{48}_{22}Ti$

24. The last electron in potassium is located in the fourth energy level because the $4s$ orbital is at a lower energy level than the $3d$ orbital. Also the properties of potassium are similar to the other elements in Group IA.

26. Noble gases each have filled s and p orbitals in the outermost energy level.

28. The elements in a group have the same number of outer energy level electrons. They are located vertically on the periodic table.

30. Valence shell electrons
    (a) 5   (c) 6   (e) 3

32. All of these elements have an s^2d^{10} electron configuration in their outermost energy levels.

34. (a) and (f)   (e) and (h)

36. 7, 33 since they are in the same periodic group.

38. (a) nonmetal, I   (c) metal, Mo

40. Period 4, Group 3B

42. Group 3A contains 3 valence electrons. Group 3B contains 2 electrons in the outermost level and 1 electron in an inner d orbital. Group A elements are representative, while Group B elements are transition elements.

44. (a) valence energy level 2, 1 valence electron
    (c) valence energy level 3, 4 valence electrons
    (e) valence energy level 2, 2 valence electrons

46. (a) 7A, Halogens   (e) 8A, Noble Gases
    (c) 1A, Alkali Metals

48. (a) Any orbital can hold a maximum of two electrons.
    (c) The third principal energy level can hold two electrons in $3s$, six electrons in $3p$, and ten electrons in $3d$ for a total of eighteen electrons.
    (e) An f sublevel can hold a maximum of fourteen electrons.

52. (a) Ne   (c) F
    (b) Ge   (d) N

54. Transition elements are found in Groups 1B–8B, lanthanides and actinides.

56. Elements number 8, 16, 34, 52, 84 all have 6 electrons in their outer shell.

58. (a) sublevel p (c) sublevel f
    (b) sublevel d

60. If 36 is a noble gas, 35 would be in periodic Group 7A and 37 would be in periodic Group 1A.

62. (a) $[Rn]7s^25f^{14}6d^{10}7p^5$
    (c) F, Cl, Br, I, At

Chapter 11

Review Questions

1. smallest Cl, Mg, Na, K, Rb largest.

3. When a third electron is removed from beryllium, it must come from a very stable electron structure corresponding to that of the noble gas, helium. In addition, the third electron must be removed from a +2 beryllium ion, which increases the difficulty of removing it.

5. The first ionization energy decreases from top to bottom because the outermost electrons in the successive noble gases are farther away from the nucleus and are more shielded by additional inner electron energy levels.

7. The first electron removed from a sodium atom is the one valence electron, which is shielded from most of its nuclear charge by the electrons of lower levels. To remove a second electron from the sodium ion requires breaking into the noble gas structure. This requires much more energy than that necessary to remove the first electron, because the Na^+ is already positive.

8. (a) K > Na   (e) Zr > Ti
   (c) O > F

10. Atomic size increases down the column, since each successive element has an additional energy level that contains electrons located farther from the nucleus.

12. Metals are less electronegative than nonmetals. Therefore, metals lose electrons more easily than nonmetals. So, metals will transfer electrons to nonmetals, leaving the metals with a positive charge and the nonmetals with a negative charge.

14. A polar bond is a bond between two atoms with very different electronegativities. A polar molecule is a dipole due to unequal electrical charge between bonded atoms resulting from unequal sharing of electrons.

16. A Lewis structure is a visual representation of the arrangement of atoms and electrons in a molecule or an ion. It shows how the atoms in a molecule are bonded together.

18. The dots in a Lewis structure represent nonbonding pairs of electrons. The lines represent bonding pairs of electrons.

22. Valence electrons are the electrons found in the outermost energy level of an atom.

24. An aluminum ion has a +3 charge because it has lost 3 electrons in acquiring a noble-gas electron structure.

Exercises

2. (a) Fe^{2+} (e) Rubidium ion.
   (c) Chloride ion.

4.   +   −   +   −
   (a) H   Cl   (e) Mg   H
   (c) C   Cl

6. (a) covalent   (c) covalent

8. (a) $F + 1 e^- \longrightarrow F^-$
   (b) $Ca \longrightarrow Ca^{2+} + 2 e^-$

10. $Ca\!:\, + \,:\!\ddot{O}\!: \longrightarrow CaO$

   $Na\cdot + \,:\!\ddot{B}r\!: \longrightarrow NaBr$

12. Si(4)   N(5)   P(5)   O(6)   Cl(7)

14. (a) Chloride ion, none
    (b) Nitrogen atom, gain 3 e^- or lose 5 e^-
    (c) Potassium atom, lose 1 e^-

16. (a) Nonpolar covalent compound; O_2.
    (c) Polar covalent compound; H_2O.

18. (a) SbH_3, Sb_2O_3 (c) HCl, Cl_2O_7

20. $BeBr_2$, beryllium bromide
    $MgBr_2$, magnesium bromide
    $SrBr_2$, strontium bromide
    $BaBr_2$, barium bromide
    $RaBr_2$, radium bromide

22. (a) $Ga\!:$ (b) $[Ga]^{3+}$ (c) $[Ca]^{2+}$

24. (a) covalent   (c) covalent
    (b) ionic   (d) covalent

26. (a) covalent   (c) ionic
    (b) covalent

28. (a) $:\!\ddot{O}::\ddot{O}\!:$ (b) $:\!\ddot{B}r\!:\!\ddot{B}r\!:$ (c) $:\!\ddot{I}\!:\!\ddot{I}\!:$

30. (a) $:\!\ddot{S}\!:\!H$ (c) $H\!:\!\ddot{N}\!:\!H$
       $\overset{|}{H}$ $\overset{|}{H}$

32. (a) $\left[:\!\ddot{I}\!:\right]^-$ (e) $\left[:\!\ddot{O}\!:\!N\!::\!O\!:\right]^-$
                                                    $\overset{|}{:\!\ddot{O}\!:}$

    (c) $\left[:\!\ddot{O}\!:\!C\!::\!\ddot{O}\!:\right]^{2-}$
             $\overset{|}{:\!\ddot{O}\!:}$

34. (a) F_2, nonpolar   (c) NH_3, polar
    (b) CO_2, nonpolar

36. (a) 3 electron pairs, trigonal planar
    (b) 4 electron pairs, tetrahedral
    (c) 4 electron pairs, tetrahedral

38. (a) tetrahedral   (c) tetrahedral
    (b) pyramidal

40. (a) tetrahedral   (b) bent   (c) bent

42. potassium

43. N_2H_4 14 e^- $:\!N\!:\!N\!:$ HN_3 16 e^- $H\!-\!\ddot{N}\!=\!N\!=\!\ddot{N}$
        $\overset{\displaystyle H\, H}{\underset{\displaystyle H\, H}{|\quad|}}$

46. (a) Hg   (c) N   (e) Au

47. (a) Zn   (c) N

51. $SnBr_2$, $GeBr_2$

54. This structure shown is incorrect since the bond is ionic. It should be represented as:

$$\left[Na\right]^+\left[:\!\ddot{O}\!:\right]^{2-}\left[Na\right]^+$$

59. (a) 105°   (c) 109.5°

63. Empirical formula is SO_3

   $:\!\ddot{O}\!:$
   $\overset{|}{:\!\ddot{O}\!-\!S\!=\!\ddot{O}\!:}$

Chapter 12

Review Questions

2. The pressure of a gas is the force that gas particles exert on the walls of a container. It depends on the temperature, the number of molecules of the gas, and the volume of the container.

4. The major components of dry air are nitrogen and oxygen.

6. The molecules of H_2 at 100°C are moving faster. Temperature is a measure of average kinetic energy. At higher temperatures, the molecules will have more kinetic energy.

7. 1 atm corresponds to 4 L.

13. The order of increasing molecular velocities is the order of decreasing molar masses.

   increasing molecular velocity $\longrightarrow$
   $\overline{\text{Rn}, F_2, N_2, CH_4, \text{He}, H_2}$
   $\longleftarrow$ increasing molar mass

   At the same temperature the kinetic energies of the gases are the same and equal to $1/2\, mv^2$. For the kinetic energies to be the same, the velocities must increase as the molar masses decrease.

15. Gases are described by the following parameters:
    (a) pressure   (c) temperature
    (b) volume   (d) number of moles

18. Charles' law: $V_1/T_1 = V_2/T_2$,
    ideal gas equation: $PV = nRT$
    Rearrange the ideal gas equation to: $V/T = nR/P$
    If you have an equal number of moles of two gases at the same pressure the right side of the rearranged ideal gas equation will be the same for both. You can set V/T for the first gas equal to V/T for the second gas (Charles' law) because the right side of both equations will cancel.

21. Equal volumes of H_2 and O_2 at the same T and P:
    (a) have equal number of molecules (Avogadro's law)
    (c) moles O_2 = moles H_2
    (e) rate H_2 = 4 times the rate of O_2 (Graham's Law of Effusion)

24. We refer gases to STP because some reference point is needed to relate volume to moles. A temperature and pressure must be specified to determine the moles of gas in a given volume, and 0°C and 760 torr are convenient reference points.

27. Heating a mole of N_2 gas at constant pressure has the following effects:
  (a) Density will decrease. Heating the gas at constant pressure will increase its volume. The mass does not change, so the increased volume results in a lower density.
  (c) Average kinetic energy of the molecules increases. This is a basic assumption of the Kinetic Molecular Theory.
  (e) Number of N_2 molecules remains unchanged. Heating does not alter the number of molecules present, except if extremely high temperatures were attained. Then, the N_2 molecules might dissociate into N atoms resulting in fewer N_2 molecules.

Exercises

2. (a) 715 torr   (c) 95.3 kPa
      (b) 953 bar
4. (a) 0.082 atm   (c) 0.296 atm
6. (a) 241 mL   (b) 577 mL
8. 3.3 atm
10. (a) 6.17 L   (b) 8.35 L
12. 7.8×10^2 mL
14. 33.4 L
16. −150°C
18. 681 torr
20. 1.450×10^3 torr
22. 1.19 L C_3H_8
24. 0.156 mol N_2
26. (a) 67.0 L SO_3 (c) 7.96 L Cl_2
       (b) 170 L C_2H_6
28. 1.78 g C_3H_6
30. 2.69×10^{22} molecules CH_4
32. 2.1 g CH_4
34. (a) 9.91 g/L Rn   (c) 2.50 g/L C_4H_8
       (b) 1.96 g/L CO_2
36. (a) 3.165 g/L Cl_2 (b) 1.46 g/L Cl_2
38. 19 L Kr
40. 72.8 L
42. 0.13 mol N_2
44. (a) 1.65×10^4 mL O_2 (b) 0.0201 mol H_2O_2
46. (a) 36 L O_2 (c) 12 L H_2O
       (b) 210 g CO_2
48. 1.12×10^3 L CO_2
51. (a) the pressure will be cut in half
       (c) the pressure will be cut in half
54. 22.4 L
56. SF_6 has the greatest density.
57. (a) CH_3
       (b) C_2H_6

$$(c) \quad H-\underset{\underset{H}{|}}{\overset{\overset{H}{|}}{C}}-\underset{\underset{H}{|}}{\overset{\overset{H}{|}}{C}}-H$$

60. 2.3×10^7 K
63. 327°C
65. 65 atm 211 K (−62°C)
67. 3.55 atm
70. 1.03×10^4 mm (33.8 ft)
71. 330 mol O_2
73. 43.2 g/mol (molar mass)
75. (a) 8.4 L H_2 (c) 8.48 g/L CO_2
77. 279 L H_2 at STP
78. Helium effuses 2.64 times faster than nitrogen.
80. C_4H_8

Chapter 13

Review Questions

2. H_2S, H_2Se, and H_2Te are gases at 0°C
4. Liquids take the shape of the container they are in. Gases also exhibit this property.

6.

10. about 70°C,
13. In Figure 13.1, it would be case (b) in which the atmosphere would reach saturation. The vapor pressure of water is the same in both (a) and (b), but since (a) is an open container the vapor escapes into the atmosphere and doesn't reach saturation.
15. The vapor pressure observed in (c) would remain unchanged. The presence of more water in (b) does not change the magnitude of the vapor pressure of the water. The temperature of the water determines the magnitude of the vapor pressure.
17. (a) At a pressure of 500 torr, water boils at 88°C.
       (b) The normal boiling point of ethyl alcohol is 78°C.
       (c) At a pressure of 0.50 atm (380 torr), ethyl ether boils at 16°C.
20. For water, to have its maximum density, the temperature must be 4°C, and the pressure sufficient to keep it liquid. $d = 1.0$ g/mL
30. (a) mercury, acetic acid, water, toluene, benzene, carbon tetrachloride, methyl alcohol, bromine
       (b) Highest boiling point is mercury; lowest is bromine.
33. Vapor pressure varies with temperature. The temperature at which the vapor pressure of a liquid equals the prevailing pressure is the boiling point of the liquid.
36. Ammonia would have a higher vapor pressure than SO_2 at −40°C because it has a lower boiling point (NH_3 is more volatile than SO_2).
40. 34.6°C, the boiling point of ethyl ether. (See Table 13.2)

Exercises

2. SO_2, SO_3, N_2O_5

4. [KOH, K_2O] [$Ba(OH)_2$, BaO] [$Ca(OH)_2$CaO]

6. (a) $Li_2O + H_2O \longrightarrow 2\,LiOH$
   (c) $Ba + 2\,H_2O \longrightarrow Ba(OH)_2 + H_2$
   (e) $SO_3 + H_2O \longrightarrow H_2SO_4$

8. (a) magnesium ammonium phosphate hexahydrate
   (b) iron(II) sulfate heptahydrate
   (c) tin(IV) chloride pentahydrate

10. (a) Distilled water has been vaporized by boiling and recondensed.
   (b) Natural waters are generally not pure, but contain dissoved minerals and suspended matter and can even contain harmful bacteria.

12. (a) HF will hydrogen bond; hydrogen is bonded to fluorine
   (c) H_2O_2 will hydrogen bond; hydrogen is bonded to oxygen.
   (e) H_2 will not hydrogen bond; hydrogen is not bonded to fluorine, oxygen, or nitrogen.

14.

16. Water forming beaded droplets is an example of cohesive forces. The water molecules have stronger attractive forces for other water molecules then they do for the surface.

18. 0.262 mol $FeI_2 \cdot 4\,H_2O$

20. 1.05 mol H_2O

22. 48.67% H_2O

24. $FePO_4 \cdot 4\,H_2O$

26. 5.5×10^4 J

28. 42 g of steam needed; not enough steam (35 g)

30. The system will be at 0°C. It will be a mixture of ice and water.

32. (a) 18.0 g H_2O (c) 18.0 g H_2O
   (b) 36.0 g H_2O

34. Steam molecules will cause a more severe burn. Steam molecules contain more energy at 100°C than water molecules at 100°C due to the energy absorbed during the vaporization state (heat of vaporization).

36. When one leaves the swimming pool, water starts to evaporate from the skin of the body. Part of the energy needed for evaporation is absorbed from the skin, resulting in the cool feeling.

38. During phase changes (ice melting to liquid water or liquid water evaporating to steam), all the heat energy is used to cause the phase change. Once the phase change is complete, the heat energy is once again used to increase the temperature of the substance.

40. 75°C at 270 torr pressure

42. $MgSO_4 \cdot 7\,H_2O$ $Na_2HPO_4 \cdot 12\,H_2O$

44. chlorine

46. When organic pollutants in water are oxidized by dissolved oxygen, there may not be sufficient dissolved oxygen to sustain marine life.

48. Na_2 zeolite(s) + $Mg^{2+}(aq) \longrightarrow$
         Mg zeolite(s) + 2 $Na^+(aq)$

51. 6.78×10^5 J

52. 40.7 kJ/mol

54. 2.30×10^6 J

56. 40.2 g H_2O

58. 1.00 mole of water vapor at STP has a volume of 22.4 L (gas)

60. 3.9×10^3 J

63. 2.71×10^4 J

Chapter 14

Review Questions

2. From Table 14.3, approximately 4.5 g of NaF would be soluble in 100 g of water at 50°C.

3. From Figure 14.3, solubilities in water at 25°C are:
   (a) KCl   35g/100 g H_2O
   (b) $KClO_3$ 9g/100 g H_2O
   (c) KNO_3 39g/100 g H_2O

6. KNO_3

8. 6×10^2 cm^2

9. The solution would be unsaturated at 40°C and 50°C.

12. The solution level in the thistle tube will fall.

16. The ions or molecules of a dissolved solute do not settle out because the individual particles are so small that the force of molecular collisions is large compared with the force of gravity.

24. For a given mass of solute, the smaller the particles, the faster the dissolution of the solute. This is due to the smaller particles having a greater surface area exposed to the dissolving action of the solvent.

28. The solution contains 16 moles of HNO_3 per liter of solution.

30. The champagne would spray out of the bottle.

34. A lettuce leaf immersed in salad dressing containing salt and vinegar will become limp and wilted as a result of osmosis.

36. (a) $1\,M$ NaOH $\qquad\qquad$ 1 L
$\quad\;$ (b) $0.6\,M\,Ba(OH)_2$ $\quad$ 0.83 L
$\quad\;$ (c) $2\,M$ KOH $\qquad\quad$ 0.50 L
$\quad\;$ (d) $1.5\,M\,Ca(OH)_2$ $\quad$ 0.33 L

42. The molarity of a 5 molal solution is less than $5\,M$.

Exercises

2. Reasonably soluble: (c) $CaCl_2$
$\quad$ Insoluble: $\qquad\qquad$ (a) PbI_2 (e) $BaSO_4$

4. (a) 16.7% $\qquad\qquad\qquad$ (c) 39% K_2CrO_4

6. 250. g solution

8. (a) 9.0 g $BaCl_2$ $\qquad\qquad$ (b) 66 g solvent

10. 33.6% NaCl

12. 22% C_6H_{14}

14. (a) $4.0\,M$ $\qquad\qquad\qquad$ (c) $1.97\,M\,C_6H_{12}O_6$

16. (a) $1.1\,mol\,HNO_3$ $\qquad\quad$ (c) 0.088 mol LiBr
$\quad\;$ (b) 7.5×10^{-3} mol $NaClO_3$

18. (a) 2.1×10^3 g H_2SO_4 $\quad$ (c) 1.2 g $Fe_2(SO_4)_3$
$\quad\;$ (b) 2.27 g $KMnO_4$

20. (a) 3.4×10^3 mL $\qquad\quad$ (b) 1.88×10^3 mL

22. (a) $1.2\,M$ $\qquad\qquad\qquad$ (c) $0.333\,M$
$\quad\;$ (b) $0.070\,M$

24. (a) 25 mL 18 $M\,H_2SO_4$ $\quad$ (b) 5.0 mL 15 $M\,NH_3$

26. (a) $1.2\,M$ HCl
$\quad\;$ (b) $4.21 \times 10^{-3}\,M$ HCl

28. (a) 3.6 mol Na_2SO_4 $\qquad$ (e) 38.25 mL H_2SO_4
$\quad\;$ (c) 0.625 mol NaOH

30. (a) 1.88×10^{-3} mol H_2O $\quad$ (e) 1.8 L CO_2
$\quad\;$ (c) 90.2 mL $HC_2H_3O_2$

32. (a) $5.5\,m\,C_6H_{12}O_6$ $\qquad$ (b) $0.25\,m\,I_2$

34. (a) $10.74\,m$ $\qquad\qquad\quad$ (c) 105.50°C
$\quad\;$ (b) −20.0°C

36. 163 g/mol

43. 60. g 10% NaOH solution

45. (a) 160 g sugar $\qquad\qquad$ (c) $0.516\,m$
$\quad\;$ (b) $0.47\,M$

47. $16.2\,m$ HCl

49. 455 g solution

51. (a) 4.5 g NaCl
$\quad\;$ (b) 450. mL H_2O must evaporate.

53. 210. mL solution

55. $6.72\,M\,HNO_3$

59. 12.0 g $Mg(OH)_2$ is the more effective acid neutralizer.

61. $6.2\,m\,H_2SO_4$
$\quad\;$ $5.0\,M\,H_2SO_4$

62. (a) $2.9\,m$ $\qquad\qquad\qquad$ (b) 101.5°C

64. (a) 8.04×10^3 g $C_2H_6O_2$ $\quad$ (c) −4.0°F
$\quad\;$ (b) 7.24×10^3 mL $C_2H_6O_2$

65. 47% $NaHCO_3$

66. (a) 7.7 L H_2O must be added
$\quad\;$ (b) 0.0178 mol
$\quad\;$ (c) 0.0015 mol H_2SO_4 in each mL

69. 2.84 g $Ba(OH)_2$ is formed.

70. (a) 0.011 mol Li_2CO_3
$\quad\;$ (c) 3.2×10^2 mL solution

Chapter 15

Review Questions

2. An electrolyte must be present in the solution for the bulb to glow.

4. First, the orientation of the polar water molecules about the Na^+ and Cl^- ions is different. Second, more water molecules will fit around Cl^-, since it is larger than the Na^+ ion.

6. tomato juice

8. Arrhenius: $HCl + NaOH \longrightarrow NaCl + H_2O$
$\quad\;$ Brønsted–Lowry: $HCl + KCN \longrightarrow HCN + KCl$
$\quad\;$ Lewis: $AlCl_3 + NaCl \longrightarrow AlCl_4^- + Na^+$

10. acids, bases, salts

13. In their crystalline structure, salts exist as positive and negative ions in definite geometric arrangement to each other, held together by the attraction of the opposite charges. When dissolved in water, the salt dissociates as the ions are pulled away from each other by the polar water molecules.

15. Molten NaCl conducts electricity because the ions are free to move. In the solid state, however, the ions are immobile and do not conduct electricity.

18. Ions are hydrated in solution because there is an electrical attraction between the charged ions and the polar water molecules.

20. (a) $[H^+] = [OH^-]$ $\qquad$ (c) $[OH^-] > [H^+]$
$\quad\;$ (b) $[H^+] > [OH^-]$

21. The net ionic equation for an acid–base reaction in aqueous solutions is: $H^+ + OH^- \longrightarrow H_2O$.

24. The fundamental difference is in the size of the particles. The particles in true solutions are less than 1 mm in size. The particles in colloids range from 1–1000 mm in size.

Exercises

2. Conjugate acid–base pairs:
$\quad\;$ (a) $HCl - Cl^-$; $NH_4^+ - NH_3$
$\quad\;$ (c) $H_3O^+ - H_2O$; $H_2CO_3 - HCO_3^-$

4. (a) $Fe_2O_3(s) + 6\,HBr(aq) \longrightarrow$
$$2\,FeBr_3(aq) + 3\,H_2O(l)$$
$\quad\;$ (c) $2\,NaOH(aq) + H_2CO_3(aq) \longrightarrow$
$$Na_2CO_3(aq) + 2\,H_2O(l)$$
$\quad\;$ (e) $Mg(s) + 2\,HClO_4(aq) \longrightarrow$
$$Mg(ClO_4)_2(aq) + H_2(g)$$

6. (a) $Fe_2O_3 + (6 H^+ + 6 Br^-) \longrightarrow$
$$(2 Fe^{3+} + 6 Br^-) + 3 H_2O$$
$$Fe_2O_3 + 6 H^+ \longrightarrow 2 Fe^{3+} + 3 H_2O$$

 (c) $(2 Na^+ + 2 OH^-) + H_2CO_3 \longrightarrow$
$$(2 Na^+ + CO_3^{2-}) + 2 H_2O$$
$$2 OH^- + H_2CO_3 \longrightarrow CO_3^{2-} + 2 H_2O$$

 (e) $Mg + (2 H^+ + 2 ClO_4^-) \longrightarrow$
$$(Mg^{2+} + 2 ClO_4^-) + H_2$$
$$Mg + 2 H^+ \longrightarrow Mg^{2+} + H_2$$

8. (a) $NaHCO_3$—salt   (e) $RbOH$—base
 (c) $AgNO_3$—salt

10. (a) $0.75\ M\ Zn^{2+}$, $1.5\ M\ Br^-$
 (b) $4.95\ M\ SO_4^{2-}$, $3.30\ M\ Al^{3+}$
 (c) $0.680\ M\ NH_4^+$, $0.340\ M\ SO_4^{2-}$
 (d) $0.0628\ M\ Mg^{2+}$, $0.126\ M\ ClO_3^-$

12. (a) $4.9\ g\ Zn^{2+}$, $12\ g\ Br^-$
 (b) $8.90\ g\ Al^{3+}$, $47.6\ g\ SO_4^{2-}$
 (c) $1.23\ g\ NH_4^+$, $3.28\ g\ SO_4^{2-}$
 (d) $0.153\ g\ Mg^{2+}$, $1.05\ g\ ClO_3^-$

14. (a) $[H^+] = 3.98 \times 10^{-3}$
 (b) $[H^+] = 1.0 \times 10^{-10}$
 (c) $[H^+] = 3.98 \times 10^{-7}$

16. (a) $[K^+] = 1.0\ M$, $[Ca^{2+}] = 0.5\ M$, $[Cl^-] = 2.0\ M$
 (b) No ions are present in the solution.
 (c) $0.67\ M\ Na^+$, $0.67\ M\ NO_3^-$

18. (a) $0.147\ M\ NaOH$
 (b) $0.964\ M\ NaOH$
 (c) $0.4750\ M\ NaOH$

20. (a) $H_2S(g) + Cd^{2+}(aq) \longrightarrow CdS(s) + 2 H^+(aq)$
 (b) $Zn(s) + 2H^+(aq) \longrightarrow Zn^{2+}(aq) + H_2(g)$
 (c) $Al^{3+}(aq) + PO_4^{3-}(aq) \longrightarrow AlPO_4(s)$

22. (a) $2\ M\ HCl$
 (b) $1\ M\ H_2SO_4$

24. $40.8\ mL$ of $0.245\ M\ HCl$

26. $7.0\%\ NaCl$ in the sample

28. $0.936\ L\ H_2$

30. (a) $pH = 7.0$ (c) $pH = 4.00$
 (b) $pH = 0.30$

32. (a) $pH = 4.30$ (b) $pH = 10.47$

34. (a) weak acid   (c) strong acid
 (b) strong base   (d) strong acid

35. (a) acidic   (e) neutral
 (c) basic

37. $0.260\ M\ Ca^{2+}$

38. $0.309\ M\ Ba(OH)_2$

42. Freezing point depression is directly related to the concentration of particles in the solution.

highest
freezing $C_{12}H_{22}O_{11} > HC_2H_3O_2 > HCl > CaCl_2$ lowest
point   1 mol   1 + mol   2 mol   3 mol   freezing point
            (particles in solution)

44. As the pH changes by 1 unit, the concentration of H^+ in solution changes by a factor of 10. For example, the pH of $0.10\ M\ HCl$ is 1.00, while the pH of $0.0100\ M\ HCl$ is 2.00.

45. 1.77% solute, 99.22% H_2O

46. $0.201\ M\ HCl$

48. $0.673\ g\ KOH$

50. $13.9\ L$ of $18.0\ M\ H_2SO_4$

52. $pH = 1.1$, acidic

54. (a) $2 NaOH(aq) + H_2SO_4(aq) \longrightarrow$
$$Na_2SO_4(aq) + 2 H_2O(l)$$
 (b) $1.0 \times 10^2\ mL\ NaOH$
 (c) $0.71\ g\ Na_2SO_4$

Chapter 16

Review Questions

2. The reaction is endothermic because the increased temperature increases the concentration of product (NO_2) present at equilibrium.

4. At equilibrium, the rate of the forward reaction equals the rate of the reverse reaction.

6. The sum of the pH and the pOH is 14. A solution whose pH is –1 would have a pOH of 15.

8. The order of molar solubilities is $AgC_2H_3O_2$, $PbSO_4$, $BaSO_4$, $AgCl$, $BaCrO_4$, $AgBr$, AgI, PbS.

9. (a) Ag_2CrO_4 (b) Ag_2CrO_4

12. The rate increases because the number of collisions increases due to the added reactant.

15. A catalyst speeds up the rate of a reaction by lowering the activation energy. A catalyst is not used up in the reaction.

18. The statement does not contradict Le Chatelier's Principle. The previous question deals with the case of dilution. If pure acetic acid is added to a dilute solution, the reaction will shift to the right, producing more ions in accordance with Le Chatelier's Principle. But the concentration of the un-ionized acetic acid will increase faster than the concentration of the ions, thus yielding a smaller percent ionization.

20. In pure water, H^+ and OH^- are produced in equal quantities by the ionization of the water molecules, $H_2O \rightleftharpoons H^+ + OH^-$. Since $pH = -\log[H^-]$, and $pOH = -\log[OH^-]$, they will always be identical for pure water. At 25°C, they each have the value of 7, but at higher temperatures, the degree of ionization is greater, so the pH and pOH would both be less than 7, but still equal.

Exercises

2. (a) $H_2O(l) \overset{100°C}{\rightleftharpoons} H_2O(g)$

(b) $SO_2(l) \rightleftharpoons SO_2(g)$

4. (a) $[NH_3]$, $[O_2]$, and $[N_2]$, will be increased. $[H_2O]$, will be decreased. Reaction shifts left.

(b) The addition of heat will shift the reaction to the left.

6. (a) left   I   I   D (add NH_3)
(b) left   I   I   D (increase volume)
(c) no change   N   N   N (add catalyst)
(d) ?   ?   I   I (add H_2 and N_2)

8.

Reaction	Increase temperature	Increase pressure	Add catalyst
(a)	right	left	no change
(b)	left	left	no change
(c)	left	left	no change

10. Equilibrium shift
(a) left   (b) right   (c) right

12. (a) $K_{eq} = \dfrac{[H^+][H_2PO_4^-]}{[H_3PO_4]}$ (c) $K_{eq} = \dfrac{[N_2O_5]^2}{[NO_2]^4[O_2]}$

(b) $K_{eq} = \dfrac{[CH_4][H_2S]^2}{[CS_2][H_2]^4}$

14. (a) $K_{sp} = [Mg^{2+}][CO_3^{2-}]$ (c) $K_{sp} = [Tl^{3+}][OH^-]^3$
(b) $K_{sp} = [Ca^{2+}][C_2O_4^{2-}]$ (d) $K_{sp} = [Pb^{2+}]^3[AsO_4^{3-}]^2$

16. If H^+ is increased,
(a) pH is decreased
(b) pOH is increased
(c) OH^- is decreased
(d) K_W, remains unchanged

18. (a) NH_4Cl acidic   (c) $(NH_4)_2SO_4$ acidic

20. (a) $NH_4^+(aq) + H_2O(l) \rightleftharpoons H_3O^+(aq) + NH_3(aq)$
(b) $SO_3^{2-}(aq) + H_2O(l) \rightleftharpoons OH^-(aq) + HSO_3^-(aq)$

22. (a) $OCl^-(aq) + H_2O(l) \rightleftharpoons OH^-(aq) + HOCl(aq)$
(b) $ClO_2^-(aq) + H_2O(l) \rightleftharpoons$
$\qquad\qquad\qquad\qquad OH^-(aq) + HClO_2(aq)$

24. When excess base gets into the bloodstream, it reacts with H^+ to form water. Then H_2CO_3 ionizes to replace H^+, thus maintaining the approximate pH of the solution.

26. (a) $K_a = 5.7 \times 10^{-6}M$ (c) 2.3×10^{-3}% ionized
(b) pH = 5.24

28. $K_a = 2.3 \times 10^{-5}$

30. (a) pH = 3.72   (c) pH = 4.72
(b) pH = 4.23

32. $K_a = 7.3 \times 10^{-6}$

34. $[H^+] = 3.0\ M$; pH = -0.47; pOH = 14.47; $[OH^-] = 3.3 \times 10^{-14}$

36. (a) pOH = 3.00; pH = 11.0
(b) pH = 0.903; pOH = 13.1
(c) pH = 5.74; pOH = 8.3

38. (a) $[OH^-] = 2.5 \times 10^{-6}$
(b) $[OH^-] = 1.1 \times 10^{-13}$

40. (a) $[H^+] = 2.2 \times 10^{-9}$
(b) $[H^+] = 1.4 \times 10^{-11}$

42. (a) 1.2×10^{-23} ZnS
(c) 1.81×10^{-18} Ag_3PO_4

44. (a) $1.1 \times 10^{-4}\ M$
(b) $9.2 \times 10^{-6}\ M$

46. (a) 8.9×10^{-4} g$BaCO_3$
(b) 9.3×10^{-9} g $AlPO_4$

48. Precipitation occurs.

50. 2.5×10^{-12} mol AgBr will dissolve.

52. pH = 4.74

54. Change in pH = $4.74 - 4.72 = 0.02$ units in the buffered solution. Initial pH = 4.74.

58. 4.2 mol HI

60. 0.500 mol HI, 2.73 mol H_2, 0.538 mol I_2

62. 128 times faster

64. pH = 4.6

66. Hypochlorous acid: $K_a = 3.5 \times 10^{-5}$
Propanoic acid: $K_a = 1.3 \times 10^{-5}$
Hydrocyanic acid: $K_a = 4.0 \times 10^{-10}$

68. (a) Precipitation occurs.
(b) Precipitation occurs.
(c) No precipitation occurs.

70. No precipitate of $PbCl_2$

72. $K_{eq} = 1.1 \times 10^4$

74. $8.0\ M\ NH_3$

76. $K_{sp} = 4.00 \times 10^{-28}$

78. (a) The temperature could have been cooler.
(b) The humidity in the air could have been higher.
(c) The air pressure could have been greater.

80. (a) $K_{eq} = \dfrac{[O_3]^2}{[O_2]^3}$ (c) $K_{eq} = \dfrac{[MgO][CO_2]}{[MgCO_3]}$

(b) $K_{eq} = \dfrac{[H_2O(l)]}{[H_2O(g)]}$ (d) $K_{eq} = \dfrac{[Bi_2S_3][H^+]^6}{[Bi^{3+}]^2[H_2S]^3}$

82. $K_{eq} = 450$

84. 1.1 g $CaSO_4$

Chapter 17

Review Questions

2. (a) I_2 has been oxidized.
   (b) Cl_2 has been reduced.

4. If the free element is higher on the list than the ion with which it is paired, the reaction occurs.
   (a) Yes. $Zn(s) + Cu^{2+}(aq) \longrightarrow Zn^{2+}(aq) + Cu(s)$
   (c) Yes. $Sn(s) + 2\,Ag^+(aq) \longrightarrow Sn^{2+}(aq) + 2\,Ag(s)$
   (e) Yes. $Ba(s) + FeCl_2(aq) \longrightarrow BaCl_2(aq) + Fe(s)$
   (g) Yes. $Ni(s) + Hg(NO_3)_2(aq) \longrightarrow$
   $$Ni(NO_3)_2(aq) + Hg(l)$$

6. (a) $2\,Al + Fe_2O_3 \longrightarrow Al_2O_3 + 2\,Fe + heat$
   (c) No. Fe is less reactive than Al.

10. (a) Oxidation occurs at the anode.
   $$2\,Cl^-(aq) \longrightarrow Cl_2(g) + 2\,e^-$$
   (b) Reduction occurs at the cathode.
   $$Ni^{2+}(aq) + 2\,e^- \longrightarrow Ni(s)$$
   (c) The net chemical reaction is
   $$Ni^{2+}(aq) + 2\,Cl^-(aq) \xrightarrow[\text{energy}]{\text{electrical}} Ni(s) + Cl_2(g)$$

12. (a) It would not be possible to monitor the voltage produced, but the reactions in the cell would still occur.
   (b) If the salt bridge were removed, the reaction would stop.

14. $Ca^{2+} + 2\,e^- \longrightarrow Ca$ cathode reaction, reduction
   $2\,Br^- \longrightarrow Br_2 + 2\,e^-$ anode reaction, oxidation

16. Since lead dioxide and lead(II) sulfate are insoluble, it is unnecessary to have salt bridges in the cells of a lead storage battery.

18. Reduction occurs at the cathode.

20. A salt bridge permits movement of ions in the cell. This keeps the solution neutral with respect to the charged particles (ions) in the solution.

Exercises

2. (a) $\underline{K}MnO_4$ $+7$
   (c) $N\underline{H}_3$ -3
   (e) $\underline{K}_2CrO_4$ $+6$

4. (a) $\underline{O}_2$ 0
   (c) $Fe(\underline{O}H)_3$ -2

6. (a) $SO_3^{2-} + H_2O \longrightarrow SO_4^{2-} + 2\,H^+ + 2\,e^-$
   S is oxidized
   (c) $S_2O_4^{2-} + 2\,H_2O \longrightarrow 2\,SO_3^{2-} + 4\,H^+ + 2\,e^-$
   S is oxidized

8. Equation (1):
   (a) As is oxidized, Ag^+ is reduced.
   (b) Ag^+ is the oxiding agent, AsH_3 the reducing agent.
   Equation (2):
   (a) Br is oxidized, Cl is reduced.
   (b) Cl_2 is the oxidizing agent, NaBr the reducing agent.

10. (a) incorrectly balanced
   $$3\,MnO_2(s) + 4\,Al(s) \longrightarrow 3\,Mn(s) + 2\,Al_2O_3(s)$$
   (c) correctly balanced

12. (a) $3\,Cl_2 + 6\,KOH \longrightarrow KClO_3 + 5\,KCl + 3\,H_2O$
   (c) $3\,CuO + 2\,NH_3 \longrightarrow N_2 + 3\,Cu + 3\,H_2O$
   (e) $5\,H_2O_2 + 2\,KMnO_4 + 3\,H_2SO_4 \longrightarrow$
   $$5\,O_2 + 2\,MnSO_4 + K_2SO_4 + 8\,H_2O$$

14. (a) $6\,H^+ + ClO_3^- + 6\,I^- \longrightarrow 3\,I_2 + Cl^- + 3\,H_2O$
   (c) $2\,H_2O + 2\,MnO_4^- + 5\,SO_2 \longrightarrow$
   $$4\,H^+ + 2\,Mn^{2+} + 5\,SO_4^{2-}$$
   (e) $8\,H^+ + Cr_2O_7^{2-} + 3\,H_3AsO_3 \longrightarrow$
   $$2\,Cr^{3+} + 3\,H_3AsO_4 + 4\,H_2O$$

16. (a) $H_2O + 2\,MnO_4^- + 3\,SO_3^{2-} \longrightarrow$
   $$2\,MnO_2 + 3\,SO_4^{2-} + 2\,OH^-$$
   (c) $8\,Al + 3\,NO_3^- + 18\,H_2O + 5\,OH^- \longrightarrow$
   $$3\,NH_3 + 8\,Al(OH)_4^-$$
   (e) $2\,Al + 6\,H_2O + 2\,OH^- \longrightarrow 2\,Al(OH)_4^- + 3\,H_2$

18. (a) $5\,Mo_2O_3 + 6\,MnO_4^- + 18\,H^+ \longrightarrow$
   $$10\,MoO_3 + 6\,Mn^{2+} + 9\,H_2O$$
   (b) $3\,BrO^- + 2\,Cr(OH)_4^- + 2\,OH^- \longrightarrow$
   $$3\,Br^- + 2\,CrO_4^{2-} + 5\,H_2O$$
   (c) $5\,S_2O_3^{2-} + 8\,MnO_4^- + 7\,H_2O \longrightarrow$
   $$10\,SO_4^{2-} + 8\,Mn^{2+} + 14\,OH^-$$

20. (a) $Pb + SO_4^{2-} \longrightarrow PbSO_4 + 2\,e^-$
   $PbO_2 + SO_4^{2-} + 4\,H^+ + 2\,e^- \longrightarrow PbSO_4 + 2\,H_2O$
   (b) The first reaction is oxidation (Pb^0 is oxidized to Pb^{2+}).
   The second reaction is reduction (Pb^{4+} is reduced to Pb^{2+}).
   (c) The first reaction (oxidation) occurs at the anode of the battery.

22. Zinc is a more reactive metal than copper, so when corrosion occurs, the zinc preferentially reacts. Zinc is above hydrogen in the Activity Series of Metals; copper is below hydrogen.

24. 20.2 L Cl_2

26. 66.2 mL of 0.200 M $K_2Cr_2O_7$ solution

28. 5.560 mol H_2

30. The electrons lost by the species undergoing oxidation must be gained (or attracted) by another species, which then undergoes reduction.

32. Sn^{4+} can only be an oxidizing agent.
   Sn^0 can only be a reducing agent.
   Sn^{2+} can be both oxidizing and reducing agents.

34. Equations (a) and (b) represent oxidations.

36. (a) $F_2 + 2\,Cl^- \longrightarrow 2\,F^- + Cl_2$
   (b) $Br_2 + Cl^- \longrightarrow NR$
   (c) $I_2 + Cl^- \longrightarrow NR$
   (d) $Br_2 + 2\,I^- \longrightarrow 2\,Br^- + I_2$

38. $4 Zn + NO_3^- + 10 H^+ \longrightarrow 4 Zn^{2+} + NH_4^+ + 3 H_2O$

40. (a) Pb is the anode.
(c) Pb (anode)
(e) Electrons flow from the lead through the wire to the silver.

Chapter 18

Review Questions

2. Alpha particles are much heavier.

4. Contributions to the early history of radioactivity include
(a) Henri Becquerel—discovered radioactivity.
(b) Marie and Pierre Curie—discovered polonium and radium.
(c) Wilhelm Roentgen—discovered X rays and developed the technique for producing them.
(d) Ernest Rutherford—discovered alpha and beta particles, established the link between radioactivity and transmutation, and produced the first successful man-made transmutation.
(e) Otto Hahn and Fritz Strassmann—were first to produce nuclear fission.

6. *Isotope* is used with reference to atoms of the same element that contain different masses. *Nuclide* refers to any isotope of any atom.

10. A disintegration series is a series of α and β emissions leading to production of a stable nuclide.

12. $^{232}_{90}Th \xrightarrow{-\alpha} {}^{228}_{88}Ra \xrightarrow{-\beta} {}^{228}_{89}Ac \xrightarrow{-\beta} {}^{228}_{90}Th \xrightarrow{-\alpha}$
$^{224}_{88}Ra \xrightarrow{-\alpha} {}^{220}_{86}Rn \xrightarrow{-\alpha} {}^{216}_{84}Po \xrightarrow{-\alpha}$
$^{212}_{82}Pb \xrightarrow{-\beta} {}^{212}_{83}Bi \xrightarrow{-\beta} {}^{212}_{84}Po \xrightarrow{-\alpha} {}^{208}_{82}Pb$

14. $^{211}_{83}Bi \longrightarrow {}^{4}_{2}He + {}^{207}_{81}Ti$ $^{207}_{81}Ti \longrightarrow {}^{0}_{-1}e + {}^{207}_{82}Pb$

17. The fission reaction in a nuclear reactor and in an atomic bomb are essentially the same. The difference is that the fissioning is "wild" or uncontrolled in the bomb. In a nuclear reactor, the fissioning rate is controlled by means of moderators, such as graphite, to slow the neutrons and control rods of cadmium or boron to absorb some of the neutrons.

20. When radioactive rays pass through normal matter, they cause that matter to become ionized (usually by knocking out electrons). Therefore, the radioactive rays are classified as ionizing radiation.

23. A radioactive "tracer" is a radioactive material, whose presence is traced by a Geiger counter or some other detecting device. Tracers are often injected into the human body, animals, and plants to determine chemical pathways, rates of circulation, etc.

26. The half-life of carbon-14 is 5730 years.
$$(4 \times 10^6 \text{ years})\left(\frac{1 \text{ half-life}}{5730 \text{ years}}\right) = 7 \times 10^2 \text{ half-lives}$$

700 half-lives would pass in 4 million years. Not enough C-14 would remain to allow detection with any degree of reliability. C-14 dating would not prove useful in this case.

30. In a nuclear power plant, a controlled nuclear fission reaction provides heat energy that is used to produce steam. The steam turns a turbine that generates electricity.

Exercises

2.

	Protons	Neutrons	Nucleons
(a) $^{235}_{92}U$	92	143	235
(b) $^{82}_{35}Br$	35	47	82

4. Its atomic number increases by one, and its mass number remains unchanged.

6. Equations for alpha decays:
(a) $^{192}_{78}Pt \longrightarrow {}^{4}_{2}He + {}^{188}_{76}Os$
(b) $^{210}_{84}Po \longrightarrow {}^{4}_{2}He + {}^{206}_{82}Pb$

8. (a) $^{239}_{93}Np \longrightarrow {}^{0}_{-1}e + {}^{239}_{94}Pu$
(b) $^{90}_{38}Sr \longrightarrow {}^{0}_{-1}e + {}^{90}_{39}Y$

10. (a) gamma-emission
(b) alpha-emission then beta-emission
(c) alpha-emission then gamma-emission

12. $^{30}_{15}P \longrightarrow {}^{30}_{14}Si + {}^{0}_{+1}e$

14. (a) $^{27}_{13}Al + {}^{4}_{2}He \longrightarrow {}^{30}_{15}P + {}^{1}_{0}n$
(c) $^{12}_{6}C + {}^{2}_{1}He \longrightarrow {}^{13}_{7}He + {}^{1}_{0}n$

16. The year 2064; 1/8 of Sr-90 remaining

18. $^{226}_{88}Ra$, alpha-emission; $^{222}_{86}Rn$, alpha-emission; $^{218}_{84}Po$, alpha-emission; $^{214}_{82}Pb$, beta-emission; $^{214}_{83}Bi$, beta-emission; $^{214}_{84}Po$, alpha-emission; $^{210}_{82}Pb$, beta-emission; $^{210}_{83}Bi$

20. (a) 5.4×10^{11} J/mol
(b) 0.199% mass loss

22. 10.9 days

24. 61.59% neutrons by mass. 0.021% electrons by mass

26. 11,460 years old

28. (a) 0.0424 g/mol
(b) 3.8×10^{12} J/mol

31. 0.16 mg remaining

34. (a) $^{235}_{92}U + {}^{1}_{0}n \longrightarrow {}^{143}_{54}Xe + 3 {}^{1}_{0}n + {}^{90}_{38}Sr$
(b) $^{235}_{92}U + {}^{1}_{0}n \longrightarrow {}^{102}_{39}Y + 3 {}^{1}_{0}n + {}^{131}_{53}I$
(c) $^{14}_{7}N + {}^{1}_{0}n \longrightarrow {}^{1}_{1}H + {}^{14}_{6}C$

36. $^{236}_{92}U \longrightarrow {}^{90}_{38}Sr + 3 {}^{1}_{0}n + {}^{143}_{54}Xe$

38. (a) $^{87}_{37}Rb \longrightarrow {}^{0}_{-1}e + {}^{87}_{38}Sr$
(b) $^{87}_{38}Sr \longrightarrow {}^{0}_{+1}e + {}^{87}_{37}Rb$

40. 7680 g

42. (a) 0.500 g left   (c) 0.0625 g
(b) 0.250 g left   (d) 9.77×10^{-4} g

Chapter 19

Review Questions

2. tetrahedral

4. (a) $\cdot\ddot{C}\cdot$

(b)
$$\begin{matrix} & H & \\ H\!:\!\overset{\displaystyle\cdot\cdot}{\underset{\displaystyle H}{C}}\!:\!H \end{matrix} \qquad H\!:\!C\!::\!C\!:\!H \qquad H\!:\!C\!:::\!C\!:\!H$$

methane   ethylene   acetylene

6. (a) CH_3-
   (b) CH_3CH_2-
   (c) $CH_3CH_2CH_2-$ (g) $(CH_3)_2CHCH_2-$
   (d) $(CH_3)_2CH-$ (h) $(CH_3)_3C-$
   (e) $CH_3CH_2CH_2CH_2-$
   (f) $CH_3CH_2\underset{\underset{\displaystyle CH_3}{|}}{C}HCH_3$

7. Compound (a) is an alkene and belongs to a homologous series of compounds represented by the formula C_nH_{2n}. Compounds (b) (c), and (d) are all alkanes, C_nH_{2n+2}

10. Vinyl acetylene, C_4H_4, $CH_2\!=\!CH\!-\!C\!\equiv\!CH$

14. The structure for 1-methylpentane would be $CH_3CH_2CH_2CH_2CH_2CH_3$. The name 1-methylpentane is not based on the longest carbon chain of 6 atoms. Therefore, the correct name is hexane.

Exercises

2. (a) inorganic   (e) organic
   (c) inorganic

4. Organic compounds are compounds in which carbon combines with hydrogen, oxygen, nitrogen, and/or sulfur. Each compound identified as organic above contains some combination of carbon with hydrogen, oxygen, nitrogen, and/or sulfur.

6. Names of alkyl groups
   (a) C_8H_{17} is octyl-   (b) $C_{10}H_{21}$ is decyl-

8. $CH_3CH_2CH_2CH_2CH_2CH_2CH_3$
   $CH_3CH_2CH_2CH_2\underset{\underset{\displaystyle CH_3}{|}}{C}HCH_3$

   $CH_3CH_2\underset{\underset{\displaystyle CH_3}{|}}{\overset{\overset{\displaystyle CH_3}{|}}{C}}CH_2CH_3$ $CH_3CH_2CH_2\underset{\underset{\displaystyle CH_3}{|}}{\overset{\overset{\displaystyle CH_3}{|}}{C}}CH_3$

   $CH_3CH_2CH_2\underset{\underset{\displaystyle CH_3}{|}}{C}HCH_2$ $CH_3\,CH_3CHCH_2CHCH_3$ with CH_3 and CH_3

   $CH_3CH_2\underset{\underset{\displaystyle CH_3}{|}}{C}HCHCH_3$ $CH_3\underset{\underset{\displaystyle CH_3}{|}}{\overset{\overset{\displaystyle CH_3}{|}}{C}}H\!-\!\overset{\overset{\displaystyle CH_3}{|}}{C}CH_3$

   $CH_3CH_2\underset{\underset{\displaystyle CH_2CH_3}{|}}{C}HCH_2CH_3$

10. (a) $CH_3CH_2CH_2Cl$ 1-chloropropane
    (b) $(CH_3)_3CCl$ 2-chloro-2-methylpropane
    (c) $CH_3CHClCH_2CH_3$ 2-chlorobutane

12. (a) 3,6-dimethyloctane
    (b) 3-ethyl-2-methylhexane (or 3-isopropylhexane)

14. (a) $CH_3CH_2\underset{\underset{\displaystyle CH_2CH_3}{|}}{C}HCH_2CH(CH_3)_2$

    (b) $CH_3CH_2CH_2\underset{\underset{\displaystyle C(CH_3)_3}{|}}{C}HCH_2CH_2CH_3$

    (c) $CH_3\underset{\underset{\displaystyle CH_3}{|}}{C}HCH_2\overset{\overset{\displaystyle CH_3}{|}}{C}CH_2CH_2\underset{\underset{\displaystyle CH(CH_3)_2}{|}}{C}H\!-\!\overset{\overset{\displaystyle CH_3}{|}}{C}HCH_2CH_3$
    with CH_2CH_3

16. (a) $CH_3\underset{\underset{\displaystyle CH_3}{|}}{\overset{\overset{\displaystyle CH_3}{|}}{C}}CH_2CH_3$
    2,2-dimethylbutane

    (b) $CH_3\underset{\underset{\displaystyle CH_3}{|}}{C}HCH_2\underset{\underset{\displaystyle CH_2CH_3}{|}}{C}HCH_3$
    2,4-dimethylhexane

18. (a) (4 isomers) $CH_3CH_2CH_2CH_2I$

    $CH_3\underset{\underset{\displaystyle CH_2CH_3}{|}}{C}HI$ $CH_3\underset{\underset{\displaystyle CH_3}{|}}{\overset{\overset{\displaystyle CH_3}{|}}{C}}\!-\!I$ $CH_3\underset{\underset{\displaystyle CH_3}{|}}{C}HCH_2I$

    (b) (4 isomers)
    $CH_3CH_2CHCl_2$ $CH_3CCl_2CH_3$
    $CH_3CHClCH_2Cl$ $CH_2ClCH_2CH_2Cl$

    (c) (5 isomers)
    $CH_3CH_2CHBrCl$ $CH_3CHClCH_2Br$
    $CH_3CHBrCH_2Cl$ $CH_2ClCH_2CH_2Br$
    $CH_3CClBrCH_3$

    (d) (9 isomers)
    $CH_3CH_2CH_2CHCl_2$ $CH_3CH_2CHClCH_2Cl$
    $CH_3CHClCH_2CH_2Cl$ $CH_2ClCH_2CH_2CH_2Cl$
    $CH_3CH_2CCl_2CH_3$ $CH_3CHClCHClCH_3$
    $CH_3CHCHCl_2$ with CH_3 CH_3CClCH_2Cl with CH_3
    CH_3CHCH_2Cl with CH_2Cl

20. (a) $CH_3CH\!=\!CH_2 + Br_2 \longrightarrow CH_3CHBrCH_2Br$
    (b) $CH_3CH\!=\!CH_2 + HBr \longrightarrow CH_3CHBrCH_3$

22. (a) CH_3OCH_3 ether
    (b) CH_3CHO aldehyde
    (c) CH_3COOH carboxylic acid
    (d) $HCOOCH_3$ ester

24. (a) methoxymethane   (c) ethanoic acid
     (b) ethanal   (d) methyl methanoate

26. (a) hexachloroethane CCl_3CCl_3
     (b) iodoethyne $CH\equiv CI$
     (c) 6-bromo-3-methyl $BrCH_2CH_2CH=CC\equiv CH$
        3-hexene-1-yne $\underset{\displaystyle CH_3}{|}$

     (d) 1,2-dibromoethene $CHBr=CHBr$

28. (a) 1,2-diphenylethene   (b) 3-pentene-1-yne

                             $CH\equiv CCH=CHCH_3$

                   (c) 3-phenyl-1-butyne

                       $CH\equiv CCHCH_3$

30. (a) $CH_3CH_2CHCH=CH_2$ 3-isopropyl-1-pentene
                $\underset{\displaystyle CH(CH_3)_2}{|}$

     (b) $CH_3CH_2CH=CCH_2CH_3$ 3-methyl-3-hexene
                     $\underset{\displaystyle CH_3}{|}$

32. (a) $CH_2=CH_2 + H_2O \xrightarrow{H^+} CH_2CH_2OH$ (ethanol)

     (b) $CH\equiv CH + 2Br_2 \longrightarrow CHBr_2CHBr_2$
                           (1,1,2,2-tetrabromoethane)

     (c) $CH_2=CH_2 + H_2 \xrightarrow[\text{Pt, 25°C}]{\text{1 atm}} CH_3CH_3$ (ethane)

34. (a)

para-dichlorobenzene
(1,4-dichlorobenzene)

     (d)

ethylbenzene

     (b)

para-nitrophenol

     (e)

para-methylphenol

     (c)

2,4,5-tribromotoluene

     (f)

2-methyl-
3-nitrobenzoic acid

36. (a)

phenol

     (b)

o-bromochlorobenzene

     (c)

1,3-dichloro-
5-nitrobenzene

     (d)

m-dinitrobenzene

38. (a)

t-butylbenzene

     (b)

1,1-diphenylethane

40.

1,3-dichloro-
2-bromobenzene

1,3-dichloro-
4-bromobenzene

1,3-dichloro-
5-bromobenzene

1,4-dichloro-
2-bromobenzene

1,2-dichloro-
3-bromobenzene

1,2-dichloro-
4-bromobenzene

42. (a) $CH_3CH-\overset{\displaystyle OH}{\overset{|}{C}}HCH_2CH_2CHCH_3$
            $\underset{\displaystyle CH_2CH_3}{|}$ $\underset{\displaystyle CH_2CH_3}{|}$
        3,7-dimethyl-4-nonanol   secondary

     (b) $HO-CH_2CH_2CHCH_2CH_2CH_3$
                       $\underset{\displaystyle CH_3CHCH_3}{|}$
        3-isopropyl-1-hexanol   primary

44. (a) 2,2-dimethyl-1-heptanol

$$HOCH_2\underset{\underset{CH_3}{|}}{\overset{\overset{CH_3}{|}}{C}}CH_2CH_2CH_2CH_2CH_3$$

(b) 1,3-propanediol $HOCH_2CH_2CH_2OH$

46. (a)

benzaldehyde

(b) $O=\underset{\underset{H}{|}}{C}CH_2CH_2\underset{\underset{H}{|}}{C}=O$ butanedial

(c) $CH_3\underset{\underset{OH}{|}}{C}HCH_2\overset{\overset{O}{||}}{C}-H$ 3-hydroxybutanal

48. (a) $CH_3\overset{\overset{O}{||}}{C}-\underset{\underset{CH_3}{|}}{\overset{\overset{CH_3}{|}}{C}}CH_3$ 3,3-dimethyl-2-butanone
methyl t-butyl ketone

(b) $CH_3\underset{\underset{O}{||}}{C}CH_2CH_2\underset{\underset{O}{||}}{C}CH_3$ 2,5-hexanedione

(c) $CH_3\underset{\underset{OH}{|}}{\overset{\overset{CH_3}{|}}{C}}CH_2\overset{\overset{O}{||}}{C}CH_3$ 4-hydroxy-4-methyl-2-pentanone

50. (a) $CH_3\underset{\underset{CH_2CH_3}{|}}{C}HCOOH$ 2-methylbutanoic acid

(b)

phenylacetic acid

(c) $CH_3CH_2CH_2CH_2COOH$ pentanoic acid

(d)

benzoic acid

52. Esters
(a) n-nonyl acetate
nonyl ethanoate $CH_3\overset{\overset{O}{||}}{C}-O-CH_2(CH_2)_7CH_3$

(b) ethyl benzoate

(c) methyl salicylate

54. (a) $H-\underset{\underset{O}{||}}{C}-OCH(CH_3)_2$ isopropyl methanoate

(b)

methyl benzoate

(c) $CH_3\underset{\underset{CH_3}{|}}{C}HCH_2\overset{\overset{O}{||}}{C}-OCH_2CH_3$ ethyl-3-methylbutanoate

56. (a) $2CH_3COOH + KOH \longrightarrow$
$$CH_3COO^-K^+ + H_2O$$

(b)

58. (a) $+CH_2-\underset{\underset{CN}{|}}{C}H\,)_n$ polyacrylonitrile

(b) $+CF_2-CF_2\,)_n$ teflon

60. (a) correct
(b) incorrect, should be butanone
(c) incorrect, should be 4-methyl-2-pentene
(d) incorrect, should be methyl propanoate
(e) incorrect, should be 3-methylpentane

62. (a) alkyl halide   (d) ester
(b) ketone   (e) hydrocarbon
(c) alkene

63. Alkanes are nonpolar molecules. Therefore, there is little attraction between them because there are no partial positive and partial negative charges to attract one another. As a result, it takes very little energy to cause alkanes to boil because there are no intermolecular attractive forces to be overcome.

68. (a) methyl ethyl ether $CH_3CH_2OCH_3$
(b) dimethyl ether CH_3OCH_3
(c) methyl ethyl ether $CH_3CH_2OCH_3$

70. Propanal and propanone are isomers, C_3H_6O. Butanal and butanone are isomers, C_4H_8O.

71. Carboxylic acids, IUPAC name, common name

	IUPAC	common
HCOOH	methanoic acid	formic acid
CH_3COOH	ethanoic acid	acetic acid
CH_3CH_2COOH	propanoic acid	propionic acid
$CH_3CH_2CH_2COOH$	butanoic acid	butyric acid
$CH_3CH_2CH_2CH_2COOH$	pentanoic acid	valeric acid

73. (a) $HCOOH + CH_3CH_2OH \xrightarrow[\Delta]{H^+}$

$$H\overset{\overset{O}{||}}{C}-OCH_2CH_3 + H_2O$$
ethyl formate

(c)

$$\text{C}_6\text{H}_5\text{—COOH} + \text{CH}_3\text{CH}_2\text{CH}_2\text{OH} \xrightarrow[\Delta]{\text{H}^+}$$

n-propyl benzoate

78. (a) CH_3CH_2OH

(c) $CH_3CH_2CH_2CHClCH_3$

(e) $CH_3\overset{\underset{\displaystyle CH_3}{|}}{\underset{\underset{\displaystyle CH_3}{|}}{C}}\text{—OH}$

80. (a) $CH_3\overset{\underset{\displaystyle CH_3}{|}}{CH}CHCH_3$ with CH_3

(c) $CH_3CH_2\overset{\overset{\displaystyle O}{\|}}{C}\text{—OH}$

(e) $CH_3CH_2\overset{\overset{\displaystyle O}{\|}}{C}\text{—O—}CH_2CH_2CH_2CH_3$

Chapter 20

Review Questions

2. Chromosomes are composed of proteins and nucleic acids.

6. Fatty acids in vegetable oils are more unsaturated than fatty acids in animal fats.

8. three disulfide linkages in beef insulin

10. Guanine and cytosine are mutually bonded to each other, as are adenine and thymine.

14.

Monosaccharides	Disaccharides	Polysaccharides
glucose	sucrose	cellulose
fructose	maltose	glycogen
galactose	lactose	starch
ribose		

18. Lactic acid is not a carbohydrate. It does not have an aldehyde or a ketone group.

19. The monosaccharide composition of:
   (a) Sucrose; a disaccharide made from one unit of glucose and one unit of fructose.
   (c) Lactose; a disaccharide made from one unit of galactose and one unit of glucose.

24. Carbohydrates are stored in the body as glycogen.

30. Structural formulas

Glycerol
$$\begin{array}{l} CH_2OH \\ | \\ CHOH \\ | \\ CH_2OH \end{array}$$

Palmitic acid $CH_3(CH_2)_{14}COOH$
Oleic acid $CH_3(CH_2)_7CH=CH(CH_2)_7COOH$
Stearic acid $CH_3(CH_2)_{16}COOH$
Linoleic acid
$$CH_3(CH_2)_4CH=CHCH_2CH=CH(CH_2)_7COOH$$

32. A triacylglycerol (triglyceride) is a triester of glycerol. Most animal fats are triacylglycerols.

38. The three essential fatty acids are linoleic, linolenic, and arachidonic acids. Diets lacking these fatty acids lead to impaired growth and reproduction and skin disorders, such as eczema and dermatitis.

39. The structural formula of cholesterol is

40. The ring structure common to all steroids is

42. Amino acids contain a carboxylic acid group (—COOH) and an amine group (—NH₂).

44. (a) phenylalanine, tryptophan, tyrosine
   (b) cysteine, methionine
   (c) serine, threonine, tyrosine

46. (a) $H_2N\text{—}CH_2\text{—}\overset{\overset{\displaystyle O}{\|}}{C}\text{—}NH\text{—}CH_2COOH$

(b) $H_2N\text{—}CH_2\text{—}\overset{\overset{\displaystyle O}{\|}}{C}\text{—}NH\text{—}CH_2\text{—}\overset{\overset{\displaystyle O}{\|}}{C}\text{—}NH\text{—}\underset{\underset{\displaystyle CH_3}{|}}{CH}\text{—}COOH$

(c) $(CH_3)_2CHCH_2\underset{\underset{\displaystyle NH_2}{|}}{CH}\overset{\overset{\displaystyle O}{\|}}{C}NH\underset{\underset{\underset{\displaystyle CH_2\text{—}S\text{—}CH_3}{|}}{\underset{\displaystyle CH_2}{|}}}{CH}\overset{\overset{\displaystyle O}{\|}}{C}NHCH_2\overset{\overset{\displaystyle O}{\|}}{C}NH\underset{\underset{\displaystyle CH_2OH}{|}}{CH}COOH$

48. Aspartame

$$CH_2CHCOOH \quad HOOCCH_2CHCOOH$$
$$| \qquad\qquad\qquad\quad |$$
$$NH_2 \qquad\qquad\qquad\quad NH_2$$
phenylalanine   aspartic acid

$$HOOCCH_2CH-NH-C-CHCH_2-$$
$$| \qquad\qquad\quad || \quad |$$
$$COOH \qquad\quad O \quad NH_2$$

$$CH_2CH-NH-C-CHCH_2COOH$$
$$| \qquad\qquad\quad || \quad |$$
$$COOH \qquad\quad O \quad NH_2$$

The methyl ester of this structure is aspartame.

55. The two helices of the double helix are joined together by hydrogen bonds between bases. The structure of the bases is such that one base will hydrogen bond to only one other specific base. That is, adenine is always hydrogen–bonded to thymine and cytosine is always bonded to guanine. Therefore, the hydrogen bonding requires a specific structure on the adjoining helix.

56. Complementary bases are the pairs that "fit" to each other by hydrogen bonds between the two helices of DNA. For DNA, the complementary pairs are thymine with adenine and cytosine with guanine or T— A and C—G.

60. The structural differences between DNA and RNA are:
(a) RNA exists in the form of a single-stranded helix, whereas DNA is a double helix.
(b) RNA contains the pentose sugar ribose, whereas DNA contains deoxyribose.
(c) RNA contains the base uracil, whereas DNA contains thymine.

62. The location of protein synthesis is in the ribosomes.

64. The red dotted lines in Figure 20.7 represent hydrogen bonds between complementary base pairs.

68. Polypeptides are numbered starting with the N-terminal acid.

1   2   3   4   5
try — gly —his— phe—val

72. Enzymes act as catalysts for biochemical reactions. Their function is to lower the activation energy of biochemical reactions.

74. CH_2O

76. The oxidation number of carbon changes from 0 to +4.

78. Cellulose: 3.7×10^3 monomer units
Starch: 25 monomer units
Cellulose: 1.9×10^{-6} m long
Starch: 1.3×10^{-8} m long

Answers to Putting It Together Review Exercises

Chapters 1–4

Multiple Choice: **1.** d **2.** a **3.** d **4.** b **5.** d **6.** c **7.** a **8.** d **9.** a **10.** b **11.** a **12.** d **13.** b **14.** c **15.** a **16.** d **17.** c **18.** b **19.** d **20.** c **21.** c **22.** a **23.** b **24.** c **25.** c **26.** a **27.** d **28.** d **29.** c **30.** a **31.** b **32.** a **33.** c **34.** c **35.** c **36.** d **37.** a **38.** a **39.** d **40.** d **41.** c **42.** c **43.** b **44.** b **45.** c **46.** b **47.** b

Free Response:

1. $(1.5\,\text{m})\left(\dfrac{100\,\text{cm}}{1\,\text{m}}\right)\left(\dfrac{1\,\text{in.}}{2.54\,\text{cm}}\right)\left(\dfrac{1\,\text{ft}}{12\,\text{in.}}\right) = 4.9\,\text{ft}$

$(4\,\text{m})\left(\dfrac{100\,\text{cm}}{1\,\text{m}}\right)\left(\dfrac{1\,\text{in.}}{2.54\,\text{cm}}\right)\left(\dfrac{1\,\text{ft}}{12\,\text{in.}}\right) = 13\,\text{ft}$

$(27°\text{C} \times 1.8) + 32 = 81°\text{F}$

2. Jane needs to time how long it took from starting to heat to when the butter is just melted. From this information, she can determine how much heat the pot and butter absorbed. Jane can look up the specific heat of copper. Jane should weigh the pot and measure the temperature of the pot and the temperature at which the butter just melted. This should allow Jane to calculate how much heat the pot absorbed. Then she simply has to subtract the heat the pot absorbed from the heat the stove put out to find out how much heat the butter absorbed.

3. $CaCO_3 \longrightarrow CaO + CO_2$

   75 g   42 g   X

   X = 75 g − 42 g = 33 g

   44 g CO_2 occupies 24 dm^3

   Therefore, 33 g CO_2 occupies $(33\,\text{g})\left(\dfrac{24\,\text{dm}^3}{44\,\text{g}}\right)\left(\dfrac{1\,\text{L}}{1\,\text{dm}^3}\right) = 18\,\text{L}$

4. (a), (b), and (c)   Picture (2) best represents a homogeneous mixture. Pictures (1) and (3) show heterogeneous mixtures, and picture (4) does not show a mixture, as only one species is present.
   Picture (1) likely shows a compound, as one of the components of the mixture is made up of more than one type of "ball." Picture (2) shows a component with more than one part, but the parts seem identical, and therefore it could be representing a diatomic molecule.

5. (a) Picture (3) because fluorine gas exists as a diatomic molecule.
   (b) Other elements that exist as diatomic molecules are oxygen, nitrogen, chlorine, hydrogen, bromine, and iodine.
   (c) Picture (2) could represent SO_3 gas.

6. (a) Tim's bowl should require less energy. Both bowls hold the same volume, but since snow is less dense than a solid block of ice, the mass of water in Tim's bowl is less than the mass of water in Sue's bowl. (Both bowls contain ice at 12°F.)

(b) $\dfrac{12°F - 32}{1.8} = -11°C$

Temperature change: $-11°C$ to $25°C = 36°C$

(c) temperature change: $-11°C$ to $0°C = 11°C$
specific heat of ice $= 2.059 \, J/g°C$

vol. of $H_2O = 1 \, qt = (0.946 \, L)\left(\dfrac{1000 \, mL}{L}\right) = 946 \, mL$

mass of ice $=$ mass of water $= (946 \, \cancel{mL})\left(\dfrac{1 \, g}{1 \, \cancel{mL}}\right) = 946 \, g$

heat required $= (m)(sp. \, ht.)(\Delta t)$

$= (946 \, g)\left(\dfrac{2.059 \, J}{g°C}\right)(11°C)\left(\dfrac{1 \, kJ}{1000 \, J}\right) = 21 \, kJ$

(d) Physical changes

7. (a) Let $x = $ RDA of iron

60% of $x = 11 \, mg \, Fe$

$x = \dfrac{11 \, mg \, Fe \times 100 \, \%}{60 \, \%} = 18 \, mg \, Fe$

(b) density of iron $= 7.86 \, g/mL$

$V = \dfrac{m}{d} = (11 \, mg \, Fe)\left(\dfrac{1 \, g}{1000 \, mg}\right)\left(\dfrac{1 \, mL}{7.86 \, g}\right) = 1.4 \times 10^{-3} \, mL \, Fe$

8. (a) $Ca_3(PO_4)_2$: molar mass $= 3(40.08) + 2(30.97) + 8(16.00) = 310.3$

%Ca in $Ca_3(PO_4)_2 = \dfrac{3(40.08)}{310.3}(100) = 38.7\%$

Let $x = mg \, Ca_3(PO_4)_2$

38.7% of $x = 162 \, mg \, Ca$

$x = \dfrac{(162 \, mg)(100)}{38.7} = 419 \, mg \, Ca_3(PO_4)_2$

(b) $Ca_3(PO_4)_2$ is a compound.

(c) Convert 120 mL to cups

$(120 \, mL)\left(\dfrac{1 \, L}{1000 \, mL}\right)\left(\dfrac{1.059 \, qt}{1 \, L}\right)\left(\dfrac{4 \, cups}{1 \, qt}\right) = 0.51 \, cup$

13% of $x = 0.51 \, cup$

$x = \dfrac{(0.51 \, cup)(100)}{13} = 3.9 \, cups$

9. If Alfred inspects the bottles carefully, he should be able to see whether the contents are solid (silver) or liquid (mercury). Alternatively, since mercury is more dense than silver, the bottle of mercury should weigh more than the bottle of silver (the question indicated that both bottles were of similar size and both were full). Density is mass/volume.

10. (a) Container holds a mixture of sulfur and oxygen.
(b) No. If the container were sealed, the total mass would remain the same whether a reaction took place or not. The mass of the reactants must equal the mass of the products.
(c) No. Density is mass/volume. The volume is the container volume, which does not change. Since the total mass remains constant even if a reaction has taken place, the density of the container, including its contents, remains constant. The density of each individual component within the container may have changed, but the total density of the container is constant.

Chapters 5–6

Multiple Choice: **1.** b **2.** d **3.** b **4.** d **5.** b **6.** b **7.** a **8.** b **9.** d **10.** c **11.** b **12.** d **13.** a **14.** c **15.** d

Names and Formulas: The following are correct: 1, 2, 4, 5, 6, 7, 9, 11, 12, 15, 16, 17, 18, 19, 21, 22, 25, 28, 30, 32, 33, 34, 36, 37, 38, 40.

Free Response:

1. (a) An ion is a charged atom or group of atoms. The charge can be either positive or negative.
    (b) Electrons have negligible mass compared with the mass of protons and neurons. The only difference between Ca and Ca^{2+} is two electrons. The mass of those two electrons is insignificant compared with the mass of the protons and neutrons present (and whose numbers do not change).

2. (a) Let x = abundance of heavier isotope.

    $$303.9303(x) + 300.9326(1 - x) = 303.001$$

    $$303.9303x - 300.9326x = 303.001 - 300.9326$$

    $$2.9977x = 2.068$$

    $$x = 0.6899$$

    $$1 - x = 0.3101$$

    % abundance of heavier isotope = 68.99%

    % abundance of lighter isotope = 31.01%

    (b) $^{304}_{120}Wz$, $^{301}_{120}Wz$

    (c) mass number − atomic number = 303 − 120 = 183 neutrons

3. Cl_2O_7 Cl: $17p \times 2 = 34$ protons
     O: $\underline{8p \times 7 = 56}$ protons
      90 protons in Cl_2O_7

    Since the molecule is electrically neutral, the number of electrons is equal to the number of protons, so Cl_2O_7 has 90 electrons. The number of neutrons cannot be precisely determined unless it is known which isotopes of Cl and O are in this particular molecule.

4. Phosphate has a −3 charge; therefore, the formula for the ionic compound is $M_3(PO_4)_2$.

    P has 15 protons; therefore, $M_3(PO_4)_2$ has 30 phosphorus protons.

    $$3 \text{ (number of protons in M)} = \frac{30 \times 6}{5} = 36 \text{ protons in 3 M}$$

    $$\text{number of protons in M} = \frac{36}{3} = 12 \text{ protons}$$

    from the periodic table, M is Mg.

5. (a) Iron can form cations with different charges (e.g., Fe^{2+} or Fe^{3+}). The Roman numeral indicating which cation of iron is involved is missing. This name cannot be fixed unless the particular cation of iron is specified.
    (b) $K_2Cr_2O_7$. Potassium is generally involved in ionic compounds. The naming system used was for covalent compounds. The name should be potassium dichromate. (Dichromate is the name of the $Cr_2O_7^{2-}$ anion.)
    (c) Sulfur and oxygen are both nonmetals and form a covalent compound. The number of each atom involved needs to be specified for covalent compounds. There are two oxides of sulfur—SO_2 and SO_3. Both elements are nonmetales, so the names should be sulfur dioxide and sulfur trioxide, respectively.

6. No. Each compound, SO_2 and SO_3, has a definite composition of sulfur and oxygen by mass. The law of multiple proportions says that two elements may combine in different ratios to form more than one compound.

7. (a) Electrons are not in the nucleus.
   (b) When an atom becomes an anion, its size increases.
   (c) An ion of Ca (Ca^{2+}) and an atom of Ar have the same number of electrons.

8. (a) $12\,amu \times 7.18 = 86.16\,amu$
   (b) The atom is most likely Rb or Sr. Other remote possibilities are Kr or Y.
   (c) Because of the possible presence of isotopes, the atom cannot be positively identified. The periodic table gives average masses.
   (d) M forms a +1 cation and is most likely in group 1A. The unknown atom is most likely $^{86}_{37}Rb$.

9. The presence of isotopes contradicts Dalton's theory that all atoms of the same element are identical. Also, the discovery of protons, neutrons, and electrons suggests that there are particles smaller than the atom and that the atom is not indivisible.

   Thomson proposed a model of an atom with no clearly defined nucleus.

   Rutherford passed alpha particles through gold foil and inspected the angles at which the alpha particles were deflected. From his results, he proposed the idea of an atom having a small dense nucleus.

Chapters 7–9

Multiple Choice: 1. a   2. a   3. d   **4.** d   5. a   6. b   7. a   8. b   9. b
10. d   11. a   12. d   13. b   14. c   15. c   16. d   17. d   18. b   19. b
20. a   21. b   22. c   23. d   24. b   25. b   26. a   27. d   28. c   29. c
30. b   31. c   32. c   33. b   34. d   35. b   36. d   37. c   38. b   39. c
40. d   41. a   42. c   43. c   44. b   45. b   46. a   47. d   48. b   49. b
50. a

Free Response:

1. (a) $104\,g\,O_2 = (104\,g\,O_2)\left(\dfrac{1\,mol}{32.00\,g}\right) = 3.25\,mol\,O_2$

   $$X\ +\ O_2\ \longrightarrow\ CO_2\ +\ H_2O$$
   $$3.25\,mol2\,mol2.5\,mol\quad \text{(multiply moles by 4)}$$

   $$4X + 13\,O_2 \longrightarrow 8\,CO_2 + 10\,H_2O$$

   Oxygen is balanced. By inspection, X must have 8/4 C atoms and 20/4 H atoms (2 C and 5 H).
   Empirical formula is C_2H_5.

   (b) Additional information needed is the molar mass of X.

2. (a)

   SO₂   O₂   SO₃

(b) $25 \text{ g SO}_2 \left(\dfrac{1 \text{ mol}}{64.07 \text{ g}} \right) = 0.39 \text{ mol SO}_2$

$5 \text{ g O}_2 \left(\dfrac{1 \text{ mol}}{32.00 \text{ g}} \right) = 0.16 \text{ mol O}_2$

$\text{mol ratio} = \dfrac{0.39}{0.16} = \dfrac{2.4 \text{ mol SO}_2}{1 \text{ mol O}_2}$

O_2 is the limiting reagent

(c) False. The percentages given are not mass percentages. The percent composition of S in SO_2 is $(32/64) \times 100 = 50.\% \text{ S}$. The percent composition of S in SO_3 is $(32/80) \times 100 = 40.\% \text{ S}$.

3. (a) $\%O = 100 - (63.16 + 8.77) = 28.07\% \text{ O}$

Start with 100 g compound Z

C: $(63.16 \text{ g}) \left(\dfrac{1 \text{ mol}}{12.01 \text{ g}} \right) = 5.259 \text{ mol}$ $\dfrac{5.259 \text{ mol}}{1.75 \text{ mol}} = 2.998$

H: $(8.77 \text{ g}) \left(\dfrac{1 \text{ mol}}{1.008 \text{ g}} \right) = 8.70 \text{ mol}$ $\dfrac{8.70 \text{ mol}}{1.754 \text{ mol}} = 4.96$

O: $(28.07 \text{ g}) \left(\dfrac{1 \text{ mol}}{16.00 \text{ g}} \right) = 1.754 \text{ mol}$ $\dfrac{1.754 \text{ mol}}{1.754 \text{ mol}} = 1.000$

The ratio of C:H:O is 3:5:1.
The empirical formula is C_3H_5O.
molar mass $= 114$; mass of empirical formula is 57
Therefore, the molecular formula is $C_6H_{10}O_2$.

(b) $2 C_6H_{10}O_2 + 15 O_2 \longrightarrow 12 CO_2 + 10 H_2O$

4. (a) Compound A must have a lower activation energy than compound B because B requires heat to overcome the activation energy for the reaction.

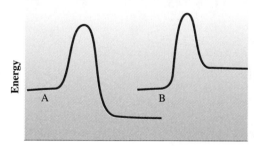

Reaction progress

(b) (i) $2 \text{ NaHCO}_3 \longrightarrow \text{Na}_2\text{CO}_3 + \text{H}_2\text{O} + \text{CO}_2$

Decomposition of 0.500 mol $NaHCO_3$ requires 85.5 kJ of heat.
If 24.0 g CO_2 is produced, then

$(24.0 \text{ g CO}_2) \left(\dfrac{1 \text{ mol CO}_2}{44.01 \text{ g}} \right) \left(\dfrac{1 \text{ mol H}_2\text{O}}{1 \text{ mol CO}_2} \right) \left(\dfrac{18.02 \text{ g}}{1 \text{ mol H}_2\text{O}} \right) = 9.83 \text{ g H}_2\text{O}$ produced

0.500 mol $NaHCO_3$ produces

$(0.500 \text{ mol NaHCO}_3) \left(\dfrac{1 \text{ mol CO}_2}{2 \text{ mol NaHCO}_3} \right) \left(\dfrac{44.01 \text{ g CO}_2}{1 \text{ mol CO}_2} \right) = 11.0 \text{ g CO}_2$

producing 11.0 g CO_2 required 85.5 kJ

producing 24.0 g CO_2 requires $\left(\dfrac{24.0 \text{ g}}{11.0 \text{ g}} \right) (85.5 \text{ kJ}) = 187 \text{ kJ}$

(ii) $NaHCO_3$ could be compound B. Since heat was absorbed for the decomposition of $NaHCO_3$, the reaction was endothermic. Decomposition of A was exothermic.

5. (a) Double-displacement reaction

(b) $2\,NH_4OH(aq) + CoSO_4(aq) \longrightarrow (NH_4)_2SO_4(aq) + Co(OH)_2(s)$

(c) 8.09 g is 25% yield

Therefore, 100% yield $= (8.09\text{ g }(NH_4)_2SO_4)\left(\dfrac{100\%}{25\%}\right) = 32.4\text{ g }(NH_4)_2SO_4$

(theoretical yield)

(d) molar mass of $(NH_4)_2SO_4 = 132.2$ g/mol

theoretical moles $(NH_4)_2SO_4 = (32.4\text{ g})\left(\dfrac{1}{132.2\text{ g/mol}}\right) = 0.245\text{ mol }(NH_4)_2SO_4$

Calculate the moles of $(NH_4)_2SO_4$ produced from 38.0 g of each reactant.

$(38.0\text{ g }NH_4OH)\left(\dfrac{1\text{ mol}}{35.05\text{ g}}\right)\left(\dfrac{1\text{ mol }(NH_4)_2SO_4}{2\text{ mol }NH_4OH}\right) = 0.542\text{ mol }(NH_4)_2SO_4$

$(38.0\text{ g }CoSO_4)\left(\dfrac{1\text{ mol}}{155.0\text{ g}}\right)\left(\dfrac{1\text{ mol }(NH_4)_2SO_4}{1\text{ mol }CoSO_4}\right) = 0.254\text{ mol }(NH_4)_2SO_4$

Limiting reactant is $CoSO_4$; NH_4OH is in excess

6. (a) $C_6H_{12}O_6 \longrightarrow 2\,C_2H_5OH + 2\,CO_2(g)$

Calculate the grams of $C_6H_{12}O_6$ that produced 11.2 g C_2H_5OH.

$(11.2\text{ g }C_2H_5OH)\left(\dfrac{1\text{ mol}}{46.07\text{ g}}\right)\left(\dfrac{1\text{ mol }C_6H_{12}O_6}{2\text{ mol }C_2H_5OH}\right)\left(\dfrac{180.1\text{ g}}{1\text{ mol}}\right) = 21.9\text{ g }C_6H_{12}O_6$

$25.0\text{ g} - 21.9\text{ g} = 3.1\text{ g }C_6H_{12}O_6$ left unreacted

Volume of CO_2 produced:

$(11.2\text{ g }C_2H_5OH)\left(\dfrac{1\text{ mol}}{46.07\text{ g}}\right)\left(\dfrac{2\text{ mol }CO_2}{2\text{ mol }C_2H_5OH}\right)\left(\dfrac{24.0\text{ L}}{\text{mol}}\right) = 5.83\text{ L}$

The assumptions made are that the conditions before and after the reaction are the same and that all reactants went to the products.

(b) theoretical yield

$= (25.0\text{ g }C_6H_{12}O_6)\left(\dfrac{1\text{ mol}}{180.1\text{ g}}\right)\left(\dfrac{2\text{ mol }C_2H_5OH}{1\text{ mol }C_6H_{12}O_6}\right)\left(\dfrac{46.07\text{ g}}{\text{mol}}\right)$

$= 12.8\text{ g }C_2H_5OH$

% yield $= \left(\dfrac{11.2\text{ g}}{12.8\text{ g}}\right)(100) = 87.5\%$

(c) decomposition reaction

7. (a) double decomposition (precipitation)

(b) lead(II) iodide (PbI_2)

(c) $Pb(NO_3)_2(aq) + 2\,KI(aq) \longrightarrow 2\,KNO_3(aq) + PbI_2(s)$

If $Pb(NO_3)_2$ is limiting, the theoretical yield is

$(25\text{ g }Pb(NO_3)_2)\left(\dfrac{1\text{ mol}}{331.2\text{ g}}\right)\left(\dfrac{1\text{ mol }PbI_2}{1\text{ mol }Pb(NO_3)_2}\right)\left(\dfrac{461.0\text{ g}}{\text{mol}}\right) = 35\text{ g }PbI_2$

If KI is limiting, the theoretical yield is

$(25\text{ g }KI)\left(\dfrac{1\text{ mol}}{166.0\text{ g}}\right)\left(\dfrac{1\text{ mol }PbI_2}{2\text{ mol }KI}\right)\left(\dfrac{461.0\text{ g}}{\text{mol}}\right) = 35\text{ g }PbI_2$

percent yield $= \left(\dfrac{7.66\text{ g}}{35\text{ g}}\right)(100) = 22\%$

8. (a) Balance the equation

$$2 XNO_3 + CaCl_2 \longrightarrow 2 XCl + Ca(NO_3)_2$$

$$(30.8 \text{ g CaCl}_2)\left(\frac{1 \text{ mol}}{111.0 \text{ g}}\right)\left(\frac{2 \text{ mol XCl}}{1 \text{ mol CaCl}_2}\right) = 0.555 \text{ mol XCl}$$

Therefore, molar mass of $XCl = \dfrac{79.6 \text{ g}}{0.555 \text{ mol}} = 143 \text{ g/mol}$

mass of $(X + Cl) = $ mass of XCl

mass of $X = 143 - 35.45 = 107.6$

$\qquad X = Ag$ (from periodic table)

(b) No. Ag is below H in the activity series.

9. (a) $2 H_2O_2 \longrightarrow 2 H_2O + O_2$

There must have been eight H_2O_2 molecules and four O_2 molecules in the flask at the start of the reaction.

(b) The reaction is exothermic.

(c) Decomposition reaction

(d) The empirical formula is OH.

Chapters 10–11

Multiple Choice: **1.** c **2.** a **3.** b **4.** a **5.** a **6.** b **7.** b **8.** b **9.** c
10. a **11.** d **12.** d **13.** a **14.** b **15.** c **16.** c **17.** b **18.** c **19.** d
20. d **21.** d **22.** a **23.** c **24.** c **25.** a **26.** b **27.** d **28.** a **29.** b
30. b **31.** a **32.** b **33.** c **34.** c **35.** d **36.** a **37.** a **38.** c **39.** d
40. b **41.** c **42.** c **43.** c **44.** a

Free Response:

1. The compound will be ionic because there is a very large difference in electronegativity between elements in Group 2A and those in Group 7A of the Periodic Table. The Lewis structure is

$$[M]^{2+} \quad \begin{matrix} [:\overset{..}{\underset{..}{X}}:]^- \\ [:\overset{..}{\underset{..}{X}}:]^- \end{matrix}$$

2. Having an even atomic number has no bearing on electrons being paired. An even atomic number means only that there is an even number of electrons. For example, carbon is atomic number six, and it has two unpaired p electrons: $1s^2 2s^2 2p_x^1 2p_y^1$.

3. False. The noble gases do not have any unpaired electrons. Their valence shell electron structure is $ns^2 np^6$ (except He).

4. The outermost electron in potassium is farther away from the nucleus than the outermost electrons in calcium, so the first ionization energy of potassium is lower than that of calcium. However, once potassium loses one electron, it achieves a noble gas electron configuration, and therefore taking a second electron away requires considerably more energy. For calcium, the second electron is still in the outermost shell and does not require as much energy to remove it.

5. The ionization energy is the energy required to *remove* an electron. A chlorine atom forms a chloride ion by *gaining* an electron to achieve a noble gas configuration.

6. The anion is Cl^-; therefore, the cation is K^+ and the noble gas is Ar. K^+ has the smallest radius, while Cl^- will have the largest. K loses an electron, and therefore, in K^+, the remaining electrons are pulled in even closer. Cl was originally larger than Ar, and gaining an electron means that, since the nuclear charge is exceeded by the number of electrons, the radius will increase relative to a Cl atom.

7. The structure shown in the question implies covalent bonds between Al and F, since the lines represent shared electrons. Solid AlF_3 is an ionic compound and therefore probably exists as an Al^{3+} ion and three F^- ions. Only valence electrons are shown in Lewis structures.

8. Carbon has four valence electrons; it needs four electrons to form a noble gas electron structure. By sharing four electrons, a carbon atom can form four covalent bonds.

9. NCl_3 is pyramidal. The presence of three pairs of electrons and a lone pair of electrons around the central atom (N) gives the molecule a tetrahedral structure and a pyramidal shape. BF_3 has three pairs of electrons and no lone pairs of electrons around the central atom (B), so both the structure and the shape of the molecule are trigonal planar.

10. The atom is Br ($35e^-$), which should form a slightly polar covalent bond with sulfur. The Lewis structure of Br is $:\overset{..}{\underset{..}{Br}}\cdot$.

Chapters 12–14

Multiple Choice: 1. b 2. a 3. b 4. c 5. a 6. d 7. d 8. b 9. a
10. b 11. c 12. a 13. c 14. c 15. d 16. a 17. c 18. a 19. a
20. d 21. b 22. c 23. a 24. a 25. d 26. d 27. a 28. b 29. c
30. a 31. a 32. c 33. c 34. c 35. a 36. b 37. b 38. c 39. d
40. a 41. c 42. d 43. b 44. c 45. c 46. c 47. a 48. c 49. d
50. b 51. a 52. c 53. d 54. b 55. b 56. a 57. d 58. c 59. b
60. b

Free Response:

1. 10.0% (m/v) has 10.0 g KCl per 100. mL of solution:

   Therefore, KCl solution contains

   $$\left(\frac{10.0\ \text{g KCl}}{100.\ \text{mL}}\right)(215\ \text{mL})\left(\frac{1\ \text{mol}}{74.55\ \text{g}}\right) = 0.288\ \text{mol KCl}$$

   NaCl solution contains $\left(\frac{1.10\ \text{mol NaCl}}{\text{L}}\right)\left(\frac{1\ \text{L}}{1000\ \text{mL}}\right)(224\ \text{mL}) = 0.246\ \text{mol NaCl}$

   The KCl solution has more particles in solution and will have the higher boiling point.

2. Mass of CO_2 in solution $= $ (molar mass)(moles) $= $ (molar mass)$\left(\dfrac{PV}{RT}\right)$

   $$= \left(\frac{44.01\ \text{g CO}_2}{\text{mol}}\right)\left(\frac{1\ \text{atm} \times 1.40\ \text{L}}{\dfrac{0.08206\ \text{L atm}}{\text{mol K}} \times 298\ \text{K}}\right)$$

   $$= 2.52\ \text{g CO}_2$$

   mass of soft drink $= (345\ \text{mL})\left(\dfrac{0.965\ \text{g}}{\text{mL}}\right) = 333\ \text{g}$

   ppm of $CO_2 = \left(\dfrac{2.52\ \text{g}}{333\ \text{g} + 2.52\ \text{g}}\right)(10^6) = 7.51 \times 10^3\ \text{ppm}$

3. 10% KOH m/v solution contains 10 g KOH in 100 mL solution.

   10% KOH by mass solution contains 10 g KOH + 90 g H_2O.

   The 10% by mass solution is the more concentrated solution and therefore would require less volume to neutralize the HCl.

4. (a) $\dfrac{0.355\ \text{mol}}{0.755\ \text{L}} = 0.470\ \text{M}$

   (b) The lower pathway represents the evaporation of water; only a phase change occurs; no new substances are formed. The upper path represents the decomposition of water. The middle path is the ionization of water.

5. Zack went to Ely, Gaye went to the Dead Sea, and Lamont was in Honolulu. Zack's b.p. was lowered, so he was in a region of lower atmospheric pressure (on a mountain). Lamont was basically at sea level, so his b.p. was about normal. Since Gaye's boiling point was raised, she was at a place of higher atmospheric pressure and therefore was possibly in a location below sea level. The Dead Sea is below sea level.

6. The particles in solids and liquids are close together (held together by inter-molecular attractions), and an increase in pressure is unable to move them significantly closer to each other. In a gas, the space between molecules is significant, and an increase in pressure is often accompanied by a decrease in volume.

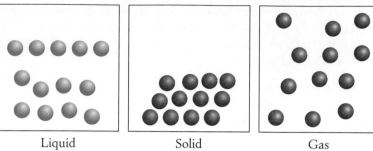

Liquid   Solid   Gas

7. (a) The CO_2 balloon will be heaviest, followed by the Ar balloon. The H_2 balloon would be the lightest. Gases at the same temperature, pressure, and volume contain the same number of moles. All balloons will contain the same number of moles of gas molecules, so when moles are converted to mass, the order from heaviest to lightest is CO_2, Ar, H_2.

  (b) Molar mass: O_2, 32.00; N_2, 28.02; Ne, 20.18

  Using equal masses of gas, we find that the balloon containing O_2 will have the lowest number of moles of gas. Since pressure is directly proportional to moles, the balloon containing O_2 will have the lowest pressure.

8. Ray probably expected to get 0.050 moles $Cu(NO_3)_2$, which is

$$(0.050 \text{ mol } Cu(NO_3)_2)\left(\frac{187.6 \text{ g}}{\text{mol}}\right) = 9.4 \text{ g } Cu(NO_3)_2$$

The fact that he got 14.775 g meant that the solid blue crystals were very likely a hydrate containing water of crystallization.

9. For most reactions to occur, molecules or ions need to collide. In the solid phase, the particles are immobile and therefore do not collide. In solution or in the gas phase, particles are more mobile and can collide to facilitate a chemical reaction.

10. $\Delta t_b = 81.48°C - 80.1°C = 1.38°C$

$$K_b = 2.53 \frac{°C \text{ kg solvent}}{\text{mol solute}}$$

$$\Delta t_b = mK_b$$

$$1.38°C = m\left(2.53 \frac{°C \text{ kg solvent}}{\text{mol solute}}\right)$$

$$0.545 \frac{\text{mol solute}}{\text{kg solvent}} = m$$

Now we convert molality to molarity.

$$\left(\frac{5.36 \text{ g solute}}{76.8 \text{ g benzene}}\right)\left(\frac{1000 \text{ g benzene}}{1 \text{ kg benzene}}\right)\left(\frac{1 \text{ kg benzene}}{0.545 \text{ mol solute}}\right) = 128. \text{ g/mol}$$

Chapters 15–17

Multiple Choice: 1. c 2. d 3. d 4. c 5. c 6. d 7. a 8. d 9. b
10. c 11. a 12. d 13. c 14. b 15. b 16. d 17. c 18. a 19. a
20. a 21. b 22. c 23. a 24. c 25. c 26. d 27. a 28. b 29. a
30. b 31. a 32. c 33. a 34. b 35. c 36. c 37. b 38. d 39. a
40. c 41. a 42. b 43. a 44. b 45. a 46. a 47. d 48. a 49. c
50. d 51. b 52. a 53. a 54. b

Balanced Equations:

55. $3\,P + 5\,HNO_3 \longrightarrow 3\,HPO_3 + 5\,NO + H_2O$

56. $2\,MnSO_4 + 5\,PbO_2 + 3\,H_2SO_4 \longrightarrow 2\,HMnO_4 + 5\,PbSO_4 + 2\,H_2O$

57. $Cr_2O_7^{2-} + 14\,H^+ + 6\,Cl^- \longrightarrow 2\,Cr^{3+} + 7\,H_2O + 3\,Cl_2$

58. $2\,MnO_4^- + 5\,AsO_3^{3-} + 6\,H^+ \longrightarrow 2\,Mn^{2+} + 5\,AsO_4^{3-} + 3\,H_2O$

59. $S^{2-} + 4\,Cl_2 + 8\,OH^- \longrightarrow SO_4^{2-} + 8\,Cl^- + 4\,H_2O$

60. $4\,Zn + NO_3^- + 6\,H_2O + 7\,OH^- \longrightarrow 4\,Zn(OH)_4^{2-} + NH_3$

61. $2\,KOH + Cl_2 \longrightarrow KCl + KClO + H_2O$

62. $4\,As + 3\,ClO_3^- + 6\,H_2O + 3\,H^+ \longrightarrow 4\,H_3AsO_3 + 3\,HClO$

63. $2\,MnO_4^- + 10\,Cl^- + 16\,H^+ \longrightarrow 2\,Mn^{2+} + 5\,Cl_2 + 8\,H_2O$

64. $Cl_2O_7 + 4\,H_2O_2 + 2\,OH^- \longrightarrow 2\,ClO_2^- + 4\,O_2 + 5\,H_2O$

Free Response:

1. $2\,Bz + 3\,Yz^{2+} \longrightarrow 2\,Bz^{3+} + 3\,Yz$

   Bz is above Yz in the activity series.

2. (a) $2\,Al(s) + 3\,Fe(NO_3)_2(aq) \longrightarrow 2\,Al(NO_3)_3(aq) + 3\,Fe(s)$

   (b) The initial solution of $Fe(NO_3)_2$ will have the lower freezing point. It has more particles in solution than the product.

3. (a) $pH = -\log[H^+] = -\log[0.10] = 1.00$

   (b) $mol\ HCl = 0.050\ L \times \dfrac{0.10\ mol}{L} = 0.0050\ mol\ HCl$

   Flask A
   $Zn(s) + 2\,HCl(aq) \longrightarrow ZnCl_2(aq) + H_2(g)$
   HCl is the limiting reactant, so no HCl will remain in the product.
   $pH = 7.0$

   Flask B
   No reaction occurs in flask B, so the pH does not change.
   $pH = 1.00$

4. (a) $2\,NaOH(aq) + H_2S(aq) \longrightarrow Na_2S(aq) + 2\,H_2O(l)$

   (b) $H_2S \rightleftharpoons 2H^+ + S^{2-}$ (aqueous solution)
   $Na_2S \longrightarrow 2\,Na^+ + S^{2-}$ (aqueous solution)

   The addition of S^{2-} to a solution of H_2S will shift the equilibrium to the left, reducing the $[H^+]$ and thereby increasing the pH (more basic).

5. (a) Yes. The K_{eq} of AgCN indicates that it is slightly soluble in water, so a precipitate will form.
   Net ionic equation: $Ag^+(aq) + CN^-(aq) \rightleftharpoons AgCN(s)$

   (b) NaCN is a salt of a weak acid and a strong base and will hydrolyze in water.
   $CN^-(aq) + H_2O(l) \rightleftharpoons HCN(aq) + OH^-(aq)$
   The solution will be basic due to increased OH^- concentration.

6. (a)   (b)

$\square$ = $PbSO_4(s)$

No reaction—contents are merely mixed.

7. (a) $2 A_3X \rightleftharpoons 2 A_2X + A_2$

$$K_{eq} = \frac{(A_2X)^2(A_2)}{(A_3X)^2} = \frac{(3)^2(6)}{(4)^2} = 3.375$$

(b) The equilibrium lies to the right. $K_{eq} > 1$

(c) Yes, it is a redox reaction because the oxidation state of A has changed. The oxidation state of A in A_2 must be 0, but the oxidation state of A in A_3X is not 0.

8. (a) $X_2 + 2 G \rightleftharpoons X_2G_2$

$$K_{eq} = \frac{(X_2G_2)}{(X_2)(G)^2} = \frac{(1)}{(3)(2)^2} = 8.33 \times 10^{-2}$$

(b) Exothermic. An increase in the amount of reactants means that the equilibrium shifted to the left.

(c) An increase in pressure will cause an equilibrium to shift by reducing the number of moles of gas in the equilibrium. If the equilibrium shifts to the right, there must be fewer moles of gas in the product than in the reactants.

9. The pH of the solution is 4.5. (Acid medium)

$$5 Fe^{2+} + MnO_4^- + 8 H^+ \longrightarrow 5 Fe^{3+} + Mn^{2+} + 4 H_2O$$

Chapters 18–20

Multiple Choice: 1. b 2. d 3. a 4. a 5. d 6. b 7. a 8. c 9. c
10. c 11. a 12. c 13. b 14. b 15. c 16. d 17. b 18. b 19. a
20. d 21. a 22. c 23. c 24. c 25. b 26. b 27. d 28. c 29. d
30. c 31. c 32. b 33. c 34. c 35. d 36. b 37. a 38. c 39. d
40. b 41. c 42. a 43. a 44. b 45. c 46. a 47. c 48. b 49. c
50. d 51. c 52. a 53. c 54. d 55. b

Free Response:

1. The nuclide lost two alpha particles and one beta particle in any order. One series could be $^{223}_{87}Fr \xrightarrow{-\alpha} {}^{219}_{85}At \xrightarrow{-\beta} {}^{219}_{86}Rn \xrightarrow{-\alpha} {}^{215}_{84}Po$. $^{215}_{84}Po$ is the resulting nuclide.

2.

Tyr-Gly-Gly-Phe

3. Prostaglandin A_2 has two alkenes, an alcohol, a ketone, and a carboxylic acid (or carboxylate) group.

4. Water interacts mainly by hydrogen bonding, and molecules that can participate significantly in hydrogen bonding generally dissolve in water. The alcohol, $CH_3CH_2CH_2OH$, is the only compound listed that can significantly participate in hydrogen bonding as both a donor and an acceptor and should be significantly soluble in water.

5. Structural isomers
   methyl propyl ether and 2-butanol; butanoic acid and methyl propanoate

6. (a) An ester can be formed from a carboxylic acid and an alcohol. Alcohol group: Ser, Tyr, or Thr. Carboxylic acid group: Asp or Glu.

(b)

(c) Cholesterol is a lipid. Lipids are water-insoluble molecules. The alcohol group in cholesterol is only a small component of a large molecule.

7. (a) The zinc ion is called an activator.

(b) RNA contains ribose sugar; DNA contains deoxyribose sugar. RNA contains the base uracil; DNA contains the base thymine. RNA is generally single stranded; DNA is generally double stranded.

(c) Yes. All living things contain small amounts of carbon-14. Some people may have also been exposed to radioactivity such as from X-rays and have higher levels of radioactivity in their bodies.

8. (a)

(b) Tyrosine contains a para substituted aromatic ring.

(c) There are five amino acids in a pentapeptide. Four moles of water are lost in preparing a pentapeptide from five amino acids.

(d) Blood sugar is glucose. When hydrolyzed, maltose increases the glucose level more than lactose. Maltose hydrolyzes to two glucose units. Lactose hydrolyzes to one glucose unit and one galactose unit. Thus, it is better to cut down on maltose.

9. $96 \, g \longrightarrow 48 \, g \longrightarrow 24 \, g \longrightarrow 12 \, g \longrightarrow 6 \, g \longrightarrow 3 \, g \longrightarrow 1.5 \, g$

Each change involves $\frac{1}{2}$ life. Six (6) half-lives are required to go from 96 g to 1.5 g.

10. (a) The process is transmutation.

(b) Transmutation produces a different element by bombarding an element with a small particle such as hydrogen, deuterium, or a neutron. Radioactive decay is the formation of a different element by the loss of an alpha or beta particle from the nucleus of a radioactive element.

(c) No. Nuclear fission involves the bombardment of a heavy element with a neutron, causing the element to split (fission) into two or more lighter elements. A large amount of energy also is formed in a nuclear fission.

GLOSSARY

A

absolute zero $-273°C$, the zero point on the Kelvin (absolute) temperature scale. *See also* Kelvin temperature scale. [2.8, 12.6]

acid (1) A substance that produces H^+ (H_3O^+) when dissolved in water. (2) A proton donor. (3) An electron pair acceptor. A substance that bonds to an electron pair. [15.1]

acid anhydride A nonmetal oxide that reacts with water to form an acid. [13.8]

acid ionization constant (K_a) The equilibrium constant for the ionization of a weak acid in water. [16.11]

activation energy The amount of energy needed to start a chemical reaction. [8.5, 16.8]

activity series of metals A listing of metallic elements in descending order of reactivity. [17.5]

actual yield The amount of product actually produced in a chemical reaction (compared with the theoretical yield). [9.6]

addition reaction In organic chemistry, a reaction in which two substances join together to produce one substance. [19.9]

alcohol An organic compound consisting of an —OH group bonded to a carbon atom in a nonaromatic hydrocarbon group; alcohols are classified as primary (1°), secondary (2°), or tertiary (3°), depending on whether the carbon atom to which the —OH group is attached is bonded to one, two, or three other carbon atoms, respectively. [19.14]

aldehyde An organic compound that contains the —CHO group. The general formula is RCHO. [19.17]

alkali metal An element (except H) from Group 1A of the periodic table. [10.4]

alkaline earth metal An element from Group 2A of the periodic table. [10.4]

alkane (saturated hydrocarbon) A compound composed of carbon and hydrogen, having only single bonds between the atoms; also known as paraffin hydrocarbon. *See also* alkene and alkyne. [19.4]

alkene (unsaturated hydrocarbon) A hydrocarbon whose molecules have at least one carbon–carbon double bond. [19.7]

alkyl group An organic group derived from an alkane by removal of one H atom. The general formula is C_nH_{2n+1} (e.g., CH_3, methyl). Alkyl groups are generally indicated by the letter R. [19.6]

alkyne (unsaturated hydrocarbon) A hydrocarbon whose molecules have at least one carbon–carbon triple bond. [19.7]

allotrope A substance existing in two or more molecular or crystalline forms (e. g., graphite and diamond are two allotropic forms of carbon). [12.17]

alpha particle (α) A particle emitted from a nucleus of an atom during radioactive decay; it consists of two protons and two neutrons with a mass of about 4 amu and a charge of +2; it is considered to be a doubly charged helium atom. [18.3]

amino acid An organic compound containing two functional groups—an amino group (NH_2) and a carboxyl group (COOH). Amino acids are the building blocks for proteins. [20.4]

amorphous A solid without shape or form. [1.7]

amphoteric (substance) A substance having properties of both an acid and a base. [15.3]

anion A negatively charged ion. *See also* ion. [3.8, 5.5, 6.2]

anode The electrode where oxidation occurs in an electrochemical reaction. [17.6]

aqueous solution A water solution. [16.3]

aromatic compound An organic compound whose molecules contain a benzene ring or that has properties resembling benzene. [19.10]

artificial radioactivity Radioactivity produced in nuclides during some types of transmutations. Artificial radioactive nuclides behave like natural radioactive elements in two ways: They disintegrate in a definite fashion, and they have a specific half-life. Sometimes called *induced radioactivity*. [18.6]

1 atmosphere The standard atmospheric pressure; that is, the pressure exerted by a column of mercury 760 mm high at a temperature of 0°C. *See also* atmospheric pressure. [12.3]

atmospheric pressure The pressure experienced by objects on Earth as a result of the layer of air surrounding our planet. A pressure of 1 atmosphere (1 atm) is the pressure that will support a column of mercury 760 mm high at 0°C. [12.3]

atom The smallest particle of an element that can enter into a chemical reaction. [3.1]

atomic mass The average relative mass of the isotopes of an element referred to the atomic mass of carbon-12. [5.9]

atomic mass number (A) The sum of the protons and neutrons in the nucleus of an atom. [5.11]

atomic mass unit (amu) A unit of mass equal to one-twelfth the mass of a carbon-12 atom. [5.11]

atomic number (Z) The number of protons in the nucleus of an atom of a given element. *See also* isotopic notation. [5.7]

atomic theory The theory that substances are composed of atoms, and that chemical reactions are explained by the properties and the interactions of these atoms. [5.2, Ch. 10]

Avogadro's law Equal volumes of different gases at the same temperature and pressure contain equal numbers of molecules. [12.11]

Avogadro's number 6.022×10^{23}; the number of formula units in 1 mole. [7.1, 9.1]

B

balanced equation A chemical equation having the same number of each kind of atom and the same electrical charge on each side of the equation. [8.2]

barometer A device used to measure atmospheric pressure. [12.3]

base A substance whose properties are due to the liberation of hydroxide (OH^-) ions into a water solution. [15.1]

basic anhydride A metal oxide that reacts with water to form a base. [13.8]

beta particle (β) A particle identical in charge (-1) and mass to an electron. [18.3]

binary compound A compound composed of two different elements. [6.4]

biochemistry The branch of chemistry concerned with chemical reactions occurring in living organisms. [20.1]

boiling point The temperature at which the vapor pressure of a liquid is equal to the pressure above the liquid. It is called the normal boiling when the pressure is 1 atmosphere. [13.5]

bond length The distance between two nuclei that are joined by a chemical bond. [13.10]

Boyle's law The relationship between pressure and volume of a gas. At constant temperature, the volume of a fixed mass of gas is inversely proportional to the pressure (PV = constant). [12.5]

Brownian movement The random motion of colloidal particles. [15.14]

buffer solution A solution that resists changes in pH when diluted or when small amounts of a strong acid or strong base are added. [16.14]

C

calorie (cal) A commonly used unit of heat energy; 1 calorie is a quantity of heat energy that will raise the temperature of 1 g of water 1°C (e.g., from 14.5 to 15.5°C). Also, 4.184 joules = 1 calorie exactly. *See also* joule. [4.6]

capillary action The spontaneous rising of a liquid in a narrow tube, which results from the cohesive forces within the liquid and the adhesive forces between the liquid and the walls of the container. [13.4]

carbohydrate A polyhydroxy aldehyde or polyhydroxy ketone, or a compound that upon hydrolysis yields a polyhydroxy aldehyde or ketone; sugars, starch, and cellulose are examples. [20.2]

carbonyl group The structure $\overset{}{>}C{=}O$. [19.17]

carboxyl group The functional group of carboxylic acids:
$$-\overset{\overset{O}{\|}}{C}-OH \quad [19.19]$$

carboxylic acid An organic compound having a carboxyl group. [19.19]

catalyst A substance that influences the rate of a reaction and can be recovered essentially unchanged at the end of the reaction. [16.8]

cathode The electrode where reduction occurs in an electrochemical reaction. [17.6]

cation A positively charged ion. *See also* ion. [3.8, 5.5]

Celsius temperature scale (°C) The temperature scale on which water freezes at 0°C and boils at 100°C at 1 atm pressure. [2.8]

chain reaction A self-sustaining nuclear or chemical reaction in which the products cause the reaction to continue or to increase in magnitude. [18.8]

Charles' law At constant pressure, the volume of a fixed mass of any gas is directly proportional to the absolute temperature (V/T = constant). [12.6]

chemical bond The attractive force that holds atoms together in a compound. [Ch. 11]

chemical change A change producing products that differ in composition from the original substances. [4.3]

chemical equation A shorthand expression showing the reactants and the products of a chemical change (for example, $2 H_2O = 2 H_2 + O_2$). [4.3, 8.1]

chemical equilibrium The state in which the rate of the forward reaction equals the rate of the reverse reaction in a chemical change. [16.3]

chemical family *See* groups or families of elements.

chemical formula A shorthand method for showing the composition of a compound, using symbols of the elements. [3.9]

chemical kinetics The study of reaction rates and reaction mechanisms. [16.2]

chemical properties The ability of a substance to form new substances either by reaction with other substances or by decomposition. [4.1]

chemistry The science of the composition, structure, properties, and reactions of matter, especially of atomic and molecular systems. [1.2]

chlorofluorocarbons (CFCs) A group of compounds made of carbon, chlorine, and fluorine. [12.17]

colligative properties Properties of a solution that depend on the number of solute particles in solution, and not on the nature of the solute (examples: vapor pressure lowering, freezing point depression, boiling point elevation). [14.7]

colloid A dispersion in which the dispersed particles are larger than the solute ions or molecules of a true solution and smaller than the particles of a mechanical suspension. [15.13]

combination reaction A direct union or combination of two substances to produce one new substance. [8.4]

combustion A chemical reaction in which heat and light are given off; generally, the process of burning or uniting a substance with oxygen. [19.4]

common-ion effect The shift of a chemical equilibrium caused by the addition of an ion common to the ions in the equilibrium. [16.12]

common names Arbitrary names that are not based on the chemical composition of compounds (examples: quick silver for mercury, laughing gas for nitrous oxide). [6.1]

compound A distinct substance composed of two or more elements combined in a definite proportion by mass. [3.8]

concentrated solution A solution containing a relatively large amount of dissolved solute. [14.6]

concentration of a solution A quantitative expression of the amount of dissolved solute in a certain quantity of solvent. [14.2]

condensation The process by which molecules in the gaseous state return to the liquid state. [13.3]

conjugate acid–base Two molecules or ions whose formulas differ by one H^+. (The acid is the species with the H^+, and the base is the species without the H^+.) [15.1]

copolymer A polymer containing two different kinds of monomer units. [19.21]

covalent bond A chemical bond formed between two atoms by sharing a pair of electrons. [11.5]

critical mass The minimum quantity of mass required to support a self-sustaining chain reaction. [18.8]

curie (Ci) A unit of radioactivity indicating the rate of decay of a radioactive substance: $1 \text{ Ci} = 3.7 \times 10^{10}$ disintegrations per second. [18.7]

D

Dalton's atomic model The first modern atomic theory to state that elements are composed of minute individual particles called *atoms*. [5.2]

Dalton's law of partial pressures The total pressure of a mixture of gases is the sum of the partial pressures exerted by each of the gases in the mixture. [12.10]

decomposition reaction A breaking down, or decomposition, of one substance into two or more different substances. [8.4]

density The mass of an object divided by its volume. [2.19]

dialysis The process by which a parchment membrane allows the passage of true solutions but prevents the passage of colloidal dispersions. [15.16]

diatomic molecules The molecules of elements that always contain two atoms. Seven elements occur as diatomic molecules: H_2, N_2, O_2, F_2, Cl_2, Br_2, and I_2. [3.17]

diffusion The property by which gases and liquids mix spontaneously because of the random motion of their particles. [12.2]

dilute solution A solution containing a relatively small amount of dissolved solute. [14.6]

dipeptide Two α-amino acids joined by a peptide linkage. [20.4]

dipole A molecule that is electrically asymmetrical, causing it to be oppositely charged at two points. [11.6]

disaccharide A carbohydrate that yields two monosaccharide units when hydrolyzed. [20.2]

dissociation The process by which a salt separates into individual ions when dissolved in water. [15.6]

DNA Deoxyribonucleic acid; a high-molar-mass polymer of nucleotides, present in all living matter, that contains the genetic code that transmits hereditary characteristics. [20.6]

double bond A covalent bond in which two pairs of electrons are shared. [11.5, 19.2]

double-displacement reaction A reaction of two compounds to produce two different compounds by exchanging the components of the reacting compounds. [8.4]

ductile A property of metals; can be drawn in wires. [3.4]

E

effusion The process by which gas molecules pass through a tiny orifice from a region of high pressure to a region of lower pressure. [12.2]

Einstein's mass–energy equation $E = mc^2$: the relationship between mass and energy. [18.12]

electrolysis The process whereby electrical energy is used to bring about a chemical change. [17.6]

electrolyte A substance whose aqueous solution conducts electricity. [15.5]

electrolytic cell An electrolysis apparatus in which electrical energy from an outside source is used to produce a chemical change. [17.6]

electron A subatomic particle that exists outside the nucleus and carries a negative electrical charge. [5.6]

electron configuration The orbital arrangement of electrons in an atom. [10.5]

electron-dot structure *See* Lewis structure.

electronegativity The relative attraction that an atom has for a pair of shared electrons in a covalent bond. [11.6]

electron shell *See* principal energy levels of electrons.

element A basic building block of matter that cannot be broken down into simpler substances by ordinary chemical changes; in 1994, there were 111 known elements. [3.1]

empirical formula A chemical formula that gives the smallest whole-number ratio of atoms in a compound—that is, the relative number of atoms of each element in the compound; also known as the simplest formula. [7.4]

endothermic reaction A chemical reaction that absorbs heat. [8.5]

energy The capacity of matter to do work. [4.5]

energy levels of electrons Areas in which electrons are located at various distances from the nucleus. [10.4]

energy sublevels The *s, p, d,* and *f* orbitals within a principal energy level occupied by electrons in an atom. [10.4]

enzyme A protein that catalyzes a biochemical reaction. [20.5]

equilibrium A dynamic state in which two or more opposing processes are taking place at the same time and at the same rate. [16.3]

equilibrium constant (K_{eq}) A constant representing the concentrations of the reactants and the products in a chemical reaction at equilibrium. [16.9]

essential amino acid An amino acid that is not synthesized by the body and therefore must be supplied in the diet. [20.4]

ester An organic compound derived from a carboxylic acid and an alcohol. The general formula is

$$R - \underset{\underset{O}{\|}}{C} - OR' \qquad [19.20]$$

ether An organic compound having two hydrocarbon groups attached to an oxygen atom. The general formula is $R - O - R'$. [19.16]

evaporation The escape of molecules from the liquid state to the gas or vapor state. [13.2]

exothermic reaction A chemical reaction in which heat is released as a product. [8.5]

F

Fahrenheit temperature scale (°F) The temperature scale on which water freezes at 32°F and boils at 212°F at 1 atm pressure. [2.8]

fats and oils Esters of fatty acids and glycerol. *See also* triacylglycerol. [20.3]

fatty acids Long-chain carboxylic acids present in lipids (fats and oils). [20.3]

formula equation A chemical equation in which all the reactants and products are written in their molecular, or normal, formula expression; also called a molecular equation. [15.12]

formula unit The atom or molecule indicated by the formula of a substance under consideration (examples: Mg, O_2, H_2O). [7.1]

free radical A neutral atom or group of atoms having one or more unpaired electrons. [12.17]

freezing or melting point The temperature at which the solid and liquid states of a substance are in equilibrium. [13.6]

frequency A measurement of the number of waves that pass a particular point per second. [10.2]

functional group An atom or group of atoms that characterizes a class of organic compounds. For example, —COOH is the functional group of carboxylic acids. [19.2]

G

galvanic cell *See* voltaic cell.

gamma ray (γ) High-energy photons emitted by radioactive nuclei; they have no electrical charge and no measurable mass. [18.3]

gas A state of matter that has no shape or definite volume so that the substance completely fills its container. [1.7]

Gay-Lussac's law At constant volume, the pressure of a fixed mass of gas is directly proportional to the absolute temperature (P/T = constant). [12.7]

Gay-Lussac's law of combining volumes (of gases) When measured at the same temperature and pressure, the ratios of the volumes of reacting gases are small whole numbers. [12.11]

genes Basic units of heredity that consist primarily of DNA and proteins and occur in the chromosomes. [20.7]

Graham's law of effusion The rates of effusion of two gases at the same temperature and pressure are inversely proportional to the square roots of their densities or molar masses. [12.2]

greenhouse effect The effect of CO_2 and other greenhouse gases to warm the atmosphere by absorbing energy near the surface of the earth, which can lead to global warming. [8.6]

ground state The lowest available enrgy level within an atom. [10.3]

groups or families (of elements) Vertical groups of elements in the periodic table (1A, 2A, and so on). Families of elements that have similar outer-orbital electron structures. [10.6]

H

half-life The time required for one-half of a specific amount of a radioactive nuclide to disintegrate; half-lives of the elements range from a fraction of a second to billions of years. [18.2]

halogenation The substitution of a halogen atom for a hydrogen atom in an organic compound. [19.3]

halogens Group 7A of the periodic table; consists of the elements fluorine, chlorine, bromine, iodine, and astatine. [10.4]

heat A form of energy associated with the motion of small particles of matter. [2.8]

heat of fusion The energy required to change 1 gram of a solid into a liquid at its melting point. [13.7]

heat of reaction The quantity of heat produced by a chemical reaction. [8.5]

heat of vaporization The amount of heat required to change 1 gram of a liquid to a vapor at its normal boiling point. [13.7]

heterogeneous Matter without a uniform composition—having two or more components or phases. [1.8]

homogeneous Matter that has uniform properties throughout. [1.8]

homologous series A series of compounds in which the members differ from one another by a regular increment. For example, each successive member of the alkane series of hydrocarbons differs by a CH_2 group. [19.4]

hydrate A solid that contains water molecules as a part of its crystalline structure. [13.9]

hydrocarbon A compound composed entirely of carbon and hydrogen. [8.5]

hydrogen bond The intermolecular force acting between molecules that contain hydrogen covalently bonded to the highly electronegative elements F, O, and N. [13.8]

hydrolysis Chemical reaction with water in which the water molecule is split into H^+ and OH^-. [16.13]

hydronium ion The result of a proton combining with a polar water molecule to form a hydrated hydrogen ion, H_3O^+. [15.1]

hypothesis A tentative explanation of certain facts to provide a basis for further experimentation. [1.5]

I

ideal gas A gas that behaves precisely according to the Kinetic Molecular Theory; also called a perfect gas. [12.2]

ideal gas law $PV = nRT$; that is, the volume of a gas varies *directly* with the number of gas molecules and the absolute temperature and *inversely* with the pressure. [12.14]

immiscible Incapable of mixing; immiscible liquids do not form a solution with one another. [14.2]

induced radioactivity *See* artificial radioactivity.

ion A positively or negatively charged atom or group of atoms. *See also* cation, anion. [3.8, 5.5]

ionic bond The chemical bond between a positively charged ion and a negatively charged ion. [11.3]

ionic compound A compound that is composed of ions (e.g., Na^+Cl^-) [3.8, 6.3]

ionization The formation of ions, which occurs as the result of a chemical reaction of certain substances with water. [15.6]

ionization energy The energy required to remove an electron from an atom, an ion, or a molecule. [11.1]

ionizing radiation Radiation with enough energy to dislocate bonding electrons and create ions when passing through matter. [18.4]

ion product constant for water (K_w) An equilibrium constant defined as the product of the H^+ ion concentration and the OH^- ion concentration, each in moles per liter. $K_w = [H^+][OH^-] = 1 \times 10^{-14}$ at 25°C. [16.10]

isomerism The phenomenon of two or more compounds having the same molecular formula but different molecular structures. [19.5]

isomers Compounds having identical molecular formulas but different structural formulas. [19.5]

isotope An atom of an element that has the same atomic number but a different atomic mass. Since their atomic numbers are identical, isotopes vary only in the number of neutrons in the nucleus. [5.10]

isotopic notation Notation for an isotope of an element where the subscript is the atomic number, the superscript is the mass number, and they are attached on the left of the symbol for the element. (For example, hydrogen-1 is notated as $_1^1H$.) *See also* atomic number, mass number. [5.10]

IUPAC International Union of Pure and Applied Chemistry, which devised (in 1921) and continually upgrades the system of nomenclature for inorganic and organic compounds. [6.1]

J

joule (J) The SI unit of energy. *See also* calorie. [4.6]

K

Kelvin temperature scale (K) Absolute temperature scale starting at absolute zero, the lowest temperature possible. Freezing and boiling points of water on this scale are 273 K and 373 K, respectively, at 1 atm pressure. *See also* absolute zero. [2.8, 12.6]

ketone An organic compound that contains a carbonyl group between two other carbon atoms. The general formula is $R_2C = O$. [19.17]

kilocalorie (kcal) 1000 cal; the kilocalorie is also known as the nutritional or large Calorie, used for measuring the energy produced by food. [4.6]

kilogram (kg) The standard unit of mass in the metric system; 1 kilogram equals 2.205 pounds. [2.9]

kilojoule (kJ) 1000 J. [4.6]

kinetic energy (KE) The energy that matter possesses due to its motion; $KE = 1/2 \, mv^2$. [4.5]

kinetic-molecular theory (KMT) A group of assumptions used to explain the behavior and properties of gases. [12.2]

L

law A statement of the occurrence of natural phenomena that occur with unvarying uniformity under the same conditions. [1.5]

law of conservation of energy Energy can be neither created nor destroyed, but it can be transformed from one form to another. [4.8]

law of conservation of mass No change is observed in the total mass of the substances involved in a chemical reaction; that is, the mass of the products equals the mass of the reactants. [4.4]

law of definite composition A compound always contains two or more elements in a definite proportion by mass. [5.3]

law of multiple proportions Atoms of two or more elements may combine in different ratios to produce more than one compound. [5.3]

Le Châtelier's principle If a stress is applied to a system in equilibrium, the system will respond in such a way as to relieve that stress and restore equilibrium under a new set of conditions. [16.4]

Lewis structure A method of indicating the covalent bonds between atoms in a molecule or an ion such that a pair of electrons (:) represents the valence electrons forming the covalent bond. [11.2]

limiting reactant A reactant that limits the amount of product formed because it is present in insufficient amount compared with the other reactants. [9.6]

linear structure In the VSEPR model, an arrangement where the pairs of electrons are arranged 180° apart for maximum separation. [11.11]

line spectrum Colored lines generated when light emitted by a gas is passed through a spectroscope. Each element possesses a unique set of line spectra. [10.3]

lipids Organic compounds found in living organisms that are water insoluble, but soluble in such fat solvents as diethyl ether, benzene, and carbon tetrachloride; examples are fats, oils, and steroids. [20.3]

liquid A state of matter in which the particles move about freely while the substance retains a definite volume; thus, liquids flow and take the shape of their containers. [1.7]

liter (L) A unit of volume commonly used in chemistry; 1 L = 1000 mL; the volume of a kilogram of water at 4°C. [2.10]

logarithm (log) The power to which 10 must be raised to give a certain number. The log of 100 is 2.0. [15.9]

M

macromolecule *See* polymer.

malleable A property of metals; can be rolled or hammered into sheets. [3.4]

mass The quantity or amount of matter that an object possesses. [2.7]

mass defect The difference between the actual mass of an atom of an isotope and the calculated mass of the protons and neutrons in the nucleus of that atom. [18.12]

mass number (A) The sum of the protons and neutrons in the nucleus of a given isotope of an atom. *See also* isotopic notation. [5.10]

mass percent solution The grams of solute in 100 g of a solution. [14.6]

matter Anything that has mass and occupies space. [1.6]

meiosis The process of cell division to form a sperm cell and an egg cell in which each cell formed contains half of the chromosomes found in the normal single cell. [20.7]

melting or freezing point *See* freezing or melting point.

meniscus The shape of the surface of a liquid when placed in a glass cylinder. It can be concave or convex. [13.4]

metal An element that is solid at room temperature and whose properties include luster, ductility, malleability, and good conductivity of heat and electricity; metals tend to lose their valence electrons and become positive ions. [3.4]

metalloid An element having properties that are intermediate between those of metals and nonmetals (for example, silicon); these elements are useful in electronics. [3.4]

meter (m) The standard unit of length in the SI and metric systems; 1 meter equals 39.37 inches. [2.5]

metric system A decimal system of measurements. *See also* SI. [2.5]

miscible Capable of mixing and forming a solution. [14.2]

mitosis Ordinary cell division in which a DNA molecule is duplicated by uncoiling to single strands and then reassembling with complementary nucleotides. Each new cell contains the normal number of chromosomes. [20.7]

mixture Matter containing two or more substances, which can be present in variable amounts; mixtures can be homogeneous (sugar water) or heterogeneous (sand and water). [1.8]

molality (*m*) An expression of the number of moles of solute dissolved in 1000 g of solvent. [14.7]

molarity (M) The number of moles of solute per liter of solution. [14.6]

molar mass The mass of Avogadro's number of atoms or molecules. The sum of the atomic masses of all the atoms in an element, compound, or ion. The mass of a mole of any formula unit. It is also known as the molecular weight. [7.1, 9.1]

molar solution A solution containing 1 mole of solute per liter of solution. [14.6]

molar volume (of a gas) The volume of 1 mol of a gas at STP equals 22.4 L/mol. [12.12]

mole The amount of a substance containing the same number of formula units (6.022×10^{23}) as there are in exactly 12 g of ^{12}C. One mole is equal to the molar mass in grams of any substance. [7.1]

molecular formula The total number of atoms of each element present in one molecule of a compound; also known as the true formula. *See also* empirical formula. [7.4]

molecule The smallest uncharged individual unit of a compound formed by the union of two or more atoms. [3.8]

mole ratio A ratio between the number of moles of any two species involved in a chemical reaction; the mole ratio is used as a conversion factor in stoichiometric calculations. [9.2]

monomer The small unit or units that undergo polymerization to form a polymer. [19.21]

monosaccharide A carbohydrate that cannot be hydrolyzed to simpler carbohydrate units (for example, simple sugars like glucose or fructose). [20.2]

N

net ionic equation A chemical equation that includes only those molecules and ions that have changed in the chemical reaction. [15.12]

neutralization The reaction of an acid and a base to form a salt plus water. [15.10]

neutron A subatomic particle that is electrically neutral and is found in the nucleus of an atom. [5.6]

noble gases A family of elements in the periodic table—helium, neon, argon, krypton, xenon, and radon—that contain a particularly stable electron structure. [10.4]

nonelectrolyte A substance whose aqueous solutions do not conduct electricity. [15.5]

nonmetal An element that has properties the opposite of metals: lack of luster, relatively low melting point and density, and generally poor conduction of heat and electricity. Nonmetals may or may not be solid at room temperature (examples: carbon, bromine, nitrogen); many are gases. They are located mainly in the upper right-hand corner of the periodic table. [3.8]

nonpolar covalent bond A covalent bond between two atoms with the same electronegativity value; thus, the electrons are shared equally between the two atoms. [11.6]

normal boiling point The temperature at which the vapor pressure of a liquid equals 1 atm or 760 torr pressure. [13.5]

nuclear binding energy The energy equivalent to the mass defect; that is, the amount of energy required to break a nucleus into its individual protons and neutrons. [18.12]

nuclear fission The splitting of a heavy nuclide into two or more intermediate-sized fragments when struck in a particular way by a neutron. As the atom is split, it releases energy and two or three more neutrons that can then cause another nuclear fission. [18.8]

nuclear fusion The uniting of two light elements to form one heavier nucleus, which is accompanied by the release of energy. [18.11]

nucleic acids Complex organic acids essential to life and found in the nucleus of living cells. They consist of thousands of units called nucleotides. Include DNA and RNA. [20.6]

nucleon A collective term for the neutrons and protons in the nucleus of an atom. [18.1]

nucleotide The building-block unit for nucleic acids. A phosphate group, a sugar residue, and a nitrogenous organic base are bonded together to form a nucleotide. [20.6]

nucleus The central part of an atom that contains all its protons and neutrons. The nucleus is very dense and has a positive electrical charge. [5.8]

nuclide A general term for any isotope of any atom. [18.1]

O

oils *See* fats and oils.

one atmosphere The standard atmospheric pressure; that is, the pressure exerted by a column of mercury 760 mm high at a temperature of 0°C. [12.3]

orbital A cloudlike region around the nucleus where electrons are located. Orbitals are considered to be energy sublevels (*s, p, d, f*) within the principal energy levels. *See also* principal energy levels. [10.3, 10.4]

orbital diagram A way of showing the arrangement of electrons in an atom, where boxes with small arrows indicating the electrons represent orbitals. [10.5]

organic chemistry The branch of chemistry that deals with carbon compounds but does not imply that these compounds must originate from some form of life. *See also* vital-force theory. [19.1]

osmosis The diffusion of water, either pure or from a dilute solution, through a semipermeable membrane into a solution of higher concentration. [14.8]

oxidation An increase in the oxidation number of an atom as a result of losing electrons. [17.2]

oxidation number A small number representing the state of oxidation of an atom. For an ion, it is the positive or negative charge on the ion; for covalently bonded atoms, it is a positive or negative number assigned to the more electronegative atom; in free elements, it is zero. [17.1]

oxidation–reduction A chemical reaction wherein electrons are transferred from one element to another; also known as redox. [17.2]

oxidation state *See* oxidation number. [17.1]

oxidizing agent A substance that causes an increase in the oxidation state of another substance. The oxidizing agent is reduced during the course of the reaction. [17.2]

ozone layer A high concentration of ozone located in the stratosphere that shields the earth from damaging ultraviolet radiation. [12.15]

P

partial pressure The pressure exerted independently by each gas in a mixture of gases. [12.10]

parts per million (ppm) A measurement of the concentration of dilute solutions now commonly used by chemists in place of mass percent. [14.6]

Pauli exclusion principle An atomic orbital can hold a maximum of two electrons, which must have opposite spins. [10.4]

peptide linkage The amide bond in a protein molecule; bonds one amino acid to another. [20.4]

percent composition of a compound The mass percent represented by each element in a compound. [7.3]

percent yield The ratio of the actual yield to the theoretical yield multiplied by 100. [9.6]

perfect gas A gas that behaves precisely according to theory; also called an *ideal gas*. [12.2]

periodic table An arrangement of the elements according to their atomic numbers. The table consists of horizontal rows, or periods, and vertical columns, or families, of elements. Each period ends with a noble gas. [3.5, 10.6]

period of elements The horizontal groupings (rows) of elements in the periodic table. [3.5, 10.6]

pH A method of expressing the H^+ concentration (acidity) of a solution; $pH = -\log[H^+]$, $pH = 7$ is a neutral solution, $pH < 7$ is acidic, and $pH > 7$ is basic. [15.9]

phase A homogeneous part of a system separated from other parts by a physical boundary. [1.8]

photon Theoretically, a tiny packet of energy that streams with others of its kind to produce a beam of light. [10.2]

physical change A change in form (such as size, shape, or physical state) without a change in composition. [4.2]

physical properties Inherent physical characteristics of a substance that can be determined without altering its composition: color, taste, odor, state of matter, density, melting point, boiling point. [4.1]

physical states of matter Solids, liquids, and gases. [1.7]

physiological saline solution A solution of 0.90% sodium chloride that is isotonic (has the same osmotic pressure) with blood plasma. [14.8]

pOH A method of expressing the basicity of a solution. $pOH = -\log[OH^-]$. $pOH = 7$ is a neutral solution, $pOH < 7$ is basic, and $pOH > 7$ is acidic. [16.11]

polar covalent bond A covalent bond between two atoms with differing electronegativity values, resulting in unequal sharing of bonding electrons. [11.5]

polyatomic ion An ion composed of more than one atom. [6.5]

polyhydroxyl alcohol An alcohol that has more than one —OH group. [19.14]

polymer (macromolecule) A natural or synthetic giant molecule formed from smaller molecules (monomers). [19.21]

polymerization The process of forming large, high-molar-mass molecules from smaller units. [19.21]

polypeptide A peptide chain containing up to 50 amino acid units. [20.4]

polysaccharide A carbohydrate that can be hydrolyzed to many monosaccharide units; cellulose, starch, and glycogen are examples. [20.2]

positron A particle with a +1 charge having the mass of an electron (a positive electron). [18.1]

potential energy (PE) Stored energy, or the energy of an object due to its relative position. [4.5]

pressure Force per unit area; expressed in many units, such as mm Hg, atm, lb/in.², torr, and pascal. [12.3]

primary alcohol An alcohol in which the carbon atom bonded to the —OH group is bonded to only one other carbon atom. [19.14]

principal energy levels of electrons Existing within the atom, these energy levels contain orbitals within which electrons are found. *See also* orbital, electron. [10.4]

product A chemical substance produced from reactants by a chemical change. [4.3]

properties The characteristics, or traits, of substances that give them their unique identities. Properties are classified as physical or chemical. [4.1]

protein A polymer consisting mainly of α-amino acids linked together; occurs in all animal and vegetable matter. [20.4]

proton A subatomic particle found in the nucleus of the atom that carries a positive electrical charge and a mass of about 1 amu. An H^+ ion is a proton. [5.6]

Q

quanta Small discrete increments of energy. From the theory proposed by physicist Max Planck that energy is emitted in energy *quanta* rather than a continuous stream. [10.3]

quantum mechanics or wave mechanics The modern theory of atomic structure based on the wave properties of matter. [10.3]

R

rad (radiation absorbed dose) A unit of absorbed radiation indicating the energy absorbed from any ionizing radiation; 1 rad = 0.01 J of energy absorbed per kilogram of matter. [18.7]

radioactive decay The process by which a radioactive element emits particles or rays and is transformed into another element. [18.2]

radioactive disintegration series The spontaneous decay of a certain radioactive nuclide by emission of alpha and beta particles from the nucleus, finally stopping at a stable isotope of lead or bismuth. [18.4]

radioactivity The spontaneous emission of radiation from the nucleus of an atom. [18.1]

rate of reaction The rate at which the reactants of a chemical reaction disappear and the products form. [16.2]

reactant A chemical substance entering into a reaction. [4.3]

redox *See* oxidation–reduction. [17.2]

reducing agent A substance that causes a decrease in the oxidation state of another substance; the reducing agent is oxidized during the course of a reaction. [17.2]

reduction A decrease in the oxidation number of an element as a result of gaining electrons. [17.2]

rem (roentgen equivalent to man) A unit of radiation-dose equivalent taking into account that the energy absorbed from different sources does not produce the same degree of biological effect. [18.7]

representative element An element in one of the A groups in the periodic table. [10.6]

resonance structure A molecule or ion that has multiple Lewis structures. *See also* Lewis structure. [11.8]

reversible chemical reaction A chemical reaction in which the products formed react to produce the original reactants. A double arrow is used to indicate that a reaction is reversible. [16.1]

RNA Ribonucleic acid; a high-molar-mass polymer of nucleotides present in all living matter. Its main function is to direct the synthesis of proteins. [20.6]

roentgen (R) A unit of exposure of gamma radiation based on the quantity of ionization produced in air. [18.7]

rounding off numbers The process by which the value of the last digit retained is determined after dropping nonsignificant digits. [2.3]

S

salts Ionic compounds of cations and anions. [Ch. 6, 15.4]

saponification The hydrolysis of a fat or an oil with an aqueous strong base (NaOH) to yield a soap. [20.3]

saturated hydrocarbon A hydrocarbon that has only single bonds between carbon atoms; classified as *alkanes*. [19.3]

saturated solution A solution containing dissolved solute in equilibrium with undissolved solute. [14.3]

scientific laws Simple statements of natural phenomena to which no exceptions are known under the given conditions. [1.5]

scientific method A method of solving problems by observation; recording and evaluating data of an experiment; formulating hypotheses and theories to explain the behavior of nature; and devising additional experiments to test the hypotheses and theories to see if they are correct. [1.5]

scientific notation A convenient way of expressing large and small numbers using powers of 10. To write a number as a power of 10 move the decimal point in the original number so that it is located after the first nonzero digit, and follow the new number by a multiplication sign and 10 with an exponent (called its *power*) that is the number of places the decimal point was moved. Example: $2468 = 2.468 \times 10^3$. [2.4]

secondary alcohol An alcohol in which the carbon atom bonded to the —OH group is bonded to two other carbon atoms. [19.14]

semipermeable membrane A membrane that allows the passage of water (solvent) molecules through it in either direction but prevents the passage of larger solute molecules or ions. [14.8]

SI An agreed-upon standard system of measurements used by scientists around the world (*Système Internationale*). *See also* metric system. [2.5]

significant figures The number of digits that are known plus one estimated digit are considered significant in a measured quantity; also called significant digits. [2.3, Appendix I]

simplest formula *See* empirical formula.

single bond A covalent bond in which one pair of electrons is shared between two atoms. [11.5]

single-displacement reaction A reaction of an element and a compound that yields a different element and a different compound. [8.4]

soap A salt of a long-carbon-chain fatty acid. [20.3]

solid A state of matter having a definite shape and a definite volume, whose particles cohere rigidly to one another, so that a solid can be independent of its container. [1.7]

solubility An amount of solute that will dissolve in a specific amount of solvent under stated conditions. [14.2]

solubility product constant (K_{sp}) The equilibrium constant for the solubility of a slightly soluble salt. [16.12]

solute The substance that is dissolved—or the least abundant component—in a solution. [14.1]

solution A system in which one or more substances are homogeneously mixed or dissolved in another substance. [14.1]

solvent The dissolving agent or the most abundant component in a solution. [14.1]

specific gravity The ratio of the density of one substance to the density of another substance taken as a standard. Water is usually the standard for liquids and solids; air, for gases. [2.9]

specific heat The quantity of heat required to change the temperature of 1 g of any substance by 1°C. [4.6]

spectator ion An ion in solution that does not undergo chemical change during a chemical reaction. [15.10]

speed (of a wave) A measurement of how fast a wave travels through space. [10.2]

spin A property of an electron that describes its appearance of spinning on an axis like a globe; the electron can spin in only two directions, and, to occupy the same orbital, two electrons must spin in opposite directions. *See also* orbital. [10.4]

standard boiling point *See* normal boiling point.

standard conditions *See* STP.

standard temperature and pressure *See* STP.

Stock (nomenclature) System A system that uses Roman numerals to name elements that form more than one type of cation. (For example: Fe^{2+}, iron(II); Fe^{3+}, iron(III).) [6.4]

stoichiometry The area of chemistry that deals with the quantitative relationships among reactants and products in a chemical reaction. [9.2]

STP (standard temperature and pressure) 0°C (273 K) and 1 atm (760 torr); also known as standard conditions. [12.8]

strong electrolyte An electrolyte that is essentially 100% ionized in aqueous solution. [15.7]

subatomic particles Particles found within the atom, mainly protons, neutrons, and electrons. [5.6]

sublimation The process of going directly from the solid state to the vapor state without becoming a liquid. [13.2]

subscript Number that appears partially below the line and to the right of a symbol of an element (example: H_2SO_4). [3.9]

substance Matter that is homogeneous and has a definite, fixed composition; substances occur in two forms—as elements and as compounds. [1.8]

substrate In biochemical reactions, the substrate is the unit acted upon by an enzyme. [20.5]

supersaturated solution A solution containing more solute than needed for a saturated solution at a particular temperature. Supersaturated solutions tend to be unstable; jarring the container or dropping in a "seed" crystal will cause crystallization of the excess solute. [14.3]

surface tension The resistance of a liquid to an increase in its surface area. [13.4]

symbol In chemistry, an abbreviation for the name of an element. [3.4]

system A body of matter under consideration. [1.8]

T

temperature A measure of the intensity of heat, or of how hot or cold a system is; the SI unit is the kelvin (K). [2.8]

tertiary alcohol An alcohol in which the carbon atom bonded to the —OH group is bonded to three other carbon atoms. [19.14]

tetrahedral structure An arrangement of the VSEPR model where four pairs of electrons are placed 109.5° degrees apart to form a tetrahedron. [11.11, 19.2]

theoretical yield The maximum amount of product that can be produced according to a balanced equation. [9.6]

theory An explanation of the general principles of certain phenomena with considerable evidence to support it; a well-established hypothesis. [1.5]

Thomson model of the atom Thomson asserted that atoms are not indivisible but are composed of smaller parts; they contain both positively and negatively charged particles—protons as well as electrons. [5.6]

titration The process of measuring the volume of one reagent required to react with a measured mass or volume of another reagent. [15.10]

torr A unit of pressure (1 torr = 1 mm Hg). [12.3]

total ionic equation An equation that shows compounds in the form in which they actually exist. Strong electrolytes are written as ions in solution, whereas nonelectrolytes, weak electrolytes, precipitates, and gases are written in the un-ionized form. [15.12]

transcription The process of forming RNA from DNA. [20.6]

transition elements The metallic elements characterized by increasing numbers of d and f electrons. These elements are located in Groups 1B–8B of the periodic table. [10.6]

transmutation The conversion of one element into another element. [18.5]

transuranium element An element that has an atomic number higher than that of uranium (>92). [18.13]

triacylglycerol (triglyceride) An ester of glycerol and three molecules of fatty acids. [20.3]

trigonal planar An arrangement of atoms in the VSEPR model where the three pairs of electrons are placed 120° apart on a flat plane. [11.11]

triple bond A covalent bond in which three pairs of electrons are shared between two atoms. [11.5, 19.2]

Tyndall effect An intense beam of light is clearly visible when passed through a colloidal dispersion, but is not visible when passed through a true solution. [15.14]

U

unsaturated hydrocarbon A hydrocarbon whose molecules contain one or more double or triple bonds between two carbon atoms; classified as *alkenes*, *alkynes*, and *aromatic compounds*. [19.3]

unsaturated solution A solution containing less solute per unit volume than its corresponding saturated solution. [14.6]

V

valence electron An electron in the outermost energy level of an atom; these electrons are the ones involved in bonding atoms together to form compounds. [10.5]

valence shell electron pair repulsion (VSEPR) theory A theory for predicting the shape of covalent compounds based on the repulsion of bonded and nonbonded pairs of valence electrons. [11.11]

vapor pressure The pressure exerted by a vapor in equilibrium with its liquid. [13.3]

vaporization *See* evaporation. [13.2]

vapor pressure curve A graph generated by plotting the temperature of a liquid on the x-axis and its vapor pressure on the y-axis. Any point on the curve represents an equilibrium between the vapor and liquid. [13.5]

vital-force theory A theory that held that organic substances could originate only from some form of living material. The theory was overthrown early in the 19th century. [19.1]

volatile (substance) A substance that evaporates readily; a liquid with a high vapor pressure and a low boiling point. [13.3]

voltaic cell A cell that produces electric current from a spontaneous chemical reaction. [17.6]

volume The amount of space occupied by matter; measured in SI units by cubic meters (m^3) but also commonly in liters and milliliters. [2.7]

volume percent (solution) The volume of solute in 100 mL of solution. [14.6]

W

water of crystallization Water molecules that are part of a crystalline structure, as in a hydrate; also called water of hydration. [13.9]

water of hydration *See* water of crystallization.

wavelength The distance between consecutive peaks and troughs in a wave; symbolized by the Greek letter lambda. [10.2]

weak electrolyte A substance that is ionized to a small extent in aqueous solution. [15.7]

weight A measure of Earth's gravitational attraction for a body (object). [2.7]

word equation A statement in words, in equation form, of the substances involved in a chemical reaction. [8.2]

PHOTO CREDITS

INDEX

A

Absolute zero, 275
Acetic acid, 66, 527
Acetone, 524, 526
Acetylene, 509
Acid anhydrides, 326
Acid ionization constant (K_a), 420–421
Acid rain, 395–396
Acid-type dry cell batteries, 455
Acids
  definitions of, 377
  electrolytes and, 380–386
  neutralization and, 389–392
  pH and, 387–388
  properties of, 374–376
  reaction with carbonates, 378
  reactions with bases, 378
  reactions with metal oxides, 378
  reactions with metals, 377
  salts and, 379, 426
Actinide series, 217
Activated charcoal, 398
Activation energy, 166, 416
Activity series of metals, 450–451
Actual yield, reaction, 190
Acute radiation damage, 488
Addition in calculations, 21–22
Addition reaction, of alkenes, 511–512
Adsorption, colloid, 398
Aeration, in water purification, 327
Agitation, rate of dissolving solids and, 344–345
Air pollution, 296–298
Alanyltyrosylglycine, 555
Alcohols
  classifications of, 518–519
  ethanol, 520
  ethylene glycol, 520
  glycerol, 520
  isopropyl alcohol, 520
  IUPAC naming system, 521–522
  methanol, 520
  naming of, 521–522
  polyhydroxy alcohol, 519
  primary, 518
  secondary, 518
  tertiary, 518
Aldehydes
  naming of, 525
  structural formulas for, 524–525
Alkali metals, 53, 215, 237
Alkaline earth metals, 53, 215
Alkaline-type dry cell batteries, 455
Alkanes
  formulas of, 500
  IUPAC naming rules, 505–507
  naming system for, 504–508

physical properties of, 500–501
as unsaturated hydrocarbon, 498
Alkenes
  addition reaction and, 511–512
  formulas for, 508–509
  IUPAC naming rules, 510
  naming system for, 510–511
  reactions of, 511
  as unsaturated hydrocarbon, 498
Alkyl groups, 504–505
Alkyl halides, 517–518
Alkynes
  IUPAC naming rules, 510
  names and formulas for, 508–509
  naming system for, 510–511
  as unsaturated hydrocarbon, 498
Allotropes, 296
Alpha particles, 473–474
Aluminum-27, 478
Amino acids, 550–557
  derived from proteins, 552–553
  digestion and, 556–557
  peptide linkage, 554
  polypeptides, 554–556
  residues, 554
  sequence determination of, 556
Amorphous solids, 6
Amphoteric hydroxides, 378
Anhydrides, 325–326
Anions, 57, 108, 111
Anodes, 452
Antarctica Toms ozone hole, 297
Aromatic acid, 527
Aromatic compounds, 498
Aromatic hydrocarbons
  characteristics of, 512–513
  disubstituted benzenes, 514
  monosubstituted benzenes, 513–514
  naming of, 513–516
  tri- and polysubstituted benzenes, 515–516
Arrhenius, Svante, 88, 374–375, 377, 381
Artificial radioactivity, 478
Artificial sweeteners, 317
Aspartame, 317
Atmospheric pressure, 268
Atomic bomb, 484–485
  characteristics of, 484
  hazards of, 484–485
Atomic clocks, 210
Atomic mass, 95–97
Atomic numbers, 93
Atomic properties
  atomic radius, 225–227
  ionization energy, 227–228
  metals and nonmetals, 225
  periodic trends in, 225–227

Atomic radius, 225–227, 233
Atomic structures, first 18 elements, 210–213
Atomic theory
  atomic mass, 95–97
  Bohr atom, 206–207
  compounds and, 86–87
  Dalton's model, 85–86
  discovery of ions, 88
  early thoughts on, 85
  electric charges, 97–98
  electromagnetic radiation, 204–205
  electron energy levels, 207–210
  electron structures, 214–218
  isotopes of elements, 94–95
  nuclear atom, 91–94
  subatomic particles, 88–91
Atoms
  Lewis structures of, 228–229
  nuclear model of, 93
  oxidation numbers of, 438
Avogadro, Amedeo, 285
Avogadro's law, 285–286
Avogadro's number, 129, 177

B

Balanced equations, 152–156, 178
Barium chloride dihydrate, 320
Barometers, 268–269
Bases
  acid reactions with, 378
  amphoteric hydroxides and, 378
  definitions of, 377
  electrolytes and, 380–386
  neutralization and, 389–392
  pH and, 387–388
  properties of, 374–376
  reactions of NaOH and KOH with metals, 378
  reactions with, 378
  salts and, 379, 426
Basic anhydrides, 325
Battery cells, 455–457
Becquerel, Antoine Henri, 91, 469
Benzenes
  disubstituted, 514
  monosubstituted, 513–514
  structure of, 512
  tri- and polysubstituted, 515–516
Benzoic acid, 527
Beta particles, 474
Binary acids, 115
Binary compounds
  acids derived from, 115
  containing metal forming two or more cations, 112–114
  containing two nonmetals, 114–115
  flow chart of, 116
  naming of, 111–112

Biochemistry
    amino acids, 550–557
    carbohydrates, 542–547
    characteristics of, 542
    DNA and genetics, 563–564
    enzymes, 557–559
    lipids, 547–550
    nucleic acids, 559–562
    proteins, 550–557
Biodiesel fuel, 499
Bohr, Niels, 206–207
Bohr atom, 206–207
Boiling point
    elevation constants, 355–357
    of liquids, 312–314
Bond angle, 324
Bond length, 323
Bond type, electronegativity and, 243
Boyle, Robert, 271
Boyle's law, 271–275
Breeder reactors, 483–484
Bronsted, J.N., 375
Bronsted-Lowry proton transfer
    theory, 375, 377, 386
Brown, Robert, 397–398
Brownian movement, 398
Buckyballs, 243
Buffer solutions, 426–429
Butane, 502
Butanone, 526

C
C-terminal residue, 554–55
Calcium precipitation, in water
    softening, 327
Calculations, significant figures in,
    20–22
Calories, 72
Capillary action, of liquids, 312
Carbohydrates, 542–547
    disaccharides, 545–546
    fructose, 544–545
    galactose, 544
    glucose, 544
    lactose, 545–546
    maltose, 545–546
    monosaccharides, 543–545
    polysaccharides, 547
    sucrose, 545–546
Carbon
    half-life for, 471
    tetrahedral structure of, 497
Carbon dioxide
    for dry cleaning, 532
    global warming and, 168–169
    oxygen exchange in cells and,
        427
    poisoning, 157
Carbon tetrachloride, 518
Carbonates, acid reactions with,
    378
Carbonyl groups, 516
Carboxylic acids, 527–528
Catalysts, effect on equilibrium,
    415–416
Cathodes, 452

Cations, 57, 107
Cellulose, 547
Celsius scale, 34–35
Cerebroside, 550
Chadwick, James, 91
Chain reaction, nuclear, 481
Charcoal, activated, 398
Charles, J.A.C., 275
Charles' law, 275–278
Chemical changes
    characteristics of, 68–70
    of common processes, 71
    energy in, 74–75
Chemical equations
    balanced, 178
    combination reactions, 159
    decomposition reactions, 159–160
    double-displacement reactions,
        161–165
    examples of, 69
    information in, 156–159
    nature of, 151–152
    single-displacement reactions,
        160–161
    types of, 159–165
    writing and balancing, 152–156
Chemical equilibrium
    acid-base properties of salt, 426
    buffer solutions, 426–429
    catalysts effect on, 415–416
    characteristics of, 408–409
    concentration effect on, 410–412
    equilibrium constants, 416–418
    ion product constant for water,
        418–420
    ionization constants, 420–422
    La Châtelier's Principle, 410
    reaction rates, 407–408
    reversible reactions, 406
    solubility product constant,
        423–425
    temperature effect on, 414–415
    volume effect on, 413–414
Chemical formulas, 59–60
Chemical kinetics, 407–408
Chemist, thinking like, 3–4
Chemistry
    nature of, 2
    purpose of studying, 2
Chernobyl accident, 483
Chloride ions, 381–382
Chlorinated fluorocarbons (CFC), 518
Chlorination, 517
Chlorine
    oxy-anions and oxy-acids of, 117
    physical properties of, 66
Chlorobenzene, 513
Chloroform, 518
Cholesterol, 550
Coal tar, 499
Colligative properties
    of electrolyte solutions, 385–386
    of solutions, 355–359
Colloids
    adsorption and, 398
    Brownian movement and, 398

characteristics of, 396–397
    dialysis and, 399
    dispersion of, 396
    properties of, 397–398
    property applications, 398–399
    Tyndall effect and, 398
Combination reactions, 159
Combined gas laws, 280–283
Combustion of alkanes, 500–501
Common ions, 425, 438
Common names, 104–105
Compounds, 56–58
    classification of, 57
    composition of, 86–87
    containing polyatomic ions,
        249–250
    ionic, 57
    Lewis structures of, 244–247
    molar mass of, 132–135
    molecular, 57
    percent composition of, 136–138
Concentrated solutions, 346
Concentration
    effect on equilibrium, 410–412
    of solutions, 340, 346–355
Condensation, 309
Copolymers, 530
Cottrell, Frederick, 398
Cottrell process, 398–399
Covalent bonds, 238–239, 438–439
Crick, Francis H.C., 556
Critical mass, 481
Crookes, Sir William, 89
Crookes tube, 89
Crystals, surface area of, 344
Cubic meters, 31
Curie, Marie, 469
Curie, Pierre, 469
Curie (Ci) unit, 479

D
Dalton's law of partial pressures,
    283–285
Dalton's model of atoms, 85–86
Davy, Humphrey, 320
De Broglie, Louis, 207
Decomposition reactions, 159–160
Demineralization, 328, 409
Density, 37–40
    of gases, 38, 288–289
    of liquids and solids, 38
    specific gravity and, 39
Deoxyribonucleic acid (DNA)
    genetics and, 563–564
    replication of, 563
    structure of, 559–561
Dialysis, 399
Dialyzing membranes, 399
Diatomic molecules, 55–56, 107
Dichlorobenzenes, 514
Diffusion, 267
Digestion, 556–557
Dilute solutions, 346
Dilution problems, 353
Dipeptides, 554
Dipoles, 240

Disinfection, in water purification, 327
Dissociation, of electrolytes, 381–382
Distallation, in water softening, 327
Disubstituted benzenes, 514
Division, 20–21
Dry air, 268
Dry cell batteries, 455

Effusion, 267
Electric charges, 97–98
Electrolysis, 69, 452–454
Electrolytes, 380–381
    colligative properties of, 385–386
    dissociation of, 381–382
    ionization of, 382–383
    strong and weak, 383–385
Electrolytic cells, 452–454
Electromagnetic radiation, 204–205
Electromagnetic fields, 473
Electromagnetic spectrum, 205
Electronegativity, 240–243
Electrons
    arrangement in noble gases, 230
    arrangement of pairs, 254
    characteristics of, 90
    configurations of, 211, 213
    covalent bonds, 238–239
    electrical charge of, 91
    relative mass of, 91
    structures, 214–218
    valence, 213
Elements, 105–109
    as diatomic molecules, 55–56
    distribution of, 49–50
    names of, 51
    in natural states, 54–55
    nature of, 48–49
    symbols of, 51–52
    transmutation of, 477–478
Empirical formulas, 139–144
Endothermic reactions, 165–167
Energy, 71–72
    in chemical changes, 74–75
    conservation of, 75–76
    kinetic energy, 71
    levels in electrons, 207–210
    potential energy, 71
Enzymes, 416, 557–559
Equations. See Chemical equations
Equilibrium. See Chemical
    equilibrium
Equilibrium constants, 416–418
Equilibrium systems, water, 323
Esterification, 528
Esters, 528–529
Ethane, 501–502
Ethanol, 519–520
Ethers, 522–524
Ethyl alcohol, 311, 313
Ethyl butyrate, 529
Ethyl chloride, 313
Ethyl ether, 311, 313
Ethylene, 509
Ethylene glycol, 520
Evaporation, 309
Exothermic reactions, 165–167, 416

F
Fahrenheit scale, 34–35
Families, in periodic table, 215
Faraday, Michael, 88, 512
Fats, 548–550
Fatty acids, 548–550
Fission products, 480
Flocculation, in water purification,
    326
Foam cars, 397
Food irradiation, 482
Formaldehyde, 524
Formalin, 524
Formic acid, 527
Formulas
    empirical (See Empirical formulas)
    equations, 392
    molecular (See Molecular
        formulas)
    percent composition from,
        136–138
    units, 132
Fossil fuels, 498–499
Frankel, Gerald, 389
Freezing point
    depression constant, 355–357
    of liquids, 314
Frequency, wave, 204
Fructose, 544–545
Functional groups, 498
Fusion reactions, 485–486

G
Galactose, 544
Galvanic cells, 454–457
Gamma rays, 474–476
Gas hydrate ice, 320
Gases, 6–8
    air pollution and, 296–298
    Avogadro's law, 285–286
    Boyle's law, 271–275
    Charles' law, 275–278
    combined gas laws, 280–283
    common materials in, 8
    Dalton's law of partial pressures,
        283–285
    density of, 38, 288–289
    diffusion of, 267
    effusion of, 267
    gas stoichiometry, 292–298
    Gay-Lussac's law, 279–280
    general properties of, 266
    Graham's law of effusion, 268
    ideal gas, 266, 290–291
    kinetic-molecular theory, 266–268
    mass-volume calculations,
        292–293
    measurement of pressure, 268–270
    mole-mass-volume relationships,
        287–288
    mole-volume calculations, 292–293
    number of molecules present,
        270–271
    physical properties of, 8
    real gases, 295–296
    temperature and, 271
    volume-volume calculations,
        294–295
Gay-Lussac, J.L., 279
Gay-Lussac's law, 279–280
Geiger counter, 478
Genes, 564
Genetic effects, of radiation, 488
Genetics, 563–564
Global warming, 168–170, 297
Glucose, 544
Glycerol, 520
Glycogen, 547
Glycosuria, 544
Glycylglycine, 554
Goldstein, Eugen, 90
Graham, Thomas, 399
Graham's law of effusion, 268
Greenhouse effect, 168–170
Greiger-Müller detecting tube,
    478–479
Ground state, 206
Groups, in periodic table, 215

H
Haber, Fritz, 413
Hahn, Otto, 480
Half-life, 471
Halogenation, 517
Halogens, 53, 160, 215
Hard water, 327
Heat
    in chemical reactions, 165–167
    of fusion, 315
    measurement of, 33, 72–74
    of reactions, 165
    of vaporization, 315
Heating curve, 315
Heredity, 563–564
Heterogeneous matter, 9
Heterogeneous mixtures, 10
Hiroshima, 485
Homogeneous matter, 9
Homogenous mixtures, 10
Homologous series, 500
Hybrid cars, 58
Hydrates, 319–321
Hydrocarbon derivatives, 516–517
Hydrocarbons, 166, 498–499
Hydrochloric acid solution, 452
Hydrofluorocarbons (HFC), 518
Hydrogen
    bonds, 316–319
    compounds, 316
    fuel cells, 58
    isotopes, 94
    line spectrum of, 206
    molecular formation of, 238
    orbital for, 207
Hydronium ion, 375
Hypotheses, 5
Hypotonic solutions, 360

I
Ice cream, 359
Ideal gas, 266, 290–291
Immiscible liquids, 339

Induced radioactivity, 478
Infectious agents, in water, 329
Inner transition elements, 217
Inorganic compounds
    common and systematic names,
        104–105
    elements and ions, 105–109
Intermolecular bonds, 318
International system. *See* Metric system
International Union of Pure and
        Applied Chemistry (IUPAC),
        105, 504–507
Ion exchange, in water softening,
        327
Ion product constant for water (K_w),
        418–420
Ionic bonds, 230–236
Ionic compounds
    acids, 118–119
    naming binary compounds, 112
    naming polyatomic ions
        compounds, 116–119
    predicting formulas of, 236–237
    writing formulas and, 109–111
Ionic equations, writing, 392–394
Ionization
    of electrolytes, 381–382
    of water, 386
Ionization constants, 420–422
Ionization energy, 227—228
Ionizing radiation, 478, 488
Ions, 105–109
    definition of, 57
    discovery of, 88
    oxidation numbers for, 438
    spectator, 390
Isoamyl acetate, 529
Isobutyl formate, 529
Isomerism, 503–504
Isomers, 503
Isopropyl alcohol, 520
Isotonic solutions, 360
Isotopes, 94–95
Isotopic notation, 94

J
Joliot-Curie, Frederic, 478
Joliot-Curie, Irene, 478
Joules, 72

K
Kekulé, August, 512
Kelvin scale, 34–35
Ketones
    naming of, 525–526
    structural formulas for, 524–525
Kevlar, 255
Kilograms, 30, 33
Kinetic energy, 71
Kinetic-molecular theory, 266–268
Kroto, Harold, 243

L
Lactose, 545–546
Lanthanide series, 217
Law of conservation of energy, 76

Law of conservation of mass, 70
Law of definite composition, 86
Law of multiple proportions, 87
Lawrence, E.O., 477
Laws, 87
Le Châtelier's Principle, 410, 415
LeChâtelier, Henri, 410
Lewis, Gilbert N., 229, 376
Lewis structures
    of alkanes, 501
    of atoms, 228–229
    complex, 247–249
    of compounds, 242–247
Lewis theory, 376, 377
Limiting reactants, 186–189
Line spectrum, 206
Linear structures, molecule, 251
Lipids, 547–550
    triacylglycerols, 548–550
    triglycerides, 548–550
Liquid crystal molecules, 255
Liquids, 6–8. *See also* Water
    boiling point of, 312–314
    changes of state, 314–316
    characteristics of, 308
    common materials in, 8
    condensation of, 309
    density of, 38
    evaporation of, 309
    freezing point of, 314
    hydrates, 319–321
    hydrogen bonds and, 316–319
    melting point and, 314
    physical properties of, 8
    sublimation of, 309
    surface tension, 311–312
    vapor pressure, 309–311
    vaporization of, 309
Liters, 31
Long-term radiation damage, 488
Lowry, T.M., 375

M
Macromolecules, 530–531
Macroscopic picture, 3
Magnesium precipitation, in water
        softening, 327
Maltose, 545–546
Mass
    conservation of, 70
    defect, 486
    mass number, 94
    measurement of, 29–30
    metric units of, 30
    percent solution, 346
    spectrometers, 95–96
Mass-mass calculations, 184–185
Mass-volume calculations, 292–293
Mass/volume percent (m/v), 348
Matter
    classification of, 8–10
    particulate nature of, 6
    physical states of, 6–8
    three states of, 308
McMillan, Edwin M., 487
Measurements, 15

mass and volume, 29–32
    scientific notation and, 15–16
    standard units in metric system, 25
    temperature, 33–37
    uncertainty and, 16–18
Meiosis, 554
Melting point, 314
Meniscus curve, 312
Messenger molecules, 284
Metal oxides, 378
Metalloids, 54
Metals
    acids reactions with, 378
    activity series for, 450–451
    atomic radii of, 233
    in periodic table, 53
    periodic trends in, 225–226
    reaction of water with, 324–326
    reactions with NaOH and KOH,
        378
Meters (m), 24
Methane
    global warming and, 168–169
    structural formula for, 501–502
Methanol
    characteristics of, 520
    structural formula of, 519
Methyl ethyl ketone (MEK), 524
Methyl salicylate, 452
Methylpropane, 502
Metric system (SI), 23–24
    common prefixes and numerical
        values, 23
    length measurement, 24
    standard units of measurement, 25
    units of mass, 30
Miscible liquids, 339
Mitosis, 564
Mixtures
    distinguishing from pure
        substances, 10
    nature of, 9–10
Molar mass
    of compounds, 132–135
    defined, 130
Molar volume, 287–288
Molarity, of solutions, 348–352, 357
Mole-mass calculations, 182–184
Mole-mole calculations, 180–182
Mole ratio, 178
Mole-volume calculations, 292–293
Molecules
    electron pair arrangement and, 254
    messenger, 284
    molecular formulas, 139–140,
        143–144
    molecular shape, 250–251
    pressure exertion by gas and,
        270–271
    relationship with moles, 177
    spherical, 243
Moles, 128–130, 177
Monomers, 530
Monosubstituted benzenes, 513–514
Multiplication in equations, 20–21
Municipal water treatment plant, 327

N

n-Octyl acetate, 529
N-terminal residue, 554–555
Nagasaki, 485
National Aeronautics and Space
    Administration (NASA), 29
Natural law, 86
Natural radioactivity, 470–474
Natural states, elements in, 54–55
Negative ions, 90
Net ionic equations, 392
Neurotransmitters, 284
Neutralization, 389–390
Neutrons, 91
Nitrogen fixation, 413–414
Noble gases, 53, 55, 215, 230
Noble metals, 55
Nonelectrolytes, 380–381
Nonmetals
    atomic radii of, 233
    in periodic table, 53–54
    periodic trends in, 225–226
    reaction of water with, 324–326
Nonpolar compounds, 241
Nonpolar covalent bonds, 240
Nonvolatile substances, 356
Normal boiling point, 313
Normal butane, 502
Nuclear atom, 91–94
    atomic numbers of elements, 93
    subatomic particle arrangement,
        92–93
Nuclear binding energy, 486
Nuclear chemistry
    acute radiation damage, 488
    alpha particles, 473–474
    artificial radioactivity, 478
    atomic bomb, 484–485
    beta particles, 474
    biological effects of radiation, 488
    discovery of radioactivity, 469–470
    gamma rays, 474–476
    genetic effects of radiation, 488
    isotopic notation and, 479
    long-term radiation damage, 488
    mass-energy relationship, 486–487
    measurement of radioactivity,
        478–479
    natural radioactivity, 470–472
    nuclear fission, 480–482
    nuclear fusion, 485–486
    nuclear power, 482–484
    radioactive disintegration series, 476
    radioactive tracers, 489
    transmutation of elements,
        477–478
    transuranium elements, 487
Nuclear fission, 480–481
Nuclear fusion, 485–486
Nuclear power, 482–484
Nuclear power plants, 283–284
Nuclear scanning, 489
Nuclear war, 485
Nucleic acids, 559–562
    DNA, 559–561
    nucleotides, 559–560

RNA, 562
    transcription, 562
Nucleons, 470
Nucleotides, 559–560
Nucleus, 93
Nuclides, 470, 472
Numerical values, in metric system, 23

O

Oils, 548–550
1 Atmosphere (atm), 269
Orbital diagrams of elements 11–18,
    213
Orbital filling, 213, 215–216
Orbitals, 207–209
Organic chemicals, in water, 329
Organic chemistry
    alcohols, 518–522
    aldehydes, 524–526
    alkanes, 500–501
    alkenes, 508–512
    alkyl halides, 517–518
    alkynes, 508–511
    aromatic hydrocarbons, 512–516
    beginnings of, 496–497
    carbon atoms, 497–498
    carboxylic acids, 527–528
    esters, 528–529
    ethers, 522–524
    functional groups, 498
    hydrocarbon derivatives, 516–517
    hydrocarbons, 498–499
    isomerism, 503–504
    ketones, 524–526
    macromolecules, 530–531
    polymers, 530–531
    structural formulas, 501–504
    vital-force theory, 495
Osmosis, 360–362
Osmotic pressure, 361–362
Oxidation number, 438–441
    of atom, 438
    for common ions, 438
    in covalently bonded substances,
        438–439
    rules for assigning, 439
Oxidation-reduction reactions
    balancing equations, 442–446
    balancing ionic redox equations,
        446–449
    characteristics of, 441–442
    electrolysis and, 452–454
    oxidizing agents and, 442
    photochromic glasses and, 447
    reducing agents and, 442
    voltaic cells and, 454–457
Oxidation states, 439
Oxy-acids, 117, 119
Oxy-anions, 117
Oxygen
    carbon dioxide exchange in cells
        and, 427
    demanding wastes, 329
    as electronegative element,
        323–324
Ozone layer, 518

P

Paraffins. *See* Alkanes
Partial pressures, 283–285
Particle size, rate of dissolving solids
    and, 344
Parts per million (PPM), 346
Pauli exclusion principle, 208
Pauling, Linus, 240
Pentane, 507
Peptides
    digestion and, 556–557
    linkage of, 554
    polypeptides, 554–556
    sequence determination of, 556
Percent composition
    of compounds, 136–138
    from experimental data, 138
Percent yield, reaction, 190
Periodic table, 214–218
    groups, 215
    groups in, 52–53
    inner transition elements, 217
    ionic charges in, 108
    metalloids in, 54
    metals in, 53
    nonmetals in, 53–54
    orbital filling and, 215–216
    outermost configurations, 218
    representative elements, 215
    rows in, 214
    transition elements, 215
pH, 386–388
    buffer solutions and, 426–429
    of common elements, 388
    definition of, 387
    HCl and NaOH and, 429
    importance of, 389
    logarithms and, 387
    meters, 389
    scale for expressing acidity, 388
    scale of acidity and basicity, 387
Phenolphthalein paint, 389
Phenyl group, 514
Phosphorus-30, 478
Photons, 205
Physical changes, 67–68, 71
Physical properties
    definition of, 66
    of water, 316, 322–323
Physiological saline solution, 360
Planck, Max, 206
Plant nutrients, in water, 329
Plaster of Paris, 321
Plutonium, 487
Polar covalent bonds, 239
Polar covalent compounds, 241
Pollutants, water, 329
Polyatomic compounds, 119
Polyatomic ions
    charges of, 117
    compounds containing, 249–250
    formulas of, 117
    naming compounds containing,
        116–119
Polyatomic molecules, 107
Polyethylene, 530–531

Polyhydroxy alcohol, 519
Polymerization, 530
Polypeptides, 554
Polysaccharides, 547
Positive ions, 90
Positron emission tomography (PET), 489
Potassium hydroxide, reaction with metals, 378
Potential energy, 71
Prefixes, in metric system, 23
Pressure, effect on solubility, 342
Pressure of gases
    measurement of, 268–270
    number of molecules present and, 270–271
Primary alcohols, 518
Principal energy levels, 207
Problem solving, 4–5, 24–27
Propane, structural formula for, 501–502
Propanone, 526
Proteins, 550–557
Protons, 90–91
Pure substances, 8
Purification, water, 326–329

Q
Quantas, 206
Quantum mechanics, 207

R
Rabner, Michael, 4
Rad (radiation absorbed dose), 479
Radiation. *See also* Nuclear chemistry; Radioactivity
    acute radiation damage, 488
    electromagnetic, 204–205
    genetic effects of, 488
    ionizing, 488
    long-term radiation damage, 488
    name origins of, 469
Radioactive decay, 470
Radioactive disintegration series, 476
Radioactive iodine (I-131), 489
Radioactive substances in water, 329
Radioactive tracers, 489
Radioactive waste production, 483
Radioactivity. *See also* Nuclear chemistry; Radiation
    alpha particles, 473–474
    artificial, 478
    beta particles, 474
    characteristics of, 476
    gamma rays, 474–476
    half-life, 471
    measurement of, 478–479
    natural, 470–474
    nuclide stability and, 472
    radiation units, 479
    radioactive decay, 470
Radium, half-life for, 471
Reactants, 70, 151
Reaction rates, 407–408

Real gases, 295–296
Red blood cells, 360
Redox. *See* Oxidation-reduction
Rem (roentgen equivalent to man), 479
Representative elements, 53, 215, 216
Resonance structure, 248
Reversible chemical reactions, 406
Ribonucleic acid (RNA), 562
Roentgen, Wilhelm Konrad, 469
Roentgen (Rg), 479
Rounding off numbers, 19–20
Rutherford, Ernest, 91–93, 469, 477

S
Salts
    acid-base properties of, 426
    characteristics of, 379
Sand filtration, in water purification, 326
Sanger, Frederick, 556
Saponification, 550
Saturated ethers, 522
Saturated fatty acids, 527
Saturated hydrocarbons. *See* Alkanes
Saturated solutions, 342
Schlatter, James, 317
Schrödinger, Erwin, 207
Scientific laws, 5
Scientific method, 5–6
Scientific notation, 15–16
Scintillation counter, 479
Screening, in water purification, 326
Secondary alcohols, 518
Sedimentation, in water purification, 326
Sediments from land erosion, 329
Semipermeable membrane, 360
Significant figures, 18–22
    addition or subtraction, 21–22
    in calculations, 20–22
    multiplication or division, 20–21
    rounding off numbers, 19–20
    rules for counting, 18–19
Silicon, 338
Single-displacement reactions, 160–161
Sodium chloride
    decomposition of, 58
    dissolution of, 340
Sodium hydroxide, 378
Sodium ions, 381–382
Solids
    characteristics of, 6
    common materials in, 8
    density of, 38
    physical properties of, 8
    rate of dissolving, 344–345
Solubility
    of alkali metal halides in water, 341
    factors related to, 340–343
    nature of, 338–339
    pressure effect on, 342
    saturated solutions, 342

solute and solvent and, 340
supersaturated solutions, 343
temperature effect on, 340–342
unsaturated solutions, 342–343
of various compounds in water, 341
Solubility product constant (K_{sp}), 423–425
Solutions. *See also* Solubility
    colligative properties of, 355–359
    common types of, 337
    concentration of, 339, 346–355
    concentration units for, 353
    dilute solutions, 346
    dilution problems, 353–355
    general properties of, 337–338
    mass percent solutions, 346–347
    mass/volume percent, 348
    molarity, 348–352
    osmosis, 360–361
    osmotic pressure, 361–362
    rate of dissolving solids, 344–345
    reaction mediums, 345–346
    solubility, 338–343
    volume percent, 348
Solvents, 337
Specific gravity, 39
Specific heat, 72–73
Spectator ions, 390
Speed, wave, 204
Spherical molecules, 243
Spin, electron, 208
Standard temperature and pressure (STP), 280
Stirring, rate of dissolving solids and, 344–345
Stock system, 112
Stoichiometry
    introduction to, 178–179
    mass-mass calculations, 184–185
    mole-mass calculations, 182–184
    mole-mole calculations, 180–182
Stoney, G.J., 88
Storage battery cells, 456
Strassmann, Fritz, 480
Strong electrolytes, 383–385
Strontium-90 isotopes, 488
Structural formulas, 501–504
Subatomic particles, 88–91
Sublevels, energy, 207–208
Sublimation, 309
Subscripts, in chemical formulas, 59
Substances
    characteristics of, 8
    distinguishing from mixtures, 10
    properties of, 66–67
Substrates, 558-55
Subtraction in equations, 21–22
Sucrose, 545–546
Sugar, physical properties of, 66
Super-iron batteries, 457
Supersaturated solutions, 343
Surface tension, liquids, 311
Sveda, Michael, 317
Systematic names, 104–105

T
Temperature
    effect on equilibrium, 414–415
    effect on solubility, 340–342
    measurement of, 17, 33–37
    pressure of gas and, 271
    rate of dissolving solids and, 344
Tertiary alcohols, 518
Tetrahedral structures
    of carbon, 497
    of molecules, 252–253
Theoretical yield, reaction, 190
Theories, 5
Thomson, J.J., 88, 89
Thomson model of the atom, 90
Three Mile Island accident, 483
Titration, 390, 394
Torricelli, E., 269
Total ionic equations, 392
Trans fats, 242
Transcription, 562
Transition elements, 53, 215, 216
Transition metals, 108
Transmutation of elements, 477–478
Transuranium elements, 487
Triacylglycerols, 548–550
Triglycerides, 548–550
Trigonal planar shape, 251
Trigonal planar structures, molecule, 251
Tyndall effect, 398

U
Ultraviolet light, 205
Un-ionized substances, 356

Unsaturated hydrocarbons, 498
Unsaturated solutions, 342–343
Uranium, half-life for, 471
Uranium oxide, 483

V
Valence electrons, 213, 215–217
Valence shell electron pair repulsion
    model (VSEPR), 251–254
Van de Graaf electrostatic generator, 477
Vapor pressure, 309–311, 314, 356
Vaporization, 309
Vasopressin, 556
Veggie Van, 499
Villard, Paul, 470
Vital-force theory, 496
Volatile substances, 311
Voltaic cells, 454–457
Volume, 31
    effect on equilibrium, 413–414
    measurement of, 31–32
    percent, 348
Volume-volume calculations,
    294–295
Volumetric flasks, 349

W
Waste production, radioactive, 483
Water. *See also* Liquids
    of crystallization, 319
    decomposition of, 57–58
    diagram of, 323
    equilibrium systems, 323
    formation of, 324

hard water, 327
of hydration, 319
ion product constant for, 418–420
ionization of, 386
maximum density of, 322
molecular structure of, 323–324
molecules in hydrates, 320–321
physical properties of, 66, 313,
    316, 322–323
pollutants of, 329
purification of, 326–329
reactions of metals with,
    324–326
reactions of nonmetals with,
    324–326
vapor pressure of, 311
Watson, James, 559
Wave mechanics, 207
Wavelength, 204
Weak electrolytes, 383–385
Weight, measurement of, 29
Wilkin, Maurice H.F., 559
Wöhler, Friedrich, 496

X
X-rays, discovery of, 469

Y
Yield calculations, 190–191

Z
Zeolite, 327
Zhang, Jim, 389
Zinc-copper voltaic cell, 454
Zinc-mercury cells, 455

Positive Ions (Cations)		Negative Ions (Anions)	
	1+	Ammonium	NH_4^+	Acetate	$C_2H_3O_2^-$

Names, Formulas, and Charges of Common Ions

Positive Ions (Cations)

Charge	Name	Formula
1+	Ammonium	NH_4^+
	Copper(I) (Cuprous)	Cu^+
	Hydrogen	H^+
	Potassium	K^+
	Silver	Ag^+
	Sodium	Na^+
2+	Barium	Ba^{2+}
	Cadmium	Cd^{2+}
	Calcium	Ca^{2+}
	Cobalt(II)	Co^{2+}
	Copper(II) (Cupric)	Cu^{2+}
	Iron(II) (Ferrous)	Fe^{2+}
	Lead(II)	Pb^{2+}
	Magnesium	Mg^{2+}
	Manganese(II)	Mn^{2+}
	Mercury(II) (Mercuric)	Hg^{2+}
	Nickel(II)	Ni^{2+}
	Tin(II) (Stannous)	Sn^{2+}
	Zinc	Zn^{2+}
3+	Aluminum	Al^{3+}
	Antimony(III)	Sb^{3+}
	Arsenic(III)	As^{3+}
	Bismuth(III)	Bi^{3+}
	Chromium(III)	Cr^{3+}
	Iron(III) (Ferric)	Fe^{3+}
	Titanium(III) (Titanous)	Ti^{3+}
4+	Manganese(IV)	Mn^{4+}
	Tin(IV) (Stannic)	Sn^{4+}
	Titanium(IV) (Titanic)	Ti^{4+}
5+	Antimony(V)	Sb^{5+}
	Arsenic(V)	As^{5+}

Negative Ions (Anions)

Charge	Name	Formula
1−	Acetate	$C_2H_3O_2^-$
	Bromate	BrO_3^-
	Bromide	Br^-
	Chlorate	ClO_3^-
	Chloride	Cl^-
	Chlorite	ClO_2^-
	Cyanide	CN^-
	Fluoride	F^-
	Hydride	H^-
	Hydrogen carbonate (Bicarbonate)	HCO_3^-
	Hydrogen sulfate (Bisulfate)	HSO_4^-
	Hydrogen sulfite (Bisulfite)	HSO_3^-
	Hydroxide	OH^-
	Hypochlorite	ClO^-
	Iodate	IO_3^-
	Iodide	I^-
	Nitrate	NO_3^-
	Nitrite	NO_2^-
	Perchlorate	ClO_4^-
	Permanganate	MnO_4^-
	Thiocyanate	SCN^-
2−	Carbonate	CO_3^{2-}
	Chromate	CrO_4^{2-}
	Dichromate	$Cr_2O_7^{2-}$
	Oxalate	$C_2O_4^{2-}$
	Oxide	O^{2-}
	Peroxide	O_2^{2-}
	Silicate	SiO_3^{2-}
	Sulfate	SO_4^{2-}
	Sulfide	S^{2-}
	Sulfite	SO_3^{2-}
3−	Arsenate	AsO_4^{3-}
	Borate	BO_3^{3-}
	Phosphate	PO_4^{3-}
	Phosphide	P^{3-}
	Phosphite	PO_3^{3-}